Elements of Pure and Applied Mathematics

Harry Lass

DOVER PUBLICATIONS, INC.
Mineola, New York

Bibliographical Note

This Dover edition, first published in 2009, is an unabridged republication of the work originally published in 1957 by the McGraw-Hill Book Company, Inc., New York.

Library of Congress Cataloging-in-Publication Data

Lass, Harry.
Elements of pure and applied mathematics / Harry Lass.
p. cm.
Reprint. Orginally published: New York : McGraw-Hill Companies, 1957.
Includes bibliographical references and index.
ISBN-13: 978-0-486-47186-0
ISBN-10: 0-486-47186-1
1. Mathematical physics. 2. Mathematical analysis. I. Title.

QA401.L35 2009
530.15—dc22

2009018678

Manufactured in the United States by Courier Corporation
47186101
www.doverpublications.com

Dedicated to
DOROTHY

PREFACE

This text is an outgrowth of various courses which the author has taught in the past sixteen years at Pomona College, California Institute of Technology, University of California at Santa Barbara, University of Illinois, and the U.C.L.A. Orange Belt Graduate Program. The contents and purposes of the text can probably best be outlined by a brief description of each chapter.

The first chapter deals with determinants, linear equations, and matrices. The summation convention of tensor analysis proves to be a valuable manipulatory aid and will be useful for future readers of tensor analysis, Riemannian geometry, and the general theory of relativity. An elementary, but fairly thorough, treatment of matrices is encountered (in Chap. 8 one finds an application of matrices to group representations—of value to the quantum physicist). Some of the topics introduced in this first chapter are linear dependence of vectors, quadratic forms, and orthogonal transformations. Problems are found at the end of each section, and to familiarize the reader with the new definitions and concepts of this chapter, more than twenty-five examples are given.

Chapter 2 covers the main essentials of vector analysis which one would expect to study in a one-semester course. More than thirty examples illustrate and extend the fundamental concepts of this chapter. Included are a discussion of the kinematics of a rigid body with one fixed point and a discussion of pursuit problems.

Chapter 3 is an extension of vector analysis and deals with the tensor calculus and some of its applications. Here one finds a brief introduction to Riemannian geometry, Lagrange's and Hamilton's equations of motion, and the important Navier-Stokes' equation of hydrodynamics. More than twenty examples are presented in this chapter.

Chapter 4 includes those elements of complex-variable theory which one would normally encounter in an elementary senior-graduate course bearing the same title. The theory and the presentation of proofs is on a more rigorous level than can be found in prevailing applied mathematics texts (the necessary tools for rigor can be found in Chap. 10, which deals with many aspects of real-variable theory). Numerous examples illustrate the theory and many of the problems are important extensions of it. The application of the Schwarz-Christoffel transformation to electrostatics is discussed.

Chapter 5 includes those topics of elementary differential equations which can be found in a junior course. There is also a discussion of differential equations in the complex domain, and included is a brief introduction to the hypergeometric function, Laguerre and Hermite polynomials, and the method of Frobenius. Existence theorems, the Wronskian, and properties of second-order linear differential equations are discussed. It is unfortunate that a greater discussion of eigenvalues, eigenfunctions, and partial differential equations is not attempted. However, a knowledge of this chapter should prove sufficient to understand the eigenvalue problems of elementary quantum mechanics. More than twenty examples illustrate the theory.

Chapter 6 includes those elements of orthogonal polynomials, Fourier series, and Fourier integrals which are so vital to the mathematician, the physicist, and the engineer. It is the author's view that an elementary knowledge of the Fourier integral, complex-variable theory, and the calculus of variations, is sufficient for the reader to gain an insight into N. Wiener's important results concerning linear predictors and filters. Included in this chapter is a discussion of the methods of Kryloff-Bogoliuboff applied to non-linear differential equations which involve the use of Fourier series.

Chapter 7 deals with the Stieltjes integral, the Laplace transform, and the calculus of variations. The reader will find that though the treatment of these topics is brief, nevertheless there is enough material in this chapter to enable him to understand the applications of these subjects to problems of physics and engineering.

Chapter 8, which embraces group theory and algebraic equations, is written chiefly for the mathematician. The physicist, however, will find it useful to peruse this chapter if he wishes to further his studies in quantum mechanics. Numerous examples illustrate the more abstract ideas associated with this chapter.

Chapter 9 deals with some elementary concepts of probability theory, an important central-limit theorem, the chi-squared distribution of statistics, and a brief discussion of game theory and the Monte Carlo method. Needless to say, it is becoming more apparent than ever that the engineer must have a thorough understanding of probability theory if he is to keep abreast of the present work that is being done in information and filter theory.

The final chapter of the text deals with many vital aspects of real-variable theory. Included are discussions of the real-number system, point-set theory, differentiation and integration, infinite series and sequences, and numerical analysis. Chapter 10 actually comprises a course on real-variable theory. Numerous references to this final chapter are given in prior chapters.

At the end of each chapter will be found a list of references. The texts listed do not necessarily give complete coverage of the field. Some of them illustrate the simpler aspects of their subjects, while others delve into the subjects at hand at a more advanced level.

The text has been written in such a way that each chapter is independent of the others although there are many cross references which tend to unify the subjects at hand. There are a considerable number of examples worked out in detail, and problems follow each section. It is not necessary that the reader complete any given chapter to gain an inkling of the fundamental concepts involved. This book is not meant to be a compendium of formulas. The discriminating reader should enjoy the rigor of analysis which is missing from most applied-mathematics texts. On the other hand, the reader who is chiefly interested in obtaining a cursory knowledge of the material will find that the text contains most of the elementary topics found in existing applied-mathematics texts. There is the added advantage that a large number of examples are treated in the text proper. There are a sufficient number of problems at the end of each section to test the reader's understanding of the theory. It is the author's opinion that this book can be used to teach almost any undergraduate mathematics course from the junior level onward, and that it can be used as an applied-mathematics text in a variety of ways which depend on the subjects to be stressed by the instructor. Although the author is highly prejudiced, he sincerely feels that this is the type of book he would have enjoyed studying as an undergraduate and graduate student. It is hoped that the author's association with the Naval Ordnance Laboratory, Corona, California; the Motorola Research Laboratory, Riverside, California; and the current association with the Jet Propulsion Laboratory of the California Institute of Technology, Pasadena, California, has given him a greater insight into the necessary methods and tools which the engineer requires along mathematical lines, and that these methods and tools have been presented adequately in the text. The author has kept alive his love of teaching through his association with the U.C.L.A. Orange Belt Graduate Program.

I should be highly ungrateful if I did not mention my extreme gratitude to John Frank for his unfailing help in proofreading the manuscript and for his very valuable suggestions.

HARRY LASS

CONTENTS

CHAPTER 1

LINEAR EQUATIONS, DETERMINANTS, AND MATRICES

1.1. Introduction. The Summation Convention. In much of the material of this chapter we shall find it expedient to adapt a summation convention first introduced by A. Einstein. Let us consider first the set of linear equations

$$\begin{aligned} a_1x + b_1y + c_1z &= d_1 \\ a_2x + b_2y + c_2z &= d_2 \\ a_3x + b_3y + c_3z &= d_3 \end{aligned} \tag{1.1}$$

We shall find it to our advantage to set $x = x^1$, $y = x^2$, $z = x^3$. The superscripts do *not* denote powers but are simply a means for distinguishing between the three quantities x, y, and z. One immediate advantage is obvious. If we were dealing with 29 variables, it would be foolish to use 29 different letters, one letter for each variable. The single letter x with a set of superscripts ranging from 1 to 29 would suffice to yield the 29 variables, written $x^1, x^2, x^3, \ldots, x^{29}$. Our reason for using superscripts rather than subscripts will soon become evident. Equations (1.1) can now be written

$$\begin{aligned} a_1x^1 + b_1x^2 + c_1x^3 &= d_1 \\ a_2x^1 + b_2x^2 + c_2x^3 &= d_2 \\ a_3x^1 + b_3x^2 + c_3x^3 &= d_3 \end{aligned} \tag{1.2}$$

Equations (1.2) still leave something to be desired, for if there were 29 such equations, our patience would be exhausted in trying to deal with the coefficients of $x^1, x^2, x^3, \ldots, x^{29}$. Let us note that in (1.2) the coefficients of x^1, x^2, x^3 may be expressed by the square array

$$\left\| \begin{matrix} a_1 & b_1 & c_1 \\ a_2 & b_2 & c_2 \\ a_3 & b_3 & c_3 \end{matrix} \right\| \tag{1.3}$$

By defining $a_1 = a_{11}$, $b_1 = a_{12}$, $c_1 = a_{13}$, $a_2 = a_{21}$, $b_2 = a_{22}$, $c_2 = a_{23}$, $a_3 = a_{31}$, $b_3 = a_{32}$, $c_3 = a_{33}$, the square array (1.3) becomes

$$\left\| \begin{matrix} a_{11} & a_{12} & a_{13} \\ a_{21} & a_{22} & a_{23} \\ a_{31} & a_{32} & a_{33} \end{matrix} \right\| \tag{1.4}$$

One advantage is immediately evident. The single element a_{ij} lies in the ith row and jth column of the square array (1.4). Equations (1.1) can now be written

$$\begin{aligned} a_{11}x^1 + a_{12}x^2 + a_{13}x^3 &= d_1 \\ a_{21}x^1 + a_{22}x^2 + a_{23}x^3 &= d_2 \\ a_{31}x^1 + a_{32}x^2 + a_{33}x^3 &= d_3 \end{aligned} \tag{1.5}$$

Using the familiar summation notation of mathematics, we rewrite (1.5) as

$$\sum_{r=1}^{3} a_{1r}x^r = d_1 \qquad \sum_{r=1}^{3} a_{2r}x^r = d_2 \qquad \sum_{r=1}^{3} a_{3r}x^r = d_3 \tag{1.6}$$

or

$$\sum_{r=1}^{3} a_{ir}x^r = d_i \qquad i = 1, 2, 3 \tag{1.7}$$

The system of equations

$$\sum_{r=1}^{n} a_{ir}x^r = d_i \qquad i = 1, 2, 3, \ldots, n \tag{1.8}$$

represents n linear equations.

A. Einstein noticed that it was superfluous to carry along the Σ sign in (1.8). We may rewrite (1.8) as

$$a_{ir}x^r = d_i \qquad i = 1, 2, \ldots, n \tag{1.9}$$

provided it is understood that whenever an index occurs exactly once both as a subscript and superscript a summation is indicated for this index over its full range of definition. In (1.9) the index r occurs both as a subscript (in a_{ir}) and as a superscript (in x^r), so that we sum on r from $r = 1$ to $r = n$. In a four-dimensional space ($x^1 = x$, $x^2 = y$, $x^3 = z$, $x^4 = ct$) summation indices range from 1 to 4. The index of summation is a dummy index since the final result is independent of the letter used. We can write

$$a_{ir}x^r \equiv a_{ij}x^j \equiv a_{i\alpha}x^\alpha \tag{1.10}$$

We may also write (1.9) as

$$a^i_r x^r = d^i \qquad i = 1, 2, \ldots, n \tag{1.11}$$

where the element a^i_j belongs to the ith row and jth column of the square array

$$\left\| \begin{matrix} a^1_1 & a^1_2 & \cdots & a^1_n \\ a^2_1 & a^2_2 & \cdots & a^2_n \\ \cdots & \cdots & \cdots & \cdots \\ a^n_1 & a^n_2 & \cdots & a^n_n \end{matrix} \right\| \tag{1.12}$$

Example 1.1. Let us consider the quantity $S = a_{\alpha\beta}x^\alpha x^\beta$, for a three-dimensional space. Since the index α occurs as both a subscript and a superscript, we sum on α from 1 to 3. This yields $S = a_{1\beta}x^1x^\beta + a_{2\beta}x^2x^\beta + a_{3\beta}x^3x^\beta$. Now each term of S is such that β is both a subscript and superscript. Summing on β from 1 to 3 as prescribed by our summation convention yields the quadratic form

$$\begin{aligned} S = \; & a_{11}x^1x^1 + a_{12}x^1x^2 + a_{13}x^1x^3 \\ & + a_{21}x^2x^1 + a_{22}x^2x^2 + a_{23}x^2x^3 \\ & + a_{31}x^3x^1 + a_{32}x^3x^2 + a_{33}x^3x^3 \end{aligned} \tag{1.13}$$

Why is $a_{\alpha\beta}x^\alpha x^\beta = a_{ij}x^ix^j = a_{rs}x^rx^s$? What does $a_{\alpha\alpha}x^\alpha x^\alpha$ mean?

Example 1.2. If $x^1, x^2, x^3, \ldots, x^n$ is a set of independent variables, then

$$\frac{\partial x^1}{\partial x^1} = \frac{\partial x^2}{\partial x^2} = \frac{\partial x^3}{\partial x^3} = \cdots = \frac{\partial x^n}{\partial x^n} = 1,$$

$\dfrac{\partial x^1}{\partial x^2} = 0$, $\dfrac{\partial x^i}{\partial x^j} = 0$ if $i \neq j$. We may write

$$\frac{\partial x^i}{\partial x^j} \equiv \delta^i_j \left.\begin{aligned} &= 1 \qquad \text{if } i = j \\ &= 0 \qquad \text{if } i \neq j \end{aligned}\right\} \tag{1.14}$$

The symbol δ^i_j is called the Kronecker[1] delta. We have

$$\delta^\alpha_\alpha = \delta^1_1 + \delta^2_2 + \cdots + \delta^n_n = n$$

Let us now assume that the quadratic form (1.13) vanishes identically for all values of the independent variables x^1, x^2, x^3, the a_{ij} to be constant. Differentiating $S = a_{\alpha\beta}x^\alpha x^\beta \equiv 0$ with respect to a given variable, say x^i, yields

$$\begin{aligned} \frac{\partial S}{\partial x^i} &= a_{\alpha\beta}x^\alpha \frac{\partial x^\beta}{\partial x^i} + a_{\alpha\beta}x^\beta \frac{\partial x^\alpha}{\partial x^i} \equiv 0 \\ &= a_{\alpha\beta}x^\alpha \delta^\beta_i + a_{\alpha\beta}x^\beta \delta^\alpha_i \equiv 0 \\ &= a_{\alpha i}x^\alpha + a_{i\beta}x^\beta \equiv 0 \qquad \text{Why?} \end{aligned}$$

Now differentiating with respect to x^j yields

$$\frac{\partial^2 S}{\partial x^j\,\partial x^i} = a_{\alpha i}\delta^\alpha_j + a_{i\beta}\delta^\beta_j \equiv 0$$

so that $a_{ji} + a_{ij} = 0$ or $a_{ij} = -a_{ji}$, $i, j = 1, 2, 3$.

Example 1.3. We define ϵ^{ij}, $i, j = 1, 2$, to have the following numerical values: Let $\epsilon^{11} = \epsilon^{22} = 0$, $\epsilon^{12} = 1$, $\epsilon^{21} = -1$. We now consider the expression

$$D = \epsilon^{ij}a^1_ia^2_j \tag{1.15}$$

Expanding (1.15) by use of our summation convention yields

$$D = \epsilon^{11}a^1_1a^2_1 + \epsilon^{12}a^1_1a^2_2 + \epsilon^{21}a^1_2a^2_1 + \epsilon^{22}a^1_2a^2_2 = a^1_1a^2_2 - a^1_2a^2_1$$

The reader who is familiar with second-order determinants quickly recognizes that

$$\epsilon^{ij}a^1_ia^2_j = \begin{vmatrix} a^1_1 & a^1_2 \\ a^2_1 & a^2_2 \end{vmatrix} \tag{1.16}$$

[1] For Leopold Kronecker (1823–1891), a world-famous German mathematician.

Example 1.4. The system of equations

$$\begin{aligned} y^1 &= y^1(x^1, x^2, \ldots, x^n) \\ y^2 &= y^2(x^1, x^2, \ldots, x^n) \\ &\cdots\cdots\cdots\cdots \\ y^n &= y^n(x^1, x^2, \ldots, x^n) \end{aligned} \tag{1.17}$$

represents a coordinate transformation from an $(x^1, x^2, \ldots, x^n)$ coordinate system to a $(y^1, y^2, \ldots, y^n)$ coordinate system. From the calculus we have

$$\begin{aligned} dy^i &= \frac{\partial y^i}{\partial x^1} dx^1 + \frac{\partial y^i}{\partial x^2} dx^2 + \cdots + \frac{\partial y^i}{\partial x^n} dx^n \qquad i = 1, 2, \ldots, n \\ &= \frac{\partial y^i}{\partial y^\alpha} dx^\alpha \end{aligned}$$

The α in the term $\frac{\partial y^i}{\partial x^\alpha}$ is to be considered as a subscript. If, furthermore, the x^i, $i = 1, 2, \ldots, n$, can be solved for the $y^1, y^2, \ldots, y^n$, and assuming differentiability of the x^i with respect to each y^i, one obtains

$$\frac{\partial y^i}{\partial y^j} \equiv \delta^i_j = \frac{\partial y^i}{\partial x^\alpha} \frac{\partial x^\alpha}{\partial y^j}$$

Differentiating this expression with respect to y^k yields

$$0 = \frac{\partial y^i}{\partial x^\alpha} \frac{\partial^2 x^\alpha}{\partial y^k \, \partial y^j} + \frac{\partial^2 y^i}{\partial x^\beta \, \partial x^\alpha} \frac{\partial x^\beta}{\partial y^k} \frac{\partial x^\alpha}{\partial y^j}$$

Multiplying both sides of this equation by $\frac{\partial x^\sigma}{\partial y^i}$ and summing on the index i yields

$$0 = \frac{\partial x^\sigma}{\partial y^i} \frac{\partial y^i}{\partial x^\alpha} \frac{\partial^2 x^\alpha}{\partial y^k \, \partial y^j} + \frac{\partial^2 y^i}{\partial x^\beta \, \partial x^\alpha} \frac{\partial x^\beta}{\partial y^k} \frac{\partial x^\alpha}{\partial y^j} \frac{\partial x^\sigma}{\partial y^i}$$

or

$$0 = \delta^\sigma_\alpha \frac{\partial^2 x^\alpha}{\partial y^k \, \partial y^j} + \frac{\partial^2 y^i}{\partial x^\beta \, \partial x^\alpha} \frac{\partial x^\beta}{\partial y^k} \frac{\partial x^\alpha}{\partial y^j} \frac{\partial x^\sigma}{\partial y^i}$$

which yields

$$\frac{\partial^2 x^\sigma}{\partial y^k \, \partial y^j} = - \frac{\partial^2 y^i}{\partial x^\beta \, \partial x^\alpha} \frac{\partial x^\alpha}{\partial y^j} \frac{\partial x^\beta}{\partial y^k} \frac{\partial x^\sigma}{\partial y^i}$$

In particular, if $y = f(x)$, then

$$\frac{d^2 x}{dy^2} = - \frac{d^2 y}{dx^2} \left(\frac{dx}{dy} \right)^3$$

Problems

1. If $y^i = a^i_\alpha x^\alpha$, $z^i = b^i_\alpha y^\alpha$, show that $z^i = b^i_\alpha a^\alpha_\beta x^\beta$.

2. If $a_{ijk} x^i x^j x^k \equiv 0$, $a_{ijk} \equiv$ constant, show that

$$a_{ijk} + a_{kij} + a_{jki} + a_{jik} + a_{kji} + a_{ikj} = 0$$

3. Show that $\delta^\alpha_\beta \frac{\partial y^i}{\partial x^\alpha} \frac{\partial x^\beta}{\partial y^j} = \delta^i_j$.

4. If $\bar{A}^i = \frac{\partial y^i}{\partial x^\alpha} A^\alpha$, show that $A^i = \frac{\partial x^i}{\partial y^\alpha} \bar{A}^\alpha$.

5. If $\bar{g}_{\alpha\beta} = g_{\mu\nu} \frac{\partial x^\mu}{\partial y^\alpha} \frac{\partial x^\nu}{\partial y^\beta}$, show that $g_{\alpha\beta} = \bar{g}_{\mu\nu} \frac{\partial y^\mu}{\partial x^\alpha} \frac{\partial y^\nu}{\partial x^\beta}$.

6. If the $g_{\mu\nu}$ of Prob. 5 are such that $g_{\mu\nu} = g_{\nu\mu}$, show that $\bar{g}_{\alpha\beta} = \bar{g}_{\beta\alpha}$.

7. We define ϵ^{ijk} as follows: The superscripts i, j, k are to take on the numerical values 1, 2, 3, not respectively, and we define $\epsilon^{123} = \epsilon^{312} = \epsilon^{231} = 1$,

$$\epsilon^{213} = \epsilon^{321} = \epsilon^{132} = -1$$

all other $\epsilon^{ijk} = 0$. Expand $\epsilon^{ijk}a_i^1a_j^2a_k^3$, and express this sum as a third-order determinant.

8. If $\varphi = \varphi(x^1, x^2, \ldots, x^n)$, $x^i = x^i(y^1, y^2, \ldots, y^n) \equiv x^i(y)$, $i = 1, 2, \ldots, n$, and if $\bar{\varphi} = \bar{\varphi}(y^1, y^2, \ldots, y^n) \equiv \varphi[x^1(y), x^2(y), \ldots, x^n(y)]$, show that

$$\frac{\partial \bar{\varphi}}{\partial y^\alpha} = \frac{\partial \varphi}{\partial x^\beta}\frac{\partial x^\beta}{\partial y^\alpha}$$

Given $\varphi_\alpha \equiv \dfrac{\partial \varphi}{\partial x^\alpha}$, show that

$$\frac{\partial \bar{\varphi}_i}{\partial y^j} - \frac{\partial \bar{\varphi}_j}{\partial y^i} = \left(\frac{\partial \varphi_\alpha}{\partial x^\beta} - \frac{\partial \varphi_\beta}{\partial x^\alpha}\right)\frac{\partial x^\alpha}{\partial y^i}\frac{\partial x^\beta}{\partial y^j}$$

9. If $a_{\alpha\beta} = -a_{\beta\alpha}$, show that $a_{\alpha\beta}x^\alpha x^\beta \equiv 0$.

1.2. Determinants. In attempting to solve (1.5) for the x^1, x^2, x^3 one is led in a very natural way to consider the square array

$$\left\|\begin{matrix} a_1^1 & a_2^1 & a_3^1 \\ a_1^2 & a_2^2 & a_3^2 \\ a_1^3 & a_2^3 & a_3^3 \end{matrix}\right\| \tag{1.18}$$

We have written $a_{ij} = a_j^i$, $i, j = 1, 2, 3$. The solution of (1.5) requires that we attach a numerical value to the matrix (as yet undefined) of elements (1.18). We do this in the following way (see Prob. 7, Sec. 1.1): We attach $3^3 = 27$ numerical values to a set of ϵ^{ijk}, $i, j, k = 1, 2, 3$. If at least two of the superscripts in ϵ^{ijk} are the same, the value of ϵ^{ijk} is zero. Thus $\epsilon^{223} = \epsilon^{131} = \epsilon^{333} = 0$, etc. If the i, j, k are all different, the value of ϵ^{ijk} is to be $+1$ or -1 according to whether it takes an even or odd number of permutations to rearrange the ijk into the natural order 123. Let us look at ϵ^{321} and hence at the arrangement 321. Permuting the integers 2 and 3 permutes 321 into 231, then permuting 3 and 1 permutes 231 into 213, and finally 213 permutes into 123 if we interchange the integers 2 and 1. Three (an odd number) permutations were required to permute 321 into 123. Thus $\epsilon^{321} = -1$. We have

$$\epsilon^{123} = \epsilon^{312} = \epsilon^{231} = +1$$

$\epsilon^{213} = \epsilon^{321} = \epsilon^{132} = -1$.

We now define

$$\begin{vmatrix} a_1^1 & a_2^1 & a_3^1 \\ a_1^2 & a_2^2 & a_3^2 \\ a_1^3 & a_2^3 & a_3^3 \end{vmatrix} \equiv \epsilon^{ijk}a_i^1a_j^2a_k^3 \tag{1.19}$$

The letters i, j, k are indices of summation. Equation (1.19) defines the determinant of the square matrix of elements (1.18). Its numerical value is given by the right-hand side of (1.19). It consists, in general, of $3! = 3 \cdot 2 \cdot 1 = 6$ terms, each term a product of three elements, one element from each row and column of (1.19). For the benefit of the student we expand (1.19) and obtain

$$\begin{aligned} \epsilon^{ijk}a_i^1a_j^2a_k^3 &= (a_1^1a_2^2a_3^3 + a_3^1a_1^2a_2^3 + a_2^1a_3^2a_1^3) \\ &\quad - (a_2^1a_1^2a_3^3 + a_3^1a_2^2a_1^3 + a_1^1a_3^2a_2^3) \end{aligned}$$

Only $3! = 6$ terms occur in the expansion of (1.19) since there are $3!$ permutations of 123. All other values of ϵ^{ijk} are zero.

We can define ϵ_{ijk} in exactly the same manner in which the ϵ^{ijk} were defined. We leave it to the reader to show that

$$\epsilon^{ijk}a_i^1a_j^2a_k^3 = \epsilon_{ijk}a_1^ia_2^ja_3^k$$

The generalization of second- and third-order determinants (the order of a determinant is the number of rows or columns of the determinant) to nth order determinants is simple. We define the $\epsilon^{i_1i_2\cdots i_n}$ to have the following numerical values: $\epsilon^{i_1i_2\cdots i_n} = 0$ if at least two of the superscripts are the same. The values of the superscripts range from 1 to n. If the $i_1, i_2, \ldots, i_n$ are distinct, the value of $\epsilon^{i_1i_2\cdots i_n}$ is to be $+1$ or -1 depending on whether an even or odd number of permutations is required to rearrange $i_1i_2 \cdots i_n$ into the natural order $123 \cdots n$. The numerical value (determinant) of the square array of elements $\|a_j^i\|$, $i, j = 1, 2, \ldots, n$, is defined as

$$\begin{aligned} |a_j^i| &= \begin{vmatrix} a_1^1 & a_2^1 & \cdots & a_n^1 \\ a_1^2 & a_2^2 & \cdots & a_n^2 \\ \cdot & \cdot & \cdots & \cdot \\ a_1^n & a_2^n & \cdots & a_n^n \end{vmatrix} \\ &= \epsilon^{i_1i_2\cdots i_n}a_{i_1}^1a_{i_2}^2 \cdots a_{i_n}^n \\ &= \epsilon_{i_1i_2\cdots i_n}a_1^{i_1}a_2^{i_2} \cdots a_n^{i_n} \end{aligned} \tag{1.20}$$

where the $\epsilon_{i_1i_2\cdots i_n}$ are defined in precisely the same manner in which the $\epsilon^{i_1i_2\cdots i_n}$ are defined. In general, (1.20) consists of $n!$ terms, each term a product of elements, one element from each row and column of $|a_j^i|$.

To facilitate writing, we shall deal with third-order determinants, but it will be obvious to the reader that any theorem derived for third-order determinants will apply to determinants of any finite order. Let us consider

$$\Delta = \begin{vmatrix} a_1^1 & a_2^1 & a_3^1 \\ a_1^2 & a_2^2 & a_3^2 \\ a_1^3 & a_2^3 & a_3^3 \end{vmatrix} = \epsilon^{ijk}a_i^1a_j^2a_k^3 = \epsilon_{ijk}a_1^ia_2^ja_3^k \tag{1.21}$$

We can obtain a new third-order determinant by interchanging the first and third row of Δ. This yields

$$\Delta' = \begin{vmatrix} a_1^3 & a_2^3 & a_3^3 \\ a_1^2 & a_2^2 & a_3^2 \\ a_1^1 & a_2^1 & a_3^1 \end{vmatrix} = \epsilon^{ijk} a_i^3 a_j^2 a_k^1 \tag{1.22}$$

But $\epsilon^{ijk} a_i^3 a_j^2 a_k^1 = \epsilon^{ijk} a_k^1 a_j^2 a_i^3 = \epsilon^{kji} a_i^1 a_j^2 a_k^3$, since i and k are dummy indices. We see that every term of (1.22) is the same as every term in (1.21) with the exception that ϵ^{ijk} is replaced by ϵ^{kji}. Since $\epsilon^{ijk} = -\epsilon^{kji}$, we conclude that $\Delta = -\Delta'$. We thus obtain the following theorems:

THEOREM 1.1. Interchanging two rows (or columns) of a determinant changes the sign of the determinant.

THEOREM 1.2. If two rows (or columns) of a determinant are the same, the value of the determinant is zero.

We note that

$$\Delta'' = \begin{vmatrix} la_1^1 & la_2^1 & la_3^1 \\ a_1^2 & a_2^2 & a_3^2 \\ a_1^3 & a_2^3 & a_3^3 \end{vmatrix} = \epsilon^{ijk}(la_i^1) a_j^2 a_k^3 = l\Delta \tag{1.23}$$

THEOREM 1.3. If a row (or column) of a determinant is multiplied by a factor l, the value of the determinant is thereby multiplied by l.

Let us now investigate the determinant

$$\begin{aligned} \Delta''' &= \begin{vmatrix} a_1^1 + la_1^3 & a_2^1 + la_2^3 & a_3^1 + la_3^3 \\ a_1^2 & a_2^2 & a_3^2 \\ a_1^3 & a_2^3 & a_3^3 \end{vmatrix} \\ &= \epsilon^{ijk}(a_i^1 + la_i^3) a_j^2 a_k^3 \\ &= \epsilon^{ijk} a_i^1 a_j^2 a_k^3 + l\epsilon^{ijk} a_i^3 a_j^2 a_k^3 \\ &= \Delta \end{aligned} \tag{1.24}$$

since $\epsilon^{ijk} a_i^3 a_j^2 a_k^3 = 0$ from Theorem (1.2). Hence we obtain the following theorem:

THEOREM 1.4. The value of a determinant remains unchanged if to the elements of any row (or column) is added a scalar multiple of the corresponding elements of another row (or column).

The theorems derived above are very useful in evaluating a determinant.

Example 1.5

$$\Delta = \begin{vmatrix} 1 & 0 & -2 & 3 \\ 4 & -2 & 6 & 2 \\ -1 & 3 & 0 & 1 \\ 3 & 0 & 0 & -1 \end{vmatrix} = 2 \begin{vmatrix} 1 & 0 & -2 & 3 \\ 2 & -1 & 3 & 1 \\ -1 & 3 & 0 & 1 \\ 3 & 0 & 0 & -1 \end{vmatrix} = 2 \begin{vmatrix} 1 & 0 & -2 & 3 \\ 0 & -1 & 7 & -5 \\ -1 & 3 & 0 & 1 \\ 3 & 0 & 0 & -1 \end{vmatrix}$$

$$\Delta = 2\begin{vmatrix} 1 & 0 & -2 & 3 \\ 0 & -1 & 7 & -5 \\ 0 & 3 & -2 & 4 \\ 3 & 0 & 0 & -1 \end{vmatrix} = 2\begin{vmatrix} 1 & 0 & -2 & 3 \\ 0 & -1 & 7 & -5 \\ 0 & 3 & -2 & 4 \\ 0 & 0 & 6 & -10 \end{vmatrix} = 2\begin{vmatrix} 1 & 0 & -2 & 3 \\ 0 & -1 & 7 & -5 \\ 0 & 0 & 19 & -11 \\ 0 & 0 & 6 & -10 \end{vmatrix}$$

$$= 2 \cdot 19 \begin{vmatrix} 1 & 0 & -2 & 3 \\ 0 & -1 & 7 & -5 \\ 0 & 0 & 1 & -\frac{11}{19} \\ 0 & 0 & 6 & -10 \end{vmatrix} = 2 \cdot 19 \begin{vmatrix} 1 & 0 & -2 & 3 \\ 0 & -1 & 7 & -5 \\ 0 & 0 & 1 & -\frac{11}{19} \\ 0 & 0 & 0 & -\frac{124}{19} \end{vmatrix}$$

$$= 2 \cdot 19 \cdot 1 \cdot (-1)1 \cdot \left(-\frac{124}{19}\right) = 248$$

The final result is obtained as follows: (1) Factor 2 from row 2; (2) multiply row 1 by -2, and add to row 2; (3) add row 1 to row 3; (4) multiply row 1 by 3, and subtract from row 4; (5) multiply row 2 by 3, and add to row 3; (6) factor out 19 from row 3; (7) multiply row 3 by -6, and add to row 4; (8) with zeros below the main diagonal, the value of the determinant is the product of the diagonal terms. Why?

The reader is urged to read Laplace's method of expanding a determinant, which can be found in various texts dealing with determinants.

Example 1.6. If the a_j^i of (1.21) are differentiable functions of a variable x, then

$$\frac{d\Delta}{dx} = \frac{d}{dx}(\epsilon^{ijk}a_i^1a_j^2a_k^3)$$
$$= \epsilon^{ijk}\frac{da_i^1}{dx}a_j^2a_k^3 + \epsilon^{ijk}a_i^1\frac{da_j^2}{dx}a_k^3 + \epsilon^{ijk}a_i^1a_j^2\frac{da_k^3}{dx}$$

$$\frac{d\Delta}{dx} = \begin{vmatrix} \frac{da_1^1}{dx} & \frac{da_2^1}{dx} & \frac{da_3^1}{dx} \\ a_1^2 & a_2^2 & a_3^2 \\ a_1^3 & a_2^3 & a_3^3 \end{vmatrix} + \begin{vmatrix} a_1^1 & a_2^1 & a_3^1 \\ \frac{da_1^2}{dx} & \frac{da_2^2}{dx} & \frac{da_3^2}{dx} \\ a_1^3 & a_2^3 & a_3^3 \end{vmatrix} + \begin{vmatrix} a_1^1 & a_2^1 & a_3^1 \\ a_1^2 & a_2^2 & a_3^2 \\ \frac{da_1^3}{dx} & \frac{da_2^3}{dx} & \frac{da_3^3}{dx} \end{vmatrix}$$

We let the reader extend this result to determinants of order n.

We now investigate the sum $S = \epsilon^{ijk}a_i^\alpha a_j^\beta a_k^\gamma$. If α, β, γ take on the values 1, 2, 3, respectively, the sum S is the determinant of the elements $\|a_j^i\|$. If α, β, γ take on the values 1, 2, 3, but not respectively, then $S = |a_j^i|$ or $S = -|a_j^i|$ depending on whether it requires an even or odd number of permutations to permute α, β, γ into 1, 2, 3 (see Theorem 1.1). In all other cases $S = 0$ from Theorem 1.2. Hence we obtain the useful identity

$$\epsilon^{ijk}a_i^\alpha a_j^\beta a_k^\gamma = |a_j^i|\epsilon^{\alpha\beta\gamma} \tag{1.25}$$

Similarly

$$\epsilon_{ijk}a_\alpha^i a_\beta^j a_\gamma^k = |a_j^i|\epsilon_{\alpha\beta\gamma} \tag{1.26}$$

We now obtain a method for multiplying two determinants of the same order. We have from (1.21), (1.26)

$$
\begin{aligned}
|a^i_j| \cdot |b^i_j| &= |a^i_j|\epsilon_{\alpha\beta\gamma} b^\alpha_1 b^\beta_2 b^\gamma_3 \\
&= \epsilon_{ijk} a^i_\alpha a^j_\beta a^k_\gamma b^\alpha_1 b^\beta_2 b^\gamma_3 \\
&= \epsilon_{ijk}(a^i_\alpha b^\alpha_1)(a^j_\beta b^\beta_2)(a^k_\gamma a^\gamma_3) \\
&= \epsilon_{ijk} c^i_1 c^j_2 c^k_3 = |c^i_j| \qquad (1.27)
\end{aligned}
$$

where $c^r_s = a^r_\alpha b^\alpha_s$, $r, s = 1, 2, 3$.

Example 1.7

$$
\begin{vmatrix} a^1_1 & a^1_2 \\ a^2_1 & a^2_2 \end{vmatrix} \cdot \begin{vmatrix} b^1_1 & b^1_2 \\ b^2_1 & b^2_2 \end{vmatrix} = \begin{vmatrix} a^1_1 b^1_1 + a^1_2 b^2_1 & a^1_1 b^1_2 + a^1_2 b^2_2 \\ a^2_1 b^1_1 + a^2_2 b^2_1 & a^2_1 b^1_2 + a^2_2 b^2_2 \end{vmatrix}
$$

$$
\begin{vmatrix} 2 & 0 & -1 \\ -1 & 2 & 0 \\ 3 & 1 & 1 \end{vmatrix} \cdot \begin{vmatrix} -2 & 0 & 0 \\ 1 & 1 & -3 \\ 0 & -1 & 1 \end{vmatrix}
$$

$$
= \begin{vmatrix} 2(-2) + 0(1) + (-1)(0) & 2(0) + 0(1) + (-1)(-1) & 2(0) + 0(-3) + (-1)(1) \\ -1(-2) + 2(1) + 0(0) & -1(0) + 2(1) + 0(-1) & -1(0) + 2(-3) + 0(1) \\ 3(-2) + 1(1) + 1(0) & 3(0) + 1(1) + 1(-1) & 3(0) + 1(-3) + 1(1) \end{vmatrix}
$$

$$
11 \cdot 4 = \begin{vmatrix} -4 & 1 & -1 \\ 4 & 2 & -6 \\ -5 & 0 & -2 \end{vmatrix} = 44
$$

Example 1.8. We define the Kronecker δ^i_j as follows [see (1.14)]:

$$
\left.\begin{aligned} \delta^i_j &= 0 \qquad \text{if } i \neq j \\ &\ 1 \qquad \text{if } i = j \end{aligned}\right\} \qquad (1.28)
$$

Thus $\delta^1_1 = \delta^2_2 = \delta^3_3 = \delta^4_4 = 1$. $\delta^1_2 = \delta^2_1 = \delta^1_4 = 0$. The determinant $|\delta^i_j|$ is the determinant

$$
|\delta^i_j| = \begin{vmatrix} 1 & 0 & 0 \\ 0 & 1 & 0 \\ 0 & 0 & 1 \end{vmatrix} = 1 \qquad i, j = 1, 2, 3.
$$

We have

$$
|a^i_j| \cdot |\delta^i_j| = |a^i_\alpha \delta^\alpha_j| = |a^i_j| = |\delta^i_j| \cdot |a^i_j|
$$

We now expand a determinant in terms of its cofactors. We have

$$
\begin{aligned}
\Delta = |a^i_j| &= \epsilon^{ijk} a^1_i a^2_j a^3_k \\
&= a^1_1(\epsilon^{1jk} a^2_j a^3_k) + a^1_2(\epsilon^{2jk} a^2_j a^3_k) + a^1_3(\epsilon^{3jk} a^2_j a^3_k) \\
&= a^1_1 A^1_1 + a^1_2 A^2_1 + a^1_3 A^3_1 \qquad (1.29)
\end{aligned}
$$

We now examine the sum $A^1_1 \equiv \epsilon^{1jk} a^2_j a^3_k$. The only terms which contribute to this sum are the terms for which j, k take on the values 2, 3, not necessarily respectively, for if j or k is set equal to 1, the value of ϵ^{1jk} is zero. We see immediately that

$$
A^1_1 = \epsilon^{1jk} a^2_j a^3_k = \begin{vmatrix} a^2_2 & a^2_3 \\ a^3_2 & a^3_3 \end{vmatrix}
$$

We call A^1_1 the cofactor of the element a^1_1. It is the determinant obtained from the elements of $\|a^i_j\|$ by removing the row and column containing the element a^1_1. Similarly

$$
A^2_1 = \epsilon^{2jk} a^2_j a^3_k = - \begin{vmatrix} a^2_1 & a^2_3 \\ a^3_1 & a^3_3 \end{vmatrix} = (-1)^{1+2} \begin{vmatrix} a^2_1 & a^2_3 \\ a^3_1 & a^3_3 \end{vmatrix}
$$

The determinant obtained from the elements of $\|a^i_j\|$ by removing the row and column containing the element a^r_s is called the minor, M^s_r, of this element. The cofactor A^s_r of the element a^r_s is such that $A^s_r = (-1)^{s+r}M^s_r$. We see that $\Delta = a^1_\alpha A^\alpha_1$. Similarly $\Delta = a^2_\alpha A^\alpha_2 = a^3_\alpha A^\alpha_3$. We note that A^i_j is the cofactor of the element a^j_i.

Let us now look at the sum $a^1_\alpha A^\alpha_2$. Here we have

$$\begin{aligned}\bar{\Delta} = a^1_\alpha A^\alpha_2 &= a^1_1(\epsilon^{i1k}a^1_i a^3_k) + a^1_2(\epsilon^{i2k}a^1_i a^3_k) + a^1_3(\epsilon^{i3k}a^1_i a^3_k) \\ &= \epsilon^{ijk}a^1_i a^1_j a^3_k \equiv 0\end{aligned}$$

since two rows of $\bar{\Delta}$ are the same.

In general, $a^i_\alpha A^\alpha_j = 0$ if $i \neq j$. Using this result along with (1.29) yields

$$\begin{aligned} a^i_\alpha A^\alpha_j &= |a^i_j|\delta^i_j \\ a^\alpha_j A^i_\alpha &= |a^i_j|\delta^i_j \end{aligned} \tag{1.30}$$

Example 1.9

$$\begin{vmatrix} 2 & -1 & 3 \\ 0 & 4 & -2 \\ 1 & 1 & 2 \end{vmatrix} = 2\begin{vmatrix} 4 & -2 \\ 1 & 2 \end{vmatrix} - (-1)\begin{vmatrix} 0 & -2 \\ 1 & 2 \end{vmatrix} + 3\begin{vmatrix} 0 & 4 \\ 1 & 1 \end{vmatrix} = 10$$

Example 1.10. We can now solve the system of equations $a^i_\alpha x^\alpha = b^i$, $i = 1, 2, \ldots, n$ for the unknowns $x^1, x^2, \ldots, x^n$ provided $|a| \neq 0$. Multiplying $a^i_\alpha x^\alpha = b^i$ by A^j_i and summing on the index i yields $a^i_\alpha A^j_i x^\alpha = b^i A^j_i$, so that $|a|\ \delta^j_\alpha x^\alpha = b^i A^j_i$ from (1.30). This in turn implies $x^j = b^i A^j_i/|a|$, $j = 1, 2, \ldots, n$, if $|a| \neq 0$. The sum $b^i A^j_i$ is also a determinant. It is that determinant obtained by replacing the elements of the jth column of $|a^i_j|$ by the elements b^i, $i = 1, 2, \ldots, n$. This is *Cramer's* rule. For the system of linear equations

$$\begin{aligned} 2x - 3y + z &= 10 \\ x + y + z &= 4 \\ x - y - z &= 0 \end{aligned}$$

we have

$$x = \frac{\begin{vmatrix} 10 & -3 & 1 \\ 4 & 1 & 1 \\ 0 & -1 & -1 \end{vmatrix}}{\begin{vmatrix} 2 & -3 & 1 \\ 1 & 1 & 1 \\ 1 & -1 & -1 \end{vmatrix}} = \frac{-16}{-8} = 2 \quad y = \frac{\begin{vmatrix} 2 & 10 & 1 \\ 1 & 4 & 1 \\ 1 & 0 & -1 \end{vmatrix}}{-8} = -1 \quad z = \frac{\begin{vmatrix} 2 & -3 & 10 \\ 1 & 1 & 4 \\ 1 & -1 & 0 \end{vmatrix}}{-8} = 3$$

Problems

1. What is the cofactor of each element of $\begin{vmatrix} a^1_1 & a^1_2 & a^1_3 \\ a^2_1 & a^2_2 & a^2_3 \\ a^3_1 & a^3_2 & a^3_3 \end{vmatrix}$?

2. From $a^i_\alpha A^\alpha_j = |a|\delta^i_j$ show that $|A^i_j| = |a|^{n-1}$.

3. If $\bar{g}_{ij} = g_{\alpha\beta}\dfrac{\partial x^\alpha}{\partial y^i}\dfrac{\partial x^\beta}{\partial y^j}$ (see Example 1.4), show that $|\bar{g}| = |g|\left|\dfrac{\partial x^i}{\partial y^j}\right|^2$.

4. If $a^i_j = a^i_j(x)$, show that $\dfrac{d|a|}{dx} = A^\alpha_\beta\dfrac{da^\beta_\alpha}{dx}$ (see Example 1.6).

5. Evaluate

$$\begin{vmatrix} 3 & -2 & 4 & 7 & 8 \\ 1 & 0 & -2 & 3 & 4 \\ -1 & 1 & 0 & 0 & 3 \\ 0 & 0 & 2 & -1 & 0 \\ 2 & 1 & -1 & 0 & 0 \end{vmatrix}$$

6. If $A^i_j = B^\alpha_\beta \dfrac{\partial x^\beta}{\partial y^i}\dfrac{\partial y^i}{\partial x^\alpha}$, show that $|A^i_j| = |B^i_j|$.

7. Prove that

$$\begin{vmatrix} 1 & 1 & 1 \\ a & b & c \\ bc & ca & ab \end{vmatrix} = (a-b)(b-c)(c-a)$$

Hint: The equation $\begin{vmatrix} 1 & 1 & 1 \\ x & b & c \\ bc & cx & bx \end{vmatrix} = 0$ is a quadratic which vanishes for $x = b$ and $x = c$.

8. Show that

$$\begin{vmatrix} 1 & -2 \\ 3 & 4 \end{vmatrix} = \begin{vmatrix} 1 & -2 & 0 \\ 3 & 4 & 0 \\ 0 & 0 & 1 \end{vmatrix}$$

Extend this result so that two determinants of different orders may be multiplied.

9. Show that $\dfrac{\partial \left|\dfrac{\partial y^i}{\partial x^j}\right|}{\partial x^k} = \left|\dfrac{\partial y_i}{\partial x^j}\right| \dfrac{\partial x^\alpha}{\partial y^\beta}\dfrac{\partial^2 y^\beta}{\partial x^k\,\partial x^\alpha}$.

10. What difficulties does one encounter in the case of a determinant of infinite order?

11. Consider $F(\lambda) = \begin{vmatrix} a^2+\lambda & ab & ac \\ ab & b^2+\lambda & bc \\ ac & bc & c^2+\lambda \end{vmatrix}$. Show that $F(0) = F'(0) = 0$.

1.3. Matrices. A matrix is defined as a rectangular array of elements (usually the elements are real or complex numbers). An algebra of matrices is developed by defining addition of matrices, multiplication of matrices, multiplication of a matrix by a scalar (real or complex number), differentiation of matrices, etc. The definitions chosen for the above-mentioned operations will be such as to make the calculus of matrices highly applicable. A matrix **A** may be denoted as follows:

$$\mathbf{A} = \left\|\begin{matrix} a^1_1 & a^1_2 & \cdots & a^1_n \\ a^2_1 & a^2_2 & \cdots & a^2_n \\ \cdots & \cdots & \cdots & \cdots \\ a^m_1 & a^m_2 & \cdots & a^m_n \end{matrix}\right\| = \|a^i_j\| \qquad \left.\begin{matrix} i = 1, 2, \ldots, m \\ j = 1, 2, \ldots, n \end{matrix}\right\} \tag{1.31}$$

If $m = n$, we say that **A** is a square matrix of order n. If **B** is the matrix of elements $\|b^i_j\|$, $i = 1, 2, \ldots, m$, $j = 1, 2, \ldots, n$, then **B** is said to be equal to **A**, written **B** = **A** or **A** = **B**, if and only if $a^i_j = b^i_j$ for the complete range of values of i and j.

Example 1.11

$$\left\|\begin{matrix} a^1_1 & a^1_2 & a^1_3 \\ a^2_1 & a^2_2 & a^2_3 \end{matrix}\right\| = \left\|\begin{matrix} 2 & 0 & -3 \\ -3 & 2 & 6 \end{matrix}\right\|$$

implies $a^1_1 = 2$, $a^1_2 = 0$, $a^1_3 = -3$, $a^2_1 = -3$, $a^2_2 = 2$, $a^2_3 = 6$.

Two matrices can be compared for equality if and only if they are comparable in the sense that they have the same number of rows and the same number of columns.

The sum of two comparable matrices **A**, **B** is defined as a new matrix **C** whose elements c^i_j are obtained by adding the corresponding elements of **A** and **B**. Thus

$$\|c^i_j\| = \mathbf{C} = \mathbf{A} + \mathbf{B} = \|a^i_j\| + \|b^i_j\| = \|a^i_j + b^i_j\| \tag{1.32}$$

We note that $\mathbf{A} + \mathbf{B} = \mathbf{B} + \mathbf{A}$.

Example 1.12

$$\begin{Vmatrix} 3 & 2 & -1 \\ 0 & 4 & -2 \end{Vmatrix} + \begin{Vmatrix} 5 & 0 & 3 \\ 1 & 0 & -1 \end{Vmatrix} = \begin{Vmatrix} 8 & 2 & 2 \\ 1 & 4 & -3 \end{Vmatrix}$$

$$\begin{Vmatrix} 3 & -2 \\ 0 & 5 \\ -1 & 4 \end{Vmatrix} + \begin{Vmatrix} 0 & 0 \\ 0 & 0 \\ 0 & 0 \end{Vmatrix} = \begin{Vmatrix} 3 & -2 \\ 0 & 5 \\ -1 & 4 \end{Vmatrix}$$

We call **A** a zero matrix if and only if each element of **A** is equal to the real number zero.

The product of a matrix **A** by a number k (real or complex) is defined as the matrix whose elements are each k times those of **A**, that is,

$$k\mathbf{A} = k\|a^i_j\| = \|ka^i_j\|$$

Example 1.13

$$4\begin{Vmatrix} 3 & 0 & -2 \\ 2 & -4 & 3 \end{Vmatrix} = \begin{Vmatrix} 12 & 0 & -8 \\ 8 & -16 & 12 \end{Vmatrix}$$

$$\begin{Vmatrix} 3 & -2 & 4 \\ -1 & 2 & 5 \end{Vmatrix} - \begin{Vmatrix} 4 & -1 & -2 \\ 0 & 1 & 3 \end{Vmatrix} = \begin{Vmatrix} -1 & -1 & 6 \\ -1 & 1 & 2 \end{Vmatrix}$$

Every matrix **A** can be associated with a negative matrix $\mathbf{B} = -\mathbf{A}$ such that $\mathbf{A} + (-\mathbf{A}) = (-\mathbf{A}) + \mathbf{A} = \mathbf{0}$ (zero matrix).

The rule for multiplying a matrix **A** by a scalar k should not be confused with the rule for multiplying a determinant by k, for in this latter case the elements of only one row or only one column are multiplied by k.

Before defining the product of two matrices let us consider the following sets of linear transformations:

$$\begin{aligned} &A: \quad z^i = a^i_j y^j \qquad i = 1, 2, \ldots, m;\ j = 1, 2, \ldots, n \\ &B: \quad y^j = b^j_k x^k \qquad k = 1, 2, \ldots, p \end{aligned}$$

Since the z's depend on the y's, which in turn depend on the x's, we can solve for the z's in terms of the x's. We write this transformation as follows:

$$AB: \quad z^i = a^i_j b^j_k x^k = c^i_k x^k \qquad c^i_k = a^i_j b^j_k$$

This suggests a method for defining multiplication of the matrices **A**, **B**.

If $\mathbf{A} = \|a^i_j\|$, $i = 1, 2, \ldots, m$, $j = 1, 2, \ldots, n$, $\mathbf{B} = \|b^i_j\|$, $i = 1, 2, \ldots, n$, $j = 1, 2, \ldots, p$, then $\mathbf{AB}$ is defined as the matrix $\mathbf{C}$ such that

$$\mathbf{C} = \mathbf{AB} = \|a^i_j\| \cdot \|b^i_j\| = \|a^i_\alpha b^\alpha_j\| = \|c^i_j\| \tag{1.33}$$

Let us note that the number of columns of the matrix $\mathbf{A}$ must equal the number of rows of $\mathbf{B}$. The matrix $\mathbf{C}$ of (1.33) is an $m \times p$ matrix. In the case of square matrices the definition for multiplication of matrices corresponds to that for multiplication of determinants [see (1.27)]. This implies that $|\mathbf{C}| = |\mathbf{A}| \cdot |\mathbf{B}|$, where $|\mathbf{C}|$ denotes the determinant of the set of elements comprising the square matrix $\mathbf{C} = \mathbf{AB}$.

Example 1.14

$$\begin{Vmatrix} 2 & -1 & 0 \\ -2 & 3 & -1 \\ 4 & -2 & 3 \end{Vmatrix} \cdot \begin{Vmatrix} 3 & 2 \\ 0 & -1 \\ 1 & 0 \end{Vmatrix}$$

$$= \begin{Vmatrix} 2(3) + (-1)0 + 0(1) & 2(2) + (-1)(-1) + 0(0) \\ -2(3) + 3(0) + (-1)(1) & -2(2) + 3(-1) + (-1)0 \\ 4(3) + (-2)0 + 3(1) & 4(2) + (-2)(-1) + 3(0) \end{Vmatrix} = \begin{Vmatrix} 6 & 5 \\ -7 & -7 \\ 15 & 10 \end{Vmatrix}$$

Example 1.15. The product of two matrices $\mathbf{A}$, $\mathbf{B}$ can yield the zero matrix with neither $\mathbf{A}$ nor $\mathbf{B}$ the zero matrix.

$$\begin{Vmatrix} 0 & 1 \\ 0 & 0 \end{Vmatrix} \cdot \begin{Vmatrix} 1 & 0 \\ 0 & 0 \end{Vmatrix} = \begin{Vmatrix} 0 & 0 \\ 0 & 0 \end{Vmatrix}$$

Example 1.16. The commutative law does not hold, in general, for multiplication of matrices.

$$\mathbf{A} \cdot \mathbf{B} = \begin{Vmatrix} 1 & 1 \\ 0 & 0 \end{Vmatrix} \cdot \begin{Vmatrix} 1 & -1 \\ 1 & 0 \end{Vmatrix} = \begin{Vmatrix} 2 & -1 \\ 0 & 0 \end{Vmatrix}$$

$$\mathbf{B} \cdot \mathbf{A} = \begin{Vmatrix} 1 & -1 \\ 1 & 0 \end{Vmatrix} \cdot \begin{Vmatrix} 1 & 1 \\ 0 & 0 \end{Vmatrix} = \begin{Vmatrix} 1 & 1 \\ 1 & 1 \end{Vmatrix}$$

Example 1.17. If $\mathbf{A}$ is a matrix whose elements are functions of a variable x, we define the derivative of $\mathbf{A}$ with respect to x, written $\dfrac{d\mathbf{A}}{dx}$, as follows:

$$\frac{d\mathbf{A}}{dx} = \left\| \frac{da^i_j}{dx} \right\| \tag{1.34}$$

Thus if $\mathbf{A} = \begin{Vmatrix} x^2 \\ \sin x \\ \ln x \end{Vmatrix}$, then $\dfrac{d\mathbf{A}}{dx} = \begin{Vmatrix} 2x \\ \cos x \\ \dfrac{1}{x} \end{Vmatrix}$.

This definition follows logically from the following considerations: If $\mathbf{A}(x) = \begin{Vmatrix} f(x) \\ g(x) \end{Vmatrix}$, then $\mathbf{A}(x + \Delta x) = \begin{Vmatrix} f(x + \Delta x) \\ g(x + \Delta x) \end{Vmatrix}$, and

$$\frac{\mathbf{A}(x+\Delta x)-\mathbf{A}(x)}{\Delta x}=\begin{Vmatrix}\dfrac{f(x+\Delta x)-f(x)}{\Delta x}\\ \dfrac{g(x+\Delta x)-g(x)}{\Delta x}\end{Vmatrix}\qquad \Delta x\neq 0$$

so that

$$\frac{d\mathbf{A}}{dx}\overset{\text{DEF}}{\equiv}\begin{Vmatrix}\lim\limits_{\Delta x\to 0}\dfrac{f(x+\Delta x)-f(x)}{\Delta x}\\ \lim\limits_{\Delta x\to 0}\dfrac{g(x+\Delta x)-g(x)}{\Delta x}\end{Vmatrix}=\begin{Vmatrix}f'(x)\\ g'(x)\end{Vmatrix}$$

Example 1.18. The transpose of a matrix $\mathbf{A}$ is the matrix, written $\mathbf{A}^T$, obtained from $\mathbf{A}$ by interchanging the rows and columns of $\mathbf{A}$. If $\mathbf{A}=\begin{Vmatrix}1 & 2 & -3\\ 4 & 1 & 5\end{Vmatrix}$, then $\mathbf{A}^T=\begin{Vmatrix}1 & 4\\ 2 & 1\\ -3 & 5\end{Vmatrix}$.

A square matrix $\mathbf{A}$ is said to be a symmetric matrix if and only if $\mathbf{A}=\mathbf{A}^T$. If $\mathbf{A}=-\mathbf{A}^T$, we say that $\mathbf{A}$ is a skew-symmetric matrix. We now exhibit a symmetric matrix $\mathbf{A}$ and a skew-symmetric matrix $\mathbf{B}$.

$$\mathbf{A}=\mathbf{A}^T=\begin{Vmatrix}2 & -1 & 4 & -2\\ -1 & 0 & 3 & 5\\ 4 & 3 & 1 & -1\\ -2 & 5 & -1 & 3\end{Vmatrix}\qquad \mathbf{B}=-\mathbf{B}^T=\begin{Vmatrix}0 & -1 & 3\\ 1 & 0 & -2\\ -3 & 2 & 0\end{Vmatrix}$$

We let the reader verify that $\frac{1}{2}(\mathbf{A}+\mathbf{A}^T)$ is a symmetric matrix if $\mathbf{A}$ is a square matrix. Let the reader first prove that $(\mathbf{A}^T)^T=\mathbf{A}$,

$$(\mathbf{A}+\mathbf{B})^T=\mathbf{A}^T+\mathbf{B}^T$$

It is easily seen that $\frac{1}{2}(\mathbf{A}-\mathbf{A}^T)$ is a skew-symmetric matrix. Any square matrix $\mathbf{A}$ can obviously be written as $\mathbf{A}=\frac{1}{2}(\mathbf{A}+\mathbf{A}^T)+\frac{1}{2}(\mathbf{A}-\mathbf{A}^T)$. Hence every square matrix can be written as the sum of a symmetric and a skew-symmetric matrix.

Example 1.19. We now prove that $(\mathbf{AB})^T=\mathbf{B}^T\mathbf{A}^T$. Let $\mathbf{A}=\|a_j^i\|$, $\mathbf{B}=\|b_j^i\|$, $\mathbf{A}^T=\|c_j^i=a_i^j\|$, $\mathbf{B}^T=\|d_j^i=b_i^j\|$. Then $\mathbf{B}^T\mathbf{A}^T=\|d_\alpha^i c_j^\alpha\|=\|b_i^\alpha a_\alpha^j\|$, $i=$ row, $j=$ column, and $\mathbf{AB}=\|a_\alpha^i b_j^\alpha\|$ so that $(\mathbf{AB})^T=\|a_\alpha^j b_i^\alpha\|=\mathbf{B}^T\mathbf{A}^T$. Q.E.D.

Let the reader now show that $(\mathbf{ABC})^T=\mathbf{C}^T\mathbf{B}^T\mathbf{A}^T$.

A square matrix with 1's down the principal diagonal and zeros elsewhere is called the identity matrix, written $\mathbf{E}=\|\delta_j^i\|$ [see (1.28)]. We easily verify that

$$\mathbf{AE}=\|a_j^i\|\cdot\|\delta_j^i\|=\|a_\alpha^i\delta_j^\alpha\|=\|a_j^i\|=\mathbf{A}=\mathbf{EA}$$

Problems

1. Show that $\mathbf{A}+\mathbf{B}=\mathbf{B}+\mathbf{A}$. Commutative law of addition.
2. Show that $\mathbf{A}+(\mathbf{B}+\mathbf{C})=(\mathbf{A}+\mathbf{B})+\mathbf{C}$. Associative law of addition.

3. Show that $\mathbf{A}(\mathbf{B} + \mathbf{C}) = \mathbf{AB} + \mathbf{AC}$, $(\mathbf{A} + \mathbf{B})\mathbf{C} = \mathbf{AC} + \mathbf{BC}$. Distributive law of multiplication with respect to addition.

4. Show that $(\mathbf{AB})\mathbf{C} = \mathbf{A}(\mathbf{BC})$. Associative law of multiplication.

5. From the distributive law show that $\mathbf{OA} = \mathbf{O}$ if $\mathbf{O} + \mathbf{B} = \mathbf{B} + \mathbf{O} = \mathbf{B}$ for all $\mathbf{B}$.

6. Show that the system of equations $y^i = a^i_\alpha x^\alpha$, $i = 1, 2, \ldots, m$, $\alpha = 1, 2, \ldots, n$, may be written in matrix form as $\mathbf{Y} = \mathbf{AX}$,

where
$$\mathbf{Y} = \begin{Vmatrix} y^1 \\ y^2 \\ \cdot \\ \cdot \\ \cdot \\ y^n \end{Vmatrix} \qquad \mathbf{X} = \begin{Vmatrix} x^1 \\ x^2 \\ \cdot \\ \cdot \\ \cdot \\ x^n \end{Vmatrix} \qquad \mathbf{A} = \begin{Vmatrix} a^1_1 & a^1_2 & \cdots & a^1_n \\ a^2_1 & a^2_2 & \cdots & a^2_n \\ \cdots & \cdots & \cdots & \cdots \\ a^m_1 & a^m_2 & \cdots & a^m_n \end{Vmatrix}$$

7. Write the system of differential equations $\dfrac{dx^i}{dt} = a^i_j x^j$, $i, j = 1, 2, \ldots, n$, in matrix form.

8. Show that

$$\begin{Vmatrix} 0 & 1 & 0 \\ 1 & 0 & 0 \\ 0 & 0 & 1 \end{Vmatrix} \cdot \begin{Vmatrix} a^1_1 & a^1_2 & a^1_3 \\ a^2_1 & a^2_2 & a^2_3 \\ a^3_1 & a^3_2 & a^3_3 \end{Vmatrix} = \begin{Vmatrix} a^2_1 & a^2_2 & a^2_3 \\ a^1_1 & a^1_2 & a^1_3 \\ a^3_1 & a^3_2 & a^3_3 \end{Vmatrix}$$

Find the square matrix $\mathbf{E}_{rs}$ such that $\mathbf{E}_{rs} \cdot \mathbf{A}$ interchanges the rth and sth rows of the square matrix $\mathbf{A}$.

9. If $\mathbf{A} = \left\Vert \dfrac{\partial y^i}{\partial x^j} \right\Vert$, $i, j = 1, 2, \ldots, n$, $\mathbf{B} = \left\Vert \dfrac{\partial x^i}{\partial y^j} \right\Vert$, show that $\mathbf{A} \cdot \mathbf{B} = \mathbf{E}$.

10. If $\mathbf{Z} = \mathbf{AY}$, $\mathbf{Y} = \mathbf{BX}$, show that $\mathbf{Z} = (\mathbf{AB})\mathbf{X}$.

11. If $\mathbf{X} = \begin{Vmatrix} x^1 \\ x^2 \\ x^3 \end{Vmatrix}$, show that $\mathbf{X}^T\mathbf{X} = \Vert (x^1)^2 + (x^2)^2 + (x^3)^2 \Vert$.

12. From the associative law show that if $\mathbf{AB} = \mathbf{E}$, then $\mathbf{BA} = \mathbf{E}$. *Hint:*

$$(\mathbf{BA})\mathbf{B} = \mathbf{B}(\mathbf{AB})$$

and assume $\mathbf{XB} = \mathbf{B}$ implies $\mathbf{X} = \mathbf{E}$.

13. If $\dfrac{d^2\mathbf{X}}{dt^2} = \mathbf{AX}$, $\mathbf{X} = \mathbf{BY}$, $\mathbf{B} = \Vert b^i_j = \text{constants}\Vert$, show that $\mathbf{B}\dfrac{d^2\mathbf{Y}}{dt^2} = \mathbf{ABY}$. If $\mathbf{B}^{-1}$ is a matrix (inverse of $\mathbf{B}$ such that $\mathbf{B}^{-1}\mathbf{B} = \mathbf{E}$), show that $\dfrac{d^2\mathbf{Y}}{dt^2} = (\mathbf{B}^{-1}\mathbf{AB})\mathbf{Y}$.

If $\mathbf{Y} = \begin{Vmatrix} y^1 \\ y^2 \\ \cdot \\ \cdot \\ \cdot \\ y^n \end{Vmatrix}$, $\mathbf{B}^{-1}\mathbf{AB} = \begin{Vmatrix} \lambda_1 & & & \\ & \lambda_2 & \text{zeros} & \\ & & \ddots & \\ \text{zeros} & & & \lambda_n \end{Vmatrix}$, show that $\dfrac{d^2y^i}{dt^2} = \lambda_i y^i$ (no sum on i), for $i = 1, 2, \ldots, n$.

14. Show that $\mathbf{A} = \mathbf{B}^T\mathbf{B}$ is a symmetric matrix. Then show that $\mathbf{B}^{-1} = \mathbf{A}^{-1}\mathbf{B}^T$ if $|\mathbf{B}| \neq 0$ (see Prob. 13 for the definition of $\mathbf{B}^{-1}$).

15. The *trace*, or *spur*, of a square matrix is defined as the sum of the terms along the principal diagonal, that is, trace $\mathbf{A} = a^\alpha_\alpha$. Show that trace $(\mathbf{AB})$ = trace $(\mathbf{BA})$. Show that trace $(\mathbf{TAS})$ = trace $\mathbf{A}$ if $\mathbf{ST} = \mathbf{E}$. Show that in general (trace $\mathbf{A}$)(trace $\mathbf{B}$) $\neq$ trace $(\mathbf{AB})$.

The Inverse of a Matrix. A square matrix $\mathbf{A}$ is said to have an inverse matrix, written $\mathbf{A}^{-1}$, if $\mathbf{A}\mathbf{A}^{-1} = \mathbf{A}^{-1}\mathbf{A} = \mathbf{E}$. Of necessity

$$|\mathbf{A}| \cdot |\mathbf{A}^{-1}| = |\mathbf{E}| = 1$$

so that $|\mathbf{A}| \neq 0$. Now let $\mathbf{A}$ be a nonsingular square matrix, $|\mathbf{A}| \neq 0$. From (1.30) we have

$$a_\alpha^i \frac{A_j^\alpha}{|a_\nu^\mu|} = \frac{A_\alpha^i}{|a_\nu^\mu|} a_j^\alpha = \delta_j^i \tag{1.35}$$

so that the matrix $\mathbf{A}^{-1}$, with elements $\dfrac{A_j^i}{|a_\nu^\mu|}$, has the property that

$$\mathbf{A}\mathbf{A}^{-1} = \mathbf{A}^{-1}\mathbf{A} = \|\delta_j^i\| = \mathbf{E}$$

[see (1.33) for the definition of multiplication of two matrices]. Recall that A_j^i is the cofactor of a_i^j.

The inverse of a matrix is unique, for if $\mathbf{B}\mathbf{A} = \mathbf{E}$, then

$$\mathbf{B}\mathbf{A}\mathbf{A}^{-1} = \mathbf{E}\mathbf{A}^{-1} = \mathbf{A}^{-1}$$

which implies $\mathbf{B}\mathbf{E} = \mathbf{B} = \mathbf{A}^{-1}$.

Example 1.20. Let $\mathbf{A} = \begin{Vmatrix} 1 & 0 & 1 \\ -1 & 0 & 2 \\ 1 & 1 & 0 \end{Vmatrix}$. $|\mathbf{A}| = -3$, $A_1^1 = \begin{vmatrix} 0 & 2 \\ 1 & 0 \end{vmatrix} = -2$,

$$A_2^1 = -\begin{vmatrix} 0 & 1 \\ 1 & 0 \end{vmatrix} = 1, \qquad A_3^1 = \begin{vmatrix} 0 & 1 \\ 0 & 2 \end{vmatrix} = 0, \qquad A_1^2 = -\begin{vmatrix} -1 & 2 \\ 1 & 0 \end{vmatrix} = 2$$

$$A_2^2 = \begin{vmatrix} 1 & 1 \\ 1 & 0 \end{vmatrix} = -1, \qquad A_3^2 = -\begin{vmatrix} 1 & 1 \\ -1 & 2 \end{vmatrix} = -3, \qquad A_1^3 = \begin{vmatrix} -1 & 0 \\ 1 & 1 \end{vmatrix} = -1$$

$$A_2^3 = -\begin{vmatrix} 1 & 0 \\ 1 & 1 \end{vmatrix} = -1, \qquad A_3^3 = \begin{vmatrix} 1 & 0 \\ -1 & 0 \end{vmatrix} = 0$$

$$\mathbf{A}^{-1} = \begin{Vmatrix} \frac{2}{3} & -\frac{1}{3} & 0 \\ -\frac{2}{3} & \frac{1}{3} & 1 \\ \frac{1}{3} & \frac{1}{3} & 0 \end{Vmatrix}$$

$$\mathbf{A}\mathbf{A}^{-1} = \mathbf{A}^{-1}\mathbf{A} = \mathbf{E}$$

Example 1.21. Let us consider the set of all nonsingular matrices ($|\mathbf{A}| \neq 0$) of order 3. This set S satisfies the following properties relative to the operation of multiplication:

M_1: $\mathbf{A}\mathbf{B} = \mathbf{C}$, $|\mathbf{C}| \neq 0$, and of third order if $\mathbf{A}$ and $\mathbf{B}$ are nonsingular of order 3. This is the closure property.

M_2: $\mathbf{A}(\mathbf{B}\mathbf{C}) = (\mathbf{A}\mathbf{B})\mathbf{C}$. The associative law.

M_3: $\mathbf{A}\mathbf{E} = \mathbf{E}\mathbf{A} = \mathbf{A}$ for all $\mathbf{A}$, where $\mathbf{E} = \begin{Vmatrix} 1 & 0 & 0 \\ 0 & 1 & 0 \\ 0 & 0 & 1 \end{Vmatrix}$. We call $\mathbf{E}$ the unit element with respect to multiplication. $\mathbf{E}$ is unique.

M_4: For every matrix **A** of S there exists a unique matrix $\mathbf{A}^{-1}$ such that

$$\mathbf{AA}^{-1} = \mathbf{A}^{-1}\mathbf{A} = \mathbf{E}$$

Any set S of elements which obey properties M_1 to M_4 relative to some operation (call it multiplication if you prefer) is called a group. Since $\mathbf{AB} \neq \mathbf{BA}$ for some **A** and **B**, the above group is non-Abelian.

If we consider the second operation, in this case addition, we also have

A_1: $\mathbf{A} + \mathbf{B} = \mathbf{B} + \mathbf{A} = \mathbf{C}$

A_2: $\mathbf{A} + (\mathbf{B} + \mathbf{C}) = (\mathbf{A} + \mathbf{B}) + \mathbf{C}$

A_3: $\mathbf{A} + \mathbf{O} = \mathbf{O} + \mathbf{A} = \mathbf{A} \qquad \mathbf{O} = \begin{Vmatrix} 0 & 0 & 0 \\ 0 & 0 & 0 \\ 0 & 0 & 0 \end{Vmatrix}$

A_4: $\mathbf{A} + (-\mathbf{A}) = (-\mathbf{A}) + \mathbf{A} = \mathbf{O}$

D_1: $\mathbf{A}(\mathbf{B} + \mathbf{C}) = \mathbf{AB} + \mathbf{AC}$

D_2: $(\mathbf{B} + \mathbf{C})\mathbf{A} = \mathbf{BA} + \mathbf{CA}$

Let the reader prove that D_1, D_2, A_3 imply $\mathbf{A} \cdot \mathbf{O} = \mathbf{OA} = \mathbf{O}$.

Although division has not been defined and will not be defined, quite a few rules of the real-number system hold for matrices. If $\mathbf{AB} = \mathbf{AC}$, $|\mathbf{A}| \neq 0$, then

$$\mathbf{A}^{-1}\mathbf{AB} = \mathbf{A}^{-1}\mathbf{AC}$$

which implies $\mathbf{B} = \mathbf{C}$. This is the law of cancellation. If $\mathbf{Y} = \mathbf{AX}$, $|\mathbf{A}| \neq 0$ (see Prob. 6, Sec. 1.2), then $\mathbf{X} = \mathbf{A}^{-1}\mathbf{Y}$. Let the reader prove this useful fact.

From the equation

$$\mathbf{ABB}^{-1}\mathbf{A}^{-1} = \mathbf{E} = (\mathbf{AB})(\mathbf{AB})^{-1} \tag{1.36}$$

we have $(\mathbf{AB})^{-1} = \mathbf{B}^{-1}\mathbf{A}^{-1}$. The inverse of a product of matrices is the product of the inverses in reverse order. Let the reader show that $(\mathbf{A}^{-1})^T = (\mathbf{A}^T)^{-1}$.

Problems

1. From the equation $\mathbf{AA}^{-1} = \mathbf{E}$, show that $\mathbf{A} = \mathbf{A}^T$ implies $\mathbf{A}^{-1} = (\mathbf{A}^{-1})^T$.
2. Let **A** be a skew-symmetric matrix of odd order, $\mathbf{A} = -\mathbf{A}^T$. Show that $|\mathbf{A}| = 0$.
3. Show that $(\mathbf{A}^{-1})^{-1} = \mathbf{A}$.
4. Show that $(\mathbf{A}^{-1}\mathbf{BA})^n = \mathbf{A}^{-1}\mathbf{B}^n\mathbf{A}$.
5. Show that $\mathbf{A}^{-1}\mathbf{B}^{-1}\mathbf{A}$ is the inverse of $\mathbf{A}^{-1}\mathbf{BA}$.
6. Show that $(\mathbf{A}^{-1}\mathbf{BA})(\mathbf{A}^{-1}\mathbf{CA}) = \mathbf{A}^{-1}(\mathbf{BC})\mathbf{A}$.
7. Find the inverse matrix of the matrix

$$\begin{Vmatrix} 1 & 2 & 3 & 4 \\ -1 & 1 & 2 & 3 \\ -2 & 4 & 1 & 2 \\ 3 & 1 & -1 & 2 \end{Vmatrix}$$

1.4. Linear Equations. Linear Dependence. We first consider the single linear homogeneous equation

$$a_1x^1 + a_2x^2 + \cdots + a_nx^n = 0 \qquad n > 1;\ a_1 \neq 0 \tag{1.37}$$

An obvious solution of (1.37) is $x^1 = x^2 = \cdots = x^n = 0$, called the trivial solution. We may obtain nontrivial solutions by choosing any values we please for $x^2, x^3, \ldots, x^n$, which in turn uniquely determine x^1. Moreover, if $x^1 = c^1, x^2 = c^2, \ldots, x^n = c^n$ is a solution of (1.37), then $x^1 = \lambda c^1, x^2 = \lambda c^2, \ldots, x^n = \lambda c^n, -\infty < \lambda < +\infty$, is also a solution of (1.37). The equation $x - 2y = 0$ has the solution $x = 2\lambda$, $y = \lambda, -\infty < \lambda < +\infty$. The equation $2x - 3y + 4z = 0$ has the solution $x = u, y = v, z = \frac{1}{4}(3v - 2u), -\infty < u < +\infty, -\infty < v < +\infty$.

Let us now consider the following system of homogeneous equations:

$$\begin{aligned} x + 2y - 4z &= 0 \\ 2x - 3y + z &= 0 \end{aligned} \tag{1.38}$$

It is obvious that $x = y = z = 0$ is a trivial solution of (1.38). We look for nontrivial solutions. From (1.38) we obtain $7y - 9z = 0$ by eliminating x, so that $z = 7\lambda$, $y = 9\lambda$, and thus $x = 10\lambda$, $-\infty < \lambda < +\infty$, is a solution of (1.38).

In dealing with a system of linear homogeneous equations we can multiply any equation by a scalar without changing the system. Moreover, we may also multiply any equation by a scalar and add the resulting equation to any other equation without changing the solution of the system. In these operations only the coefficients of the unknowns play a role. Hence one needs only to manipulate the elements of the coefficient matrix. For (1.38) we have $\left\|\begin{matrix} 1 & 2 & -4 \\ 2 & -3 & 1 \end{matrix}\right\|$, which transforms into $\left\|\begin{matrix} 1 & 2 & -4 \\ 0 & -7 & 9 \end{matrix}\right\|$ by multiplying the first row by -2 and adding the corresponding elements to the second row. We then solve $-7y + 9z = 0$ for y, z and then solve for x in the equation $x + 2y - 4z = 0$.

Example 1.22. Let us look at a system of four linear homogeneous equations in five unknowns.

$$\begin{aligned} x - y + 2z + 3u + 3v &= 0 \\ 2x + y - z + u + 4v &= 0 \\ -3x + 2y + 4z + u &= 0 \\ 4y + 4z + 2v &= 0 \end{aligned} \tag{1.39}$$

We now triagonalize the elements of the coefficient matrix

$$\left\|\begin{matrix} 1 & -1 & 2 & 3 & 3 \\ 2 & 1 & -1 & 1 & 4 \\ -3 & 2 & 4 & 1 & 0 \\ 0 & 4 & 4 & 0 & 2 \end{matrix}\right\| \tag{1.40}$$

Multiplying row one by -2 and adding to row two, and multiplying row one by 3 and adding to row three, yields

$$\left\|\begin{matrix} 1 & -1 & 2 & 3 & 3 \\ 0 & 3 & -5 & -5 & -2 \\ 0 & -1 & 10 & 10 & 9 \\ 0 & 4 & 4 & 0 & 2 \end{matrix}\right\| \tag{1.41}$$

Continuing, (1.41) transforms into

$$\left\|\begin{matrix} 1 & -1 & 2 & 3 & 3 \\ 0 & -1 & 10 & 10 & 9 \\ 0 & 0 & 1 & 1 & 1 \\ 0 & 0 & 0 & -4 & 6 \end{matrix}\right\| \tag{1.42}$$

At one stage we interchanged rows two and three.

The equivalent system of equations is

$$\begin{aligned} -4u - 6v &= 0 \\ z + u + v &= 0 \\ -y + 10z + 10u + 9v &= 0 \\ x - y + 2z + 3u + 3v &= 0 \end{aligned} \tag{1.43}$$

Letting $v = -2$ yields in turn $u = 3$, $z = -1$, $y = 2$, $x = 1$. The most general solution of (1.39) is $x = \lambda$, $y = 2\lambda$, $z = -\lambda$, $u = 3\lambda$, $v = -2\lambda$, $-\infty < \lambda < +\infty$.

From the above consideration the reader can prove by mathematical induction or otherwise that a system of n linear homogeneous equations in m unknowns always possesses a nontrivial solution if $m > n$.

We now consider the case $m = n$. The system

$$a^i_j x^j = 0 \qquad i, j = 1, 2, \ldots, n \tag{1.44}$$

has the unique trivial solution $x^1 = x^2 = \cdots = x^n = 0$ if $|a^i_j| \neq 0$ from Example 1.10. The only possibility for the existence of a nontrivial solution occurs for the case $|a^i_j| = 0$. Triagonalizing the matrix $\|a^i_j\|$ yields a new equivalent triangular matrix $\|b^i_j\|$ such that $|b^i_j| = 0$. Let the reader explain this. We thus obtain

$$\begin{vmatrix} b^1_1 & b^1_2 & & \cdots & b^1_n \\ 0 & b^2_2 & & \cdots & b^2_n \\ 0 & 0 & b^3_3 & \cdots & b^3_n \\ \cdot & \cdot & \cdot & \cdots & \cdot \\ 0 & 0 & 0 & \cdots & b^n_n \end{vmatrix} = 0$$

which implies $b^i_i = 0$ (no summation) for at least one value of i. If $b^n_n = 0$, we have reduced our original system to one containing more unknowns than equations, for which a nontrivial solution exists. If $b^n_n \neq 0$, $b^{n-1}_{n-1} = 0$, then $x^n = 0$ and again we have more unknowns than equations so that a nontrivial solution again exists. Continuing, we see that the vanishing of at least one element along the main diagonal implies the existence of a nontrivial solution.

Example 1.23. The determinant of the coefficient matrix of the system

$$\begin{aligned} x + y + z + u &= 0 \\ 2x - y + z - u &= 0 \\ 5x - y + 3z &= 0 \\ -x + 5y + z + 2u &= 0 \end{aligned} \tag{1.45}$$

vanishes. Triagonalizing the coefficient matrix yields

$$\begin{Vmatrix} 1 & 1 & 1 & 1 \\ 0 & -3 & -1 & -3 \\ 0 & 0 & 0 & 7 \\ 0 & 0 & 0 & -3 \end{Vmatrix}$$

System (1.45) becomes $-3u = 7u = 0$ so that $u = 0$, and we have $x + y + z = 0$, $-3y - z = 0$, so that $z = -3\lambda$, $y = \lambda$, $x = 2\lambda$, $-\infty < \lambda < +\infty$.

On considering the system

$$\begin{aligned} x - y &= 0 \\ 2x + y &= 0 \\ x + y &= 0 \end{aligned} \tag{1.46}$$

we see immediately that only the trivial solution $x = y = 0$ exists.

Generally speaking, a system of m linear homogeneous equations in n unknowns, $m > n$, does not possess a nontrivial solution. The reader is referred to Ferrar's text on algebra for the complete discussion of this case. The *rank* of a matrix plays an important role in discussing solutions of linear equations. For a discussion of the rank of a matrix see the above-mentioned text.

Example 1.24. The column matrix $\mathbf{X} = \begin{Vmatrix} x^1 \\ x^2 \\ \cdot \\ \cdot \\ \cdot \\ x^n \end{Vmatrix}$ will be called a vector. The number x^j is called the jth component of $\mathbf{X}$. We call the number, n, the dimensionality of the space. The determinant of the matrix $\mathbf{X}^T\mathbf{X}$ is called the square of the magnitude of the vector $\mathbf{X}$. If the x^i, $i = 1, 2, \ldots, n$, are complex numbers, we define $|\bar{\mathbf{X}}^T\mathbf{X}|$ as the square of the magnitude of $\mathbf{X}$, where $\bar{\mathbf{X}}^T = \|\bar{x}^1\bar{x}^2 \cdots \bar{x}^n\|$, $\bar{x}^i$ = conjugate complex of x^i.

The system of vectors

$$\mathbf{X}_r = \begin{Vmatrix} x_r^1 \\ x_r^2 \\ \cdot \\ \cdot \\ \cdot \\ x_r^n \end{Vmatrix} \qquad r = 1, 2, \ldots, m \tag{1.47}$$

is said to be linearly dependent if there exist scalars $\lambda^1, \lambda^2, \ldots, \lambda^m$ not all zero such that

$$\lambda^\alpha \mathbf{X}_\alpha = \mathbf{0} \tag{1.48}$$

An equivalent definition of linear dependence is the following: The set of vectors (1.47) is a linearly independent system if Eq. (1.48) implies $\lambda^1 = \lambda^2 = \cdots = \lambda^m = 0$.

We now prove the following theorem: Any m vectors in an n-dimensional space are linearly dependent if $m > n$. The proof is as follows: The system of equations $\lambda^\alpha \mathbf{X}_\alpha = 0$ is equivalent to the system of n linear homogeneous equations in the m unknowns $\lambda^1, \lambda^2, \ldots, \lambda^m$

$$x^i_\alpha \lambda^\alpha = 0 \qquad i = 1, 2, \ldots, n; \alpha = 1, 2, \ldots, m$$

Such a system always has a nontrivial solution for $m > n$. Q.E.D.

Problems

1. Solve the system

$$\begin{aligned} x - 2y + 3z - 4u &= 0 \\ 2x + y - z + u &= 0 \\ 3x - y + 2z - u &= 0 \end{aligned}$$

2. Solve the system

$$\begin{aligned} x + 2y - z - 3u &= 0 \\ 2x - y + z + 4u &= 0 \end{aligned}$$

3. Solve the system

$$\begin{aligned} 2x + y - z + u &= 0 \\ x - y - z + u &= 0 \\ x - 4y - 2z + 2u &= 0 \\ 5x + y - 3z + 3u &= 0 \end{aligned}$$

4. Determine λ so that the following system will have nontrivial solutions, and solve:

$$\begin{aligned} \lambda x &= 4x + y \\ \lambda y &= -2x + y \end{aligned}$$

5. Show that the vectors $\mathbf{X}_1 = \left\|\begin{matrix}1\\1\\1\end{matrix}\right\|$, $\mathbf{X}_2 = \left\|\begin{matrix}0\\-1\\1\end{matrix}\right\|$, $\mathbf{X}_3 = \left\|\begin{matrix}1\\-1\\0\end{matrix}\right\|$ are linearly independent. Given $\mathbf{X} = \left\|\begin{matrix}2\\-1\\3\end{matrix}\right\|$, find scalars $\lambda_1, \lambda_2, \lambda_3$ such that $\mathbf{X} = \sum_{i=1}^{3} \lambda_i \mathbf{X}_i$.

1.5. Quadratic Forms. The square of the distance between a point $P(x, y, z)$ and the origin $O(0, 0, 0)$ in a Euclidean space is given by

$$L^2 = x^2 + y^2 + z^2 \tag{1.49}$$

using a Euclidean coordinate system. The generalization of (1.49) to an n-dimensional space yields

$$L^2 = \sum_{i=1}^{n} (x^i)^2 \tag{1.50}$$

The linear transformation $x^i = a^i_\alpha y^\alpha$, $i, \alpha = 1, 2, \ldots, n$, $|a^i_j| \neq 0$, yields, from (1.50),

$$L^2 = \sum_{i=1}^{n} a^i_\alpha a^i_\beta \, y^\alpha y^\beta \tag{1.51}$$

If we desire that the y's be the components of a Euclidean coordinate system, we need

$$L^2 = \sum_{i=1}^{n} (y^i)^2 \tag{1.52}$$

Comparing (1.51), (1.52) yields

$$\sum_{i=1}^{n} a^i_\alpha a^i_\beta = \delta_{\alpha\beta} = \left.\begin{matrix} 1 & \text{if } \alpha = \beta \\ 0 & \text{if } \alpha \neq \beta \end{matrix}\right\} \tag{1.53}$$

The system of Eqs. (1.53) in matrix form becomes

$$\mathbf{A}^T\mathbf{A} = \mathbf{A}\mathbf{A}^T = \mathbf{E} \qquad \mathbf{A} = \|a^i_j\| \tag{1.54}$$

Equation (1.54) implies in turn that $\mathbf{A}^T = \mathbf{A}^{-1}$. A matrix $\mathbf{A}$ satisfying (1.54) is called an *orthogonal* matrix. If $\mathbf{A}_1$, $\mathbf{A}_2$ are any two different column vectors of $\mathbf{A}$, then $\mathbf{A}_1^T \cdot \mathbf{A}_2 = \mathbf{0}$. We say that the vectors are perpendicular.

For complex components (1.50) becomes $L^2 = \sum_{i=1}^{n} \bar{x}^i x^i$, and for the y's to be the components of an orthogonal coordinate system we find that the matrix $\mathbf{A}$ must satisfy $\bar{\mathbf{A}}^T\mathbf{A} = \mathbf{E}$ or $\bar{\mathbf{A}}^T = \mathbf{A}^{-1}$, where $\bar{\mathbf{A}} = \|\bar{a}^i_j\|$, $\bar{a}^i_j$ the complex conjugate of a^i_j. If $\bar{\mathbf{A}}^T = \mathbf{A}^{-1}$, we say that $\mathbf{A}$ is a *unitary* matrix.

We now consider the quadratic form

$$Q = a_{\alpha\beta}x^\alpha x^\beta \qquad a_{\alpha\beta} \text{ real; } \alpha, \beta = 1, 2, \ldots, n \tag{1.55}$$

and ask whether it is possible to find an orthogonal transformation $\mathbf{X} = \mathbf{BY}$ such that (1.55) reduces to the canonical form

$$Q = \sum_{i=1}^{n} \lambda_i (y^i)^2 \tag{1.56}$$

Let the student note that (1.55) may be written in matrix form as

$$Q = \mathbf{X}^T\mathbf{A}\mathbf{X} \tag{1.57}$$

noting that $\mathbf{X}^T\mathbf{A}\mathbf{X}$ is a matrix of just one element and so is written as a scalar. Under the transformation $\mathbf{X} = \mathbf{BY}$, (1.57) becomes

$$\begin{aligned} Q &= (\mathbf{BY})^T\mathbf{A}(\mathbf{BY}) \\ &= \mathbf{Y}^T(\mathbf{B}^T\mathbf{A}\mathbf{B})\mathbf{Y} \end{aligned} \tag{1.58}$$

so that (1.58) will have the form of (1.56) if and only if

$$\mathbf{B}^T\mathbf{A}\mathbf{B} = \begin{Vmatrix} \lambda_1 & 0 & 0 & \cdots & 0 \\ 0 & \lambda_2 & 0 & \cdots & 0 \\ 0 & 0 & \lambda_3 & \cdots & 0 \\ \cdots & \cdots & \cdots & \cdots & \cdots \\ 0 & 0 & 0 & \cdots & \lambda_n \end{Vmatrix} \tag{1.59}$$

that is, $\mathbf{B}^T\mathbf{A}\mathbf{B}$ must be a diagonal matrix. Our problem has been reduced to that of finding a matrix $\mathbf{B}$ satisfying (1.59), and hence satisfying

$$\begin{aligned} \mathbf{B}^{-1}\mathbf{A}\mathbf{B} &= \|\lambda_i\delta_{ij}\| \\ \text{or} \qquad \mathbf{A}\mathbf{B} &= \mathbf{B}\|\lambda_i\delta_{ij}\| \end{aligned} \tag{1.60}$$

since $\mathbf{B}^T = \mathbf{B}^{-1}$ is required if $\mathbf{X} = \mathbf{B}\mathbf{Y}$ is to be an orthogonal transformation. We may consider $\mathbf{A}$ to be a symmetric matrix, $\mathbf{A} = \mathbf{A}^T$, since $Q \equiv \mathbf{X}^T[\frac{1}{2}(\mathbf{A} + \mathbf{A}^T)]\mathbf{X} + \mathbf{X}^T[\frac{1}{2}(\mathbf{A} - \mathbf{A}^T)]\mathbf{X}$, and $\mathbf{X}^T[\frac{1}{2}(\mathbf{A} - \mathbf{A}^T)]\mathbf{X} \equiv \mathbf{0}$ (see Prob. 9, Sec. 1.1), while $\frac{1}{2}(\mathbf{A} + \mathbf{A}^T)$ is a symmetric matrix.

We now attempt to find the square matrix $\mathbf{B}$ satisfying (1.60). If $\mathbf{B} = \|b_{ij}\|$, Eq. (1.60) becomes

$$\sum_{\alpha=1}^{n} a_{i\alpha}b_{\alpha j} = \sum_{\alpha=1}^{n} b_{i\alpha}\lambda_\alpha\delta_{\alpha j} = b_{ij}\lambda_j \qquad i, j = 1, 2, \ldots, n \tag{1.61}$$

For $j = 1$ we have

$$\begin{gathered} \sum_{\alpha=1}^{n} a_{i\alpha}b_{\alpha 1} = b_{i1}\lambda_1 \\ \text{or} \qquad \mathbf{A} \cdot \begin{Vmatrix} b_{11} \\ b_{21} \\ \cdot \\ \cdot \\ \cdot \\ b_{n1} \end{Vmatrix} = \lambda_1 \begin{Vmatrix} b_{11} \\ b_{21} \\ \cdot \\ \cdot \\ \cdot \\ b_{n1} \end{Vmatrix} \end{gathered} \tag{1.62}$$

If $\mathbf{B}_1$ is the column matrix (vector), $\begin{Vmatrix} b_{11} \\ b_{21} \\ \cdot \\ \cdot \\ \cdot \\ b_{n1} \end{Vmatrix}$, (1.62) may be written

$$\begin{gathered} \mathbf{A}\mathbf{B}_1 = \lambda\mathbf{B}_1 \qquad \lambda = \lambda_1 \\ \text{or} \qquad (\mathbf{A} - \lambda\mathbf{E})\mathbf{B}_1 = \mathbf{0} \end{gathered} \tag{1.63}$$

Equation (1.63) represents a system of n linear homogeneous equations in the n unknowns comprising the column matrix $\mathbf{B}_1$. From Sec. 1.4 a necessary and sufficient condition that nontrivial solutions exist is that

$$|\mathbf{A} - \lambda\mathbf{E}| = 0$$

or

$$\begin{vmatrix} a_{11} - \lambda & a_{12} & \cdots & a_{1n} \\ a_{21} & a_{22} - \lambda & \cdots & a_{2n} \\ \cdots & \cdots & \cdots & \cdots \\ a_{n1} & a_{n2} & \cdots & a_{nn} - \lambda \end{vmatrix} = 0 \tag{1.64}$$

We call (1.64) the *characteristic* equation of the matrix $\mathbf{A}$. It is a polynomial equation in λ of degree n, so has n roots $\lambda_1, \lambda_2, \ldots, \lambda_n$, the λ_i real or complex. The roots $\lambda_1, \lambda_2, \ldots, \lambda_n$ determine the column vectors $\mathbf{B}_1, \mathbf{B}_2, \ldots, \mathbf{B}_n$, which in turn comprise the matrix $\mathbf{B}$, that is,

$$\mathbf{B} = \|\mathbf{B}_1 \quad \mathbf{B}_2 \quad \cdots \quad \mathbf{B}_n\| \qquad \text{a square matrix}$$

The solution $\mathbf{B}_1$ of $\mathbf{AB}_1 = \lambda_1\mathbf{B}_1$ is called an *eigenfunction, eigenvector,* or *characteristic vector.* λ_1 is called the *eigenvalue,* or *latent root,* or *characteristic root* corresponding to the eigenfunction $\mathbf{B}_1$. If $\mathbf{B}_1$ is a solution of (1.63), so is $\mathbf{B}_1$/length of $\mathbf{B}_1$, a vector of unit length. If the $\mathbf{B}_i$, $i = 1, 2, \ldots, n$, are unit vectors, it is easy to prove that the matrix $\mathbf{B}$ is an orthogonal matrix. Let

$$\begin{aligned} \mathbf{AB}_1 &= \lambda_1\mathbf{B}_1 \\ \mathbf{AB}_2 &= \lambda_2\mathbf{B}_2 \end{aligned} \qquad \lambda_1 \neq \lambda_2 \tag{1.65}$$

Then $(\mathbf{AB}_1)^T \equiv \mathbf{B}_1^T\mathbf{A}^T = \mathbf{B}_1^T\mathbf{A} = \lambda_1\mathbf{B}_1^T$ so that $\mathbf{B}_1^T\mathbf{AB}_2 = \lambda_1\mathbf{B}_1^T\mathbf{B}_2$. Moreover $\mathbf{B}_1^T\mathbf{AB}_2 = \lambda_2\mathbf{B}_1^T\mathbf{B}_2$, so that $(\lambda_1 - \lambda_2)\mathbf{B}_1^T\mathbf{B}_2 = 0$ and $\mathbf{B}_1^T\mathbf{B}_2 = 0$. If the λ_i, $i = 1, 2, \ldots, n$, are all different, the matrix $\mathbf{B}$ is an orthogonal matrix.

Example 1.25. We now find the linear orthogonal transformation which transforms the quadratic form $Q = 7x^2 + 7y^2 + 7z^2 + 6xy + 8yz$ into canonical form. We have

$$Q = \|x \quad y \quad z\| \begin{Vmatrix} 7 & 3 & 0 \\ 3 & 7 & 4 \\ 0 & 4 & 7 \end{Vmatrix} \begin{Vmatrix} x \\ y \\ z \end{Vmatrix}$$

The characteristic equation is

$$\begin{vmatrix} 7 - \lambda & 3 & 0 \\ 3 & 7 - \lambda & 4 \\ 0 & 4 & 7 - \lambda \end{vmatrix} = 0$$

which reduces to $(7 - \lambda)(\lambda - 12)(\lambda - 2) = 0$ so that $\lambda_1 = 7$, $\lambda_2 = 12$, $\lambda_3 = 2$. For $\lambda_1 = 7$ Eq. (1.63) becomes $3b_2 = 0$, $3b_1 + 4b_3 = 0$, $4b_2 = 0$. A unit vector $\mathbf{B}_1$ whose components b_1, b_2, b_3 satisfy these equations is $\mathbf{B}_1 = \begin{Vmatrix} \frac{4}{5} \\ 0 \\ -\frac{3}{5} \end{Vmatrix}$. For $\lambda_2 = 12$ Eq. (1.63) becomes $-5b_1 + 3b_2 = 0, 3b_1 - 5b_2 + 4b_3 = 0, 4b_2 - 5b_3 = 0$, so that $\mathbf{B}_2 = \begin{Vmatrix} \frac{3}{5\sqrt{2}} \\ \frac{1}{\sqrt{2}} \\ \frac{4}{5\sqrt{2}} \end{Vmatrix}$.

For $\lambda_3 = 2$ we obtain $\mathbf{B}_3 = \begin{Vmatrix} -\frac{3}{5\sqrt{2}} \\ \frac{1}{\sqrt{2}} \\ -\frac{4}{5\sqrt{2}} \end{Vmatrix}$. Thus

$$\mathbf{B} = \begin{Vmatrix} \frac{4}{5} & \frac{3}{5\sqrt{2}} & -\frac{3}{5\sqrt{2}} \\ 0 & \frac{1}{\sqrt{2}} & \frac{1}{\sqrt{2}} \\ -\frac{3}{5} & \frac{4}{5\sqrt{2}} & -\frac{4}{5\sqrt{2}} \end{Vmatrix} \quad \text{and} \quad \begin{Vmatrix} x \\ y \\ z \end{Vmatrix} = \begin{Vmatrix} \frac{4}{5} & \frac{3}{5\sqrt{2}} & -\frac{3}{5\sqrt{2}} \\ 0 & \frac{1}{\sqrt{2}} & \frac{1}{\sqrt{2}} \\ -\frac{3}{5} & \frac{4}{5\sqrt{2}} & -\frac{4}{5\sqrt{2}} \end{Vmatrix} \begin{Vmatrix} u \\ v \\ w \end{Vmatrix}$$

so that

$$\begin{aligned} x &= \frac{1}{5\sqrt{2}}(4\sqrt{2}\,u + 3v - 3w) \\ y &= \frac{1}{5\sqrt{2}}(5v + 5w) \\ z &= \frac{1}{5\sqrt{2}}(-3\sqrt{2}\,u + 4v - 4w) \end{aligned} \tag{1.66}$$

The quadratic form $Q = 7x^2 + 7y^2 + 7z^2 + 6xy + 8yz$ becomes

$$Q = 7u^2 + 12v^2 + 2w^2$$

under the linear orthogonal transformation (1.66).

Each root λ_i, $i = 1, 2, \ldots, n$, of $|\mathbf{A} - \lambda\mathbf{E}| = 0$ determined a column vector of the orthogonal matrix $\mathbf{B}$. If multiple roots of $|\mathbf{A} - \lambda\mathbf{E}| = 0$ occur, it appears at first glance that we cannot complete the full matrix $\mathbf{B}$. However, we can show that an orthogonal matrix $\mathbf{F}$ exists such that the quadratic form $Q = \mathbf{X}^T\mathbf{A}\mathbf{X}$, $\mathbf{A} = \mathbf{A}^T$ is canonical in form for the transformation $\mathbf{X} = \mathbf{F}\mathbf{W}$.

First let us note the following pertinent facts: If $\mathbf{X} = \mathbf{B}\mathbf{Y}$, $\mathbf{Y} = \mathbf{C}\mathbf{Z}$ are orthogonal transformations, then $\mathbf{X} = (\mathbf{B}\mathbf{C})\mathbf{Z}$ is an orthogonal transformation. We have $(\mathbf{B}\mathbf{C})^T = \mathbf{C}^T\mathbf{B}^T = \mathbf{C}^{-1}\mathbf{B}^{-1} = (\mathbf{B}\mathbf{C})^{-1}$, so that $\mathbf{B}\mathbf{C}$ is an orthogonal matrix. In other words, the matrix product of orthogonal matrices is an orthogonal matrix. Next we note that, if $\mathbf{B}$ is an orthogonal matrix in a k-dimensional space, $k < n$, then

$$\mathbf{C} = \left\| \begin{array}{ccccccc} 1 & 0 & 0 & \cdots & & & 0 \\ 0 & 1 & 0 & \cdots & & & 0 \\ \cdot & \cdot & \cdot & \cdot & \cdot & \cdot & \cdot \\ \cdot & \cdot & \cdot & \cdot & \cdot & \cdot & \cdot \\ \cdot & \cdot & \cdot & \cdot & \cdot & \cdot & \cdot \\ 0 & \cdots & 1 & 0 & & \cdots & 0 \\ 0 & \cdots & 0 & & & & \\ \cdot & \cdots & & & & & \\ \cdot & \cdots & & & \mathbf{B} & & \\ \cdot & \cdots & & & & & \\ 0 & 0 \cdots & 0 & & & & \end{array} \right\|$$

is an orthogonal matrix for the n-dimensional space. The reader can easily verify that **C** has the necessary properties for an orthogonal matrix. Finally the roots of $|\mathbf{A} - \lambda\mathbf{E}| = 0$ are the roots of $|\mathbf{B}^{-1}\mathbf{A}\mathbf{B} - \lambda\mathbf{E}| = 0$, and conversely. From $\mathbf{B}^{-1}\mathbf{A}\mathbf{B} - \lambda\mathbf{E} \equiv \mathbf{B}^{-1}(\mathbf{A} - \lambda\mathbf{E})\mathbf{B}$ we have

$$\begin{aligned} |\mathbf{B}^{-1}\mathbf{A}\mathbf{B} - \lambda\mathbf{E}| &= |\mathbf{B}^{-1}(\mathbf{A} - \lambda\mathbf{E})\mathbf{B}| \\ &= |\mathbf{B}^{-1}|\,|\mathbf{A} - \lambda\mathbf{E}|\,|\mathbf{B}| \\ &= |\mathbf{B}^{-1}|\,|\mathbf{B}|\,|\mathbf{A} - \lambda\mathbf{E}| \\ &= |\mathbf{B}^{-1}\mathbf{B}|\,|\mathbf{A} - \lambda\mathbf{E}| \\ &= |\mathbf{A} - \lambda\mathbf{E}| \end{aligned}$$

which proves our statement.

Now let λ_1 be any root of $|\mathbf{A} - \lambda\mathbf{E}| = 0$. It is immaterial whether λ_1 is a multiple root. We solve $(\mathbf{A} - \lambda_1\mathbf{E})\mathbf{B}_1 = 0$ for $\mathbf{B}_1$, $\mathbf{B}_1$ a unit vector. We now obtain an orthogonal matrix **B** with $\mathbf{B}_1$ as its first column. This can be done as follows (the method does not yield a unique answer): Let $\mathbf{B}_2 = \|b_{i2}\|$ be the elements of the second column of **B**. In order that $\mathbf{B}_2$ be orthogonal to $\mathbf{B}_1$, we need $\sum_{i=1}^{n} b_{i1}b_{i2} = 0$. This is a single homogeneous equation in the unknowns b_{i2}, $i = 1, 2, \ldots, n$. We know that we can find a nontrivial solution which can be normalized so that $\sum_{i=1}^{n} b_{i2}^2 = 1$. To obtain the third column, we need $\sum_{i=1}^{n} b_{i1}b_{i3} = 0$, $\sum_{i=1}^{n} b_{i2}b_{i3} = 0$. A normalized nontrivial solution exists for $n > 2$. By continuing this process we construct an orthogonal matrix **B**. The final column of **B** involves a system of $n - 1$ equations in n unknowns for which a nontrivial solution exists. From the construction of **B** it follows that $\mathbf{B}^{-1}\mathbf{A}\mathbf{B}$ has λ_1 as the element of the first row and first column and has zeros elsewhere in the first column. From

$$(\mathbf{B}^{-1}\mathbf{A}\mathbf{B})^T = \mathbf{B}^T\mathbf{A}^T(\mathbf{B}^{-1})^T = \mathbf{B}^{-1}\mathbf{A}(\mathbf{B}^T)^{-1} = \mathbf{B}^{-1}\mathbf{A}(\mathbf{B}^{-1})^{-1} = \mathbf{B}^{-1}\mathbf{A}\mathbf{B}$$

we note that $\mathbf{B}^{-1}\mathbf{A}\mathbf{B}$ is a symmetric matrix. We have used the facts that $\mathbf{A} = \mathbf{A}^T$, $\mathbf{B}^T = \mathbf{B}^{-1}$. Hence under the orthogonal transformation $\mathbf{X} = \mathbf{B}\mathbf{Y}$, Q becomes

$$Q = \mathbf{X}^T\mathbf{A}\mathbf{X} = \mathbf{Y}^T(\mathbf{B}^{-1}\mathbf{A}\mathbf{B})\mathbf{Y}$$

$$= \|y^1 \quad y^2 \quad \cdots \quad y^n\| \begin{Vmatrix} \lambda_1 & 0 \quad 0 \quad \cdots \quad 0 \\ 0 & \\ 0 & \\ \cdot & \mathbf{C} \\ \cdot & \\ \cdot & \\ 0 & \end{Vmatrix} \begin{Vmatrix} y^1 \\ y^2 \\ \cdot \\ \cdot \\ \cdot \\ y^n \end{Vmatrix}$$

$$= \lambda_1(y^1)^2 + \sum_{\alpha,\beta=2}^{n} c_{\alpha\beta}y^\alpha y^\beta = \sum_{\alpha,\beta=1}^{n} c'_{\alpha\beta}y^\alpha y^\beta$$

The characteristic roots of the matrix $\|c'_{\alpha\beta}\| = \mathbf{B}^{-1}\mathbf{A}\mathbf{B}$ satisfy

$$\begin{vmatrix} \lambda_1 - \lambda & 0 & 0 & \cdots & 0 \\ 0 & c_{22} - \lambda & c_{23} & \cdots & c_{2n} \\ 0 & c_{32} & c_{33} - \lambda & \cdots & c_{3n} \\ \cdot & \cdot & \cdot & & \\ \cdot & \cdot & & \cdot & \\ \cdot & \cdot & & & \cdot \\ 0 & c_{n2} & & & c_{nn} - \lambda \end{vmatrix} = 0$$

Hence the remaining roots of $|\mathbf{A} - \lambda\mathbf{E}| = 0$ satisfy

$$\begin{vmatrix} c_{22} - \lambda & c_{23} & \cdots & c_{2n} \\ c_{32} & c_{33} - \lambda & \cdots & c_{3n} \\ \cdot & \cdot & & \\ \cdot & & \cdot & \\ \cdot & & & \cdot \\ c_{n2} & & & c_{nn} - \lambda \end{vmatrix} = 0$$

By the same procedure we can reduce $\sum_{\alpha,\beta=2}^{n} c_{\alpha\beta}y^\alpha y^\beta$ to the form

$$\lambda_2(z^2)^2 + \sum_{\alpha,\beta=3}^{n} d_{\alpha\beta}z^\alpha z^\beta$$

so that Q becomes $Q = \lambda_1(z^1)^2 + \lambda_2(z^2)^2 + \sum_{\alpha,\beta=3}^{n} d_{\alpha\beta}z^\alpha z^\beta$ with $\lambda_1 = \lambda_2$ if λ_1

is a repeated root. Continuing this process reduces Q to canonical form. Q.E.D.

1.6. Positive-definite Quadratic Forms. Inverse of a Matrix. Let us consider the quadratic form

$$Q = \mathbf{X}^T(\mathbf{S}^T\mathbf{S})\mathbf{X} \tag{1.67}$$

where $\mathbf{S}$ is a triangular matrix, $\mathbf{S} = \|s_j^i\|$, $s_j^i = 0$ for $i > j$. From (1.67) we have $Q = (\mathbf{X}^T\mathbf{S}^T)(\mathbf{S}\mathbf{X}) = (\mathbf{S}\mathbf{X})^T(\mathbf{S}\mathbf{X}) = \sum_{i=1}^{n} (s_\alpha^i x^\alpha)^2$. It is obvious that $Q > 0$ unless $s_\alpha^i x^\alpha = 0$, $i = 1, 2, \ldots, n$. If $|s_j^i| \neq 0$, we know that this system of equations has only the trivial solution

$$x^1 = x^2 = \cdots = x^n = 0$$

Thus the special quadratic form (1.67) for $|\mathbf{S}| \neq 0$ has the property $Q > 0$ unless $x^1 = x^2 = \cdots = x^n = 0$. If $Q = \mathbf{X}^T\mathbf{A}\mathbf{X} > 0$ unless $\mathbf{X} = \mathbf{0}$, we say that Q is a *positive-definite* quadratic form. For such a form it can be shown that

$$|a_{11}| > 0 \qquad \begin{vmatrix} a_{11} & a_{12} \\ a_{21} & a_{22} \end{vmatrix} > 0, \qquad \begin{vmatrix} a_{11} & a_{12} & a_{13} \\ a_{21} & a_{22} & a_{23} \\ a_{31} & a_{32} & a_{33} \end{vmatrix} > 0 \qquad \cdots$$

$$|a_j^i| = |\mathbf{A}| > 0 \tag{1.68}$$

(see "Algebra," by W. L. Ferrar, Oxford University Press). Conversely, (1.68) implies that Q is positive-definite.

The above considerations suggest that, if Q is a positive-definite form, $Q = \mathbf{X}^T\mathbf{A}\mathbf{X}$, $\mathbf{A} = \mathbf{A}^T$, then there exists a triangular matrix $\mathbf{S}$ such that $\mathbf{S}^T\mathbf{S} = \mathbf{A}$. We now show that this is true. The equation $\mathbf{S}^T\mathbf{S} = \mathbf{A}$ implies

$$\sum_{\alpha=1}^{n} s_{\alpha i}s_{\alpha j} = a_{ij} \qquad i, j = 1, 2, \ldots, n \tag{1.69}$$

For $i = j = 1$ we have $s_{11}^2 = a_{11}$ so that $s_{11} = (a_{11})^{\frac{1}{2}}$. For $i = 1$, $j > 1$ we obtain $s_{11}s_{1j} = a_{1j}$ since $s_{21} = s_{31} = \cdots = s_{n1} = 0$, so that

$$s_{1j} = \frac{a_{1j}}{s_{11}} = a_{1j}(a_{11})^{-\frac{1}{2}}$$

$j = 2, 3, \ldots, n$. For $i = j = 2$ we have $(s_{12})^2 + (s_{22})^2 = a_{22}$ so that

$$s_{22} = (a_{22} - s_{12}^2)^{\frac{1}{2}} = \left(a_{22} - \frac{a_{12}^2}{a_{11}}\right)^{\frac{1}{2}} = \left(\frac{\begin{vmatrix} a_{11} & a_{12} \\ a_{21} & a_{22} \end{vmatrix}}{|a_{11}|}\right)^{\frac{1}{2}}$$

Remember that $a_{ij} = a_{ji}$. For $i = 2$, $j > 2$ we have $s_{12}s_{1j} + s_{22}s_{2j} = a_{2j}$

and $s_{2j} = (a_{2j} - s_{12}s_{1j})/s_{22}$. Continuing this process, one obtains the set of equations

$$\begin{aligned}
s_{11} &= (a_{11})^{\frac{1}{2}} \\
s_{1j} &= \frac{a_{1j}}{a_{11}^{\frac{1}{2}}} \qquad j > 1 \\
s_{22} &= (a_{22} - s_{12}^2)^{\frac{1}{2}} \\
s_{2j} &= \frac{a_{2j} - s_{12}s_{1j}}{s_{22}} \qquad j > 2 \\
&\cdots\cdots\cdots\cdots\cdots \\
s_{kk} &= [a_{kk} - (s_{1k}^2 + s_{2k}^2 + \cdots + s_{k-1,k}^2)]^{\frac{1}{2}} \\
s_{kl} &= \frac{a_{kl} - (s_{1k}s_{1l} + s_{2k}s_{2l} + \cdots + s_{k-1,k}s_{k-1,l})}{s_{kk}} \qquad l > k \\
&\cdots\cdots\cdots\cdots\cdots\cdots\cdots\cdots\cdots\cdots
\end{aligned} \tag{1.70}$$

Given the numbers a_{ij}, the elements s_{ij} can be calculated step by step from (1.70). Let the reader show that

$$s_{33}^2 = \frac{\begin{vmatrix} a_{11} & a_{12} & a_{13} \\ a_{21} & a_{22} & a_{23} \\ a_{31} & a_{32} & a_{33} \end{vmatrix}}{\begin{vmatrix} a_{11} & a_{12} \\ a_{21} & a_{22} \end{vmatrix}} \tag{1.71}$$

The above method for determining **S** can be used to find the inverse matrix of **A** provided $\mathbf{X}^T\mathbf{A}\mathbf{X}$ is positive-definite. Since $\mathbf{A} = \mathbf{S}^T\mathbf{S}$, we have $\mathbf{A}^{-1} = \mathbf{S}^{-1}(\mathbf{S}^T)^{-1} = \mathbf{S}^{-1}(\mathbf{S}^{-1})^T$. To find $\mathbf{S}^{-1}$ from **S**, we proceed as follows: Let $\mathbf{T} = \mathbf{S}^{-1}$, $t_{ij} = 0$, $i > j$. From $\mathbf{TS} = \mathbf{E}$ we obtain

$$\sum_{\alpha=1}^{n} t_{i\alpha}s_{\alpha j} = \delta_{\alpha j} \qquad i, j = 1, 2, \ldots, n \tag{1.72}$$

Equation (1.72) enables one to solve for the t_{ij}. Thus $t_{11}s_{11} = 1$, or $t_{11} = 1/s_{11}$. For $i = 1$, $j = 2$ we have $t_{11}s_{12} + t_{12}s_{22} = \delta_{12} = 0$ so that $t_{12} = -t_{11}s_{12}/s_{22}$. Continuing, we can compute $t_{13}, \ldots, t_{1n}, t_{22}, t_{23}, \ldots, t_{nn}$.

Example 1.26. We invert the matrix

$$\mathbf{A} = \begin{Vmatrix} 1 & 2 & 1 \\ 2 & 5 & 3 \\ 1 & 3 & 4 \end{Vmatrix}$$

From (1.70) $s_{11} = 1$, $s_{12} = 2$, $s_{13} = 1$, $s_{23} = 1$, $s_{33} = \sqrt{2}$. From (1.72) $t_{11} = 1$, $t_{12} = -2$, $t_{13} = -(t_{11}s_{13} + t_{12}s_{23})/s_{33} = 1/\sqrt{2}$, $t_{22} = 1$, $t_{23} = -1/\sqrt{2}$, $t_{33} = 1/\sqrt{2}$. Thus

$$S^{-1} = \begin{Vmatrix} 1 & -2 & \frac{1}{\sqrt{2}} \\ 0 & 1 & -\frac{1}{\sqrt{2}} \\ 0 & 0 & \frac{1}{\sqrt{2}} \end{Vmatrix}$$

and

$$A^{-1} = S^{-1}(S^{-1})^T = \begin{Vmatrix} 1 & -2 & \frac{1}{\sqrt{2}} \\ 0 & 1 & -\frac{1}{\sqrt{2}} \\ 0 & 0 & \frac{1}{\sqrt{2}} \end{Vmatrix} \cdot \begin{Vmatrix} 1 & 0 & 0 \\ -2 & 1 & 0 \\ \frac{1}{\sqrt{2}} & -\frac{1}{\sqrt{2}} & \frac{1}{\sqrt{2}} \end{Vmatrix} = \frac{1}{2}\begin{Vmatrix} 11 & -5 & 1 \\ -5 & 3 & -1 \\ 1 & -1 & 1 \end{Vmatrix}$$

We now invert the matrix **A** by another method. Let us first note that the matrix

$$E_{12} = \begin{Vmatrix} 1 & k & 0 \\ 0 & 1 & 0 \\ 0 & 0 & 1 \end{Vmatrix}$$

has the property that $E_{12}B$ is a new matrix which can be obtained by multiplying the second row of **B** by k and adding these elements to the corresponding elements of the first row of **B**. Thus

$$\begin{Vmatrix} 1 & k & 0 \\ 0 & 1 & 0 \\ 0 & 0 & 1 \end{Vmatrix} \cdot \begin{Vmatrix} b_{11} & b_{12} & b_{13} \\ b_{21} & b_{22} & b_{23} \\ b_{31} & b_{32} & b_{33} \end{Vmatrix} = \begin{Vmatrix} b_{11} + kb_{21} & b_{12} + kb_{22} & b_{13} + kb_{23} \\ b_{21} & b_{22} & b_{23} \\ b_{31} & b_{32} & b_{33} \end{Vmatrix}$$

Let the reader show that placing k in the rth row and sth column of the unit matrix **E** produces a matrix E_{rs} such that $E_{rs}B$ is a matrix identical to **B** except that the elements $b_{r\alpha}$, $\alpha = 1, 2, \ldots, n$, are replaced by $b_{r\alpha} + kb_{s\alpha}$, $r \neq s$. Notice that $|E_{rs}| = 1$ for $r \neq s$. Now let **A** be a square matrix such that $|A| \neq 0$. We consider the matrix $C = \|A, E\|$. For the **A** in Example 1.26

$$C = \begin{Vmatrix} 1 & 2 & 1 & 1 & 0 & 0 \\ 2 & 5 & 3 & 0 & 1 & 0 \\ 1 & 3 & 4 & 0 & 0 & 1 \end{Vmatrix} \tag{1.73}$$

We manipulate **C** by operations on the rows until the first three rows and columns of **C** become **E**. These operations are equivalent to multiplying **C** on the left by matrices of the type E_{rs} discussed above. Let **B** be the product of all the E_{rs}. Then

$$BC = \|BA, BE\| = \|E, B\|$$

Hence $BA = E$, and $B = A^{-1}$. We obtain **B** from $\|E, B\|$. For example, starting with the **C** of (1.73), we multiply the first row by -2 and add to the second row, and subtract the first row from the third row. This yields

$$\begin{Vmatrix} 1 & 2 & 1 & 1 & 0 & 0 \\ 0 & 1 & 1 & -2 & 1 & 0 \\ 0 & 1 & 3 & -1 & 0 & 1 \end{Vmatrix}$$

We can easily obtain zeros in the first and third row of the second column. We now have

$$\begin{Vmatrix} 1 & 0 & -1 & 5 & -2 & 0 \\ 0 & 1 & 1 & -2 & 1 & 0 \\ 0 & 0 & 2 & 1 & -1 & 1 \end{Vmatrix}$$

Adding $\frac{1}{2}$ the third row to the first row and subtracting $\frac{1}{2}$ the third row from the second row yields

$$\begin{Vmatrix} 1 & 0 & 0 & \frac{11}{2} & -\frac{5}{2} & \frac{1}{2} \\ 0 & 1 & 0 & -\frac{5}{2} & \frac{3}{2} & -\frac{1}{2} \\ 0 & 0 & 2 & 1 & -1 & 1 \end{Vmatrix}$$

Factoring 2 from the third row, multiplication by $\begin{Vmatrix} 1 & 0 & 0 \\ 0 & 1 & 0 \\ 0 & 0 & \frac{1}{2} \end{Vmatrix}$ yields the inverse matrix

$$\mathbf{A}^{-1} = \frac{1}{2}\begin{Vmatrix} 11 & -5 & 1 \\ -5 & 3 & -1 \\ 1 & -1 & 1 \end{Vmatrix}$$

1.7. Differential Equations. We now consider the system of differential equations

$$\frac{d^2x^i}{dt^2} = a^i_\alpha x^\alpha \qquad i, \alpha = 1, 2, \ldots, n;\ a^i_j = a^j_i = \text{constants}$$

which can be written in the matrix form

$$\frac{d^2\mathbf{X}}{dt^2} = \mathbf{AX} \qquad \mathbf{A} = \mathbf{A}^T \tag{1.74}$$

We look for a linear transformation which will simplify (1.74). Let $\mathbf{X} = \mathbf{BY}$, so that (1.74) becomes

$$\frac{d^2\mathbf{Y}}{dt^2} = (\mathbf{B}^{-1}\mathbf{AB})\mathbf{Y} \tag{1.75}$$

From previous considerations we have shown that it is possible to find a matrix $\mathbf{B}$ such that $\mathbf{B}^{-1}\mathbf{AB}$ is diagonal (nondiagonal terms are zero). For this matrix $\mathbf{B}$ we have

$$\frac{d^2y^i}{dt^2} = \lambda_i y^i \qquad \text{no summation on } i \tag{1.76}$$

since

$$\mathbf{B}^{-1}\mathbf{AB} = \begin{Vmatrix} \lambda_1 & 0 & 0 & \cdots & 0 \\ 0 & \lambda_2 & 0 & \cdots & 0 \\ 0 & 0 & \lambda_3 & \cdots & 0 \\ \cdot & \cdot & \cdot & \cdots & \cdot \\ 0 & 0 & 0 & \cdots & \lambda_n \end{Vmatrix}$$

The solution of (1.76) is

$$y^i = C^i e^{\sqrt{\lambda_i}\, t} + D^i e^{-\sqrt{\lambda_i}\, t} \qquad i = 1, 2, \ldots, n \tag{1.77}$$

From $\mathbf{X} = \mathbf{BY}$ we can solve for $x^i(t)$, $i = 1, 2, \ldots, n$. The y^i, $i = 1, 2, \ldots, n$, are called normal coordinates.

Example 1.27. Let us consider two particles of masses m_1, m_2, respectively, moving in a one-dimensional continuum, coupled in such a way that equal and opposite forces proportional to their distance apart act on the particles. The differential equations of motion are

$$\begin{aligned} m_1 \frac{d^2x_1}{dt^2} &= -a(x_1 - x_2) \\ m_2 \frac{d^2x_2}{dt^2} &= a(x_1 - x_2) \end{aligned} \tag{1.78}$$

For convenience, let $y_1 = \sqrt{m_1}\, x_1$, $y_2 = \sqrt{m_2}\, x_2$, $\omega_1^2 = a/m_1$, $\omega_2^2 = a/m_2$,

$$\frac{a}{(m_1 m_2)^{\frac{1}{2}}} = k,$$

so that (1.78) becomes

$$\begin{aligned} \frac{d^2y_1}{dt^2} &= -\omega_1^2 y_1 + k y_2 \\ \frac{d^2y_2}{dt^2} &= k y_1 - \omega_2^2 y_2 \end{aligned}$$

or

$$\frac{d^2\mathbf{Y}}{dt^2} = \frac{d^2}{dt^2}\begin{Vmatrix} y_1 \\ y_2 \end{Vmatrix} = \begin{Vmatrix} -\omega_1^2 & k \\ k & -\omega_2^2 \end{Vmatrix} \cdot \begin{Vmatrix} y_1 \\ y_2 \end{Vmatrix} = \mathbf{AY}$$

The characteristic equation is

$$\begin{vmatrix} -\omega_1^2 - \lambda & k \\ k & -\omega_2^2 - \lambda \end{vmatrix} = 0$$

so that $\lambda_1 = \frac{1}{2}\{(\omega_1^2 + \omega_2^2) + [(\omega_1 - \omega_2)^2 + 4k^2]^{\frac{1}{2}}\}$. Let the reader find $y_1(t)$, $y_2(t)$, and hence $x_1(t)$, $x_2(t)$.

1.8. Subdivision of a Matrix. A matrix $\mathbf{A}$ may be subdivided into rectangular arrays, each array in turn being thought of as a matrix. For example,

$$\mathbf{A} = \begin{Vmatrix} 2 & -1 & 3 & 4 \\ 0 & 1 & -1 & 2 \\ 3 & 2 & 1 & 5 \\ -1 & 4 & 6 & -2 \\ 0 & -2 & 1 & 1 \end{Vmatrix} = \left\Vert \begin{array}{cc:cc} 2 & -1 & 3 & 4 \\ 0 & 1 & -1 & 2 \\ \hdashline 3 & 2 & 1 & 5 \\ -1 & 4 & 6 & -2 \\ 0 & -2 & 1 & 1 \end{array} \right\Vert$$

$$= \begin{Vmatrix} \mathbf{A}_1 & \mathbf{A}_2 \\ \mathbf{A}_3 & \mathbf{A}_4 \end{Vmatrix}$$

where

$$\mathbf{A}_1 = \begin{Vmatrix} 2 & -1 \\ 0 & 1 \end{Vmatrix} \qquad \mathbf{A}_2 = \begin{vmatrix} 3 & 4 \\ -1 & 2 \end{vmatrix} \qquad \mathbf{A}_3 = \begin{Vmatrix} 3 & 2 \\ -1 & 4 \\ 0 & -2 \end{Vmatrix} \qquad \mathbf{A}_4 = \begin{Vmatrix} 1 & 5 \\ 6 & -2 \\ 1 & 1 \end{Vmatrix}$$

The linear system

$$y^i = a^i_\alpha x^\alpha \qquad i, \alpha = 1, 2, \ldots, n$$

may be written

$$y^i = \sum_{\alpha=1}^{k} a^i_\alpha x^\alpha + \sum_{\alpha=k+1}^{n} a^i_\alpha x^\alpha \qquad i = 1, 2, 3, \ldots, k, k+1, \ldots, n$$

In matrix form

$$\left\| \begin{array}{c} y^1 \\ y^2 \\ \cdot \\ \cdot \\ \cdot \\ y^k \\ \hdashline y^{k+1} \\ \cdot \\ \cdot \\ \cdot \\ y^n \end{array} \right\| = \left\| \begin{array}{cccc:ccc} a^1_1 & a^1_2 & \cdots & a^1_k & a^1_{k+1} & \cdots & a^1_n \\ a^2_1 & a^2_2 & \cdots & a^2_k & a^2_{k+1} & \cdots & a^2_n \\ \cdots & \cdots & \cdots & \cdots & \cdots & \cdots & \cdots \\ \cdots & \cdots & \cdots & \cdots & \cdots & \cdots & \cdots \\ \cdots & \cdots & \cdots & \cdots & \cdots & \cdots & \cdots \\ a^k_1 & a^k_2 & \cdots & a^k_k & a^k_{k+1} & \cdots & a^k_n \\ \hdashline a^{k+1}_1 & a^{k+1}_2 & \cdots & a^{k+1}_k & a^{k+1}_{k+1} & \cdots & a^{k+1}_n \\ \cdots & \cdots & \cdots & \cdots & \cdots & \cdots & \cdots \\ \cdots & \cdots & \cdots & \cdots & \cdots & \cdots & \cdots \\ \cdots & \cdots & \cdots & \cdots & \cdots & \cdots & \cdots \\ a^n_1 & a^n_2 & \cdots & a^n_k & a^n_{k+1} & \cdots & a^n_n \end{array} \right\| \cdot \left\| \begin{array}{c} x^1 \\ x^2 \\ \cdot \\ \cdot \\ \cdot \\ x^k \\ \hdashline x^{k+1} \\ \cdot \\ \cdot \\ \cdot \\ x^n \end{array} \right\|$$

or

$$\left\| \begin{array}{c} \mathbf{Y}_1 \\ \mathbf{Y}_2 \end{array} \right\| = \left\| \begin{array}{cc} \mathbf{A}_1 & \mathbf{A}_2 \\ \mathbf{A}_3 & \mathbf{A}_4 \end{array} \right\| \cdot \left\| \begin{array}{c} \mathbf{X}_1 \\ \mathbf{X}_2 \end{array} \right\|$$

Hence

$$\begin{aligned} \mathbf{Y}_1 &= \mathbf{A}_1\mathbf{X}_1 + \mathbf{A}_2\mathbf{X}_2 \\ \mathbf{Y}_2 &= \mathbf{A}_3\mathbf{X}_1 + \mathbf{A}_4\mathbf{X}_2 \end{aligned} \tag{1.79}$$

If we are not interested in solving for $x^{k+1}, x^{k+2}, \ldots, x^n$, we can eliminate $\mathbf{X}_2$ from (1.79). We obtain

$$\mathbf{X}_2 = \mathbf{A}_4^{-1}(\mathbf{Y}_2 - \mathbf{A}_3\mathbf{X}_1) \qquad |\mathbf{A}_4| \neq 0$$

so that

$$\mathbf{X}_1 = (\mathbf{A}_1 - \mathbf{A}_2\mathbf{A}_4^{-1}\mathbf{A}_3)^{-1}(\mathbf{Y}_1 - \mathbf{A}_2\mathbf{A}_4^{-1}\mathbf{Y}_2)$$

In a mesh circuit the y^i are the impressed voltages, the a^i_j are the impedances, and the x^i are the currents.

1.9. Conclusion. In the calculus the solution of $\dfrac{dx}{dt} = ax$, $a =$ constant, is $x = x_0e^{at}$. Can we generalize this for the system $\dfrac{d\mathbf{X}}{dt} = \mathbf{AX}$? We see immediately that one would be led to consider matrices of the form $e^{\mathbf{A}t}$. How should we go about defining such a matrix? From the calculus $e^x = \sum_{n=0}^{\infty} x^n/n!$. This suggests that if $\mathbf{B}$ is a square matrix we define

$$e^{\mathbf{B}} = \mathbf{E} + \mathbf{B} + \frac{1}{2!}\mathbf{B}^2 + \cdots + \frac{1}{n!}\mathbf{B}^n + \cdots \tag{1.80}$$

This poses a new problem. What do we mean by the sum of an infinite number of matrices? We define

$$\mathbf{B}_r \equiv \mathbf{E} + \mathbf{B} + \frac{1}{2!}\mathbf{B}^2 + \cdots + \frac{1}{r!}\mathbf{B}^r$$

as the rth partial sum of (1.80). If $\lim_{r\to\infty} \mathbf{B}_r$ exists in the sense that each element of $\mathbf{B}_r$ converges, we define

$$e^{\mathbf{B}} = \lim_{r\to\infty} \mathbf{B}_r$$

If $\mathbf{B}$ is a square matrix of order n, whose terms b^i_j are uniformly bounded, that is, $|b^i_j| < A =$ constant, for $i, j = 1, 2, \ldots, n$, then $\lim_{r\to\infty} \mathbf{B}_r$ exists. The proof is as follows: Consider the matrix $\mathbf{B}^2$. Its elements are of the form $b^i_\alpha b^\alpha_j$. Thus each element of $\mathbf{B}^2$ in absolute value is less than nA^2. The elements of $\mathbf{B}^3$ are of the form $b^i_\alpha b^\alpha_\beta b^\beta_j$ which are bounded by n^2A^3. The elements of $\mathbf{B}^k$ are bounded by $n^{k-1}A^k$. Hence every element of $\mathbf{B}_r - \mathbf{E}$ is bounded by the series

$$\sum_{k=1}^{r} \frac{1}{k!} n^{k-1}A^k = \frac{1}{n}\sum_{k=1}^{r} \frac{(nA)^k}{k!}$$

Thus each term of $\mathbf{B}_r$ converges since each term of $\mathbf{B}_r - \mathbf{E}$ is a series bounded in absolute value term by term by the elements of the series $1/n \sum_{k=1}^{r} (nA)^k/k!$ which converges to $(1/n)e^{nA}$ as $r \to \infty$. Let the reader define $\cos \mathbf{B}$ and $\sin \mathbf{B}$. Is $\sin^2 \mathbf{B} + \cos^2 \mathbf{B} = \mathbf{E}$? Let the reader also show that

$$\frac{d}{dt} e^{\mathbf{B}t} = \mathbf{B}e^{\mathbf{B}t}$$

for a constant matrix $\mathbf{B}$.

Problems

1. Show that the roots of (1.64) are real if a_{ij} is real and $a_{ij} = a_{ji}$. *Hint:* Consider $\mathbf{AX} = \lambda\mathbf{X}$, assume $\lambda = \lambda_1 + i\lambda_2$, $\mathbf{X} = \mathbf{X}_1 + i\mathbf{X}_2$, and show that $\lambda_2 = 0$.

2. Consider the quadratic form $Q = a_{\alpha\beta}\bar{x}^\alpha x^\beta$, the x^i, a^α_β complex. We may write $Q = \bar{\mathbf{X}}^T\mathbf{AX}$. If $\mathbf{A} = \bar{\mathbf{A}}^T$, show that Q is real by showing that $Q = \bar{Q}$. A matrix $\mathbf{A}$ such that $\mathbf{A} = \bar{\mathbf{A}}^T$ is called a *Hermitian* matrix.

3. If $\mathbf{A}$ and $\mathbf{B}$ are Hermitian (see Prob. 2), show that $\mathbf{AB} + \mathbf{BA}$ and $i(\mathbf{AB} - \mathbf{BA})$ are Hermitian.

4. Show that the roots of (1.64) are real if $\mathbf{A}$ is Hermitian.

5. Find an orthogonal transformation which reduces

$$Q = x^2 + y^2 + z^2 + 2xz + 4\sqrt{2}\,yz$$

to canonical form.

6. Show that the characteristic roots of the matrix **A** are the same as those of the matrix $\mathbf{B}^{-1}\mathbf{AB}$.

7. Write the system

$$m\frac{d^2x}{dt^2} + He\frac{dy}{dt} = Ee$$
$$m\frac{d^2y}{dt^2} - He\frac{dx}{dt} = 0$$

in the matrix form

$$\frac{d^2\mathbf{X}}{dt^2} + \frac{He}{m}\mathbf{A}\frac{d\mathbf{X}}{dt} = \mathbf{B}$$

where $\mathbf{X} = \begin{Vmatrix} x \\ y \end{Vmatrix} \qquad \mathbf{A} = \begin{Vmatrix} 0 & 1 \\ -1 & 0 \end{Vmatrix} \qquad \mathbf{B} = \begin{Vmatrix} \frac{Ee}{m} \\ 0 \end{Vmatrix}$

Let $\mathbf{X} = \mathbf{CY}$, $|\mathbf{C}| \neq 0$, and find **C** so that the system becomes

$$\frac{d^2\mathbf{Y}}{dt^2} + \frac{He}{m}\begin{Vmatrix} i & 0 \\ 0 & -i \end{Vmatrix}\mathbf{Y} = \frac{i}{2}\begin{Vmatrix} -i & -1 \\ -i & 1 \end{Vmatrix}\begin{Vmatrix} \frac{Ee}{m} \\ 0 \end{Vmatrix}$$

Integrate this system of equations.

8. Solve the system

$$\begin{aligned} x_1 + x_2 + x_3 + x_4 &= 0 \\ x_1 - x_2 + x_3 - x_4 &= 0 \\ 2x_1 + 3x_2 - x_3 + x_4 &= 7 \\ 3x_1 - x_2 + x_3 - x_4 &= 2 \end{aligned}$$

for x_1, x_2 by first eliminating x_3, x_4.

9. Consider the system of differential equations

$$\frac{d^2\mathbf{X}}{dt^2} + \mathbf{A}\frac{d\mathbf{X}}{dt} + \mathbf{BX} = \mathbf{0}$$

where **A** and **B** are constant matrices. Let

$$\mathbf{X} = e^{wt}\mathbf{C} \qquad \mathbf{C} \text{ constant}$$

and show that w satisfies $|w^2\delta^i_j + wa^i_j + b^i_j| = 0$ if **C** is not the zero vector.

10. The characteristic equation of **A** is the determinant $|\mathbf{A} - \lambda\mathbf{E}| = 0$. This is a polynomial in λ, written

$$\lambda^n + b_1\lambda^{n-1} + b_2\lambda^{n-2} + \cdots + b_n = 0$$

See Dickson, "Modern Algebraic Theories," for a proof that **A** satisfies its characteristic equation, that is,

$$\mathbf{A}^n + b_1\mathbf{A}^{n-1} + b_2\mathbf{A}^{n-2} + \cdots + b_n\mathbf{E} = \mathbf{0}$$

This is the *Hamilton-Cayley* theorem.

REFERENCES

Aitken, A. C.: "Determinants and Matrices," Oliver & Boyd, Ltd., London, 1942.

Albert, A. A.: "Introduction to Algebraic Theories," University of Chicago Press, Chicago, 1941.

Birkhoff, G., and S. MacLane: "A Survey of Modern Algebra," The Macmillan Company, New York, 1941.

Ferrar, W. L.: "Algebra," Oxford University Press, New York, 1941.

Michal, A. D.: "Matrix and Tensor Calculus," John Wiley & Sons, Inc., New York, 1947.

Veblen, O.: "Invariants of Quadratic Differential Forms," Cambridge University Press, New York, 1933.

CHAPTER 2

VECTOR ANALYSIS

2.1. Introduction. Elementary vector analysis is a study of directed line segments. The reader is well aware that displacements, velocities, accelerations, forces, etc., require for their description a direction as well as a magnitude. One cannot completely describe the motion of a particle by simply stating that a 2-lb force acts upon it. The direction of the applied force must be stipulated with reference to a particular coordinate system. In much the same manner the knowledge that a particle has a speed of 3 fps relative to a given observer does not yield all the pertinent information as regards the motion of the particle with respect to the observer. One must know the direction of motion of the particle.

A vector, by definition, is a directed line segment. Any physical quantity which can be represented by a vector will also be designated as a vector. The length of a vector when compared with a unit of length will be called the magnitude of the vector. The magnitude of a vector is thus a scalar. A scalar differs from a vector in that no direction is associated with a scalar. Speed, temperature, volume, etc., including elements of the real-number system, are examples of scalars.

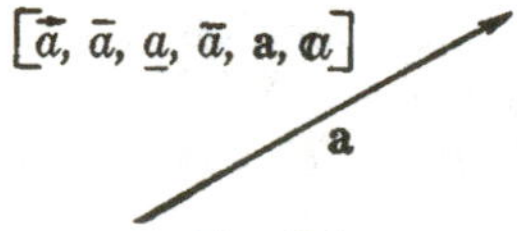

Fig. 2.1

Vectors will be represented by arrows (see Fig. 2.1), and **boldface** type will be used to distinguish a vector from a scalar. The student will have to adopt his own notation for describing a vector in writing.

A vector of length 1 is called a unit vector. There are an infinity of unit vectors since the direction of a unit vector is arbitrary. If a represents the length of a vector, we shall write $a \equiv |\mathbf{a}|$. If $|\mathbf{a}| = 0$, we say that $\mathbf{a}$ is a zero vector, $\mathbf{a} = \mathbf{0}$.

2.2. Equality of Vectors. Two vectors will be defined to be equal if, and only if, they are parallel, have the same sense of direction, and are of equal magnitude. The starting points of the vectors are immaterial. This does not imply that two forces which are equal will produce the same physical result. Our definition of equality is purely a mathematical definition. We write $\mathbf{a} = \mathbf{b}$ if the vectors are equal. Moreover, we imply further that if $\mathbf{a} = \mathbf{b}$ we may replace $\mathbf{a}$ in any vector equation by $\mathbf{b}$

and, conversely, we may replace **b** by **a**. Figure 2.2 shows two vectors **a**, **b** which are equal to each other.

FIG. 2.2

2.3. Multiplication of a Vector by a Scalar. If we multiply a vector **a** by a real number x, we define the product $x\mathbf{a}$ to be a new vector parallel to **a**; the magnitude of $x\mathbf{a}$ is $|x|$ times the magnitude of **a**. If $x > 0$, $x\mathbf{a}$ is parallel to and has the same direction as **a**. If $x < 0$, $x\mathbf{a}$ is parallel to and has the reverse direction of **a** (see Fig. 2.3).

We note that

$$x(y\mathbf{a}) = (xy)\mathbf{a} = xy\mathbf{a}$$
$$0\mathbf{a} = \mathbf{0}$$

It is immediately evident that two vectors are parallel if, and only if, one of them can be written as a scalar multiple of the other.

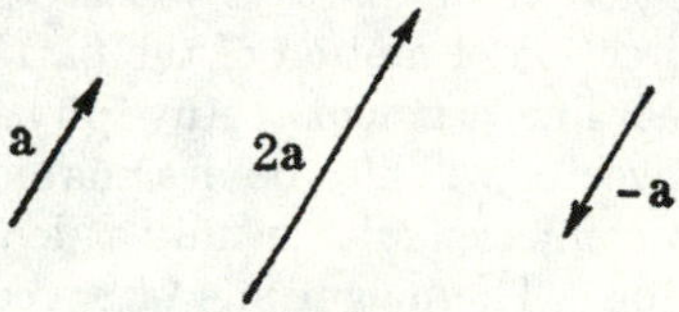

FIG. 2.3

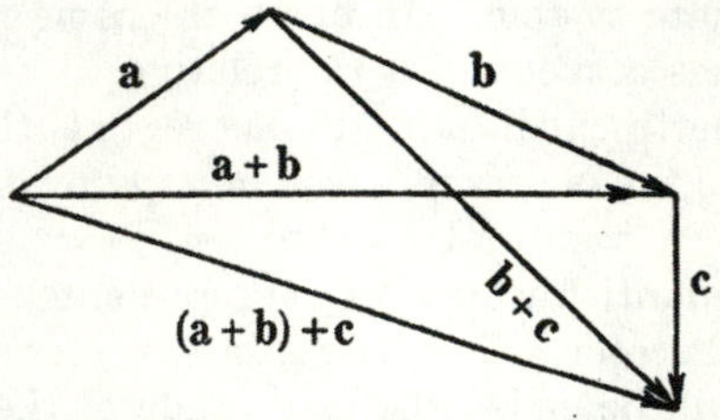

FIG. 2.4

2.4. Addition of Vectors. Let us suppose we have two vectors given, say **a** and **b**. The vector sum, written $\mathbf{a} + \mathbf{b}$, is defined as follows: A triangle is constructed with **a** and **b** forming two sides of the triangle. The vector drawn from the starting point of **a** to the arrow of **b** is defined as the vector sum $\mathbf{a} + \mathbf{b}$ (see Fig. 2.4).

From Euclidean geometry we note that

$$\mathbf{a} + \mathbf{b} = \mathbf{b} + \mathbf{a} \tag{2.1}$$
$$(\mathbf{a} + \mathbf{b}) + \mathbf{c} = \mathbf{a} + (\mathbf{b} + \mathbf{c}) \tag{2.2}$$
$$x(\mathbf{a} + \mathbf{b}) = x\mathbf{a} + x\mathbf{b} \tag{2.3}$$

Furthermore, $\mathbf{a} + \mathbf{0} = \mathbf{a}$, $(x + y)\mathbf{a} = x\mathbf{a} + y\mathbf{a}$, and if $\mathbf{a} = \mathbf{b}$, $\mathbf{c} = \mathbf{d}$, then $\mathbf{a} + \mathbf{c} = \mathbf{b} + \mathbf{d}$. The reader should give geometric proofs of the above statements.

Subtraction of vectors can be reduced to addition by defining

$$\mathbf{a} - \mathbf{b} \equiv \mathbf{a} + (-\mathbf{b})$$

An equivalent definition of $\mathbf{a} - \mathbf{b}$ is the following: We look for the vector $\mathbf{c}$ such that $\mathbf{c} + \mathbf{b} = \mathbf{a}$. $\mathbf{c}$ is defined as the vector $\mathbf{a} - \mathbf{b}$ (see Fig. 2.5).

2.5. Applications to Geometry. Let $\mathbf{a}$, $\mathbf{b}$, $\mathbf{c}$, be vectors with a common origin O whose end points A, B, C, respectively, lie on a straight line.

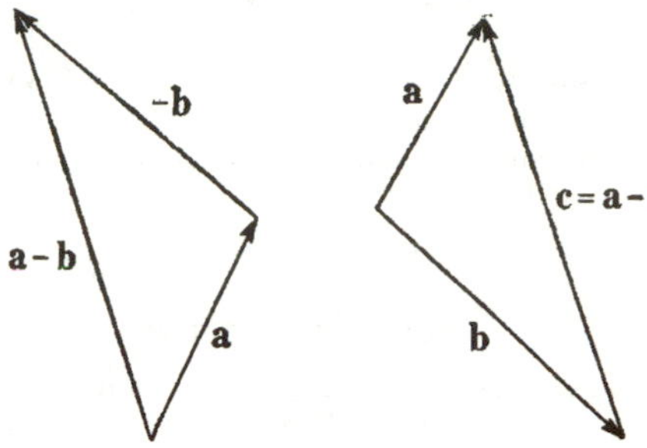

FIG. 2.5

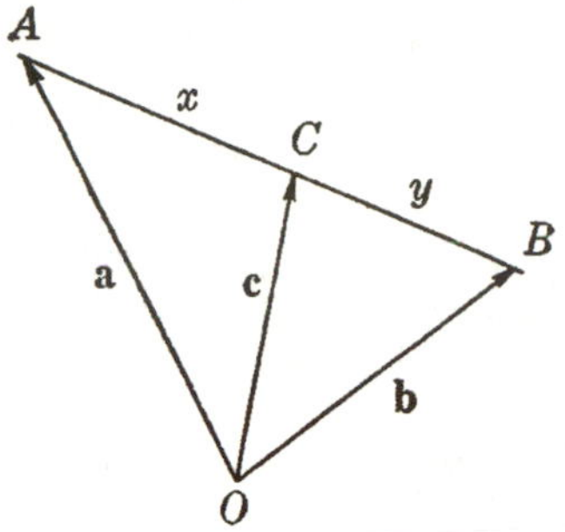

FIG. 2.6

Let C divide AB in the ratio $x:y$, $x + y = 1$ (see Fig. 2.6). We propose to determine $\mathbf{c}$ as a linear combination of $\mathbf{a}$ and $\mathbf{b}$. It is evident that $\mathbf{c} = \mathbf{a} + \overrightarrow{AC}$. But $\overrightarrow{AC} = x\overrightarrow{AB} = x(\mathbf{b} - \mathbf{a})$. Hence

$$\mathbf{c} = (1 - x)\mathbf{a} + x\mathbf{b} = y\mathbf{a} + x\mathbf{b} \tag{2.4}$$

Conversely, let $\mathbf{c} = y\mathbf{a} + x\mathbf{b}$, $x + y = 1$. If $\mathbf{a}$, $\mathbf{b}$, $\mathbf{c}$ have a common origin O, we show that their end points lie on a straight line. We write $\mathbf{c} = (1 - x)\mathbf{a} + x\mathbf{b}$ so that $\mathbf{c} = \mathbf{a} + x(\mathbf{b} - \mathbf{a})$. Since $x(\mathbf{b} - \mathbf{a})$ is a vector parallel to the vector $\mathbf{b} - \mathbf{a} = \overrightarrow{AB}$, we note from the definition of vector addition that the end point C lies on the line joining A to B. Q.E.D.

Equation (2.4) is very useful in solving geometric problems involving ratios of line segments.

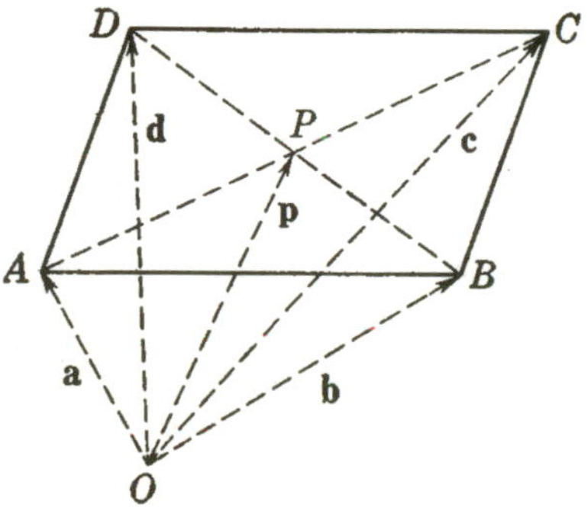

FIG. 2.7

Example 2.1. The diagonals of a parallelogram bisect each other. Let $ABCD$ be any parallelogram, and let O be any point in space (see Fig. 2.7). The statement $\mathbf{c} - \mathbf{d} = \mathbf{b} - \mathbf{a}$ defines the parallelogram $ABCD$. Hence

$$\tfrac{1}{2}(\mathbf{a} + \mathbf{c}) = \frac{\mathbf{a} + \mathbf{c}}{2} = \frac{\mathbf{b} + \mathbf{d}}{2} = \tfrac{1}{2}(\mathbf{b} + \mathbf{d})$$

The vector $(\mathbf{a} + \mathbf{c})/2$ with origin at O has its end point on the line joining AC [see Eq. (2.4)]. The vector $(\mathbf{b} + \mathbf{d})/2$ with origin at O has its end point on the line joining BD. There is only one vector from O whose end point lies on these two lines, namely, the vector $\mathbf{p}$. Hence P bisects AC and BD.

Example 2.2. In Fig. 2.8, D divides CB in the ratio 3:1; E divides AB in the ratio 3:2. How does P divide CE, AD?

Imagine vectors **a**, **b**, **c**, **d**, **e**, **p** drawn from a fixed point O to the points A, B, C, D, E, P, respectively. From Eq. (2.4) we have

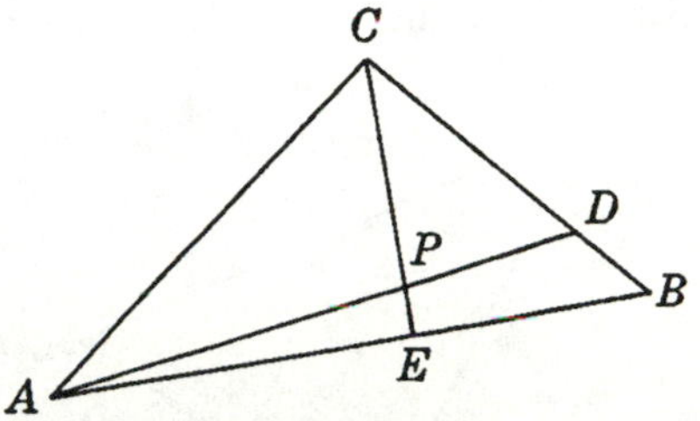

FIG. 2.8

$$\mathbf{d} = \frac{\mathbf{c} + 3\mathbf{b}}{4} \qquad \mathbf{e} = \frac{2\mathbf{a} + 3\mathbf{b}}{5}$$

Since **p** depends linearly on **a** and **d**, and also on **c** and **e**, we eliminate the vector **b** from the above two equations. This yields

$$4\mathbf{d} + 2\mathbf{a} = 5\mathbf{e} + \mathbf{c}$$

or

$$\tfrac{4}{6}\mathbf{d} + \tfrac{2}{6}\mathbf{a} = \tfrac{5}{6}\mathbf{e} + \tfrac{1}{6}\mathbf{c}$$

The vector $\frac{4}{6}\mathbf{d} + \frac{2}{6}\mathbf{a}$ with origin at O must have its end point lying on the line AD. Similarly $\frac{5}{6}\mathbf{e} + \frac{1}{6}\mathbf{c}$ has its end point lying on the line EC. These two vectors have been shown to be equal. Then

$$\mathbf{p} = \tfrac{4}{6}\mathbf{d} + \tfrac{2}{6}\mathbf{a} = \tfrac{5}{6}\mathbf{e} + \tfrac{1}{6}\mathbf{c} \qquad \text{Why?}$$

Thus P divides AD in the ratio 2:1 and divides CE in the ratio 5:1.

Problems

1. Interpret $\dfrac{\mathbf{a}}{|\mathbf{a}|}$.

2. **a**, **b**, **c** are consecutive vectors forming a triangle. What is their vector sum? Generalize this result.

3. **a** and **b** are consecutive vectors of a parallelogram. Express the diagonal vectors in terms of **a** and **b**.

4. **a** and **b** are not parallel. If $x\mathbf{a} + y\mathbf{b} = l\mathbf{a} + m\mathbf{b}$, show that $x = l$, $y = m$.

5. Show graphically that $|\mathbf{a}| + |\mathbf{b}| \geqq |\mathbf{a} + \mathbf{b}|$, $|\mathbf{a} - \mathbf{b}| \geqq \big|\,|\mathbf{a}| - |\mathbf{b}|\,\big|$.

6. Show that the midpoints of the lines which join the midpoints of the opposite sides of a quadrilateral coincide. The four sides of the quadrilateral are not necessarily coplanar.

7. Show that the medians of a triangle meet at a point P which divides each median in the ratio 1:2.

8. Vectors are drawn from the center of a regular polygon to its vertices. Show that the vector sum is zero.

9. **a**, **b**, **c**, **d** are vectors with a common origin. Find a necessary and sufficient condition that their end points lie in a plane.

10. Show that, if two triangles in space are so situated that the three points of intersection of corresponding sides lie on a line, then the lines joining the corresponding vertices pass through a common point, and conversely. This is Desargues' theorem.

2.6. Coordinate Systems. The reader is already familiar with the Euclidean space of three dimensions encountered in the analytic geometry and the calculus. The cartesian coordinate system is frequently used for describing the position of a point in this space. The reader, no doubt, also is acquainted with other coordinate systems, e.g., cylindrical coordinates, spherical coordinates.

We let **i**, **j**, **k** be the three unit vectors along the positive x, y, and z axes respectively. If **r** is the vector from the origin to a point $P(x, y, z)$, then (see Fig. 2.9)

$$\mathbf{r} = x\mathbf{i} + y\mathbf{j} + z\mathbf{k} \tag{2.5}$$

The numbers x, y, z are called the components of the vector **r**. Note that they represent the projections of the vector **r** on the x, y, and z axes. **r** is called the position vector of the point P.

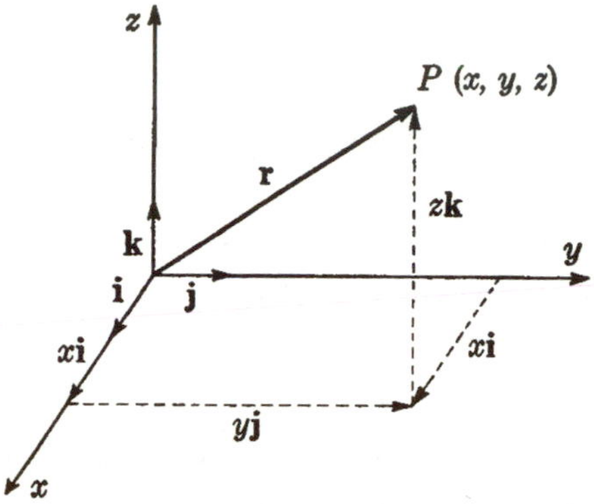

FIG. 2.9

By translating the origin of any vector **A** to the origin of our cartesian coordinate system it can be easily seen that

$$\mathbf{A} = A_1\mathbf{i} + A_2\mathbf{j} + A_3\mathbf{k} \tag{2.6}$$

A_1, A_2, A_3 are the projections of **A** on the x, y, and z axes, respectively. They are called the components of **A**.

Let us now consider the motion of a fluid covering all of space. At any point $P(x, y, z)$ the fluid will have a velocity **V** with components u, v, w. Thus

$$\mathbf{V} = u(x, y, z, t)\mathbf{i} + v(x, y, z, t)\mathbf{j} + w(x, y, z, t)\mathbf{k} \tag{2.7}$$

The velocity components u, v, w will, in general, depend on the point $P(x, y, z)$ and the time t. The most general vector encountered will be of the form given by (2.7). Equation (2.7) describes a vector field. If we fix t, we obtain an instantaneous view of the vector field. Every point in space has a vector associated with it. As time goes on, the vector field changes. A special case of (2.7) is the vector field

$$\mathbf{V} = u(x, y, z)\mathbf{i} + v(x, y, z)\mathbf{j} + w(x, y, z)\mathbf{k} \tag{2.8}$$

Equation (2.8) describes a *steady-state* vector field. The vector field is independent of the time, but the components depend on the coordinates of the point $P(x, y, z)$.

The simplest vector field occurs when the components of **V** are constant throughout all of space. A vector field of this type is said to be *uniform*.

Example 2.3. A particle of mass m is placed at the origin. The force of attraction which the mass would exert on a unit mass placed at the point $P(x, y, z)$ is

$$\mathbf{F} = -\frac{Gm\mathbf{r}}{r^3} = -Gm\frac{x\mathbf{i} + y\mathbf{j} + z\mathbf{k}}{(x^2 + y^2 + z^2)^{\frac{3}{2}}}$$

This is Newton's law of attraction. Note that **F** is a steady-state vector field.

It is easily verified that if

$$\mathbf{A} = A_1\mathbf{i} + A_2\mathbf{j} + A_3\mathbf{k}$$
$$\mathbf{B} = B_1\mathbf{i} + B_2\mathbf{j} + B_3\mathbf{k}$$

then

$$\mathbf{A} + \mathbf{B} = (A_1 + B_1)\mathbf{i} + (A_2 + B_2)\mathbf{j} + (A_3 + B_3)\mathbf{k}$$
$$x\mathbf{A} + y\mathbf{B} = (xA_1 + yB_1)\mathbf{i} + (xA_2 + yB_2)\mathbf{j} + (xA_3 + yB_3)\mathbf{k}$$

2.7. Scalar, or Dot, Product. We define the scalar, or dot, product of two vectors by the identity

$$\mathbf{a} \cdot \mathbf{b} \equiv |\mathbf{a}|\,|\mathbf{b}| \cos \theta \tag{2.9}$$

where θ is the angle between the two vectors when they are drawn from a common origin. Since $\cos \theta = \cos(-\theta)$, there is no ambiguity as to how θ is chosen.

From (2.9) it follows that

$$\begin{aligned} \mathbf{a} \cdot \mathbf{b} &= \mathbf{b} \cdot \mathbf{a} \\ x\mathbf{a} \cdot y\mathbf{b} &= xy\mathbf{a} \cdot \mathbf{b} \\ \mathbf{a} \cdot \mathbf{a} &= |\mathbf{a}|^2 = \mathbf{a}^2 \\ \mathbf{a} = \mathbf{b}, \mathbf{c} = \mathbf{d} \quad &\text{implies} \quad \mathbf{a} \cdot \mathbf{c} = \mathbf{b} \cdot \mathbf{d} \end{aligned} \tag{2.10}$$

If **a** is perpendicular to **b**, then $\mathbf{a} \cdot \mathbf{b} = 0$. Conversely, if $\mathbf{a} \cdot \mathbf{b} = 0$, $|\mathbf{a}| \neq 0$, $|\mathbf{b}| \neq 0$, then **a** is perpendicular to **b**.

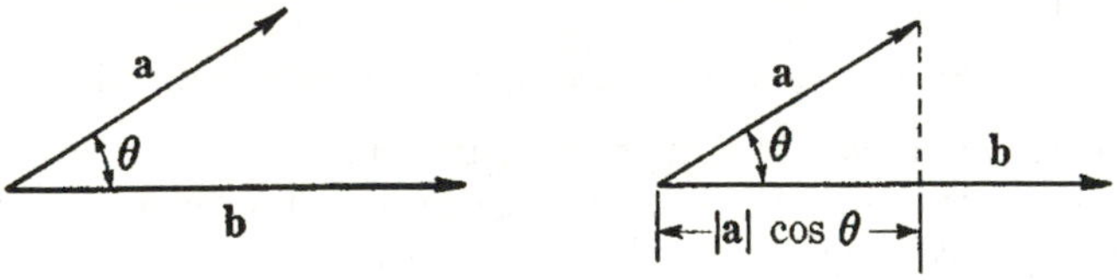

FIG. 2.10

Now $\mathbf{a} \cdot \mathbf{b}$ is equal to the projection of **a** onto **b** multiplied by the length of **b** (see Fig. 2.10). Thus

$$\mathbf{a} \cdot \mathbf{b} = (\text{proj}_b\, \mathbf{a})|\mathbf{b}| = (\text{proj}_a\, \mathbf{b})|\mathbf{a}|$$

With this in mind we proceed to prove the distributive law,

$$\mathbf{a} \cdot (\mathbf{b} + \mathbf{c}) = \mathbf{a} \cdot \mathbf{b} + \mathbf{a} \cdot \mathbf{c} \tag{2.11}$$

From Fig. 2.11 it is apparent that

$$\begin{aligned}\mathbf{a}\cdot(\mathbf{b}+\mathbf{c}) &= [\text{proj}_a\,(\mathbf{b}+\mathbf{c})]|\mathbf{a}| \\ &= (\text{proj}_a\,\mathbf{b} + \text{proj}_a\,\mathbf{c})|\mathbf{a}| \\ &= (\text{proj}_a\,\mathbf{b})|\mathbf{a}| + (\text{proj}_a\,\mathbf{c})|\mathbf{a}| \\ &= \mathbf{b}\cdot\mathbf{a} + \mathbf{c}\cdot\mathbf{a} \\ &= \mathbf{a}\cdot\mathbf{b} + \mathbf{a}\cdot\mathbf{c}\end{aligned}$$

A repeated application of (2.11) yields

$$\begin{aligned}(\mathbf{a}+\mathbf{b})\cdot(\mathbf{c}+\mathbf{d}) &= \mathbf{a}\cdot(\mathbf{c}+\mathbf{d}) + \mathbf{b}\cdot(\mathbf{c}+\mathbf{d}) \\ &= \mathbf{a}\cdot\mathbf{c} + \mathbf{a}\cdot\mathbf{d} + \mathbf{b}\cdot\mathbf{c} + \mathbf{b}\cdot\mathbf{d}\end{aligned}$$

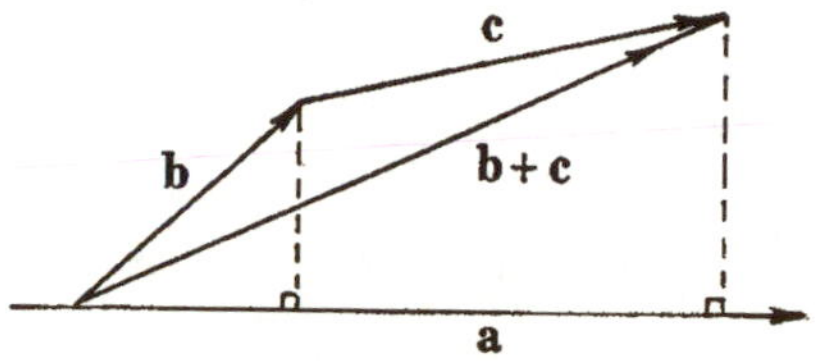

FIG. 2.11

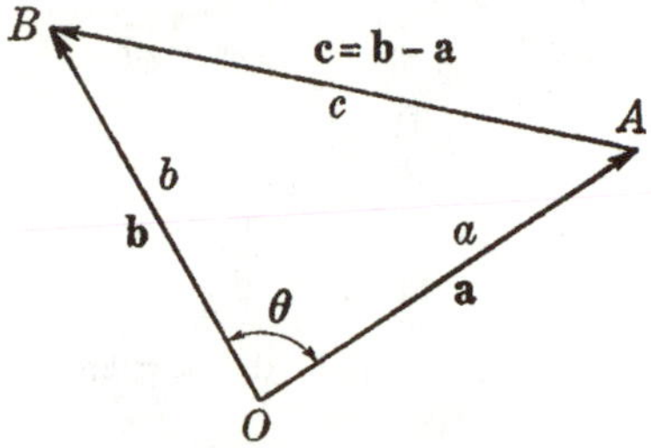

FIG. 2.12

Example 2.4. *Cosine Law of Trigonometry* (Fig. 2.12)

$$\begin{aligned}\mathbf{c} &= \mathbf{b} - \mathbf{a} \\ \mathbf{c}\cdot\mathbf{c} &= (\mathbf{b}-\mathbf{a})\cdot(\mathbf{b}-\mathbf{a}) \\ &= \mathbf{b}\cdot\mathbf{b} + \mathbf{a}\cdot\mathbf{a} - 2\mathbf{a}\cdot\mathbf{b} \\ c^2 &= b^2 + a^2 - 2ab\cos\theta\end{aligned}$$

Example 2.5. $\mathbf{i}\cdot\mathbf{i} = \mathbf{j}\cdot\mathbf{j} = \mathbf{k}\cdot\mathbf{k} = 1$, $\mathbf{i}\cdot\mathbf{j} = \mathbf{j}\cdot\mathbf{k} = \mathbf{k}\cdot\mathbf{i} = 0$. Hence if

$$\mathbf{A} = A_1\mathbf{i} + A_2\mathbf{j} + A_3\mathbf{k}$$

$\mathbf{B} = B_1\mathbf{i} + B_2\mathbf{j} + B_3\mathbf{k}$, then

$$\mathbf{A}\cdot\mathbf{B} = A_1B_1 + A_2B_2 + A_3B_3 \tag{2.12}$$

Formula (2.12) is very useful. It should be memorized.

Example 2.6. Let $Ox^1x^2x^3$ and $O\bar{x}^1\bar{x}^2\bar{x}^3$ be orthogonal rectangular cartesian coordinate systems with common origin O. We now use the superscript convention of Chap. 1. The unit vectors along the x^j axes, $j = 1, 2, 3$, are designated by $\mathbf{i}_j$. Similarly $\bar{\mathbf{i}}_j$, $j = 1, 2, 3$, is the set of unit vectors along the $\bar{x}^j$ axes. Let a_α^β be the cosine of the angle between the vectors $\mathbf{i}_\alpha$, $\bar{\mathbf{i}}_\beta$, $\alpha, \beta = 1, 2, 3$. The projections of $\mathbf{i}_1$ on the $\bar{x}^1$, $\bar{x}^2$, $\bar{x}^3$ axes are a_1^1, a_1^2, a_1^3, respectively. Hence $\mathbf{i}_1 = a_1^1\bar{\mathbf{i}}_1 + a_1^2\bar{\mathbf{i}}_2 + a_1^3\bar{\mathbf{i}}_3 = a_1^\alpha\bar{\mathbf{i}}_\alpha$. Let the reader show that

$$\mathbf{i}_\alpha = a_\alpha^\beta\bar{\mathbf{i}}_\beta \tag{2.13}$$

We have (Fig. 2.13) $\mathbf{r} = \bar{\mathbf{r}}$, so that $x^\alpha\mathbf{i}_\alpha = \bar{x}^\alpha\bar{\mathbf{i}}_\alpha$. From (2.13) $a_\alpha^\beta x^\alpha\bar{\mathbf{i}}_\beta = \bar{x}^\beta\bar{\mathbf{i}}_\beta$ so that

$$\bar{x}^\beta = a_\alpha^\beta x^\alpha \qquad \beta = 1, 2, 3 \tag{2.14}$$

Equation (2.14) represents the coordinate transformation (linear) between our two coordinate systems. In matrix form, $\bar{\mathbf{X}} = \mathbf{AX}$. From $\sum_{i=1}^{3} x^i x^i = \sum_{i=1}^{3} \bar{x}^i \bar{x}^i$ it follows that $\mathbf{X}^T\mathbf{X} = \bar{\mathbf{X}}^T\bar{\mathbf{X}}$, which in turn implies $\mathbf{A}^T\mathbf{A} = \mathbf{E}$ or $\mathbf{A}^T = \mathbf{A}^{-1}$.

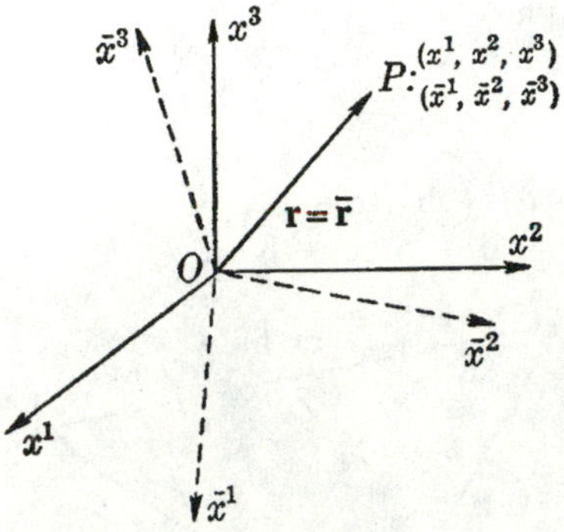

FIG. 2.13

If U^1, U^2, U^3 are the components of a vector when referred to the x coordinate system, and if $\bar{U}^1$, $\bar{U}^2$, $\bar{U}^3$ are the components of the same vector when referred to the $\bar{x}$ coordinate system, one obtains

$$\bar{U}^\beta = a^\beta_\alpha U^\alpha \tag{2.15}$$

This result is obtained in exactly the same manner in which Eq. (2.14) was derived. From (2.14), $\frac{\partial \bar{x}^\beta}{\partial x^\alpha} = a^\beta_\alpha$ so that (2.15) may be written

$$\bar{U}^\beta = \frac{\partial \bar{x}^\beta}{\partial x^\alpha} U^\alpha \tag{2.16}$$

Equation (2.16) will be the starting point for the definition of a contravariant vector field (see Chap. 3).

Problems

1. Add and subtract the vectors $\mathbf{a} = 2\mathbf{i} - 3\mathbf{j} + 5\mathbf{k}$, $\mathbf{b} = -2\mathbf{i} + 2\mathbf{j} + 2\mathbf{k}$. Show that the vectors are perpendicular.

2. Find the angle between the vectors $\mathbf{a} = 2\mathbf{i} - 3\mathbf{j} + \mathbf{k}$, $\mathbf{b} = 3\mathbf{i} - \mathbf{j} - 2\mathbf{k}$.

3. Let $\mathbf{a}$ and $\mathbf{b}$ be unit vectors in the xy plane making angles α and β with the x axis. Show that $\mathbf{a} = \cos\alpha\,\mathbf{i} + \sin\alpha\,\mathbf{j}$, $\mathbf{b} = \cos\beta\,\mathbf{i} + \sin\beta\,\mathbf{j}$, and prove that

$$\cos(\alpha - \beta) = \cos\alpha\cos\beta + \sin\alpha\sin\beta$$

4. Show that the equation of a sphere with center at $P_0(x_0, y_0, z_0)$ and radius a is $(x - x_0)^2 + (y - y_0)^2 + (z - z_0)^2 = a^2$.

5. Show that the equation of the plane passing through the point $P_0(x_0, y_0, z_0)$ normal to the vector $A\mathbf{i} + B\mathbf{j} + C\mathbf{k}$ is

$$A(x - x_0) + B(y - y_0) + C(z - z_0) = 0$$

6. Show that the equation of a straight line through the point $P_0(x_0, y_0, z_0)$ parallel to the vector $l\mathbf{i} + m\mathbf{j} + n\mathbf{k}$ is

$$x = x_0 + \lambda l \qquad y = y_0 + \lambda m \qquad z = z_0 + \lambda n \qquad -\infty < \lambda < \infty$$

7. Prove that the sum of the squares of the diagonals of a parallelogram is equal to the sum of the squares of its sides.

8. Show that the shortest distance from the point $P_0(x_0, y_0, z_0)$ to the plane

$$Ax + By + Cz + D = 0$$

is

$$\left| \frac{Ax_0 + By_0 + Cz_0 + D}{(A^2 + B^2 + C^2)^{\frac{1}{2}}} \right|$$

9. For Eq. (2.14), $\sum_{\beta=1}^{3} x^\beta x^\beta = \sum_{\beta=1}^{3} \bar{x}^\beta \bar{x}^\beta$. Show that this implies

$$\sum_{\beta=1}^{3} a_\alpha^\beta a_\gamma^\beta \begin{array}{ll} = 0 & \text{if } \alpha \neq \gamma \\ = 1 & \text{if } \alpha = \gamma \end{array}$$

10. For the a_β^α of Prob. 9 show that if $\bar{U}^\alpha = a_\beta^\alpha U^\beta$, $\bar{V}^\alpha = a_\beta^\alpha V^\beta$, then

$$\sum_{\alpha=1}^{3} \bar{U}^\alpha \bar{V}^\alpha = \sum_{\alpha=1}^{3} U^\alpha V^\alpha$$

2.8. Vector, or Cross, Product. One can construct a vector **c** from two given vectors **a**, **b** as follows: Let **a** and **b** be translated so that they have a common origin, and let them form the sides of a parallelogram of area $A = |\mathbf{a}|\,|\mathbf{b}| \sin \theta$ (see Fig. 2.14). We define **c** to be perpendicular to the plane of this parallelogram with magnitude equal to the area of the parallelogram. The direction of **c** is obtained by rotating **a** into **b** (angle of rotation less than 180°) and considering the motion of a right-hand screw. The vector **c** thus obtained is defined as the vector, or cross, product of **a** and **b**, written

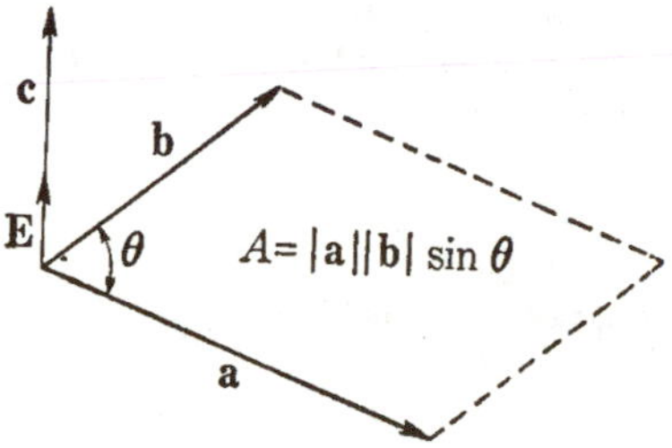

FIG. 2.14

$$\mathbf{c} = \mathbf{a} \times \mathbf{b} = |\mathbf{a}|\,|\mathbf{b}| \sin \theta\, \mathbf{E} \tag{2.17}$$

with $|\mathbf{E}| = 1$, $\mathbf{E} \cdot \mathbf{a} = \mathbf{E} \cdot \mathbf{b} = 0$.

The vector product occurs frequently in mechanics and electricity, but for the present we discuss its algebraic behavior. It follows that

$$\mathbf{a} \times \mathbf{b} = -\mathbf{b} \times \mathbf{a}$$

so that the vector product is not commutative. If **a** and **b** are parallel, $\mathbf{a} \times \mathbf{b} = \mathbf{0}$. Conversely, if $\mathbf{a} \times \mathbf{b} = \mathbf{0}$, then **a** and **b** are parallel provided $|\mathbf{a}| \neq 0$, $|\mathbf{b}| \neq 0$. In particular $\mathbf{a} \times \mathbf{a} = \mathbf{0}$.

The distributive law, $\mathbf{a} \times (\mathbf{b} + \mathbf{c}) = \mathbf{a} \times \mathbf{b} + \mathbf{a} \times \mathbf{c}$, can be shown to hold as follows: Let

$$\mathbf{u} \equiv \mathbf{a} \times (\mathbf{b} + \mathbf{c}) - \mathbf{a} \times \mathbf{b} - \mathbf{a} \times \mathbf{c}$$

We attempt to show that $\mathbf{u} \equiv \mathbf{0}$. Since the distributive law for scalar multiplication holds, we have

$$\mathbf{v} \cdot \mathbf{u} = \mathbf{v} \cdot \mathbf{a} \times (\mathbf{b} + \mathbf{c}) - \mathbf{v} \cdot (\mathbf{a} \times \mathbf{b}) - \mathbf{v} \cdot (\mathbf{a} \times \mathbf{c}) \tag{2.18}$$

for arbitrary $\mathbf{v}$. In the next paragraph it will be shown that

$$\mathbf{a} \cdot (\mathbf{b} \times \mathbf{c}) = (\mathbf{a} \times \mathbf{b}) \cdot \mathbf{c}$$

Thus (2.18) may be written

$$\begin{aligned} \mathbf{v} \cdot \mathbf{u} &= (\mathbf{v} \times \mathbf{a}) \cdot (\mathbf{b} + \mathbf{c}) - (\mathbf{v} \times \mathbf{a}) \cdot \mathbf{b} - (\mathbf{v} \times \mathbf{a}) \cdot \mathbf{c} \\ &= (\mathbf{v} \times \mathbf{a}) \cdot \mathbf{b} + (\mathbf{v} \times \mathbf{a}) \cdot \mathbf{c} - (\mathbf{v} \times \mathbf{a}) \cdot \mathbf{b} - (\mathbf{v} \times \mathbf{a}) \cdot \mathbf{c} \\ &= 0 \end{aligned}$$

This implies that $\mathbf{u} = \mathbf{0}$ or $\mathbf{v} \perp \mathbf{u}$. Since $\mathbf{v}$ can be chosen arbitrarily, and hence picked not perpendicular to $\mathbf{u}$, it follows that $\mathbf{u} = \mathbf{0}$ so that

$$\mathbf{a} \times (\mathbf{b} + \mathbf{c}) = \mathbf{a} \times \mathbf{b} + \mathbf{a} \times \mathbf{c} \tag{2.19}$$

Example 2.7. One sees that $\mathbf{i} \times \mathbf{i} = \mathbf{0}, \mathbf{j} \times \mathbf{j} = \mathbf{0}, \mathbf{k} \times \mathbf{k} = \mathbf{0}, \mathbf{i} \times \mathbf{j} = \mathbf{k}, \mathbf{j} \times \mathbf{k} = \mathbf{i}$, $\mathbf{k} \times \mathbf{i} = \mathbf{j}$. For the vectors $\mathbf{a} = a_1\mathbf{i} + a_2\mathbf{j} + a_3\mathbf{k}$, $\mathbf{b} = b_1\mathbf{i} + b_2\mathbf{j} + b_3\mathbf{k}$ we obtain $\mathbf{a} \times \mathbf{b} = (a_2b_3 - a_3b_2)\mathbf{i} + (a_3b_1 - a_1b_3)\mathbf{j} + (a_1b_2 - a_2b_1)\mathbf{k}$ from the distributive law. Symbollically

$$\mathbf{a} \times \mathbf{b} = \begin{vmatrix} \mathbf{i} & \mathbf{j} & \mathbf{k} \\ a_1 & a_2 & a_3 \\ b_1 & b_2 & b_3 \end{vmatrix} \tag{2.20}$$

Equation (2.20) is to be expanded by the ordinary method of determinants.

Example 2.8

$$\mathbf{a} = \mathbf{i} - 3\mathbf{j} + 2\mathbf{k} \qquad \mathbf{b} = 4\mathbf{i} + \mathbf{j} - 3\mathbf{k}$$

$$\mathbf{a} \times \mathbf{b} = \begin{vmatrix} \mathbf{i} & \mathbf{j} & \mathbf{k} \\ 1 & -3 & 2 \\ 4 & 1 & -3 \end{vmatrix} = 7\mathbf{i} + 11\mathbf{j} + 13\mathbf{k}$$

$$(\mathbf{a} \times \mathbf{b}) \cdot \mathbf{a} = 7 - 33 + 26 = 0 \qquad (\mathbf{a} \times \mathbf{b}) \cdot \mathbf{b} = 28 + 11 - 39 = 0$$

Example 2.9. *Rotation of a Particle.* Assume that a particle is rotating about a fixed line L with angular speed ω. We assume that the shortest distance of the particle from L remains constant. Let us define the angular velocity of the particle as the vector, $\boldsymbol{\omega}$, whose direction is along L and whose length is ω. We choose the direction of $\boldsymbol{\omega}$ in the usual sense of a right-hand-screw advance (see Fig. 2.15). Let $\mathbf{r}$ be the position vector of P with origin O on the line L. It is a simple matter for the reader to show that the velocity of P, say $\mathbf{V}$, is parallel to, and has the same magnitude as, $\boldsymbol{\omega} \times \mathbf{r}$. Thus $\mathbf{V} = \boldsymbol{\omega} \times \mathbf{r}$.

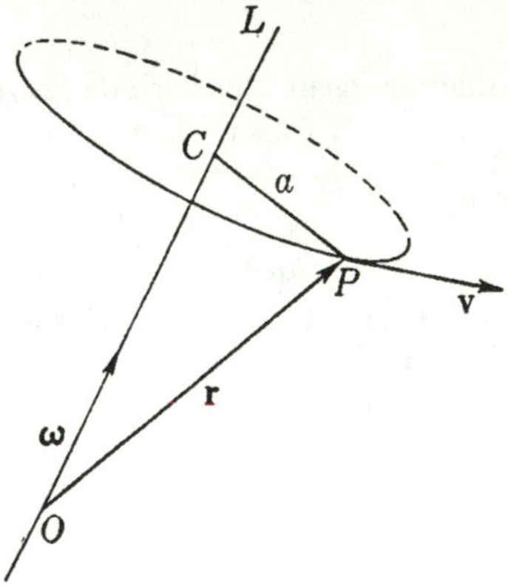

FIG. 2.15

Example 2.10. *Motion of a Rigid Body with One Fixed Point.* Let $Oxyz$ be a fixed coordinate system, and $O\bar{x}\bar{y}\bar{z}$ a coordinate system attached to the rigid body whose fixed point is the origin O. Let P be a point of the rigid body. As time progresses, the coordinates $\bar{x}, \bar{y}, \bar{z}$ remain constant since the $O\bar{x}\bar{y}\bar{z}$ coordinate system is rigidly attached to the moving frame. From (2.14) we have $\bar{x}^i = a^i_j x^j$. Hence

$$\frac{d\bar{x}^i}{dt} = 0 = \frac{da^i_j}{dt} x^j + a^i_j \frac{dx^j}{dt}$$

so that $\frac{dx^k}{dt} = -A^k_i \frac{da^i_j}{dt} x^j$, $A^k_i a^i_j = \delta^k_j$, or $\|A^i_j\| = \|a^i_j\|^{-1}$. If we define $\omega^k_j = -A^k_i \frac{da^i_j}{dt}$, we have

$$\frac{dx^k}{dt} = \omega^k_j x^j \tag{2.21}$$

However, $\sum_{k=1}^{3} x^k x^k$ represents the invariant distance from O to P so that $\sum_{k=1}^{3} x^k \frac{dx^k}{dt} = 0$ and $\sum_{k=1}^{3} \omega^k_j x^j x^k = 0$. From Example 1.2 it follows that $\omega^k_j = -\omega^j_k$. Equation (2.21) can now be written as

$$\begin{aligned}
\frac{dx^1}{dt} &= \omega^1_2 x^2 + \omega^1_3 x^3 = \omega^1_3 x^3 - \omega^2_1 x^2 \\
\frac{dx^2}{dt} &= \omega^2_1 x^1 + \omega^2_3 x^3 = \omega^2_1 x^1 - \omega^3_2 x^3 \\
\frac{dx^3}{dt} &= \omega^3_1 x^1 + \omega^3_2 x^2 = \omega^3_2 x^2 - \omega^1_3 x^1
\end{aligned}$$

so that $\mathbf{v} = \frac{dx^1}{dt}\mathbf{i} + \frac{dx^2}{dt}\mathbf{j} + \frac{dx^3}{dt}\mathbf{k} = \boldsymbol{\omega} \times \mathbf{r}$, $\boldsymbol{\omega} = \omega^3_2\mathbf{i} + \omega^1_3\mathbf{j} + \omega^2_1\mathbf{k}$. It follows from the result of Example 2.9 that the motion of a rigid body with one point fixed can be characterized as follows: There exists an angular velocity vector $\boldsymbol{\omega}$ whose components, in general, change with time, such that at any instant the motion of the rigid body is one of pure rotation with angular velocity $\boldsymbol{\omega}$. This property is very important in the study of the motion of a gyroscope. It can easily be shown that the most general rigid-body motion consists of a translation plus a pure rotation.

2.9. Multiple Scalar and Vector Products. The triple scalar product $\mathbf{a} \cdot (\mathbf{b} \times \mathbf{c})$ has a simple geometric interpretation. This scalar represents the volume of the parallelepiped formed by the coterminous sides **a**, **b**, **c**, since

$$\begin{aligned}
\mathbf{a} \cdot (\mathbf{b} \times \mathbf{c}) &= |\mathbf{a}|\,|\mathbf{b}|\,|\mathbf{c}| \sin\theta \cos\alpha \\
&= hA = \text{volume}
\end{aligned}$$

where A is the area of the parallelogram with sides **b** and **c** and h is the altitude of the parallelepiped (see Fig. 2.16). It is easy to see that $(\mathbf{a} \times \mathbf{b}) \cdot \mathbf{c}$ represents the same volume. Hence it is permissible to interchange the dot and cross in the triple scalar product. Since there can be no confusion as to the meaning of $\mathbf{a} \cdot (\mathbf{b} \times \mathbf{c})$, it is usually written as $(\mathbf{abc})$. The expression $\mathbf{a} \times (\mathbf{b} \cdot \mathbf{c})$ is meaningless. Why? We let the reader prove that

$$\mathbf{a} \cdot \mathbf{b} \times \mathbf{c} = (\mathbf{abc}) = \begin{vmatrix} a_1 & a_2 & a_3 \\ b_1 & b_2 & b_3 \\ c_1 & c_2 & c_3 \end{vmatrix} \tag{2.22}$$

From determinant theory, or otherwise, it follows that

$$(\mathbf{abc}) = (\mathbf{cab}) = (\mathbf{bca}) = -(\mathbf{bac}) = -(\mathbf{cba}) = -(\mathbf{acb})$$

It is easy to show that a necessary and sufficient condition that **a**, **b**, **c** lie in the same plane when they have a common origin is that $(\mathbf{abc}) = 0$. In particular, $(\mathbf{aac}) = 0$.

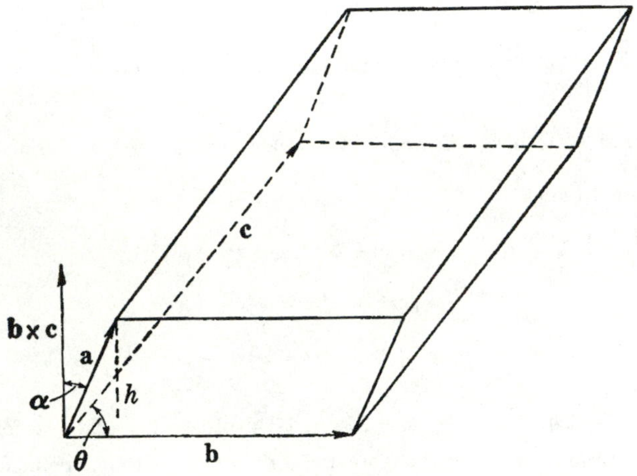

FIG. 2.16

The triple vector product $\mathbf{a} \times (\mathbf{b} \times \mathbf{c})$ is an important vector. It is certainly a vector, call it **V**, since it is the vector product of **a** and $\mathbf{b} \times \mathbf{c}$. We know that **V** is perpendicular to the vector $\mathbf{b} \times \mathbf{c}$. However, $\mathbf{b} \times \mathbf{c}$ is perpendicular to the plane of **b** and **c** so that **V** lies in the plane of **b** and **c**. If **b** and **c** are not parallel, then $\mathbf{V} = \lambda\mathbf{b} + \mu\mathbf{c}$. If **b** and **c** are parallel, $\mathbf{V} = \mathbf{0}$. Since $\mathbf{V} \cdot \mathbf{a} = 0$, we have $\lambda(\mathbf{b} \cdot \mathbf{a}) + \mu(\mathbf{c} \cdot \mathbf{a}) = 0$ so that

$$\mathbf{V} = \lambda_1[(\mathbf{c} \cdot \mathbf{a})\mathbf{b} - (\mathbf{b} \cdot \mathbf{a})\mathbf{c}]$$

It can be shown that $\lambda_1 \equiv 1$ so that

$$\mathbf{a} \times (\mathbf{b} \times \mathbf{c}) = (\mathbf{a} \cdot \mathbf{c})\mathbf{b} - (\mathbf{a} \cdot \mathbf{b})\mathbf{c} \tag{2.23}$$

The expansion (2.23) of $\mathbf{a} \times (\mathbf{b} \times \mathbf{c})$ is often referred to as the rule of the middle factor. Similarly

$$(\mathbf{a} \times \mathbf{b}) \times \mathbf{c} = (\mathbf{a} \cdot \mathbf{c})\mathbf{b} - (\mathbf{b} \cdot \mathbf{c})\mathbf{a} \tag{2.24}$$

More complicated products can be simplified by use of the triple products. For example, we can expand $(\mathbf{a} \times \mathbf{b}) \times (\mathbf{c} \times \mathbf{d})$ by considering $\mathbf{a} \times \mathbf{b}$ as a single vector and applying (2.23).

$$\begin{aligned}(\mathbf{a} \times \mathbf{b}) \times (\mathbf{c} \times \mathbf{d}) &= (\mathbf{a} \times \mathbf{b} \cdot \mathbf{d})\mathbf{c} - (\mathbf{a} \times \mathbf{b} \cdot \mathbf{c})\mathbf{d} \\ &= (\mathbf{abd})\mathbf{c} - (\mathbf{abc})\mathbf{d}\end{aligned}$$

$$\text{Also} \qquad (\mathbf{a} \times \mathbf{b}) \cdot (\mathbf{c} \times \mathbf{d}) = [(\mathbf{a} \times \mathbf{b}) \times \mathbf{c}] \cdot \mathbf{d}$$
$$= [(\mathbf{a} \cdot \mathbf{c})\mathbf{b} - (\mathbf{b} \cdot \mathbf{c})\mathbf{a}] \cdot \mathbf{d}$$
$$= (\mathbf{a} \cdot \mathbf{c})(\mathbf{b} \cdot \mathbf{d}) - (\mathbf{b} \cdot \mathbf{c})(\mathbf{a} \cdot \mathbf{d})$$
$$= \begin{vmatrix} \mathbf{a} \cdot \mathbf{c} & \mathbf{a} \cdot \mathbf{d} \\ \mathbf{b} \cdot \mathbf{c} & \mathbf{b} \cdot \mathbf{d} \end{vmatrix}$$

Example 2.11. Consider the spherical triangle ABC (sides are arcs of great circles) (see Fig. 2.17). Let the sphere be of radius 1. Now

$$(\mathbf{a} \times \mathbf{b}) \cdot (\mathbf{a} \times \mathbf{c}) = \mathbf{b} \cdot \mathbf{c} - (\mathbf{a} \cdot \mathbf{b})(\mathbf{a} \cdot \mathbf{c})$$

since $\mathbf{a} \cdot \mathbf{a} = 1$. The angle between $\mathbf{a} \times \mathbf{b}$ and $\mathbf{a} \times \mathbf{c}$ is the same as the dihedral angle A between the planes OAC and OAB, since $\mathbf{a} \times \mathbf{b}$ is perpendicular to the plane of OAB and since $\mathbf{a} \times \mathbf{c}$ is perpendicular to the plane of OAC. Hence

$$\sin \gamma \sin \beta \cos A = \cos \alpha - \cos \gamma \cos \beta$$

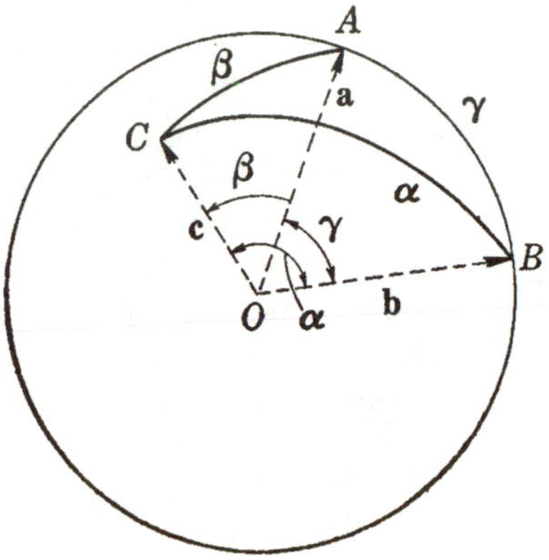

FIG. 2.17

Problems

1. Show by two methods that the vectors $\mathbf{a} = 3\mathbf{i} - \mathbf{j} + 2\mathbf{k}$, $\mathbf{b} = -12\mathbf{i} + 4\mathbf{j} - 8\mathbf{k}$ are parallel.

2. Find a unit vector perpendicular to the vectors $\mathbf{a} = \mathbf{i} - \mathbf{j} + 2\mathbf{k}$, $\mathbf{b} = 3\mathbf{i} + \mathbf{j} - \mathbf{k}$.

3. If $\mathbf{a}, \mathbf{b}, \mathbf{c}, \mathbf{d}$ have a common origin, interpret the equation $(\mathbf{a} \times \mathbf{b}) \cdot (\mathbf{c} \times \mathbf{d}) = 0$.

4. Write a vector equation which specifies that the plane through $\mathbf{a}$ and $\mathbf{b}$ is parallel to the plane through $\mathbf{c}$ and $\mathbf{d}$.

5. Show that $\mathbf{d} \times (\mathbf{a} \times \mathbf{b}) \cdot (\mathbf{a} \times \mathbf{c}) = (\mathbf{abc})(\mathbf{a} \cdot \mathbf{d})$.

6. Show that $(\mathbf{a} \times \mathbf{b}) \cdot (\mathbf{b} \times \mathbf{c}) \times (\mathbf{c} \times \mathbf{a}) = (\mathbf{abc})^2$.

7. Show that $\mathbf{a} \times (\mathbf{b} \times \mathbf{c}) + \mathbf{b} \times (\mathbf{c} \times \mathbf{a}) + \mathbf{c} \times (\mathbf{a} \times \mathbf{b}) = 0$.

8. Find an expression for the shortest distance from the end point of the vector $\mathbf{r}_1$, to the plane passing through the end points of the vectors $\mathbf{r}_2$, $\mathbf{r}_3$, $\mathbf{r}_4$. All four vectors have a common origin O.

9. Assume $(\mathbf{abc}) \neq 0$. Let $\mathbf{d} = x\mathbf{a} + y\mathbf{b} + z\mathbf{c}$. Show that

$$x = \frac{(\mathbf{dbc})}{(\mathbf{abc})} \qquad y = \frac{(\mathbf{adc})}{(\mathbf{abc})} \qquad z = \frac{(\mathbf{abd})}{(\mathbf{abc})}$$

10. If $(\mathbf{abc}) \neq 0$, show that

$$\mathbf{d} = \frac{\mathbf{c} \cdot \mathbf{d}}{(\mathbf{abc})} \mathbf{a} \times \mathbf{b} + \frac{\mathbf{a} \cdot \mathbf{d}}{(\mathbf{abc})} \mathbf{b} \times \mathbf{c} + \frac{\mathbf{b} \cdot \mathbf{d}}{(\mathbf{abc})} \mathbf{c} \times \mathbf{a}$$

11. Consider the system of equations

$$\begin{aligned} a_1x + b_1y + c_1z &= d_1 \\ a_2x + b_2y + c_2z &= d_2 \\ a_3x + b_3y + c_3z &= d_3 \end{aligned} \qquad (2.25)$$

Let $\mathbf{a} = a_1\mathbf{i} + a_2\mathbf{j} + a_3\mathbf{k}$, etc., and show that (2.25) can be written as

$$\mathbf{a}x + \mathbf{b}y + \mathbf{c}z = \mathbf{d}$$

Show that $x = \frac{(\mathbf{dbc})}{(\mathbf{abc})}$, etc., $(\mathbf{abc}) \neq 0$.

2.10. Differentiation of Vectors. Let us consider the vector field

$$\mathbf{u} = A_1(x, y, z, t)\mathbf{i} + A_2(x, y, z, t)\mathbf{j} + A_3(x, y, z, t)\mathbf{k} \tag{2.26}$$

At any point $P(x, y, z)$ and at any time t, (2.26) defines a vector. If we keep P fixed, the vector $\mathbf{u}$ can still change because of the time dependence of its components A_1, A_2, A_3. If we keep the time fixed, we note that the vector at $P(x, y, z)$ will, in general, differ from the vector at $Q(x + dx, y + dy, z + dz)$. Now, in the calculus, the student has learned how to find the change (or differential) of a single function of x, y, z, t. What difficulties do we encounter in the case of a vector? Actually none, since we easily note that $\mathbf{u}$ will change (in magnitude and/or direction) if and only if the components of $\mathbf{u}$ change. The vectors $\mathbf{i}$, $\mathbf{j}$, $\mathbf{k}$ are assumed fixed in space throughout the discussion. We are thus led to the following definition for the differential of a vector:

$$d\mathbf{u} = dA_1\,\mathbf{i} + dA_2\,\mathbf{j} + dA_3\,\mathbf{k} \tag{2.27}$$

where $$dA_i = \frac{\partial A_i}{\partial x}\,dx + \frac{\partial A_i}{\partial y}\,dy + \frac{\partial A_i}{\partial z}\,dz + \frac{\partial A_i}{\partial t}\,dt \qquad i = 1, 2, 3$$

If x, y, z are functions of t, then

$$\frac{d\mathbf{u}}{dt} = \frac{dA_1}{dt}\,\mathbf{i} + \frac{dA_2}{dt}\,\mathbf{j} + \frac{dA_3}{dt}\,\mathbf{k} \tag{2.28}$$

where $$\frac{dA_i}{dt} = \frac{\partial A_i}{\partial x}\frac{dx}{dt} + \frac{\partial A_i}{\partial y}\frac{dy}{dt} + \frac{\partial A_i}{\partial z}\frac{dz}{dt} + \frac{\partial A_i}{\partial t} \qquad i = 1, 2, 3 \tag{2.29}$$

In particular, let $\mathbf{r} = x\mathbf{i} + y\mathbf{j} + z\mathbf{k}$ be the position vector of a moving particle $P(x, y, z)$. Then

$$\mathbf{v} \equiv \frac{d\mathbf{r}}{dt} = \frac{dx}{dt}\,\mathbf{i} + \frac{dy}{dt}\,\mathbf{j} + \frac{dz}{dt}\,\mathbf{k} \tag{2.30}$$

and $$\mathbf{a} \equiv \frac{d\mathbf{v}}{dt} = \frac{d^2\mathbf{r}}{dt^2} = \frac{d^2x}{dt^2}\,\mathbf{i} + \frac{d^2y}{dt^2}\,\mathbf{j} + \frac{d^2z}{dt^2}\,\mathbf{k} \tag{2.31}$$

Equations (2.30) and (2.31) are, by definition, the velocity and acceleration of the particle.

If a vector $\mathbf{u}$ depends on a single variable t, we can define

$$\frac{d\mathbf{u}}{dt} = \lim_{\Delta t \to 0} \frac{\mathbf{u}(t + \Delta t) - \mathbf{u}(t)}{\Delta t} \tag{2.32}$$

(see Fig. 2.18).

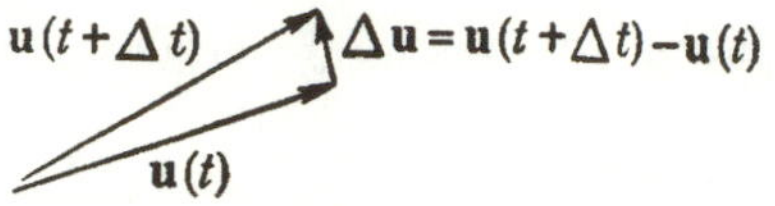

FIG. 2.18

It is easy to verify that (2.32) is equivalent to (2.28). If

$$\mathbf{u} = \mathbf{u}(x, y, z, \ldots)$$

then

$$\frac{\partial \mathbf{u}}{\partial x} = \lim_{\Delta x \to 0} \frac{\mathbf{u}(x + \Delta x, y, z, \ldots) - \mathbf{u}(x, y, z, \ldots)}{\Delta x}$$

or

$$\frac{\partial \mathbf{u}}{\partial x} = \frac{\partial A_1}{\partial x}\mathbf{i} + \frac{\partial A_2}{\partial x}\mathbf{j} + \frac{\partial A_3}{\partial x}\mathbf{k} \tag{2.33}$$

2.11. Differentiation Rules. Consider

$$\begin{aligned}\varphi(t) &= \mathbf{u}(t) \cdot \mathbf{v}(t)\\ \varphi(t + \Delta t) - \varphi(t) &= \mathbf{u}(t + \Delta t) \cdot \mathbf{v}(t + \Delta t) - \mathbf{u}(t) \cdot \mathbf{v}(t)\\ &= (\mathbf{u}(t) + \Delta\mathbf{u}) \cdot (\mathbf{v}(t) + \Delta\mathbf{v}) - \mathbf{u}(t) \cdot \mathbf{v}(t)\\ &= \mathbf{u} \cdot \Delta\mathbf{v} + \mathbf{v} \cdot \Delta\mathbf{u} + \Delta\mathbf{u} \cdot \Delta\mathbf{v}\end{aligned}$$

Hence

$$\frac{\varphi(t + \Delta t) - \varphi(t)}{\Delta t} = \mathbf{u} \cdot \frac{\Delta\mathbf{v}}{\Delta t} + \mathbf{v} \cdot \frac{\Delta\mathbf{u}}{\Delta t} + \Delta\mathbf{u} \cdot \frac{\Delta\mathbf{v}}{\Delta t}$$

and

$$\frac{d\varphi}{dt} = \lim_{\Delta t \to 0} \frac{\varphi(t + \Delta t) - \varphi(t)}{\Delta t} = \mathbf{u} \cdot \frac{d\mathbf{v}}{dt} + \mathbf{v} \cdot \frac{d\mathbf{u}}{dt}$$

or

$$\frac{d}{dt}(\mathbf{u} \cdot \mathbf{v}) = \mathbf{u} \cdot \frac{d\mathbf{v}}{dt} + \mathbf{v} \cdot \frac{d\mathbf{u}}{dt} \tag{2.34}$$

Equation (2.34) also can be easily obtained by writing **u** and **v** in component form. Thus $\varphi = \sum_{i=1}^{3} u_i v_i$, $\frac{d\varphi}{dt} = \sum_{i=1}^{3} u_i \frac{dv_i}{dt} + \sum_{i=1}^{3} v_i \frac{du_i}{dt}$,

$$\frac{d\varphi}{dt} = \mathbf{u} \cdot \frac{d\mathbf{v}}{dt} + \mathbf{v} \cdot \frac{d\mathbf{u}}{dt}$$

Similarly

$$\frac{d}{dt}(\mathbf{u} \times \mathbf{v}) = \mathbf{u} \times \frac{d\mathbf{v}}{dt} + \frac{d\mathbf{u}}{dt} \times \mathbf{v} \tag{2.35}$$

$$\frac{d}{dt}(f\mathbf{u}) = f\frac{d\mathbf{u}}{dt} + \frac{df}{dt}\mathbf{u} \tag{2.36}$$

Example 2.12. Let **u** be a vector of magnitude u. Then $\mathbf{u} \cdot \mathbf{u} = u^2$ so that

$$\mathbf{u} \cdot \frac{d\mathbf{u}}{dt} = u\frac{du}{dt} \tag{2.37}$$

This is a very useful result. In particular, if the magnitude of **u** remains constant, $\frac{du}{dt} = 0$ and $\mathbf{u} \cdot \frac{d\mathbf{u}}{dt} = 0$. This implies, in general, that $\frac{d\mathbf{u}}{dt}$ is perpendicular to **u**, if

$|\mathbf{u}| =$ constant and $\dfrac{d\mathbf{u}}{dt} \neq \mathbf{0}$. The reader should give a geometric proof of this statement.

Example 2.13. *Motion in a Plane.* Let $\mathbf{r}$ be the position vector of a point P moving in a plane having polar coordinates (r, θ) (see Fig. 2.19). Now $\mathbf{r} = r\mathbf{R}$, $\mathbf{R}$ a unit vector. Since $\mathbf{R} = \cos\theta\,\mathbf{i} + \sin\theta\,\mathbf{j}$, we have $\mathbf{P} \equiv \dfrac{d\mathbf{R}}{d\theta} = -\sin\theta\,\mathbf{i} + \cos\theta\,\mathbf{j}$. Why is $\mathbf{P}$ perpendicular to $\mathbf{R}$? Notice that $\mathbf{P}$ is also a unit vector. Differentiating $\mathbf{r} = r\mathbf{R}$ yields

$$\begin{aligned} \mathbf{v} = \frac{d\mathbf{r}}{dt} &= \frac{dr}{dt}\mathbf{R} + r\frac{d\mathbf{R}}{dt} = \frac{dr}{dt}\mathbf{R} + r\frac{d\mathbf{R}}{d\theta}\frac{d\theta}{dt} \\ &= \frac{dr}{dt}\mathbf{R} + r\frac{d\theta}{dt}\mathbf{P} \end{aligned}$$

and
$$\mathbf{a} = \frac{d\mathbf{v}}{dt} = \frac{d^2r}{dt^2}\mathbf{R} + \frac{dr}{dt}\frac{d\mathbf{R}}{d\theta}\frac{d\theta}{dt} + \frac{dr}{dt}\frac{d\theta}{dt}\mathbf{P} + r\frac{d^2\theta}{dt^2}\mathbf{P} + r\frac{d\theta}{dt}\frac{d\mathbf{P}}{d\theta}\frac{d\theta}{dt}$$

so that the acceleration of the particle is

$$\mathbf{a} = \left[\frac{d^2r}{dt^2} - r\left(\frac{d\theta}{dt}\right)^2\right]\mathbf{R} + \frac{1}{r}\frac{d}{dt}\left(r^2\frac{d\theta}{dt}\right)\mathbf{P} \tag{2.38}$$

since $\dfrac{d\mathbf{P}}{d\theta} = -\cos\theta\,\mathbf{i} - \sin\theta\,\mathbf{j} = -\mathbf{R}$. If the particle moves under the action of a central force field, $\mathbf{f} = f\mathbf{R}$, then $\dfrac{d}{dt}\left(r^2\dfrac{d\theta}{dt}\right) = 0$ so that $\frac{1}{2}r^2\dfrac{d\theta}{dt} = h =$ constant. The sectoral area swept out by the particle in time dt is $dA = \frac{1}{2}r^2\,d\theta$, so that $\dfrac{dA}{dt} = h$, and equal areas are swept out in equal periods of time. This is Kepler's first law of planetary motion.

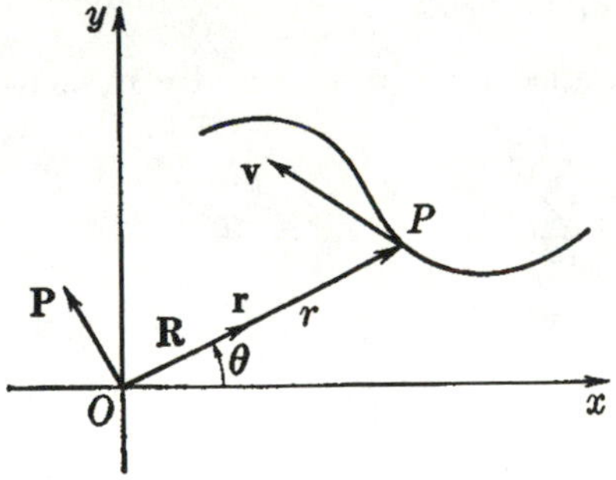

FIG. 2.19

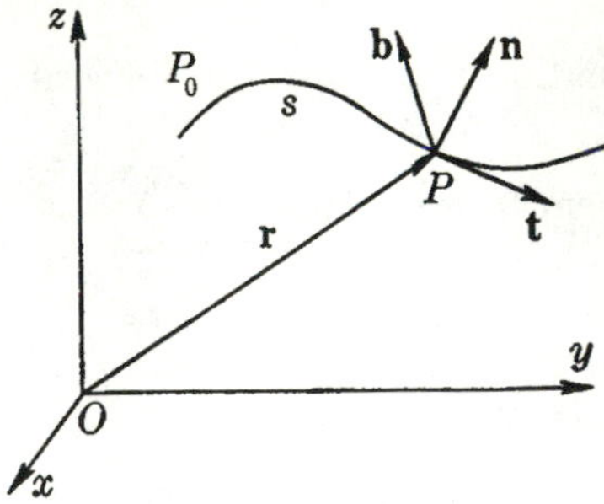

FIG. 2.20

Example 2.14. *Frenet-Serret Formulas.* A three-dimensional curve, Γ, in a Euclidean space can be represented by the locus of the end point of the position vector given by

$$\mathbf{r}(t) = x(t)\mathbf{i} + y(t)\mathbf{j} + z(t)\mathbf{k} \tag{2.39}$$

where t is a parameter ranging over a set of values $t_0 \leqq t \leqq t_1$. If s is arc length along the space curve, then $\dfrac{d\mathbf{r}}{ds}\cdot\dfrac{d\mathbf{r}}{ds} = \dfrac{dx^2 + dy^2 + dz^2}{ds^2} = 1$ from the calculus. It is natural to define $\dfrac{d\mathbf{r}}{ds}$ as the unit *tangent* vector to the space curve Γ (see Fig. 2.20). Since

$\mathbf{t} = \frac{d\mathbf{r}}{ds}$ is a unit vector, $\frac{d\mathbf{t}}{ds}$ is perpendicular to $\mathbf{t}$. Moreover, $\frac{d\mathbf{t}}{ds}$ tells us how fast the direction of $\mathbf{t}$ is changing with respect to arc length s. Hence we define the curvature, κ, of the space curve Γ by $\kappa^2 = \frac{d\mathbf{t}}{ds} \cdot \frac{d\mathbf{t}}{ds}$. κ^2, in general, is a function of s, and hence κ varies from point to point. The principal *normal* vector to the space curve Γ is defined to be the unit vector, $\mathbf{n}$, parallel to $\frac{d\mathbf{t}}{ds}$. Thus

$$\frac{d\mathbf{t}}{ds} = \kappa\mathbf{n} \tag{2.40}$$

The reciprocal of the curvature is called the radius of curvature, $\rho = 1/\kappa$. At any point P on Γ we now have two vectors $\mathbf{t}$ and $\mathbf{n}$ at right angles to each other. This enables us to set up a local coordinate system at P by defining a third vector at right angles to $\mathbf{t}$ and $\mathbf{n}$. We define as the *binormal* the vector $\mathbf{b} \equiv \mathbf{t} \times \mathbf{n}$. The three fundamental vectors $\mathbf{t}$, $\mathbf{n}$, $\mathbf{b}$ form a trihedral at P; any vector associated with the space curve Γ can be written as a linear combination of $\mathbf{t}$, $\mathbf{n}$, $\mathbf{b}$.

Let us now evaluate $\frac{d\mathbf{b}}{ds}$ and $\frac{d\mathbf{n}}{ds}$. From $\mathbf{b} \cdot \mathbf{t} = 0$ we obtain $\frac{d\mathbf{b}}{ds} \cdot \mathbf{t} + \mathbf{b} \cdot \frac{d\mathbf{t}}{ds} = 0$ or $\frac{d\mathbf{b}}{ds} \cdot \mathbf{t} = 0$ since $\mathbf{b} \cdot \mathbf{n} = 0$. Hence $\frac{d\mathbf{b}}{ds}$ is perpendicular to $\mathbf{t}$. Since $\frac{d\mathbf{b}}{ds}$ is also perpendicular to $\mathbf{b}$ ($\mathbf{b}$ is a unit vector), we see that $\frac{d\mathbf{b}}{ds}$ must be parallel to $\mathbf{n}$. Consequently $\frac{d\mathbf{b}}{ds} = \tau\mathbf{n}$, where τ by definition is the magnitude of $\frac{d\mathbf{b}}{ds}$. τ is called the torsion of the curve Γ. To obtain $\frac{d\mathbf{n}}{ds}$, we note that $\mathbf{n} = \mathbf{b} \times \mathbf{t}$ so that

$$\frac{d\mathbf{n}}{ds} = \mathbf{b} \times \frac{d\mathbf{t}}{ds} + \frac{d\mathbf{b}}{ds} \times \mathbf{t} = \mathbf{b} \times \kappa\mathbf{n} + \tau\mathbf{n} \times \mathbf{t} = -\kappa\mathbf{t} - \tau\mathbf{b}$$

The famous Frenet-Serret formulas are

$$\begin{aligned} \frac{d\mathbf{t}}{ds} &= \kappa\mathbf{n} \\ \frac{d\mathbf{n}}{ds} &= -(\kappa\mathbf{t} + \tau\mathbf{b}) \\ \frac{d\mathbf{b}}{ds} &= \tau\mathbf{n} \end{aligned} \tag{2.41}$$

As an example of (2.41) we consider the circular helix given by

$$\mathbf{r} = a \cos t\, \mathbf{i} + a \sin t\, \mathbf{j} + bt\mathbf{k}$$

We have $\mathbf{t} = \frac{d\mathbf{r}}{ds} = (-a \sin t\, \mathbf{i} + a \cos t\, \mathbf{j} + b\mathbf{k}) \frac{dt}{ds}$. From $\mathbf{t} \cdot \mathbf{t} = 1$ we obtain

$$1 = (a^2 + b^2) \left(\frac{dt}{ds}\right)^2$$

so that

$$\mathbf{t} = (-a \sin t\, \mathbf{i} + a \cos t\, \mathbf{j} + b\mathbf{k})(a^2 + b^2)^{-\frac{1}{2}}$$

Thus $\kappa\mathbf{n} = \frac{d\mathbf{t}}{ds} = (-a \cos t\, \mathbf{i} - a \sin t\, \mathbf{j})(a^2 + b^2)^{-1}$, and $\kappa = a(a^2 + b^2)^{-1}$, since $\kappa = \left|\frac{d\mathbf{t}}{ds}\right|$.

From $\mathbf{b} = \mathbf{t} \times \mathbf{n}$ let the reader show that $\mathbf{b} = (b \sin t\, \mathbf{i} - b \cos t\, \mathbf{j} + a\mathbf{k})(a^2 + b^2)^{-\frac{1}{2}}$, $\dfrac{d\mathbf{b}}{ds} = \tau\mathbf{n} = (b \cos t\, \mathbf{i} + b \sin t\, \mathbf{j})(a^2 + b^2)^{-1}$, $\tau = b(a^2 + b^2)^{-1}$.

Example 2.15. *Pursuit Problems.* Let us consider the problem of a missile M pursuing a target T, the motion taking place in the xy plane. $\mathbf{r}_T$ and $\mathbf{V}_T$ are the position and velocity vectors of the target; $\mathbf{r}_M$ and $\mathbf{V}_M$ are the position and velocity vectors of the missile; θ, φ, ψ are defined by Fig. 2.21.

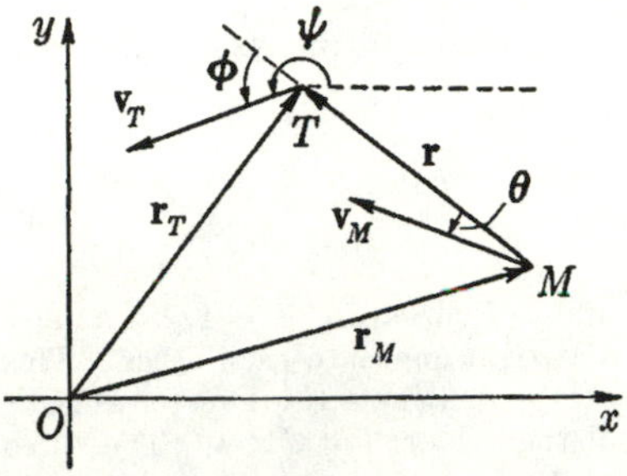

FIG. 2.21

Let $\mathbf{r} = \mathbf{r}_T - \mathbf{r}_M$ so that $\dfrac{d\mathbf{r}}{dt} = \mathbf{V}_T - \mathbf{V}_M$ and $\mathbf{r} \cdot \dfrac{d\mathbf{r}}{dt} = \mathbf{r} \cdot \mathbf{V}_T - \mathbf{r} \cdot \mathbf{V}_M$. Thus $r\dfrac{dr}{dt} = rV_T \cos \varphi - rV_M \cos \theta$, and

$$\frac{dr}{dt} = V_T \cos \varphi - V_M \cos \theta \tag{2.42}$$

Differentiating the identity $\mathbf{r} \cdot \mathbf{V}_T = rV_T \cos \varphi$ yields

$$\mathbf{r} \cdot \frac{d\mathbf{V}_T}{dt} + (\mathbf{V}_T - \mathbf{V}_M) \cdot \mathbf{V}_T = V_T \frac{d}{dt}(r \cos \varphi) + r \cos \varphi \frac{dV_T}{dt} \tag{2.43}$$

If $\mathbf{t}$ is the unit tangent vector to the curve Γ traversed by the target, then $\mathbf{V}_T = V_T\mathbf{t}$ and $\dfrac{d\mathbf{V}_T}{dt} = \dfrac{dV_T}{dt}\mathbf{t} + V_T\dfrac{d\mathbf{t}}{dt}$ so that $\dfrac{d\mathbf{V}_T}{dt} = \dfrac{dV_T}{dt}\mathbf{t} + V_T\dfrac{d\psi}{dt}\mathbf{n}$ since $\dfrac{d\mathbf{t}}{dt} = \left|\dfrac{d\mathbf{t}}{dt}\right|\mathbf{n}$ (see Example 2.14). Equation (2.43) becomes

$$r\frac{dV_T}{dt}\cos \varphi - rV_T\frac{d\psi}{dt}\sin \varphi + V_T^2 - V_MV_T \cos(\theta - \varphi) = V_T\frac{d}{dt}(r \cos \varphi) + r \cos \varphi \frac{dV_T}{dt} \tag{2.44}$$

For the special case of constant target speed, $\dfrac{dV_T}{dt} = 0$, (2.44) becomes

$$-rV_T\frac{d\psi}{dt}\sin \varphi + V_T^2 - V_MV_T \cos(\varphi - \theta) = V_T\frac{d}{dt}(r \cos \varphi) \tag{2.45}$$

Let us apply (2.42), (2.45) to the dog-rabbit problem. At $t = 0$ the rabbit starts at the origin and runs along the positive y axis with constant speed V_T. A dog starts at $(a, 0)$ at $t = 0$ and pursues the rabbit in such a manner (direct pursuit) that $\theta \equiv 0$ throughout the motion. The constant speed of the dog is V_M. We have

$$\begin{aligned}\frac{dr}{dt} &= V_T \cos \varphi - V_M \\ V_T^2 - V_MV_T \cos \varphi &= V_T\frac{d}{dt}(r \cos \varphi)\end{aligned} \tag{2.46}$$

since $\psi \equiv \pi/2$. Equations (2.46) can be written as

$$V_T\frac{d}{dt}(r \cos \varphi) + V_M\frac{dr}{dt} = V_T^2 - V_M^2$$

Integration yields $V_Tr \cos \varphi + V_Mr = (V_T^2 - V_M^2)t + V_Ma$, since $r = a$, $\varphi = -\pi/2$, at $t = 0$. The rabbit is caught when $r = 0$, or at $t = V_Ma(V_M^2 - V_T^2)^{-1}$.

Problems

1. Prove (2.35), (2.36).

2. Show that $\frac{d}{dt}\left(\mathbf{r} \times \frac{d\mathbf{r}}{dt}\right) = \mathbf{r} \times \frac{d^2\mathbf{r}}{dt^2}$.

3. $\mathbf{r} = \mathbf{a} \cos \omega t + \mathbf{b} \sin \omega t$; $\mathbf{a}$, $\mathbf{b}$, ω are constants. Prove that $\mathbf{r} \times \frac{d\mathbf{r}}{dt} = \omega \mathbf{a} \times \mathbf{b}$ and $\frac{d^2\mathbf{r}}{dt^2} + \omega^2 \mathbf{r} = 0$.

4. $\mathbf{R}$ is a unit vector in the direction $\mathbf{r}$, $r = |\mathbf{r}|$. Show that $\mathbf{R} \times d\mathbf{R} = \frac{\mathbf{r} \times d\mathbf{r}}{r^2}$.

5. If $\frac{d\mathbf{a}}{dt} = \boldsymbol{\omega} \times \mathbf{a}$, $\frac{d\mathbf{b}}{dt} = \boldsymbol{\omega} \times \mathbf{b}$, show that $\frac{d}{dt}(\mathbf{a} \times \mathbf{b}) = \boldsymbol{\omega} \times (\mathbf{a} \times \mathbf{b})$.

6. If $\mathbf{r} = \mathbf{a}e^{\omega t} + \mathbf{b}e^{-\omega t}$, $\mathbf{a}$, $\mathbf{b}$, ω constants, show that $\frac{d^2\mathbf{r}}{dt^2} - \omega^2 \mathbf{r} = 0$.

7. Consider the differential equation

$$\frac{d^2\mathbf{u}}{dt^2} + 2A \frac{d\mathbf{u}}{dt} + B\mathbf{u} = 0 \tag{2.47}$$

A, B constants. Assume a solution of the form $\mathbf{u}(t) = \mathbf{C}e^{wt}$, $\mathbf{C}$ a constant vector, w a constant scalar. Show that $\mathbf{u}(t) = \mathbf{C}_1 e^{w_1 t} + \mathbf{C}_2 e^{w_2 t}$ is a solution of (2.47) where w_1, w_2 are roots of $w^2 + 2Aw + B = 0$. What if $w_1 = w_2$?

8. Find a vector $\mathbf{u}$ which satisfies $\frac{d^3\mathbf{u}}{dt^3} - \frac{d^2\mathbf{u}}{dt^2} - 2\frac{d\mathbf{u}}{dt} = 0$ such that $\mathbf{u} = \mathbf{i}$, $\frac{d\mathbf{u}}{dt} = \mathbf{j}$, $\frac{d^2\mathbf{u}}{dt^2} = \mathbf{k}$ for $t = 0$.

9. For the space curve $x = 3t - t^3$, $y = 3t^2$, $z = 3t + t^3$ show that

$$\kappa = \tau = \tfrac{1}{3}(1 + t^2)^{-2}$$

10. Show that $\frac{d^3\mathbf{r}}{ds^3} = -\kappa^2 \mathbf{t} + \frac{d\kappa}{ds}\mathbf{n} - \kappa\tau\mathbf{b}$.

11. Four particles on the corners of a square (sides $= b$) begin to move toward each other in a clockwise fashion in a direct-pursuit course. Each has constant speed V. Show that they move a distance b before contact takes place.

12. In navigational pursuit $V_M \sin \theta = V_T \sin \varphi$. Interpret this result geometrically, assuming θ, φ constants.

13. A target moves on the circumference of a circle with constant speed V. A missile starts at the center of this circle and pursues the target. The speed of the missile is also V. The pursuit is such that the center of the circle, the missile, and the target are collinear. Show that the target moves one-fourth of the circumference up to the moment of capture.

2.12. The Gradient. Let $\varphi(x, y, z)$ be any differentiable space function. From the calculus

$$d\varphi = \frac{\partial \varphi}{\partial x} dx + \frac{\partial \varphi}{\partial y} dy + \frac{\partial \varphi}{\partial z} dz \tag{2.48}$$

The right-hand side of (2.48) suggests that the scalar product of two vectors might be involved. If $\mathbf{r} = x\mathbf{i} + y\mathbf{j} + z\mathbf{k}$ is the position vector of the point $P(x, y, z)$, then $d\mathbf{r} = dx\,\mathbf{i} + dy\,\mathbf{j} + dz\,\mathbf{k}$. Hence, to express $d\varphi$

as a scalar product, one need only define the vector with components $\frac{\partial \varphi}{\partial x}, \frac{\partial \varphi}{\partial y}, \frac{\partial \varphi}{\partial z}$. This vector is called the gradient of $\varphi(x, y, z)$, written del $\varphi \equiv \nabla\varphi$. We define

$$\nabla\varphi \equiv \frac{\partial \varphi}{\partial x}\mathbf{i} + \frac{\partial \varphi}{\partial y}\mathbf{j} + \frac{\partial \varphi}{\partial z}\mathbf{k} \tag{2.49}$$

so that

$$d\varphi = d\mathbf{r} \cdot \nabla\varphi \tag{2.50}$$

The reader should recall from the calculus that $d\varphi$ represents the change in φ as we move from $P(x, y, z)$ to $Q(x + dx, y + dy, z + dz)$, except for infinitesimals of higher order. Equation (2.50) states that this change in φ can be obtained by evaluating the gradient of φ at P and computing the scalar product of $\nabla\varphi$ and $d\mathbf{r}$, $d\mathbf{r}$ being the vector from P to Q.

We now give a geometrical interpretation of $\nabla\varphi$. From $\varphi(x, y, z)$ we can form a family of surfaces $\varphi(x, y, z) =$ constant. The surface S given by $\varphi(x, y, z) = \varphi(x_0, y_0, z_0)$ contains the point $P(x_0, y_0, z_0)$. $\varphi(x, y, z)$ has the constant value $\varphi(x_0, y_0, z_0)$ if we remain on this particular surface S. Now let Q be any point on S near P. Since $d\varphi = 0$, we have, from (2.50), $d\mathbf{r} \cdot \nabla\varphi = 0$. Hence $\nabla\varphi$ is perpendicular to $d\mathbf{r}$. Thus $\nabla\varphi$, at P, is normal to all possible tangents to the surface at P so that $\nabla\varphi$ necessarily must be normal to the surface $\varphi(x, y, z) = \varphi(x_0, y_0, z_0)$ at $P(x_0, y_0, z_0)$. This is a highly important result and should be thoroughly understood by the reader.

Let us now return to (2.50). If $ds = |d\mathbf{r}|$, (2.50) states that

$$\frac{d\varphi}{ds} = \frac{d\mathbf{r}}{ds} \cdot \nabla\varphi = \mathbf{u} \cdot \nabla\varphi \tag{2.50'}$$

where $|\mathbf{u}| = 1$. Hence $\frac{d\varphi}{ds} = |\nabla\varphi| \cos\theta$, where θ is the angle between $\mathbf{u}$ and $\nabla\varphi$. It is obvious that $\frac{d\varphi}{ds}$ has its maximum when $\theta = 0$,

$$\left(\frac{d\varphi}{ds}\right)_{\max} = |\nabla\varphi|$$

The greatest change in $\varphi(x, y, z)$ at $P(x_0, y_0, z_0)$ occurs in the direction of $\nabla\varphi$, that is, the greatest change in φ occurs when we move normal to the surface $\varphi(x, y, z) = \varphi(x_0, y_0, z_0)$. This is to be expected. Let the reader show that $\nabla(\varphi_1 + \varphi_2) = \nabla\varphi_1 + \nabla\varphi_2$.

Example 2.16. Let us find a unit vector perpendicular to the surface

$$x^2 - xy + yz = 1$$

at the point $P(1, 1, 1)$. Here $\varphi(x, y, z) = x^2 - xy + yz$, and

$$\nabla\varphi = (2x - y)\mathbf{i} + (z - x)\mathbf{j} + y\mathbf{k}$$
$$\nabla\varphi = \mathbf{i} + \mathbf{k} \text{ at } P(1, 1, 1)$$
$$\mathbf{N} = \frac{\nabla\varphi}{|\nabla\varphi|} = \frac{\mathbf{i} + \mathbf{k}}{\sqrt{2}}$$

Example 2.17. By direct computation $\nabla r = \mathbf{r}/r$ for $r = (x^2 + y^2 + z^2)^{\frac{1}{2}}$. We obtain this result by a different method. The surface r = constant is a sphere with center at the origin. Since ∇r is perpendicular to the sphere, $\nabla r = f\mathbf{r}$. Now

$$dr = \nabla r \cdot d\mathbf{r} = f\mathbf{r} \cdot d\mathbf{r} = fr\,dr$$

so that $f = 1/r$. Q.E.D.

Example 2.18. Consider $\nabla f(u)$, $u = u(x, y, z)$. We have

$$\nabla f(u) = \frac{\partial f}{\partial x}\mathbf{i} + \frac{\partial f}{\partial y}\mathbf{j} + \frac{\partial f}{\partial z}\mathbf{k} = f'(u)\frac{\partial u}{\partial x}\mathbf{i} + f'(u)\frac{\partial u}{\partial y}\mathbf{j} + f'(u)\frac{\partial u}{\partial z}\mathbf{k}$$

where $f'(u) = \dfrac{df}{du}\cdot$ Hence $\nabla f(u) = f'(u)\nabla u$.

Example 2.19. The operator del, $\nabla \equiv \mathbf{i}\dfrac{\partial}{\partial x} + \mathbf{j}\dfrac{\partial}{\partial y} + \mathbf{k}\dfrac{\partial}{\partial z}$, is a useful concept. It is helpful to keep in mind that ∇ acts both as a differential operator and as a vector in some sense.

Thus $\nabla\varphi = \left(\mathbf{i}\dfrac{\partial}{\partial x} + \mathbf{j}\dfrac{\partial}{\partial y} + \mathbf{k}\dfrac{\partial}{\partial z}\right)\varphi = \mathbf{i}\dfrac{\partial\varphi}{\partial x} + \mathbf{j}\dfrac{\partial\varphi}{\partial y} + \mathbf{k}\dfrac{\partial\varphi}{\partial z}\cdot$ It is easy to show that $\nabla(C_1\varphi_1 + C_2\varphi_2) = C_1\nabla\varphi_1 + C_2\nabla\varphi_2$ if C_1, C_2 are constants. Let the reader show that

$$\nabla(\varphi_1\varphi_2) = \varphi_1\nabla\varphi_2 + \varphi_2\nabla\varphi_1 \tag{2.51}$$

Notice how (2.51) conforms to the rule of calculus for the derivative of a product.

Problems

1. Find the equation of the tangent plane to the surface $xy - z = 1$ at the point $(2, 1, 1)$.

2. Show that $\nabla(\mathbf{a} \cdot \mathbf{r}) = \mathbf{a}$, where $\mathbf{a}$ is a constant vector.

3. If $r = (x^2 + y^2 + z^2)^{\frac{1}{2}}$, show that $\nabla r^n = nr^{n-2}\mathbf{r}$.

4. If $\varphi = (\mathbf{r} \times \mathbf{a}) \cdot (\mathbf{r} \times \mathbf{b})$, show that $\nabla\varphi = \mathbf{b} \times (\mathbf{r} \times \mathbf{a}) + \mathbf{a} \times (\mathbf{r} \times \mathbf{b})$ when $\mathbf{a}$ and $\mathbf{b}$ are constant vectors.

5. Show that the ellipse $r_1 + r_2 = c_1$ and the hyperbola $r_1 - r_2 = c_2$ intersect at right angles when they have the same foci.

6. Find the change of $\varphi = x^2y + yz^2 - xz$ in the direction normal to the surface $yx^2 + xy^2 + z^2y = 3$ at the point $P(1, 1, 1)$.

7. Prove (2.51).

8. If $f = f(u_1, u_2, \ldots, u_n)$, $u_k = u_k(x, y, z)$, $k = 1, 2, \ldots, n$, show that

$$\nabla f = \sum_{k=1}^{n} \frac{\partial f}{\partial u_k}\nabla u_k$$

9. If $\varphi = \varphi(x, y, z, t)$, show that $\dfrac{d\varphi}{dt} = \dfrac{\partial\varphi}{\partial t} + \dfrac{d\mathbf{r}}{dt}\cdot\nabla\varphi$.

10. The equation of an ellipse is $r_1 + r_2 =$ constant. Why is $\nabla(r_1 + r_2) \cdot \mathbf{T} = 0$ if $\mathbf{T}$ is a unit tangent to the ellipse at the point P? $\nabla(r_1 + r_2)$ is computed at P. From $\nabla(r_1 + r_2) = \nabla r_1 + \nabla r_2$, $\nabla r_1 = \frac{\mathbf{r}_1}{r_1}$, $\nabla r_2 = \frac{\mathbf{r}_2}{r_2}$ give a geometric interpretation to

$$\nabla(r_1 + r_2) \cdot \mathbf{T} = 0$$

2.13. The Divergence of a Vector. Let us consider the motion of a fluid of density $\rho(x, y, z, t)$, the velocity of the fluid at any point being given as $\mathbf{V} = \mathbf{V}(x, y, z, t)$. Let

$$\mathbf{f} = \rho\mathbf{V} = X\mathbf{i} + Y\mathbf{j} + Z\mathbf{k}$$

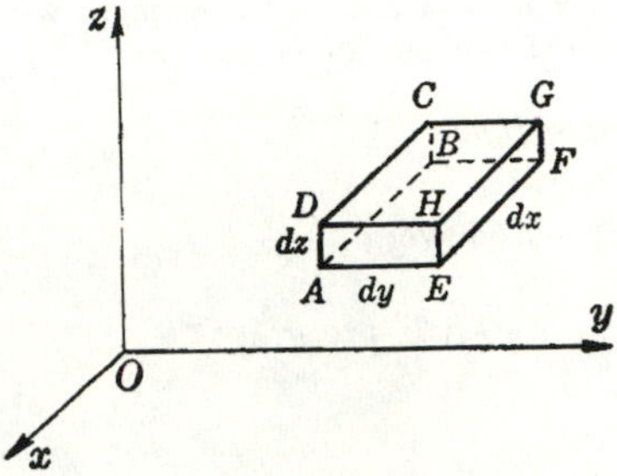

FIG. 2.22

We now concentrate on the flow of fluid through a small parallelogram $ABCDEFGH$ (Fig. 2.22) of dimensions dx, dy, dz. At time t let us calculate the amount of fluid entering the box through the face $ABCD$. The x and z components of the velocity contribute nothing to the flow through $ABCD$. Now $Y(x, y, z, t)$ has the dimension $\frac{M}{L^2T}$, $M =$ mass, $L =$ length, $T =$ time. Thus $Y\,dx\,dz$ has the dimension MT^{-1} and denotes the gain of mass per unit time by the box because of flow through the face $ABCD$. Similarly $\left(Y + \frac{\partial Y}{\partial y}\,dy\right)dx\,dz$ represents the loss of mass per unit time because of flow through the face $EFGH$ at the same time t. The loss of mass per unit time is thus $\frac{\partial Y}{\partial y}\,dx\,dy\,dz$. If we also take into consideration the other faces of the box, we find that the total loss of mass per unit time is

$$\left(\frac{\partial X}{\partial x} + \frac{\partial Y}{\partial y} + \frac{\partial Z}{\partial z}\right) dx\,dy\,dz \tag{2.52}$$

Hence $\frac{\partial X}{\partial x} + \frac{\partial Y}{\partial y} + \frac{\partial Z}{\partial z}$ is the loss of mass per unit time per unit volume.

The scalar $\frac{\partial X}{\partial x} + \frac{\partial Y}{\partial y} + \frac{\partial Z}{\partial z}$ is called the divergence of the vector field $\mathbf{f}$, written

$$\operatorname{div} \mathbf{f} = \frac{\partial X}{\partial x} + \frac{\partial Y}{\partial y} + \frac{\partial Z}{\partial z} \tag{2.53}$$

Returning to the operator $\nabla = \mathbf{i}\frac{\partial}{\partial x} + \mathbf{j}\frac{\partial}{\partial y} + \mathbf{k}\frac{\partial}{\partial z}$, we note that

$$\nabla \cdot \mathbf{f} = \left(\mathbf{i}\frac{\partial}{\partial x} + \mathbf{j}\frac{\partial}{\partial y} + \mathbf{k}\frac{\partial}{\partial z}\right) \cdot (X\mathbf{i} + Y\mathbf{j} + Z\mathbf{k})$$
$$= \frac{\partial X}{\partial x} + \frac{\partial Y}{\partial y} + \frac{\partial Z}{\partial z} = \operatorname{div} \mathbf{f} \tag{2.54}$$

provided we interpret ∇ as both a vector and a differential operator. Let the reader show that

$$\nabla \cdot (\varphi \mathbf{f}) = \varphi \nabla \cdot \mathbf{f} + \nabla \varphi \cdot \mathbf{f} \tag{2.55}$$

Example 2.20. For $\mathbf{r} = x\mathbf{i} + y\mathbf{j} + z\mathbf{k}$, $\nabla \cdot \mathbf{r} = 3$. For $\mathbf{f} = r^{-3}\mathbf{r}$,

$$\nabla \cdot \mathbf{f} = r^{-3}\nabla \cdot \mathbf{r} + \nabla r^{-3} \cdot \mathbf{r} = 3r^{-3} - 3r^{-4}\nabla r \cdot \mathbf{r} = 3r^{-3} - 3r^{-5}\mathbf{r} \cdot \mathbf{r} = 0$$

(see Example 2.17).

Example 2.21. What is the divergence of a gradient?

$$\nabla \cdot (\nabla \varphi) = \nabla \cdot \left(\frac{\partial \varphi}{\partial x}\mathbf{i} + \frac{\partial \varphi}{\partial y}\mathbf{j} + \frac{\partial \varphi}{\partial z}\mathbf{k}\right)$$
$$= \frac{\partial^2 \varphi}{\partial x^2} + \frac{\partial^2 \varphi}{\partial y^2} + \frac{\partial^2 \varphi}{\partial z^2}$$

This important scalar is called the *Laplacian* of $\varphi(x, y, z)$.

$$\operatorname{Lap} \varphi = \nabla \cdot (\nabla \varphi) = \nabla^2 \varphi = \frac{\partial^2 \varphi}{\partial x^2} + \frac{\partial^2 \varphi}{\partial y^2} + \frac{\partial^2 \varphi}{\partial z^2} \tag{2.56}$$

2.14. The Curl of a Vector. Let v_i, $i = 1, 2, 3$, be the components of the velocity of a fluid in an $x^1x^2x^3$ Euclidean coordinate system. The differential change in the components of $\mathbf{v}$ is given by

$$dv_i = \frac{\partial v_i}{\partial x^j}\,dx^j + \frac{\partial v_i}{\partial t}\,dt$$
$$= \frac{1}{2}\left(\frac{\partial v_i}{\partial x^j} + \frac{\partial v_j}{\partial x^i}\right)dx^j + \frac{1}{2}\left(\frac{\partial v_i}{\partial x^j} - \frac{\partial v_j}{\partial x^i}\right)dx^j + \frac{\partial v_i}{\partial t}\,dt \tag{2.57}$$

The terms $s_{ij} \equiv \dfrac{\partial v_i}{\partial x^j} - \dfrac{\partial v_j}{\partial x^i}$ $i, j = 1, 2, 3$, now occupy our attention. The s_{ij} are the elements of a skew-symmetric matrix. As a result there are three important elements, listed as

$$t_1 = \frac{\partial v_3}{\partial x^2} - \frac{\partial v_2}{\partial x^3}$$
$$t_2 = \frac{\partial v_1}{\partial x^3} - \frac{\partial v_3}{\partial x^2} \tag{2.58}$$
$$t_3 = \frac{\partial v_2}{\partial x^1} - \frac{\partial v_1}{\partial x^2}$$

The vector $t_1\mathbf{i} + t_2\mathbf{j} + t_3\mathbf{k}$ is defined and called the curl of $\mathbf{v}$. Using the

∇ operator, we note that the curl of $\mathbf{v}$ may be written

$$\text{curl}\ \mathbf{v} = \nabla \times \mathbf{v} = \begin{vmatrix} \mathbf{i} & \mathbf{j} & \mathbf{k} \\ \dfrac{\partial}{\partial x} & \dfrac{\partial}{\partial y} & \dfrac{\partial}{\partial z} \\ v_1 & v_2 & v_3 \end{vmatrix} \tag{2.59}$$

Example 2.22. Let $\mathbf{f} = x^2yz\mathbf{i} - 2xyz^2\mathbf{j} + y^2z\mathbf{k}$.

$$\begin{aligned} \nabla \times \mathbf{f} &= \begin{vmatrix} \mathbf{i} & \mathbf{j} & \mathbf{k} \\ \dfrac{\partial}{\partial x} & \dfrac{\partial}{\partial y} & \dfrac{\partial}{\partial z} \\ x^2yz & -2xyz^2 & y^2z \end{vmatrix} \\ &= (2yz + 4xyz)\mathbf{i} + x^2y\mathbf{j} - (2yz^2 + x^2z)\mathbf{k} \end{aligned}$$

Example 2.23. Consider $\nabla \times (\varphi\mathbf{f})$. We have, for $\mathbf{f} = u\mathbf{i} + v\mathbf{j} + w\mathbf{k}$,

$$\begin{aligned} \nabla \times (\varphi\mathbf{f}) &= \begin{vmatrix} \mathbf{i} & \mathbf{j} & \mathbf{k} \\ \dfrac{\partial}{\partial x} & \dfrac{\partial}{\partial y} & \dfrac{\partial}{\partial z} \\ \varphi u & \varphi v & \varphi w \end{vmatrix} \\ &= \varphi \begin{vmatrix} \mathbf{i} & \mathbf{j} & \mathbf{k} \\ \dfrac{\partial}{\partial x} & \dfrac{\partial}{\partial y} & \dfrac{\partial}{\partial z} \\ u & v & w \end{vmatrix} + \begin{vmatrix} \mathbf{i} & \mathbf{j} & \mathbf{k} \\ \dfrac{\partial \varphi}{\partial x} & \dfrac{\partial \varphi}{\partial y} & \dfrac{\partial \varphi}{\partial z} \\ u & v & w \end{vmatrix} \\ &= \varphi\nabla \times \mathbf{f} + (\nabla\varphi) \times \mathbf{f} \end{aligned} \tag{2.60}$$

To obtain the curl of $\varphi\mathbf{f}$, keep φ fixed, and let ∇ operate on $\mathbf{f}$, yielding $\varphi\nabla \times \mathbf{f}$; then keep $\mathbf{f}$ fixed, and allow ∇ to operate on φ, yielding $\nabla\varphi$, and complete the vector product $(\nabla\varphi) \times \mathbf{f}$. The sum of these operations yields $\nabla \times (\varphi\mathbf{f})$.

Example 2.24. The curl of a gradient is zero.

$$\begin{aligned} \nabla \times (\nabla\varphi) &= \begin{vmatrix} \mathbf{i} & \mathbf{j} & \mathbf{k} \\ \dfrac{\partial}{\partial x} & \dfrac{\partial}{\partial y} & \dfrac{\partial}{\partial z} \\ \dfrac{\partial \varphi}{\partial x} & \dfrac{\partial \varphi}{\partial y} & \dfrac{\partial \varphi}{\partial z} \end{vmatrix} \\ &= \mathbf{i}\left(\frac{\partial^2\varphi}{\partial y\,\partial z} - \frac{\partial^2\varphi}{\partial z\,\partial y}\right) + \mathbf{j}\left(\frac{\partial^2\varphi}{\partial z\,\partial x} - \frac{\partial^2\varphi}{\partial x\,\partial z}\right) + \mathbf{k}\left(\frac{\partial^2\varphi}{\partial x\,\partial y} - \frac{\partial^2\varphi}{\partial y\,\partial x}\right) \\ &= 0 \end{aligned}$$

provided φ has continuous mixed second derivatives.

2.15. Further Properties of ∇ Operator. We define the product $\mathbf{u} \cdot \nabla$ to be the scalar differential operator

$$u_x \frac{\partial}{\partial x} + u_y \frac{\partial}{\partial y} + u_z \frac{\partial}{\partial z} \tag{2.61}$$

We then have

$$(\mathbf{u} \cdot \nabla)\mathbf{v} = u_x \frac{\partial \mathbf{v}}{\partial x} + u_y \frac{\partial \mathbf{v}}{\partial y} + u_z \frac{\partial \mathbf{v}}{\partial z}$$

Let us now investigate $\mathbf{u} \times (\nabla \times \mathbf{v})$. Assuming that we can expand this term by the rule of the middle factor [remember that the expansion $\mathbf{a} \times (\mathbf{b} \times \mathbf{c}) = (\mathbf{a} \cdot \mathbf{c})\mathbf{b} - (\mathbf{a} \cdot \mathbf{b})\mathbf{c}$ holds true only for vectors; ∇, strictly speaking, is not a vector], we have

$$\mathbf{u} \times (\nabla \times \mathbf{v}) = \nabla_v(\mathbf{u} \cdot \mathbf{v}) - (\mathbf{u} \cdot \nabla)\mathbf{v} \tag{2.62}$$

The subscript v in the term $\nabla_v(\mathbf{u} \cdot \mathbf{v})$ means that ∇_v operates only on the components of $\mathbf{v}$. Thus

$$\begin{aligned}\nabla_v(\mathbf{u} \cdot \mathbf{v}) &= \nabla_v(u_x v_x + u_y v_y + u_z v_z) \\ &= \left(u_x \frac{\partial v_x}{\partial x} + u_y \frac{\partial v_y}{\partial x} + u_z \frac{\partial v_z}{\partial x}\right)\mathbf{i} + \cdots\end{aligned}$$

Interchanging the role of $\mathbf{u}$ and $\mathbf{v}$ yields

$$\mathbf{v} \times (\nabla \times \mathbf{u}) = \nabla_u(\mathbf{u} \cdot \mathbf{v}) - (\mathbf{v} \cdot \nabla)\mathbf{u} \tag{2.63}$$

Adding (2.62) and (2.63) results in

$$\nabla_u(\mathbf{u} \cdot \mathbf{v}) + \nabla_v(\mathbf{u} \cdot \mathbf{v}) = \mathbf{u} \times (\nabla \times \mathbf{v}) + \mathbf{v} \times (\nabla \times \mathbf{u}) + (\mathbf{u} \cdot \nabla)\mathbf{v} + (\mathbf{v} \cdot \nabla)\mathbf{u}$$

and

$$\nabla(\mathbf{u} \cdot \mathbf{v}) = \mathbf{u} \times (\nabla \times \mathbf{v}) + \mathbf{v} \times (\nabla \times \mathbf{u}) + (\mathbf{u} \cdot \nabla)\mathbf{v} + (\mathbf{v} \cdot \nabla)\mathbf{u} \tag{2.64}$$

The above analysis in no way constitutes a proof of (2.64). We leave it to the reader to verify (2.64) by direct expansion. The same remarks hold for the following examples:

$$\begin{aligned}\nabla \times (\mathbf{u} \times \mathbf{v}) &= \nabla_u \times (\mathbf{u} \times \mathbf{v}) + \nabla_v \times (\mathbf{u} \times \mathbf{v}) \\ &= (\mathbf{v} \cdot \nabla)\mathbf{u} - \mathbf{v}(\nabla \cdot \mathbf{u}) + (\mathbf{u} \cdot \nabla)\mathbf{v} - (\mathbf{u} \cdot \nabla)\mathbf{v}\end{aligned} \tag{2.65}$$

$$\begin{aligned}\nabla \cdot (\mathbf{u} \times \mathbf{v}) &= \nabla_u \cdot (\mathbf{u} \times \mathbf{v}) + \nabla_v \cdot (\mathbf{u} \times \mathbf{v}) \\ &= (\nabla \times \mathbf{u}) \cdot \mathbf{v} - \nabla_v \cdot (\mathbf{v} \times \mathbf{u}) \\ &= (\nabla \times \mathbf{u}) \cdot \mathbf{v} - (\nabla \times \mathbf{v}) \cdot \mathbf{u}\end{aligned} \tag{2.66}$$

We now list some important identities:

(1) $\nabla(uv) = u\nabla v + v\nabla u$
(2) $\nabla \cdot (\varphi\mathbf{u}) = \varphi\nabla \cdot \mathbf{u} + (\nabla\varphi) \cdot \mathbf{u}$
(3) $\nabla \times (\varphi\mathbf{u}) = \varphi\nabla \times \mathbf{u} + (\nabla\varphi) \times \mathbf{u}$
(4) $\nabla \times (\nabla\varphi) = 0$
(5) $\nabla \cdot (\nabla \times \mathbf{u}) = 0$
(6) $\nabla \cdot (\mathbf{u} \times \mathbf{v}) = (\nabla \times \mathbf{u}) \cdot \mathbf{v} - (\nabla \times \mathbf{v}) \cdot \mathbf{u}$
(7) $\nabla \times (\mathbf{u} \times \mathbf{v}) = (\mathbf{v} \cdot \nabla)\mathbf{u} + (\mathbf{u} \cdot \nabla)\mathbf{v} - \mathbf{v}(\nabla \cdot \mathbf{u}) - \mathbf{u}(\nabla \cdot \mathbf{v})$
(8) $\nabla(\mathbf{u} \cdot \mathbf{v}) = \mathbf{u} \times (\nabla \times \mathbf{v}) + \mathbf{v} \times (\nabla \times \mathbf{u}) + (\mathbf{u} \cdot \nabla)\mathbf{v} + (\mathbf{v} \cdot \nabla)\mathbf{u}$
(9) $\nabla \times (\nabla \times \mathbf{u}) = \nabla(\nabla \cdot \mathbf{u}) - \nabla^2\mathbf{u}$
(10) $(\mathbf{u} \cdot \nabla)\mathbf{r} = \mathbf{u}$
(11) $\nabla \cdot \mathbf{r} = 3$

(12) $\nabla \times \mathbf{r} = 0$

(13) $d\varphi = d\mathbf{r} \cdot \nabla\varphi + \frac{\partial \varphi}{dt}\, dt$

(14) $d\mathbf{f} = (d\mathbf{r} \cdot \nabla)\mathbf{f} + \frac{\partial \mathbf{f}}{\partial t}\, dt$

(15) $\nabla \cdot (r^{-3}\mathbf{r}) = 0$

Problems

1. Show that the divergence of a curl is zero.

2. Find the divergence and curl of $x\mathbf{i} - y\mathbf{j}/x + y$; of $x \cos z\, \mathbf{i} + y \ln x\, \mathbf{j} - z^2\mathbf{k}$.

3. If $\mathbf{A} = ax\mathbf{i} + by\mathbf{j} + cz\mathbf{k}$, show that $\nabla(\mathbf{A} \cdot \mathbf{r}) = 2\mathbf{A}$.

4. Show that $\nabla^2(1/r) = 0$, $r = (x^2 + y^2 + z^2)^{\frac{1}{2}}$.

5. Show that $\nabla \times [f(r)\mathbf{r}] = 0$.

6. If $\mathbf{f} = u\nabla v$ show that $\mathbf{f} \cdot \nabla \times \mathbf{f} = 0$, u not constant.

7. Show that $(\mathbf{v} \cdot \nabla)\mathbf{v} = \frac{1}{2}\nabla\mathbf{v}^2 - \mathbf{v} \times (\nabla \times \mathbf{v})$.

8. If $\mathbf{A}$ is a constant unit vector show that $\mathbf{A} \cdot [\nabla(\mathbf{A} \cdot \mathbf{v}) - \nabla \times (\mathbf{v} \times \mathbf{A})] = \nabla \cdot \mathbf{v}$.

9. If $\mathbf{w}$ is a constant vector, $\mathbf{r} = x\mathbf{i} + y\mathbf{j} + z\mathbf{k}$, show that $\nabla \times (\mathbf{w} \times \mathbf{r}) = 2\mathbf{w}$.

10. Show that $\nabla \cdot (\mathbf{u} \times \mathbf{v}) = (\nabla \times \mathbf{u}) \cdot \mathbf{v} - (\nabla \times \mathbf{v}) \cdot \mathbf{u}$.

11. Show that $d\mathbf{f} = (d\mathbf{r} \cdot \nabla)\mathbf{f} + \frac{\partial \mathbf{f}}{\partial t}\, dt$ if $\mathbf{f} = \mathbf{f}(x, y, z, t)$.

12. Show that $\nabla \times (\nabla \times \mathbf{u}) = \nabla(\nabla \cdot \mathbf{u}) - \nabla^2\mathbf{u}$.

13. Let $u = u(x, y, z)$, $v = v(x, y, z)$, and assume $\nabla u \times \nabla v = 0$. Let $d\mathbf{r}$ be the vector from P to Q; P and Q are points on the surface $u(x, y, z) =$ constant. From $dv = d\mathbf{r} \cdot \nabla v$, show that $dv = 0$ and hence that v remains constant when u is constant. This implies that $v = f(u)$ or $F(u, v) = 0$. Show conversely that if u and v satisfy a relationship $F(u, v) = 0$ then $\nabla u \times \nabla v = 0$. *Hint:* $\nabla F(u, v) = 0 = \frac{\partial F}{\partial u} \nabla u + \frac{\partial F}{\partial v} \nabla v$. We assume that $\frac{\partial F}{\partial u}$ and $\frac{\partial F}{\partial v}$ do not both vanish identically.

14. Prove that a necessary and sufficient condition that u, v, w satisfy an equation $F(u, v, w) = 0$ is that $\nabla u \cdot \nabla v \times \nabla w = 0$, or

$$J\left(\frac{u, v, w}{x, y, z}\right) \equiv \begin{vmatrix} \frac{\partial u}{\partial x} & \frac{\partial u}{\partial y} & \frac{\partial u}{\partial z} \\ \frac{\partial v}{\partial x} & \frac{\partial v}{\partial y} & \frac{\partial v}{\partial z} \\ \frac{\partial w}{\partial x} & \frac{\partial w}{\partial y} & \frac{\partial w}{\partial z} \end{vmatrix} = 0$$

This determinant is called the *Jacobian* of (u, v, w) with respect to (x, y, z).

15. Let $\mathbf{A} = \nabla \times (\varphi\mathbf{i})$, $\nabla^2\varphi = 0$, $\varphi = X(x)Y(y)Z(z)$. Show that $\mathbf{A} \cdot \nabla \times \mathbf{A} = 0$.

16. Show that $(\nabla \times \mathbf{f}) \cdot (\mathbf{u} \times \mathbf{v}) = [(\mathbf{u} \cdot \nabla)\mathbf{f}] \cdot \mathbf{v} - [(\mathbf{v} \cdot \nabla)\mathbf{f}] \cdot \mathbf{u}$.

2.16. Orthogonal Curvilinear Coordinates. Up to the present moment we have expressed the formulas for the gradient, divergence, curl, and Laplacian in the familiar rectangular cartesian coordinate system. It is often quite necessary to express the above quantities in other coordinate systems. For example, if one were to solve $\nabla^2 V = 0$ subject to the boundary condition that $V =$ constant on the sphere $x^2 + y^2 + z^2 = a^2$, one would find it of great aid to express $\nabla^2 V$ in spherical coordinates.

The boundary conditions of a physical problem dictate to a great extent the coordinate system to be used.

Let us now consider a spherical coordinate system (see Fig. 2.23). The relationships between x, y, z and r, θ, φ are

$$
\begin{aligned}
r &= (x^2 + y^2 + z^2)^{\frac{1}{2}} \\
\theta &= \cos^{-1} \frac{z}{(x^2 + y^2 + z^2)^{\frac{1}{2}}} \\
\varphi &= \tan^{-1} \frac{y}{x}
\end{aligned}
\tag{2.67}
$$

and

$$
\begin{aligned}
x &= r \sin \theta \cos \varphi \\
y &= r \sin \theta \sin \varphi \\
z &= r \cos \theta
\end{aligned}
\tag{2.68}
$$

Let us note the following pertinent facts: Through any point $P(x, y, z)$, other than the origin, there pass the sphere $r =$ constant, the cone $\theta =$ constant, and the plane $\varphi =$ constant. These surfaces intersect in pairs which yield three curves through P. The intersection of the sphere and the cone is a circle. Along this curve only φ can change, since r and θ are constant. This curve is called appropriately the φ curve. At P we construct a unit vector, $\mathbf{e}_\varphi$, tangent to the φ curve. The direction of $\mathbf{e}_\varphi$ is chosen in the direction of positive increase in φ. Similarly one obtains $\mathbf{e}_r$ and $\mathbf{e}_\theta$. The reader can easily verify that these unit tangents form an orthogonal trihedral at P such that $\mathbf{e}_r \times \mathbf{e}_\theta = \mathbf{e}_\varphi$. The vectors $\mathbf{e}_r$, $\mathbf{e}_\theta$, $\mathbf{e}_\varphi$ form a basis for spherical coordinates in exactly the same manner that $\mathbf{i}$, $\mathbf{j}$, $\mathbf{k}$ form a basis for rectangular coordinates. Any vector at P may be written $\mathbf{f} = f_r\mathbf{e}_r + f_\theta\mathbf{e}_\theta + f_\varphi\mathbf{e}_\varphi$, where f_r, f_θ, f_φ are the projections of $\mathbf{f}$ on the vectors $\mathbf{e}_r$, $\mathbf{e}_\theta$, $\mathbf{e}_\varphi$, respectively. Unlike $\mathbf{i}$, $\mathbf{j}$, $\mathbf{k}$ the vectors $\mathbf{e}_r$, $\mathbf{e}_\theta$, $\mathbf{e}_\varphi$ change directions as we move from point to point. Thus we may expect to find more complicated formulas arising when the gradient, divergence, etc., are computed in spherical coordinates.

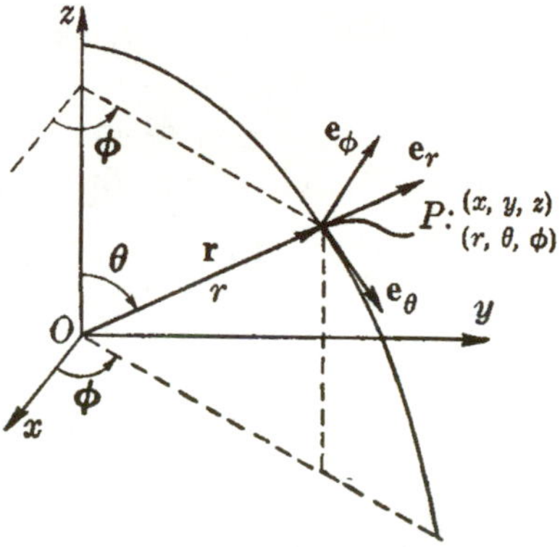

FIG. 2.23

Since ∇r is perpendicular to the surface $r =$ constant, ∇r is parallel to $\mathbf{e}_r$. Similarly $\nabla\theta$ is parallel to $\mathbf{e}_\varphi$; $\nabla\varphi$ is parallel to $\mathbf{e}_\varphi$. Thus

$$
\begin{aligned}
\mathbf{e}_r &= h_r \nabla r \\
\mathbf{e}_\theta &= h_\theta \nabla \theta \\
\mathbf{e}_\varphi &= h_\varphi \nabla \varphi
\end{aligned}
\tag{2.69}
$$

Let $d\mathbf{r}$ be a vector of length ds parallel to $\mathbf{e}_r$. Then $dr = d\mathbf{r} \cdot \nabla r$ [see (2.50)], so that $h_r\, dr = d\mathbf{r} \cdot h_r \nabla r = d\mathbf{r} \cdot \mathbf{e}_r = ds$. Since $dr = ds$, we have $h_r = 1$. Now let $d\mathbf{r}$ be a vector of length ds parallel to $\mathbf{e}_\theta$. We have $d\theta = d\mathbf{r} \cdot \nabla\theta$, $h_\theta\, d\theta = d\mathbf{r} \cdot h_\theta \nabla\theta = d\mathbf{r} \cdot \mathbf{e}_\theta = ds$. It is seen that h_θ is that factor which must be multiplied into $d\theta$ to yield arc length. Thus $h_\theta = r$, and similarly $h_\varphi = r \sin\theta$. In spherical coordinates

$$ds^2 = dr^2 + r^2\, d\theta^2 + r^2 \sin^2\theta\, d\varphi^2$$

The positive square roots of the coefficients of dr^2, $d\theta^2$, $d\varphi^2$ yield h_r, h_θ, h_φ, respectively.

We can now write

$$\begin{aligned} \mathbf{e}_r &= \nabla r = \mathbf{e}_\theta \times \mathbf{e}_\varphi = r^2 \sin\theta \nabla\theta \times \nabla\varphi \\ \mathbf{e}_\theta &= r\nabla\theta = \mathbf{e}_\varphi \times \mathbf{e}_r = r \sin\theta \nabla\varphi \times \nabla r \\ \mathbf{e}_\varphi &= r \sin\theta \nabla\varphi = \mathbf{e}_r \times \mathbf{e}_\theta = r\nabla r \times \nabla\theta \end{aligned} \tag{2.70}$$

The differential of volume is

$$\begin{aligned} dV = ds_1\, ds_2\, ds_3 &= h_r h_\theta h_\varphi\, dr\, d\theta\, d\varphi \\ &= r^2 \sin\theta\, dr\, d\theta\, d\varphi \end{aligned}$$

It is easy to show that $\nabla r \cdot \nabla\theta \times \nabla\varphi = (h_r h_\theta h_\varphi)^{-1} = (r^2 \sin\theta)^{-1}$.

Now we consider the gradient of $f(r, \theta, \varphi)$. We have

$$\begin{aligned} \nabla f &= \frac{\partial f}{\partial r} \nabla r + \frac{\partial f}{\partial \theta} \nabla\theta + \frac{\partial f}{\partial \varphi} \nabla\varphi \\ &= \frac{1}{h_r} \frac{\partial f}{\partial r} \mathbf{e}_r + \frac{1}{h_\theta} \frac{\partial f}{\partial \theta} \mathbf{e}_\theta + \frac{1}{h_\varphi} \frac{\partial f}{\partial \varphi} \mathbf{e}_\varphi \\ &= \frac{\partial f}{\partial r} \mathbf{e}_r + \frac{1}{r} \frac{\partial f}{\partial \theta} \mathbf{e}_\theta + \frac{1}{r \sin\theta} \frac{\partial f}{\partial \varphi} \mathbf{e}_\varphi \end{aligned} \tag{2.71}$$

Equation (2.71) is the gradient of a scalar in spherical coordinates.

To compute the divergence of $\mathbf{f} = f_r \mathbf{e}_r + f_\theta \mathbf{e}_\theta + f_\varphi \mathbf{e}_\varphi$, we write

$$\mathbf{f} = f_r r^2 \sin\theta \nabla\theta \times \nabla\varphi + f_\theta r \sin\theta \nabla\varphi \times \nabla r + f_\varphi r \nabla r \times \nabla\varphi$$

from (2.70). Then

$$\nabla \cdot \mathbf{f} = \nabla(f_r r^2 \sin\theta) \cdot \nabla\theta \times \nabla\varphi + \nabla(f_\theta r \sin\theta) \cdot \nabla\varphi \times \nabla r + \nabla(f_\varphi r) \cdot \nabla r \times \nabla\varphi$$

since $\nabla \cdot (\nabla\theta \times \nabla\varphi) = 0$, etc. [see formula for $\nabla \cdot (\mathbf{u} \times \mathbf{v})$]. But

$$\begin{aligned} \nabla(f_r r^2 \sin\theta) \cdot \nabla\theta \times \nabla\varphi &= \frac{\partial}{\partial \theta} (f_r r^2 \sin\theta) \nabla\theta \cdot \nabla\theta \times \nabla\varphi \\ &\quad + \frac{\partial (f_r r^2 \sin\theta)}{\partial \varphi} \nabla\varphi \cdot \nabla\theta \times \nabla\varphi \\ &\quad + \frac{\partial (f_r r^2 \sin\theta)}{\partial r} \nabla r \cdot \nabla\theta \times \nabla\varphi \\ &= \frac{1}{h_r h_\theta h_\varphi} \frac{\partial}{\partial r} (f_r r^2 \sin\theta) = \frac{1}{r^2 \sin\theta} \frac{\partial}{\partial r} (f_r r^2 \sin\theta) \end{aligned}$$

We thus obtain

$$\nabla \cdot \mathbf{f} = \frac{1}{r^2 \sin \theta} \left[\frac{\partial}{\partial r} (f_r r^2 \sin \theta) + \frac{\partial}{\partial \theta} (f_\theta r \sin \theta) + \frac{\partial}{\partial \varphi} (f_\varphi r) \right] \quad (2.72)$$

Equation (2.72) is the divergence of **f** in spherical coordinates.

If we apply (2.71) and (2.72) to $\mathbf{f} = \nabla V$, we obtain

$$\nabla^2 V = \frac{1}{r^2 \sin \theta} \left[\frac{\partial}{\partial r} \left(r^2 \sin \theta \frac{\partial V}{\partial r} \right) + \frac{\partial}{\partial \theta} \left(\sin \theta \frac{\partial V}{\partial \theta} \right) + \frac{\partial}{\partial \varphi} \left(\frac{1}{\sin \theta} \frac{\partial V}{\partial \varphi} \right) \right] \quad (2.73)$$

Equation (2.73) is the Laplacian of V in spherical coordinates.

To obtain the curl of **f**, let $\mathbf{f} = f_r h_r \nabla r + f_\theta h_\theta \nabla \theta + f_\varphi h_\varphi \nabla \varphi$. It can be easily shown that

$$\nabla \times \mathbf{f} = \frac{1}{h_r h_\theta h_\varphi} \begin{vmatrix} h_r \mathbf{e}_r & h_\theta \mathbf{e}_\theta & h_\varphi \mathbf{e}_\varphi \\ \frac{\partial}{\partial r} & \frac{\partial}{\partial \theta} & \frac{\partial}{\partial \varphi} \\ h_r f_r & h_\theta f_\theta & h_\varphi f_\varphi \end{vmatrix} \quad (2.74)$$

For the general orthogonal curvilinear coordinate system (u_1, u_2, u_3) where $ds^2 = h_1^2 \, du_1^2 + h_2^2 \, du_2^2 + h_3^2 \, du_3^2$ we list without proof

$$\nabla f = \frac{1}{h_1} \frac{\partial f}{\partial u_1} \mathbf{e}_1 + \frac{1}{h_2} \frac{\partial f}{\partial u_2} \mathbf{e}_2 + \frac{1}{h_3} \frac{\partial f}{\partial u_3} \mathbf{e}_3$$

$$\nabla \cdot \mathbf{f} = \frac{1}{h_1 h_2 h_3} \left[\frac{\partial}{\partial u_1} (h_2 h_3 f_1) + \frac{\partial}{\partial u_2} (h_3 h_1 f_2) + \frac{\partial}{\partial u_3} (h_1 h_2 f_3) \right]$$

$$\nabla \times \mathbf{f} = \frac{1}{h_1 h_2 h_3} \begin{vmatrix} h_1 \mathbf{e}_1 & h_2 \mathbf{e}_2 & h_3 \mathbf{e}_3 \\ \frac{\partial}{\partial u_1} & \frac{\partial}{\partial u_2} & \frac{\partial}{\partial u_3} \\ h_1 f_1 & h_2 f_2 & h_3 f_3 \end{vmatrix} \quad (2.75)$$

$$\nabla^2 f = \frac{1}{h_1 h_2 h_3} \left[\frac{\partial}{\partial u_1} \left(\frac{h_2 h_3}{h_1} \frac{\partial f}{\partial u_1} \right) + \frac{\partial}{\partial u_2} \left(\frac{h_3 h_1}{h_2} \frac{\partial f}{\partial u_2} \right) + \frac{\partial}{\partial u_3} \left(\frac{h_1 h_2}{h_3} \frac{\partial f}{\partial u_3} \right) \right]$$

It would be a good exercise for the student to derive (2.75).

Problems

1. For $x = r \sin \theta \cos \varphi$, $y = r \sin \theta \sin \varphi$, $z = r \cos \theta$ show that the form

$$ds^2 = dx^2 + dy^2 + dz^2$$

becomes $ds^2 = dr^2 + r^2 \, d\theta^2 + r^2 \sin^2 \theta \, d\varphi^2$.

2. For $x = r \cos \varphi$, $y = r \sin \varphi$, $z = z$ show that $ds^2 = dr^2 + r^2 \, d\varphi^2 + dz^2$.

3. Express ∇f, $\nabla \cdot \mathbf{f}$, $\nabla \times \mathbf{f}$ in cylindrical coordinates and show that

$$\nabla^2 f = \frac{1}{r} \left[\frac{\partial}{\partial r} \left(r \frac{\partial f}{\partial r} \right) + \frac{\partial}{\partial \varphi} \left(\frac{1}{r} \frac{\partial f}{\partial \varphi} \right) + \frac{\partial}{\partial z} \left(r \frac{\partial f}{\partial f} \right) \right]$$

4. Solve $\nabla^2 V = 0$ in spherical coordinates if $V = V(r)$.
5. Solve $\nabla^2 V = 0$ in cylindrical coordinates if $V = V(r)$.
6. Show that $\nabla \times [f(r)\mathbf{r}] = 0$, r a spherical coordinate.
7. By making use of $\nabla^2\mathbf{V} = \nabla(\nabla \cdot \mathbf{V}) - \nabla \times (\nabla \times \mathbf{V})$, find $\nabla^2\mathbf{V}$ for $\mathbf{V} = v(r)\mathbf{e}_r$, $\mathbf{V}$ purely radial in spherical coordinates.
8. Express $\nabla^2\mathbf{V}$, for $\mathbf{V} = f(r)\mathbf{e}_r + \varphi(z)\mathbf{e}_z$, in cylindrical coordinates.
9. Consider the equation

$$(\lambda + \mu)\nabla(\nabla \cdot \mathbf{s}) + \mu\nabla^2\mathbf{s} = \rho \frac{\partial^2 \mathbf{s}}{\partial t^2}$$

λ, μ, ρ constants. Assume $\mathbf{s} = e^{ipt}\mathbf{s}_1$, p constant, $i = \sqrt{-1}$, and show that

$$(\lambda + \mu)\nabla(\nabla \cdot \mathbf{s}_1) + (\mu + \rho p^2)\mathbf{s}_1 = 0$$

Next show that $[\nabla^2 + (\mu + \rho p^2)/\lambda + \mu](\nabla \cdot \mathbf{s}_1) = 0$, $\lambda + \mu \neq 0$.
10. If $\mathbf{A} = \nabla \times (x\mathbf{r})$, $\nabla^2\psi = 0$, $\psi = R(r)\Theta(\theta)\Phi(\varphi)$, show that $\mathbf{A} \cdot \nabla \times \mathbf{A} = 0$. Is it necessary that $\psi = R(r)\Theta(\theta)\Phi(\varphi)$?
11. If $ds^2 = \sum_{i=1}^{3} dx^i\, dx^i$, $x^i = x^i(y^1, y^2, y^3)$, $i = 1, 2, 3$, show that

$$ds^2 = g_{\alpha\beta}\, dy^\alpha\, dy^\beta$$

where

$$g_{\alpha\beta} = \sum_{i=1}^{3} \frac{\partial x^i}{\partial y^\alpha} \frac{\partial x^i}{\partial y^\beta}$$

12. Derive (2.74).
13. Derive (2.75).

2.17. The Line Integral. We start with a vector field

$$\mathbf{f} = X(x, y, z)\mathbf{i} + Y(x, y, z)\mathbf{j} + Z(x, y, z)\mathbf{k}$$

Let $\mathbf{r}(t) = x(t)\mathbf{i} + y(t)\mathbf{j} + z(t)\mathbf{k}$, $a \leq t \leq b$, be a space curve Γ joining the points A and B with position vectors $\mathbf{r}(a)$, $\mathbf{r}(b)$. One may subdivide Γ into n parts by subdividing t into

$$a = t_0 < t_1 < t_2 < \cdots t_j < \cdots < t_n = b$$

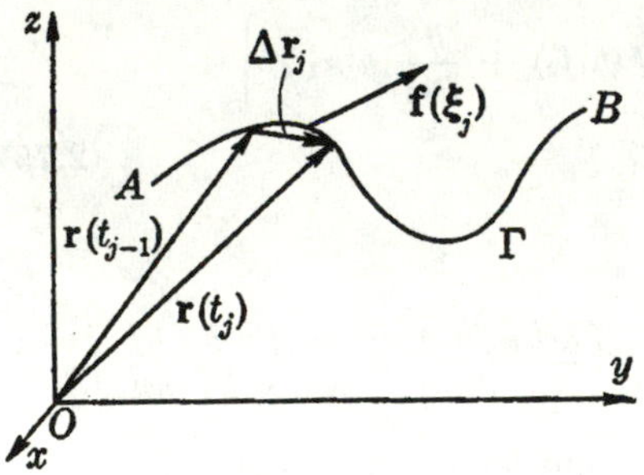

FIG. 2.24

(see Fig. 2.24). Let $\Delta\mathbf{r}_j = \mathbf{r}(t_j) - \mathbf{r}(t_{j-1})$, and let ξ_j be any number such that $t_{j-1} \leq \xi_j \leq t_j$. We can compute $x(\xi_j)$, $y(\xi_j)$, $z(\xi_j)$, which yields the vector $\mathbf{f}(\xi_j)$. One then forms the sum

$$S_n = \sum_{j=1}^{n} \mathbf{f}(\xi_j) \cdot \Delta\mathbf{r}_j \tag{2.76}$$

If $\lim_{n \to \infty} S_n$ exists independent of how the ξ_j are chosen provided maximum

$|\Delta \mathbf{r}_j| \to 0$ as $n \to \infty$, we define this limit to be the line integral of **f** along the curve Γ from A to B.

As in the calculus the limit is written

$$\int_A^B \mathbf{f} \cdot d\mathbf{r} = \int_\Gamma \mathbf{f} \cdot d\mathbf{r} \tag{2.77}$$

If **f** is continuous along Γ and if Γ has continuous turning tangents, that is, $\dfrac{d\mathbf{r}}{ds}$ is continuous, then (2.77) exists from Riemann integration theory and can be written

$$\int_A^B \mathbf{f} \cdot d\mathbf{r} = \int_a^b \left\{ X[x(t), y(t), z(t)] \frac{dx(t)}{dt} + Y[x(t), y(t), z(t)] \frac{dy(t)}{dt} + Z[x(t), y(t), z(t)] \frac{dz}{dt} \right\} dt \tag{2.77'}$$

We use (2.77′) as a means of evaluating the line integral. There will be some vector fields for which the line integral from A to B will be independent of the curve Γ joining A and B. Such vector fields are said to be *conservative.* If **f** is a force field, (2.77) defines the work done by the force field as one moves a unit particle (mass or charge) from A to B. We now work out a few examples and then take up the case of conservative vector fields.

Example 2.25. Let $\mathbf{f} = xy\mathbf{i} + z\mathbf{j} - xyz\mathbf{k}$, and let the path of integration be the curve $x = t$, $y = t^2$, $z = t$, the integration performed from the origin O to the point $P(1, 1, 1)$, the range of t being given by $0 \leq t \leq 1$. Along the curve, $\mathbf{f} = t^3\mathbf{i} + t\mathbf{j} - t^4\mathbf{k}$, and $d\mathbf{r} = \dfrac{d\mathbf{r}}{dt}\,dt = (\mathbf{i} + 2t\mathbf{j} + \mathbf{k})\,dt$ so that $\mathbf{f} \cdot d\mathbf{r} = (t^3 + 2t^2 - t^4)\,dt$. Hence

$$\int_0^P \mathbf{f} \cdot d\mathbf{r} = \int_0^1 (t^3 + 2t^2 - t^4)\,dt = \tfrac{43}{60}$$

If we choose the straight-line path $x = t$, $y = t$, $z = t$ from the origin to the point $P(1, 1, 1)$, we obtain $\int_0^P \mathbf{f} \cdot d\mathbf{r} = \int_0^1 (t^2 + t - t^3)\,dt = \tfrac{7}{12}$. It is seen that the vector field **f** is not conservative.

Example 2.26. Let $\mathbf{f} = x^2\mathbf{i} + y^3\mathbf{j}$, and let the path of integration be the parabola $y = x^2$ from $(0, 0)$ to $(1, 1)$. Let $x = t$ so that $y = t^2$, $0 \leq t \leq 1$, and

$$\mathbf{r}(t) = x\mathbf{i} + y\mathbf{j} = t\mathbf{i} + t^2\mathbf{j}$$

$$\int_{(0,0)}^{(1,1)} \mathbf{f} \cdot d\mathbf{r} = \int_0^1 (t^2\mathbf{i} + t^6\mathbf{j}) \cdot (\mathbf{i} + 2t\mathbf{j})\,dt = \int_0^1 (t^2 + 2t^7)\,dt = \tfrac{7}{12}$$

For the same **f** let us compute the line integral by moving along the x axis from $x = 0$ to $x = 1$ and then moving along the line $x = 1$ from $y = 0$ to $y = 1$. Although the continuous curve does not have a continuous tangent at the point $(1, 0)$, we need not be concerned since one point of discontinuity does not affect the Riemann integral provided $\frac{d\mathbf{r}}{dt}$ is bounded in the neighborhood of the discontinuity. We have

$$\int_{(0,0)}^{(1,1)} \mathbf{f} \cdot d\mathbf{r} = \int_{(0,0)}^{(1,0)} \mathbf{f} \cdot d\mathbf{r} + \int_{(1,0)}^{(1,1)} \mathbf{f} \cdot d\mathbf{r}$$

Along the first part of the curve $x = t$, $y = 0$, $dx = dt$, $dy = 0$. Along the second part of the curve $x = 1$, $dx = 0$, $y = t$, $dy = dt$, and

$$\int_{(0,0)}^{(1,1)} \mathbf{f} \cdot d\mathbf{r} = \int_0^1 t^2\, dt + \int_0^1 t^3\, dt = \tfrac{7}{12}$$

We become suspicious and guess that $\mathbf{f}$ is conservative. Notice that

$$\mathbf{f} = \nabla\left(\frac{x^3}{3} + \frac{y^4}{4} + \text{constant}\right) = \nabla\varphi \qquad \varphi = \frac{x^3}{3} + \frac{y^4}{4} + \text{constant}$$

Hence $\mathbf{f} \cdot d\mathbf{r} = \nabla\varphi \cdot d\mathbf{r} = d\varphi$ so that

$$\int_A^B \mathbf{f} \cdot d\mathbf{r} = \int_A^B d\varphi = \varphi \Big|_A^B = \varphi(B) - \varphi(A)$$

Since φ is single-valued, the value of $\int_A^B \mathbf{f} \cdot d\mathbf{r}$ depends only upon the upper and lower limits and is independent of the path of integration from A to B.

Example 2.27. We have just seen from Example 2.26 that if $\mathbf{f} = \nabla\varphi$, φ single-valued, then $\mathbf{f}$ is a conservative vector field. Conversely, let us assume that $\int \mathbf{f} \cdot d\mathbf{r}$ is independent of the path. We show that $\mathbf{f}$ is the gradient of a scalar. Define

$$\varphi(x, y, z) = \int_{P_0(x_0,y_0,z_0)}^{P(x,y,z)} \mathbf{f} \cdot d\mathbf{r}$$

The value of φ depends only on the upper limit (we keep P_0 fixed). Then

$$\varphi(x + \Delta x, y, z) = \int_{P(x_0,y_0,z_0)}^{Q(x+\Delta x,y,z)} \mathbf{f} \cdot d\mathbf{r}$$

and $$\varphi(x + \Delta x, y, z) - \varphi(x, y, z) = \int_{P(x,y,z)}^{Q(x+\Delta x,y,z)} X(x, y, z)\, dx + Y(x, y, z)\, dy + Z(x, y, z)\, dz$$

We choose the straight line path from P to Q as our curve of integration. Then $dy = dz = 0$, and

$$\varphi(x + \Delta x, y, z) - \varphi(x, y, z) = \int_x^{x+\Delta x} X(x, y, z)\, dx$$

Applying the theorem of the mean for integrals yields

$$\varphi(x + \Delta x, y, z) - \varphi(x, y, z) = X(\xi, y, z)\, \Delta x \qquad x \leqq \xi \leqq x + \Delta x$$

so that $$\frac{\partial\varphi}{\partial x} = \lim_{\Delta x \to 0} \frac{\varphi(x + \Delta x, y, z) - \varphi(x, y, z)}{\Delta x} = \lim_{\Delta x \to 0} X(\xi, y, z) = X(x, y, z)$$

assuming X continuous. Similarly $Y = \frac{\partial\varphi}{\partial y}$, $Z = \frac{\partial\varphi}{\partial z}$, so that

$$\mathbf{f} = X\mathbf{i} + Y\mathbf{j} + Z\mathbf{k} = \frac{\partial\varphi}{\partial x}\mathbf{i} + \frac{\partial\varphi}{\partial y}\mathbf{j} + \frac{\partial\varphi}{\partial z}\mathbf{k} \qquad \text{Q.E.D.}$$

A quick test to determine whether **f** is conservative is the following: Note that, if $\mathbf{f} = \nabla\varphi$, then $\nabla \times \mathbf{f} = \mathbf{0}$. Conversely, assume $\nabla \times \mathbf{f} = \mathbf{0}$. Then for $\mathbf{f} = X\mathbf{i} + Y\mathbf{j} + Z\mathbf{k}$ we have

$$\begin{aligned} \frac{\partial X}{\partial y} &= \frac{\partial Y}{\partial x} \\ \frac{\partial Y}{\partial z} &= \frac{\partial Z}{\partial y} \\ \frac{\partial Z}{\partial x} &= \frac{\partial X}{\partial z} \end{aligned} \tag{2.78}$$

Let

$$\varphi(x, y, z) = \int_{x_0}^{x} X(x, y, z)\,dx + \int_{y_0}^{y} Y(x_0, y, z)\,dy + \int_{z_0}^{z} Z(x_0, y_0, z)\,dz \tag{2.79}$$

We now show that $\mathbf{f} = \nabla\varphi$. From the calculus

$$\begin{aligned} \frac{\partial\varphi}{\partial x} &= X(x, y, z) \\ \frac{\partial\varphi}{\partial y} &= \int_{x_0}^{x} \frac{\partial X}{\partial y}\,dx + Y(x_0, y, z) = \int_{x_0}^{x} \frac{\partial Y}{\partial x}\,dx + Y(x_0, y, z) \\ &= Y(x, y, z) - Y(x_0, y, z) + Y(x_0, y, z) \\ &= Y(x, y, z) \\ \frac{\partial\varphi}{\partial z} &= \int_{x_0}^{x} \frac{\partial X}{\partial z}\,dx + \int_{y_0}^{y} \frac{\partial Y}{\partial z}\,dy + Z(x_0, y_0, z) \\ &= \int_{x_0}^{x} \frac{\partial Z}{\partial x}\,dx + \int_{y_0}^{y} \frac{\partial Z}{\partial y}\,dy + Z(x_0, y_0, z) \\ &= Z(x, y, z) - Z(x_0, y, z) + Z(x_0, y, z) - Z(x_0, y_0, z) + Z(x_0, y_0, z) \\ &= Z(x, y, z) \end{aligned}$$

Hence $$\mathbf{f} = X\mathbf{i} + Y\mathbf{j} + Z\mathbf{k} = \frac{\partial\varphi}{\partial x}\mathbf{i} + \frac{\partial\varphi}{\partial y}\mathbf{j} + \frac{\partial\varphi}{\partial z}\mathbf{k} = \nabla\varphi$$

The constants x_0, y_0, z_0 can be chosen arbitrarily.

We have proved that a necessary and sufficient condition that **f** be the gradient of a scalar is that $\nabla \times \mathbf{f} = \mathbf{0}$. Thus for **f** conservative we have $\mathbf{f} = \nabla\varphi$ or $\nabla \times \mathbf{f} = \mathbf{0}$. We also say that such an **f** is an irrotational vector field.

Example 2.28. It is easy to show that $\mathbf{f} = 2xye^z\mathbf{i} + x^2e^z\mathbf{j} + x^2ye^z\mathbf{k}$ is irrotational. Then

$$\begin{aligned} \varphi(x, y, z) &= \int_0^x 2xye^z\,dx + \int_0^y 0^2e^z\,dy + \int_0^z 0^2 \cdot 0 \cdot e^z\,dz \\ &= x^2ye^z \end{aligned}$$

and $$\mathbf{f} = \nabla(x^2ye^z + \text{constant})$$

Problems

1. Given $\mathbf{f} = e^x\mathbf{i} - xy\mathbf{j}$, evaluate $\int \mathbf{f} \cdot d\mathbf{r}$ along the curve $y = x^3$ from the origin to the point (2, 8).

2. $\mathbf{f} = (y + \sin z)\mathbf{i} + x\mathbf{j} + x \cos z\mathbf{k}$. Show that $\mathbf{f}$ is conservative, and find φ so that $\mathbf{f} = \nabla\varphi$.

3. Let $\mathbf{f} = -y\mathbf{i} + x\mathbf{j}$. Evaluate $\int \mathbf{f} \cdot d\mathbf{r}$ around the circle with center at the origin and radius a.

4. Show that if $\mathbf{f} = \nabla\varphi$, φ single-valued, the line integral around a closed path, written $\oint \mathbf{f} \cdot d\mathbf{r}$, vanishes. Prove the converse.

5. Let $\mathbf{f} = (-y\mathbf{i} + x\mathbf{j})/x^2 + y^2$. Show that $\mathbf{f} = \nabla \tan^{-1}(y/x)$ and integrate $\mathbf{f}$ around any closed path surrounding the origin, and show that for this path

$$\oint \mathbf{f} \cdot d\mathbf{r} = 2\pi$$

Why does this integral not vanish? See Prob. 4. Notice that $\mathbf{f}$ is not defined at the origin. The curve of integration contains the origin in its interior. This will be important in complex-variable theory.

6. If $\mathbf{A}$ is a constant vector, why is it true that $\oint \mathbf{A} \cdot d\mathbf{r} = 0$?

2.18. Stokes's Theorem. We begin by studying the locus of the end points of the vector

$$\mathbf{r} = x(u, v)\mathbf{i} + y(u, v)\mathbf{j} + z(u, v)\mathbf{k} \tag{2.80}$$

where u and v range over a continuous set of values and x, y, z are assumed to have continuous partial derivatives in u and v. For a fixed $v = v_0$ the end points of $\mathbf{r}$ trace a space curve as we let u vary continuously. For each v a space curve exists, and if we let u vary, we obtain a locus of space curves which collectively form a surface. The curves obtained by setting $v =$ constant are called the u curves, and the curves obtained by setting $u =$ constant are called the v curves. We thus have a two-parameter family of curves forming the surface.

A simple example will illustrate what we have been talking about. Consider

$$\mathbf{r} = r \sin\theta \cos\varphi\, \mathbf{i} + r \sin\theta \sin\varphi\, \mathbf{j} + r\cos\theta\, \mathbf{k} \qquad \left.\begin{array}{l} r = \text{constant} \\ 0 \leqq \varphi \leqq 2\pi \\ 0 \leqq \theta \leqq \pi \end{array}\right\} \tag{2.81}$$

We use θ and φ instead of u and v. Let us notice that $\mathbf{r} \cdot \mathbf{r} = r^2 =$ constant, so that the end points of the vector $\mathbf{r}$ lie on a sphere of radius r. For a fixed $\theta = \theta_0$ the z component of $\mathbf{r}$, namely, $r \cos\theta_0$, is constant. For $0 \leqq \varphi \leqq 2\pi$ the end points of $\mathbf{r}$ trace out a circle of latitude. The φ curves are thus circles of latitude. It is easy to show that the θ curves are the meridians of longitude. We can show that the θ curves intersect the φ curves orthogonally. The expression $\frac{\partial\mathbf{r}}{\partial\theta}$ represents a vector tangent to a θ curve, while $\frac{\partial\mathbf{r}}{\partial\varphi}$ represents a vector tangent to a φ curve. From

$$\frac{\partial \mathbf{r}}{\partial \theta} = r \cos \theta \cos \varphi \, \mathbf{i} + r \cos \theta \sin \varphi \, \mathbf{j} - r \sin \theta \, \mathbf{k}$$

$$\frac{\partial \mathbf{r}}{\partial \varphi} = -r \sin \theta \sin \varphi \, \mathbf{i} + r \sin \theta \cos \varphi \, \mathbf{j}$$

we have
$$\frac{\partial \mathbf{r}}{\partial \theta} \cdot \frac{\partial \mathbf{r}}{\partial \varphi} = 0. \quad \text{Q.E.D.}$$

If we move from a point $P(u, v)$ to the point $Q(u + du, v + dv)$, P and Q on the surface, then $d\mathbf{r} = \overrightarrow{PQ} = \frac{\partial \mathbf{r}}{\partial u} du + \frac{\partial \mathbf{r}}{\partial v} dv$. Hence

$$ds^2 = d\mathbf{r} \cdot d\mathbf{r} = \frac{\partial \mathbf{r}}{\partial u} \cdot \frac{\partial \mathbf{r}}{\partial u} du^2 + 2 \frac{\partial \mathbf{r}}{\partial u} \cdot \frac{\partial \mathbf{r}}{\partial v} du \, dv + \frac{\partial \mathbf{r}}{\partial v} \cdot \frac{\partial \mathbf{r}}{\partial v} dv^2$$

where ds is arc length. For the sphere $ds^2 = r^2 \, d\theta^2 + r^2 \sin^2 \theta \, d\varphi^2$.

Let us now consider a surface of the type given by (2.80) bounded by a rectifiable curve Γ that lies on the surface (see Fig. 2.25).

The vector $\frac{\partial \mathbf{r}}{\partial u} \times \frac{\partial \mathbf{r}}{\partial v}$ is perpendicular to the surface since $\frac{\partial \mathbf{r}}{\partial u}$ and $\frac{\partial \mathbf{r}}{\partial v}$ are tangent to the surface. As we move along the curve Γ, keeping our head in the same direction as $\frac{\partial \mathbf{r}}{\partial u} \times \frac{\partial \mathbf{r}}{\partial v}$, we keep track of the area to our left. It is this surface, S, with which we keep in touch. Γ will be called the boundary of S. We neglect the rest of the surface $\mathbf{r}(u, v)$. We now consider a mesh on the surface formed by a collection of parametric curves (the u and v curves). The mesh will be taken fine enough so that, if (u, v) are the coordinates of A, then $(u + du, v)$, $(u + du, v + dv)$, $(u, v + dv)$ are the coordinates of B, C, D, respectively (Fig. 2.25). Now consider

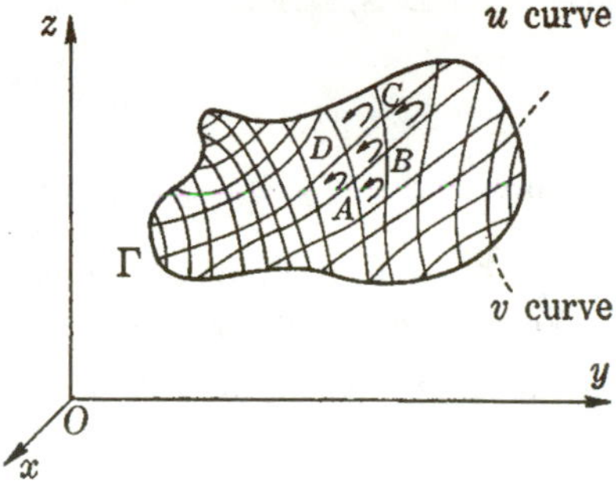

FIG. 2.25

$$\oint_{ABCD} \mathbf{f} \cdot d\mathbf{r}$$

The value of $\mathbf{f}$ at A is $\mathbf{f}(u, v)$; at B it is $\mathbf{f}(u + du, v)$; at C it is $\mathbf{f}(u + du, v + dv)$; at D it is $\mathbf{f}(u, v + dv)$. Now

$$\begin{aligned} \mathbf{f}(u + du, v) &= \mathbf{f}(u, v) + d\mathbf{f}_u \\ &= \mathbf{f}(u, v) + \frac{\partial \mathbf{f}}{\partial u} du \\ &= \mathbf{f}(u, v) + du \left(\frac{\partial \mathbf{r}}{\partial u} \cdot \nabla \right) \mathbf{f} \end{aligned}$$

except for infinitesimals of higher order. Similarly

$$\mathbf{f}(u, v + dv) = \mathbf{f}(u, v) + dv\left(\frac{\partial \mathbf{r}}{\partial v}\cdot\nabla\right)\mathbf{f}$$

Hence, but for infinitesimals of higher order,

$$\begin{aligned}\oint_{ABCD} \mathbf{f}\cdot d\mathbf{r} &= \mathbf{f}\cdot\frac{\partial \mathbf{r}}{\partial u}\,du + \left[\mathbf{f} + du\left(\frac{\partial \mathbf{r}}{\partial u}\cdot\nabla\right)\mathbf{f}\right]\cdot\frac{\partial \mathbf{r}}{\partial v}\,dv \\ &\quad - \mathbf{f}\cdot\frac{\partial \mathbf{r}}{\partial v}\,dv - \left[\mathbf{f} + dv\left(\frac{\partial \mathbf{r}}{\partial v}\cdot\nabla\right)\mathbf{f}\right]\cdot\frac{\partial \mathbf{r}}{\partial u}\,du \\ &= \left\{\left[\left(\frac{\partial \mathbf{r}}{\partial u}\cdot\nabla\right)\mathbf{f}\right]\cdot\frac{\partial \mathbf{r}}{\partial v} - \left[\left(\frac{\partial \mathbf{r}}{\partial v}\cdot\nabla\right)\mathbf{f}\right]\cdot\frac{\partial \mathbf{r}}{\partial u}\right\}du\,dv \\ &= (\nabla\times\mathbf{f})\cdot\frac{\partial \mathbf{r}}{\partial u}\times\frac{\partial \mathbf{r}}{\partial v}\,du\,dv\end{aligned}$$

(see Prob. 16, Sec. 2.15).

The vector $\frac{\partial \mathbf{r}}{\partial u}\,du \times \frac{\partial \mathbf{r}}{\partial v}\,dv$ is normal to the surface S. Its magnitude is the area of the sector $ABCD$, except for infinitesimals of higher order, since $ABCD$ is, strictly speaking, not a parallelogram. We define

$$d\boldsymbol{\sigma} \equiv \frac{\partial \mathbf{r}}{\partial u}\times\frac{\partial \mathbf{r}}{\partial v}\,du\,dv \tag{2.82}$$

so that

$$\oint_{ABCD} \mathbf{f}\cdot d\mathbf{r} = \nabla\times\mathbf{f}\cdot d\boldsymbol{\sigma}$$

except for infinitesimals of higher order.

We now sum over the entire network. Interior line integrals cancel out in pairs, leaving only $\oint_\Gamma \mathbf{f}\cdot d\mathbf{r}$. Also

$$\sum_S (\nabla\times\mathbf{f})\cdot d\boldsymbol{\sigma} \to \iint_S (\nabla\times\mathbf{f})\cdot d\boldsymbol{\sigma}$$

as the mesh gets finer and finer. We thus have Stokes's theorem

$$\oint_\Gamma \mathbf{f}\cdot d\mathbf{r} = \iint_S (\nabla\times\mathbf{f})\cdot d\boldsymbol{\sigma} \tag{2.83}$$

Comments

1. Since Γ may not be a parametric curve, (2.82) may not hold for a mesh circuit containing Γ as part of its boundary. This is true, but fortunately we need not worry about the inequality. The line integrals cancel out in pairs no matter what subdivision we use, and for a fine network the contributions of those areas next to Γ contribute little to $\iint_S \nabla\times\mathbf{f}\cdot d\boldsymbol{\sigma}$. The limiting process takes care of this apparent negligence.

2. Stokes's theorem has been proved for a surface of the type $\mathbf{r}(u, v)$ [see (2.80)]. The theorem is easily seen to be true if we have a finite number of these surfaces connected continuously (edges). The case of an infinite number of edges requires further consideration.

3. Stokes's theorem is also true for a surface containing a conical point where no $d\boldsymbol{\sigma}$ can be defined. We just neglect to integrate over a small area covering this point. Since the area can be made arbitrarily small, it cannot affect the integral.

4. The reader is referred to the text of Kellogg, "Foundations of Potential Theory," for a more rigorous proof of Stokes's theorem.

5. The tremendous importance of Stokes's theorem cannot be overemphasized. It relates a line integral to a surface integral, and conversely.

6. In order to apply Stokes's theorem it is necessary that $\nabla \times \mathbf{f}$ exist and be integrable over the surface S.

Examples of Stokes's Theorem

Example 2.29. Let $\mathbf{f} = -y\mathbf{i} + x\mathbf{j}$, and let us evaluate $\oint_\Gamma \mathbf{f} \cdot d\mathbf{r}$, Γ any rectifiable curve in the xy plane. Applying Stokes's theorem, we have

$$\oint_\Gamma \mathbf{f} \cdot d\mathbf{r} = \iint_S (\nabla \times \mathbf{f}) \cdot \mathbf{k}\, dy\, dx = \iint_S 2\mathbf{k} \cdot \mathbf{k}\, dy\, dx = 2A$$

Thus the area A, bounded by Γ, is

$$A = \tfrac{1}{2}\oint x\, dy - y\, dx \tag{2.84}$$

For the ellipse $x = a \cos t$, $y = b \sin t$, $0 \leqq t \leqq 2\pi$, we have $dx = -a \sin t\, dt$, $dy = b \cos t\, dt$, and $A = \frac{1}{2}\int_0^{2\pi} ab(\cos^2 t + \sin^2 t)\, dt = \pi ab$.

Example 2.30. If $\nabla \times \mathbf{f} = \mathbf{0}$ everywhere, it follows from Stokes's theorem that $\oint \mathbf{f} \cdot d\mathbf{r} = 0$ around every closed path. Conversely, assume $\oint \mathbf{f} \cdot d\mathbf{r} = 0$ around every closed path, and assume $\nabla \times \mathbf{f}$ is continuous. If $\nabla \times \mathbf{f} \not\equiv \mathbf{0}$, then $\nabla \times \mathbf{f} \neq \mathbf{0}$ at some point P. From continuity we have $\nabla \times \mathbf{f} \neq \mathbf{0}$ in some neighborhood of P and $\nabla \times \mathbf{f}$ nearly parallel to $(\nabla \times \mathbf{f})_P$ in this neighborhood. Choose a small plane surface S through P with boundary Γ in this neighborhood of P. The normal to the plane is chosen parallel to $(\nabla \times \mathbf{f})_P$. Then $\oint \mathbf{f} \cdot d\mathbf{r} = \iint_S \nabla \times \mathbf{f} \cdot d\boldsymbol{\sigma} > 0$, a contradiction.

An irrotational field is characterized by any of the three conditions

$$\begin{aligned} &\text{(i)}\ \ \mathbf{f} = \nabla\varphi \\ &\text{(ii)}\ \ \nabla \times \mathbf{f} = 0 \\ &\text{(iii)}\ \ \oint \mathbf{f} \cdot d\mathbf{r} = 0 \qquad \text{for every closed path} \end{aligned} \tag{2.85}$$

Any of these conditions implies the other two.

Example 2.31. Let $\mathbf{f} = f(x, y, z)\mathbf{a}$, where $\mathbf{a}$ is any constant vector. Applying Stokes's theorem yields

$$\oint f\mathbf{a} \cdot d\mathbf{r} = \iint_S \nabla \times (f\mathbf{a}) \cdot d\boldsymbol{\sigma}$$

$$\mathbf{a} \cdot \oint f\, d\mathbf{r} = \iint_S \nabla f \times \mathbf{a} \cdot d\boldsymbol{\sigma} = \mathbf{a} \cdot \iint_S d\boldsymbol{\sigma} \times \nabla f$$

so that $\mathbf{a} \cdot \left(\oint f \, d\mathbf{r} - \iint_S d\boldsymbol{\sigma} \times \nabla f \right) = 0$. Since $\mathbf{a}$ is arbitrary, it follows that

$$\oint f \, d\mathbf{r} = \iint_S d\boldsymbol{\sigma} \times \nabla f \tag{2.86}$$

It can be shown that

$$\oint d\mathbf{r} * \mathbf{f} = \iint_S (d\boldsymbol{\sigma} \times \nabla) * \mathbf{f} \tag{2.87}$$

The asterisk can denote scalar, vector, or ordinary multiplication. In the latter case $\mathbf{f}$ becomes the scalar f.

Problems

1. Show that $\oint \mathbf{r} \cdot d\mathbf{r} = 0$ by two methods.

2. Show that $\oint \mathbf{a} \times \mathbf{r} \cdot d\mathbf{r} = 2\mathbf{a} \cdot \iint_S d\boldsymbol{\sigma}$ if $\mathbf{a}$ is constant.

3. By Stokes's theorem show that $\nabla \times \nabla\varphi = 0$.

4. Prove that $\oint u\nabla v \cdot d\mathbf{r} = \iint_S \nabla u \times \nabla v \cdot d\boldsymbol{\sigma}$.

5. Prove that $\oint u\nabla v \cdot d\mathbf{r} = -\oint v\nabla u \cdot d\mathbf{r}$.

6. If $\mathbf{f} = \cos y \, \mathbf{i} + x(1 + \sin y)\mathbf{j}$, find the value of $\oint \mathbf{f} \cdot d\mathbf{r}$ around the circle with center at the origin and radius r.

7. If $\oint \mathbf{E} \cdot d\mathbf{r} = -\frac{1}{c}\frac{\partial}{\partial t}\iint_S \mathbf{B} \cdot d\boldsymbol{\sigma}$ for all surfaces S show that $\nabla \times \mathbf{E} = -\frac{1}{c}\frac{\partial \mathbf{B}}{\partial t}$.

8. Show that $\iint_S \nabla \times \mathbf{f} \cdot d\boldsymbol{\sigma} = 0$ if S is a closed surface.

9. $\mathbf{f} = (x^2 - y^2)\mathbf{i} + 2xy\mathbf{j}$. Find $\oint \mathbf{f} \cdot d\mathbf{r}$ around the square with vertices at (0, 0), (1, 0), (1, 1), (0, 1). Do this by two methods.

10. If a vector is normal to a surface at every point, show that its curl is zero or is tangent to the surface at each point.

11. Let $\mathbf{f} = \mathbf{a} \times \mathbf{g}$, $\mathbf{a}$ any constant vector. Apply Stokes's theorem to $\mathbf{f}$, and show that $\oint d\mathbf{r} \times \mathbf{g} = \iint_S (d\boldsymbol{\sigma} \times \nabla) \times \mathbf{g}$.

12. Assume $\nabla \times \mathbf{f} \neq \mathbf{0}$. Show that, if a scalar $\mu(x, y, z)$ exists such that

$$\nabla \times (\mu\mathbf{f}) = \mathbf{0}$$

then $\mathbf{f} \cdot \nabla \times \mathbf{f} = 0$.

13. Show that $|\oint d\mathbf{r} \times \mathbf{r}|$ taken around a closed curve in the xy plane is twice the area enclosed by the curve.

14. Let $\mathbf{f} = X(x, y)\mathbf{i} + Y(x, y)\mathbf{j}$, and let S be the area bounded by the closed curves Γ_1 and Γ_2 lying on the xy plane, Γ_1 interior to Γ_2. Show that

$$\iint_S \mathbf{f} \cdot d\boldsymbol{\sigma} = \oint_{\Gamma_2} \mathbf{f} \cdot d\mathbf{r} - \oint_{\Gamma_1} \mathbf{f} \cdot d\mathbf{r}$$

Both line integrals are taken in the counterclockwise sense.

2.19. The Divergence Theorem (Gauss). Let S be a closed surface containing the volume V in its interior. We assume S has a well-defined normal almost everywhere. We now subdivide the volume into many elementary volumes. From Sec. 2.13 we note that except for infinitesimals of higher order

$$\iint_{\Delta S} \mathbf{f} \cdot d\mathbf{\sigma} = \nabla \cdot \mathbf{f}\, \Delta\tau$$

where ΔS is the entire surface bounding the elementary volume $\Delta\tau$. If we sum over all volumes and pass to the limit as the maximum $\Delta\tau \to 0$, we obtain

$$\iint_S \mathbf{f} \cdot d\mathbf{\sigma} = \iiint_V \nabla \cdot \mathbf{f}\, d\tau \tag{2.88}$$

Equation (2.88) is the divergence theorem of Gauss. It relates a surface integral to a volume integral. It has tremendous applications to the various fields of science. In the derivation of (2.88) use has been made of the fact that for each internal $d\mathbf{\sigma}$ there is a $-d\mathbf{\sigma}$, so that all interior surface integrals cancel in pairs, leaving only the boundary surface S as a contributing factor.

Equation (2.88) may be interpreted as follows: Any vector field $\mathbf{f}$ may be looked upon as representing the flow of a fluid, $\mathbf{f} = \rho\mathbf{v}$. From Sec. 2.13 $\nabla \cdot \mathbf{f}$ represents the loss of fluid per unit volume per unit time. The total loss of fluid per unit time throughout V is $\iiint_V \nabla \cdot \mathbf{f}\, d\tau$. Now if $\mathbf{f}$ and $\nabla \cdot \mathbf{f}$ are continuous in V, there cannot be any sources or sinks in V which would create or destroy matter. Consequently the total loss of fluid per unit time must be due to the fluid leaving the surface S. We might station a great many observers on the boundary S, let each observer measure the outward flow of fluid, and then sum up their recorded data. At a point on the surface with normal vector area $d\mathbf{\sigma}$ the component of the velocity normal to the surface is $\mathbf{v} \cdot \mathbf{N}$, and $\rho\mathbf{v} \cdot d\mathbf{\sigma} = \mathbf{f} \cdot d\mathbf{\sigma}$ represents the outward flow of mass per unit time. The total loss of mass per unit time is $\iint_S \mathbf{f} \cdot d\mathbf{\sigma}$, and thus (2.88) is obtained. For a more detailed and rigorous proof of Gauss's theorem, see Kellogg, "Foundations of Potential Theory."

Example 2.32. Let $\mathbf{E} = q\mathbf{r}/r^3$, and let V be a region surrounding the origin with boundary surface S. We wish to compute $\iint_S \mathbf{E} \cdot d\mathbf{\sigma}$. We cannot apply the divergence theorem to the region V since $\nabla \cdot \mathbf{E}$ is discontinuous at $r = 0$. We overcome this difficulty by surrounding the origin by a small sphere Σ of radius ϵ with center at the origin (see Fig. 2.26). The divergence theorem can be applied to the region V' (V less interior of Σ sphere) with boundaries S and Σ. Thus

$$\iint_S \mathbf{E} \cdot d\mathbf{d} + \iint_\Sigma \mathbf{E} \cdot d\mathbf{d} = \iiint_{V'} \nabla \cdot \left(\frac{q\mathbf{r}}{r^3}\right) d\tau$$

We have seen that $\nabla \cdot (q\mathbf{r}/r^3) = 0$, $r \neq 0$, so that

$$\iint_S \mathbf{E} \cdot d\mathbf{d} = -\iint_\Sigma \frac{q\mathbf{r}}{r^3} \cdot d\mathbf{d}$$

Now on Σ, $r = \epsilon$, $d\mathbf{d} = (-\mathbf{r}/r)\, dS$, so that $(q\mathbf{r}/r^3) \cdot d\mathbf{d} = (-q/\epsilon^2)\, dS$ and

$$\iint_\Sigma \frac{q\mathbf{r}}{r^3} \cdot d\mathbf{d} = -\frac{q}{\epsilon^2} \iint dS = -4\pi q$$

Hence

$$\iint_S \mathbf{E} \cdot d\mathbf{d} = 4\pi q$$

$\mathbf{E} = q\mathbf{r}/r^3$ is the electrostatic field due to a point charge q at the origin. For a con-

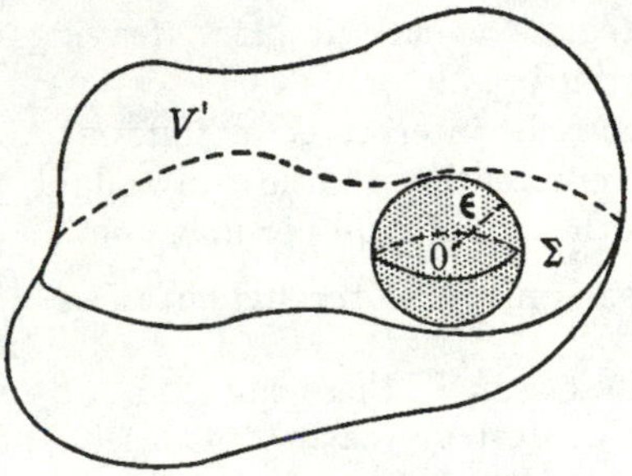

Fig. 2.26

tinuous distribution of charge of density ρ in S it can be shown that

$$\iint_S \mathbf{E} \cdot d\mathbf{d} = 4\pi \iiint_V \rho \, d\tau$$

Assuming that the divergence theorem can be applied (see Kellogg's "Foundations of Potential Theory" for proof of this fact), then

$$\iiint_V \nabla \cdot \mathbf{E}\, d\tau = 4\pi \iiint_V \rho\, d\tau \tag{2.89}$$

Since (2.89) holds for all volumes we must have $\nabla \cdot \mathbf{E} = 4\pi\rho$, provided $\nabla \cdot \mathbf{E} - 4\pi\rho$ is continuous. In a uniform dielectric medium $\nabla \cdot \mathbf{D} = 4\pi\rho$, $\mathbf{D} = \kappa\mathbf{E}$. For magnetism one has $\nabla \cdot \mathbf{B} = 0$. The equations $\nabla \cdot \mathbf{D} = 4\pi\rho$, $\nabla \cdot \mathbf{B} = 0$ comprise two of Maxwell's equations. It can easily be shown that $\mathbf{E} = -\nabla V$ for an inverse-square force. In empty space $\rho = 0$ so that $\nabla^2 V = 0$. A great deal of electrostatic theory deals with the solution of Laplace's equation, $\nabla^2 V = 0$.

Example 2.33. *Green's Theorem.* We apply (2.88) to $\mathbf{f}_1 = u\nabla v$ and $\mathbf{f}_2 = v\nabla u$ and obtain

$$\begin{aligned}\iiint_V \nabla\cdot(u\nabla v)\,d\tau &= \iiint_V (u\nabla^2 v + \nabla u\cdot\nabla v)\,d\tau = \iint_S u\nabla v\cdot d\boldsymbol{\sigma}\\ \iiint_V \nabla\cdot(v\nabla u)\,d\tau &= \iiint_V (v\nabla^2 u + \nabla v\cdot\nabla u)\,d\tau = \iint_S v\nabla u\cdot d\boldsymbol{\sigma}\end{aligned} \tag{2.90}$$

Subtracting, we obtain

$$\iiint_V (u\nabla^2 v - v\nabla^2 u)\,d\tau = \iint_S (u\nabla v - v\nabla u)\cdot d\boldsymbol{\sigma} \tag{2.91}$$

Equations (2.90) and (2.91) are Green's formulas.

Example 2.34. *A Uniqueness Theorem.* Let φ and ψ satisfy Laplace's equation inside a region R, and let $\varphi = \psi$ on the boundary S of R. We show that $\varphi \equiv \psi$. Let $\theta = \varphi - \psi$. Hence $\nabla^2\theta = \nabla^2\varphi - \nabla^2\psi = 0$ in R, and $\theta \equiv 0$ on S. Applying (2.90) with $u = v = \theta$ yields

$$\iiint_R (\theta\nabla^2\theta + \nabla\theta\cdot\nabla\theta)\,d\tau = \iint_S \theta\nabla\theta\cdot d\boldsymbol{\sigma}$$

so that $\iiint_R (\nabla\theta\cdot\nabla\theta)\,d\tau = 0$. Since $\nabla\theta$ is continuous ($\nabla^2\theta$ is assumed to exist), we have $\nabla\theta \equiv 0$ in R and θ = constant. Thus $\varphi - \psi$ = constant. Assuming $\varphi - \psi$ is continuous as we approach the boundary, we must have $\varphi \equiv \psi$, since $\varphi = \psi$ on the boundary.

Example 2.35. Let $\mathbf{f} = f(x, y, z)\mathbf{a}$, where $\mathbf{a}$ is any constant vector. Applying (2.88) yields

$$\mathbf{a}\cdot\iint_S f\,d\boldsymbol{\sigma} = \iiint_V \nabla\cdot(f\mathbf{a})\,d\tau = \mathbf{a}\cdot\iiint_V \nabla f\,d\tau$$

Hence

$$\iint_S f\,d\boldsymbol{\sigma} = \iiint_V \nabla f\,d\tau$$

We leave it to the reader to show that

$$\iint_S d\boldsymbol{\sigma} * \mathbf{f} = \iiint_R (\nabla * \mathbf{f})\,d\tau \tag{2.92}$$

Example 2.36. A vector field $\mathbf{f}$ whose flux $\left(\oiint_S \mathbf{f}\cdot d\boldsymbol{\sigma}\right)$ over every closed surface vanishes is called a solenoidal field. From (2.88) it follows that $\nabla\cdot\mathbf{f} = 0$. Now assume $\mathbf{f} = \nabla\times\mathbf{g}$, so that $\nabla\cdot\mathbf{f} = \nabla\cdot(\nabla\times\mathbf{g}) = 0$. Thus the curl of a vector is a solenoidal vector. Is the converse true? If $\nabla\cdot\mathbf{f} = 0$, can we write $\mathbf{f} = \nabla\times\mathbf{g}$? The answer is "yes," and we call $\mathbf{g}$ the vector potential of $\mathbf{f}$. Notice that $\mathbf{g}$ cannot be unique for $\nabla\times(\mathbf{g} + \nabla\varphi) = \nabla\times\mathbf{g}$. We now exhibit a method for determining $\mathbf{g}$. Let $\mathbf{f} = X\mathbf{i} + Y\mathbf{j} + Z\mathbf{k}$, and assume $\mathbf{g} = \alpha\mathbf{i} + \beta\mathbf{j} + \gamma\mathbf{k}$. We wish to determine $\mathbf{g}$ so that $\nabla\times\mathbf{g} = \mathbf{f}$ provided $\nabla\cdot\mathbf{f} = 0$. Thus α, β, γ must satisfy

$$\begin{aligned}\frac{\partial \gamma}{\partial y} - \frac{\partial \beta}{\partial z} &= X \\ \frac{\partial \alpha}{\partial z} - \frac{\partial \gamma}{\partial x} &= Y \\ \frac{\partial \beta}{\partial x} - \frac{\partial \alpha}{\partial y} &= Z\end{aligned} \tag{2.93}$$

We are to determine α, β, γ. Assume $\alpha = 0$. Then

$$\begin{aligned}\frac{\partial \gamma}{\partial y} - \frac{\partial \beta}{\partial z} &= X \\ -\frac{\partial \gamma}{\partial x} &= Y \\ \frac{\partial \beta}{\partial x} &= Z\end{aligned}$$

Of necessity,

$$\beta(x, y, z) = \int_{x_0}^{x} Z(x, y, z)\,dx + \sigma(y, z)$$

$$\gamma(x, y, z) = -\int_{x_0}^{x} Y(x, y, x)\,dx + \tau(y, z)$$

Hence
$$\begin{aligned}\frac{\partial \gamma}{\partial y} - \frac{\partial \beta}{\partial z} &= -\int_{x_0}^{x} \left(\frac{\partial Y}{\partial y} + \frac{\partial Z}{\partial z}\right) dx + \frac{\partial \tau}{\partial y} - \frac{\partial \sigma}{\partial z} \\ &= \int_{x_0}^{x} \frac{\partial X}{\partial x}\,dx + \frac{\partial \tau}{\partial y} - \frac{\partial \sigma}{\partial z}\end{aligned}$$

since $\nabla \cdot \mathbf{f} = 0$. Therefore

$$\frac{\partial \gamma}{\partial y} - \frac{\partial \beta}{\partial z} = X(x, y, z) - X(x_0, y, z) + \frac{\partial \tau}{\partial y} - \frac{\partial \sigma}{\partial z}$$

We need only choose σ and τ so that $\frac{\partial \tau}{\partial y} - \frac{\partial \sigma}{\partial z} = X(x_0, y, z)$. Let $\sigma = 0$ and

$$\tau(y, z) = \int_{y_0}^{y} X(x_0, y, z)\,dy$$

x_0 and y_0 are constants of integration. Hence

$$\mathbf{g} = \mathbf{j} \int_{x_0}^{x} Z(x, y, z)\,dx + \mathbf{k}\left[\tau(y, z) - \int_{x_0}^{x} Y(x, y, z)\,dx\right] + \nabla\varphi \tag{2.94}$$

where $\tau(y, z)$ is defined above and φ is arbitrary.

For example, let $\mathbf{f} = x^2\mathbf{i} - xy\mathbf{j} - xz\mathbf{k}$ so that $\nabla \cdot \mathbf{f} = 0$. Choosing $x_0 = y_0 = 0$ yields $\tau(y, z) = \int_0^y 0\,dy = 0$, $\int_{x_0}^{x} Z\,dx = -z \int_0^x x\,dx = -(zx^2/2)$,

$$\int_0^x Y\,dx = -y \int_0^x x\,dx = -\left(\frac{yx^2}{2}\right)$$

and $\mathbf{g} = -(zx^2/2)\mathbf{j} + (yx^2/2)\mathbf{k} + \nabla\varphi$. It is to verify that $\mathbf{f} = \nabla \times \mathbf{g}$.

Example 2.37. *Integration of Laplace's Equation.* Let S be the surface of a region R for which $\nabla^2\varphi = 0$. Let P be any point of R, and let r be the distance from P to any point Q in R or on S. We make use of Green's formula

$$\iiint_R (\varphi\nabla^2\psi - \psi\nabla^2\varphi)\,d\tau = \iint_S (\varphi\nabla\psi - \psi\nabla\varphi) \cdot d\mathbf{\sigma}$$

We choose $\psi = 1/r$, which yields a discontinuity inside R, namely, at P, where $r = 0$. In order to overcome this difficulty, we proceed as in Example 2.32. Surround P by a sphere Σ of radius ϵ. Using the fact that $\nabla^2\varphi = \nabla^2\psi = 0$ in R' (R less the Σ sphere) yields

$$\iint_S \left(\varphi\nabla\frac{1}{r} - \frac{1}{r}\nabla\varphi\right)\cdot d\boldsymbol{\sigma} = \iint_\Sigma \left(\frac{1}{r}\nabla\varphi - \varphi\nabla\frac{1}{r}\right)\cdot d\boldsymbol{\sigma}$$

We leave it to the reader to show that, as $\epsilon \to 0$, $\iint_\Sigma (1/r)\nabla\varphi \cdot d\boldsymbol{\sigma} \to 0$ and $\iint_\Sigma \varphi\nabla(1/r)\cdot d\boldsymbol{\sigma} \to -4\pi\varphi(P)$. Hence

$$\varphi(P) = \frac{1}{4\pi}\iint_S \left(\frac{1}{r}\nabla\varphi - \varphi\nabla\frac{1}{r}\right)\cdot d\boldsymbol{\sigma} \tag{2.95}$$

This formula states that the value of φ at any point P in R is determined by the value of φ and $\nabla\varphi\cdot\mathbf{N} = \dfrac{\partial\varphi}{\partial n}$ on the surface S, where $\mathbf{N}$ is the unit vector normal to S.

Problems

1. If $\mathbf{f} = x\mathbf{i} - y\mathbf{j} + (z^2 - 1)\mathbf{k}$, find the value of $\iint_S \mathbf{f}\cdot d\boldsymbol{\sigma}$ over the closed surface bounded by the planes $z = 0$, $z = 1$, and the cylinder $x^2 + y^2 = 1$.

2. Show that $x\mathbf{i} + y\mathbf{j}/x^2 + y^2$ is solenoidal.

3. Prove that $\iint_S d\boldsymbol{\sigma} = 0$ if S is a closed surface.

4. Prove that $\iint_S d\boldsymbol{\sigma} \times \mathbf{f} = \iiint_V \nabla \times \mathbf{f}\, d\tau$.

5. Show that $\iiint_V \mathbf{f}\cdot\nabla\varphi\, d\tau = \iint_S \varphi\mathbf{f}\cdot d\boldsymbol{\sigma} - \iiint_V \varphi\nabla\cdot\mathbf{f}\, d\tau$.

6. If $\mathbf{w} = \frac{1}{2}\nabla \times \mathbf{v}$, $\mathbf{v} = \nabla \times \mathbf{u}$, show that

$$\tfrac{1}{2}\iiint_R \mathbf{v}^2\, d\tau = \tfrac{1}{2}\iint_S (\mathbf{u}\times\mathbf{v})\cdot d\boldsymbol{\sigma} + \iiint_R \mathbf{u}\cdot\mathbf{w}\, d\tau$$

7. If $\mathbf{v} = \nabla\varphi$, $\nabla^2\varphi = 0$, show that for a closed surface $\iiint_R \mathbf{v}^2\, d\tau = \iint_S \varphi\mathbf{v}\cdot d\boldsymbol{\sigma}$.

8. If $\mathbf{f}_1$ and $\mathbf{f}_2$ are irrotational, show that $\mathbf{f}_1 \times \mathbf{f}_2$ is solenoidal.

9. Find a vector $\mathbf{g}$ such that $yz\mathbf{i} - zx\mathbf{j} + (x^2 + y^2)\mathbf{k} = \nabla \times \mathbf{g}$.

10. Find a vector $\mathbf{g}$ such that $\mathbf{r}/r^3 = \nabla \times \mathbf{g}$.

11. If $-\iint_S \rho\mathbf{v}\cdot d\boldsymbol{\sigma} = \iiint_R \dfrac{\partial\rho}{\partial t}\, d\tau$ for all surfaces, show that $\dfrac{\partial\rho}{\partial t} + \nabla\cdot(\rho\mathbf{v}) = 0$.

12. Find a vector $\mathbf{f}$ such that $\nabla\cdot\mathbf{f} = 2x + y - 1$, $\nabla \times \mathbf{f} = z\mathbf{i}$.

REFERENCES

Brand, L.: "Vector and Tensor Analysis," John Wiley & Sons, Inc., New York, 1947.

Kellogg, O. D.: "Foundations of Potential Theory," John Murray, London, 1929.

Lass, H.: "Vector and Tensor Analysis," McGraw-Hill Book Company, Inc., New York, 1950.

Phillips, H. B.: "Vector Analysis," John Wiley & Sons, Inc., New York, 1933.

Rutherford, D. E.: "Vector Methods," Oliver & Boyd, Ltd., London, 1944.

Weatherburn, C. E.: "Elementary Vector Analysis," George Bell & Sons, Ltd., London, 1921.

———: "Advanced Vector Analysis," George Bell & Sons, Ltd., London, 1944.

CHAPTER 3

TENSOR ANALYSIS

3.1. Introduction. In this chapter we wish to generalize the notion of a vector. In Chap. 2 the concept of a vector was highly geometric since we looked upon a vector as a directed line segment. This spatial concept of a vector is easily understood for a space of one, two, or three dimensions. To extend the idea of a vector to a space of dimension higher than 3 (whatever that may be) becomes rather difficult if we hold to the simple idea that a vector is to be a directed line segment. To avoid this difficulty, we look for an algebraic viewpoint of a vector. This can be done in the following manner: In Euclidean coordinates the vector $\mathbf{A}$ can be written $\mathbf{A} = A_1\mathbf{i} + A_2\mathbf{j} + A_3\mathbf{k}$. We can represent the vector $\mathbf{A}$ by the number triple (A_1, A_2, A_3) and write $\mathbf{A} = (A_1, A_2, A_3)$. The unit vectors $\mathbf{i}, \mathbf{j}, \mathbf{k}$ can be represented by the triples (1, 0, 0), (0, 1, 0), (0, 0, 1), respectively. We define addition of number triples and multiplication of a number triple by a real number α as follows:

$$\begin{aligned}(A_1, A_2, A_3) + (B_1, B_2, B_3) &= (A_1 + B_1, A_2 + B_2, A_3 + B_3)\\ \alpha(A_1, A_2, A_3) &= (\alpha A_1, \alpha A_2, \alpha A_3)\end{aligned} \tag{3.1}$$

Equations (3.1) define a linear vector space. We note that

$$(A_1, A_2, A_3) = A_1(1, 0, 0) + A_2(0, 1, 0) + A_3(0, 0, 1) \tag{3.2}$$

The elements A_1, A_2, A_3 of $\mathbf{A}$ are called the components of the number triple.

Throughout this chapter we shall use the summation convention, and at times, for convenience, the superscript convention, of Chap. 1. To continue our discussion of vectors, let $\mathbf{A} = (A_1, A_2, A_3)$, $\mathbf{B} = (B^1, B^2, B^3)$ be two vectors represented as triples. We define the scalar, or inner, product of $\mathbf{A}$ and $\mathbf{B}$ by

$$\mathbf{A} \cdot \mathbf{B} = A_\alpha B^\alpha \tag{3.3}$$

The square of the norm (or length) of the vector $\mathbf{A}$ is defined to be $L^2 = A_\alpha A^\alpha$, $A_i = A^i$, $i = 1, 2, 3$. If $L^2 = 1$, $\mathbf{A}$ is a unit vector. The cosine of the angle between two vectors is defined by

$$\cos\theta = \frac{A_\alpha B^\alpha}{(A_i A^i B_j B^j)^{\frac{1}{2}}} \tag{3.4}$$

It is not difficult to show that $|\cos\theta| \leqq 1$. We must show that

$$A_i A^i B_j B^j \geqq (A_\alpha B^\alpha)^2 \tag{3.5}$$

Let us consider

$$y = \sum_{\alpha=1}^{3} (A_\alpha x - B_\alpha)^2 \geqq 0 \tag{3.6}$$

for x real.

Now $y = A_i A^i x^2 - 2A_\alpha B^\alpha x + B_j B^j$ represents a parabola. Since $y \geqq 0$ for all real x, $y = 0$ has no real roots or two equal real roots. Hence (3.5), the Schwarz-Cauchy inequality, holds. Let the reader show that, if the equality sign holds in (3.5), then $A_\alpha = \lambda B_\alpha$, $\alpha = 1, 2, 3$. If $\lambda > 0$, $\cos\theta = 1$. If $\lambda < 0$, $\cos\theta = -1$. If $\cos\theta = 0$, that is,

$$A_\alpha B^\alpha = 0$$

we say that **A** and **B** are orthogonal.

We can define the vector, or cross, product of **A** and **B** algebraically as follows: Let A^i, B^i, $i = 1, 2, 3$, be the components of **A** and **B**, respectively. Let C_k, $k = 1, 2, 3$, be the components of the number triple $\mathbf{C} = (C_1, C_2, C_3)$, where

$$C_k = \epsilon_{ijk} A^i B^j \tag{3.7}$$

The epsilons of (3.7) were defined in Sec. 1.2. We note that

$$\begin{aligned} C_1 &= \epsilon_{ij1} A^i B^j = \epsilon_{231} A^2 B^3 + \epsilon_{321} A^3 B^2 \\ &= A^2 B^3 - A^3 B^2 \\ C_2 &= A^3 B^1 - A^1 B^3 \\ C_3 &= A^1 B^2 - A^2 B^1 \end{aligned} \tag{3.8}$$

Let us consider the scalar product of **A** and **C**. We have

$$\begin{aligned} \mathbf{A}\cdot\mathbf{C} &= A^k C_k = \epsilon_{ijk} A^i A^k B^j \\ &= \epsilon_{kji} A^k A^i B^j \end{aligned}$$

But $\epsilon_{kji} = -\epsilon_{ijk}$ so that $\mathbf{A}\cdot\mathbf{C} = 0$. Similarly $\mathbf{B}\cdot\mathbf{C} = 0$.

If $\mathbf{A} = (A_1(t), A_2(t), A_3(t))$, then $\frac{d\mathbf{A}}{dt}$ is defined to be the triple $\left(\frac{dA_1}{dt}, \frac{dA_2}{dt}, \frac{dA_3}{dt}\right)$. If $\varphi(x^1, x^2, x^3)$ is a scalar function of $x = x^1$, $y = x^2$, $z = x^3$, the gradient of φ is defined to be the triple

$$\operatorname{grad}\varphi = \nabla\varphi = \left(\frac{\partial\varphi}{\partial x^1}, \frac{\partial\varphi}{\partial x^2}, \frac{\partial\varphi}{\partial x^3}\right) \tag{3.9}$$

It is easy to define the divergence and curl of a vector in terms of number triples. Let the reader show that

$$B^k = \epsilon^{ijk}\left(\frac{\partial A_i}{\partial x^j} - \frac{\partial A_j}{\partial x^i}\right) \tag{3.10}$$

yields the conventional $\nabla \times \mathbf{A}$. The divergence of $\mathbf{A}$ is the scalar

$$\nabla \cdot \mathbf{A} = \frac{\partial A^\alpha}{\partial x^\alpha} \tag{3.11}$$

The definitions above refer to Euclidean coordinates.

In the above presentation geometry has been omitted. Everything depends on the rules for manipulating the number triples. One need only define an n-dimensional vector as an n-tuple

$$\mathbf{A} = (A_1, A_2, A_3, \ldots, A_n) \tag{3.12}$$

The definitions for manipulating triples are easily extended to the case of n-tuples. There is some difficulty in connection with the vector product and the curl. This will be discussed later.

Let us hope that the reader does not feel that it is absolutely necessary to visualize a vector in a four-dimensional space in order to speak of such a vector. He may feel that an abstract idea can have no place in the realm of science. This is not the case. No one can visualize a four-dimensional space. Yet the general theory of relativity is essentially a theory of a four-dimensional Riemannian geometry.

One further generalization before we take up tensor formalism. Let S be a deformable body which is in a given state of rest. Let $P(x^1, x^2, x^3)$ be any point of this body. Now assume that the body is displaced from its position to a new position. Let $s_i(x^1, x^2, x^3)$, $i = 1, 2, 3$, represent the displacement vector of the point P. The displacement vector at a nearby point $Q(x^1 + dx^1, x^2 + dx^2, x^3 + dx^3)$ is

$$s_i(x^1, x^2, x^3) + \frac{\partial s_i}{\partial x^j}\, dx^j \tag{3.13}$$

except for infinitesimals of higher order. We can write

$$\frac{\partial s_i}{\partial x^j} \equiv \frac{1}{2}\left(\frac{\partial s_i}{\partial x^j} + \frac{\partial s_j}{\partial x^i}\right) + \frac{1}{2}\left(\frac{\partial s_i}{\partial x^j} - \frac{\partial s_j}{\partial x^i}\right) \tag{3.14}$$

The nine terms of $\frac{1}{2}\left(\frac{\partial s_i}{\partial x^j} + \frac{\partial s_j}{\partial x^i}\right)$, $i, j = 1, 2, 3$, are highly important in deformation theory. It is convenient to represent these nine terms as the elements of a 3 by 3 matrix. A simple generalization tells us that we may wish to speak of a large collection of elements represented by

$$\begin{gathered} T^{ij\cdots k}_{ab\cdots c}(x^1, x^2, \ldots, x^n) \\ i, j, \ldots, k, a, b, \ldots, c = 1, 2, 3, \ldots, n \end{gathered} \tag{3.15}$$

This is our first introduction to tensors. Before we speak of tensors, we must say a word about coordinate systems and coordinate transformations.

The totality of n-tuples $(x^1, x^2, \ldots, x^n)$ forms the arithmetic n-space, the x^i real, $i = 1, 2, \ldots, n$. By a space of n dimensions we mean any set of objects which can be put into one-to-one correspondence with the arithmetic n-space. Thus

$$P \leftrightarrow (x^1, x^2, \ldots, x^n) \tag{3.16}$$

The correspondence (3.16) attaches an n-tuple to each point P of our space. We look upon (3.16) as a coordinate system imposed on the space of elements P. We now consider the n equations

$$y^i = y^i(x^1, x^2, \ldots, x^n) \qquad i = 1, 2, \ldots, n \tag{3.17}$$

and assume that we can solve for the x^i so that

$$x^i = x^i(y^1, y^2, \ldots, y^n) \qquad i = 1, 2, \ldots, n \tag{3.18}$$

We assume that (3.17) and (3.18) are single-valued. The reader may read that excellent text, "Mathematical Analysis," by Goursat-Hedrick on the conditions imposed upon (3.17) in order that (3.18) exists. The n-space of which P is a point can also be put into one-to-one correspondence with the set of n-tuples of the form $(y^1, y^2, \ldots, y^n)$, so that a new coordinate system has been imposed on our n-space. The point P has not changed, but we have a new method for attaching coordinates to the elements P of our n-space. It is for this reason that (3.17) is called a transformation of coordinates.

3.2. Contravariant Vectors. We consider the arithmetic n space and define a space curve, Γ, in this V_n by

$$\Gamma: \quad \begin{aligned} &x^i = x^i(t) \qquad i = 1, 2, \ldots, n \\ &\alpha \leqq t \leqq \beta \end{aligned} \tag{3.19}$$

In a 3-space the components of a vector tangent to the space curve are $\dfrac{dx^1}{dt}, \dfrac{dx^2}{dt}, \dfrac{dx^3}{dt}$. Generalizing, we define a tangent vector to the space curve (3.19) as the n-tuple $\left(\dfrac{dx^1}{dt}, \dfrac{dx^2}{dt}, \ldots, \dfrac{dx^n}{dt}\right)$, written

$$\frac{dx^i}{dt} \qquad i = 1, 2, \ldots, n \tag{3.20}$$

The elements of (3.20) are the components of a tangent vector to the space curve (3.19). Now let us consider an allowable (one-to-one and single-valued) coordinate transformation of the type given by (3.17). We immediately see that

$$y^i = y^i(x^1, x^2, \ldots, x^n) = y^i[x^1(t), x^2(t), \ldots, x^n(t)] = y^i(t) \quad (3.21)$$

$i = 1, 2, \ldots, n$. Equation (3.21) represents the space curve Γ as described by the y coordinate system. An observer using the y coordinate system will say that the components of the tangent vector to Γ are given by

$$\frac{dy^i}{dt} \qquad i = 1, 2, \ldots, n \quad (3.22)$$

Needless to say, the x coordinate system describing Γ is no more important than the y coordinate system used to describe the same curve. Remember that the points of P have not changed: only the description of these points has changed. We cannot say that $\left(\frac{dx^1}{dt}, \frac{dx^2}{dt}, \ldots, \frac{dx^n}{dt}\right)$ is the tangent vector any more than we can say that $\left(\frac{dy^1}{dt}, \frac{dy^2}{dt}, \ldots, \frac{dy^n}{dt}\right)$ is the tangent vector. If we were to consider all allowable coordinate transformations, we would obtain the whole class of tangent elements, each element claiming to be a tangent vector for that particular coordinate system. It is the abstract collection of all these elements which is said to be the tangent vector. We now ask what relationship exists between the components of the tangent vector in the x coordinate system and the components of the tangent vector in the y coordinate system. We easily answer this question since

$$\frac{dy^i}{dt} = \frac{\partial y^i}{\partial x^\alpha}\frac{dx^\alpha}{dt} \qquad i = 1, 2, \ldots, n \quad (3.23)$$

We note that, in general, $\frac{dy^i}{dt}$ depends on every $\frac{dx^\alpha}{dt}$ since the index α is summed from 1 to n. We leave it to the reader to show that

$$\frac{dx^i}{dt} = \frac{\partial x^i}{\partial y^\alpha}\frac{dy^\alpha}{dt} \quad (3.24)$$

We now make the following generalization: The numbers $A^i(x^1, x^2, \ldots, x^n)$, $i = 1, 2, \ldots, n$, which transform according to the law

$$\bar{A}^i(\bar{x}^1, \bar{x}^2, \ldots, \bar{x}^n) = \frac{\partial \bar{x}^i}{\partial x^\alpha} A^\alpha(x^1, x^2, \ldots, x^n) \qquad i = 1, 2, \ldots, n \quad (3.25)$$

under the coordinate transformation

$$\bar{x}^i = \bar{x}^i(x^1, x^2, \ldots, x^n) \qquad i = 1, 2, \ldots, n \tag{3.26}$$

are said to be the components of a contravariant vector. The contravariant vector is not just the set of components in one coordinate system but is rather the abstract quantity which is represented in each coordinate system x by the set of components $A^i(x)$.

One can manufacture as many contravariant vector fields as one desires. Let $(A^1(x), A^2(x), \ldots, A^n(x))$ be any n-tuple in an

$$x = (x^1, x^2, \ldots, x^n)$$

coordinate system. In any $\bar{x}$ coordinate system related to the x coordinate system by (3.26), define the $\bar{A}^i$, $i = 1, 2, \ldots, n$, as in (3.25). We have constructed a contravariant vector field by this device. If the components of a contravariant vector are known in one coordinate system, then the components are known in all other allowable coordinate systems, by (3.25). A coordinate transformation does not yield a new vector; it merely changes the components of the same vector. We say that a contravariant vector is an invariant under a coordinate transformation. An object of any sort which is not changed by transformations of coordinates is called an invariant. If the reader is confused, let him remember that a point is an invariant. The point does not change under a coordinate transformation; the description of the point changes.

The law of transformation for a contravariant vector is transitive. Let

$$\bar{A}^i(\bar{x}) = \frac{\partial \bar{x}^i}{\partial x^\alpha} A^\alpha(x) \qquad \bar{\bar{A}}^i(\bar{\bar{x}}) = \frac{\partial \bar{\bar{x}}^i}{\partial \bar{x}^\alpha} \bar{A}^\alpha(\bar{x})$$

Then

$$\bar{\bar{A}}^i(\bar{\bar{x}}) = \frac{\partial \bar{\bar{x}}^i}{\partial \bar{x}^\alpha} \frac{\partial \bar{x}^\alpha}{\partial x^\beta} A^\beta(x) = \frac{\partial \bar{\bar{x}}^i}{\partial x^\beta} A^\beta(x)$$

which proves our statement.

Example 3.1. Let X, Y, Z be the components of a contravariant vector in a Euclidean space for which $ds^2 = dx^2 + dy^2 + dz^2$. The components of this vector in a cylindrical coordinate system are

$$\begin{aligned}
R &= \frac{\partial r}{\partial x} X + \frac{\partial r}{\partial y} Y + \frac{\partial r}{\partial z} Z = \cos\theta X + \sin\theta Y \\
\Theta &= \frac{\partial \theta}{\partial x} X + \frac{\partial \theta}{\partial y} Y + \frac{\partial \theta}{\partial z} Z = -\frac{\sin\theta}{r} X + \frac{\cos\theta}{r} Y \\
Z_1 &= \frac{\partial z}{\partial x} X + \frac{\partial z}{\partial y} Y + \frac{\partial z}{\partial z} Z = Z
\end{aligned}$$

where $r = (x^2 + y^2)^{\frac{1}{2}}$, $\theta = \tan^{-1}(y/x)$, $z = z$. Notice that the dimension of Θ is not the same as the dimensions of X, Y, and Z. The quantity $r\Theta$ has the correct dimensions. $(R, r\Theta, Z)$ are the physical components of the vector as distinguished from the vector components (R, Θ, Z). R, $r\Theta$, and Z are the projections of the vector

$$\mathbf{A} = X\mathbf{i} + Y\mathbf{j} + Z\mathbf{k}$$

on the unit vectors $\mathbf{e}_r$, $\mathbf{e}_\theta$, $\mathbf{e}_z = \mathbf{k}$, respectively.

Problems

1. Show that, if the components of a contravariant vector vanish in one coordinate system, they vanish in all coordinate systems.

2. If A^i and B^i are contravariant vector fields (the A^i and B^i, $i = 1, 2, \ldots, n$, are really the components of the vector fields), show that $C^i = A^i \pm B^i$ is also a contravariant vector field.

3. What can be said of two contravariant vectors whose components are equal in one coordinate system?

4. If X, Y, Z are the components given in Example 3.1, find the components in a spherical coordinate system. By what must Θ and Φ be multiplied to yield the physical components?

5. Let A^i and B^i be the components of two contravariant vector fields. Define $C^{ij}(x) = A^i(x)B^j(x)$, $\bar{C}^{ij}(\bar{x}) = \bar{A}^i(\bar{x})\bar{B}^j(\bar{x})$. Show that

$$\bar{C}^{ij}(\bar{x}) = \frac{\partial \bar{x}^i}{\partial x^\alpha} \frac{\partial \bar{x}^j}{\partial x^\beta} C^{\alpha\beta}(x)$$

6. Referring to Prob. 5, show that

$$C^{ij}(x) = \frac{\partial x^i}{\partial \bar{x}^\alpha} \frac{\partial x^j}{\partial \bar{x}^\beta} \bar{C}^{\alpha\beta}(\bar{x})$$

3.3. Covariant Vectors. We consider the scalar point function $\varphi(x^1, x^2, \ldots, x^n)$, which is assumed to be an absolute invariant in that under the allowable coordinate transformation $\bar{x}^i = \bar{x}^i(x^1, x^2, \ldots, x^n)$, $i = 1, 2, \ldots, n$,

$$\begin{aligned}\bar{\varphi}(\bar{x}^1, \bar{x}^2, \ldots, \bar{x}^n) &= \varphi(x^1, x^2, \ldots, x^n) \\ &= \varphi(x^1(\bar{x}), x^2(\bar{x}), \ldots, x^n(\bar{x}))\end{aligned} \tag{3.27}$$

where $x^i(\bar{x}) = x^i(\bar{x}^1, \bar{x}^2, \ldots, \bar{x}^n)$, $i = 1, 2, \ldots, n$, is the inverse transformation of (3.26). We form the n-tuple

$$\left(\frac{\partial \varphi}{\partial x^1}, \frac{\partial \varphi}{\partial x^2}, \ldots, \frac{\partial \varphi}{\partial x^n}\right) \tag{3.28}$$

which is an obvious generalization of the gradient of φ. Differentiating (3.27) yields

$$\frac{\partial \bar{\varphi}}{\partial \bar{x}^i} = \frac{\partial \varphi}{\partial x^\alpha} \frac{\partial x^\alpha}{\partial \bar{x}^i} = \frac{\partial x^\alpha}{\partial \bar{x}^i} \frac{\partial \varphi}{\partial x^\alpha} \tag{3.29}$$

Equation (3.29) relates the components of grad φ in the x coordinate system with the components of grad $(\bar{\varphi} = \varphi)$ in the $\bar{x}$ coordinate system.

More generally, the numbers $A_i(x^1, x^2, \ldots, x^n)$, $i = 1, 2, \ldots, n$ which transform according to the law

$$\bar{A}_i(\bar{x}^1, \bar{x}^2, \ldots, \bar{x}^n) = \frac{\partial x^\alpha}{\partial \bar{x}^i} A_\alpha(x^1, x^2, \ldots, x^n) \qquad i = 1, 2, \ldots, n \tag{3.30}$$

are said to be the components of a covariant vector field. The remarks of Sec. 3.2 concerning contravariant vectors apply here as well. One may ask what the difference is between a covariant and a contravariant vector. It is the law of transformation. Compare (3.25) with (3.30)! The reason that no such distinction was made in Chap. 2 will be answered in Sec. 3.7.

3.4. Scalar Product of Two Vectors. Let $A^i(x)$ and $B_i(x)$ be the components of a contravariant vector and a covariant vector. We consider the sum $A^\alpha B_\alpha$. What is the form of $A^\alpha B_\alpha$ in the $\bar{x}$ coordinate system? The letter $\bar{x}$ is an abbreviation for the set $(\bar{x}^1, \bar{x}^2, \ldots, \bar{x}^n)$. Now

$$A^\alpha = \bar{A}^\beta \frac{\partial x^\alpha}{\partial \bar{x}^\beta} \qquad B_\alpha = \bar{B}_\gamma \frac{\partial \bar{x}^\gamma}{\partial x^\alpha}$$

so that

$$\begin{aligned} A^\alpha B_\alpha &= \bar{A}^\beta \bar{B}_\gamma \frac{\partial x^\alpha}{\partial \bar{x}^\beta} \frac{\partial \bar{x}^\gamma}{\partial x^\alpha} \\ &= \bar{A}^\beta \bar{B}_\gamma \frac{\partial \bar{x}^\gamma}{\partial \bar{x}^\beta} = \bar{A}^\beta \bar{B}_\gamma \delta^\gamma_\beta \\ &= \bar{A}^\beta \bar{B}_\beta = \bar{A}^\alpha \bar{B}_\alpha \end{aligned}$$

Hence the form of $A^\alpha B_\alpha$ remains invariant under a coordinate transformation. This scalar invariant is called the scalar, dot, or inner product of the two vectors **A** and **B**.

Problems

1. If $\bar{A}_i = A_\alpha \dfrac{\partial x^\alpha}{\partial \bar{x}^i}$, show that $A_i = \bar{A}_\alpha \dfrac{\partial \bar{x}^\alpha}{\partial x^i}$.

2. If φ and ψ are scalar invariants show that

$$\begin{aligned} \text{grad } (\varphi\psi) &= \varphi \text{ grad } \psi + \psi \text{ grad } \varphi \\ \text{grad } F(\varphi) &= F'(\varphi) \text{ grad } \varphi \end{aligned}$$

3. If A_i and B_i are the components of two covariant vector fields, show that $C_{ij} = A_i B_j$ transforms according to the law

$$\bar{C}_{ij}(\bar{x}) = C_{\alpha\beta}(x) \frac{\partial x^\alpha}{\partial \bar{x}^i} \frac{\partial x^\beta}{\partial \bar{x}^j} \qquad \bar{C}_{\alpha\beta} = \bar{A}_\alpha \bar{B}_\beta$$

4. Show that $C^i_j = A^i B_j$ transforms according to the law $\bar{C}^i_j(\bar{x}) = C^\alpha_\beta(x) \dfrac{\partial x^\beta}{\partial \bar{x}^j} \dfrac{\partial \bar{x}^i}{\partial x^\alpha}$. A^i and B_i are the components of a contravariant vector and covariant vector, respectively.

5. Assume $A^i(x)B_i(x) = \bar{A}^i(\bar{x})\bar{B}_i(\bar{x})$ for all contravariant vector fields A^i. Show that B_i is a covariant vector field.

6. Let $\bar{s}_i = s_\alpha \dfrac{\partial x^\alpha}{\partial \bar{x}^i}$. Show that

$$\frac{\partial \bar{s}_i}{\partial \bar{x}^j} - \frac{\partial \bar{s}_j}{\partial \bar{x}^i} = \left(\frac{\partial s_\alpha}{\partial x^\beta} - \frac{\partial s_\beta}{\partial x^\alpha}\right)\frac{\partial x^\alpha}{\partial \bar{x}^i}\frac{\partial x^\beta}{\partial \bar{x}^j}$$

3.5. Tensors. The contravariant and covariant vectors defined above are special cases of differential invariants called tensors. The components of the tensor are of the form

$$T^{a_1a_2\cdots a_r}_{b_1b_2\cdots b_s}(x^1, x^2, \ldots, x^n) \tag{3.31}$$

where the indices $a_1, a_2, \ldots, a_r, b_1, b_2, \ldots, b_s$ run through the values $1, 2, \ldots, n$ and the components transform according to the rule

$$\bar{T}^{a_1a_2\cdots a_r}_{b_1b_2\cdots b_s}(\bar{x}) = \left|\frac{\partial x}{\partial \bar{x}}\right|^N T^{\alpha_1\alpha_2\cdots\alpha_r}_{\beta_1\beta_2\cdots\beta_s}(x)\frac{\partial \bar{x}^{a_1}}{\partial x^{\alpha_1}}\cdots\frac{\partial \bar{x}^{a_r}}{\partial x^{\alpha_r}}\frac{\partial x^{\beta_1}}{\partial \bar{x}^{b_1}}\cdots\frac{\partial x^{\beta_s}}{\partial \bar{x}^{b_s}} \tag{3.32}$$

The exponent N of the Jacobian $\left|\dfrac{\partial x}{\partial \bar{x}}\right|$ is called the weight of the tensor. If $N = 0$, we say that the tensor field is absolute; otherwise the tensor field is relative of weight N. For $N = 1$ we have a tensor density. The tensor of (3.32) is said to be contravariant of order r and covariant of order s. If $s = 0$ (no subscripts), the tensor is purely contravariant, and if $r = 0$ (no superscripts), the tensor is purely covariant; otherwise we have a mixed tensor. The vectors of Secs. 3.2 and 3.3 are absolute tensors. If no indices occur, we are speaking of a scalar.

At times we shall call $T^{\alpha_1\alpha_2\cdots\alpha_r}_{\beta_1\beta_2\cdots\beta_s}(x)$ a tensor, although strictly speaking the various T's are the components of the tensor in the x coordinate system.

Two tensors are said to be of the same kind if the tensors have the same number of covariant indices, the same number of contravariant indices, and the same weight. Let the reader show that the sum of two tensors of the same kind is again a tensor of the same kind.

We can construct further tensors as follows:

(*a*) The sum and difference of two tensors of the same kind are again a tensor of the same kind.

(*b*) The product of two tensors is a tensor. We show this for a special case. Let

$$\bar{T}^i_j = \left|\frac{\partial x}{\partial \bar{x}}\right|^2 T^\alpha_\beta \frac{\partial x^\beta}{\partial \bar{x}^j}\frac{\partial \bar{x}^i}{\partial x^\alpha}$$

$$\bar{S}^{kl} = \left|\frac{\partial x}{\partial \bar{x}}\right|^3 S^{\sigma\tau}\frac{\partial \bar{x}^k}{\partial x^\sigma}\frac{\partial \bar{x}^l}{\partial x^\tau}$$

so that
$$(\bar{T}^i_j\bar{S}^{kl}) = \left|\frac{\partial x}{\partial \bar{x}}\right|^5 (T^\alpha_\beta S^{\sigma\tau})\frac{\partial x^\beta}{\partial \bar{x}^j}\frac{\partial \bar{x}^i}{\partial x^\alpha}\frac{\partial \bar{x}^k}{\partial x^\sigma}\frac{\partial \bar{x}^l}{\partial x^\tau}$$

(*c*) *Contraction.* Let us consider the absolute mixed tensor A_j^i. We have

$$\bar{A}_j^i(\bar{x}) = A_\beta^\alpha(x) \frac{\partial x^\beta}{\partial \bar{x}^j} \frac{\partial \bar{x}^i}{\partial x^\alpha} \tag{3.33}$$

The elements of A_j^i may be looked upon as the elements of a matrix. Let us assume that we are interested in the sum of the diagonal terms, A_i^i. In (3.33) let $j = i$, and sum. We obtain

$$\begin{aligned} \bar{A}_i^i(\bar{x}) &= A_\beta^\alpha(x) \frac{\partial x^\beta}{\partial \bar{x}^i} \frac{\partial \bar{x}^i}{\partial x^\alpha} = A_\beta^\alpha \frac{\partial x^\beta}{\partial x^\alpha} = A_\beta^\alpha \delta_\alpha^\beta \\ &= A_\alpha^\alpha(x) = A_i^i(x) \end{aligned}$$

Hence $A_i^i(x)$ is a scalar invariant (invariant in both form and numerical value). From the mixed tensor we obtained a scalar by equating a superscript to a subscript and summing. Let us extend this result to the absolute tensor A_{klm}^{ij}. We have

$$\bar{A}_{klm}^{ij} = A_{\rho\sigma\tau}^{\alpha\beta} \frac{\partial \bar{x}^i}{\partial x^\alpha} \frac{\partial \bar{x}^j}{\partial x^\beta} \frac{\partial x^\rho}{\partial \bar{x}^k} \frac{\partial x^\sigma}{\partial \bar{x}^l} \frac{\partial x^\tau}{\partial \bar{x}^m}$$

Now let $l = j$, and sum. We obtain

$$\begin{aligned} \bar{A}_{kjm}^{ij} &= A_{\rho\sigma\tau}^{\alpha\beta} \frac{\partial \bar{x}^i}{\partial x^\alpha} \frac{\partial \bar{x}^j}{\partial x^\beta} \frac{\partial x^\rho}{\partial \bar{x}^k} \frac{\partial x^\sigma}{\partial \bar{x}^j} \frac{\partial x^\tau}{\partial \bar{x}^m} \\ &= A_{\rho\sigma\tau}^{\alpha\beta} \frac{\partial \bar{x}^i}{\partial x^\alpha} \frac{\partial x^\rho}{\partial \bar{x}^k} \frac{\partial x^\tau}{\partial \bar{x}^m} \frac{\partial x^\sigma}{\partial \bar{x}^j} \frac{\partial \bar{x}^j}{\partial x^\beta} \\ &= A_{\rho\sigma\tau}^{\alpha\beta} \frac{\partial \bar{x}^i}{\partial x^\alpha} \frac{\partial x^\rho}{\partial \bar{x}^k} \frac{\partial x^\tau}{\partial \bar{x}^m} \frac{\partial x^\sigma}{\partial x^\beta} \\ &= A_{\rho\sigma\tau}^{\alpha\beta} \frac{\partial \bar{x}^i}{\partial x^\alpha} \frac{\partial x^\rho}{\partial \bar{x}^k} \frac{\partial x^\tau}{\partial \bar{x}^m} \delta_\beta^\sigma \\ &= A_{\rho\sigma\tau}^{\alpha\sigma} \frac{\partial \bar{x}^i}{\partial x^\alpha} \frac{\partial x^\rho}{\partial \bar{x}^k} \frac{\partial x^\tau}{\partial \bar{x}^m} \end{aligned}$$

so that $A_{\rho\sigma\tau}^{\alpha\sigma}$ are the elements of an absolute mixed tensor. We may write $B_{\rho\tau}^\alpha = A_{\rho\sigma\tau}^{\alpha\sigma}$ since the index σ is summed from 1 to n and hence disappears. Notice that the contravariant and covariant orders of $B_{\rho\tau}^\alpha$ are one less than those of A_{klm}^{ij}. It is not difficult to see why this method of producing a new tensor from a given one is called the method of contraction.

(*d*) *Quotient Law.* One may wish to show that the elements $B_{ij}(x)$ are the elements of a tensor. Assume that it is known that A^iB_{jk} is a tensor for arbitrary contravariant vectors A^i. Then

$$\begin{aligned} \bar{A}^i\bar{B}_{jk} &= A^\alpha B_{\beta\gamma} \frac{\partial \bar{x}^i}{\partial x^\alpha} \frac{\partial x^\beta}{\partial \bar{x}^j} \frac{\partial x^\gamma}{\partial \bar{x}^k} \\ &= \bar{A}^i B_{\beta\gamma} \frac{\partial x^\beta}{\partial \bar{x}^j} \frac{\partial x^\gamma}{\partial \bar{x}^k} \end{aligned}$$

so that $\bar{A}^i\left(\bar{B}_{jk} - B_{\beta\gamma}\frac{\partial x^\beta}{\partial \bar{x}^j}\frac{\partial x^\gamma}{\partial \bar{x}^k}\right) = 0$. Since $\bar{A}^i$ is arbitrary, we must have

$$\bar{B}_{jk} = B_{\beta\gamma}\frac{\partial x^\beta}{\partial \bar{x}^j}\frac{\partial x^\gamma}{\partial \bar{x}^k}$$

This is the desired result. If $A^i B^{j\cdots}_{k\cdots}$ is a tensor for arbitrary A^i, then $B^{j\cdots}_{k\cdots}$ is a tensor. This is the quotient law.

Example 3.2. The Kronecker delta δ^i_j (see Sec. 1.1) is a mixed absolute tensor, for

$$\delta^i_j \frac{\partial x^j}{\partial \bar{x}^\alpha}\frac{\partial \bar{x}^\beta}{\partial x^i} = \frac{\partial x^i}{\partial \bar{x}^\alpha}\frac{\partial \bar{x}^\beta}{\partial x^i} = \frac{\partial \bar{x}^\beta}{\partial \bar{x}^\alpha} = \bar{\delta}^\beta_\alpha$$

If $\varphi(x^1, x^2, \ldots, x^n) = \varphi(x)$ is an absolute scalar, $\varphi(x) = \bar{\varphi}(\bar{x})$, then $\varphi\delta^i_j$ is an absolute mixed tensor whose components in the x coordinate system equal the corresponding components in the $\bar{x}$ coordinate system. Conversely, let $A^i_j(x)$ be an *isotropic* tensor, that is, $\bar{A}^i_j(\bar{x}(x)) = A^i_j(x)$. Then

$$A^i_j = A^\alpha_\beta \frac{\partial \bar{x}^i}{\partial x^\alpha}\frac{\partial x^\beta}{\partial \bar{x}^j}$$

$$A^i_j \frac{\partial \bar{x}^j}{\partial x^\sigma} = A^\alpha_\sigma \frac{\partial \bar{x}^i}{\partial x^\alpha}$$

$$\frac{\partial \bar{x}_j}{\partial x^\alpha}(\delta^\alpha_\sigma A^i_j - A^\alpha_\sigma \delta^i_j) = 0$$

Since $\frac{\partial \bar{x}^j}{\partial x^\alpha}$ can be chosen arbitrarily, it follows that

$$\delta^\alpha_\sigma A^i_j = A^\alpha_\sigma \delta^i_j$$

for all i, j, α, σ. Choosing $i \neq j$ and $\alpha = \sigma$, it follows that $A^i_j = 0$, $i \neq j$. Now choose $i = j$, $\alpha = \sigma$, no summation intended. Then $A^\alpha_\alpha = A^i_i$, no summation occurring. Hence the diagonal elements of A^i_j are equal to some scalar φ, so that $A^i_j = \varphi\delta^i_j$. Let the reader show that if A^{ij}_{kl} is isotropic then

$$A^{ij}_{kl} = \varphi(x)\delta^i_k\delta^j_l + f(x)\delta^i_l\delta^j_k \tag{3.34}$$

Example 3.3. Let $g_{ij}(x)$ be the components of an absolute covariant tensor so that

$$\bar{g}_{ij} = g_{\alpha\beta}\frac{\partial x^\alpha}{\partial \bar{x}^i}\frac{\partial x^\beta}{\partial \bar{x}^j}$$

and

$$\bar{g}_{ij}\frac{\partial \bar{x}^j}{\partial x^\sigma} = g_{\alpha\sigma}\frac{\partial x^\alpha}{\partial \bar{x}^i}$$

If $g_{\alpha\sigma} = g_{\sigma\alpha}$, we have, upon taking determinants, that

$$|\bar{g}_{ij}|\left|\frac{\partial \bar{x}}{\partial x}\right| = |g|\left|\frac{\partial x}{\partial \bar{x}}\right|$$

and

$$|\bar{g}_{ij}|^{\frac{1}{2}} = |g_{ij}|^{\frac{1}{2}}\left|\frac{\partial x}{\partial \bar{x}}\right|$$

Now if A^i are the components of an absolute contravariant vector, then $\bar{A}^i = A^\alpha\frac{\partial \bar{x}^i}{\partial x^\alpha}$, and

$$\bar{B}^i = |\bar{g}|^{\frac{1}{2}}\bar{A}^i = \left|\frac{\partial x}{\partial \bar{x}}\right| |g|^{\frac{1}{2}}A^\alpha\frac{\partial \bar{x}^i}{\partial x^\alpha}$$

$$= \left|\frac{\partial x}{\partial \bar{x}}\right| B^\alpha\frac{\partial \bar{x}^i}{\partial x^\alpha}$$

Thus B^i is a vector density. This method affords a means of changing absolute tensors into relative tensors.

Example 3.4. Assume $g_{\alpha\beta}\,dx^\alpha\,dx^\beta$ is an absolute scalar invariant, that is,

$$\bar{g}_{\alpha\beta}\,d\bar{x}^\alpha\,d\bar{x}^\beta = g_{\alpha\beta}\,dx^\alpha\,dx^\beta$$

Moreover we assume that $g_{\alpha\beta} = g_{\beta\alpha}$. From $d\bar{x}^\alpha = \frac{\partial \bar{x}^\alpha}{\partial x^\mu}\,dx^\mu$ we have

$$\bar{g}_{\alpha\beta}\frac{\partial \bar{x}^\alpha}{\partial x^\mu}\frac{\partial \bar{x}^\beta}{\partial x^\nu}\,dx^\mu\,dx^\nu = g_{\mu\nu}\,dx^\mu\,dx^\nu$$

or

$$\left(\bar{g}_{\alpha\beta}\frac{\partial \bar{x}^\alpha}{\partial x^\mu}\frac{\partial \bar{x}^\beta}{\partial x^\nu} - g_{\mu\nu}\right)dx^\mu\,dx^\nu = 0$$

for arbitrary dx^μ. It follows from Example 1.2, that

$$g_{\mu\nu} = \bar{g}_{\alpha\beta}\frac{\partial \bar{x}^\alpha}{\partial x^\mu}\frac{\partial \bar{x}^\beta}{\partial x^\nu}$$

Hence the $g_{\alpha\beta}(x)$ are the components of an absolute covariant tensor of rank 2 (two subscripts).

Example 3.5. Let A_i and B_i be absolute covariant vectors. Let the reader show that

$$C_{ij} \equiv A_iB_j - A_jB_i$$

is an absolute covariant tensor of rank 2 and that $C_{ij} = -C_{ji}$. For a three-dimensional space

$$\|C_{ij}\| = \begin{Vmatrix} 0 & A_1B_2 - A_2B_1 & A_1B_3 - A_3B_1 \\ -(A_1B_2 - A_2B_1) & 0 & A_2B_3 - A_3B_2 \\ -(A_1B_3 - A_3B_1) & -(A_2B_3 - A_3B_2) & 0 \end{Vmatrix}$$

The nonvanishing terms correspond to the components of the vector product.

Problems

1. If the components of a tensor are zero in one coordinate system, show that the components are zero in all coordinate systems.

2. If $g_{\alpha\beta} = g_{\beta\alpha}$ and $\bar{g}_{ij} = g_{\alpha\beta}\frac{\partial x^\alpha}{\partial \bar{x}^i}\frac{\partial x^\beta}{\partial \bar{x}^j}$, show that $\bar{g}_{ij} = \bar{g}_{ji}$.

3. From $g_{\alpha\beta} \equiv \frac{1}{2}(g_{\alpha\beta} + g_{\beta\alpha}) + \frac{1}{2}(g_{\alpha\beta} - g_{\beta\alpha})$, show that

$$g_{\alpha\beta}\,dx^\alpha\,dx^\beta = \tfrac{1}{2}(g_{\alpha\beta} + g_{\beta\alpha})\,dx^\alpha\,dx^\beta$$

4. If A^{ij}_{kl} is an absolute tensor, show that A^{ij}_{ij} is an absolute scalar.

5. If $A_{\alpha\beta}$ is an absolute tensor, and if $A^{\alpha\beta}A_{\beta\gamma} = \delta^\alpha_\gamma$, show that $A^{\alpha\beta}$ is an absolute tensor. The two tensors are said to be reciprocal.

6. Show that the cofactors of the determinant $|a_{ij}|$ are the components of a relative tensor of weight 2 if a_{ij} is an absolute covariant tensor.

7. If A^i is an absolute contravariant vector, show that $\frac{\partial A^i}{\partial x^j}$ is not a mixed tensor.

8. Assume A^{ij}_{kl} is an isotropic tensor. Show that (3.34) holds.

9. Use (1.26) and $|\bar{g}_{ij}| = |g_{ij}|\left|\frac{\partial x}{\partial \bar{x}}\right|^2$ to show that $|g_{ij}|^{\frac{1}{2}}\epsilon_{ijk}$ is an absolute tensor.

3.6. The Line Element. For the Euclidean space of three dimensions we have

$$ds^2 = dx^2 + dy^2 + dz^2 \tag{3.35}$$

The simple generalization to a Euclidean n-space yields

$$\begin{aligned} ds^2 &= (dx^1)^2 + (dx^2)^2 + \cdots + (dx^n)^2 \\ &= \delta_{\alpha\beta}\, dx^\alpha\, dx^\beta \qquad \delta_{\alpha\beta} = 0 \text{ if } \alpha \neq \beta;\ \delta_{\alpha\beta} = 1 \text{ if } \alpha = \beta \end{aligned} \tag{3.36}$$

If we apply a coordinate transformation, $x^i = x^i(\bar{x}^1, \bar{x}^2, \ldots, \bar{x}^n)$, $i = 1, 2, \ldots, n$, we have $dx^\alpha = \dfrac{\partial x^\alpha}{\partial \bar{x}^\mu}\, d\bar{x}^\mu$, $dx^\beta = \dfrac{\partial x^\beta}{\partial \bar{x}^\nu}\, d\bar{x}^\nu$, so that (3.36) takes the form

$$\begin{aligned} ds^2 &= \delta_{\alpha\beta} \frac{\partial x^\alpha}{\partial \bar{x}^\mu} \frac{\partial x^\beta}{\partial \bar{x}^\nu}\, d\bar{x}^\mu\, d\bar{x}^\nu \\ &= g_{\mu\nu}\, d\bar{x}^\mu\, d\bar{x}^\nu \end{aligned} \tag{3.37}$$

where $\bar{g}_{\mu\nu} = \delta_{\alpha\beta} \dfrac{\partial x^\alpha}{\partial \bar{x}^\mu} \dfrac{\partial x^\beta}{\partial \bar{x}^\nu} = \displaystyle\sum_{\alpha=1}^{3} \dfrac{\partial x^\alpha}{\partial \bar{x}^\mu} \dfrac{\partial x^\alpha}{\partial \bar{x}^\nu}$. Riemann considered the general quadratic form

$$ds^2 = g_{\alpha\beta}(x)\, dx^\alpha\, dx^\beta \tag{3.38}$$

This quadratic form (the line element ds^2) is called a Riemannian metric. The $g_{\alpha\beta}(x)$ are the components of the metric tensor (see Example 3.4). A space characterized by the metric (3.38) is called a Riemannian space.

Theorems regarding this Riemannian space yield a Riemannian geometry. Given the form (3.38), it does not follow that a coordinate transformation exists such that $ds^2 = \delta_{\alpha\beta}\, dy^\alpha\, dy^\beta$. If there is a coordinate transformation $x^i = x^i(y^1, y^2, \ldots, y^n)$, such that $ds^2 = \delta_{\alpha\beta}\, dy^\alpha\, dy^\beta$, we say that the Riemannian space is Euclidean. The y coordinate system is said to be a *Euclidean* coordinate system. Any coordinate system for which the g_{ij} are constants is called a *cartesian* coordinate system (after Descartes).

We can choose the metric tensor symmetric, for

$$g_{ij} \equiv \tfrac{1}{2}(g_{ij} + g_{ji}) + \tfrac{1}{2}(g_{ij} - g_{ji})$$

and $\frac{1}{2}(g_{ij} - g_{ji})\, dx^i\, dx^j \equiv 0$. The terms $\frac{1}{2}(g_{ij} + g_{ji})$ are symmetric. We assume that the quadratic form is positive-definite (see Sec. 1.6).

Example 3.6. In a three-dimensional Euclidean space using Euclidean coordinates one has

$$ds^2 = (dx^1)^2 + (dx^2)^2 + (dx^3)^2$$

so that

$$\|g_{ij}\| = \begin{Vmatrix} 1 & 0 & 0 \\ 0 & 1 & 0 \\ 0 & 0 & 1 \end{Vmatrix}$$

For spherical coordinates

$$\begin{aligned} x^1 &= r \sin\theta \cos\varphi = y^1 \sin y^2 \cos y^3 \\ x^2 &= r \sin\theta \sin\varphi = y^1 \sin y^2 \sin y^3 \\ x^3 &= r \cos\theta = y^1 \cos y^2 \end{aligned}$$

The $\bar{g}_{ij}$ of the spherical coordinates system can be obtained from

$$\bar{g}_{ij}(y^1, y^2, y^3) = g_{\alpha\beta}(x(y)) \frac{\partial x^\alpha}{\partial y^i} \frac{\partial x^\beta}{\partial y^j}$$

Hence

$$\begin{aligned} \bar{g}_{11} &= \left(\frac{\partial x^1}{\partial y^1}\right)^2 + \left(\frac{\partial x^2}{\partial y^1}\right)^2 + \left(\frac{\partial x^3}{\partial y^1}\right)^2 = 1 \\ \bar{g}_{22} &= (y^1)^2 \\ \bar{g}_{33} &= (y^1)^2(\sin y^2)^2 \\ \bar{g}_{ij} &= 0 \qquad \text{for } i \neq j \end{aligned}$$

We obtain

$$\begin{aligned} ds^2 &= (dy^1)^2 + (y^1)^2(dy^2)^2 + (y^1 \sin y^2)^2(dy^3)^2 \\ &= dr^2 + r^2\, d\theta^2 + r^2 \sin^2\theta\, d\varphi^2 \end{aligned}$$

This spherical coordinate system is not cartesian since $\bar{g}_{22} \neq$ constant.

Example 3.7. We define g^{ij} as the reciprocal tensor to g_{ij}, that is, $g^{ij}g_{jk} = \delta^i_k$ (see Prob. 5, Sec. 3.5). The g^{ij} are the signed minors of the g_{ji} divided by the determinant of the g_{ij}. The g^{ij} are the elements of the inverse matrix of the matrix $\|g_{ij}\|$. For the spherical coordinates of Example 3.6 we have

$$\|g_{ij}\| = \left\|\begin{matrix} 1 & 0 & 0 \\ 0 & r^2 & 0 \\ 0 & 0 & r^2\sin^2\theta \end{matrix}\right\| \qquad \|g^{ij}\| = \left\|\begin{matrix} 1 & 0 & 0 \\ 0 & \dfrac{1}{r^2} & 0 \\ 0 & 0 & \dfrac{1}{r^2\sin^2\theta} \end{matrix}\right\|$$

Example 3.8. We define the length L of a vector A^i in a Riemannian space by the quadratic form

$$L^2 = g_{\alpha\beta}A^\alpha A^\beta \tag{3.39}$$

The associated vector of A^i is the covariant vector

$$A_i = g_{i\alpha}A^\alpha$$

It is easily seen that $A^i = g^{i\beta}A_\beta$ so that

$$L^2 = g_{\alpha\beta}A^\alpha A^\beta = g^{\alpha\beta}A_\alpha A_\beta \tag{3.40}$$

The cosine of the angle between two vectors A^i, B^i is defined by

$$\cos\theta = \frac{g_{ij}A^iB^j}{(g_{\alpha\beta}A^\alpha A^\beta)^{\frac{1}{2}}(g_{\mu\nu}A^\mu A^\nu)^{\frac{1}{2}}} \tag{3.41}$$

Let the reader show that $|\cos\theta| \leqq 1$.

Problems

1. Prove (3.40).
2. Show that $|\cos\theta| \leqq 1$, $\cos\theta$ defined by (3.41).

3. For paraboloidal coordinates

$$x^1 = y^1y^2 \cos y^3$$
$$x^2 = y^1y^2 \sin y^3$$
$$x^3 = \tfrac{1}{2}[(y^1)^2 - (y^2)^2]$$

Show that

$$ds^2 = (dx^1)^2 + (dx^2)^2 + (dx^3)^2 = [(y^1)^2 + (y^2)^2][(dy^1)^2 + (dy^2)^2] + (y^1y^2)^2(dy^3)^2$$

4. Consider the hypersurface $x^i = x^i(u^1, u^2)$ embedded in a Riemannian 3-space. If we keep u^1 fixed, $u^1 = u_0^1$, we obtain the space curve $x^i = x^i(u_0^1, u^2)$, called the u_2 curve. Similarly $x^i = x^i(u^1, u_0^2)$ represents a u_1 curve on this surface. These curves are called the coordinate curves of the surface. Show that the metric for the surface is $ds^2 = h_{ij}\, du^i\, du^j$, where $h_{ij} = g_{\alpha\beta} \dfrac{\partial x^\alpha}{\partial u^i} \dfrac{\partial x^\beta}{\partial u^j}$. Show that the coordinate curves intersect orthogonally if $h_{12} = 0$.

5. The equation $\varphi(x^1, x^2, \ldots, x^n) = 0$ determines a hypersurface of a V_n. Show that the $\dfrac{\partial \varphi}{\partial x^i}$ are the components of a covariant vector normal to the surface.

6. Show that $\dfrac{dx^i}{ds}$ is a unit tangent vector to the space curve $x^i(s)$, $ds^2 = g_{\alpha\beta}\, dx^\alpha\, dx^\beta$.

7. Show that the unit vectors tangent to the u^1 and u^2 curves of Prob. 4 are given by

$$h_{11}^{-\frac{1}{2}} \frac{\partial x^i}{\partial u^1}, \quad h_{22}^{-\frac{1}{2}} \frac{\partial x^i}{\partial u^2}$$

8. If θ is the angle between the coordinate curves of Prob. 4, show that

$$\cos \theta = (h_{11}h_{22})^{-\frac{1}{2}} h_{12}$$

3.7. Geodesics in a Riemannian Space. If a space curve in a Riemannian space is given by $x^i = x^i(t)$, we can compute the distance between two points of the curve by the formula

$$s = \int_{t_0}^{t_1} \left(g_{\alpha\beta}(x(t)) \frac{dx^\alpha}{dt} \frac{dx^\beta}{dt} \right)^{\frac{1}{2}} dt \tag{3.42}$$

The geodesic is defined as that particular curve $x^i(t)$ joining $x^i(t_0)$ and $x^i(t_1)$ which extremalizes (3.42). The problem of determining the geodesic reduces to a problem in the calculus of variations. We apply the Euler-Lagrange formula to (3.42) (see Sec. 7.6). The differential equation of Euler-Lagrange is

$$\frac{d}{dt}\left(\frac{\partial f}{\partial \dot{x}^i}\right) - \frac{\partial f}{\partial x^i} = 0 \tag{3.43}$$

where $f = (g_{\alpha\beta}\dot{x}^\alpha \dot{x}^\beta)^{\frac{1}{2}}$. Now

$$\frac{\partial f}{\partial x^i} = \frac{1}{2f} \frac{\partial g_{\alpha\beta}}{\partial x^i} \dot{x}^\alpha \dot{x}^\beta$$
$$\frac{\partial f}{\partial \dot{x}^i} = \frac{1}{2f} (g_{\alpha i}\dot{x}^\alpha + g_{i\beta}\dot{x}^\beta)$$

If instead of using t as a parameter we switch to arc length s, then $t \equiv s$ and $f = 1$. Hence

$$\frac{\partial f}{\partial x^i} = \frac{1}{2}\frac{\partial g_{\alpha\beta}}{\partial x^i}\dot{x}^\alpha \dot{x}^\beta$$

$$\frac{d}{dt}\left(\frac{\partial f}{\partial \dot{x}^i}\right) = \frac{1}{2}(g_{\alpha i}\ddot{x}^\alpha + g_{i\beta}\ddot{x}^\beta) + \frac{1}{2}\left(\frac{\partial g_{\alpha i}}{\partial x^\beta} + \frac{\partial g_{i\beta}}{\partial x^\alpha}\right)\dot{x}^\alpha \dot{x}^\beta$$

so that (using the fact that $g_{ij} = g_{ji}$) (3.43) becomes

$$g_{i\alpha}\ddot{x}^\alpha + \frac{1}{2}\left(\frac{\partial g_{\alpha i}}{\partial x^\beta} + \frac{\partial g_{i\beta}}{\partial x^\alpha} - \frac{\partial g_{\alpha\beta}}{\partial x^i}\right)\dot{x}^\alpha \dot{x}^\beta = 0$$

Multiplying by $g^{\tau i}$ and summing on i yields

$$\frac{d^2x^\tau}{ds^2} + \tfrac{1}{2}g^{\tau i}\left(\frac{\partial g_{\alpha i}}{\partial x^\beta} + \frac{\partial g_{i\beta}}{\partial x^\alpha} - \frac{\partial g_{\alpha\beta}}{\partial x^i}\right)\frac{dx^\alpha}{ds}\frac{dx^\beta}{ds} = 0 \tag{3.44}$$

or

$$\frac{d^2x^\tau}{ds^2} + \Gamma^\tau_{\alpha\beta}\frac{dx^\alpha}{ds}\frac{dx^\beta}{ds} = 0 \qquad \tau = 1, 2, \ldots, n$$

where

$$\Gamma^\tau_{\alpha\beta} = \tfrac{1}{2}g^{\tau i}\left(\frac{\partial g_{\alpha i}}{\partial x^\beta} + \frac{\partial g_{i\beta}}{\partial x^\alpha} - \frac{\partial g_{\alpha\beta}}{\partial x^i}\right) \tag{3.44'}$$

Equations (3.44) are the second-order differential equations of the geodesics or paths. The functions $\Gamma^\tau_{\alpha\beta}$ are called the Christoffel symbols of the second kind.

Example 3.9. For a Euclidean space using cartesian coordinates we have $g_{ij} \equiv$ constant so that $\frac{\partial g_{ij}}{\partial x^k} = 0$ and (3.44′) yields $\Gamma^\tau_{\alpha\beta} = 0$. Hence the geodesics are given by $\frac{d^2x^\tau}{ds^2} = 0$, or $x^\tau = a^\tau s + b^\tau$, linear paths.

Example 3.10. Assume that we live on the surface of a right helicoid immersed in a Euclidean 3-space, $ds^2 = (dx^1)^2 + [(x^1)^2 + c^2](dx^2)^2$. We have

$$\|g_{ij}\| = \begin{Vmatrix} 1 & 0 \\ 0 & (x^1)^2 + c^2 \end{Vmatrix} \qquad \|g^{ij}\| = \begin{Vmatrix} 1 & 0 \\ 0 & \dfrac{1}{(x^1)^2 + c^2} \end{Vmatrix}$$

Applying (3.44′), we have

$$\Gamma^2_{12} = \tfrac{1}{2}g^{2i}\left(\frac{\partial g_{1i}}{\partial x^2} + \frac{\partial g_{i2}}{\partial x^1} - \frac{\partial g_{12}}{\partial x^i}\right)$$

Since $g^{2i} = 0$ unless $i = 2$, we have

$$\Gamma^2_{12} = \tfrac{1}{2}g^{22}\left(\frac{\partial g_{12}}{\partial x^2} + \frac{\partial g_{22}}{\partial x^1} - \frac{\partial g_{12}}{\partial x^2}\right)$$
$$= \frac{1}{2[(x^1)^2 + c^2]}2x^1 = \frac{x^1}{(x^1)^2 + c^2}$$

Similarly,

$$\Gamma^1_{11} = 0,\ \Gamma^2_{11} = 0,\ \Gamma^1_{12} = \Gamma^1_{21} = 0,\ \Gamma^1_{22} = -x^1,\ \Gamma^2_{22} = 0,\ \Gamma^2_{12} = \Gamma^2_{21} = \frac{x^1}{[(x^1)^2 + c^2]}$$

The differential equations of the geodesics are

$$\frac{d^2x^1}{ds^2} + \Gamma^1_{11}\left(\frac{dx^1}{ds}\right)^2 + 2\Gamma^1_{12}\frac{dx^1}{ds}\frac{dx^2}{ds} + \Gamma^1_{22}\left(\frac{dx^2}{ds}\right)^2 = 0$$
$$\frac{d^2x^2}{ds^2} + \Gamma^2_{11}\left(\frac{dx^1}{ds}\right)^2 + 2\Gamma^2_{12}\frac{dx^1}{ds}\frac{dx^2}{ds} + \Gamma^2_{22}\left(\frac{dx^2}{ds}\right)^2 = 0$$

or

$$\frac{d^2x^1}{ds^2} - x^1\left(\frac{dx^2}{ds}\right)^2 = 0$$
$$\frac{d^2x^2}{ds^2} + \frac{2x^1}{(x^1)^2 + c^2}\frac{dx^1}{ds}\frac{dx^2}{ds} = 0 \tag{3.44''}$$

Integrating the second of these equations yields

$$\frac{dx^2}{ds}[(x^1)^2 + c^2] = \text{constant} = A$$

Let the reader show that $\left(\frac{dx^1}{ds}\right)^2 + \frac{A}{(x^1)^2 + c^2} = \text{constant.}$

Problems

1. Derive the $\Gamma^\alpha_{\beta\gamma}$ of Example 3.10.

2. Find the differential equations of the geodesics for the line element

$$ds^2 = (dx^1)^2 + \sin^2 x^1 (dx^2)^2$$

Integrate these equations with initial conditions $x^1 = \theta_0$, $x^2 = \varphi_0$, $\frac{dx^1}{ds} = 1$, $\frac{dx^2}{ds} = 0$ at $s = 0$.

3. Show that $\Gamma^\tau_{\alpha\beta} = \Gamma^\tau_{\beta\alpha}$.

4. Obtain the Christoffel symbols and the differential equations of the geodesics for the surface

$$x^1 = u^1 \cos u^2$$
$$x^2 = u^1 \sin u^2$$
$$x^3 = 0$$

The surface is the plane $x^3 = 0$, and the coordinates are polar coordinates

$$ds^2 = (du^1)^2 + (u^1)^2(du^2)^2$$

5. From (3.44′) show that

$$\frac{\partial g_{\alpha\beta}}{\partial x^\mu} = g_{\sigma\beta}\Gamma^\sigma_{\alpha\mu} + g_{\alpha\sigma}\Gamma^\sigma_{\beta\mu}$$

6. Let $ds^2 = E\,du^2 + 2F\,du\,dv + G\,dv^2$. Calculate $|g_{ij}|$, $\|g^{ij}\|$, Γ^i_{jk}.

7. If $\bar\Gamma^i_{jk}(\bar x) = \Gamma^\alpha_{\beta\gamma}(x)\frac{\partial x^\beta}{\partial \bar x^j}\frac{\partial x^\gamma}{\partial \bar x^k}\frac{\partial \bar x^i}{\partial x^\alpha} + \frac{\partial^2 x^\sigma}{\partial \bar x^j \partial \bar x^k}\frac{\partial \bar x^i}{\partial x^\sigma}$, show that $\Gamma^\alpha_{\beta\gamma} - \Gamma^\alpha_{\gamma\beta}$ is a tensor.

8. Obtain the Christoffel symbols for a Euclidean space using cylindrical and spherical coordinates.

9. The Christoffel symbols of the first kind are $\{i, jk\} = g_{i\sigma}\Gamma^\sigma_{jk}$. Show that $\Gamma^\tau_{jk} = g^{i\tau}\{i, jk\}$.

3.8. Law of Transformation for the Christoffel Symbols. Let the equations of the geodesics be given by

$$\frac{d^2x^i}{ds^2} + \Gamma^i_{jk}(x)\frac{dx^j}{ds}\frac{dx^k}{ds} = 0 \tag{3.45}$$

and

$$\frac{d^2\bar{x}^i}{ds^2} + \bar{\Gamma}^i_{jk}(\bar{x})\frac{d\bar{x}^j}{ds}\frac{d\bar{x}^k}{ds} = 0 \tag{3.46}$$

for the two coordinate systems x^i, $\bar{x}^i$ in a Riemannian space. We now find a relationship between the Γ^i_{jk} and the $\bar{\Gamma}^i_{jk}$. From $\bar{x}^i = \bar{x}^i(x)$ we have

$$\frac{d\bar{x}^i}{ds} = \frac{\partial \bar{x}^i}{\partial x^\alpha}\frac{dx^\alpha}{ds} \qquad \frac{d^2\bar{x}^i}{ds^2} = \frac{\partial^2\bar{x}^i}{\partial x^\beta\,\partial x^\alpha}\frac{dx^\alpha}{ds}\frac{dx^\beta}{ds} + \frac{\partial\bar{x}^i}{\partial x^\alpha}\frac{d^2x^\alpha}{ds^2}$$

Substituting into (3.46) yields

$$\frac{\partial\bar{x}^i}{\partial x^\alpha}\frac{d^2x^\alpha}{ds^2} + \frac{\partial^2\bar{x}^i}{\partial x^\beta\,\partial x^\alpha}\frac{dx^\alpha}{ds}\frac{dx^\beta}{ds} + \bar{\Gamma}^i_{jk}\frac{\partial\bar{x}^j}{\partial x^\alpha}\frac{\partial\bar{x}^k}{\partial x^\beta}\frac{dx^\alpha}{ds}\frac{dx^\beta}{ds} = 0$$

Multiply this equation by $\dfrac{\partial x^\sigma}{\partial\bar{x}^i}$ and one obtains (after summing on i)

$$\frac{d^2x^\sigma}{ds^2} + \left(\bar{\Gamma}^i_{jk}\frac{\partial\bar{x}^j}{\partial x^\alpha}\frac{\partial\bar{x}^k}{\partial x^\beta}\frac{\partial x^\sigma}{\partial\bar{x}^i} + \frac{\partial^2\bar{x}^i}{\partial x^\beta\,\partial x^\alpha}\frac{\partial x^\sigma}{\partial\bar{x}^i}\right)\frac{dx^\alpha}{ds}\frac{dx^\beta}{ds} = 0$$

Comparing with (3.45), we have

$$\Gamma^i_{jk} = \bar{\Gamma}^\alpha_{\beta\gamma}\frac{\partial\bar{x}^\beta}{\partial x^j}\frac{\partial\bar{x}^\gamma}{\partial x^k}\frac{\partial x^i}{\partial\bar{x}^\alpha} + \frac{\partial^2\bar{x}^\sigma}{\partial x^j\,\partial x^k}\frac{\partial x^i}{\partial\bar{x}^\sigma} \tag{3.47}$$

Equation (3.47) is the law of transformation for the Christoffel symbols. Note that the Γ^i_{jk} are not the components of a mixed tensor so that $\Gamma^i_{jk}(x)$ may be zero in one coordinate system but not in all coordinate systems. Let the reader show that

$$\bar{\Gamma}^i_{jk} = \Gamma^\alpha_{\beta\gamma}\frac{\partial x^\beta}{\partial\bar{x}^j}\frac{\partial x^\gamma}{\partial\bar{x}^k}\frac{\partial\bar{x}^i}{\partial x^\alpha} + \frac{\partial^2 x^\sigma}{\partial\bar{x}^j\,\partial\bar{x}^k}\frac{\partial\bar{x}^i}{\partial x^\sigma} \tag{3.48}$$

Example 3.11. From Prob. 4, Sec. 1.2, and the definition of the g^{ij} we have

$$\frac{\partial|g_{ij}|}{\partial x^\mu} = |g_{ij}|g^{\alpha\beta}\frac{\partial g_{\alpha\beta}}{\partial x^\mu}$$

From Prob. 5, Sec. 3.7,

$$\frac{\partial g_{\alpha\beta}}{\partial x^\mu} = g_{\sigma\beta}\Gamma^\sigma_{\alpha\mu} + g_{\sigma\alpha}\Gamma^\sigma_{\beta\mu}$$

so that

$$\begin{aligned}\frac{\partial \ln|g|}{\partial x^\mu} &= g^{\alpha\beta}g_{\sigma\beta}\Gamma^\sigma_{\alpha\mu} + g^{\alpha\beta}g_{\alpha\sigma}\Gamma^\sigma_{\beta\mu}\\ &= \delta^\alpha_\sigma\Gamma^\sigma_{\alpha\mu} + \delta^\beta_\sigma\Gamma^\sigma_{\beta\mu}\\ &= \Gamma^\alpha_{\alpha\mu} + \Gamma^\beta_{\beta\mu} = 2\Gamma^\alpha_{\alpha\mu}\end{aligned}$$

Hence

$$\frac{\partial \ln|g_{ij}|^{\frac{1}{2}}}{\partial x^\mu} = \Gamma^\alpha_{\alpha\mu} \tag{3.49}$$

Example 3.12. Let us consider a Euclidean space using Euclidean coordinates. In this coordinate system $\Gamma^i_{jk}(x) = 0$. From (3.48)

$$\bar{\Gamma}^i_{jk}(\bar{x}) = \frac{\partial^2 x^\sigma}{\partial \bar{x}^j \, \partial \bar{x}^k} \frac{\partial \bar{x}^i}{\partial x^\sigma}$$

for the $\bar{x}$ coordinate system. If the $\bar{\Gamma}^i_{jk}$ also vanish, then $\dfrac{\partial^2 x^\sigma}{\partial \bar{x}^j \, \partial \bar{x}^k} = 0$ of necessity, so that

$$x^\sigma = a^\sigma_\alpha \bar{x}^\alpha + b^\sigma$$

where a^σ_α, b^σ are constants of integration. Hence the coordinate transformation between two cartesian coordinate systems is linear. If the transformation is orthogonal,

$$\sum_{\sigma=1}^{n} a^\sigma_\alpha a^\sigma_\beta = \delta_{\alpha\beta}$$

[see (1.53)]. For orthogonal transformations $\bar{g}_{ij} = g_{\alpha\beta} \dfrac{\partial x^\alpha}{\partial \bar{x}^i} \dfrac{\partial x^\beta}{\partial \bar{x}^j}$ reduces to

$$\delta_{ij} = \delta_{\alpha\beta} \frac{\partial x^\alpha}{\partial \bar{x}^i} \frac{\partial x^\beta}{\partial \bar{x}^j}$$

so that

$$\delta_{ij} \frac{\partial \bar{x}^i}{\partial x^\mu} = \delta_{\alpha\beta} \frac{\partial x^\alpha}{\partial \bar{x}^i} \frac{\partial \bar{x}^i}{\partial x^\mu} \frac{\partial x^\beta}{\partial \bar{x}^j}$$

$$\frac{\partial \bar{x}^j}{\partial x^\mu} = \delta_{\alpha\beta} \delta^\alpha_\mu \frac{\partial x^\beta}{\partial \bar{x}^j} = \frac{\partial x^\mu}{\partial \bar{x}^j} \tag{3.50}$$

Now let us compare the laws of transformation for covariant and contravariant vectors. We have

$$\bar{A}^i = A^\alpha \frac{\partial \bar{x}^i}{\partial x^\alpha} \qquad \bar{A}_i = A_\alpha \frac{\partial x^\alpha}{\partial x^i} \tag{3.51}$$

Making use of (3.50) yields

$$\bar{A}^i = \sum_{\alpha=1}^{n} A^\alpha \frac{\partial x^\alpha}{\partial \bar{x}^i} \tag{3.52}$$

so that orthogonal transformations affect contravariant vectors in exactly the same way covariant vectors are affected. This is why there was no distinction made between these two types of vectors in the elementary treatment of vectors.

Problems

1. Prove (3.48).
2. By differentiating the identity $g^{i\alpha} g_{\alpha j} = \delta^i_j$ show that

$$\frac{\partial g^{ik}}{\partial x^j} = -g^{hk} \Gamma^i_{hj} - g^{hi} \Gamma^k_{hj}$$

3. Show that the law of transformation for the Christoffel symbols is transitive.
4. Derive the law of transformation for the Christoffel symbols from

$$\bar{g}_{ij} = g_{\alpha\beta} \frac{\partial x^\alpha}{\partial \bar{x}^i} \frac{\partial x^\beta}{\partial \bar{x}^j}$$

5. Define $g^*_{ij}(x) = \mu(x)g_{ij}(x)$. Show that

$$\Gamma^{*i}_{jk}(x) = \Gamma^i_{jk}(x) + \varphi_k\delta^i_j + \varphi_j\delta^i_k - g^{i\sigma}g_{jk}\varphi_\sigma$$

where $\varphi_\sigma = \frac{1}{2}\frac{\partial \ln \mu}{\partial x^\sigma}$ and Γ^{*i}_{jk}, Γ^i_{jk} are the Christoffel symbols for the g^*_{ij}, g_{ij}, respectively.

3.9. Covariant Differentiation. Let us differentiate the absolute covariant vector given by the transformation

$$\bar{A}_j = A_\mu \frac{\partial x^\mu}{\partial \bar{x}^j}$$

with respect to an absolute scalar $t = \bar{t}$. We obtain

$$\frac{d\bar{A}_j}{d\bar{t}} = \frac{dA_\mu}{dt}\frac{\partial x^\mu}{\partial \bar{x}^j} + A_\mu \frac{\partial^2 x^\mu}{\partial \bar{x}^k\,\partial \bar{x}^j}\frac{d\bar{x}^k}{d\bar{t}} \tag{3.53}$$

since $t = \bar{t}$. It is at once apparent that $\frac{dA^\alpha}{dt}$ is not a covariant vector. We wish to determine a vector (covariant) which will reduce to the ordinary derivative in a Euclidean space using Euclidean coordinates. We accomplish this in the following manner by making use of the transformation law for the Christoffel symbols: Multiply (3.48) by $\bar{A}_i \frac{d\bar{x}^k}{d\bar{t}} = A_\mu \frac{\partial x^\mu}{\partial \bar{x}^i}\frac{d\bar{x}^k}{d\bar{t}}$ to obtain

$$\begin{aligned}\bar{\Gamma}^i_{jk}\bar{A}_i \frac{d\bar{x}^k}{d\bar{t}} &= \Gamma^\alpha_{\beta\gamma} A_\mu \frac{\partial x^\mu}{\partial \bar{x}^i}\frac{\partial x^\beta}{\partial \bar{x}^j}\frac{\partial x^\gamma}{\partial \bar{x}^k}\frac{\partial \bar{x}^i}{\partial x^\alpha}\frac{d\bar{x}^k}{d\bar{t}} + A_\mu \frac{\partial x^\mu}{\partial \bar{x}^i}\frac{\partial \bar{x}^i}{\partial x^\sigma}\frac{\partial^2 x^\sigma}{\partial \bar{x}^j\,\partial \bar{x}^k}\frac{d\bar{x}^k}{d\bar{t}}\\ &= \Gamma^\alpha_{\mu\gamma}A_\alpha \frac{\partial x^\mu}{\partial \bar{x}^j}\frac{dx^\gamma}{dt} + A_\mu \frac{\partial^2 x^\mu}{\partial \bar{x}^j\,\partial \bar{x}^k}\frac{d\bar{x}^k}{d\bar{t}}\end{aligned}$$

Subtracting from (3.53) yields

$$\frac{d\bar{A}_j}{d\bar{t}} - \bar{\Gamma}^i_{jk}\bar{A}_i \frac{d\bar{x}^k}{d\bar{t}} = \left(\frac{dA_\mu}{dt} - \Gamma^\alpha_{\mu\gamma}A_\alpha \frac{dx^\gamma}{dt}\right)\frac{\partial x^\mu}{\partial \bar{x}^j} \tag{3.54}$$

Hence $\frac{dA_\mu}{dt} - \Gamma^\alpha_{\mu\gamma}A_\alpha \frac{dx^\gamma}{dt}$ is a covariant vector.

For Euclidean coordinates $\Gamma^\alpha_{\mu\gamma} \equiv 0$ so that this covariant vector becomes the ordinary derivative $\frac{dA_\mu}{dt}$. We call $\frac{dA_\mu}{dt} - \Gamma^\alpha_{\mu\gamma}A_\alpha \frac{dx^\gamma}{dt}$ the intrinsic derivative of A_μ with respect to t. Its value depends on the direction we choose since it involves $\frac{dx^\gamma}{dt}$. We write

$$\frac{\delta A_\mu}{\delta t} \equiv \left(\frac{dA_\mu}{dt} - \Gamma^\alpha_{\mu\gamma}A_\alpha \frac{dx^\gamma}{dt}\right) \tag{3.55}$$

Since

$$\frac{\delta A_\mu}{\delta t} = \left(\frac{\partial A_\mu}{\partial x^\gamma} - \Gamma^\alpha_{\mu\gamma}A_\alpha\right)\frac{dx^\gamma}{dt}$$

is a vector for arbitrary $\frac{dx^\gamma}{dt}$, it follows from the quotient law that

$$\frac{\partial A_\mu}{\partial x^\gamma} - \Gamma^\alpha_{\mu\gamma} A_\alpha \tag{3.56}$$

is a covariant tensor of rank 2. We call (3.56) the **covariant derivative** of A_μ, and we write

$$A_{\mu,\gamma} \equiv \frac{\partial A_\mu}{\partial x^\gamma} - \Gamma^\alpha_{\mu\gamma} A_\alpha \tag{3.57}$$

The comma in $A_{\mu,\gamma}$ denotes **covariant differentiation.**

We now consider a scalar of weight N,

$$\bar{A} = \left|\frac{\partial x}{\partial \bar{x}}\right|^N A$$

We have

$$\frac{\partial \bar{A}}{\partial \bar{x}^j} = \left|\frac{\partial x}{\partial \bar{x}}\right|^N \frac{\partial A}{\partial x^\alpha}\frac{\partial x^\alpha}{\partial \bar{x}^j} + N\left|\frac{\partial x}{\partial \bar{x}}\right|^{N-1} \frac{\partial\left|\frac{\partial x}{\partial \bar{x}}\right|}{\partial \bar{x}^j} A$$

From Prob. 9, Sec. 1.2, we have

$$\frac{\partial\left|\frac{\partial x}{\partial \bar{x}}\right|}{\partial \bar{x}^j} = \left|\frac{\partial x}{\partial \bar{x}}\right| \frac{\partial \bar{x}^\alpha}{\partial x^\beta}\frac{\partial^2 x^\beta}{\partial \bar{x}^j\,\partial \bar{x}^\alpha}$$

so that

$$\frac{\partial \bar{A}}{\partial \bar{x}^j} = \left|\frac{\partial x}{\partial \bar{x}}\right|^N \frac{\partial A}{\partial x^\alpha}\frac{\partial x^\alpha}{\partial \bar{x}^j} + N\left|\frac{\partial x}{\partial \bar{x}}\right|^N \frac{\partial \bar{x}^\alpha}{\partial x^\beta}\frac{\partial^2 x^\beta}{\partial \bar{x}^j\,\partial \bar{x}^\alpha} A \tag{3.58}$$

Multiplying $\bar{\Gamma}^\alpha_{j\alpha} = \Gamma^\sigma_{\alpha\sigma}\frac{\partial x^\alpha}{\partial \bar{x}^j} + \frac{\partial^2 x^\sigma}{\partial \bar{x}^\alpha\,\partial \bar{x}^j}\frac{\partial \bar{x}^\alpha}{\partial x^\sigma}$ by $N\bar{A} = N\left|\frac{\partial x}{\partial \bar{x}}\right|^N A$ and subtracting from (3.58) yields

$$\frac{\partial \bar{A}}{\partial \bar{x}^j} - N\bar{A}\bar{\Gamma}^\sigma_{j\sigma} = \left|\frac{\partial x}{\partial \bar{x}}\right|^N \left(\frac{\partial A}{\partial x^\alpha} - NA\Gamma^\sigma_{\alpha\sigma}\right)\frac{\partial x^\alpha}{\partial \bar{x}^j}$$

Hence the invariant (in form), $\frac{\partial A}{\partial x^\alpha} - NA\Gamma^\sigma_{\alpha\sigma}$, is a covariant vector of weight N. We write

$$A_{,\alpha} \equiv \frac{\partial A}{\partial x^\alpha} - NA\Gamma^\sigma_{\alpha\sigma} \tag{3.59}$$

We call $A_{,\alpha}$ the covariant derivative of A. The comma in $A_{,\alpha}$ denotes covariant differentiation. If A is an absolute scalar, $N = 0$ and

$$A_{,\alpha} = \frac{\partial A}{\partial x^\alpha}$$

the gradient of A.

In general, it can be shown that if $T^{\alpha_1\alpha_2\cdots\alpha_r}_{\beta_1\beta_2\cdots\beta_s}(x)$ is a relative tensor of weight N, then

$$\begin{aligned} T^{\alpha_1\alpha_2\cdots\alpha_r}_{\beta_1\beta_2\cdots\beta_s,m} \equiv \frac{\partial T^{\alpha_1\alpha_2\cdots\alpha_r}_{\beta_1\beta_2\cdots\beta_s}}{\partial x^m} &+ T^{\mu\alpha_2\cdots\alpha_r}_{\beta_1\beta_2\cdots\beta_s}\Gamma^{\alpha_1}_{\mu m} + \cdots + T^{\alpha_1\alpha_2\cdots\mu}_{\beta_1\beta_2\cdots\beta_s}\Gamma^{\alpha_r}_{\mu m} \\ &- T^{\alpha_1\alpha_2\cdots\alpha_r}_{\mu\beta_2\cdots\beta_s}\Gamma^{\mu}_{\beta_1 m} - \cdots - T^{\alpha_1\alpha_2\cdots\alpha_r}_{\beta_1\beta_2\cdots\mu}\Gamma^{\mu}_{\beta_s m} \\ &- NT^{\alpha_1\alpha_2\cdots\alpha_r}_{\beta_1\beta_2\cdots\beta_s}\Gamma^{\mu}_{\mu m} \qquad (3.60) \end{aligned}$$

is a relative tensor of weight N, of covariant rank one greater than $T^{\alpha_1\alpha_2\cdots\alpha_r}_{\beta_1\beta_2\cdots\beta_s}$. $T^{\alpha_1\alpha_2\cdots\alpha_r}_{\beta_1\beta_2\cdots\beta_s,m}$ is called the covariant derivative of $T^{\alpha_1\alpha_2\cdots\alpha_r}_{\beta_1\beta_2\cdots\beta_s}$. Since the covariant derivative is a tensor, successive covariant differentiations can be applied.

Example 3.13. We apply (3.60) to the metric tensor g_{ij}. We have

$$g_{ij,k} = \frac{\partial g_{ij}}{\partial x^k} - g_{\mu j}\Gamma^{\mu}_{ik} - g_{i\mu}\Gamma^{\mu}_{jk} = 0$$

from Prob. 5, Sec. 3.7.

Example 3.14. *Curl of a Vector.* Let A_i be an absolute covariant vector. We have

$$A_{i,j} = \frac{\partial A_i}{\partial x^j} - A_\alpha\Gamma^{\alpha}_{ij} \qquad \text{and} \qquad A_{j,i} = \frac{\partial A_j}{\partial x^i} - A_\alpha\Gamma^{\alpha}_{ji}$$

so that $A_{i,j} - A_{j,i} \equiv \dfrac{\partial A_i}{\partial x^j} - \dfrac{\partial A_j}{\partial x^i}$ is a tensor. It is called the curl of A and is a covariant tensor of rank 2. Strictly speaking, the curl is not a vector but a tensor of rank 2. In a three-dimensional space, however, the curl may be looked upon as a vector [see (3.10)]. If $A_i = \dfrac{\partial\varphi}{\partial x^i}$, then curl $A_i = 0$. Conversely, if curl $A_i = 0$, then $A_i = \dfrac{\partial\varphi}{\partial x^i}$, where [see (2.79)]

$$\begin{aligned} \varphi(x^1, x^2, \ldots, x^n) = &\int_{x_0^1}^{x^1} A_1(x^1, x^2, \ldots, x^n)\,dx^1 \\ &+ \int_{x_0^2}^{x^2} A_2(x_0^1, x^2, \ldots, x^n)\,dx^2 \\ &+ \cdots + \int_{x_0^n}^{x^n} A_n(x_0^1, x_0^2, \ldots, x_0^{n-1}, x^n)\,dx^n \qquad (3.61) \end{aligned}$$

The constants $x_0^1, x_0^2, \ldots, x_0^n$ can be chosen arbitrarily.

Example 3.15. *Divergence of a Vector.* The divergence of an absolute contravariant vector is defined as the contraction of its covariant derivative. Hence

$$\text{div } A^i = A^{\alpha}_{,\alpha} = \frac{\partial A_\alpha}{\partial x^\alpha} + A^\alpha\Gamma^{i}_{\alpha i}$$

From (3.49), $\Gamma^{i}_{\alpha i} = \dfrac{\partial \ln |g|^{\frac{1}{2}}}{\partial x^\alpha}$, $|g| = |g_{ij}|$, so that

$$\begin{aligned} \text{div } A^i &= \frac{\partial A_\alpha}{\partial x^\alpha} + |g|^{-\frac{1}{2}}A^\alpha\frac{\partial |g|^{\frac{1}{2}}}{\partial x^\alpha} \\ &= \frac{1}{|g|^{\frac{1}{2}}}\frac{\partial}{\partial x^\alpha}(|g|^{\frac{1}{2}}A^\alpha) \end{aligned} \qquad (3.62)$$

If we wish to obtain the divergence of A_i, we consider the associated vector $A^i = g^{i\alpha}A_\alpha$. The A^α of (3.62) are the vector components of **A**. To keep div A^i dimensionally correct, we replace the vector components of A^i by the physical components of A^i. For spherical coordinates

$$|g|^{\frac{1}{2}} = \begin{vmatrix} 1 & 0 & 0 \\ 0 & r^2 & 0 \\ 0 & 0 & r^2 \sin^2\theta \end{vmatrix}^{\frac{1}{2}} = r^2 \sin\theta$$

so that $\operatorname{div} A^i = \dfrac{1}{r^2\sin\theta}\left[\dfrac{\partial}{\partial r}(r^2\sin\theta A^r) + \dfrac{\partial}{\partial\theta}(r^2\sin\theta A^\theta) + \dfrac{\partial}{\partial\varphi}(r^2\sin\theta A^\varphi)\right]$ and changing to physical components (see Prob. 4, Sec. 3.2)

$$\operatorname{div} A^i = \frac{1}{r^2\sin\theta}\left[\frac{\partial}{\partial r}(r^2\sin\theta A^r) + \frac{\partial}{\partial\theta}(r\sin\theta A^\theta) + \frac{\partial}{\partial\varphi}(rA^\varphi)\right]$$

Example 3.16. *The Laplacian of a Scalar Invariant.* The Laplacian of the scalar invariant φ is defined as the divergence of the gradient of φ. We consider the associated vector of the gradient of φ, namely, $g^{\alpha\beta}\dfrac{\partial\varphi}{\partial x^\beta}$. Applying (3.62) to $g^{\alpha\beta}\dfrac{\partial\varphi}{\partial x^\beta}$ yields the Laplacian

$$\operatorname{Lap}\varphi \equiv \nabla^2\varphi = \frac{1}{|g|^{\frac{1}{2}}}\frac{\partial}{\partial x^\alpha}\left(|g|^{\frac{1}{2}}g^{\alpha\beta}\frac{\partial\varphi}{\partial x^\beta}\right) \tag{3.63}$$

In spherical coordinates

$$|g_{ij}|^{\frac{1}{2}} = r^2\sin\theta \qquad \|g^{ij}\| = \begin{Vmatrix} 1 & 0 & 0 \\ 0 & \frac{1}{r^2} & 0 \\ 0 & 0 & \frac{1}{r^2\sin^2\theta} \end{Vmatrix}$$

$$\nabla^2 V = \frac{1}{r^2\sin\theta}\left[\frac{\partial}{\partial r}\left(r^2\sin\theta\frac{\partial V}{\partial r}\right) + \frac{\partial}{\partial\theta}\left(\sin\theta\frac{\partial V}{\partial\theta}\right) + \frac{\partial}{\partial\varphi}\left(\frac{1}{\sin\theta}\frac{\partial V}{\partial\varphi}\right)\right]$$

Problems

1. Starting with $\bar{A}^i = A^\alpha\dfrac{\partial\bar{x}^i}{\partial x^\alpha}$, show that $A^i_{,j} \equiv \dfrac{\partial A^i}{\partial x^j} + A^\alpha\Gamma^i_{\alpha j}$ is a mixed tensor without recourse to (3.60).

2. Show that $(A^iB_j)_{,k} = A^iB_{j,k} + A^i_{,k}B_j$.

3. Show that $(g_{i\alpha}A^\alpha)_{,j} = g_{i\alpha}A^\alpha_{,j}$.

4. Show that $|g_{ij}|_{,k} = 0$.

5. Show that $\delta^i_{j,k} = 0$.

6. Find the Laplacian of V in cylindrical coordinates.

7. Show that $A^r_{s,t} = \dfrac{\partial A^r_s}{\partial x^t} + \Gamma^r_{\mu t}A^\mu_s - \Gamma^\mu_{st}A^r_\mu$ for an absolute mixed tensor A^r_s without recourse to (3.60).

8. Show that $A^\alpha_{i,\alpha} = \dfrac{1}{|g|^{\frac{1}{2}}}\dfrac{\partial}{\partial x^\alpha}(|g|^{\frac{1}{2}}A^\alpha_i) - A^\alpha_\beta\Gamma^\beta_{i\alpha}$.

9. Write out the form of $A^\alpha_{,rs}$ (two covariant differentiations).

10. If $A_i(x, t)$ is a covariant vector, show that $\dfrac{\delta A_i}{\delta t} = \dfrac{\partial A_i}{\partial t} + A_{i,j}\dfrac{dx^j}{dt}$. Show that $\dfrac{\partial v^i}{\partial t} + v^i_{,j}v^j$ is the acceleration vector if v^i is the velocity vector.

3.10. Geodesic Coordinates. Since the Christoffel symbols Γ^i_{jk} are not the components of a tensor, it may be possible to find a coordinate system in the neighborhood of $x^i = q^i$ such that $\Gamma^i_{jk}(q) = 0$. We now show that this can be done. Let

$$\bar{x}^i = (x^i - q^i) + \tfrac{1}{2}\Gamma^i_{\alpha\beta}(q)(x^\alpha - q^\alpha)(x^\beta - q^\beta) \tag{3.64}$$

so that $\left.\dfrac{\partial \bar{x}^i}{\partial x^j}\right|_q = \delta^i_j$ and $\left|\dfrac{\partial \bar{x}^i}{\partial x^j}\right|_q = 1$. Hence (3.64) is nonsingular in a neighborhood of $x = q$. The point $x^i = q^i$ corresponds to the point $\bar{x}^i = 0$, that is, the point $x = q$ now becomes the origin of the $\bar{x}$ coordinate system. Differentiating (3.64) yields

$$\delta^i_j = \frac{\partial \bar{x}^i}{\partial \bar{x}^j} = \frac{\partial x^i}{\partial \bar{x}^j} + \Gamma^i_{\alpha\beta}(q)(x^\alpha - q^\alpha)\frac{\partial x^\beta}{\partial \bar{x}^j} \tag{3.65}$$

since $\Gamma^i_{\alpha\beta}(q) = \Gamma^i_{\beta\alpha}(q)$. Thus $\delta^i_j = \left.\dfrac{\partial x^i}{\partial \bar{x}^j}\right|_q$. Differentiating (3.65) with respect to $\bar{x}^k$ yields

$$0 = \frac{\partial^2 x^i}{\partial \bar{x}^k\,\partial \bar{x}^j} + \Gamma^i_{\alpha\beta}(q)\frac{\partial x^\alpha}{\partial \bar{x}^k}\frac{\partial x^\beta}{\partial \bar{x}^j} + \Gamma^i_{\alpha\beta}(q)(x^\alpha - q^\alpha)\frac{\partial^2 x^\beta}{\partial \bar{x}^k\,\partial \bar{x}^j}$$

so that

$$\begin{aligned}\left.\frac{\partial^2 x^i}{\partial \bar{x}^k\,\partial \bar{x}^j}\right|_q &= -\Gamma^i_{\alpha\beta}(q)\left.\frac{\partial x^\alpha}{\partial \bar{x}^k}\right|_q\left.\frac{\partial x^\beta}{\partial \bar{x}^j}\right|_q \\ &= -\Gamma^i_{\alpha\beta}(q)\delta^\alpha_k\delta^\beta_j = -\Gamma^i_{kj}(q) = -\Gamma^i_{jk}(q)\end{aligned}$$

Now

$$\bar{\Gamma}^i_{jk}(\bar{x}) = \Gamma^\alpha_{\beta\gamma}(x)\frac{\partial x^\beta}{\partial \bar{x}^j}\frac{\partial x^\gamma}{\partial \bar{x}^k}\frac{\partial \bar{x}^i}{\partial x^\alpha} + \frac{\partial^2 x^\sigma}{\partial \bar{x}^j\,\partial \bar{x}^k}\frac{\partial \bar{x}^i}{\partial x^\sigma}$$

and evaluating at $x^i = q^i$, $\bar{x}^i = 0$, yields

$$\begin{aligned}\bar{\Gamma}^i_{jk}(0) &= \Gamma^\alpha_{\beta\gamma}(q)\delta^\beta_j\delta^\gamma_k\delta^i_\alpha - \Gamma^\sigma_{jk}(q)\delta^i_\sigma \\ &= \Gamma^i_{jk}(q) - \Gamma^i_{jk}(q) \equiv 0. \quad \text{Q.E.D.}\end{aligned}$$

Any system of coordinates for which $(\Gamma^i_{jk})_P = 0$ at a point P is called a geodesic coordinate system. In such a system of coordinates the covariant derivative, when evaluated at P, becomes the ordinary derivative when evaluated at P. For example,

$$\begin{aligned}(A^i_{,j})_P &= \left(\frac{\partial A^i}{\partial x^j}\right)_P + (\Gamma^i_{\alpha j})_P(A^\alpha)_P \\ &= \left(\frac{\partial A^i}{\partial x^j}\right)_P\end{aligned}$$

since $(\Gamma^i_{\alpha j})_P = 0$, if the x^i are the coordinates of a geodesic coordinate system.

We now show that

$$\begin{aligned}(A^i + B^i)_{,j} &= A^i_{;j} + B^i_{;j} \\ (A^iB^j)_{,k} &= A^iB^j_{,k} + A^i_{,k}B^j\end{aligned} \tag{3.66}$$

System (3.66) is true in geodesic coordinates at any point P. But if two tensors are equal to each other (components are equal) in one coordinate system, they are equal to each other in all coordinate systems. Hence (3.66) holds for all coordinate systems at any point P.

Equation (3.64) yields one geodesic coordinate system. There are infinitely many such systems since we could have added

$$A^i(q)\varphi_{\alpha\beta\gamma}(q)(x^\alpha - q^\alpha)(x^\beta - q^\beta)(x^\gamma - q^\gamma)$$

to the right-hand side of (3.64) and still have obtained $\bar{\Gamma}^i_{jk}(0) = 0$.

A special type of geodesic coordinate system is the following: Let us consider the family of geodesics passing through the point P, $x^i = x^i_0$. Each $\xi^i = \left.\dfrac{dx^i}{ds}\right|_P$ determines a unique geodesic passing through P. This follows since the differential equations of the geodesics are of second order. Suitable restrictions (say, analyticity) on the $\Gamma^i_{jk}(x)$ guarantee a unique solution of the second-order system of differential equations when the initial conditions are proposed. The ξ^i are the components of the tangent vector of the geodesic through P. We now move along the geodesic (determined by ξ^i) a distance s. This determines a unique point Q. Conversely, if Q is near P, there is a unique geodesic passing through P and Q which determines a unique ξ^i and s. We define

$$\bar{x}^i \equiv \xi^i s \tag{3.67}$$

The $\bar{x}^i$ are called Riemannian coordinates. A simple example of Riemannian coordinates occurs in a two-dimensional Euclidean space. There is a unique geodesic through the origin with slope $m = \tan\theta$. The r coordinate of polar coordinates corresponds to the s of Riemannian coordinates. Thus ($\xi^1 = \cos\theta$, $\xi^2 = \sin\theta$), $r = s$, and $x^1 = r\cos\theta$, $x^2 = r\sin\theta$. In this case the Riemannian coordinate system $(\bar{x}^1, \bar{x}^2)$ corresponds to the Euclidean coordinate system (x, y).

The differential equations of the geodesics in Riemannian coordinates are

$$\frac{d^2\bar{x}^i}{ds^2} + \bar{\Gamma}^i_{jk}(\bar{x})\frac{d\bar{x}^j}{ds}\frac{d\bar{x}^k}{ds} = 0$$

But $\dfrac{d\bar{x}^i}{ds} = \xi^i$, $\dfrac{d^2\bar{x}^i}{ds^2} = 0$, so that

$$\bar{\Gamma}^i_{jk}(\bar{x})\xi^j\xi^k = 0 \tag{3.68}$$

Since (3.68) holds at the point P (the origin of the Riemannian coordinate system, $\bar{x}^i = 0$) for arbitrary ξ^i, it follows that $\bar{\Gamma}^i_{jk}(0) = 0$. Hence a Riemannian coordinate system is a geodesic coordinate system.

3.11. The Curvature Tensor. We consider the absolute vector V^i. Its covariant derivative yields the mixed tensor

$$V^i_{,j} = \frac{\partial V^i}{\partial x^j} + V^\alpha \Gamma^i_{\alpha j}$$

On again differentiating covariantly we obtain

$$\begin{aligned} V^i_{,jk} &= \frac{\partial V^i_{,j}}{\partial x^k} + V^\alpha_{,j}\Gamma^i_{\alpha k} - V^i_{,\alpha}\Gamma^\alpha_{jk} \\ &= \frac{\partial^2 V^i}{\partial x^k\,\partial x^j} + \frac{\partial V^\alpha}{\partial x^k}\Gamma^i_{\alpha j} + V^\alpha \frac{\partial \Gamma^i_{\alpha j}}{\partial x^k} + \left(\frac{\partial V^\alpha}{\partial x^j} + V^\beta \Gamma^\alpha_{\beta j}\right)\Gamma^i_{\alpha k} \\ &\qquad - \left(\frac{\partial V^i}{\partial x^\alpha} + V^\beta \Gamma^i_{\beta\alpha}\right)\Gamma^\alpha_{jk} \end{aligned}$$

Interchanging k and j and subtracting yields

$$V^i_{,jk} - V^i_{,kj} = V^\alpha R^i_{\alpha jk} \tag{3.69}$$

where

$$R^i_{\alpha jk} = \frac{\partial \Gamma^i_{\alpha j}}{\partial x^k} - \frac{\partial \Gamma^i_{\alpha k}}{\partial x^j} + \Gamma^\beta_{\alpha j}\Gamma^i_{\beta k} - \Gamma^\beta_{\alpha k}\Gamma^i_{\beta j} \tag{3.70}$$

A necessary and sufficient condition that $V^i_{,jk} = V^i_{,kj}$ is that $R^i_{\alpha jk} = 0$. It follows from (3.69) and the quotient law that $R^i_{\alpha jk}$ is a tensor. The tensor $R^i_{\alpha jk}$ depends only on the metric tensor of the space. It is called the curvature tensor. Its name and importance will become apparent in the next two paragraphs. The contracted curvature tensor

$$R_{ij} \equiv R^\alpha_{i\alpha j} = \frac{\partial \Gamma^\alpha_{i\alpha}}{\partial x^j} - \frac{\partial \Gamma^\alpha_{ij}}{\partial x^\alpha} + \Gamma^\beta_{i\alpha}\Gamma^\alpha_{\beta j} - \Gamma^\beta_{ij}\Gamma^\alpha_{\beta\alpha} \tag{3.71}$$

is called the Ricci tensor and plays a most important role in the general theory of relativity.

The scalar invariant $R = g^{ij}R_{ij}$ is called the scalar curvature. The tensor

$$R_{hijk} = g_{h\alpha}R^\alpha_{ijk} \tag{3.72}$$

is called the Riemann-Christoffel, or covariant curvature, tensor.

Problems

1. Show that, for Riemannian coordinates, $\bar{\Gamma}^i_{jk}(\bar{x})\bar{x}^j\bar{x}^k = 0$.

2. Show that $(A^i_j + B^i_j)_{,k} = A^i_{j,k} + B^i_{j,k}$.

3. Show that $R_{ij} = R_{ji}$.

4. Show that $R^i_{\alpha kj,\sigma} + R^i_{\alpha\sigma j,k} + R^i_{\alpha k\sigma,j} = 0$. This is the Bianchi identity. *Hint:* Use geodesic coordinates.

5. Show that $R_{hijk} = -R_{ihjk} = -R_{hikj}$, $R_{iijk} = 0$, $R_{ijkk} = 0$.

6. If $R_{ij} = kg_{ij}$, show that $R = nk$, n the dimension of the space.

3.12. Euclidean, or Flat, Space. If a space is Euclidean, there exists a coordinate system for which the g_{ij} are constants, so that $\Gamma^i_{jk}(x) \equiv 0$, $\dfrac{\partial \Gamma^i_{jk}}{\partial x^l} = 0$. It immediately follows that the curvature tensor R^i_{jkl} vanishes (all its components are zero). We now show the converse. If the curvature tensor is zero, the space is Euclidean. Let us first note that if an x coordinate system exists such that the $g_{ij}(x)$ are constants then $\Gamma^i_{jk}(x) = 0$, and conversely. If there is an x coordinate system for which $\Gamma^i_{jk}(x) = 0$, then

$$\bar{\Gamma}^i_{jk}(y) = \frac{\partial^2 x^\alpha}{\partial y^j \, \partial y^k} \frac{\partial y^i}{\partial x^\alpha} \tag{3.73}$$

and conversely, if (3.73) holds, then $\Gamma^i_{jk}(x) = 0$. Our problem reduces to the following: Given the Christoffel symbols $\bar{\Gamma}^i_{jk}(y)$ in any y coordinate system, can we find a set of x^i, $i = 1, 2, 3, \ldots, n$, satisfying (3.73)?

The system of second-order differential equations (3.73) can be written

$$\frac{\partial^2 x^\sigma}{\partial y^j \, \partial y^k} = \frac{\partial x^\sigma}{\partial y^i} \bar{\Gamma}^i_{jk}(y) \tag{3.74}$$

We reduce (3.74) to a system of first-order differential equations. Let

$$\frac{\partial x^\sigma}{\partial y^i} \equiv u^\sigma_i \tag{3.75}$$

so that

$$\frac{\partial u^\sigma_k}{\partial y^j} = u^\sigma_i \bar{\Gamma}^i_{jk}(y) \tag{3.76}$$

For each σ we have the first-order system of differential equations given by (3.75) and (3.76). These equations are special cases of the more general system

$$\frac{\partial z^k}{\partial y^j} = F^k_j(z^1, z^2, \ldots, z^n, z^{n+1}, y^1, y^2, \ldots, y^n) \tag{3.77}$$

with $k = 1, 2, \ldots, n + 1$; $j = 1, 2, \ldots, n$. If we let $z^1 = x^\sigma$, $z^2 = u^\sigma_1, \ldots, z^{n+1} = u^\sigma_n$, (3.77) reduces to (3.75), (3.76).

If a solution of (3.77) exists, then of necessity (assuming differentiability and continuity of the second mixed derivatives)

$$\frac{\partial F^k_j}{\partial y^l} + \frac{\partial F^k_j}{\partial z^\mu}\frac{\partial z^\mu}{\partial y^l} = \frac{\partial F^k_l}{\partial y^j} + \frac{\partial F^k_l}{\partial z^\mu}\frac{\partial z^\mu}{\partial y^j}$$

or

$$\frac{\partial F^k_j}{\partial y^l} + \frac{\partial F^k_j}{\partial z^\mu} F^\mu_l = \frac{\partial F^k_l}{\partial y^j} + \frac{\partial F^k_l}{\partial z^\mu} F^\mu_j \tag{3.78}$$

If the F^k_j are analytic, it can be shown that the integrability conditions (3.78) are also sufficient that (3.77) has a unique solution satisfying the initial conditions $z^k = z^k_0$ at $y^i = y^i_0$. The reader is referred to the elegant proof found in Gaston Darboux, "Leçons systèmes orthogonaux et les coordonnées curvilignes," pp. 325–336, Gauthier-Villars, Paris, 1910.

The integrability conditions (3.78) when referred to (3.75), (3.76) become

$$\begin{aligned} \Gamma^\alpha_{jk} u^\sigma_\alpha &= \Gamma^\alpha_{kj} u^\sigma_\alpha \\ R^\alpha_{jkl} u^\sigma_\alpha &= 0 \end{aligned} \tag{3.79}$$

The first equation of (3.79) is satisfied from the symmetry of the Γ^i_{jk}, and the second is satisfied if $R^\alpha_{jkl} = 0$. Hence a necessary and sufficient condition that a Riemannian space be Euclidean is that the curvature tensor vanish.

3.13. Parallel Displacement of a Vector. Consider an absolute contravariant vector A^i in a Euclidean space. Using cartesian coordinates yields $\Gamma^i_{jk}(x) = 0$. We assume further that the A^i are constant. From

$$\bar{A}^i(\bar{x}) = A^\alpha \frac{\partial \bar{x}^i}{\partial x^\alpha}$$

we have

$$\begin{aligned} d\bar{A}^i &= A^\alpha \frac{\partial^2 \bar{x}^i}{\partial x^\beta\, \partial x^\alpha} \frac{\partial x^\beta}{\partial \bar{x}^\gamma}\, d\bar{x}^\gamma \\ &= \bar{A}^\sigma \frac{\partial^2 \bar{x}^i}{\partial x^\beta\, \partial x^\alpha} \frac{\partial x^\beta}{\partial \bar{x}^\gamma} \frac{\partial x^\alpha}{\partial \bar{x}^\sigma}\, d\bar{x}^\gamma \end{aligned}$$

Moreover

$$\begin{aligned} \bar{\Gamma}^i_{\gamma\sigma} &= \frac{\partial^2 x^\alpha}{\partial \bar{x}^\gamma\, \partial \bar{x}^\sigma} \frac{\partial \bar{x}^i}{\partial x^\alpha} \qquad \text{since } \Gamma^i_{jk}(x) = 0 \\ &= - \frac{\partial^2 \bar{x}^i}{\partial x^\beta\, \partial x^\alpha} \frac{\partial x^\beta}{\partial \bar{x}^\gamma} \frac{\partial x^\alpha}{\partial \bar{x}^\sigma} \end{aligned}$$

from Example 1.4. Hence

$$d\bar{A}^i = -\bar{A}^\sigma \bar{\Gamma}^i_{\sigma\gamma}\, d\bar{x}^\gamma \tag{3.80}$$

Now the A^i, being constant in a Euclidean space, can be looked upon as yielding a uniform or parallel vector field. Equation (3.80) describes how the components of this parallel vector field change in various coordinate systems. Since, generally, a Riemannian space is not Euclidean, we can use (3.80) to define parallelism of a vector field.

We say that A^i is parallelly displaced with respect to the Riemannian V_n along the curve $x^i = x^i(s)$ if

$$\frac{dA^i}{ds} = -A^\alpha \Gamma^i_{\alpha\beta} \frac{dx^\beta}{ds} \tag{3.81}$$

or

$$\frac{\delta A^i}{\delta s} = 0$$

We say that the vector suffers a parallel displacement along the curve. If a vector suffers a parallel displacement along all curves, then from $\frac{dA^i}{ds} = \frac{\partial A^i}{\partial x^\beta}\frac{dx^\beta}{ds}$ it follows that

$$\frac{\partial A^i}{\partial x^\beta} = -A^\alpha \Gamma^i_{\alpha\beta}$$

or $A^i_{,\beta} = 0$.

Example 3.17. Let us consider two unit vectors A^i, B^i which undergo parallel displacement along a curve. We have

$$\cos\theta = g_{\alpha\beta}A^\alpha B^\beta$$

and

$$\frac{\delta(\cos\theta)}{\delta s} = g_{\alpha\beta}\frac{\delta A^\alpha}{\delta s}B^\beta + g_{\alpha\beta}A^\alpha\frac{\delta B^\beta}{\delta s} = 0$$

since $g_{\alpha\beta,\gamma} = 0$, $\frac{\delta A^\alpha}{\delta s} = 0$, $\frac{\delta B^\beta}{\delta s} = 0$. Hence, if two vectors of constant magnitude undergo parallel displacements along a curve, they are inclined at a constant angle.

Two vectors at a point are said to be parallel if their corresponding components are proportional. If A^i is a vector of constant magnitude, the vector $B^i = \varphi A^i$, $\varphi =$ scalar, is parallel to A^i. If A^i is also parallel with respect to the V_n along the curve $x^i = x^i(s)$, we have $\frac{\delta A^i}{\delta s} = 0$. Now

$$\begin{aligned}\frac{\delta B^i}{\delta s} &= \varphi\frac{\delta A^i}{\delta s} + \frac{d\varphi}{ds}A^i = \frac{d\varphi}{ds}A^i \\ &= \frac{1}{\varphi}\frac{d\varphi}{ds}B^i\end{aligned}$$

We desire B^i to be parallel with respect to the V_n along the curve. Hence a vector of variable magnitude must satisfy an equation of the type

$$\frac{\delta B^i}{\delta s} = f(s)B^i \tag{3.82}$$

if it is to be parallelly displaced along the curve.

Example 3.18. The curvature tensor arises under the following considerations: Let $x^i = x^i(t)$, $0 \leqq t \leqq 1$, be an infinitesimal closed path. The change in the components of a contravariant vector on being parallelly displaced along this closed path is

$$\Delta A^i = -\oint \Gamma^i_{\alpha\beta}A^\alpha\, dx^\beta$$

If we expand A^α, $\Gamma^i_{\alpha\beta}$ in a Taylor series about $x^i_0 = x^i(0)$ and neglect infinitesimals of higher order, it can be shown that

$$\Delta A^i = \tfrac{1}{2}(R^i_{\alpha\beta\gamma})_0(A^\alpha)_0\oint x^\gamma\, dx^\beta - x^\beta\, dx^\gamma \tag{3.83}$$

where $R^i_{\alpha\beta\gamma}$ is the curvature tensor of Secs. 3.11, 3.12.

Problems

1. Show that the Riemannian space for which $ds^2 = d\theta^2 + \sin^2 \theta \, d\varphi^2$ is not Euclidean.

2. Derive (3.79).

3. Show that the unit tangent vector to a geodesic suffers a parallel displacement along the geodesic.

4. If B^i satisfies (3.82), show by letting $B^i = \psi A^i$ that it is possible to find ψ so that $\frac{\delta A^i}{\delta s} = 0$.

5. Derive (3.83).

3.14. Lagrange's Equations of Motion. Let L be an absolute scalar invariant of the space coordinates $(x^1, x^2, \ldots, x^n)$, their time derivatives $(\dot{x}_1, \dot{x}_2, \ldots, \dot{x}_n)$, and the invariant time t, $t = \bar{t}$. Hence

$$L(x^1, x^2, \ldots, x^n, \dot{x}^1, \dot{x}^2, \ldots, \dot{x}^n, t) = \bar{L}(\bar{x}^1, \bar{x}^2, \ldots, \bar{x}^n, \dot{\bar{x}}^1, \ldots, \dot{\bar{x}}^n, \bar{t})$$

under the transformations

$$\begin{aligned} x^i &= x^i(\bar{x}^1, \bar{x}^2, \ldots, \bar{x}^n) \qquad i = 1, 2, \ldots, n \\ t &= \bar{t} \end{aligned} \tag{3.84}$$

From (3.84)

$$\dot{x}^i = \frac{\partial x^i}{\partial \bar{x}^\alpha} \dot{\bar{x}}^\alpha = F^i(\bar{x}, \dot{\bar{x}})$$

$$\frac{\partial \dot{x}^i}{\partial \bar{x}^j} = \frac{\partial^2 x^i}{\partial \bar{x}^j \, \partial \bar{x}^\alpha} \dot{\bar{x}}^\alpha$$

$$\frac{\partial \dot{x}^i}{\partial \dot{\bar{x}}^j} = \frac{\partial x^i}{\partial \bar{x}^\alpha} \frac{\partial \dot{\bar{x}}^\alpha}{\partial \dot{\bar{x}}^j} = \frac{\partial x^i}{\partial \bar{x}^\alpha} \delta^\alpha_j = \frac{\partial x^i}{\partial \bar{x}^j}$$

when it is assumed that the $\bar{x}^i$ and $\dot{\bar{x}}^i$ are independent variables as far as $\bar{L}$ is concerned. Now

$$\begin{aligned} \frac{\partial \bar{L}}{\partial \bar{x}^i} &= \frac{\partial L}{\partial x^\alpha} \frac{\partial x^\alpha}{\partial \bar{x}^i} + \frac{\partial L}{\partial \dot{x}^\alpha} \frac{\partial \dot{x}^\alpha}{\partial \bar{x}^i} \\ &= \frac{\partial L}{\partial x^\alpha} \frac{\partial x^\alpha}{\partial \bar{x}^i} + \frac{\partial L}{\partial \dot{x}^\alpha} \frac{\partial^2 x^\alpha}{\partial \bar{x}^i \, \partial \bar{x}^\beta} \dot{\bar{x}}^\beta \end{aligned} \tag{3.85}$$

Also

$$\frac{\partial \bar{L}}{\partial \dot{\bar{x}}^i} = \frac{\partial L}{\partial \dot{x}^\alpha} \frac{\partial \dot{x}^\alpha}{\partial \dot{\bar{x}}^i} = \frac{\partial L}{\partial \dot{x}^\alpha} \frac{\partial x^\alpha}{\partial \bar{x}^i}$$

so that

$$\begin{aligned} \frac{d}{d\bar{t}}\left(\frac{\partial \bar{L}}{\partial \dot{\bar{x}}^i}\right) &= \frac{d}{dt}\left(\frac{\partial \bar{L}}{\partial \dot{\bar{x}}^i}\right) \\ &= \frac{d}{dt}\left(\frac{\partial L}{\partial \dot{x}^\alpha}\right) \frac{\partial x^\alpha}{\partial \bar{x}^i} + \frac{\partial L}{\partial \dot{x}^\alpha} \frac{\partial^2 x^\alpha}{\partial \bar{x}^\beta \, \partial \bar{x}^i} \dot{\bar{x}}^\beta \end{aligned} \tag{3.86}$$

Subtracting (3.85) from (3.86) yields

$$\frac{d}{d\bar{t}}\left(\frac{\partial \bar{L}}{\partial \dot{\bar{x}}^i}\right) - \frac{\partial \bar{L}}{\partial \bar{x}^i} = \frac{\partial x^\alpha}{\partial \bar{x}^i} \left[\frac{d}{dt}\left(\frac{\partial L}{\partial \dot{x}^\alpha}\right) - \frac{\partial L}{\partial x^\alpha}\right] \tag{3.87}$$

Equation (3.87) implies immediately that

$$\frac{d}{dt}\left(\frac{\partial L}{\partial \dot{x}^\alpha}\right) - \frac{\partial L}{\partial x^\alpha}$$

is an absolute covariant tensor.

Let us consider a system of n particles, the mass m_i, $i = 1, 2, \ldots, n$, located at the point x_i^α, $\alpha = 1, 2, 3$. We assume that the coordinates are Euclidean and that Newtonian mechanics apply. Let

$$V(x_1^1, x_1^2, x_1^3, x_2^1, x_2^2, x_2^3, \ldots, x_n^1, x_n^2, x_n^3)$$

be the potential function such that

$$F_s^r \equiv -\frac{\partial V}{\partial x_s^r}$$

represents the rth component of the force applied to particle s. In Newtonian mechanics $F_s^r = m_s \dfrac{d^2x_s^r}{dt^2}$ for Euclidean coordinates. The kinetic energy of the system is defined as

$$T = \sum_{i=1}^{n} \tfrac{1}{2}m_i\left(\frac{ds_i}{dt}\right)^2 = \sum_{i=1}^{n} \tfrac{1}{2}m_i g_{\alpha\beta}\dot{x}_i^\alpha \dot{x}_i^\beta$$

where $g_{\alpha\beta} = \delta_{\alpha\beta}$. The *Lagrangian* of the system is defined by

$$\begin{aligned} L &= T - V \\ &= \sum_{i=1}^{n} \tfrac{1}{2}m_i g_{\alpha\beta}\dot{x}_i^\alpha \dot{x}_i^\beta - V \end{aligned}$$

Thus

$$\frac{\partial L}{\partial \dot{x}_s^r} = m_s g_{\alpha r}\dot{x}_s^\alpha = m_s \dot{x}_s^r$$

$$\frac{d}{dt}\left(\frac{\partial L}{\partial \dot{x}_s^r}\right) = m_s \ddot{x}_s^r$$

$$\frac{\partial L}{\partial x_s^r} = -\frac{\partial V}{\partial x_s^r}$$

so that

$$\frac{d}{dt}\left(\frac{\partial L}{\partial \dot{x}_s^r}\right) - \frac{\partial L}{\partial x_s^r} = m_s \ddot{x}_s^r + \frac{\partial V}{\partial x_s^r} = 0$$

for Newtonian mechanics. Since $\dfrac{d}{dt}\left(\dfrac{\partial L}{\partial \dot{x}_s^r}\right) - \dfrac{\partial L}{\partial x_s^r}$ is a vector, it vanishes in all coordinate systems. We replace the x_s^r by any system of coordinates $(q^1, q^2, \ldots, q^n)$ which completely specifies the configuration of the system of particles. Lagrange's equations of motion are

$$\frac{d}{dt}\left(\frac{\partial L}{\partial \dot{q}^r}\right) - \frac{\partial L}{\partial q^r} = 0 \qquad r = 1, 2, \ldots, n \tag{3.88}$$

Example 3.19. We consider the motion of a particle acted upon by the force field $\mathbf{F} = -\nabla V$. We use cylindrical coordinates.

$$L = T - V = \tfrac{1}{2}m(\dot{r}^2 + r^2\dot{\theta}^2 + \dot{z}^2) - V$$

so that

$$\frac{\partial L}{\partial r} = m(r\dot{\theta}^2) - \frac{\partial V}{\partial r}$$

$$\frac{d}{dt}\left(\frac{\partial L}{\partial \dot{r}}\right) = m\ddot{r}$$

and one of Lagrange's equations of motion is

$$m\ddot{r} - mr\dot{\theta}^2 + \frac{\partial V}{\partial r} = 0$$

Since $-\dfrac{\partial V}{\partial r}$ represents the radial force, the term $\ddot{r} - r\dot{\theta}^2$ must be the radial acceleration.

If no potential function exists, or if it is difficult to obtain the potential function, we can modify Lagrange's equations as follows: Since the kinetic energy is a scalar invariant, we have that

$$Q_r \equiv \frac{d}{dt}\left(\frac{\partial T}{\partial \dot{x}^r}\right) - \frac{\partial T}{\partial x^r} \tag{3.89}$$

is a covariant vector. Let the reader show that if the x^i are Euclidean coordinates, then Q_r is the Newtonian force. If f_r is the force in a y coordinate system, then

$$Q_r = f_\alpha \frac{\partial y^\alpha}{\partial x^r} \qquad Q_r\, dx^r = f_\alpha \frac{\partial y^\alpha}{\partial x^r}\, dx^r = f_\alpha\, dy^\alpha$$

so that $Q_r\, dx^r$ is a scalar invariant and represents the differential of work dW, and

$$Q_i = \frac{\partial W}{\partial x^i}$$

We obtain Q_i by allowing x^i to vary by an amount Δx^i, keeping $x^1, x^2, \ldots, x^{i-1}, x^{i+1}, \ldots, x^n$ fixed, calculate the work ΔW_i done by the forces acting on the system, and compute

$$Q_i = \lim_{\Delta x^i \to 0} \frac{\Delta W_i}{\Delta x^i} \qquad i \text{ not summed}$$

Example 3.20. A hoop rotates about a vertical diameter with constant angular speed ω. A bead is free to slide on the wire hoop, and there is no friction between hoop and bead. Gravity acts on the bead of mass m. We set up the equations of motion of the bead, using spherical coordinates. The force of the hoop on the bead is given by $\mathbf{F} = R\mathbf{e}_r + \Phi\mathbf{e}_\varphi$, R, Φ unknown, and the force of gravity is

$$\mathbf{G} = -mg\cos\theta\mathbf{e}_r + mg\sin\theta\mathbf{e}_\theta$$

Hence

$$\begin{aligned} Q_r &= R - mg\cos\theta \\ Q_\theta &= mgr\sin\theta \\ Q_\varphi &= \Phi r\sin\theta \end{aligned}$$

From $T = \frac{1}{2}m(\dot{r}^2 + r^2\dot{\theta}^2 + r^2\sin^2\theta\dot{\varphi}^2)$, we have

$$\begin{aligned} \frac{\partial T}{\partial r} &= m(r\dot{\theta}^2 + r\sin^2\theta\,\dot{\varphi}^2) & \frac{d}{dt}\left(\frac{\partial T}{\partial \dot{r}}\right) &= m\ddot{r} \\ \frac{\partial T}{\partial \theta} &= mr^2\sin\theta\cos\theta\,\dot{\varphi}^2 & \frac{d}{dt}\left(\frac{\partial T}{\partial \dot{\theta}}\right) &= \frac{d}{dt}(mr^2\dot{\theta}) \\ \frac{\partial T}{\partial \varphi} &= 0 & \frac{d}{dt}\left(\frac{\partial T}{\partial \dot{\varphi}}\right) &= \frac{d}{dt}(mr^2\sin^2\theta\,\dot{\varphi}) \end{aligned}$$

Equation (3.89) yields

$$\begin{aligned} \frac{d^2r}{dt^2} - (r\dot{\theta}^2 + r\sin^2\theta\,\dot{\varphi}^2) &= \frac{R}{m} - g\cos\theta \\ \frac{d}{dt}(r^2\dot{\theta}) - r^2\sin\theta\cos\theta\,\dot{\varphi}^2 &= gr\sin\theta \\ \frac{d}{dt}(r^2\sin^2\theta\,\dot{\varphi}) &= \frac{\Phi}{m}r\sin\theta \end{aligned} \tag{3.90}$$

The geometry of the configuration yields $r = \text{constant} = r_0$, $\dot{r} = \ddot{r} = 0$, $\dot{\varphi} = \omega$, $\ddot{\varphi} = 0$. The solution of the problem consists in finding $\theta(t)$ by integrating the second equation of (3.90). Once this is done, R and Φ can be obtained from the other two equations. Let the reader show that

$$\tfrac{1}{2}r_0^2\left(\frac{d\theta}{dt}\right)^2 - \tfrac{1}{2}r_0^2\omega^2\sin^2\theta = -gr_0\cos\theta + \text{constant} \tag{3.91}$$

3.15. Hamilton's Equations of Motion. From the Lagrangian $L(x, \dot{x}, t)$ we define

$$p_i \equiv \frac{\partial L}{\partial \dot{x}^i} \qquad i = 1, 2, \ldots, n \tag{3.92}$$

We show that p_i is a covariant vector, called the generalized momentum vector. Since $L = \bar{L}$, we have

$$\begin{aligned} \frac{\partial \bar{L}}{\partial \dot{\bar{x}}^i} &= \frac{\partial L}{\partial \dot{x}^\alpha}\frac{\partial \dot{x}^\alpha}{\partial \dot{\bar{x}}^i} = \frac{\partial L}{\partial \dot{x}^\alpha}\frac{\partial x^\alpha}{\partial \bar{x}^i} \\ \bar{p}_i &= p_\alpha\frac{\partial x^\alpha}{\partial \bar{x}^i} \end{aligned}$$

which proves our statement. We shall now use q^i instead of x^i to represent the coordinate system. The *Hamiltonian* is defined by

$$H = p_\alpha\dot{q}^\alpha - L(q, \dot{q}, t) \tag{3.93}$$

From $p_i = \dfrac{\partial L}{\partial \dot{q}^i} \equiv F_i(q, \dot{q}, t)$ we assume that we can solve for the $\dot{q}^i$, $i = 1, 2, \ldots, n$, in terms of the p_i, q_i, and time. Thus the Hamiltonian H now becomes a function of the p's, q's, and time,

$$H = H(p, q, t)$$

Hence
$$\begin{aligned}\frac{\partial H}{\partial q^i} &= p_\alpha \frac{\partial \dot{q}^\alpha}{\partial q^i} - \frac{\partial L}{\partial q^i} - \frac{\partial L}{\partial \dot{q}^\alpha}\frac{\partial \dot{q}^\alpha}{\partial q^i} \\ &= -\frac{\partial L}{\partial q^i}\end{aligned}$$

since $\dfrac{\partial L}{\partial \dot{q}^\alpha} = p_\alpha$. From Lagrange's equations of motion,

$$\frac{\partial L}{\partial q^i} = \frac{d}{dt}\left(\frac{\partial L}{\partial \dot{q}^i}\right) = \frac{dp_i}{dt}$$

so that
$$\frac{dp_i}{dt} = -\frac{\partial H}{\partial q^i}$$

Also
$$\begin{aligned}\frac{\partial H}{\partial p_i} &= \dot{q}^i + p_\alpha \frac{\partial \dot{q}^\alpha}{\partial p_i} - \frac{\partial L}{\partial \dot{q}^\alpha}\frac{\partial \dot{q}^\alpha}{\partial p_i} \\ &= \dot{q}^i = \frac{dq^i}{dt}\end{aligned}$$

from (3.92) and (3.93).

Hamilton's equations of motion are

$$\begin{aligned}\frac{dq^i}{dt} &= \frac{\partial H}{\partial p_i} \\ \frac{dp_i}{dt} &= -\frac{\partial H}{\partial q^i}\end{aligned} \tag{3.94}$$

Whereas Lagrange's equations of motion are, in general, a system of n second-order differential equations, Hamilton's equations are a system of $2n$ first-order differential equations.

Example **3.21.** Referring to Example 3.19,

$$\begin{aligned}p_r &= \frac{\partial L}{\partial \dot{r}} = m\dot{r} & \dot{q}_r &= \dot{r} = \frac{p_r}{m} \\ p_\theta &= \frac{\partial L}{\partial \dot{\theta}} = mr^2\dot{\theta} & \dot{q}_\theta &= \dot{\theta} = \frac{p_\theta}{mr^2} \\ p_z &= \frac{\partial L}{\partial \dot{z}} = m\dot{z} & \dot{q}_z &= \dot{z} = \frac{p_z}{m}\end{aligned}$$

and
$$\begin{aligned}H &= p_\alpha \dot{q}^\alpha - L \\ &= \frac{p_r^2}{m} + \frac{p_\theta^2}{mr^2} + \frac{p_z^2}{m} - \frac{m}{2}\left(\frac{p_r^2}{m^2} + \frac{p_\theta^2}{m^2r^2} + \frac{p_z^2}{m^2}\right) + V(r, \theta, z, t) \\ &= \frac{1}{2m}\left(p_r^2 + \frac{p_\theta^2}{r^2} + p_z^2\right) + V(r, \theta, z, t)\end{aligned}$$

Hamilton's equations are

$$\frac{dr}{dt} = \frac{\partial H}{\partial p_r} = \frac{p_r}{m} \qquad \frac{dp_r}{dt} = -\frac{\partial H}{\partial r} = \frac{p_\theta^2}{mr^3} - \frac{\partial V}{\partial r}$$

$$\frac{d\theta}{dt} = \frac{p_\theta}{mr^2} \qquad \frac{dp_\theta}{dt} = -\frac{\partial V}{\partial \theta}$$

$$\frac{dz}{dt} = \frac{p_z}{m} \qquad \frac{dp_z}{dt} = -\frac{\partial V}{\partial z}$$

Hence

$$m\frac{d^2r}{dt^2} = \frac{p_\theta^2}{mr^3} - \frac{\partial V}{\partial r} = mr\left(\frac{d\theta}{dt}\right)^2 - \frac{\partial V}{\partial r}$$

$$\frac{m}{r}\frac{d}{dt}\left(r^2\frac{d\theta}{dt}\right) = -\frac{1}{r}\frac{\partial V}{\partial \theta}$$

$$m\frac{d^2z}{dt^2} = -\frac{\partial V}{\partial z}$$

These are Newton's equations of motion for a particle using cylindrical coordinates.

Problems

1. Find the components of the acceleration vector in spherical coordinates and in cylindrical coordinates for a particle.

2. A particle slides in a frictionless tube which rotates in a horizontal plane with constant angular speed ω. Neglecting gravity, find the reaction between the tube and the particle.

3. If $T = a_{\alpha\beta}(q^1, q^2, \ldots, q^n)\dot{q}^\alpha\dot{q}^\beta$, show that $2T = \dfrac{\partial T}{\partial \dot{q}^\alpha}\dot{q}^\alpha$.

4. Set up Hamilton's equations for a particle moving in one dimension under a Hooke's law force, $F = -k^2x$.

5. Integrate (3.91) under the assumption $\theta \approx 180°$.

3.16. Euler's Equations of Motion. We discuss the motion of a rigid body with one fixed point. The reader is urged to read Example 2.10 the results of which we shall use. The x coordinate system will be fixed in space, and the $\bar{x}$ coordinate system will be rigidly attached to the moving body. We shall use the Σ sign to represent an integration over the complete rigid body.

In vector notation the angular-momentum vector is defined to be

$$\mathbf{H} = \Sigma\mathbf{r} \times m\mathbf{v}$$

and in tensor notation

$$H_k = \Sigma m\epsilon_{ijk}x^i\dot{x}^j \tag{3.95}$$

From (3.95)

$$\frac{dH_k}{dt} = \sum m\epsilon_{ijk}x^i\ddot{x}^j + \sum m\epsilon_{ijk}\dot{x}^i\dot{x}^j$$

$$= \sum m\epsilon_{ijk}x^i\ddot{x}^j \tag{3.96}$$

since $\epsilon_{ijk}\dot{x}^i\dot{x}^j = \epsilon_{jik}\dot{x}^j\dot{x}^i = -\epsilon_{ijk}\dot{x}^j\dot{x}^i = -\epsilon_{ijk}\dot{x}^i\dot{x}^j = 0$.

The moment or torque vector is defined as

$$\mathbf{L} = \sum \mathbf{r} \times \mathbf{f} = \sum \mathbf{r} \times m \frac{d^2\mathbf{r}}{dt^2}$$
$$L_k = \sum m\epsilon_{ijk} x^i \ddot{x}^j \tag{3.97}$$

Hence
$$\frac{dH_k}{dt} = L_k \tag{3.98}$$

Now
$$H_k = \Sigma m \epsilon_{ijk} x^i \omega_l^j x^l$$

since $\dot{x}^j = \omega_l^j x^l$ [see (2.21)]. Since ϵ_{ijk} and $\omega_l^j(t)$ are independent of the space coordinates, we write

$$H_k = \epsilon_{ijk}\omega_l^j \Sigma m x^i x^l$$

We define $I^{il} \equiv \Sigma m x^i x^l$ as the inertia tensor. Since $x^i = A_\alpha^i \bar{x}^\alpha$, we have

$$I^{il} = A_\alpha^i A_\beta^l \Sigma m \bar{x}^\alpha \bar{x}^\beta = A_\alpha^i A_\beta^l \bar{I}^{\alpha\beta}$$
$$= \bar{I}^{\alpha\beta} \frac{\partial x^i}{\partial \bar{x}^\alpha} \frac{\partial x^l}{\partial \bar{x}^\beta}$$

The components of the inertia tensor relative to the moving frame are constants since the frame is rigidly attached to the rigid body. This is not the case for the I^{il} since the a_α^i are, in general, functions of time. Remember that the a_α^i are the direction cosines, which vary with time. Thus it will be useful to deal with the $\bar{x}$ coordinate system. In this frame

$$\bar{H}_k = \bar{\epsilon}_{ijk} \bar{I}^{il} \bar{\omega}_l^j \tag{3.99}$$

The $\bar{\epsilon}_{ijk}$ transform like the components of an absolute tensor for orthogonal transformations since $\left| \dfrac{\partial \bar{x}^i}{\partial x^j} \right| = |a_j^i| = 1$ [see (1.25)]. Moreover

$$\bar{H}_k = \frac{\partial x^\alpha}{\partial \bar{x}^k} H_\alpha = A_k^\alpha H_\alpha$$
$$\frac{d\bar{H}_k}{dt} = A_k^\alpha \frac{dH_\alpha}{dt} + \frac{dA_k^\alpha}{dt} H_\alpha$$
$$= \bar{L}_k + \frac{dA_k^\alpha}{dt} H_\alpha$$

From $A_j^\alpha a_\alpha^i = \delta_j^i$, $\omega_j^k = -A_i^k \dfrac{da_j^i}{dt}$ it follows that

$$\frac{dA_k^\alpha}{dt} = A_i^\alpha A_k^\beta a_\gamma^i \omega_\beta^\gamma$$

so that
$$\frac{dA_k^\alpha}{dt} H_\alpha = (A_i^\alpha H_\alpha)(A_k^\beta a_\gamma^i \omega_\beta^\gamma) = \bar{H}_i \bar{\omega}_k^i$$

Hence [using (3.99)]

$$\bar{\epsilon}_{ijk}\bar{I}^{il}\frac{d\bar{\omega}_l^j}{dt} - \bar{\epsilon}_{\alpha ji}\bar{I}^{\alpha l}\bar{\omega}_l^j\bar{\omega}_k^i = \bar{L}_k \tag{3.100}$$

Equation (3.100) is one form of Euler's equations of motion. Since I^{il} is a quadratic form, we can find an orthogonal transformation which diagonalizes this tensor or matrix (see Sec. 1.5). Let us now use this new coordinate system (it is also fixed in the moving body). We omit the bars of (3.100) for convenience. We have

$$I^{il} = \begin{Vmatrix} I_1 & 0 & 0 \\ 0 & I_2 & 0 \\ 0 & 0 & I_3 \end{Vmatrix}$$

It is easy to see that $I_1 + I_2 = A_z$, the moment of inertia about the z axis, etc. The xyz coordinate system is fixed in the moving body. Equation (3.100) becomes, for $k = 1$,

$$\epsilon_{ij1}I^{il}\frac{d\omega_l^j}{dt} - \epsilon_{\alpha ji}I^{\alpha l}\omega_l^j\omega_1^i = L_1$$

$$\epsilon_{231}I^{22}\frac{d\omega_2^3}{dt} + \epsilon_{321}I^{33}\frac{d\omega_3^2}{dt} - \epsilon_{132}I^{11}\omega_1^3\omega_1^2 - \epsilon_{312}I^{33}\omega_3^1\omega_1^2 - \epsilon_{123}I^{11}\omega_1^2\omega_1^3$$
$$- \epsilon_{213}I^{22}\omega_2^1\omega_1^3 = L_1$$

$$(I^{22} + I^{33})\frac{d\omega_2^3}{dt} + [(I^{22} + I^{11}) - (I^{33} + I^{11})]\omega_3^1\omega_1^2 = L_1$$

or

$$A_x\frac{d\omega_x}{dt} + (A_z - A_y)\omega_y\omega_z = L_x$$

Similarly

$$A_y\frac{d\omega_y}{dt} + (A_x - A_z)\omega_z\omega_x = L_y \tag{3.101}$$

$$A_z\frac{d\omega_z}{dt} + (A_y - A_x)\omega_x\omega_y = L_z$$

Problems

1. For a free body, $\mathbf{L} = \mathbf{0}$, show that

$$A_x\omega_x^2 + A_y\omega_y^2 + A_z\omega_z^2 = \text{constant}$$

and

$$A_x^2\omega_x^2 + A_y^2\omega_y^2 + A_z^2\omega_z^2 = \text{constant}$$

follow from (3.101).

2. Assume that, at $t = 0$ for a free body, $\omega_x = \omega_0$, $\omega_y = \omega_z = 0$. Describe the motion for $t > 0$.

3. Solve for ω_x, ω_y, ω_z for the free body with $A_x = A_y$.

4. Show that the body of Prob. 3 precesses with constant angular speed about the angular-momentum vector.

5. Derive the second and third equations of (3.101).

3.17. The Navier-Stokes Equations of Motion for a Viscous Fluid. Let us first consider the motion of a fluid in the neighborhood of a point

$P(x)$ of the fluid. Let the velocity of the fluid at P be given by u^i, or $u_i = g_{i\alpha}u^\alpha$. The velocity at a nearby point $Q(x + dx)$ is, except for infinitesimals of higher order,

$$\begin{aligned} u_i(x + dx) &= u_i(x) + du_i \\ &= u_i(x) + \frac{\partial u_i}{\partial x^\alpha}\, dx^\alpha \\ &= u_i(x) + \frac{1}{2}\left(\frac{\partial u_i}{\partial x^\alpha} - \frac{\partial u_\alpha}{\partial x^i}\right) dx^\alpha + \frac{1}{2}\left(\frac{\partial u_i}{\partial x^\alpha} + \frac{\partial u_\alpha}{\partial x^i}\right) dx^\alpha \end{aligned}$$

The partial derivatives are evaluated at the point P. Strictly speaking, du_i is not a vector, so that we should be concerned with the intrinsic differential δu_i. The above can be written

$$u_i(x + dx) = u_i(x) + \tfrac{1}{2}(u_{i,\alpha} - u_{\alpha,i})\, dx^\alpha + \tfrac{1}{2}(u_{i,\alpha} + u_{\alpha,i})\, dx^\alpha \tag{3.102}$$

We now analyze (3.102), which states that the velocity of Q is the sum of three terms:

1. $u_i(x)$, the velocity of P. Q is carried along with P, so that $u_i(x)$ is a translational effect.

2. The term $\frac{1}{2}(u_{i,\alpha} - u_{\alpha,i})\, dx^\alpha$ corresponds to a rotation with angular velocity $\boldsymbol{\omega} = \frac{1}{2}\nabla \times \mathbf{u}$. A sphere in the neighborhood of P with center at P would be translated and rotated under the action of terms 1 and 2. It follows that terms 1 and 2 are rigid-body motions.

3. Hence the term $\frac{1}{2}(u_{i,\alpha} + u_{\alpha,i})\, dx^\alpha$ must be responsible if any deformations of the fluid take place.

We define

$$s_{ij} = \tfrac{1}{2}(u_{i,j} + u_{j,i}) \tag{3.103}$$

as the strain tensor. Let $dx^j \equiv x^j$ temporarily, that is, let P be the origin of our coordinate system, x^j the coordinates of the nearby point Q. We can write $s_{ij}x^i = \frac{1}{2}\nabla(s_{ij}x^ix^j)$ using Euclidean coordinates. However, $s_{ij}x^ix^j$ is a quadratic form which can be reduced to

$$\bar{\varphi} = \bar{s}_{11}(\bar{x}^1)^2 + \bar{s}_{22}(\bar{x}^2)^2 + \bar{s}_{33}(\bar{x}^3)^2$$

under an orthogonal coordinate transformation. In the $\bar{x}$ system, $\frac{1}{2}\nabla\bar{\varphi} = \bar{s}_{11}\bar{x}^1\mathbf{i} + \bar{s}_{22}\bar{x}^2\mathbf{j} + \bar{s}_{33}\bar{x}^3\mathbf{k}$. Thus, along with the rigid-body motions of terms 1 and 2, occurs a velocity whose components in the $\bar{x}^1$, $\bar{x}^2$, $\bar{x}^3$ directions are proportional to $\bar{x}^1$, $\bar{x}^2$, $\bar{x}^3$, respectively. The sphere surrounding P tends to grow into an ellipsoid whose principal axes are the $\bar{x}^1$, $\bar{x}^2$, $\bar{x}^3$ axes.

Terms 1, 2, and 3 characterize the motion of a fluid completely. In order to discuss the dynamics of a fluid, one must consider the stress tensor t^{ij}. We refer to Fig. 2.22. The face $ABCD$ is in contact with a part of the fluid. This part exerts a force on the face. This force per unit area has components t^{xy}, t^{yy}, t^{zy}. The y refers to the fact that the normal to the face $ABCD$ points in the y direction. By considering the other two principal faces one is led to the stress tensor t^{ij}, $i, j = 1, 2, 3$. The total force on a closed surface S is given by

$$\iint_S t^{ij} N_j \, d\sigma \tag{3.104}$$

where N_j is the unit normal vector to the surface area $d\sigma$. Let the reader show that (3.104) can be written as

$$\iiint_R t^{ij}_{,j} \, d\tau \tag{3.105}$$

so that $t^{ij}_{,j}$ represents the force per unit volume due to the stress tensor t^{ij}. The equality of (3.104) and (3.105) is essentially the divergence theorem of vector analysis.

In order to obtain the Navier-Stokes equations of motion, we make the fundamental assumption that the components of the stress tensor be proportional to the components of the strain tensor. Thus

$$t^i_j = a^{i\alpha}_{j\beta} s^\beta_\alpha \tag{3.106}$$

We further assume that $a^{i\alpha}_{j\beta}$ is an isotropic tensor so that

$$a^{i\alpha}_{j\beta} = k(x)\delta^i_j\delta^\alpha_\beta + l(x)\delta^i_\beta\delta^\alpha_j \tag{3.107}$$

(see Example 3.2).

Combining (3.106) and (3.107) yields

$$\begin{aligned} t^i_j &= k(x)\delta^i_j\delta^\alpha_\beta s^\beta_\alpha + l(x)\delta^i_\beta\delta^\alpha_j s^\beta_\alpha \\ &= k(x)\delta^i_j s^\alpha_\alpha + l(x) s^i_j \end{aligned} \tag{3.108}$$

and

$$t^i_i = 3ks^i_i + ls^i_i = (3k + l)s^i_i$$

In hydrostatics

$$t^i_j = \begin{vmatrix} -p & 0 & 0 \\ 0 & -p & 0 \\ 0 & 0 & -p \end{vmatrix}$$

so that $t^i_i = -3p$. The pressure p is defined to be $p = -\frac{1}{3}t^i_i$ for the general case. Thus

$$-3p = (3k + l)s^i_i \qquad ks^i_i = -p - \frac{l}{3}s^i_i$$

Equation (3.108) becomes

$$t^i_j = -p\delta^i_j - \frac{l}{3}\,\delta^i_j s^\alpha_\alpha + l s^i_j$$

and

$$t^i_{j,i} = -p_{,j} - \frac{l}{3}\,s^\alpha_{\alpha,j} + l s^i_{j,i}$$

or

$$t^i_{j,i} = -p_{,j} - \frac{l}{3}\,g^{\alpha k}s_{\alpha k,j} + l g^{ik}s_{jk,i} \tag{3.109}$$

From $s_{jk,i} = \frac{1}{2}u_{j,ki} + \frac{1}{2}u_{k,ji}$ and $l = 2\nu$, (3.109) becomes

$$F_j \equiv t^i_{j,i} = -p_{,j} + \tfrac{1}{3}\nu g^{\alpha k}u_{\alpha,kj} + \nu g^{ki}u_{j,ki} \tag{3.110}$$

In vector form

$$\mathbf{F} = -\nabla p + \frac{\nu}{3}\nabla(\nabla\cdot\mathbf{u}) + \nu\nabla^2\mathbf{u} \tag{3.111}$$

where $\mathbf{F}$ is the force per unit volume due to the stress tensor t^i_j. Let f be the external force per unit mass so that $(\mathbf{F} + \rho\mathbf{f})\,d\tau$ is the total force acting on the element $d\tau$. Newton's second law of motion states that

$$\frac{d}{dt}(\rho\,d\tau\,\mathbf{u}) = (\mathbf{F} + \rho\mathbf{f})\,d\tau$$

or

$$\rho\,d\tau\,\frac{d\mathbf{u}}{dt} = (\mathbf{F} + \rho\mathbf{f})\,d\tau$$

since $\rho\,d\tau$ is a constant during the motion. Hence

$$\rho\,\frac{d\mathbf{u}}{dt} = \rho\mathbf{f} - \nabla p + \frac{\nu}{3}\nabla(\nabla\cdot\mathbf{u}) + \nu\nabla^2\mathbf{u} \tag{3.112}$$

where ν is the viscosity of the fluid. Equation (3.112) is the Navier-Stokes equation of motion for a viscous fluid.

Problems

1. For an incompressible fluid show that

$$\rho\,\frac{d\mathbf{u}}{dt} = \rho\mathbf{f} - \nabla p + \nu\nabla^2\mathbf{u}$$

2. Consider the steady state of flow of an incompressible fluid through a cylindrical tube of radius a. Let $\mathbf{u} = u(r)\mathbf{k}$, $r^2 = x^2 + y^2$. Show that $p = p(z)$, $\nu\nabla^2 u = \dfrac{\partial p}{\partial z}$. Then show that $u = \dfrac{A}{4\nu}(r^2 - a^2)$, $\dfrac{dp}{dz} = A = \text{constant}$.

3. Solve for the steady-state motion of an incompressible viscous fluid between two parallel plates (infinite in extent), one of the plates being fixed, the other moving at a constant velocity, the distance between the two plates remaining constant.

4. Find the steady-state motion of an incompressible viscous fluid surrounding a sphere which is rotating about a diameter with constant angular velocity. No external forces exist.

REFERENCES

Brand, L.: "Vector and Tensor Analysis," John Wiley & Sons, Inc., New York, 1947.

Brillouin, L.: "Les Tenseurs," Dover Publications, New York, 1946.

Lass, H.: "Vector and Tensor Analysis," McGraw-Hill Book Company, Inc., New York, 1950.

McConnell, A. J.: "Applications of the Absolute Differential Calculus," Blackie & Son, Ltd., Glasgow, 1931.

Michal, A. D.: "Matrix and Tensor Calculus," John Wiley & Sons, Inc., New York, 1947.

Thomas, T. Y.: "Differential Invariants of Generalized Spaces," Cambridge University Press, New York, 1934.

Veblen, O.: "Invariants of Quadratic Differential Forms," Cambridge University Press, New York, 1933.

Weatherburn, C. E.: "Riemannian Geometry," Cambridge University Press, New York, 1942.

CHAPTER 4

COMPLEX-VARIABLE THEORY

4.1. Introduction. The reader is already familiar with some aspects of complex numbers. We enter now into a discussion of some of the simpler properties of complex numbers. In order to attach a solution to the equation $x^2 + 1 = 0$, the mathematician is forced to invent a new number, i, such that $i^2 + 1 = 0$, or $i = \sqrt{-1}$. We say that i is an imaginary number in order to distinguish it from elements of the real-number field (see Chap. 10 for a discussion of this field). The solution of the quadratic equation $ax^2 + bx + c = 0$, $a \neq 0$, requires a discussion of complex numbers of the form $\alpha + \beta i$, α and β real. The set of all such complex numbers subject to certain rules and operations listed below is called the complex-number field, an extension of the real-number field. We note that the complex numbers are to satisfy the following set of rules or postulates with respect to the operations of addition and multiplication:

1. Addition is closed, that is, the sum of two complex numbers is a complex number.

$$(a + bi) + (c + di) = (a + c) + (b + d)i$$

2. Addition obeys the associative law. If $z_1 = a_1 + b_1i$, $z_2 = a_2 + b_2i$, $z_3 = a_3 + b_3i$, then

$$z_1 + (z_2 + z_3) = (z_1 + z_2) + z_3$$

3. The unique zero element exists for addition.

$$0 = 0 + 0 \cdot i \qquad \text{and} \qquad z + 0 = 0 + z = z \qquad \text{for any complex number } z$$

4. Every complex number z has a unique negative, written $-z$, such that $z + (-z) = (-z) + z = 0$. If $z = a + bi$, then

$$-z = (-a) + (-b)i = -a - bi$$

5. Addition is commutative.

$$z_1 + z_2 = z_2 + z_1$$

6. Multiplication is defined as follows: If $z_1 = a_1 + b_1i$, $z_2 = a_2 + b_2i$, then

$$z_1z_2 = (a_1a_2 - b_1b_2) + (a_1b_2 + a_2b_1)i$$

The product of two complex numbers is again a complex number. This is the closure property for multiplication.

7. Multiplication obeys the associative law.

$$z_1(z_2z_3) = (z_1z_2)z_3 = z_1z_2z_3$$

8. The unique element $1 = 1 + 0 \cdot i$ exists for multiplication.

$$1 \cdot z = z \cdot 1 = z \qquad \text{for all } z$$

9. Every nonzero complex number z has a unique inverse, written z^{-1} or $1/z$, such that $zz^{-1} = z^{-1}z = 1$. If $z = a + bi$, $a^2 + b^2 \neq 0$, then

$$z^{-1} = \frac{a}{a^2 + b^2} - \frac{b}{a^2 + b^2} i$$

10. Multiplication is commutative.

$$z_1z_2 = z_2z_1$$

11. The distributive law holds with respect to addition.

$$z_1(z_2 + z_3) = z_1z_2 + z_1z_3 \qquad (z_1 + z_2)z_3 = z_1z_3 + z_2z_3$$

4.2. The Argand Plane. The complex number $z = x + iy$ admits of a very simple geometric representation. We may consider z as a vector whose origin is the origin of the Euclidean xy plane of analytic geometry and whose terminus is the point P with abscissa x and ordinate y. The mapping of complex numbers in this manner yields the Argand z plane. Addition of complex numbers obeys the parallelogram law of addition of vectors. We call x the real part of z, written $x = \text{Rl } z$, and we call y the imaginary component of z, written $y = \text{Im } z$ (see Fig. 4.1).

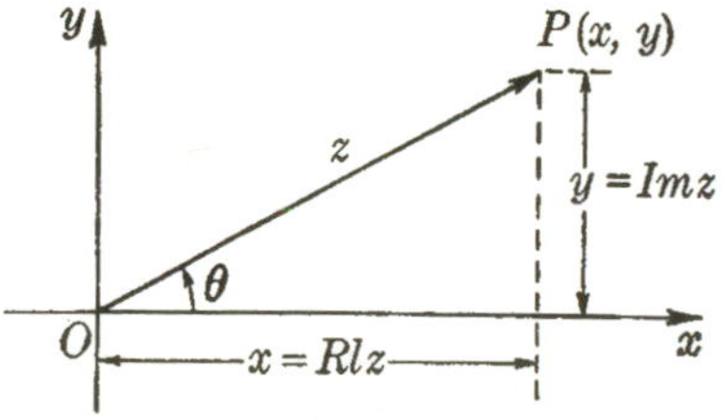

FIG. 4.1

The length of the vector representing $z = x + iy$ is called the modulus of z, written mod $z \equiv |z| = (x^2 + y^2)^{\frac{1}{2}}$. The argument of the complex number z is the angle between the real x axis and the vector z, measured in the counterclockwise sense. The argument of z is not single-valued, for if $\theta = \arg z$, then $\theta \pm 2\pi n$ is also the argument of z for any integer n. We define the principal value of arg z by the inequality $-\pi < \text{Arg } z \leqq \pi$.

Example 4.1. If $z = 1 + i$, then $|z| = \sqrt{2}$, and Arg $z = \pi/4$. If $z = -1$, then $|z| = 1$, Arg $z = \pi$. If $z = -i$, then $|z| = 1$, Arg $z = -\pi/2$.

Example 4.2. If we use polar coordinates, we may write

$$z = x + iy = r(\cos\theta + i \sin\theta)$$

$|z| = r$, $\arg z = \theta$. If $z_1 = r_1(\cos\theta_1 + i\sin\theta_1)$, $z_2 = r_2(\cos\theta_2 + i\sin\theta_2)$, we leave it to the reader to show that

$$z_1 z_2 = r_1 r_2[\cos(\theta_1 + \theta_2) + i\sin(\theta_1 + \theta_2)]$$
$$\frac{z_1}{z_2} = \frac{r_1}{r_2}[\cos(\theta_1 - \theta_2) + i\sin(\theta_1 - \theta_2)] \qquad r_2 \neq 0$$

We leave it to the reader to attach a simple geometric interpretation to multiplication and division of complex numbers.

Example 4.3. The reader should verify that

$$\begin{aligned} \text{Rl}\,(z_1 + z_2) &= \text{Rl}\,z_1 + \text{Rl}\,z_2 \\ \text{Im}\,(z_1 + z_2) &= \text{Im}\,z_1 + \text{Im}\,z_2 \\ |z_1 z_2| &= |z_1| \cdot |z_2| \\ \arg z_1 z_2 &= \arg z_1 + \arg z_2 \pm 2\pi n \qquad n \text{ an integer} \\ \arg\frac{z_1}{z_2} &= \arg z_1 - \arg z_2 \pm 2\pi n \\ -|z| &\leqq \text{Rl}\,z \leqq |z| \\ -|z| &\leqq \text{Im}\,z \leqq |z| \end{aligned}$$

If $z = x + iy$, we define the complex conjugate of z by the equation $\bar{z} = x - iy$. The conjugate of a complex number is obtained by replacing i by $-i$. Obviously

$$\begin{aligned} \overline{z_1 + z_2} &= \bar{z}_1 + \bar{z}_2 \\ z + \bar{z} &= 2\,\text{Rl}\,z \\ z - \bar{z} &= 2i\,\text{Im}\,z \\ z\bar{z} &= x^2 + y^2 = |z|^2 \end{aligned}$$

From $|z_1 + z_2|^2 = (z_1 + z_2)(\bar{z}_1 + \bar{z}_2)$ we let the reader deduce that $|z_1 + z_2| \leqq |z_1| + |z_2|$, and from this that $|z_1 - z_2| \geqq \big|\,|z_1| - |z_2|\,\big|$ for all z_1, z_2.

4.3. Simple Mappings. Henceforth the complex number z will stand always for the complex number $x + iy$. We now examine the complex number $w = z^2 = (x + iy)^2 = x^2 - y^2 + 2xyi$. It will be highly beneficial to construct a new Argand plane, called the w plane, with

$$w = u + iv$$

u and v Euclidean coordinates. The equation $w = z^2$ may be looked upon as a mapping of the complex numbers of the z plane into complex numbers of the w plane. The transformation $w = z^2$ maps the point P with coordinates (x, y) of the z plane into the point Q of the w plane whose coordinates are $u = x^2 - y^2$, $v = 2xy$. A curve in the xy plane will, in general, map into a curve in the uv plane. For example, the transformation $w = z^2$ maps the straight line $x = t$, $y = mt$, $-\infty < t < \infty$, into the straight-line segment $u = (1 - m^2)t$, $v = 2mt$, $0 \leqq t < \infty$.

The hyperbolas $x^2 - y^2 = \text{constant} = c$ map into the straight lines $u = c$. The hyperbolas $xy = c$ map into the straight lines $v = 2c$.

Example 4.4. We now examine the transformation $w = z + 1/z$, $z \neq 0$. We have $z = r(\cos \theta + i \sin \theta)$, $w = u + iv$, so that

$$u + iv = r(\cos \theta + i \sin \theta) + \frac{1}{r}(\cos \theta - i \sin \theta)$$

and

$$u = \left(r + \frac{1}{r}\right) \cos \theta$$

$$v = \left(r - \frac{1}{r}\right) \sin \theta$$

From $\cos^2 \theta + \sin^2 \theta = 1$ we have

$$\frac{u^2}{(r + 1/r)^2} + \frac{v^2}{(r - 1/r)^2} = 1 \qquad r \neq 1$$

The circles $r = a \neq 1$ map into the ellipses

$$\frac{u^2}{(a + 1/a)^2} + \frac{v^2}{(a - 1/a)^2} = 1$$

of the uv plane. Into what curve does the circle $r = 1$ map?

Problems

1. If $a + bi = c + di$, a, b, c, d real, show that $a = c$, $b = d$.

2. Show that $0 \cdot z = 0$ for all z by use of the distributive law and the definition of the zero element.

3. From $|z_1 + z_2|^2 = (z_1 + z_2)(\bar{z}_1 + \bar{z}_2)$ show that $|z_1 + z_2| \leqq |z_1| + |z_2|$.

4. Show that $(\cos \theta + i \sin \theta)^n = \cos (n\theta) + i \sin (n\theta)$, n an integer. This is a formula of De Moivre. Obtain an identity involving $\cos 4\theta$. Solve for the roots of $z^3 = 1$, $z^4 = 1$, $z^5 = 1$.

5. Examine the transformation $w = z - 1/z$ in the manner of Example 4.4.

6. Examine the transformations $w = az$, a complex, $w = z + b$, b complex, $w = 1/z$. Examine the important bilinear transformation $w = (az + b)/(cz + d)$, a, b, c, d complex.

4.4. Definition of a Complex Function. Let $Z = \{z\}$ be a set of complex numbers defined in some manner. Now assume that by some rule or set of rules we can set up a correspondence such that for every point z of Z there corresponds a unique complex number w, and let $W = \{w\}$ be the totality of complex numbers obtained in this manner by exhausting the set Z. We thus have a mapping $Z \to W$ which defines a complex function of z over the set Z. The correspondence between the element z and the element w is usually written

$$w = f(z) \tag{4.1}$$

It is customary to consider $f(z)$ as the complex function of z. We can write $f(z) = u(x, y) + iv(x, y)$, since, given z, we are given x and y, which in turn yield the real and imaginary parts of w, called $u(x, y)$ and $v(x, y)$, respectively.

Example 4.5. Let Z be the set of complex numbers z such that $|z| > 1$. The correspondence $z \rightarrow w$ such that $w = z/(|z| - 1)$ defines a complex function of z for the domain of definition of z. In this case

$$u(x, y) = \frac{x}{\sqrt{x^2 + y^2} - 1} \qquad v(x, y) = \frac{y}{\sqrt{x^2 + y^2} - 1}$$

with

$$x^2 + y^2 > 1$$

Example 4.6. Let Z be the finite set of complex numbers $z = 1$ and $z = 2$. Let $z = 1$ correspond to $w = 5$, $z = 2$ correspond to $w = 1 + i$. This mapping defines a complex function of z defined over the set Z.

Example 4.7. Let Z be the set of real integers, and let z of Z correspond to the constant $w = i$ for all z of Z. Note that in this case more than one element of Z corresponds to the same value w. A complex function can be a many-to-one mapping. Remember, however, that only one w corresponds to each z. This is what we mean by a single-valued function of z.

4.5. Continuous Functions. We define continuity of a single-valued complex function in the following manner: Let the points z of Z be mapped into the points w of W, and consider the two Argand planes, the z plane and the w plane. We say that $f(z)$ is continuous at $z = z_0$, z_0 in Z, if the following holds: Consider any circle C of nonzero radius with center at $w_0 = f(z_0)$. We must be able to determine a circle C' of nonzero radius with center at z_0 such that every point z of Z interior to C' maps into a point w of W interior to C. Analytically this means that, given any $\epsilon > 0$ (the radius of the circle C), there exists a $\delta > 0$ (the radius of the circle C') such that

$$|f(z) - f(z_0)| < \epsilon$$

for all z of Z such that $|z - z_0| < \delta$. This is the usual ϵ, δ definition of continuity of real-variable theory. If $f(z)$ is continuous at every point z of Z, we say that $f(z)$ is continuous over Z.

If z_i, $i = 1, 2, 3, \ldots$, is an infinite sequence of Z tending to z_0 as a limit, z_0 in Z, then the above definition of continuity implies that

$$\lim_{z \rightarrow z_0} f(z) = f(z_0)$$

Let the reader verify this.

Example 4.8. Let $f(z) = z$ for all z. We easily note that this function is continuous for all z since we need only pick $\delta = \epsilon$ for every choice of $\epsilon > 0$.

Example 4.9. Let $f(z)$ and $g(z)$ be defined over the same set $Z = \{z\}$, and assume that $f(z)$ and $g(z)$ are continuous at the point z_0 in Z. We now show that $f(z) + g(z)$ is continuous at z_0. Choose any $\epsilon > 0$, and consider $\epsilon/2 > 0$. Since $f(z)$ is continuous at z_0 there exists a $\delta_1 > 0$ such that $|f(z) - f(z_0)| < \epsilon/2$ for $|z - z_0| < \delta_1$, z in Z. A similar statement holds for $g(z)$, with δ_2 replacing δ_1. The reader should be able to show that $|(f(z) + g(z)) - (f(z_0) + g(z_0))| < \epsilon$ for $|z - z_0| < \delta$, z in Z, δ the smaller of δ_1, δ_2.

Example 4.10. Let the reader prove the following statements: Let $f(z)$ and $g(z)$ be defined over Z, and assume $f(z)$ and $g(z)$ continuous at z_0 in Z. Then

1. $f(z) - g(z)$ is continuous at z_0.
2. $f(z)g(z)$ is continuous at z_0.
3. $f(z)/g(z)$ is continuous at z_0 provided $g(z_0) \neq 0$.

Problems

1. Prove the statements of Example 4.10.

2. Show that $f(z) = a_0z^n + a_1z^{n-1} + \cdots + a_n$, n a positive integer, is continuous for all z, $z \neq \pm\infty$.

3. Show that the $f(z)$ of Examples 4.6 and 4.7 are continuous over their domain of definition.

4. Let $f(z) = (z^3 - 1)/(z - 1)$, $z \neq 1$, $f(1) = 3$. Show that $f(z)$ is continuous at $z = 1$.

5. Let $f(z)$ be continuous over a closed and bounded set. Show that $f(z)$ is uniformly continuous (see Sec. 10.12 for the definition of uniform continuity).

6. Let $f(z) = 1/(1 - z)$, $z \neq 1$, $f(1) = 5$. Show that $f(z)$ is discontinuous at $z = 1$.

7. Show that $f(z) = x \sin(1/x) + iy$, $x \neq 0$, $f(0) = 0$, is continuous at the origin $(0, 0)$.

4.6. Differentiability. Let $f(z)$ be defined for the set $Z = \{z\}$. Let z_0 be a point of Z, and let $z_1, z_2, z_3, \ldots, z_n, \ldots$ be any infinite sequence of elements of Z which tend to z_0 as a limit. None of the z_i, $i = 1, 2, 3, \ldots$, is to be equal to z_0. We say that $f(z)$ is differentiable at z_0 if

$$\lim_{z_n \to z_0} \frac{f(z_n) - f(z_0)}{z_n - z_0} \tag{4.2}$$

exists independent of the sequential approach to z_0. We can state differentiability in an equivalent manner. $f(z)$ is differentiable at z_0 if there exists a constant, written $f'(z_0)$, so that for any $\epsilon > 0$ there exists a $\delta > 0$ such that

$$\left| \frac{f(z) - f(z_0)}{z - z_0} - f'(z_0) \right| < \epsilon$$

whenever $|z - z_0| < \delta$ for z in Z, $z \neq z_0$.

In the cases we shall be interested in, z_0 will be an interior point of Z (see Sec. 10.7) for the definition of an interior point. In this case, (4.2) becomes

$$f'(z_0) = \lim_{\Delta z \to 0} \frac{f(z_0 + \Delta z) - f(z_0)}{\Delta z} \tag{4.3}$$

independent of the approach to zero of Δz. Let us consider

$$f(z) = u(x, y) + iv(x, y)$$

and investigate the conditions that will be imposed on $u(x, y)$ and $v(x, y)$

in order that

$$\lim_{\Delta z \to 0} \frac{f(z + \Delta z) - f(z)}{\Delta z}$$

shall exist independent of the approach to zero of Δz. We have

$$\begin{aligned} \Delta z &= \Delta x + i\,\Delta y \\ f(z + \Delta z) &= u(x + \Delta x, y + \Delta y) + iv(x + \Delta x, y + \Delta y) \end{aligned}$$

so that
$$\begin{aligned} \frac{f(z + \Delta z) - f(z)}{\Delta z} &= \frac{u(x + \Delta x, y + \Delta y) - u(x, y)}{\Delta x + i\,\Delta y} \\ &\quad + i\,\frac{v(x + \Delta x, y + \Delta y) - v(x, y)}{\Delta x + i\,\Delta y} \end{aligned} \tag{4.4}$$

First we let $\Delta z \to 0$ by keeping y constant, that is, $\Delta z = \Delta x$, $\Delta y = 0$. Then (4.4) becomes

$$\frac{f(z + \Delta z) - f(z)}{\Delta z} = \frac{u(x + \Delta x, y) - u(x, y)}{\Delta x} + i\,\frac{v(x + \Delta x, y) - v(x, y)}{\Delta x}$$

and
$$\lim_{\Delta z \to 0} \frac{f(z + \Delta z) - f(z)}{\Delta z} = \frac{\partial u}{\partial x} + i\,\frac{\partial v}{\partial x}$$

provided $\dfrac{\partial u}{\partial x}$ and $\dfrac{\partial v}{\partial x}$ exist. Now we let $\Delta z \to 0$ by keeping x constant, that is, $\Delta x = 0$, $\Delta z = i\,\Delta y$. Then

$$\begin{aligned} \lim_{\Delta z \to 0} \frac{f(z + \Delta z) - f(z)}{\Delta z} &= \frac{1}{i}\frac{\partial u}{\partial y} + \frac{\partial v}{\partial y} \\ &= \frac{\partial v}{\partial y} - i\,\frac{\partial u}{\partial y} \end{aligned}$$

provided $\dfrac{\partial u}{\partial y}$ and $\dfrac{\partial v}{\partial y}$ exist. Assuming differentiability of $f(z)$ yields

$$\frac{\partial u}{\partial x} + i\,\frac{\partial v}{\partial x} = \frac{\partial v}{\partial y} - i\,\frac{\partial u}{\partial y}$$

so that
$$\begin{aligned} \frac{\partial u}{\partial x} &= \frac{\partial v}{\partial y} \\ \frac{\partial v}{\partial x} &= -\frac{\partial u}{\partial y} \end{aligned} \tag{4.5}$$

The *Cauchy-Riemann* conditions (4.5) are necessary if $f(z)$ is to be differentiable. We have, however, neglected the infinity of other methods by which Δz may approach zero. But it will turn out that the Cauchy-Riemann equations (4.5) will be sufficient for differentiability provided we further assume that the partial derivatives of (4.5) are continuous. Let us proceed to prove this statement. The right-hand side of (4.4) may be written

$$\frac{u(x + \Delta x, y + \Delta y) - u(x, y + \Delta y) + u(x, y + \Delta y) - u(x, y)}{\Delta x + i\,\Delta y}$$
$$+ i\,\frac{v(x + \Delta x, y + \Delta y) - v(x, y + \Delta y) + v(x, y + \Delta y) - v(x, y)}{\Delta x + i\,\Delta y}$$
$$\equiv \frac{\Delta x\,(\partial u/\partial x + \xi_1) + \Delta y\,(\partial u/\partial y + \xi_2)}{\Delta x + i\,\Delta y}$$
$$+ i\,\frac{\Delta x\,(\partial v/\partial x + \xi_3) + \Delta y\,(\partial v/\partial y + \xi_4)}{\Delta x + i\,\Delta y}$$

where ξ_1, ξ_2, ξ_3, ξ_4 tend to zero as $\Delta z \to 0$. We have applied the theorem of the mean of the differential calculus. The $\xi_i \to 0$ if we assume continuity of the partial derivatives. Making use of the Cauchy-Riemann equations yields

$$\frac{f(z + \Delta z) - f(z)}{\Delta z} = \frac{\partial u}{\partial x} + i\,\frac{\partial v}{\partial x} + \frac{\xi_1\,\Delta x + \xi_2\,\Delta y + i(\xi_3\,\Delta x + \xi_4\,\Delta y)}{\Delta x + i\,\Delta y}$$

We leave it to the reader to show that

$$\lim_{\Delta x + i\,\Delta y \to 0} \frac{\xi_1\,\Delta x + \xi_2\,\Delta y + i(\xi_3\,\Delta x + \xi_4\,\Delta y)}{\Delta x + i\,\Delta y} = 0$$

Hence $\lim_{\Delta z \to 0} \frac{f(z + \Delta z) - f(z)}{\Delta z}$ exists and equals $\frac{\partial u}{\partial x} + i\,\frac{\partial v}{\partial x}$ independent of the manner of approach to zero of Δz. We have proved Theorem 4.1.

THEOREM 4.1. $f(z) = u + iv$ is differentiable if u and v satisfy the Cauchy-Riemann equations and if, furthermore, the four first partial derivatives of u and v with respect to x and y are continuous.

The following example shows that we cannot, in general, discard the continuity of the partial derivatives. Let

$$f(z) = \frac{x^3(1 + i) - y^3(1 - i)}{x^2 + y^2} \qquad z \neq 0 \qquad f(0) = 0$$

Here
$$u(x, y) = \frac{x^3 - y^3}{x^2 + y^2} \qquad u(0, 0) = 0$$
$$v(x, y) = \frac{x^3 + y^3}{x^2 + y^2} \qquad v(0, 0) = 0$$

Moreover
$$\left.\frac{\partial u}{\partial x}\right|_{0,0} = \lim_{x \to 0} \frac{u(x, 0) - u(0, 0)}{x - 0} = \lim_{x \to 0} \frac{x}{x} = 1$$

Similarly $\frac{\partial u}{\partial y} = -1$, $\frac{\partial v}{\partial x} = 1$, $\frac{\partial y}{\partial v} = 1$, at $z = 0$, so that the Cauchy-Riemann equations hold at $z = 0$. However, $f'(0)$ depends on the approach to the origin, for let $y = mx$, so that

$$f(z) = \frac{x(1 + i) - xm^3(1 - i)}{1 + m^2}$$

along $y = mx$, and

$$\lim_{z \to 0} \frac{f(z) - f(0)}{z - 0} = \lim_{x \to 0} \frac{(1 + i) - m^3(1 - i)}{(1 + m^2)(1 + mi)} \frac{x}{x} = \frac{(1 + i) - m^3(1 - i)}{(1 + m^2)(1 + mi)}$$

which depends on m. Let the reader show that $\frac{\partial u}{\partial x}$ is discontinuous at the origin by showing that $\lim\limits_{\substack{x \to 0 \\ y \to 0}} \frac{\partial u}{\partial x}$ does not exist.

DEFINITION 4.1. Let $f(z)$ be differentiable at $z = z_0$. If, furthermore, there exists a circle with center at z_0 such that $f(z)$ is differentiable at every interior point of that circle, we say that $f(z)$ is regular, or analytic, at z_0. Analyticity at a point is stronger than mere differentiability at that point (see Prob. 10 of this section). If $f(z)$ is not analytic at z_0, we say that z_0 is a singular point of $f(z)$.

Example 4.11. We show that $f(z) = z^2$ is differentiable everywhere. Since $f(z) = (x + iy)^2 = x^2 - y^2 + 2xyi$, we have $u = x^2 - y^2$, $v = 2xy$. Hence

$$\frac{\partial u}{\partial x} = 2x = \frac{\partial v}{\partial y}, \frac{\partial u}{\partial y} = -2y = -\frac{\partial v}{\partial x}$$

so that the Cauchy-Riemann equations hold. Moreover it is obvious that the partial derivatives are continuous. Thus $f'(z) = \frac{\partial u}{\partial x} + i\frac{\partial v}{\partial x} = 2x + 2yi = 2(x + iy) = 2z$. We could have shown that $f'(z)$ exists and equals $2z$ by applying (4.3).

$$\lim_{\Delta z \to 0} \frac{(z + \Delta z)^2 - z^2}{\Delta z} = \lim_{\Delta z \to 0} \frac{2z\,\Delta z + (\Delta z)^2}{\Delta z} = 2z$$

Let the reader show that if $f(z) = z^n$, n an integer, then $f'(z) = nz^{n-1}$.

Example 4.12. The reader should verify that if $f'(z_0)$, $g'(z_0)$ exist, then

$$(1)\quad \frac{d}{dz}[f(z) + g(z)]\Big|_{z=z_0} = f'(z_0) + g'(z_0)$$

$$(2)\quad \frac{d}{dz}[f(z)g(z)]\Big|_{z=z_0} = f(z_0)g'(z_0) + f'(z_0)g(z_0)$$

$$(3)\quad \frac{d}{dz}\left[\frac{f(z)}{g(z)}\right]\Big|_{z=z_0} = \frac{g(z_0)f'(z_0) - f(z_0)g'(z_0)}{[g(z_0)]^2} \qquad g(z_0) \neq 0$$

We assume that $f(z)$ and $g(z)$ have the same domain of definition.

Problems

1. If u and v satisfy (4.5) and if their second partials exist, show that $\nabla^2 u = \nabla^2 v = 0$. Also show that

$$\frac{\partial u}{\partial x}\frac{\partial u}{\partial y} + \frac{\partial v}{\partial x}\frac{\partial v}{\partial y} = 0$$

Give a geometric interpretation of this last equation. Remember that

$$\nabla u = \frac{\partial u}{\partial x}\mathbf{i} + \frac{\partial u}{\partial y}\mathbf{j}$$

is a vector normal to the curve u = constant.

The fact that u and v must satisfy Laplace's equation proves to be useful in the application of complex-variable theory to electricity and hydrodynamics.

2. Let $f(z) = u + iv$ be differentiable, and assume u is given such that $\nabla^2 u = 0$. Show that $v(x, y) = \int_{x_0}^{x} -\frac{\partial u(x, y)}{\partial y}\,dx + \int_{y_0}^{y} \frac{\partial u(x_0, y)}{\partial x}\,dy$. Let

$$f(z) = x^3 - 3xy^2 + iv(x, y)$$

be differentiable. Find $v(x, y)$. Note that $\nabla^2(x^3 - 3xy^2) = 0$.

3. Show that $f(z) = x - iy$ is not differentiable.

4. Show that $f(z) = x^3 - 3xy^2 + i(3x^2y - y^3)$ is analytic everywhere. Then show that $f(z) = z^3$. If $f(x + iy) = u(x, y) + iv(x, y)$, show that $f(z) = u(z, 0) + iv(z, 0)$.

5. Let $\sin z = \sin x \cosh y + i \cos x \sinh y$. Show that $\sin z$ is analytic everywhere. Is this definition of $\sin z$ consistent for z real? Define $\cos z \equiv \frac{d \sin z}{dz}$, and show that $\sin^2 z + \cos^2 z = 1$.

6. If $f'(z) = 0$ for all z, show that $f(z)$ = constant.

7. Prove the statements in Example 4.12.

8. If $f(z) = a_n z^n + a_{n-1}x^{n-1} + \cdots + a_0 = \sum_{k=0}^{n} a_k z^k$, show that $f'(z) = \sum_{k=1}^{n} k a_k z^{k-1}$.

9. If $f(z) = \sum_{n=0}^{\infty} a_n z^n$ converges for $0 \leqq |z| < R$, show that $f'(z) = \sum_{n=1}^{\infty} n a_n z^{n-1}$ for $0 \leqq |z| < R$.

10. Show that $f(z) = x^2y^2$ is differentiable only at the origin.

11. Assume $f'(a)$ and $g'(a)$ exist, $g'(a) \neq 0$. If $f(a) = g(a) = 0$, show that

$$\lim_{z \to a} \frac{f(z)}{g(z)} = \frac{f'(a)}{g'(a)},$$

4.7. The Definite Integral. The reader is urged to read first those sections of Chap. 10 concerning the uniform continuity of a continuous function, rectifiable Jordan curves, the Riemann integral, and Cauchy sequences. Let Γ be a rectifiable Jordan curve[1] joining the points $z = \alpha$, $z = \beta$. Let $f(z)$ be a continuous complex function defined on Γ. We are not concerned with the definition of $f(z)$ elsewhere. Neither do we introduce the differentiability of $f(z)$. The Riemann integral of $f(z)$ over Γ is defined as follows: Subdivide Γ into n parts in any manner whatsoever. Call the points of the subdivision $\alpha = z_0, z_1, z_2, \ldots, z_k,$

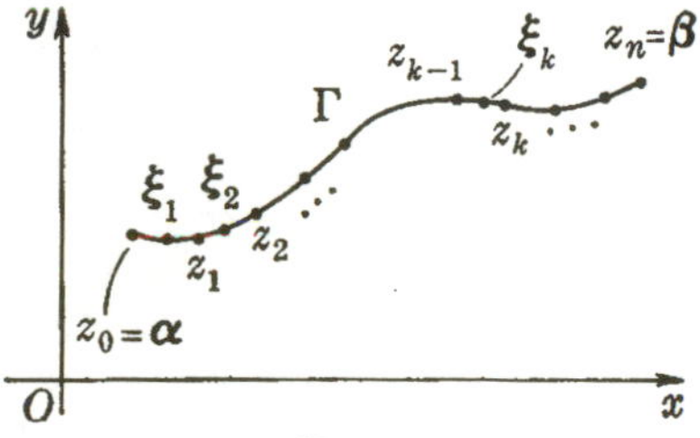

FIG. 4.2

[1] Called a simple curve.

$z_{k+1}, \ldots, z_n = \beta$ (see Fig. 4.2). On each of the paths $z_0z_1, z_1z_2, \ldots, z_kz_{k+1}, \ldots, z_{n-1}z_n$ choose a point $\xi_1, \xi_2, \ldots, \xi_k, \xi_{k+1}, \ldots, \xi_n$, and form the partial sum

$$J_n = \sum_{k=1}^{n} f(\xi_k)(z_k - z_{k-1}) \tag{4.6}$$

If $\lim_{n\to\infty} J_n = J$ exists and is unique independent of the choice of the ξ_k and independent of the method of subdividing Γ provided only that, as n tends to infinity, the maximum of the arc lengths from z_{k-1} to z_k, $k = 1, 2, \ldots, n$, tends to zero, we say that $f(z)$ is Riemann-integrable over Γ and write

$$\begin{aligned} J &= \int_\alpha^\beta f(z)\,dz \qquad \text{over } \Gamma \\ &= \int_{\alpha(\Gamma)}^{\beta} f(z)\,dz = \int_\Gamma f(z)\,dz \end{aligned} \tag{4.7}$$

This definition agrees with the definition of the Riemann integral of real-variable theory if Γ is a section of the real axis, Γ: $\alpha \leqq x \leqq \beta$, and if $f(z)$ is a real function of x.

We now show that, if $f(z)$ is continuous on the rectifiable Jordan arc given by $x = f(t)$, $y = \varphi(t)$, $t_0 \leqq t \leqq t_1$, then the Riemann integral of $f(z)$ over Γ exists. The proof proceeds as follows: Let us first look at any arc of Γ joining $z = a$ to $z = b$ and consider

$$S = f(\xi)(b - a)$$

where ξ is any point on the arc joining $z = a$ and $z = b$. A further subdivision of this arc into $z_0 = a, z_1, z_2, \ldots, z_n = b$ yields the partial sum defined as in (4.6),

$$S_n = \sum_{k=1}^{n} f(\xi_k)(z_k - z_{k-1})$$

Now $S = f(\xi)(b - a) = f(\xi) \sum_{k=1}^{n} (z_k - z_{k-1}) = \sum_{k=1}^{n} f(\xi)(z_k - z_{k-1})$

so that
$$S - S_n = \sum_{k=1}^{n} (f(\xi) - f(\xi_k))(z_k - z_{k-1})$$

If furthermore the maximum variation of $f(z)$ on the arc joining $z = a$, $z = b$ is σ, then

$$|S - S_n| \leqq \sum_{k=1}^{n} |f(\xi) - f(\xi_k)|\,|z_k - z_{k-1}| \leqq \sigma \sum_{k=1}^{n} |z_k - z_{k-1}| \leqq \sigma L \tag{4.8}$$

where L is the length of arc from $z = a$ to $z = b$. Why is

$$\sum_{k=1}^{n} |z_k - z_{k-1}| \leqq L?$$

This result will be important in what follows. Now we use the property that $f(z)$ is uniformly continuous on Γ. Choose a subdivision of Γ so that the maximum variation of $f(z)$ on any segmental arc of the subdivision is less than $\frac{1}{2}$, and obtain J_1 [see (4.6)]. Now impose a finer subdivision on the previous subdivision such that the maximum variation of $f(z)$ on any segmental arc of the new subdivision is less than $1/2^2$, and form a J_2 for this subdivision. Continue this process. For J_n the subdivisions are so fine that the maximum variation of $f(z)$ on any segmental arc is less than $1/2^n$. Moreover the maximum subdivision tends to zero in size. We obtain the sequence of complex numbers

$$J_1, J_2, J_3, \ldots, J_n, \ldots \tag{4.9}$$

We now show that $\lim_{n\to\infty} J_n$ exists. Choose any $\epsilon > 0$. We can find an integer n_0 such that $1/2^n < \epsilon/L$ for $n \geqq n_0$, where L is the length of arc of Γ. Now for $m \geqq n \geqq n_0$ we have

$$|J_m - J_n| \leqq \frac{1}{2^n} L < \epsilon$$

using the result of (4.8). Hence the sequence (4.9) is a Cauchy sequence and

$$\lim_{n\to\infty} J_n = J$$

We must now show that J is the same limit for any other choice of the ξ_k. For the new choice of the ξ_k we obtain the sequence

$$J_1', J_2', \ldots, J_n', \ldots$$

and by exactly the same reasoning as above we have

$$\lim_{n\to\infty} J_n' = J'$$

But $|J_n - J_n'| < (1/2^n)L$, and we leave it to the reader to show that this implies $J = J'$.

The final step is to show that the same limit J occurs for any other method of subdividing provided the maximum length of the subdivisions tends to zero. For any other sequence $K_1, K_2, \ldots, K_n, \ldots$ of partial sums of the form (4.6) we can superimpose the subdivisions which yield

K_n on the subdivisions of J_n of (4.9). Then

$$|J_n - K_n| < \frac{L}{2^n}$$

from (4.8). We leave it to the reader to show that $\lim_{n\to\infty} K_n = \lim_{n\to\infty} J_n$. This concludes the proof. The reader is referred to that excellent text by Knopp, "Theory of Functions," Dover Publications, 1945, for a much clearer expository of the Riemann integral.

For the reader who has trouble in understanding what has been attempted let us note that if $f(z) = u(x, y) + iv(x, y)$ and if Γ is given by $x = x(t)$, $y = y(t)$, $t_0 \leqq t \leqq t_1$, then $dz = dx + i\,dy$,

$$f(z)\,dz = u(x, y)\,dx - v(x, y)\,dy + i[u(x, y)\,dy + v(x, y)\,dx]$$

so that it would be logical to define

$$\int_\Gamma f(z)\,dz = \int_\Gamma u(x, y)\,dx - v(x, y)\,dy + i\left[\int_\Gamma u(x, y)\,dy + v(x, y)\,dx\right] \tag{4.10}$$

where $\int_\Gamma u(x, y)\,dx$, etc., are the ordinary line integrals of real-variable theory. It is not very difficult to show that this definition and existence of the line integral agrees with that discussed above for a continuous $f(z)$ and hence a continuous $u(x, y)$, $v(x, y)$. In particular if the Jordan curve is regular in the sense that $\frac{dx}{dt}$ and $\frac{dy}{dt}$ are continuous, then (4.10) becomes

$$\int_\Gamma f(z)\,dz = \int_{t_0}^{t_1}\left[u(x(t), y(t))\,\frac{dx(t)}{dt} - v(x(t), y(t))\,\frac{dy(t)}{dt}\right]dt + i\int_{t_0}^{t_1}\left[u(x(t), y(t))\,\frac{dy(t)}{dt} + v(x(t), y(t))\,\frac{dx(t)}{dt}\right]dt \tag{4.11}$$

Example 4.13. We evaluate $\int_\Gamma f(z)\,dz$, where $f(z) = z$ and the curve Γ is given as the straight line joining the origin and the point (3, 4). Γ may be written $x = 3t$, $y = 4t$, $0 \leqq t \leqq 1$. Along Γ

$$f(z) = z = x + iy = (3 + 4i)t$$

Moreover the vector z to the curve is $z = (3 + 4i)t$ so that $dz = (3 + 4i)\,dt$ and $f(z)\,dz = (3 + 4i)^2 t\,dt$.

$$\int_\Gamma f(z)\,dz = \int_0^1 (3 + 4i)^2 t\,dt = \tfrac{1}{2}(3 + 4i)^2$$

We could save ourselves all this work if we knew that

$$\int_\alpha^\beta z\,dz = \tfrac{1}{2}(\beta^2 - \alpha^2)$$

for any simple path Γ from $z = \alpha$ to $z = \beta$. Let the reader show that this is true.
Hint: subdivide Γ into $\alpha = z_0, z_1, z_2, \ldots, z_n = \beta$, and consider

$$J_n = \sum_{k=1}^{n} z_{k-1}(z_k - z_{k-1})$$

$$J_n' = \sum_{k=1}^{n} z_k(z_k - z_{k-1})$$

Show that $J_n + J_n' = \beta^2 - \alpha^2$, and let $n \to \infty$.

In this example we notice that $\int_\Gamma z\,dz$ depends only on the end points of Γ and not on Γ itself. The significance of this will become apparent in the next section.

Example 4.14. We evaluate $\int_\Gamma f(z)\,dz$ for $f(z) = x - iy$ with Γ the straight line from (0, 0) to (1, 1). Here $x = t$, $y = t$, $0 \leqq t \leqq 1$, $z = t + it$, $dz = (1 + i)\,dt$ so that

$$\int_\Gamma f(z)\,dz = \int_0^1 (1 - i)(1 + i)t\,dt = 1$$

For the same $f(z)$ let us take Γ as the sum of the two straight-line segments, one from (0, 0) to (1, 0), the second from (1, 0) to (1, 1). For the first path $x = t$, $y = 0$, $0 \leqq t \leqq 1$, so that $dz = dt$ and $f(z) = t$ and $\int_{\Gamma_1} f(z)\,dz = \frac{1}{2}$. Along the second path $x = 1$, $dx = 0$, $y = t$, $dz = i\,dt$, $0 \leqq t \leqq 1$, and $f(z) = 1 - it$ so that

$$\int_{\Gamma_2} f(z)\,dz = \int_0^1 (1 - it)i\,dt = \tfrac{1}{2} + i$$

Hence

$$\int_\Gamma f(z)\,dz = \int_{\Gamma_1} f(z)\,dz + \int_{\Gamma_2} f(z)\,dz = 1 + i$$

Thus, for $f(z) = x - iy$, the line integral is not independent of the path. One may guess that the nonanalyticity of $f(z) = x - iy$ may be the answer. The next section will verify this fact.

Example 4.15. The function $f(z) = 1/z$ is continuous and analytic everywhere except at the origin. Let us compute $\oint f(z)\,dz$, where Γ is a circle of radius a with center at the origin. It is easy to see that $z = a(\cos\theta + i\sin\theta)$, $0 \leqq \theta \leqq 2\pi$, represents the circle. Hence $dz = a(-\sin\theta + i\cos\theta)\,d\theta$, and

$$\begin{aligned}\oint \frac{-\sin\theta + i\cos\theta}{\cos\theta + i\sin\theta}\,d\theta &= \int_0^{2\pi} (-\sin\theta + i\cos\theta)(\cos\theta - i\sin\theta)\,d\theta \\ &= \int_0^{2\pi} i\,d\theta = 2\pi i\end{aligned}$$

Even though $f(z)$ is analytic on Γ, $\oint f(z)\,dz \neq 0$, in this case. We shall see that the singularity of $f(z)$ at the interior point $z = 0$ accounts for this fact. Notice that the value of the integral is not dependent on the radius of Γ. Indeed, it will be shown that for any closed rectifiable path Γ, with the origin in its interior, one has

$$\oint_\Gamma \frac{dz}{z} = 2\pi i$$

Problems

1. Show that $\int_\alpha^\beta z^2\,dz = \frac{1}{3}(\beta^3 - \alpha^3)$.

2. Show that $\oint (z - z_0)^m\,dz = \begin{matrix} 0 & \text{if } m \neq -1 \\ 2\pi i & \text{if } m = -1 \end{matrix}$

m a positive or negative integer, if the integration is performed around the circle Γ with radius a and center z_0. The integration is performed in the counterclockwise sense.

3. Define $\int_\Gamma |f(z)|\,|dz|$, and show that

$$\left| \int_\Gamma f(z)\,dz \right| \leqq \int_\Gamma |f(z)|\,|dz|$$

4. Show that $\underset{(\Gamma)}{\int_\alpha^\beta} f(z)\,dz = -\underset{(\Gamma)}{\int_\beta^\alpha} f(z)\,dz$.

5. If $|f(z)| \leqq M$ along Γ and L is the length of Γ, show that

$$\left| \int_\Gamma f(z)\,dz \right| \leqq LM$$

6. Why is it that $\int_\Gamma cf(z)\,dz = c\int_\Gamma f(z)\,dz$ for c constant?

7. Why is it that $\int_\Gamma f_1(z)\,dz + \int_\Gamma f_2(z)\,dz = \int_\Gamma [f_1(z) + f_2(z)]\,dz$?

8. Evaluate $\int_\Gamma (z + |z|)\,dz$, where Γ is the straight line from (0, 0) to (1, 1). Do the same for the other path discussed in Example 4.14.

9. Evaluate $\int_1^{z_0} \frac{dz}{z}$ along the path Γ consisting of Γ_1 and Γ_2, where Γ_1 is the straight line from (1, 0) to ($|z_0|$, 0) and Γ_2 is the arc of the circle from ($|z_0|$, 0) to z_0 with center at the origin. Is the value of the integral single-valued?

4.8. Cauchy's Integral Theorem. The fundamental theorem of complex-variable theory is due to Cauchy. There are various forms of this theorem. We present now a proof of one form.

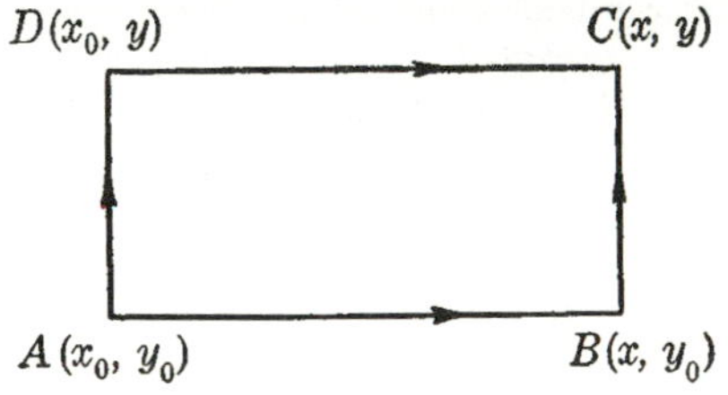

FIG. 4.3

Let S be a simply connected open region such that the partial derivatives of $u(x, y)$ and $v(x, y)$ are continuous and satisfy the Cauchy-Riemann equations,

$$f(z) = u(x, y) + iv(x, y)$$

at every point of S. In Sec. 4.6 we saw that these conditions were sufficient for $f(z)$ to be analytic in S. We shall show that, if Γ is any closed simple curve inside S, then

$$\oint_{\Gamma} f(z)\, dz = 0 \tag{4.12}$$

The statement concerning (4.12) is Cauchy's integral theorem. We prove first that $\oint f(z)\, dz = 0$ for the rectangle C (see Fig. 4.3) contained entirely in S.

We can obtain a single-valued complex function of the two real variables x and y by defining

$$F(x, y) = \int_{x_0}^{x} f(t + iy_0)\, dt + i \int_{y_0}^{y} f(x + it)\, dt$$

$F(x, y)$ is the sum of the integrals of $f(z)$ along the straight lines AB and BC. Similarly $G(x, y)$ is obtained by integrating $f(z)$ along AD and DC. The integration of $f(z)$ around the rectangle C in the counterclockwise sense is

$$\oint_C f(z)\, dz = F(x, y) - G(x, y)$$

where $$G(x, y) = i \int_{y_0}^{y} f(x_0 + it)\, dt + \int_{x_0}^{x} f(t + iy)\, dt$$

If we can show that $F(x, y) = G(x, y)$, then $\oint_C f(z)\, dz = 0$.

Let the reader show that

$$\frac{\partial F}{\partial x} = f(x + iy_0) + i \int_{y_0}^{y} \frac{\partial f(x + it)}{\partial x}\, dt \tag{4.13}$$

We need

$$\frac{d}{dx} \int_{x_0}^{x} f(t + iy_0)\, dt = f(x + iy_0)$$
$$\frac{d}{dx} \int_{y_0}^{y} f(x + it)\, dt = \int_{y_0}^{y} \frac{\partial f(x + it)}{\partial x}\, dt$$

in order to obtain (4.13). These statements are proved in Chap. 10 for f real. The student need only write $f = u + iv$ and apply the theorems of real-variable theory to u and v separately. The continuity of $\dfrac{\partial f}{\partial x}$ is used to perform the differentiation underneath the integral. Also we have

$$\begin{aligned} \frac{\partial F}{\partial y} &= if(x + iy) = if(z) \\ \frac{\partial G}{\partial x} &= f(x + iy) = f(z) \\ \frac{\partial G}{\partial y} &= if(x_0 + iy) + \int_{x_0}^{x} \frac{\partial f(t + iy)}{\partial y}\, dt \end{aligned} \tag{4.14}$$

As yet, we have not made use of the Cauchy-Riemann equations. Thus

$$f'(z) = \frac{\partial f}{\partial x} = \frac{\partial u}{\partial x} + i\frac{\partial v}{\partial x} = \frac{\partial v}{\partial y} - i\frac{\partial u}{\partial y} = -i\left(\frac{\partial u}{\partial y} + i\frac{\partial v}{\partial y}\right) = -i\frac{\partial f}{\partial y}$$

Hence $$\frac{\partial F}{\partial x} = f(x + iy_0) + \int_{y_0}^{y} \frac{\partial f(x + it)}{\partial t}\, dt = f(x + iy) = \frac{\partial G}{\partial x} \tag{4.15}$$

Similarly $\dfrac{\partial F}{\partial y} = \dfrac{\partial G}{\partial y}$, and hence $dF = dG$, $F = G + c$. Since $F = G = 0$ at (x_0, y_0), the constant c is zero. This proves that $\oint_C f(z)\, dz = 0$.

We now consider $\oint_{\Gamma_0} f(z)\, dz$, where Γ_0 is any closed rectifiable Jordon curve lying inside the rectangle $ABCD$. Parts of the curve may be segments of the sides of the rectangle (see Fig. 4.4). Let $P(x, y)$ be any point on Γ_0, and define $F(x, y)$ at P in exactly the same manner in which F was defined at $C(x, y)$. From (4.14) and (4.15)

$$\begin{aligned} dF &= \frac{\partial F}{\partial x}\, dx + \frac{\partial F}{\partial y}\, dy = f(z)\, dx + if(z)\, dy \\ &= f(z)(dx + i\, dy) = f(z)\, dz \end{aligned}$$

Hence $$\oint_{\Gamma_0} f(z)\, dz = \oint_{\Gamma_0} dF = \oint_{\Gamma_0} dU + i\oint_{\Gamma_0} dV = 0$$

where $F = U + iV$. Certainly $\oint dU = \oint dV = 0$.

The final part of the proof uses the following reasoning: Let Γ be any closed simple curve in S. Γ is a closed set of points. Now S was assumed to be an open set. If we adjoin to S its boundary points, we obtain $\bar{S}$, a closed set, consisting of S and the boundary of S, say, $\bar{\Gamma}$. There will be a minimum distance between Γ and $\bar{\Gamma}$ not equal to zero. We can therefore impose a fine enough mesh on S (the mesh consists of rectangles) so that the rectangles which contain the points of Γ will be entirely in S.

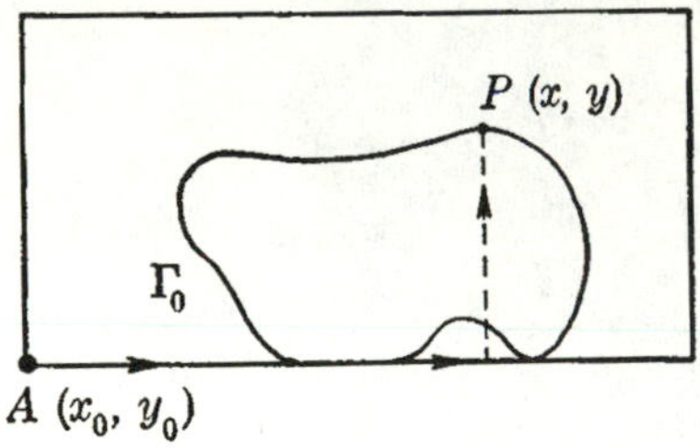

FIG. 4.4

The integration over all rectangles interior to Γ plus the integration of $f(z)$ over those boundaries which include Γ vanish from the above results. However, all integrations over the rectangles interior to Γ vanish in pairs, leaving

$$\oint_\Gamma f(z)\, dz = 0 \tag{4.16}$$

The condition that the partial derivatives of u and v be continuous can actually be omitted. The proof that $\oint_C f(z)\, dz = 0$ for the rectangle

C under the condition that $f(z)$ be analytic in S can be shown as follows: First assume $\oint_C f(z)\,dz \neq 0$. Subdivide the region R (C and its interior), into four new regions (see Fig. 4.5) by halving the sides of the rectangles. Now $\oint_C f(z)\,dz$ equals the sum of the integrals of $f(z)$ over the boundaries of the four new regions since the internal integrations cancel each other in pairs. Hence at least one of the four integrals does not vanish. We choose that boundary C_1 for which $\left|\oint_{C_1} f(z)\,dz\right|$ has the largest value. Again we subdivide, choose C_2, and continue this process indefinitely. We obtain a sequence of regions $R, R_1, R_2, \ldots, R_n, \ldots$, with boundaries $C, C_1, C_2, C_3, \ldots, C_n, \ldots$ such that $\oint_{C_n} f(z)\,dz \neq 0$. The regions R_n, $n = 1, 2, \ldots$ are closed and bounded sets, and the diameter of R_n tends to zero as $n \to \infty$. From the theorem of nested sets (see Chap. 10) there exists a unique point P which belongs to every R_n, $n = 1, 2, 3, \ldots$. Let P be the point z_0. Obviously z_0 is interior to R, or z_0 lies on C. Hence $f(z)$ is differentiable at $z = z_0$ so that

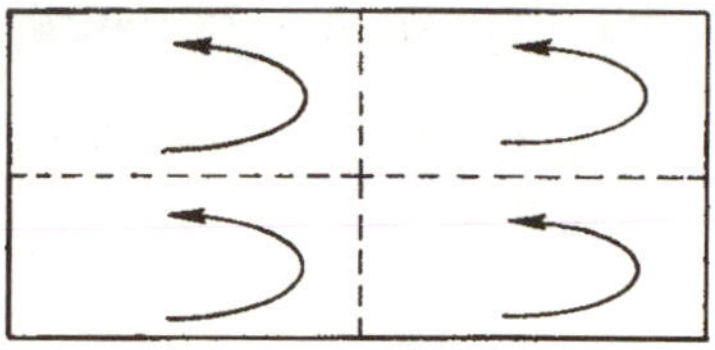

FIG. 4.5

$$f(z) = f(z_0) + f'(z_0)(z - z_0) + \epsilon(z, z_0)(z - z_0)$$

where $\epsilon(z, z_0)$ tends to zero as z tends to z_0. Hence, given any $\epsilon_0 > 0$, we can pick a region R_n with boundary C_n such that $|\epsilon(z, z_0)| < \epsilon_0$ for all z on C_n. Now

$$\left|\oint_C f(z)\,dz\right| \leqq 4\left|\oint_{C_1} f(z)\,dz\right| \leqq 4^2\left|\oint_{C_2} f(z)\,dz\right| \leqq \cdots \leqq 4^n\left|\oint_{C_n} f(z)\,dz\right|$$

But $$\oint_{C_n} f(z)\,dz = \oint_{C_n} f(z_0)\,dz + \int_{C_n} f'(z_0)(z - z_0)\,dz + \int_{C_n} \epsilon(z_0, z)(z - z_0)\,dz$$

so that $$\oint_{C_n} f(z)\,dz = \int_{C_n} \epsilon(z_0, z)(z - z_0)\,dz$$

Remember that $\oint dz = 0$, $\oint z\,dz = 0$. For any positive ϵ_0 we choose n sufficiently large so that $|\epsilon(z, z_0)| < \epsilon_0$ for all z on C_n. If l_n is the length of the diagonal of the rectangle C_n, then

$$\left|\oint_{C_n} f(z)\,dz\right| < \epsilon_0 \oint_{C_n} |z - z_0|\,|dz| \leqq \epsilon_0 4l_n^2 = 4\epsilon_0\left(\frac{L}{2^n}\right)^2$$

since $l_n = L/2^n$, where L is the length of the diagonal of C. Hence

$$\left|\oint_C f(z)\,dz\right| < 4^n\epsilon_0\frac{4L^2}{4^n} = 4\epsilon_0 L^2$$

Since ϵ_0 can be chosen arbitrarily small, the constant $\oint_C f(z)\,dz$ must be zero. Q.E.D. The proof of Cauchy's theorem follows then in exactly the same manner as demonstrated above. This proof is due to Bliss in the *American Mathematical Society Colloquium Publications*, vol. 16.

A very strong statement of Cauchy's theorem is as follows: Let Γ be a simple closed path such that $f(z)$ is analytic in the interior R of Γ and such that $f(z)$ is continuous on Γ. Then

$$\oint f(z)\,dz = 0$$

The proof is not trivial and is omitted here. Continuity of $f(z)$ on Γ in this case means $\lim\limits_{z\to\zeta} f(z) = f(\zeta)$, ζ on Γ, z on Γ or in R.

We now state some immediate consequences of Cauchy's integral theorem:

A. Let $f(z)$ be analytic in a simply connected region R. For $z = \alpha$, $z = \beta$, in R,

$$\int_\alpha^\beta f(z)\,dz$$

is independent of the simple path chosen from $z = \alpha$ to $z = \beta$ provided the path lies entirely in R. Let the reader verify this statement.

B. Let $f(z)$ be analytic in a region R bounded by two simple closed paths C_1, C_2 (see Fig. 4.6). Then $\oint_{C_1} f(z)\,dz = \oint_{C_2} f(z)\,dz$ provided both integrations are done in a clockwise sense or a counterclockwise

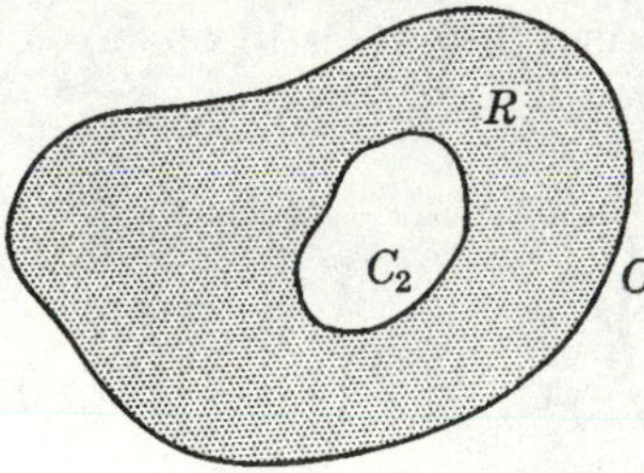

FIG. 4.6

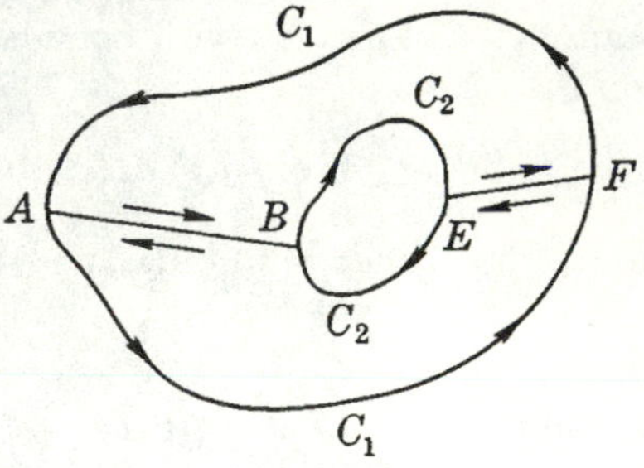

FIG. 4.7

sense. The proof is as follows: Construct the paths AB and EF (see Fig. 4.7). From Cauchy's theorem

$$\oint_{ABC_2EFC_1A} f(z)\,dz = 0 \qquad \oint_{AC_1FEC_2BA} f(z)\,dz = 0$$

Adding yields

$$\oint_{C_1\circlearrowleft} f(z)\,dz + \oint_{C_2\circlearrowright} f(z)\,dz = 0$$

and

$$\oint_{C_1\circlearrowleft} f(z)\,dz = \oint_{C_2\circlearrowleft} f(z)\,dz$$

Notice that nothing need be known about $f(z)$ outside of C_1 or inside of C_2. We have assumed $f(z)$ continuous on C_1 and C_2.

C. For Fig. 4.8 let the reader show that

$$\oint_{\Gamma_0} f(z)\,dz = \oint_{\Gamma_1} f(z)\,dz + \oint_{\Gamma_2} f(z)\,dz$$

$f(z)$ is analytic inside Γ_0 and outside Γ_1 and Γ_2 and is continuous on Γ_0, Γ_1, Γ_2. Generalize this statement for the curves $\Gamma_0, \Gamma_1, \Gamma_2, \ldots, \Gamma_n$.

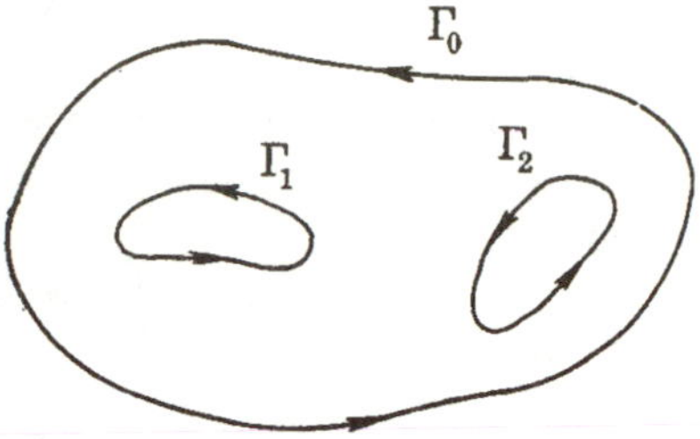

FIG. 4.8

D. Let $f(z)$ be analytic inside a simply connected open region R, and consider

$$F(z) = \int_{z_0}^{z} f(t)\,dt$$

where z_0 and z are in R. The path of integration is omitted since the integral is independent of the path (see **A**). Hence

$$F(z + \Delta z) - F(z) = \int_{z_0}^{z+\Delta z} f(t)\,dt + \int_{z}^{z_0} f(t)\,dt = \int_{z}^{z+\Delta z} f(t)\,dt$$

We choose the straight-line path from z to $z + \Delta z$ as the path of integration. Then $t = z + \mu\,\Delta z$, $0 \leqq \mu \leqq 1$, along this path, $dt = d\mu\,\Delta z$, and

$$\frac{F(z + \Delta z) - F(z)}{\Delta z} = \int_0^1 f(z + \mu\,\Delta z)\,d\mu$$

Since $f(z)$ is analytic at z,

$$\lim_{\Delta z \to 0} \frac{f(z + \mu\,\Delta z) - f(z)}{\mu\,\Delta z} = f'(z)$$

and

$$f(z + \mu\,\Delta z) = f(z) + \mu\,\Delta z f'(z) + \epsilon\mu\,\Delta z \tag{4.16'}$$

where $|\epsilon| \to 0$ as $\Delta z \to 0$. Hence

$$\frac{F(z + \Delta z) - F(z)}{\Delta z} = f(z)\int_0^1 d\mu + \Delta z f'(z)\int_0^1 \mu\,d\mu + \Delta z\int_0^1 \epsilon\mu\,d\mu$$

It is obvious that

$$\lim_{\Delta z \to 0} \frac{F(z + \Delta z) - F(z)}{\Delta z} = f(z)$$

so that $F(z)$ is analytic in R.

E. Let $f(z)$ be continuous inside a simply connected open region R, and assume

$$F(z) = \int_{z_0}^{z} f(t)\,dt$$

is independent of the path from z_0 to all z, the path lying entirely in R. We show that $F(z)$ is analytic in R. The variable t is simply the complex variable of integration. The reader can prove this statement easily enough by proceeding as in **D**. Continuity implies that

$$f(z + \mu\,\Delta z) = f(z) + \eta \tag{4.16''}$$

where $\eta \to 0$ as $\Delta z \to 0$. In the proof of **D** it was not necessary to use the analyticity of $f(z)$ twice. We could have used (4.16″) in place of (4.16′).

F. The fundamental theorem of the integral calculus applies equally well to the theory of complex variables. Assuming the conditions stated in **D** yields

$$F(z) = \int_{z_0}^{z} f(t)\,dt \qquad F'(z) = f(z)$$

Let $G(z)$ be any function such that $G'(z) = f(z)$. Then

$$\frac{d}{dz}\,(F(z) - G(z)) = 0$$

and $F(z) = G(z) - C$. Hence $G(z) - C = \int_{z_0}^{z} f(t)\,dt$,

$$G(z_0) - C = \int_{z_0}^{z_0} f(t)\,dt = 0$$

so that

$$G(z) - G(z_0) = \int_{z_0}^{z} f(t)\,dt$$

As a simple example, $\displaystyle\int_\alpha^\beta z^2\,dz = \tfrac{1}{3}(\beta^3 - \alpha^3)$ since $\dfrac{d}{dz}\,(\tfrac{1}{3}z^3) = z^2$.

Problems

1. Show that $\displaystyle\oint \frac{dz}{z - z_0} = 2\pi i$ for any simple closed path Γ, z_0 in the interior of Γ. Show that $\displaystyle\oint \frac{dz}{z - z_0} = 0$ if z_0 is exterior to Γ.

2. Evaluate $\displaystyle\oint \frac{dz}{z^2 - z}$ for any simple closed curve Γ enclosing the circle $r = 1$. *Hint:* Use **C** above, and write $1/(z^2 - z) = 1/(z - 1) - 1/z$.

3. Construct a function $f(z)$ such that $\oint f(z)\,dz = 0$ for all simple closed paths, $f(z)$ not analytic. Does this contradict Cauchy's theorem?

4. Let $f(z, t)$ be a complex function of the complex variable z and the real variable t. Assume $f(z, t)$ and $\dfrac{\partial f(z, t)}{\partial t}$ analytic in z for $t_0 \leq t \leq t_1$, and consider

$$F(t) = \int_\alpha^\beta f(z, t)\,dz$$
$$G(t) = \int_\alpha^\beta \frac{\partial f(z, t)}{\partial t}\,dz$$

If, furthermore, $\frac{\partial f}{\partial t}(z, t)$ is continuous in z and t, show that

$$\int_{t_0}^{t} G(u)\,du = F(t) - F(t_0)$$

so that

$$\frac{d}{dt}\int_{\alpha}^{\beta} f(z, t)\,dt = \int_{\alpha}^{\beta} \frac{\partial f(z, t)}{\partial t}\,dz$$

5. Let $f(z)$ and $g(z)$ be analytic in a simply connected region R. From

$$\frac{d}{dz}(f(z)g(z)) = f(z)g'(z) + f'(z)g(z)$$

show that

$$\int_{\alpha}^{\beta} f(z)g'(z)\,dz = f(\beta)g(\beta) - f(\alpha)g(\alpha) - \int_{\alpha}^{\beta} g(z)f'(z)\,dz$$

The path of integration from $z = \alpha$ to $z = \beta$ lies in R.

4.9. Cauchy's Integral Formula. A truly fundamental consequence of Cauchy's integral theorem of Sec. 4.8 is the following formula due to Cauchy: Let $f(z)$ be analytic in the simply connected open region R, and let Γ be a simple closed curve in R. Then

$$f(a) = \frac{1}{2\pi i}\oint_{\Gamma} \frac{f(z)\,dz}{z - a} \tag{4.17}$$

if $z = a$ is an interior point of Γ. The sense of integration around Γ is such that as we move around Γ the region containing $z = a$ lies to our left. The proof is as follows: From Sec. 4.8B we can replace the curve of integration Γ by any circle Γ_0 with center at $z = a$, Γ_0 interior to Γ.

Then
$$\begin{aligned}\frac{1}{2\pi i}\oint_{\Gamma} \frac{f(z)\,dz}{z - a} &= \frac{1}{2\pi i}\oint_{\Gamma_0} \frac{f(z)\,dz}{z - a}\\ &= \frac{1}{2\pi i}\int_0^{2\pi} \frac{f(z)b(-\sin\theta + i\cos\theta)\,d\theta}{b(\cos\theta + i\sin\theta)}\\ &= \frac{1}{2\pi}\int_0^{2\pi} f[a + b(\cos\theta + i\sin\theta)]\,d\theta \end{aligned} \tag{4.17'}$$

since $z = a + b(\cos\theta + i\sin\theta)$, $0 \leqq \theta \leqq 2\pi$, is the equation of Γ_0, b the radius of Γ_0. Since $f(z)$ is continuous at $z = a$, we have

$$f[a + b(\cos\theta + i\sin\theta)] = f(a) + \eta$$

where $\eta \to 0$ as $b \to 0$. Hence

$$\int_0^{2\pi} f[a + b(\cos\theta + i\sin\theta)]\,d\theta = 2\pi f(a) + \int_0^{2\pi} \eta\,d\theta$$

and

$$\lim_{b\to 0} \frac{1}{2\pi}\int_0^{2\pi} f[a + b(\cos\theta + i\sin\theta)]\,d\theta = f(a)$$

Since the left-hand side of (4.17′) is independent of b, (4.17) must result. We can now observe one important consequence of analyticity. To evaluate the right-hand side of (4.17), one needs only to know the value of $f(z)$ on the boundary Γ. After the integration is performed, the value of $f(z)$ is known in the interior of Γ. Thus, if $f(z)$ is known to be analytic inside Γ, if $f(z)$ is continuous on Γ so that Cauchy's theorem holds, then the values of $f(z)$ interior to Γ can be determined if we know only the values of $f(z)$ on Γ. Analyticity is, indeed, a powerful condition.

Since $f'(a)$ is known to exist, we may hope that $f'(a)$ can be obtained from (4.17) by differentiating underneath the integral. If this were possible, we would obtain

$$f'(a) = \frac{1}{2\pi i} \oint_\Gamma \frac{f(z)\,dz}{(z-a)^2} \tag{4.18}$$

Let us prove that (4.18) is correct. We have

$$f(a+h) = \frac{1}{2\pi i} \oint \frac{f(z)\,dz}{z-(a+h)}$$

$$f(a+h) - f(a) = \frac{1}{2\pi i} \oint \left[\frac{1}{z-(a+h)} - \frac{1}{z-a}\right] f(z)\,dz$$

$$\frac{f(a+h)-f(a)}{h} = \frac{1}{2\pi i} \oint \frac{f(z)\,dz}{(z-a)(z-a-h)}$$

Since $f'(a) = \lim_{h\to 0} \dfrac{f(a+h)-f(a)}{h}$, we need only show that

$$\lim_{h\to 0} \oint \frac{f(z)\,dz}{(z-a)(z-a-h)} = \oint \frac{f(z)\,dz}{(z-a)^2} \tag{4.19}$$

to obtain (4.18). Consider

$$\oint \frac{f(z)\,dz}{(z-a)(z-a-h)} - \oint \frac{f(z)\,dz}{(z-a)^2} \equiv h \oint \frac{f(z)\,dz}{(z-a)^2(z-a-h)}$$

For h sufficiently small it is easy to see that $\left|\dfrac{f(z)}{(z-a)(z-a-h)}\right|$ is uniformly bounded for all z on Γ, so that as $h \to 0$ (4.19) holds, which in turn yields (4.18). By mathematical induction let the reader show that

$$f^{(n)}(a) = \frac{n!}{2\pi i} \oint \frac{f(z)\,dz}{(z-a)^{n+1}} \qquad n = 0, 1, 2, 3, \ldots \tag{4.20}$$

Equation (4.20) embraces Cauchy's integral formula for $f(a)$ and all its derivatives. It is important to note that analyticity of $f(z)$ is strong enough to guarantee the existence of all derivatives of $f(z)$. In real-variable theory the existence of $f'(x)$ in a neighborhood of $x = a$ in no way yields any information about the existence of further derivatives of $f(x)$ at $x = a$.

Problems

1. Let $f(z)$ be analytic within a circle C of radius R and center a, $f(z)$ continuous on C. Let M be an upper bound of $|f(z)|$ on C. Show that

$$|f^{(n)}(a)| \leqq \frac{Mn!}{R^n}$$

2. If $f(z)$ is analytic for all z it is said to be an entire function. Use the results of Prob. 1 to show that an entire bounded function must be a constant. $f(z)$ is said to be bounded if a constant M exists such that $|f(z)| < M$ for all z. Hence prove the first theorem of Liouville that a nonconstant entire function assumes arbitrarily large values (in absolute value) outside every circle with center at $z = 0$.

3. Consider the polynomial

$$f(z) = a_0 z^n + a_1 z^{n-1} + a_2 z^{n-2} + \cdots + a_n$$

Show that

$$|f(z)| \geqq |z|^n \left(|a_0| - \frac{|a_1|}{|z|} - \frac{|a_2|}{|z|^2} - \cdots - \frac{|a_n|}{|z|^n} \right)$$

Hence show that for any positive M there exists an R such that, for $|z| > R$, $|f(z)| > M$.

4. Let $f(z)$ be the polynomial of Prob. 3. Assume $f(z) \neq 0$ for all z, and consider $g(z) = 1/f(z)$. Why is $g(z)$ an entire function? Use the result of Prob. 3 to show that $g(z)$ is a bounded entire function. Since $g(z) \not\equiv$ constant, use the result of Prob. 2 to deduce that a z_0 exists such that $f(z_0) = 0$. This is the fundamental theorem of algebra (Gauss). Deduce that $f(z)$ has n zeros.

5. Prove Morera's theorem: If $f(z)$ is continuous in a simply connected region R, and if $\oint f(z)\,dz = 0$ for every simple closed path in R, then $f(z)$ is analytic in R.

6. Prove (4.20) for $n = 2$.

7. Let $\varphi_n(z)$, $n = 1, 2, 3, \ldots$, be a sequence of functions analytic inside and on the simple closed curve Γ which converges uniformly to $\varphi(z)$. Show that $\varphi(z)$ is analytic inside Γ.

4.10. Taylor's Expansion. In real-variable theory certain functions could be written as infinite series, called Taylor's series or expansion. For example,

$$e^x = 1 + x + \frac{x^2}{2!} + \cdots + \frac{x^n}{n!} + \cdots = \sum_{n=0}^{\infty} \frac{x^n}{n!}$$

$$\sin x = x - \frac{x^3}{3!} + \frac{x^5}{5!} - \cdots + \frac{(-1)^n x^{2n+1}}{(2n+1)!} + \cdots$$

$$= \sum_{n=0}^{\infty} \frac{(-1)^n x^{2n+1}}{(2n+1)!}$$

$$\ln(1 + x) = x - \frac{x^2}{2} + \frac{x^3}{3} + \cdots + \frac{(-1)^{n-1} x^n}{n} + \cdots$$

$$= \sum_{n=1}^{\infty} \frac{(-1)^{n-1} x^n}{n}$$

The function $f(x) = e^{-1/x^2}$, $x \neq 0$, $f(0) = 0$, is such that

$$f(0) = f'(0) = f''(0) = \cdots = f^{(n)}(0) = \cdots = 0$$

However, this function has no Taylor-series expansion about $x = 0$. We shall find, however, that if $f(z)$ is analytic at $z = a$, a Taylor-series expansion will exist for $f(z)$ at $z = a$. Remember that analyticity of $f(z)$ at $z = a$ means differentiability of $f(z)$ for all points in some neighborhood of $z = a$. Before proving this result the student should review the definitions and theorems involving infinite series (see Chap. 10). The definitions of convergence and uniform convergence of series, and the theorems regarding term-by-term integration and differentiation of a series, apply equally well for an infinite series whose terms are complex.

We now proceed to the development of Taylor's theorem. Let $f(z)$ be a function analytic in an open region R, and let z_0 be an interior point of R. There exists a unique power series $\sum_{n=0}^{\infty} a_n(z - z_0)^n$ such that

$$
\begin{aligned}
f(z) &= \sum_{n=0}^{\infty} a_n(z - z_0)^n \\
a_n &= \frac{1}{n!} f^{(n)}(z_0)
\end{aligned}
\tag{4.21}
$$

for all z in some neighborhood of $z = z_0$. The series (4.21) for $f(z)$ converges to $f(z)$ for all z inside a circle C surrounding $z = z_0$, and the circumference of C contains those singularities of $f(z)$ which are closest to $z = z_0$. A point z_1 is said to be a singular point of $f(z)$ if $f(z)$ is not analytic at $z = z_1$.

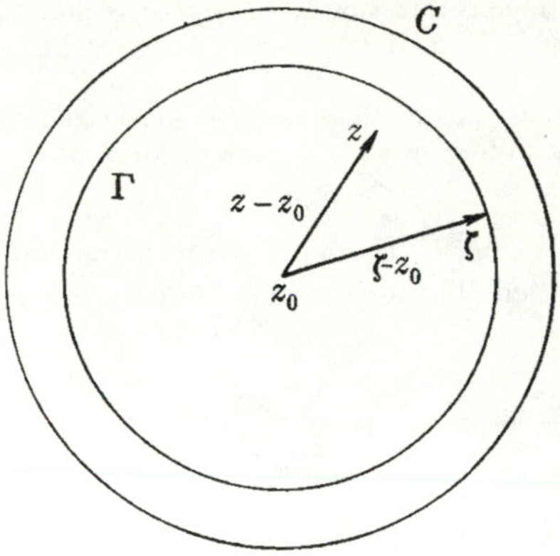

FIG. 4.9

We proceed to the proof. Let C be the circle mentioned above. C has the property that there is at least one point on C at which $f(z)$ is not analytic. Moreover $f(z)$ is analytic at every interior point of C. Now let z be any interior point of C, and construct a circle Γ with center z_0 containing z in its interior, Γ interior to C (see Fig. 4.9).

Let ζ be any point on Γ. From Cauchy's integral formula

$$
\begin{aligned}
f(z) &= \frac{1}{2\pi i} \oint_\Gamma \frac{f(\zeta)\, d\zeta}{\zeta - z} \\
&= \frac{1}{2\pi i} \oint_\Gamma \frac{f(\zeta)\, d\zeta}{(\zeta - z_0) - (z - z_0)} \\
&= \frac{1}{2\pi i} \oint \frac{f(\zeta)\, d\zeta}{(\zeta - z_0)[1 - (z - z_0)/(\zeta - z_0)]}
\end{aligned}
\tag{4.22}
$$

Since $\left|\frac{z - z_0}{\zeta - z_0}\right| < K < 1$ (see Fig. 4.9), we have

$$\frac{1}{1 - (z - z_0)/(\zeta - z_0)} = \sum_{n=0}^{\infty} \left(\frac{z - z_0}{\zeta - z_0}\right)^n$$

Equation (4.22) becomes

$$f(z) = \frac{1}{2\pi i} \oint_\Gamma \sum_{n=0}^{\infty} \frac{(z - z_0)^n f(\zeta)}{(\zeta - z_0)^{n+1}} \, d\zeta$$

The uniform convergence of the series $\sum_{n=0}^{\infty} \frac{(z - z_0)^n f(\zeta)}{(\zeta - z_0)^{n+1}}$, ζ the variable of integration, z_0 and z fixed, enables us to interchange integration and summation. Hence

$$\begin{aligned} f(z) &= \frac{1}{2\pi i} \sum_{n=0}^{\infty} (z - z_0)^n \oint \frac{f(\zeta)\, d\zeta}{(\zeta - z_0)^{n+1}} \\ &= \sum_{n=0}^{\infty} a_n (z - z_0)^n \\ &= \sum_{n=0}^{\infty} \frac{f^{(n)}(z_0)}{n!} (z - z_0)^n \end{aligned}$$

where $a_n = \frac{1}{2\pi i} \oint \frac{f(\zeta)\, d\zeta}{(\zeta - z_0)^{n+1}} = \frac{f^{(n)}(z_0)}{n!}$ see (4.20) Q.E.D.

To show uniqueness, assume

$$f(z) = \sum_{n=0}^{\infty} a_n (z - z_0)^n = \sum_{n=0}^{\infty} b_n (z - z_0)^n$$

for all z inside C. For $z = z_0$ we have $a_0 = b_0$, so that

$$\sum_{n=1}^{\infty} a_n (z - z_0)^n = \sum_{n=1}^{\infty} b_n (z - z_0)^n \tag{4.23}$$

for all z inside C. But (4.23) implies

$$(z - z_0) \sum_{n=1}^{\infty} a_n (z - z_0)^{n-1} = (z - z_0) \sum_{n=1}^{\infty} b_n (z - z_0)^{n-1}$$

for all z inside C, which in turn implies

$$\sum_{n=1}^{\infty} a_n(z - z_0)^{n-1} = \sum_{n=1}^{\infty} b_n(z - z_0)^{n-1}$$

for all z inside C with the possible exception of $z = z_0$.

Hence $$\lim_{z \to z_0} \sum_{n=1}^{\infty} a_n(z - z_0)^{n-1} = \lim_{z \to z_0} \sum_{n=1}^{\infty} b_n(z - z_0)^{n-1}$$

so that $a_1 = b_1$. By mathematical induction the reader can show that $a_n = b_n$, $n = 0, 1, 2, \ldots$. Q.E.D.

Example 4.16. The function $f(z)$ defined by

$$f(z) \equiv e^x \cos y + ie^x \sin y$$

is easily seen to be analytic everywhere. We have

$$f'(z) = \frac{\partial u}{\partial x} + i\frac{\partial v}{\partial x} = e^x \cos y + ie^x \sin y = f(z)$$

so that by mathematical induction $f^{(n)}(z) = f(z)$ and $f^{(n)}(0) = f(0) = 1$. Hence

$$f(z) = \sum_{n=0}^{\infty} \frac{z^n}{n!} = e^x(\cos y + i \sin y)$$

We define $f(z)$ to be $e^z \equiv e^x e^{iy}$. We note that, if z is real, $f(z)$ reduces to e^x. By direct multiplication of the series representing e^z and e^{z_1} it can be shown that $e^z e^{z_1} = e^{z+z_1}$. An easier way is the following: Let the reader show that $\frac{d}{dz} e^{a-z} = -e^{a-z}$ for any constant a. Then

$$\frac{d}{dz}(e^z e^{a-z}) = -e^z e^{a-z} + e^z e^{a-z} \equiv 0$$

so that $e^z e^{a-z}$ is independent of z. Let $z = 0$, and we have $e^z e^{a-z} = e^a$. Now let $a = z + z_1$ so that $e^z e^{z_1} = e^{z+z_1}$. From this result we have $e^{iy} = \cos y + i \sin y$, y real. Let the reader show that this result of Euler's also holds for y complex.

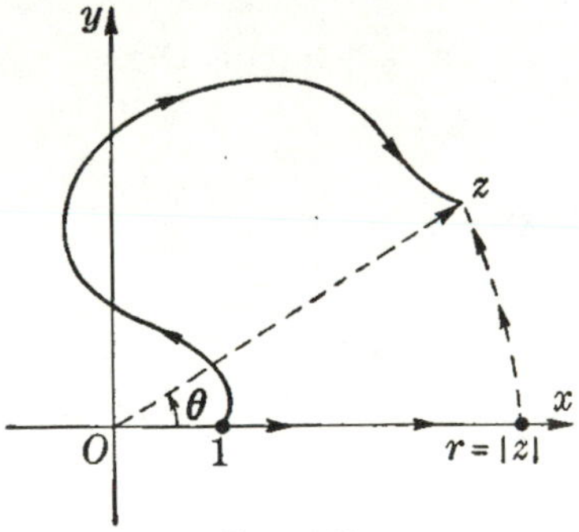

FIG. 4.10

Example 4.17. Let us define Ln z as follows. Let z be any complex number other than $z = 0$, $z = r(\cos \theta + i \sin \theta) = re^{i\theta}$ from Example 4.16, $-\pi < \theta \leq \pi$. We evaluate $\text{Ln } z \equiv \int_1^z \frac{dt}{t}$ along any curve Γ not passing through the origin or crossing the negative x axis (see Fig. 4.10). We can replace Γ by the path from $x = 1$ to $x = r$ followed by the arc of the circle with radius r and center at $z = 0$ until we reach z.

This yields

$$\begin{aligned}\text{Ln } z &= \int_1^r \frac{dx}{x} + \int_0^\theta \frac{ire^{i\varphi}\,d\varphi}{re^{i\varphi}} \\ &= \ln r + i\theta \\ &= \ln |z| + i \text{ Arg } z\end{aligned}$$

Ln z is single-valued and analytic everywhere except at $z = 0$, since $\dfrac{d \text{ Ln } z}{dz} = \dfrac{1}{z}$. For z real and positive, Ln z becomes the ordinary $\log_e x$. It is easy to show that

$$\frac{d^2 \text{ Ln } z}{dz^2} = -\frac{1}{z^2}, \ldots, \frac{d^n \text{ Ln } z}{dz^n} = \frac{(n-1)!}{z^n}(-1)^{n+1}$$

so that

$$\text{Ln } z = (z-1) - \frac{(z-1)^2}{2} + \frac{(z-1)^3}{3} \cdots$$

Why does the series expansion for Ln z converge only for $|z - 1| < 1$?

If we had not imposed the condition $-\pi < \theta \leqq \pi$, then Ln z would not have been single-valued. Let the reader show that for this case

$$\ln z = \int_1^z \frac{dt}{t} = \ln |z| + i(\text{Arg } z + 2\pi n)$$

where n is the number of times the path of integration Γ encircles the origin. If the integration is performed in a clockwise fashion, n is negative; otherwise n is a positive integer. Let the reader show that

$$\begin{aligned}\text{Ln }(-1) &= \pi i \\ \ln z_1 z_2 &= \ln z_1 + \ln z_2 + 2\pi i n \\ \text{Ln } x_1 x_2 &= \text{Ln } x_1 + \text{Ln } x_2 \qquad x_1 > 0,\ x_2 > 0\end{aligned}$$

Ln z is called the principal value of ln z. We imagine a cut exists along the negative x axis which forbids us from crossing the negative x axis. We call the point $z = 0$ a branch point of ln z. If we wish to pass from Ln z to ln z, we need only imagine that as we approach the cut from the top half of the z plane we have the ability to slide under the cut into a new z plane. In this new Riemann surface, or sheet, we have

$$\ln^{(1)} z = \text{Ln } z + 2\pi i$$

If we now swing around the origin and slide into a new surface, we have

$$\ln^{(2)} z = \text{Ln } z + 4\pi i$$

This process can be extended indefinitely. On each Riemann surface

$$\ln^{(k)} z = \text{Ln } z + 2\pi k i$$

is a single-valued function.

Example 4.18. The function $f(z) = 1/(1 - z)$, $z \neq 1$, $f(1)$ defined in any way we please, has for its only singularity the point $z = 1$. The Taylor-series expansion of $f(z)$ about $z = 0$ must converge for $|z| < 1$. Indeed

$$f(z) = \sum_{n=0}^{\infty} \frac{z^n}{n} \tag{4.24}$$

converges for $|z| < 1$ and can be used to represent $1/(1 - z)$ for $|z| < 1$. Let the reader show that

$$a_n = \frac{1}{n!}\frac{d^n}{dz^n}\left(\frac{1}{1-z}\right)\bigg|_{z=0} = 1 \qquad n = 0, 1, 2, \ldots$$

The series expansion of $1/(1-z)$ about the point $z = i/2$ should converge for

$$\left|z - \frac{i}{2}\right| < \frac{\sqrt{5}}{2}$$

since the point $z = i/2$ is $\sqrt{5}/2$ units from $z = 1$. We write

$$\frac{1}{1-z} = \frac{1}{1 - \frac{i}{2} - \left(z - \frac{i}{2}\right)}$$

$$= \frac{1}{1-\frac{i}{2}}\frac{1}{1 - \frac{z - i/2}{1 - i/2}}$$

$$= \frac{1}{1 - i/2}\sum_{n=0}^{\infty}\left(\frac{z - i/2}{1 - i/2}\right)^n \qquad (4.25)$$

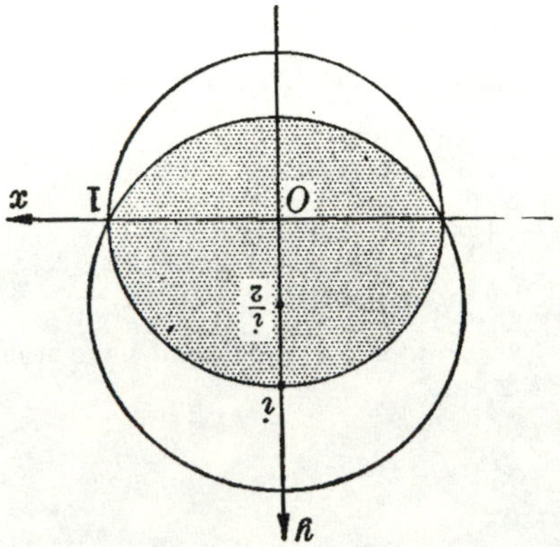

FIG. 4.11

The circles of convergence, $|z| = 1$ and $\left|z - \frac{i}{2}\right| = \frac{\sqrt{5}}{2}$, overlap (see Fig. 4.11). In this shaded region of overlapping, (4.24) and (4.25) converge to the same value of $1/(1-z)$. Equations (4.24) and (4.25) are said to be analytic continuations of each other.

Problems

1. Obtain the Taylor-series expansion of $\sinh z \equiv (e^z - e^{-z})/2$ about $z = 0$. The same for $\cosh z \equiv (e^z + e^{-z})/2$ about $z = 0$. Show that $\cosh^2 z = 1 + \sinh^2 z$, $\frac{d}{dz}\sinh z = \cosh z$, $\frac{d}{dz}\cosh z = \sinh z$.

2. Show that $\operatorname{Ln}(1+z) = \sum_{n=1}^{\infty}\frac{(-1)^{n-1}z^n}{n}$.

3. Show that $\sum_{n=0}^{\infty}(z^n/n!)$ converges for all z. Multiply the series $\sum_{n=0}^{\infty}(z^n/n!)$ and $\sum_{n=0}^{\infty}(z_1^n/n!)$ to obtain $e^z e^{z_1} = e^{z+z_1}$.

4. If one defines $\tan^{-1} z$ by $\tan^{-1} z = \int_0^z \frac{dt}{1+t^2}$, what difficulties occur? How can one make $\tan^{-1} z$ a single-valued function?

5. Show that $e^{2\pi n i} = 1$ if n is an integer. If $e^{z+\alpha} = e^z$, show that $\alpha = 2\pi n i$, n an integer.

6. Define $w = \sqrt{z}$ as that function w such that $w^2 = z$. If $z = re^{i\theta}$, $w = \rho e^{i\varphi}$, show that $\rho = \sqrt{r}$, $\varphi = \pi/2$. Since $z = re^{i(\theta+2\pi)}$, show that $\varphi = \theta/2 + \pi$ also. For $z \neq 0$, show that w is double-valued. Construct a Riemann surface so that w is single-valued. Is the origin a branch point?

7. If $w = \operatorname{Ln} z$, show that $z = e^w$. *Hint:* $\operatorname{Ln} z = \ln|z| + i\theta$, $e^w = e^{\ln|z| + i\theta}$.

8. Define $w = z^\alpha$, α complex, by the equation $w = z^\alpha = e^{\alpha \ln z}$. How can one make w single-valued? Show that for this case $\dfrac{dw}{dz} = \alpha z^{\alpha-1}$.

9. Let $f(z)$ be analytic for $|z| < R$, $f(z) = \sum_{n=0}^{\infty} a_n z^n$. Show that, for $r < R$,

$$\frac{1}{2\pi}\int_0^{2\pi} |f(re^{i\theta})|^2\, d\theta = \sum_{n=0}^{\infty} |a_n|^2 r^{2n}$$

10. Let $f(z)$ be analytic in a simply connected region R bounded by a simple closed path Γ, $f(z)$ continuous on Γ, so that Cauchy's theorem applies. Let z_0 be an interior point of R so that $f(z) = \sum_{n=0}^{\infty} a_n(z - z_0)^n$. Let C be a circle with center at z_0 and radius r, C inside Γ. Show that $|a_0|^2 + |a_1|^2 r^2 + |a_2|^2 r^4 + \cdots \leqq |a_0|^2$ if it is assumed that $|f(z_0)| \geqq |f(z)|$ for all z in R. Hence prove the maximum-modulus theorem, which states that $|f(z_0)| \geqq |f(z)|$ for all z in R implies z_0 on Γ if $f(z) \not\equiv$ constant.

4.11. An Identity Theorem. Analytic Continuation. We have seen that, if a function $f(z)$ is analytic at a point z_0, a Taylor-series expansion exists. If one desires, then, one could define the class of analytic functions as the totality of series of the form $\sum_{n=0}^{\infty} a_n(z - z_0)^n$ with nonzero radii of convergence. Some of the series would be analytic continuations of each other. Thus one could start with a particular analytic function $f(z) = \sum_{n=0}^{\infty} a_n(z - z_0)^n$ which converges for all z such that $|z - z_0| < r \neq 0$. Now choose a point, z_1, inside this circle. Since $f(z)$ is analytic at z_1, we can find a series expansion for $f(z)$ in the form $\sum_{n=0}^{\infty} b_n(z - z_1)^n$ which converges for all z such that

$$|z - z_1| < r_1 \neq 0$$

This new region of convergence may extend beyond the original circle of convergence (see Fig. 4.12).

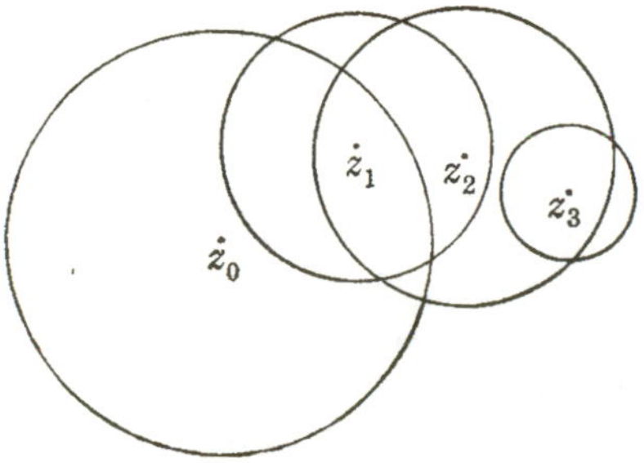

FIG. 4.12

This process can be continued. Each series is an analytic continuation of its predecessor, and conversely. All of them represent the original $f(z)$, which has now been extended to other portions of the z plane. One might naturally ask, if a point $z = \zeta$ is reached by two different paths of analytic continuation, do the two series representations thus obtained converge to the same value in their common region of convergence? We cannot answer this question until we prove Theorem 4.2.

THEOREM 4.2. Let $f(z)$ and $g(z)$ be analytic in a simply connected open region R. Assume $f(z) = g(z)$ for a sequence of points $z_1, z_2, \ldots, z_n, \ldots$ having z_0 as a limit point, z_i, $i = 0, 1, 2, \ldots, n, \ldots$, in R. Then $f(z) \equiv g(z)$ in R.

The proof proceeds as follows: First note that $f(z_n) = g(z_n)$ so that $f(z_0) = \lim_{n\to\infty} f(z_n) = \lim_{n\to\infty} g(z_n) = g(z_0)$. Moreover, since $f(z)$ and $g(z)$ are analytic at z_0,

$$f(z) = \sum_{k=0}^{\infty} a_k(z - z_0)^k$$

$$g(z) = \sum_{k=0}^{\infty} b_k(z - z_0)^k$$

and $f(z_0) = g(z_0)$ implies $a_0 = b_0$. Hence

$$(z_n - z_0) \sum_{k=1}^{\infty} a_k(z_n - z_0)^{k-1} = (z_n - z_0) \sum_{k=1}^{\infty} b_k(z_n - z_0)^{k-1}$$

for $n = 1, 2, 3, \ldots$. This implies

$$\sum_{k=1}^{\infty} a_k(z_n - z_0)^{k-1} = \sum_{k=1}^{\infty} b_k(z_n - z_0)^{k-1}$$

for $n = 1, 2, 3, \ldots$. Hence

$$\lim_{n\to\infty} \sum_{k=1}^{\infty} a_k(z_n - z_0)^{k-1} = \lim \sum_{k=1}^{\infty} b_k(z_n - z_0)^{k-1}$$

which implies $a_1 = b_1$. By mathematical induction the reader can show that $a_n = b_n$, $n = 0, 1, 2, \ldots$. This shows that $f(z) \equiv g(z)$ for some neighborhood of z_0. This neighborhood (a circle) extends up to the nearest singularity of $f(z)$ or $g(z)$. Now let ζ be any point of R, and construct a simple path Γ_0 from z_0 to ζ lying entirely in R. Let Γ be the boundary of R (see Fig. 4.13). Call the shortest distance from Γ_0 to Γ, ρ. The radius of convergence of $f(z)$ about $z = z_0$ is $\geqq \rho$. Why? Now choose a point z_1 on Γ_0 interior to the circle of convergence of $f(z)$ and $g(z)$ about z_0. Since $f(z)$ and $g(z)$ are identical in a neighborhood of z_1, we easily prove in exactly the same manner as above that $f(z) \equiv g(z)$ for some circle of convergence about z_1 whose radius is greater than or equal to ρ. Let the reader show that $z = \zeta$ can be reached in a finite number of steps. When ζ is interior to one of the circles of convergence,

$f(\zeta) = g(\zeta)$, which proves the theorem. It is important to note that at each point $z_0, z_1, z_2, \ldots, z_k$ the actual given function $f(z)$ and its derivatives are used to obtain the Taylor series expansion of $f(z)$. Analytic continuation is not used since we are not at all sure that the value of

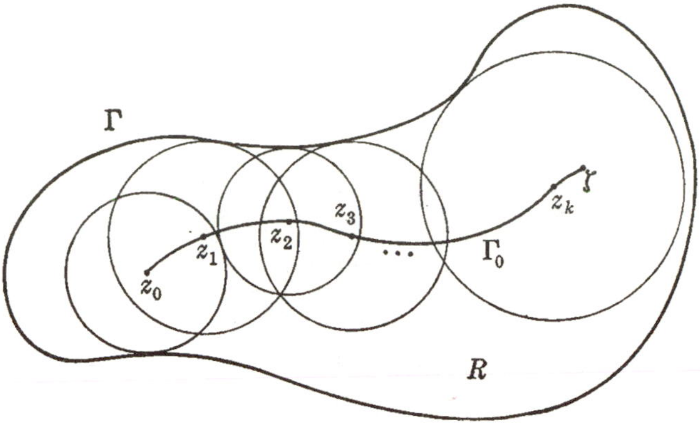

Fig. 4.13

$f(z)$ at $z = \zeta$ would be equal to the value at $z = \zeta$ obtained by analytic continuation. That this is true for a simply connected region requires some proof. This is essentially the monodromy theorem, whose proof we omit. The formal statement of the monodromy theorem is this:

Let R be a simply connected region, and let $f(z) = \sum_{n=0}^{\infty} a_n(z - z_0)^n$ have a nonzero radius of convergence. If $f(z)$ can be continued analytically from z along every path in R, then this continuation gives rise to a function which is single-valued and analytic in R.

We now state some consequences of the identity theorem:

(*a*) The identity theorem holds for a connected region.

(*b*) Let $f(z)$ be analytic at $z = z_0$. Let $z_1, z_2, z_3, \ldots, z_n, \ldots$ be a sequence of points which tend to z_0 as a limit such that $f(z_n) = 0$, $n = 1, 2, 3, \ldots$. Then $f(z) \equiv 0$ in a neighborhood of $z = z_0$.

(*c*) Let $f(z)$ be analytic at z_0, $f(z) \not\equiv$ constant. A neighborhood, N, of $z = z_0$ can be found such that $f(z_1) \neq f(z_0)$ for z_1 in N, $z_1 \neq z_0$.

(*d*) Let $f(z)$ and $g(z)$ be analytic in an open connected region R. Assume $f^{(n)}(z_0) = g^{(n)}(z_0)$, $n = 0, 1, 2, \ldots$, z_0 in R. It can be shown that $f(z) \equiv g(z)$ in R.

Problems

1. Prove (*a*).
2. Prove (*b*).
3. Prove (*c*).
4. Prove (*d*).

5. Consider $f(z) = \sin \dfrac{1}{1-z}$, $|z| < 1$. Show that $f(z)$ has an infinite number of zeros in the region $|z| < 1$. Does this contradict (*b*)?

6. Let $\varphi(x)$ be a real-valued function of the real variable x, $a \leqq x \leqq b$. Show that, if it is at all possible to continue $\varphi(x)$ analytically into the z plane, the continuation is unique. *Hint:* Assume $f(z)$ and $g(z)$ analytic for z real, $a \leqq z \leqq b$, $f(x) = \varphi(x) = g(x)$.

7. Use the results of Prob. 6 to show that

$$f(z) = \sum_{n=0}^{\infty} \frac{z^n}{n!}$$

is the only possible definition of e^z.

8. Consider $z^\alpha = e^{\alpha \operatorname{Ln} z}$. Define Ω as the operator of continuation by starting at z and returning to z, the path of continuation encircling the origin. Show that

$$\Omega z^\alpha = e^{2\pi i \alpha} z^\alpha$$

Hint: $\operatorname{Ln} z = \ln |z| + i\theta$. After encircling the origin $\operatorname{Ln} z = \ln |z| + i\theta + 2\pi i$.

9. Let $f(z)$ be analytic at $z = z_0$ so that $f(z) = \sum_{n=0}^{\infty} a_n(z - z_0)^n$ converges for $|z - z_0| < R$. Show that $f(z)$ has at least one singular point on the circle $z = z_0 + Re^{i\theta}$, $0 \leqq \theta \leqq 2\pi$. *Hint:* Assume $f(z)$ analytic at each point of the circle, obtain a circle of convergence at each point, apply the Heine-Borel theorem, and extend the radius of convergence, a contradiction.

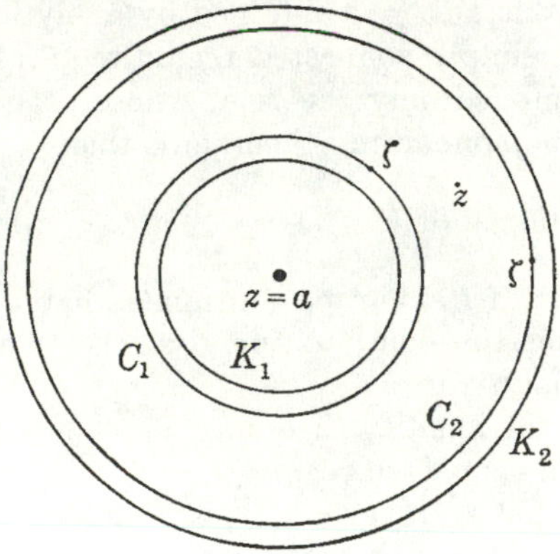

FIG. 4.14

4.12. Laurent's Expansion. A generalization of Taylor series is due to Laurent. Let $f(z)$ be analytic in an annular ring bounded by the two circles K_1, K_2 with common center $z = a$. Nothing is said of $f(z)$ inside K_1 or outside K_2 (see Fig. 4.14). Now let z be any point in the annular ring, and construct circles C_1 and C_2 with centers at $z = a$ such that z is interior to the annular ring lying between C_1 and C_2. Moreover C_1 and C_2 lie in the annular ring between K_1 and K_2. From Cauchy's integral formula and theorem

$$f(z) = \frac{1}{2\pi i} \oint_{C_2} \frac{f(\zeta)\, d\zeta}{\zeta - z} - \frac{1}{2\pi i} \oint_{C_1} \frac{f(\zeta)\, d\zeta}{\zeta - z} \tag{4.26}$$

This result can be obtained easily by using the method found in Sec. 4.8B. Now

$$\oint_{C_2} \frac{f(\zeta)\, d\zeta}{\zeta - z} = \oint_{C_2} \frac{f(\zeta)\, d\zeta}{(\zeta - a) - (z - a)} = \oint \frac{f(\zeta)\, d\zeta}{(\zeta - a)[1 - (z - a)/(\zeta - a)]}$$

Since $\left|\frac{z-a}{\zeta-a}\right| < r < 1$ for ζ on C_2, we write

$$\frac{1}{1-(z-a)/(\zeta-a)} = \sum_{n=0}^{\infty}\left(\frac{z-a}{\zeta-a}\right)^n$$

and interchange the order of integration and summation because of the uniform convergence of the series. This yields

$$\frac{1}{2\pi i}\oint_{C_2}\frac{f(\zeta)\,d\zeta}{\zeta-z} = \sum_{n=0}^{\infty} a_n(z-a)^n$$
$$a_n = \frac{1}{2\pi i}\oint_{C_2}\frac{f(\zeta)\,d\zeta}{(\zeta-a)^{n+1}} \qquad n = 0, 1, 2, \ldots \tag{4.27}$$

For the second integral

$$-\oint_{C_1}\frac{f(\zeta)\,d\zeta}{\zeta-z} = -\oint_{C_1}\frac{f(\zeta)\,d\zeta}{(\zeta-a)-(z-a)}$$
$$= \oint_{C_1}\frac{f(\zeta)\,d\zeta}{(z-a)[1-(\zeta-a)/(z-a)]}$$

Since $\left|\frac{\zeta-a}{z-a}\right| < s < 1$ for ζ on C_1, we write

$$\frac{1}{1-(\zeta-a)/(z-a)} = \sum_{n=0}^{\infty}\left(\frac{\zeta-a}{z-a}\right)^n$$

Interchanging integration and summation yields

$$-\frac{1}{2\pi i}\oint_{C_1}\frac{f(\zeta)\,d\zeta}{\zeta-z} = \sum_{n=1}^{\infty} a_n(z-a)^{-n}$$
$$a_n = \frac{1}{2\pi i}\oint_{C_1}(\zeta-a)^{n-1}f(\zeta)\,d\zeta \qquad n = 1, 2, 3, \ldots \tag{4.28}$$

Hence

$$f(z) = \sum_{n=0}^{\infty} a_n(z-a)^n + \sum_{n=1}^{\infty} a_{-n}(z-a)^{-n}$$
$$= S_1(z-a) + S_2(z-a)$$

where

$$S_1(z-a) = \sum_{n=0}^{\infty} a_n(z-a)^n \qquad \sum_{n=1}^{\infty} a_{-n}(z-a)^{-n} = S_2(z-a)$$

Since $(z-a)^{-(n+1)}f(z)$ and $(z-a)^{n-1}f(z)$ are analytic in the annular ring between K_1 and K_2, we can choose any path of integration to calculate the a_n, $n = 0, \pm1, \pm2, \ldots$, provided the path Γ encircles the point $z = a$

exactly once. Hence we can write

$$f(z) = \sum_{n=-\infty}^{\infty} a_n(z-a)^n$$
$$a_n = \frac{1}{2\pi i}\oint_\Gamma \frac{f(\zeta)\,d\zeta}{(\zeta - a)^{n+1}} \qquad n = 0, \pm 1, \pm 2, \ldots \tag{4.29}$$

Equation (4.29) is the Laurent expansion of $f(z)$ valid for all z in the annular ring between K_1 and K_2.

We leave it as an exercise for the reader to show that $S_1(z)$ converges for all z inside K_2 and that $S_2(z)$ converges for all z outside K_1. The common region of convergence is the above-mentioned annular ring.

Example 4.19. Consider $f(z) = 1/(z-1)(z-2)$. Certainly $f(z)$ is analytic for all z such that $1 < |z| < 2$. Let us find the Laurent expansion for $f(z)$ in this region. It is not necessary to find the a_n of (4.29). We wish to write $f(z)$ as the sum of two series, one series converging for $|z| < 2$, the other for $|z| > 1$. We write

$$\frac{1}{(z-1)(z-2)} = \frac{1}{z-2} - \frac{1}{z-1} = -\frac{1}{2}\frac{1}{1-z/2} - \frac{1}{z}\frac{1}{1-1/z}$$
$$= -\sum_{n=0}^{\infty}\frac{z^n}{2^{n+1}} - \sum_{n=1}^{\infty}\frac{1}{z^n}$$

The first series converges for $|z| < 2$ and the second for $|z| > 1$. Let us check this answer by using (4.29) to find the a_n.

$$a_n = \frac{1}{2\pi i}\oint_\Gamma \frac{d\zeta}{(\zeta-1)(\zeta-2)\zeta^{n+1}}$$

where Γ is any simple path enclosing the origin and lying between the circles $|z| = 1$, $|z| = 2$. We write, for $n \geqq 0$,

$$\frac{1}{(\zeta-1)(\zeta-2)\zeta^{n+1}} = \frac{A}{\zeta-1} + \frac{B}{\zeta-2} + \frac{C}{\zeta} + \frac{D}{\zeta^2} + \cdots + \frac{E}{\zeta^{n+1}} \tag{4.30}$$

Hence
$$\frac{1}{2\pi i}\oint \frac{d\zeta}{(\zeta-1)(\zeta-2)\zeta^{n+1}} = A + C \qquad \text{Why?}$$

To find A, multiply (4.30) by $\zeta - 1$, and then let $\zeta \to 1$. We obtain $-1 = A$. It is easy to see that $B = 1/2^{n+1}$. To find C, multiply (4.30) by ζ, and let $\zeta \to \infty$. We have $0 = A + B + C$ so that $C = 1 - 1/2^{n+1}$. Hence

$$a_n = \frac{1}{2\pi i}\oint \frac{d\zeta}{(\zeta-1)(\zeta-2)\zeta^{n+1}} = -\frac{1}{2^{n+1}} \qquad n \geqq 0$$

Let the reader show that $a_n = -1$ for $n < 0$.

4.13. Singular Points. We had previously stated that z_0 is a singular point of $f(z)$ if $f(z)$ is not analytic at z_0. If, moreover, $f(z)$ is analytic for some neighborhood of z_0 with the exception of z_0, we say that z_0 is an isolated singular point. In this case the Laurent expansion of $f(z)$ converges for $0 < |z - z_0| < r$, where r is the distance from z_0 to the nearest singular point of $f(z)$ other than z_0 itself. We distinguish now between 3 types of functions which have isolated singular points.

Case 1. The series $f(z) = \sum_{n=0}^{\infty} a_n(z - z_0)^n$ is such that $a_n = 0$, for all $n < 0$. We need only redefine $f(z)$ at z_0 to be a_0 and $f(z)$ becomes analytic at $z = z_0$. A singularity of this type is said to be a removable singularity. Thus $f(z) = \sum_{n=0}^{\infty} (z^n/n!)$ for $z \neq 0$, $f(0) = 2$ can be made analytic at $z = 0$ by defining $f(0) = \lim_{z \to 0} \sum_{n=0}^{\infty} (z^n/n!) = 1$.

Case 2. All but a finite number of the a_n, $n < 0$, vanish. In this case we say that z_0 is a pole of $f(z)$. We write

$$f(z) = \frac{a_{-n}}{(z - z_0)^n} + \cdots + \frac{a_{-1}}{z - z_0} + \sum_{k=0}^{\infty} a_k(z - z_0)^k \qquad a_{-n} \neq 0$$

and z_0 is said to be a pole of order n. If

$$f(z) = \frac{a_{-1}}{z - z_0} + \sum_{k=0}^{\infty} a_k(z - z_0)^k$$

$z = z_0$ is said to be a simple pole of $f(z)$.

The reader can easily verify that $|f(z)|$ becomes unbounded as $z \to z_0$ if z_0 is a pole. A pole is likewise called a nonessential singular point.

Case 3. An infinite number of the a_n, $n < 0$, do not vanish. In this case we say that z_0 is an essential singular point. For example,

$$e^{1/z} = 1 + \frac{1}{z} + \frac{1}{2!z^2} + \cdots + \frac{1}{n!z^n} + \cdots$$

has an essential singularity at $z = 0$. An important property of an essential singularity is Theorem 4.3, due to Picard, which we state without proof.

THEOREM 4.3. In any neighborhood of an essential singularity a single-valued function takes on every value, with one possible exception, an infinity of times.

Let us consider $e^{1/z}$ at $z = 0$ as an example. Are there an infinite number of z in any neighborhood of $z = 0$ for which $e^{1/z} = e^{50}$? The answer is "yes"! Since $e^{2\pi ni} = 1$, we have $e^{1/z+2\pi ni} = e^{50}$; so the equality holds if $1/z_n + 2\pi ni = 50$, or $z_n = 1/(50 - 2\pi ni)$. Let $n = -1, -2, -3$, etc., and we have an infinite sequence of z_n which tend to $z = 0$ as a limit and such that $e^{1/z_n} = e^{50}$. Let the reader show that the same applies to $e^{1/z} = i$. The exceptional value stated in Picard's theorem is zero. There is no z in any neighborhood of $z = 0$ for which $e^{1/z} = 0$. Let the reader show this.

We shall see in Chap. 5 that the point at infinity will play an important role in the development of differential equations in the complex domain. First we say that the point of infinity is an isolated singular point if $f(z)$ is analytic for all z such that $|z| > R > 0$. We then replace z in $f(z)$ by $1/t$ and investigate $f(1/t)$ at $t = 0$. The nature of $f(1/t)$ at $t = 0$ is defined to be the nature of $f(z)$ at $z = \infty$.

Example 4.20. (*a*) $f(z) = 1/z$, $f(1/t) = t$, so that, if we define $f(\infty) = 0$, $f(z)$ is analytic at $z = \infty$. (*b*) $f(z) = z - 1/z$, $f(1/t) = 1/t - t$, which has a simple pole at $t = 0$. We say, therefore, that $f(z)$ has a simple pole at infinity. (*c*) $f(z) = \sum_{n=0}^{\infty} \frac{z^n}{n!}$, $f(1/t) = \sum_{n=0}^{\infty} \frac{1}{n!t^n}$. Since $f(1/t)$ has an essential singularity at $t = 0$ we say that $f(z)$ has an essential singularity at $z = \infty$.

Problems

1. Find the Laurent expansion of $f(z) = 1/(z^2 + 1)(z - 2)$ for $1 < |z| < 2$.

2. Find the Laurent expansion of $f(z) = \dfrac{1}{z^2(z - 1)(z + 2)}$ for $0 < |z| < 1$; for $1 < |z| < 2$; for $|z| > 2$.

3. Show that the coefficients in the Laurent expansion (4.29) are unique.

4. Find all the roots of $e^z = i$.

5. Find a function $f(z)$ which has a simple pole at $z = 0$, $z = 1$, and $z = \infty$.

6. Define $\cosh z = (e^z + e^{-z})/2$. Show that

$$\cosh\left(z + \frac{1}{z}\right) = a_0 + \sum_{n=1}^{\infty} a_n\left(z^n + \frac{1}{z^n}\right)$$

$$a_n = \frac{1}{2\pi}\int_0^{2\pi} \cos n\theta \cosh (2 \cos \theta)\, d\theta$$

for $|z| > 0$.

7. Let $f(z)$ have a simple pole at $z = z_0$, $f(z) = \dfrac{a_{-1}}{z - z_0} + \sum_{n=0}^{\infty} a_n(z - z_0)^n$. We call a_{-1} the residue of $f(z)$ at $z = z_0$. Show that $a_{-1} = \lim_{z\to z_0} (z - z_0)f(z)$.

8. Let $f(z)$ have a pole of order k at $z = z_0$. Show that the residue a_{-1} is given by

$$a_{-1} = \frac{1}{(k-1)!}\frac{d^{k-1}}{dz^{k-1}}[(z - z_0)^k f(z)]\bigg|_{z=z_0}$$

9. Find the residues of $f(z) = z^4/(c^2 + z^2)^4$ at its poles.

10. Find the residue of $f(z) = 1 + z + z^2$ at $z = \infty$.

11. Let $f(z, t) = e^{(t/2)(z-1/z)}$. Show that the Laurent expansion of $f(z, t)$ for $|z| > 0$ is

$$f(z, t) = \sum_{n=-\infty}^{\infty} J_n(t) z^n$$

where $$J_n(t) = (-1)^n J_n(-t) = \frac{1}{2\pi} \int_0^{2\pi} \cos (n\theta - t \sin \theta) \, d\theta$$

12. Let $f(z)$ be analytic in the infinite strip given by $-a < \operatorname{Im} z < a$, $a > 0$. Assume $f(z) = f(z + 2\pi)$. By use of Laurent's theorem show that

$$f(z) = \sum_{n=-\infty}^{\infty} c_n e^{inz}$$

$$c_n = \frac{1}{2\pi} \int_0^{2\pi} f(z) e^{-inz} \, dz$$

13. Let $f(z)$ be an entire function (analytic everywhere) with a pole of order n at $z = \infty$. Obtain the Laurent expansion of $f(z)$ for $0 \leqq |z| < \infty$, and show that

$$f(z) = a_0 + a_1 z + \cdots + a_n z^n$$

14. Let $f(z)$ be analytic everywhere with the exception of a finite number of poles at $z_1, z_2, \ldots, z_k$ and a pole at $z = \infty$. The order of the pole at z_i is α_i, $i = 1, 2, \ldots, n$. Consider

$$\varphi(z) = (z - z_1)^{\alpha_1}(z - z_2)^{\alpha_2} \cdots (z - z_k)^{\alpha_k} f(z)$$

and show that $f(z)$ is a rational function, the quotient of two polynomials. The result of Prob. 13 is useful.

15. Let $p(z)$ have simple poles at the finite points $z = \xi, \eta, \zeta$. Also assume that $2z - z^2 p(z)$ is analytic at $z = \infty$. Consider

$$P(z) = (z - \xi)(z - \eta)(z - \zeta) p(z)$$

and show that $P(z)$ has a pole at most of the order 2 at $z = \infty$. Hence show that

$$p(z) = \frac{A}{z - \xi} + \frac{B}{z - \eta} + \frac{C}{z - \zeta}$$

where A, B, C are constants.

4.14. Residue Theorem. Contour Integration. Let $z = z_0$ be an isolated singular point of $f(z)$. For $f(z)$ single-valued and analytic in the region $0 < |z - z_0| < R$ we have the Laurent expansion

$$f(z) = \sum_{n=-\infty}^{\infty} a_n (z - z_0)^n$$

Let Γ be a simple closed path encircling $z = z_0$ lying in the region $0 < |z - z_0| < R$. Then

$$\oint f(z)\,dz = \oint_\Gamma \sum_{n=-\infty}^{\infty} a_n(z - z_0)^n\,dz$$
$$= \sum_{n=-\infty}^{\infty} a_n \oint_\Gamma (z - z_0)^n\,dz$$
$$= 2\pi i a_{-1}$$

since $\oint_\Gamma (z - z_0)^n\,dz = 0$ for $n \neq -1$, and $\oint_\Gamma \dfrac{dz}{z - z_0} = 2\pi i$. The interchange of integration and summation can be justified. We leave this as an exercise for the reader. The constant a_{-1} is called the residue of $f(z)$ at $z = z_0$.

Let the reader prove Theorem 4.4.

THEOREM 4.4. Let R be a simply connected open set, and let Γ be a simple closed path in R. Let $f(z)$ be single-valued and analytic in R with the exception of a finite number of isolated singular points. Then

$$\oint_\Gamma f(z)\,dz = 2\pi i \cdot (\text{sum of residues of } f(z) \text{ inside } \Gamma) \tag{4.31}$$

The residue theorem (4.31) is highly useful in the evaluation of real definite integrals. We discuss some examples.

Example 4.21. Let us evaluate $\int_{-\infty}^{\infty} \dfrac{dx}{1 + x^2}$. We deal naturally with

$$f(z) = \frac{1}{1 + z^2}$$

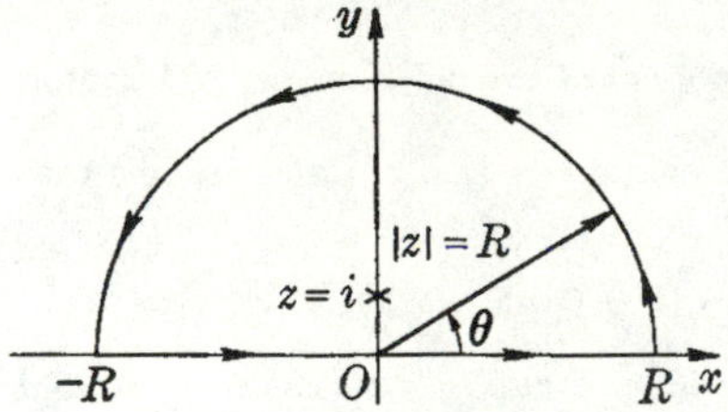

FIG. 4.15

Now $f(z)$ has a simple pole at $z = i$. To apply the residue theorem, we look for a path containing $z = i$ in its interior. At the same time we desire that part of the simple closed path be part of the real axis. We choose as Γ the straight-line segment extending from $x = -R$ to $x = R$ and the upper semicircle $|z| = R > 1$ (see Fig. 4.15). Since $f(z) = 1/(1 + z^2)$ has a simple pole at $z = i$, the residue of $f(z)$ at $z = i$ is $\lim_{z \to i} \dfrac{z - i}{1 + z^2} = \lim_{z \to i} \dfrac{z - i}{(z - i)(z + i)} = \dfrac{1}{2i}$ (see Prob. 7, Sec. 4.13). Hence

$$\int_{-R}^{R} \frac{dx}{1 + x^2} + \int_0^\pi \frac{Rie^{i\theta}\,d\theta}{1 + R^2 e^{2i\theta}} = 2\pi i \frac{1}{2i} = \pi \tag{4.32}$$

Now
$$\left| \int_0^\pi \frac{Rie^{i\theta}\,d\theta}{1 + R^2e^{2i\theta}} \right| \leqq R \int_0^\pi \frac{d\theta}{|1 + R^2e^{2i\theta}|} \leqq \frac{R\pi}{R^2 - 1}$$

since $|R^2e^{2i\theta} + 1| \geqq |R^2e^{2i\theta}| - 1 = R^2 - 1$, $R > 1$. If we allow R to become infinite, (4.32) becomes

$$\int_{-\infty}^{\infty} \frac{dx}{1 + x^2} = \pi$$

since $\lim_{R \to \infty} R\pi/(R^2 - 1) = 0$.

The method used above is known as contour integration. This method can be used to advantage to evaluate integrals of the type $\int_{-\infty}^{\infty} f(x)\,dx$. The choice of the closed contour of integration is not always apparent. One does not always pick a semicircle as part of the contour.

Example 4.22. Let us evaluate $\int_{-\infty}^{\infty} \frac{x \sin x}{x^2 + a^2}\,dx$, a positive. Instead of dealing with $(z \sin z)/(z^2 + a^2)$ we consider $f(z) = ze^{iz}/(z^2 + a^2)$. Remember that

$$e^{ix} = \cos x + i \sin x$$

which introduces the term $\sin x$. e^{iz} is much easier to handle than $\sin z$. As the contour of integration let us use that of Example 4.21 with $R > a$, since $f(z)$ has a simple pole at $z = ai$. We apply the residue theorem and obtain

$$\int_{-R}^{R} \frac{xe^{ix}}{x^2 + a^2}\,dx + \int_C \frac{ze^{iz}\,dz}{z^2 + a^2} = 2\pi i \left(\frac{aie^{i(ai)}}{2ai}\right) = i\pi e^{-a}$$

where C is the upper semicircle of Fig. 4.15.

On C, $z = Re^{i\theta}$, $dz = Rie^{i\theta}\,d\theta$, so that $\frac{ze^{iz}\,dz}{z^2 + a^2} = \frac{R^2 e^{iRe^{i\theta}} ie^{i\theta}\,d\theta}{R^2 + a^2}$. We see that it will be difficult to determine $\lim_{R \to \infty} \int_C \frac{ze^{iz}\,dz}{z^2 + a^2}$. We abandon the semicircle and choose as our contour of integration the rectangle with vertices at $(-R, 0)$, $(R, 0)$, $(R, R + Ri)$, $(-R, -R + Ri)$. Then

$$\int_{-R}^{R} \frac{xe^{ix}}{x^2 + a^2}\,dx + \int_0^R \frac{(R + iy)e^{i(R+iy)}i\,dy}{(R + iy)^2 + a^2} + \int_R^{-R} \frac{(x + Ri)e^{i(x+Ri)}\,dx}{(x + Ri)^2 + a^2}$$
$$+ \int_R^0 \frac{(-R + iy)e^{i(-R+iy)}i\,dy}{(-R + iy)^2 + a^2} = \pi ie^{-a}$$

Let the reader show that all integrals tend to zero as $R \to \infty$ except $\int_{-R}^{R} \frac{xe^{ix}\,dx}{x^2 + a^2}$ so that

$$\int_{-\infty}^{\infty} \frac{xe^{ix}\,dx}{x^2 + a^2} = \pi e^{-a} i$$

and

$$\int_{-\infty}^{\infty} \frac{x \sin x\,dx}{x^2 + a^2} = \pi e^{-a}$$

by equating the imaginary parts.

Example 4.23. As an illustration of the residue theorem consider

$$f(z) = (z - z_0)^n \sum_{k=0}^{\infty} a_k(z - z_0)^k \qquad a_0 \neq 0$$

We say that $z = z_0$ is an nth-order zero of $f(z)$. We can also write $f(z) = (z - z_0)^n \varphi(z)$, $\varphi(z_0) \neq 0$, $\varphi(z)$ analytic in a neighborhood of z_0. Let the reader show that there exists a neighborhood of z_0 such that $\varphi(z) \neq 0$ for all z in this neighborhood. Then

$$\begin{aligned} f'(z) &= n(z - z_0)^{n-1}\varphi(z) + (z - z_0)^n \varphi'(z) \\ &= (z - z_0)^{n-1}[n\varphi(z) + (z - z_0)\varphi'(z)] \end{aligned}$$

We see that $z = z_0$ is an $(n - 1)$st-order root of $f'(z) = 0$. Moreover

$$g(z) \equiv \frac{f'(z)}{f(z)} = \frac{n\varphi(z) + (z - z_0)\varphi'(z)}{(z - z_0)\varphi(z)} = \frac{n}{z - z_0} + \frac{\varphi'(z)}{\varphi(z)}$$

so that $z = z_0$ is a simple pole of $g(z)$, since $\varphi(z_0) \neq 0$. Hence for any simple closed path Γ surrounding z_0 for which $\varphi(z) \neq 0$ for all z inside Γ we have

$$\oint \frac{f'(z)}{f(z)}\,dz = 2\pi i n$$

Similarly, let $f(z) = \dfrac{a_{-m}}{(z - z_0)^m} + \cdots + \dfrac{a_{-1}}{z - z_0} + \sum_{k=0}^{\infty} a_k(z - z_0)^k$ so that

$$f(z) = \frac{\varphi(z)}{(z - z_0)^m}$$

$\varphi(z)$ analytic in a neighborhood of $z = z_0$, $\varphi(z_0) = a_{-m} \neq 0$. Let the reader show that $f'(z)/f(z) = -m/(z - z_0) + \varphi'(z)/\varphi(z)$. m is called the order of the pole at $z = z_0$. The reader should now be able to prove Theorem 4.5.

THEOREM 4.5. Let $f(z)$ be analytic in a simply connected set R with the possible exception of isolated singular points. Let C be a simple closed path in R enclosing a finite number of these isolated singular points which are poles, $f(z) \neq 0$ on C. Then

$$\frac{1}{2\pi i}\oint \frac{f'(z)}{f(z)}\,dz = N - P \tag{4.33}$$

where N is the number of zeros of $f(z)$ inside C, the order of each zero counted in determining N, and P is the number of poles, the order of each pole counted in determining P.

Problems

1. Show that $\displaystyle\int_{-\infty}^{\infty} \frac{x^2\,dx}{(x^2 + a^2)^3} = \frac{\pi}{8a^3}$, Rl $a > 0$.

2. Evaluate $\displaystyle\int_{-\infty}^{\infty} \frac{x^2\,dx}{x^4 + 5x^2 + 4}$.

3. Integrate e^{iz}/z around the rectangle with vertices at $z = \pm R, \pm R \pm Ri$ indented at the origin, and show that $\displaystyle\int_{-\infty}^{\infty} \frac{\sin x}{x}\,dx = \pi$, see Fig. 4.16.

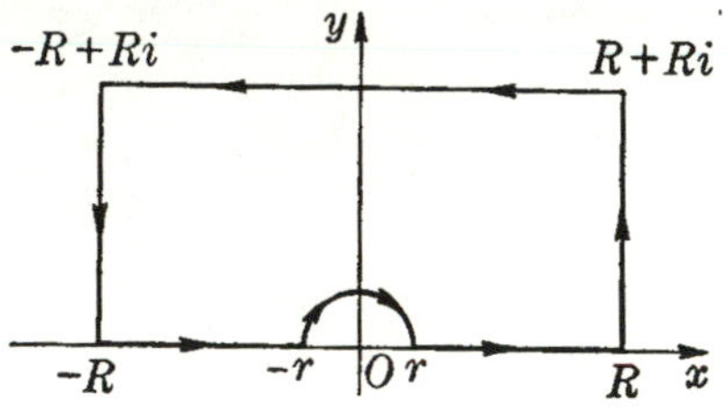

FIG. 4.16

4. Prove (4.33).

5. Let $f(z) = a_0 z^n + a_1 z^{n-1} + \cdots + a_n$, and let C be a circle $|z| = r$ such that $|f(z)| > 1$ for $|z| > r$. Show that $\dfrac{1}{2\pi i}\oint \dfrac{f'(z)}{f(z)}\,dz = n$. From (4.33) state a theorem concerning $f(z) = 0$.

6. Consider $z = e^{i\theta} = \cos\theta + i\sin\theta$. Show that $\cos\theta = \frac{1}{2}(z + 1/z)$, $\sin\theta = (1/2i)(z - 1/z)$.

Use this result to evaluate $\int_0^{2\pi} \frac{d\theta}{2 + \cos\theta}$

7. Integrate e^{-z^2} around the rectangle whose sides are $x = -R$, $x = R$, $y = 0$, $y = b > 0$, and show that

$$\int_{-\infty}^{\infty} e^{-x^2} \cos 2bx \, dx = e^{-b^2} \int_{-\infty}^{\infty} e^{-x^2} \, dx$$

Then show that

$$\int_{-\infty}^{\infty} e^{-x^2} \, dx = \left(\int_{-\infty}^{\infty} db \int_{-\infty}^{\infty} e^{-x^2} \cos 2bx \, dx \right)^{\frac{1}{2}} = \sqrt{\pi}$$

from (6.78).

8. For $0 < a < 2$ show that $\int_0^{\infty} \frac{x^{a-1}}{1 + x^2} \, dx = \pi \left(2 \sin \frac{\pi a}{2} \right)^{-1}$.

9. Show that $\int_0^{\infty} \frac{\ln (1 + x)^2}{1 + x^2} \, dx = \pi \ln 2$.

10. For $a > 0$ show that $\int_0^{\infty} \frac{dx}{x^4 + a^4} = \frac{\pi}{2\sqrt{2}\, a^3}$.

11. For $a > 0$, $b > 0$ show that

$$\int_{-\infty}^{\infty} \frac{dx}{(x^2 + a^2)(x^2 + b^2)^2} = \frac{\pi(a + 2b)}{2ab^3(a + b)^2}$$

12. For $a > b > 0$ show that

$$\int_{-\infty}^{\infty} \frac{\cos x \, dx}{(x^2 + a^2)(x^2 + b^2)} = \frac{\pi}{a^2 - b^2} \left(\frac{e^{-b}}{b} - \frac{e^{-a}}{a} \right)$$

4.15. The Schwarz-Christoffel Transformation. Let us consider a closed polygon in the w plane. We assume that the polygon does not intersect itself (see Fig. 4.17).

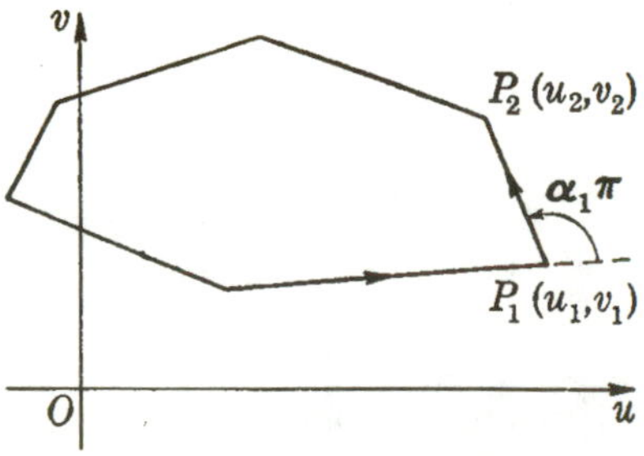

FIG. 4.17

We wish to find a function $z = F(w)$ or $w = f(z)$ which maps the polygon into the real axis of the z plane. Let P_1 be mapped into $(x_1, 0)$, P_2 into $(x_2, 0)$, etc. As we move along the polygon in the w plane, we move to the right along the x axis in the z plane. Now if such a transformation exists, then $dw = f'(z)\,dz$ and $\arg dw = \arg f'(z) + \arg dz$. Along the x axis, $\arg dz = \arg dx = 0$, so that

$$\arg dw = \arg f'(z)$$

along the polygon. Hence

$$\Delta \arg dw = \Delta \arg f'(z)$$

along the polygon, where Δ represents an abrupt change in the value of arg dw. This occurs, for example, at P_1. As we turn the corner at P_1, arg dw changes abruptly by an amount $\alpha_1\pi$, $-1 < \alpha < +1$. Now consider $f'(z) = (z - x_1)^{-\alpha_1}$, α_1 real, so that $\arg f'(z) = -\alpha_1 \arg (z - x_1)$. Thus

$$\begin{aligned}\Delta \arg f'(z) &= -\alpha_1 \Delta \arg (z - x_1)\\ &= -\alpha_1[\arg_{z>x_1} (z - x_1) - \arg_{z<x_1} (z - x_1)]\\ &= -\alpha_1(0 - \pi) = \pi\alpha_1\end{aligned}$$

This suggests that the transformation we are looking for satisfies

$$\frac{dw}{dz} = f'(z) = A(z - x_1)^{-\alpha_1}(z - x_2)^{-\alpha_2} \cdots (z - x_n)^{-\alpha_n} \tag{4.34}$$

The reader can easily verify that

$$\Delta \arg dw = \Delta \arg f'(z) = \pi\alpha_k \qquad k = 1, 2, \ldots, n \tag{4.35}$$

It can be shown rigorously that

$$w(z) = A \int_0^z (z - x_1)^{-\alpha_1}(z - x_2)^{-\alpha_2} \cdots (z - x_n)^{-\alpha_n}\, dz + B \tag{4.36}$$

is the required transformation. A and B are constants in the Schwarz-Christoffel transformation given by (4.36). It can also be shown that the interior of the polygon maps into the upper half plane of the z axis. Since $\sum_{i=1}^{n} \alpha_i = 2$, the reader can easily verify that $w(\infty)$ and $w(-\infty)$ exist. By integrating (4.34) around the closed path of Fig. 4.18, allowing the radii of the small semicircles to tend to zero, and allowing the radius of the large semicircle to become infinite, the reader can verify that $w(-\infty) = w(\infty)$, not neglecting the fact that $-1 < \alpha_i < +1$, $i = 1, 2, \ldots, n$.

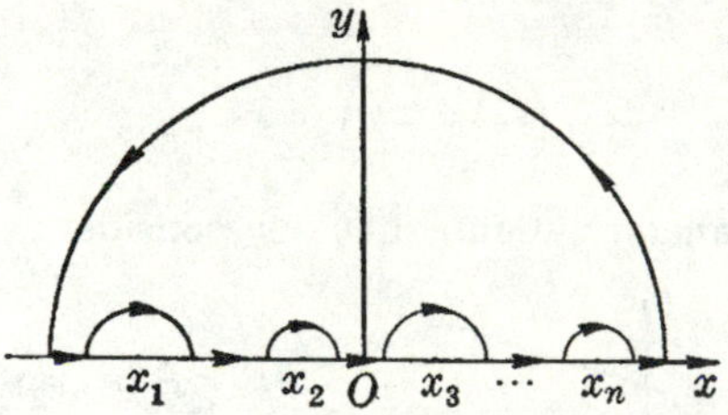

FIG. 4.18

If we desire that x_n be the point at infinity, we define $z = x_n - 1/\zeta$. When $z = x_n$, $\zeta = \infty$. Moreover $dz = (1/\zeta^2)\, d\zeta$ so that (4.36) becomes

$$\begin{aligned}w(z) &= A \int_0^z \left(x_n - x_1 - \frac{1}{\zeta}\right)^{-\alpha_1} \cdots \left(x_n - x_n - \frac{1}{\zeta}\right)^{-\alpha_n} \frac{d\zeta}{\zeta^2} + B\\ &= A_1 \int_0^z (\zeta - a_1)^{-\alpha_1}(\zeta - a_2)^{-\alpha_2} \cdots (\zeta - a_{n-1})^{-\alpha_{n-1}} \zeta^{\sum_{i=1}^{n} \alpha_i} \frac{d\zeta}{\zeta^2} + B\\ &= A_1 \int_0^z (z - a_1)^{-\alpha_1}(z - a_2)^{-\alpha_2} \cdots (z - a_{n-1})^{-\alpha_{n-1}}\, dz + B \qquad (4.37)\end{aligned}$$

since $\sum_{i=1}^{n} \alpha_i = 2$. The a_i, $i = 1, 2, \ldots, n-1$ are constants. Equation (4.37) is exactly of the form of (4.36) with the term $x_n = \infty$ omitted.

Example 4.24. We consider the polygon of Fig. 4.19.

Two of the vertices of the polygon are at $-\pi/2 + \infty i$, $\pi/2 + \infty i$. Let us map P into $(-a, 0)$ and Q into $(a, 0)$ of the z plane. At P, $\alpha_1 = \frac{1}{2}$, at Q, $\alpha_2 = \frac{1}{2}$, so that (4.36) can be written

$$w(z) = A \int_0^z (z+a)^{-\frac{1}{2}}(z-a)^{-\frac{1}{2}}\,dz + B$$
$$= A \int_0^z \frac{dz}{\sqrt{z^2 - a^2}} + B$$

The transformation $z = a \sin u$, $dz = a \cos u\, du$, $\sqrt{z^2 - a^2} = ai \cos u$, yields

$$w = \frac{A}{i} \sin^{-1} \frac{z}{a} + B$$

From the conditions $(z = -a, w = -\pi/2)$, $(z = a, w = \pi/2)$ we have

$$w = \sin^{-1} \frac{z}{a} \qquad z = a \sin w \qquad (4.38)$$

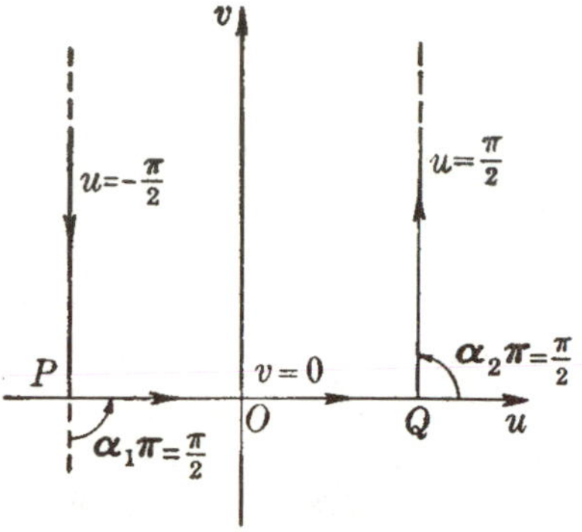

FIG. 4.19

If we let $w = \pi/2 + iR$, then $z = a \cos (iR) = a \cosh R$. As $R \to \infty$, $z \to \infty$. Similarly if we let $w = -\pi/2 + iR$, we see that $z \to \infty$ as $R \to \infty$. The transformation (4.38) unfolds the polygon. For $z = x + iy$, $w = u + iv$, (4.38) becomes

$$x + iy = a \sin (u + iv) = a \sin u \cosh v + ia \cos u \sinh v$$
$$x = a \sin u \cosh v$$
$$y = a \cos u \sinh v$$

Hence $x^2/(a^2 \sin^2 u) - y^2/(a^2 \cos^2 u) = 1$ so that the straight lines $u = u_0 \neq \left.\begin{matrix}0\\ \pi/2\end{matrix}\right\}$ map into hyperbolas in the z plane. Similarly the straight lines $v = v_0$ map into ellipses in the z plane.

Example 4.25. The Schwarz-Christoffel transformation is very useful in solving certain problems in two-dimensional electrostatic theory. First let us note that for an analytic function, $w = f(z) = u(z, y) + iv(x, y)$, we have $\nabla^2 u = 0$ and $\nabla^2 v = 0$, (see Prob. 1, Sec. 4.6). If $v(x, y)$ is the electrostatic potential, then $\nabla^2 v = 0$. The lines of force, $u(x, y) = \text{constant}$, are at right angles to the equipotential curves, $v(x, y) = \text{constant}$. But for the analytic function, $w = u + iv$, we know that the curves $u(x, y) = \text{constant}$ intersect the curves $v(x, y) = \text{constant}$ at right angles (see Prob. 1, Sec. 4.6). Thus, to solve a two-dimensional electrostatic problem, we need only find $w = f(z) = u + iv$ such that $v(x, y)$ satisfies the electrostatic boundary conditions. Once this is done, $\mathbf{E} = -\nabla v$ yields the electric field. Moreover

$$|\mathbf{E}| = \left|\left(\frac{\partial v}{\partial x}\right)^2 + \left(\frac{\partial v}{\partial y}\right)^2\right|^{\frac{1}{2}} = \left|\left(\frac{\partial u}{\partial x}\right)^2 + \left(\frac{\partial v}{\partial x}\right)^2\right| = \left|\frac{dw}{dz}\right|$$

Now let us consider an infinite line charge q whose projection in the xy plane is the origin. It is easy to show that $\nabla^2 u = 0$, $u = u(r)$, $r = (x^2 + y^2)^{\frac{1}{2}}$, has the solution

$u(r) = 2q \ln r$. If we consider $w = 2qi \ln z$, we have

$$u + iv = 2qi \ln (re^{i\theta}) = i2q \ln r - 2q\theta$$

so that $v = 2q \ln r$ is the required potential. If the charge were placed at z_0, then $w = 2qi \ln (z - z_0)$. A similar line charge, $-q$, placed at $\bar{z}_0$ yields

$$w = -2qi \ln (z - \bar{z}_0)$$

The field due to both charges can be obtained from

$$w = 2qi \ln (z - z_0) - 2qi \ln (z - \bar{z}_0)$$

or

$$w = 2qi \ln \frac{z - z_0}{z - \bar{z}_0} \tag{4.39}$$

If we consider a point on the x axis, $z = x$, then

$$w = u + iv = 2qi \ln \frac{x - z_0}{x - \bar{z}_0}$$

Let the reader show that $2qi \ln \dfrac{x - z_0}{x - \bar{z}_0}$ is real, so that $v = 0$ for the x axis. Hence (4.39) will yield the electric field for an infinite grounded plane (the x axis) due to an infinite line charge placed at z_0. Remember that the imaginary part of w of (4.39) satisfies Laplace's equation and satisfies the boundary condition $v = 0$ when $y = 0$.

Now we consider a more complicated example where we shall make use of the Schwarz-Christoffel transformation. Consider two semi-infinite grounded planes intersecting at an angle φ. We find $w(z)$ for an infinite line charge q placed at z_0 (see Fig. 4.20). First we find the transformation which maps the polygon AOB into the line $v(x, y) = 0$, which is the u axis. We need only apply the Schwarz-Christoffel transformation (4.36) with z and w interchanged. We map $z = 0$ into $w = 0$. α, at $z = 0$, is easily seen to be $(\pi - \varphi)/\pi = 1 - \varphi/\pi$, so that $dz = A_0 w^{\varphi/\pi - 1}\, dw$ and $z = Aw^{\varphi/\pi}$, $w = Bz^{\pi/\varphi}$.

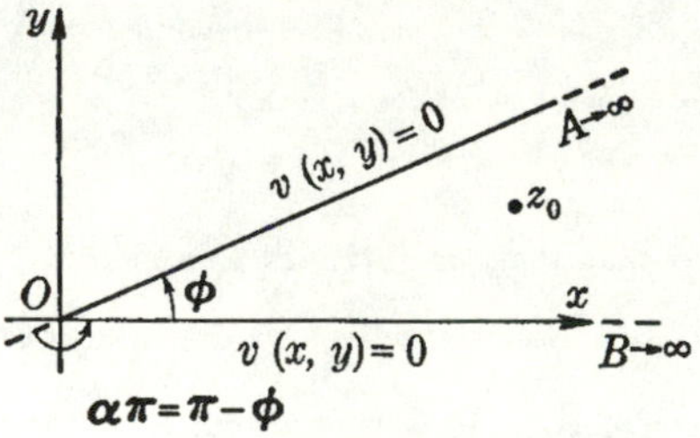

FIG. 4.20

The charge at z_0 maps into $w_0 = Bz_0^{\pi/\varphi}$. The complex function for an infinite grounded plane with charge at w_0 has been worked out above. It is

$$W = U + iV = 2qi \ln \frac{w - w_0}{w - \bar{w}_0}$$

so that

$$W = U + iV = 2qi \ln \frac{z^{\pi/\varphi} - z_0^{\pi/\varphi}}{z^{\pi/\varphi} - \bar{z}_0^{\pi/\varphi}} \tag{4.40}$$

$V(x, y)$ is the required potential function.

As a special case let $\varphi = \pi$, $z_0 = r_0 i$. Equation (4.40) becomes

$$U + iV = 2qi \ln \frac{z - r_0 i}{z + r_0 i}$$
$$= 2qi[\ln (z - r_0 i) - \ln (z + r_0 i)]$$
$$= 2qi[\ln |z - ir_0| + i \arg (z - ir_0)]$$
$$-2qi[\ln |z + ir_0| + i \arg |z + ir_0|]$$

Hence
$$V = 2q[\ln |z - ir_0| - \ln |z + ir_0|]$$
$$= q \ln \frac{x^2 + (y - r_0)^2}{x^2 + (y + r_0)^2}$$

For $y \equiv 0$, $V = q \ln 1 = 0$.

Problems

1. What electrostatic problem is solved by the transformation (4.38)?

2. Map the rectangle of Fig. 4.21 with two vertices at infinity into the real axis of the z plane. Show that $w = (a/\pi) \cosh^{-1} z$.

3. Find the electric field due to a charge q placed midway between two infinite grounded planes (see Fig. 4.22).

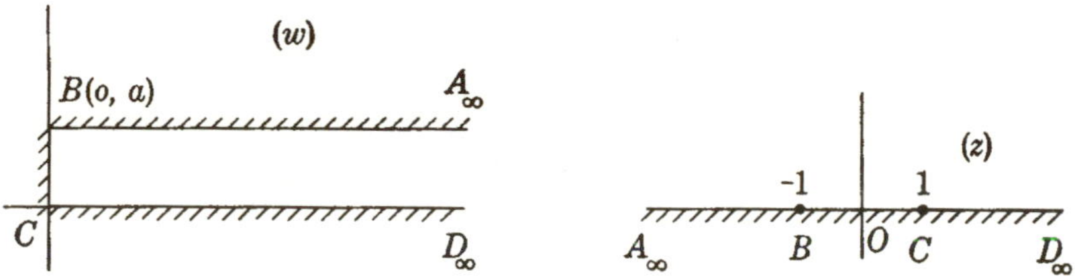

FIG. 4.21

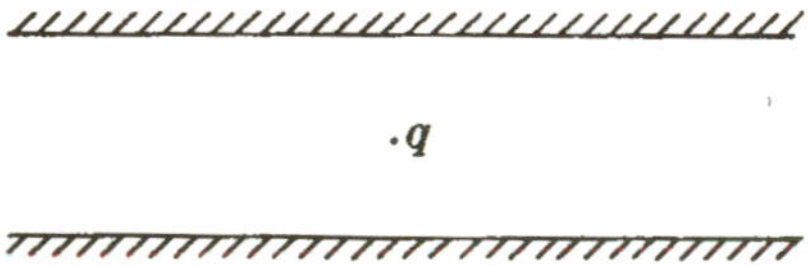

FIG. 4.22

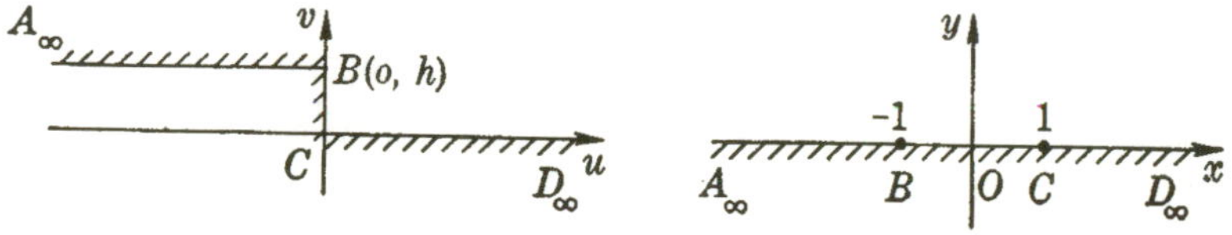

FIG. 4.23

4. Map the polygon of Fig. 4.23 into the real axis of the z plane, and show that $w = (h/\pi)(\sqrt{z^2 - 1} + \cosh^{-1} z)$.

5. Consider a closed curve C given parametrically by $x = f(t)$, $y = \varphi(t)$, $a \leqq t \leqq b$. Consider the transformation $z = x + iy = f(w) + i\varphi(w)$. Show that under this transformation the closed curve C of the z plane maps into the real axis of the w plane.

4.16. The Gamma Function. The Beta Function. It was Euler who first obtained a function of the real variable x which is continuous for x positive and which reduces to $n!$ when $x = n$, a positive integer. We consider

$$\Gamma(x) = \int_0^\infty e^{-t}t^{x-1}\,dt \tag{4.41}$$

The existence of the integral depends on the behavior of the integrand at $t = 0$ and at $t = \infty$. We first write

$$\Gamma(x) = \int_0^1 e^{-t}t^{x-1}\,dt + \int_1^\infty e^{-t}t^{x-1}\,dt$$

For $x > 0$ we have that $e^{-t}t^{x-1}$ behaves like t^{x-1} for t near zero. It is well known that $\int_0^1 t^{x-1}\,dt$ exists for $x > 0$ since

$$\lim_{\epsilon\to 0}\int_\epsilon^1 t^{x-1}\,dt = \lim_{\epsilon\to 0}\left(\frac{1}{x} - \frac{\epsilon^x}{x}\right) = \frac{1}{x} \qquad \begin{matrix}\epsilon > 0\\ x > 0\end{matrix}$$

To investigate the second integral, we note that

$$e^{-t}t^{x-1} = e^{-t/2}(e^{-t/2}t^{x-1}) \qquad \lim_{t\to\infty} e^{-t/2}t^{x-1} = 0$$

Hence for $t \geqq T$ there exists a constant A such that $|e^{-t}t^{x-1}| < Ae^{-t/2}$. Since $\int_T^\infty Ae^{-t/2}\,dt$ exists, of necessity $\int_1^\infty e^{-t}t^{x-1}\,dt$ exists. We now integrate (4.41) by parts and obtain

$$\begin{aligned}\Gamma(x) &= e^{-t}\frac{t^x}{x}\Big|_{t=0}^{t=\infty} + \int_0^\infty e^{-t}\frac{t^x}{x}\,dt \qquad x > 0\\ &= \frac{1}{x}\int_0^\infty e^{-t}t^x\,dt = \frac{1}{x}\Gamma(x+1)\end{aligned} \tag{4.42}$$

Thus $\Gamma(x + 1) = x\Gamma(x)$. If x is a positive integer, $x = n$, we have $\Gamma(n + 1) = n\Gamma(n) = n(n - 1)\Gamma(n - 1)$ and, by repeated applications, we obtain

$$\Gamma(n+1) = n(n-1)(n-2)\cdots 2\cdot 1\Gamma(1) = n! \tag{4.43}$$

since $\Gamma(1) = \int_0^\infty e^{-t}\,dt = 1$.

The gamma function of a complex variable z is defined by

$$\Gamma(z) = \int_0^\infty e^{-t}t^{z-1}\,dt \tag{4.44}$$

t^{z-1} is defined as $t^{z-1} = e^{\ln t^{z-1}} = e^{(z-1)\,\mathrm{Ln}\,t}$, where $\mathrm{Ln}\,t$ is the principal value of $\ln t$. By the same reasoning as above one deduces that $\Gamma(z)$ exists for $\mathrm{Rl}\,z > 0$, and $\Gamma(z + 1) = z\Gamma(z)$.

We can write

$$\begin{aligned}\Gamma(z) &= \int_0^1 e^{-t}t^{z-1}\,dt + \int_1^\infty e^{-t}t^{z-1}\,dt \\ &= \int_0^1 t^{z-1}\sum_{k=0}^{\infty}\frac{(-1)^k t^k}{k!}\,dt + \int_1^\infty e^{-t}t^{z-1}\,dt \\ &= \int_0^1 \sum_{k=0}^{\infty}\frac{(-1)^k t^{k+z-1}}{k!}\,dt + \int_1^\infty e^{-t}t^{z-1}\,dt\end{aligned}$$

Since $\sum_{k=0}^{\infty}\frac{(-1)^k t^{k+z-1}}{k!}$ converges uniformly on (0, 1) for Rl $z > 0$, we can interchange the order of integration and summation. Hence

$$\begin{aligned}\Gamma(z) &= \sum_{k=0}^{\infty}\int_0^1 \frac{(-1)^k t^{k+z-1}}{k!}\,dt + \int_1^\infty e^{-t}t^{z-1}\,dt \\ &= \sum_{k=0}^{\infty}\frac{(-1)^k}{k!(z+k)} + \int_1^\infty e^{-t}t^{z-1}\,dt \qquad (4.45)\end{aligned}$$

Now $\int_1^\infty e^{-t}t^{z-1}\,dt$ exists for all z, and $\sum_{k=0}^{\infty}\frac{(-1)^k}{k!(z+k)}$ converges for all z other than $z = 0, -1, -2, \ldots, -n, \ldots$. Hence $\Gamma(z)$ defined by (4.45) is analytic everywhere except at $z = -n$, $n = 0, 1, 2, \ldots$. At $z = -n$, $\Gamma(z)$ has a simple pole. $\Gamma(z)$ defined by (4.45) is the analytic continuation of $\Gamma(z)$ as defined by (4.44).

The function $\varphi(z) = \Gamma(z)\Gamma(-z)$ has simple poles at $z = 0, \pm 1, \pm 2, \pm 3, \ldots$ and is analytic elsewhere. The function $1/(\sin \pi z)$ has simple poles at $z = 0, \pm 1, \pm 2, \ldots$ and is analytic elsewhere. It can be shown that

$$\Gamma(z)\Gamma(1-z) = \frac{\pi}{\sin \pi z} \qquad (4.46)$$

Now the right-hand side of (4.46) never vanishes. Hence, if a z_0 exists such that $\Gamma(z_0) = 0$, then, of necessity, $\Gamma(1 - z_0)$ could not be finite so that $1 - z_0$ is a pole of $\Gamma(z)$. Hence $1 - z_0 = -n$ and $z_0 = 1 + n$, $\Gamma(z_0) = \Gamma(1 + n) = n! \neq 0$, a contradiction. Thus $\Gamma(z)$ has no zeros.

We state without proof Legendre's duplication formula,

$$\sqrt{\pi}\,\Gamma(2z) = 2^{2z-1}\Gamma(z)\Gamma(z + \tfrac{1}{2}) \qquad (4.47)$$

An important result is Stirling's approximation to $n!$ for n large. It is

$$n! \approx \sqrt{2\pi n}\left(\frac{n}{e}\right)^n \tag{4.48}$$

This result will be obtained in Sec. 10.25 by use of the Euler-Maclaurin sum formula.

The beta function defined by

$$B(p, q) = \int_0^1 t^{p-1}(1-t)^{q-1}\, dt \tag{4.49}$$

is closely related to the gamma function. The convergence of (4.49) exists if Rl $p > 0$ and Rl $q > 0$. We choose

$$t^{p-1} = e^{(p-1)\,\mathrm{Ln}\,t} \qquad (1-t)^{q-1} = e^{(q-1)\,\mathrm{Ln}\,(t-1)}$$

One can show with some labor that

$$B(p, q) = \frac{\Gamma(p)\Gamma(q)}{\Gamma(p+q)} \tag{4.50}$$

The substitution $u = (1-t)/t$ or $t = 1/(1+u)$ in (4.49) yields

$$B(p, q) = \int_0^\infty \frac{u^{p-1}\, du}{(1+u)^{p+q}}$$

Thus $$\Gamma(z)\Gamma(1-z) = \int_0^\infty \frac{u^{z-1}}{1+u}\, du \qquad 0 < \mathrm{Rl}\, z < 1$$

The substitution $u = e^t$ yields

$$\Gamma(z)\Gamma(1-z) = \int_{-\infty}^{\infty} \frac{e^{zt}}{1+e^t}\, dt$$

To evaluate $\int_{-\infty}^{\infty} \frac{e^{zt}}{1+e^t}\, dt$ one applies the theorem of residues to the closed rectangle C with vertices at $t = -R$, $t = S$, $t = S + 2\pi i$,

$$t = -R + 2\pi i$$

R and S real and positive, t the complex variable of integration. The pole of $e^{zt}/(1+e^t)$ inside this contour exists at $t = \pi i$. The residue of $e^{zt}/(1+e^t)$ at $t = \pi i$ is

$$\lim_{t\to\pi i} \frac{(t-\pi i)e^{zt}}{1+e^t} = e^{\pi i z} \lim_{t\to\pi i} \frac{t-\pi i}{1+e^t} = \frac{e^{\pi i z}}{e^{\pi i}} = -e^{\pi i z}$$

By letting R and S tend to infinity and using the fact that $0 < \mathrm{Rl}\, z < 1$ one can obtain $\Gamma(z)\Gamma(1-z) = \pi/\sin \pi z$ [see (4.46)]. By analytic continuation it follows that (4.46) holds throughout the domain of definition of $1/\sin \pi z$.

Problems

1. Show that $\Gamma(\frac{1}{2}) = \sqrt{\pi}$.
2. Prove that $(2n)! = 2^{2n} n! \Gamma(n + \frac{1}{2}) \pi^{-\frac{1}{2}}$.
3. From (4.45) show that $\lim_{z \to -n} (z + n)\Gamma(z) = (-1)^n/n!$, n a positive integer.
4. Show that

$$\int_0^{\pi/2} \sin^{2p-1} \theta \cos^{2q-1} \theta \, d\theta = \tfrac{1}{2} B(p, q)$$

for Rl $p > 0$, Rl $q > 0$.

5. Integrate $t^{z-1}e^{-t}$ around the complete boundary of a quadrant of a circle indented at the origin, and show that

$$\int_0^\infty t^{z-1} e^{-it} \, dt = e^{-\pi z i/2} \Gamma(z)$$

for $0 <$ Rl $z < 1$.

6. Show that

$$\int_0^\infty e^{-x^4} \, dx = \tfrac{1}{4} \Gamma(\tfrac{1}{4})$$

REFERENCES

Ahlfors, L. V.: "Complex Analysis," McGraw-Hill Book Company, Inc., New York, 1953.

Churchill, R. V.: "Introduction to Complex Variables and Applications," McGraw-Hill Book Company, Inc., New York, 1948.

Copson, E. T.: "Theory of Functions of a Complex Variable," Oxford University Press, New York, 1935.

Knopp, K.: "Theory of Functions," Dover Publications, New York, 1945.

MacRobert, T. M.: "Functions of a Complex Variable," St. Martin's Press, Inc., New York, 1938.

Phillips, E. G.: "Functions of a Complex Variable," Interscience Publishers, Inc., New York, 1943.

Titchmarsh, E. C.: "The Theory of Functions," Oxford University Press, New York, 1932.

Whittaker, E. T., and G. N. Watson: "A Course of Modern Analysis," The Macmillan Company, New York, 1944.

CHAPTER 5

DIFFERENTIAL EQUATIONS

5.1. General Remarks. A differential equation is any equation involving derivatives of a dependent variable with respect to one or more independent variables. Thus

$$\frac{dy}{dx} = y - x \tag{5.1}$$

$$\frac{d^2y}{dx^2} + 4\frac{dy}{dx} + 5y = \sin x \tag{5.2}$$

$$\frac{\partial^2\varphi}{\partial x^2} + \frac{\partial^2\varphi}{\partial y^2} = 0 \tag{5.3}$$

$$\left(\frac{dy}{dx} + y\right)^2 + x = e^y \tag{5.4}$$

are classified as differential equations. Equation (5.3) is called a partial differential equation for obvious reasons. The others are called ordinary differential equations. One may have a system of differential equations involving more than one dependent variable [see (5.5)],

$$\begin{aligned} \frac{d^2x}{dt^2} + \frac{dy}{dt} &= t \\ \frac{dx}{dt} - y &= 0 \end{aligned} \tag{5.5}$$

In addition to their own mathematical interest differential equations are particularly important since the scientist attempts to describe the behavior of certain aspects of the universe in terms of differential equations. We list a few of this type.

$$m\frac{d^2x}{dt^2} + k\frac{dx}{dt} + lx = f(x) \tag{5.6}$$

$$L\frac{d^2x}{dt^2} + R\frac{dx}{dt} + \frac{1}{C}x = A\cos\omega t \tag{5.7}$$

$$\frac{\partial^2 y}{\partial x^2} = \frac{1}{c^2}\frac{\partial^2 y}{\partial t^2} \tag{5.8}$$

$$\frac{d^2u}{dx^2} + \frac{8\pi^2 m}{h^2}[E - V(x)]u = 0 \tag{5.9}$$

$$R_{ij} = 0 \tag{5.10}$$

The order of the highest derivative occurring in a differential equation is called the order of the differential equation.

5.2. Solution of a Differential Equation. Initial and Boundary Conditions. Let a differential equation be given involving the dependent variable φ and the independent variables x and y. Any function $\varphi(x, y)$ which satisfies the differential equation is called a solution of the differential equation. For example, it is easy to prove that $\varphi(x, y) = e^x \sin y$ satisfies $\nabla^2\varphi = \frac{\partial^2\varphi}{\partial x^2} + \frac{\partial^2\varphi}{\partial y^2} = 0$. We say that $e^x \sin y$ is a solution of $\nabla^2\varphi = 0$. It is important to realize that a differential equation has, in general, infinitely many solutions. For example, $y'' = 0$ admits any function $y = ax + b$ as a solution, where a and b are constants which can be chosen arbitrarily. To specify a particular solution, either initial conditions like

$$y = 2,\ y' = 1 \qquad \text{when } x = 3$$

or boundary conditions like

$$y = 2 \text{ when } x = 3 \qquad \text{and} \qquad y = 4 \text{ when } x = -1$$

must be given in addition to the differential equation. In the first case $y = x - 1$ is the solution, and in the second case $y = -\frac{1}{2}x + \frac{7}{2}$ is the solution. A full study of a differential equation implies the determination of the most general solution of the equation (involving arbitrary elements which may or may not be constants) and a discussion of how many additional conditions must be imposed in order to fix uniquely the arbitrary elements entering in the general solution. Without attempting an exact statement or proof, at the moment, we state the fact that "in general" the most general solution of an ordinary differential equation of order n contains exactly n arbitrary constants to be uniquely determined by n initial conditions.

Problems

1. Verify that the following functions are solutions of the corresponding differential equations:

(a) $y = x^2 - e^x + 1$ $\qquad y''' + xy'' - y + \frac{y}{x}e^x = \frac{1}{x}e^x(1 - e^x)$

(b) $u = x^2 - y$ $\qquad \frac{\partial^2 u}{\partial x^2} + x\frac{\partial u}{\partial x}\frac{\partial u}{\partial y} = 2(1 - x^2)$

(c) $y + x + 1 = 0$ $\qquad (y - x)y' - (y^2 - x^2) = 0$

2. Verify that $y = ae^x + be^{2x}$ is a solution of the differential equation

$$y'' - 3y' + 2y = 0$$

Determine the particular solution which satisfies the initial conditions $y = 0$, $y' = 3$ for $x = 0$.

3. Find the solution of $y' + 3xe^x = 0$ which satisfies the initial condition $y = 1$ when $x = 1$.

5.3. The Differential Equation of a Family of Curves. The family of straight lines in the plane is characterized by the equation $y = ax + b$, where a and b are arbitrary constants. To fix these constants means to select one member of the family, that is, to fix our attention on one straight line among all the others. The differential equation $y'' = 0$ is called the differential equation of this family of straight lines because every function of the form $y = ax + b$ satisfies this equation and, conversely, every solution of $y'' = 0$ is a member of the family $y = ax + b$. The differential equation $y'' = 0$ characterizes the family as a whole without specific reference to the particular members.

More generally, a family of curves can be described by

$$y = f(x, a_1, a_2, \ldots, a_n)$$

or implicitly by $F(x, y, a_1, a_2, \ldots, a_n) = 0$, in which n arbitrary constants appear. The differential equation of the family is obtained by successively differentiating n times and eliminating the constants between the resulting $n + 1$ relations. The differential equation that results is of order n.

Example 5.1. To find the differential equation of the family of parabolas

$$y = ax + bx^2$$

we differentiate twice to obtain

$$y' = a + 2bx$$
$$y'' = 2b$$

The last equation is solved for b, and the result is substituted into the previous equation. This equation is solved for a, and the expressions for a and b are substituted into $y = ax + bx^2$. The result is the differential equation

$$y = xy' - \tfrac{1}{2}x^2y''$$

The elimination of the constants a and b can also be obtained by considering the equations

$$\begin{aligned} xa + x^2b + (-y)1 &= 0 \\ a + 2xb + (-y')1 &= 0 \\ 2b + (-y'')1 &= 0 \end{aligned}$$

as a system of homogeneous linear equations in a, b, 1. The solution $(a, b, 1)$ is nontrivial, and hence the determinant of the coefficients vanishes.

$$\begin{vmatrix} x & x^2 & -y \\ 1 & 2x & -y' \\ 0 & 2 & -y'' \end{vmatrix} = 0$$

Expansion about the third column yields the result above.

Problems

Find the differential equations whose solutions are the families

1. $y = C_1e^x + C_2e^{-x}$

2. $y = C_1 \cos 2x + C_2 \sin 2x$

3. $y = e^{2x}(C_1 + C_2x)$

4. Find the differential equation of all circles which have their centers on the y axis.

5. Find the differential equation of all circles in the plane.

6. Find the differential equation of all parabolas whose principal axes are parallel to the x axis.

7. Find the differential equation of all straight lines whose intercepts total 1.

5.4. Ordinary Differential Equations of the First Order and First Degree. Let y be the dependent variable, and let x be the independent variable. The most general equation of the first order is any equation involving x, y, and y'. If y' enters in the equation only linearly, that is, only in the first power, the equation is said to be of the first degree. Such an equation can be written in the form

$$\frac{dy}{dx} = f(x, y) = -\frac{M(x, y)}{N(x, y)} \tag{5.11}$$

We shall see later that under suitable restrictions on $f(x, y)$ there always exists a unique solution $y = \varphi(x)$, such that $y_0 = \varphi(x_0)$ and

$$\frac{d\varphi(x)}{dx} = f(x, \varphi(x))$$

The discussion of (5.11) will be restricted to the simplest cases at present:

(*a*) $f(x, y) = \alpha(x)\beta(y)$.

(*b*) $f(x, y)$ is homogeneous of degree zero, that is, $f(tx, ty) = t^n f(x, y)$, $n = 0$.

(*c*) Exact equations.

(*d*) Integrating factors.

(*e*) $f(x, y) = -p(x)y + q(x)$.

Case (a): Separation of Variables. If

$$\frac{dy}{dx} = -\frac{M(x)}{N(y)} \tag{5.12}$$

then $M(x)\,dx + N(y)\,dy = 0$. Consider

$$F(x, y) = \int^x M(x)\,dx + \int^y N(y)\,dy = \text{constant} \tag{5.13}$$

The integrals in (5.13) are indefinite (no lower limit). We wish to show that y as a function of x given implicitly by (5.13) satisfies (5.12). We have

$$\frac{\partial F}{\partial x} + \frac{\partial F}{\partial y}\frac{dy}{dx} = 0 \qquad \frac{dy}{dx} = -\frac{\partial F/\partial x}{\partial F/\partial y} \qquad \frac{\partial F}{\partial y} \neq 0$$

But $\dfrac{\partial F}{\partial x} = M(x)$, $\dfrac{\partial F}{\partial y} = N(y)$, so that $\dfrac{dy}{dx} = -\dfrac{M(x)}{N(y)}$. Q.E.D. Conversely, let $y = y(x)$ be any solution of (5.12), so that

$$\frac{dy(x)}{dx} = -\frac{M(x)}{N(y(x))} \tag{5.14}$$

Consider $F(x, y(x)) = \int^x M(x)\,dx + \int^{y(x)} N(y)\,dy$. We have

$$\frac{dF}{dx} = M(x) + N(y(x))\,\frac{dy(x)}{dx} \equiv 0$$

from (5.14) so that $F(x, y(x)) =$ constant. Hence any solution of (5.12) can be obtained from (5.13) by solving implicitly for y in terms of x.

Example 5.2. We solve $y\,dx + (1 + x^2)\tan^{-1} x\,dy = 0$. Separating the variables, we have

$$\frac{dx}{(1 + x^2)\tan^{-1} x} + \frac{dy}{y} = 0$$

Integration yields $\ln \tan^{-1} x + \ln y = C_1 = \ln C_2$ so that $y \tan^{-1} x = C_2$ or

$$y = C_2(\tan^{-1} x)^{-1}$$

Problems

Solve:

1. $dx - \sqrt{a^2 - x^2}\,dy = 0$

2. $y^2 \cos \sqrt{x}\,dx + 2\sqrt{x}\,e^{1/y}\,dy = 0$

3. $x(y^2 - 1)\,dx + y(x^2 + 1)\,dy = 0$

4. $xy\,\dfrac{dy}{dx} = 1 + y^2$

5. For what curves is the portion of the tangent between the axes bisected at the point of contact?

6. Find the general equation of all curves for which the tangent makes a constant angle φ with the radius vector.

7. Find the function which is equal to zero when $x = 1$, and to 1 when $x = 4$, and whose rate of change is inversely proportional to its value.

8. Find $r(x)$ if $r\,\dfrac{dr}{dx} = \sqrt{a^4 - r^4}$.

9. Find the family of curves intersecting the family of parabolas $y^2 = 4px$ at right angles for all p. These curves are called the orthogonal trajectories of the system of parabolas.

10. The area bounded by the x axis, the arc of a curve, a fixed ordinate, and a variable ordinate is proportional to the arc between these ordinates. Find the equation of the curve.

11. Assume that a drop (sphere) of liquid evaporates at a rate proportional to its area of surface. Find the radius of the drop as a function of the time.

12. Brine containing 2 lb of salt per gallon runs at the rate of 3 gpm into a 10-gal tank initially filled with fresh water. The mixture is stirred uniformly and flows out at the same rate. Find the amount of the salt in the tank at the end of 1 hr.

13. It begins to snow some time before noon and continues to snow at a constant rate throughout the day. At noon a machine begins to shovel at a constant rate. By 1400 two blocks of snow have been cleared, and by 1600 one more block of snow is cleared. What time before noon did it begin to snow? Assume width of street a constant.

Case (*b*). We say that $f(x, y)$ is homogeneous in x and y of degree n if $f(tx, ty) = t^n f(x, y)$. For example, $f(x, y) = x^2 + y^2$ is homogeneous of degree 2 since $f(tx, ty) = t^2x^2 + t^2y^2 = t^2(x^2 + y^2) = t^2 f(x, y)$. We now consider

$$\frac{dy}{dx} = f(x, y) \tag{5.15}$$

where $f(tx, ty) = t^n f(x, y)$. We let $y = tx$, so that $\dfrac{dy}{dx} = t + x\dfrac{dt}{dx}$. Moreover $f(x, tx) = x^n f(1, t)$ so that (5.15) becomes

$$t + x\frac{dt}{dx} = x^n f(1, t) \tag{5.16}$$

If $n = 0$, that is, if $f(x, y)$ is homogeneous of degree zero, then

$$\frac{dt}{f(1, t) - t} = \frac{dx}{x} \qquad f(1, t) - t \neq 0 \tag{5.17}$$

and the variables have been separated. We can solve for $t = t(x)$, and $y = xt(x)$ satisfies (5.15). In particular, if

$$\frac{dy}{dx} = \frac{M(x, y)}{N(x, y)}$$

and $M(x, y)$, $N(x, y)$ are homogeneous of the same degree, then M/N is homogeneous of degree zero, so that the substitution $y = tx$ yields an equation in t and x with variables separable.

Example 5.3. Consider

$$\frac{dy}{dx} = \frac{x + y}{x - y}$$

$f(x, y) = (x + y)/(x - y)$ is homogeneous of degree zero. Let $y = tx$, so that

$$t + x\frac{dt}{dx} = \frac{x + tx}{x - tx} = \frac{1 + t}{1 - t}$$
$$\frac{1 - t}{1 + t^2}\,dt = \frac{dx}{x}$$
$$\tan^{-1} t - \tfrac{1}{2}\ln(1 + t^2) = \ln x + \ln C$$
$$\tan^{-1}\frac{y}{x} - \tfrac{1}{2}\ln\left(1 + \frac{y^2}{x^2}\right) = \ln Cx$$

Problems

1. $\dfrac{dy}{dx} = \dfrac{y^2}{x^2 - xy}$
2. $\dfrac{dy}{dx} = \dfrac{y + \sqrt{x^2 + y^2}}{x}$
3. $\dfrac{dy}{dx} = \dfrac{x + y}{x}$

4. $\dfrac{dy}{dx} = \dfrac{x + y + 1}{x - y + 2}$. *Hint:* Let $x = \bar{x} + a$, $y = \bar{y} + b$, and find a and b so that

$$\frac{d\bar{y}}{d\bar{x}} = \frac{\bar{x} + \bar{y}}{\bar{x} - \bar{y}}$$

5. Discuss (5.15) if $f(1, t) = t$ in (5.17).

Case (c): Exact Equations. We consider

$$\frac{dy}{dx} = -\frac{M(x, y)}{N(x, y)}$$

or

$$M(x, y)\,dx + N(x, y)\,dy = 0 \tag{5.18}$$

We say that Eq. (5.18) is exact if there exists a function $\varphi(x, y)$ such that

$$d\varphi(x, y) = M(x, y)\,dx + N(x, y)\,dy \tag{5.19}$$

If (5.18) is exact, then $d\varphi = 0$ and $\varphi(x, y) =$ constant is a solution of (5.18). From the calculus

$$d\varphi = \frac{\partial\varphi}{\partial x}\,dx + \frac{\partial\varphi}{\partial y}\,dy \tag{5.20}$$

so that for (5.18) to be exact we must have

$$\begin{aligned} \frac{\partial\varphi}{\partial x} &= M(x, y) \\ \frac{\partial\varphi}{\partial y} &= N(x, y) \end{aligned} \tag{5.21}$$

Further differentiation yields

$$\frac{\partial^2\varphi}{\partial y\,\partial x} = \frac{\partial M}{\partial y} \qquad \frac{\partial^2\varphi}{\partial x\,\partial y} = \frac{\partial N}{\partial x} \tag{5.22}$$

If we assume continuity of the second mixed partials, of necessity

$$\frac{\partial M}{\partial y} = \frac{\partial N}{\partial x} \tag{5.23}$$

Equation (5.23) is a necessary condition that (5.18) be exact.

Conversely, assume $\dfrac{\partial M}{\partial y} = \dfrac{\partial N}{\partial x}$. We show that a function $\varphi(x, y)$ exists such that $d\varphi = M\,dx + N\,dy$. Consider

$$\varphi(x, y) = \int_{x_0}^{x} M(x, y)\,dx + \int_{y_0}^{y} N(x_0, y)\,dy \tag{5.24}$$

x_0 and y_0 are arbitrary constants. Then

$$\frac{\partial \varphi}{\partial x} = M(x, y)$$

$$\frac{\partial \varphi}{\partial y} = \int_{x_0}^{x} \frac{\partial M}{\partial y}\, dx + N(x_0, y)$$

$$= \int_{x_0}^{x} \frac{\partial N}{\partial x}\, dx + N(x_0, y)$$

$$= N(x, y) - N(x_0, y) + N(x_0, y)$$

$$= N(x, y)$$

Hence (5.24) yields the required function $\varphi(x, y)$.

If this seems familiar, there is a good reason, since the same material was covered in Chap. 2, Vector Analysis. Let $\mathbf{f} = M(x, y)\mathbf{i} + N(x, y)\mathbf{j}$. Then $\mathbf{f} \cdot d\mathbf{r} = M\, dx + N\, dy$. If $\mathbf{f} = \nabla\varphi$, then $d\varphi = M\, dx + N\, dy$. But $\mathbf{f} = \nabla\varphi$ implies $\nabla \times \nabla\varphi = \mathbf{0}$, which is statement (5.23).

Example 5.4. Consider

$$(\ln x + \ln y + 1)\, dx + \frac{x}{y}\, dy = 0 \tag{5.25}$$

$M = \ln x + \ln y + 1$, $\dfrac{\partial M}{\partial y} = \dfrac{1}{y}$, $N = \dfrac{x}{y}$, $\dfrac{\partial N}{\partial x} = \dfrac{1}{y} = \dfrac{\partial M}{\partial y}$, so that (5.25) is exact. Applying (5.24) yields

$$\varphi(x, y) = \int_1^x (\ln x + \ln y + 1)\, dx + \int_1^y \frac{1}{y}\, dy = \text{constant}$$

Integration yields

$$\varphi(x, y) = x \ln x \Big|_1^x - x \Big|_1^x + x \ln y \Big|_{x=1}^{x=x} + x \Big|_1^x + \ln y \Big|_{y=1}^{y=y} = c$$

or

$$x \ln x - x + 1 + x \ln y - \ln y + x - 1 + \ln y = c$$

$$x \ln (xy) = c$$

We chose the lower limits to be 1 rather than zero since $\ln x$ is not defined at $x = 0$. The reader can check easily that

$$d[x \ln (xy)] = (\ln x + \ln y + 1)\, dx + \frac{x}{y}\, dy$$

Problems

1. $(xe^x + 2 \cos y)\, dx - 2x \sin y\, dy = 0$
2. $\dfrac{y}{(x + y)^2}\, dx - \dfrac{x\, dy}{(x + y)^2} = 0$
3. $\sin y\, e^{x \sin y}\, dx + (xe^{x \sin y} \cos y - 2 \cos y \sin y)\, dy = 0$

Case (d): Integrating Factors. It may readily turn out that (5.18) is not exact, that is, $\dfrac{\partial M}{\partial y} \neq \dfrac{\partial N}{\partial x}$. Let us assume that this is so. Perhaps one can find a function $\mu(x, y)$ such that

$$\mu M\, dx + \mu N\, dy = 0 \tag{5.26}$$

is now exact. We call $\mu(x, y)$ an integrating factor of (5.18). The solution of (5.26) is also a solution of (5.18) since in both cases $\frac{dy}{dx} = -\frac{M}{N}$. In general it is very difficult to find an integrating factor. There are two cases, however, for which an integrating factor can be found. Let us determine the condition on M and N such that $\mu = \mu(y)$ is an integrating factor. Of necessity

$$\frac{\partial}{\partial y}(\mu(y)M) = \frac{\partial}{\partial x}(\mu(y)N)$$

and

$$\frac{d\mu}{dy}M + \mu\frac{\partial M}{\partial y} = \mu\frac{\partial N}{\partial x}$$

so that

$$\frac{1}{\mu}\frac{d\mu}{dy} = \frac{\partial N/\partial x - \partial M/\partial y}{M} \tag{5.27}$$

If $\dfrac{\partial N/\partial x - \partial M/\partial y}{M}$ is a function of y only, then (5.27) can be solved for y, for in this case

$$\mu(y) = \exp\int\frac{\dfrac{\partial N}{\partial x} - \dfrac{\partial M}{\partial y}}{M}\,dy$$

Example 5.5. Consider

$$-y\,dx + x\,dy = 0 \tag{5.28}$$

We have $\frac{\partial M}{\partial y} = -1$, $\frac{\partial N}{\partial x} = 1$, so that (5.28) is not exact. However,

$$\frac{\partial N/\partial x - \partial M/\partial y}{M} = -\frac{2}{y}$$

so that an integrating factor exists, namely,

$$\mu(y) = \exp\left(-\int\frac{2}{y}\,dy\right) = e^{-2\ln y} = e^{\ln y^{-2}} = \frac{1}{y^2}$$

Multiplying (5.28) by $1/y^2$ yields

$$-\frac{1}{y}\,dx + \frac{x}{y^2}\,dy = 0 \tag{5.29}$$

The reader can easily verify that (5.29) is exact. The solution of (5.29) is x/y = constant, which is also a solution of (5.28).

For an integrating factor of the type $\mu(x)$, one needs of necessity

$$\frac{\partial M/\partial y - \partial N/\partial x}{N} \equiv \varphi(x) \tag{5.30}$$

Problems

1. $2xy\,dx + (y^2 - 3x^2)\,dy = 0$
2. $(x - xy^2)\,dx + x^2y\,dy = 0$

3. Prove the statement concerning (5.30).

4. $\left(3 - \frac{y}{x^2}\right) dx + \left(\frac{2y}{x^2} - \frac{1}{x}\right) dy = 0$

Case (e): Linear Equations of the First Order. The equation

$$\frac{dy}{dx} + p(x)y = q(x) \tag{5.31}$$

is called a linear differential equation of the first order. The expression "linear" refers to the fact that both y and y' occur linearly in (5.31). The functions $p(x)$ and $q(x)$ need not be linear in x. If $q(x) \equiv 0$, (5.31) is said to be homogeneous; otherwise it is inhomogeneous.

The solution of the homogeneous equation

$$\frac{dy}{dx} + p(x)y = 0$$

is trivial since the equation is separable. Hence

$$y = A \exp\left[-\int p(x)\, dx\right] \tag{5.32}$$

where A is a constant of integration, $\int p(x)\, dx$ is an indefinite integral. Now we attempt to use (5.32) in order to find a solution of (5.31). We introduce a new variable $z(x)$ by the equation

$$y = z \exp\left[-\int p(x)\, dx\right]$$

or

$$z = y \exp \int p(x)\, dx$$

where $y(x)$ is a solution of (5.31). Then

$$\frac{dz}{dx} = \frac{dy}{dx} \exp \int p(x)\, dx + yp \exp \int p(x)\, dx$$

But $\frac{dy}{dx} = q(x) - p(x)y$ from (5.31) so that

$$\begin{aligned} \frac{dz}{dx} &= (q - py) \exp \int p\, dx + py \exp \int p(x)\, dx \\ &= q(x) \exp \int p(x)\, dx \end{aligned}$$

Hence

$$z(x) = \int q(x) \left(\exp \int p(x)\, dx\right) dx + C$$

so that

$$y(x) = \exp\left[-\int p(x)\, dx\right]\left[\int q(x)\left(\exp \int p(x)\, dx\right) dx + C\right] \tag{5.33}$$

The reader can easily show that (5.33) is a solution of (5.31).

Example 5.6. Consider

$$\frac{dy}{dx} + y \cot x = x^2 \csc x$$

Here $p(x) = \cot x$, $q(x) = x^2 \csc x$ so that (5.33) becomes

$$\begin{aligned} y(x) &= \exp\left(-\int \cot x\, dx\right)\left[\int x^2 \csc x \left(\exp \int \cot x\, dx\right) dx + C\right] \\ &= e^{-\ln \sin x}\left(\int x^2 \csc x e^{\ln \sin x}\, dx + C\right) \\ &= \frac{1}{\sin x}\left(\int x^2\, dx + C\right) \\ &= \frac{1}{\sin x}\left(\frac{x^3}{3} + C\right) \end{aligned}$$

Example 5.7. We now investigate (5.33) in more detail. We have

$$\begin{aligned} y(x) &= C \exp\,[-\textstyle\int p(x)\, dx] + \exp\,[-\textstyle\int p(x)\, dx][\textstyle\int q(x)\,(\exp \textstyle\int p(x)\, dx)\, dx] \\ &= y_1(x) + y_2(x) \end{aligned}$$

We notice that $y_1(x) = C \exp\,[-\int p(x)\, dx]$ is the general solution of the homogeneous equation $\frac{dy}{dx} + p(x)y = 0$. Moreover $y_2(x)$ is a particular solution of (5.31) obtained from (5.33) by choosing $C = 0$ In other words we have Theorem 5.1.

THEOREM 5.1. The most general solution of the inhomogeneous equation is the sum of a particular solution plus the general solution of the corresponding homogeneous equation.

This important result is valid also for linear equations of higher order and is discussed later. In simple cases a particular solution can be found by inspection. For example, consider

$$\frac{dy}{dx} + y = x \tag{5.34}$$

We look for a particular solution of the form $y = ax + b$. Then

$$a + ax + b = x$$

holds if $a + b = 0$, $a = 1$ so that $b = -1$. Hence $y = x - 1$ is a particular solution of (5.34). This method of obtaining a particular solution is called the method of undetermined coefficients. The general solution of the corresponding homogeneous equation, $\frac{dy}{dx} + y = 0$, is $y = ce^{-x}$, so that

$$y = ce^{-x} + x - 1$$

is the general solution of (5.34).

Problems

1. $\frac{dy}{dx} + xy = x$
2. $\frac{dx}{dy} - y^2x = 1 - y^3$

3. $x\frac{dy}{dx} - y - x\cos x = 0$

4. Change $\frac{dy}{dx} + p(x)y = q(x)y^n$ to a linear equation by the substitution $z = y^{1-n}$, $n \neq 1$. What if $n = 1$?

Review Problems

1. $\frac{dy}{dx} + \frac{2x}{x^2+1}y = x$

2. A body falls because of gravity. There is a retarding force of friction proportional to the speed of the body, say, kv. If the body starts from rest, find the distance fallen as a function of the time.

3. $(x + y)^2\,dx + x^2\,dy = 0$

4. $[\sin(xy) + xy\cos(xy)]\,dx + x^2\cos(xy)\,dy = 0$

5. $e^{-x^2y}(2xy - 2x^3y^2)\,dx + e^{-x^2y}(3x^2 - x^4y)\,dy = 0$

6. Brine containing 2 lb of salt per gallon runs at the rate of 3 gpm into a 10-gal tank initially filled and containing 15 lb of salt. The mixture is stirred uniformly and flows out at the same rate into another 10-gal tank initially filled with pure water. This mixture is also uniformly stirred and is emptied at the rate of 3 gpm. Find the amount of brine in the second tank at any time t.

5.5. An Existence Theorem. We consider the equation

$$\frac{dy}{dx} = f(x, y) \tag{5.35}$$

We wish to show that if $f(x, y)$ is suitably restricted in a neighborhood of the point $P_0(x_0, y_0)$ there exists a unique function $y = y(x)$ which satisfies (5.35) such that $y_0 = y(x_0)$. The restriction we impose on $f(x, y)$ is the following: Assume that a constant $M \neq \infty$ exists such that for all points $P(x, y)$, $Q(x, z)$ in some neighborhood of $P_0(x_0, y_0)$ we have

$$|f(x, z) - f(x, y)| < M|z - y| \tag{5.36}$$

A function $f(x, y)$ satisfying (5.36) is said to obey the Lipschitz condition. An immediate consequence of (5.36) is the following: If $\frac{\partial f}{\partial y}$ exists in a neighborhood of $P_0(x_0, y_0)$, then $\frac{\partial f}{\partial y}$ is uniformly bounded in that neighborhood, for [from (5.36)]

$$\left|\frac{\partial f}{\partial y}\right| = \lim_{z \to y}\left|\frac{f(x, z) - f(x, y)}{z - y}\right| < M$$

Conversely, if $\frac{\partial f}{\partial y}$ is continuous for a closed neighborhood of $P_0(x_0, y_0)$, then $f(x, y)$ satisfies the Lipschitz criterion for that neighborhood. Remember that a continuous function on a closed and bounded set is uniformly bounded over the set. We assume further that $f(x, y)$ is continuous so that the integrals of (5.37) exist.

The proof of the existence of a solution of (5.35) is based upon Picard's method of successive approximations. Define the sequence of functions $y_1(x), y_2(x), \ldots, y_n(x), \ldots$ as follows:

$$
\begin{aligned}
y_1(x) &= \int_{x_0}^{x} f(x, y_0)\, dx + y_0 \\
y_2(x) &= \int_{x_0}^{x} f(x, y_1(x))\, dx + y_0 \\
y_3(x) &= \int_{x_0}^{x} f(x, y_2(x))\, dx + y_0 \\
&\cdots\cdots\cdots\cdots\cdots\cdots \\
y_{n-1}(x) &= \int_{x_0}^{x} f(x, y_{n-2}(x))\, dx + y_0 \\
y_n(x) &= \int_{x_0}^{x} f(x, y_{n-1}(x))\, dx + y_0 \\
&\cdots\cdots\cdots\cdots\cdots\cdots
\end{aligned}
\tag{5.37}
$$

The $y_i(x)$, $i = 1, 2, \ldots, n, \ldots$, are obtained in a very natural manner. In $f(x, y)$ we replace y by the initial constant y_0, integrate, and obtain $y_1(x)$ such that $y_1(x_0) = y_0$. We now replace y in $f(x, y)$ by $y_1(x)$, integrate, and obtain $y_2(x)$. This process is continued indefinitely. The next endeavor is to show that the sequence thus obtained converges to a function $y(x)$ which we hope is the solution of (5.35). Let us note that

$$y_n(x) \equiv y_0 + (y_1 - y_0) + (y_2 - y_1) + \cdots + (y_n - y_{n-1})$$

Hence the investigation of the convergence of the sequence hinges on the convergence of the series

$$y_0 + (y_1 - y_0) + (y_2 - y_1) + \cdots + (y_n - y_{n-1}) + \cdots \tag{5.38}$$

Let K be an upper bound of $f(x, y)$ in the neighborhood of $P_0(x_0, y_0)$ for which the Lipschitz condition holds. Then from (5.37)

$$|y_1(x) - y_0| = \left| \int_{x_0}^{x} f(x, y_0)\, dx \right| \le \left| \int_{x_0}^{x} K\, dx \right| = K|x - x_0| \tag{5.39}$$

Also from (5.37)

$$
\begin{aligned}
|y_2(x) - y_1(x)| &= \left| \int_{x_0}^{x} [f(x, y_1) - f(x, y_0)]\, dx \right| \\
&< \left| \int_{x_0}^{x} M|y_1 - y_0|\, dx \right|
\end{aligned}
$$

from the Lipschitz condition. Applying (5.39) yields

$$
\begin{aligned}
|y_2(x) - y_1(x)| &< MK \left| \int_{x_0}^{x} |x - x_0|\, dx \right| \\
&< MK \frac{|x - x_0|^2}{2!}
\end{aligned}
$$

Let the reader show that

$$|y_3(x) - y_2(x)| < KM^2 \frac{|x - x_0|^3}{3!}$$

and by mathematical induction that

$$|y_n(x) - y_{n-1}(x)| < KM^{n-1} \frac{|x - x_0|^n}{n!} \tag{5.40}$$

Hence each term of the series (5.38) is bounded in absolute value by the terms of the converging series

$$y_0 + \sum_{n=1}^{\infty} \frac{K}{M} M^n \frac{|x - x_0|^n}{n!} = \frac{K}{M} (e^{M|x-x_0|} - 1) + y_0$$

Since the series for $e^{M|x-x_0|}$ converges uniformly in any closed and bounded set, the series (5.38) converges uniformly for those x in the region for which the Lipschitz condition holds. We have

$$\lim_{n \to \infty} y_n(x) = y(x)$$

and the convergence is uniform. From

$$y_n(x) = \int_{x_0}^{x} f(x, y_{n-1}(x))\, dx + y_0$$

we have

$$\begin{aligned} y(x) = \lim_{n \to \infty} y_n(x) &= \lim_{n \to \infty} \int_{x_0}^{x} f(x, y_{n-1}(x))\, dx + y_0 \\ &= \int_{x_0}^{x} \lim_{n \to \infty} f(x, y_{n-1}(x))\, dx + y_0 \end{aligned}$$

because of uniform convergence. The Lipschitz condition (5.36) also guarantees continuity of $f(x, y)$ with respect to y, for

$$\lim_{z \to y} |f(x, z) - f(x, y)| = 0$$

since $M|z - y| \to 0$ as $z \to y$. Hence

$$y(x) = \int_{x_0}^{x} f(x, y(x))\, dx + y_0$$

and

$$\frac{dy}{dx} = f(x, y(x)) \qquad y(x_0) = y_0$$

so that $y(x)$ satisfies (5.35) and the initial condition. Q.E.D.

We must now show that $y(x)$ is unique. Let $z(x)$ be a solution of (5.35) such that $z(x_0) = y_0$. Then

$$\frac{dz}{dx} = f(x, z(x)) \qquad z(x_0) = y_0$$

and integration yields

$$z(x) = \int_{x_0}^{x} f(x, z(x))\, dx + y_0$$

Hence

$$z(x) - y(x) = \int_{x_0}^{x} [f(x, z(x)) - f(x, y(x))]\, dx$$
$$|z(x) - y(x)| < M \left| \int_{x_0}^{x} |z(x) - y(z)|\, dx \right| \tag{5.41}$$

by applying the Lipschitz condition. Let L be an upper bound of $|z(x) - y(x)|$. Then

$$|z(x) - y(x)| < ML|x - x_0|$$

and, applying (5.41) successively, we obtain

$$|z(x) - y(x)| < M^2L \frac{|x - x_0|^2}{2!}$$
$$\cdots\cdots\cdots\cdots$$
$$|z(x) - y(x)| < M^nL \frac{|x - x_0|^n}{n!}$$
$$\cdots\cdots\cdots\cdots$$

Since $\dfrac{M^n|x - x_0|^n}{n!}$ is the nth term of $\displaystyle\sum_{n=0}^{\infty} \frac{M^n|x - x_0|^n}{n!}$ which is known to converge, we have

$$\lim_{n\to\infty} \frac{LM^n|x - x_0|^n}{n!} = 0$$

for any fixed x. Hence the difference between $z(x)$ and $y(x)$ can be made as small as we please. This means that $z(x) \equiv y(x)$. Q.E.D.

Problems

1. Show that $f(x, y) = xy$ satisfies the Lipschitz condition for $|x| < A < \infty$.

2. Show that (5.40) holds.

3. Show that $f(x, y) = x \sin y$ does not satisfy the Lipschitz condition for all x.

4. Consider $\dfrac{dy}{dx} = y$, $y(0) = 1$. Obtain the sequence (5.37), and show that

$$\lim_{n\to\infty} y_n(x) = e^x$$

5. Consider $\dfrac{dy}{dx} = x + y^2$, $y(0) = 0$. Obtain $y_1(x)$, $y_2(x)$, $y_3(x)$, $y_4(x)$ from (5.37).

6. Consider the system

$$\frac{dy}{dx} = f(x, y, z)$$
$$\frac{dz}{dx} = g(x, y, z) \tag{5.42}$$

Impose suitable restrictions on $f(x, y, z)$ and $g(x, y, z)$ in a neighborhood of the point $P_0(x_0, y_0, z_0)$, and obtain a sequence of functions $y_1(x)$, $y_2(x)$, . . . , $y_n(x)$, . . . , and a sequence of functions $z_1(x), z_2(x), \ldots, z_n(x), \ldots$, such that $\lim_{n\to\infty} y_n(x) = y(x)$, $\lim_{n\to\infty} z_n(x) = z(x)$, where $y(x)$ and $z(x)$ satisfy (5.42). Show that $y(x)$ and $z(x)$ are unique if $y(x_0) = y_0$, $z(x_0) = z_0$.

5.6. Linear Dependence. The Wronskian. A system of functions $y_1(x), y_2(x), \ldots, y_n(x)$, $a \leqq x \leqq b$, are said to be linearly dependent over the interval (a, b) if there exists a set of constants $c_1, c_2, \ldots, c_n$, not all zero, such that

$$\sum_{i=1}^{n} c_i y_i(x) \equiv 0 \tag{5.43}$$

for all x on (a, b). Otherwise we say that the $y(x)$ are linearly independent. Equation (5.43) implies that at least one of the functions can be written as a linear combination of the others. Thus, if $c_1 \neq 0$, then

$$y_1 = -\frac{1}{c_1}\sum_{i=2}^{n} c_i y_i(x)$$

Linear dependence will be important in the study of linear differential equations. Let us find a criterion for linear dependence. Assume the $y_i(x)$ of (5.43) differentiable $n - 1$ times. Then

$$\sum_{i=1}^{n} c_i y_i^{(j)}(x) = 0 \qquad j = 0, 1, 2, \ldots, n-1 \tag{5.44}$$

by successive differentiations. The system (5.44) may be looked upon as a system of n linear homogeneous equations in the unknowns c_i, $i = 1, 2, \ldots, n$. Since a nontrivial solution exists (remember not all c_i vanish), we must have

$$|y_i^{(j)}(x)| = 0$$

or

$$W(y_1, y_2, \ldots, y_n) = \begin{vmatrix} y_1(x) & y_2(x) & \cdots & y_n(x) \\ y_1'(x) & y_2'(x) & \cdots & y_n'(x) \\ \cdots & \cdots & \cdots & \cdots \\ y_1^{(n-1)}(x) & y_2^{(n-1)}(x) & \cdots & y_n^{(n-1)}(x) \end{vmatrix} = 0 \tag{5.45}$$

Determinant (5.45) is a necessary condition that the $y_i(x)$ be linearly dependent on $a \leqq x \leqq b$. This important determinant is called the Wronskian of $y_1, y_2, \ldots, y_n$. If $W(y_1, y_2, \ldots, y_n) \neq 0$, the y_i are linearly independent.

Let us now investigate the converse. We take first an easy case. Let

$$W(y_1, y_2) = \begin{vmatrix} y_1 & y_2 \\ y_1' & y_2' \end{vmatrix} = 0$$

for $a \leqq x \leqq b$. Assume $y_1(x) \neq 0$ for $a \leqq x \leqq b$. The same result would be obtained if we assumed $y_2(x) \neq 0$ for $a \leqq x \leqq b$. Then

$$y_1 y_2' - y_1' y_2 = 0 \qquad \frac{y_1 y_2' - y_2 y_1'}{y_1^2} = 0$$

and $\dfrac{d}{dx}\left(\dfrac{y_2}{y_1}\right) = 0$ so that $y_2 = cy_1$, which shows that y_1 and y_2 are linearly dependent.

The proof for the general case is not so easy. We assume that (5.45) holds for $a \leqq x \leqq b$. Furthermore we assume continuity of the derivatives, along with the assumption that at least one of the minors of the last row of (5.45) does not vanish for $a \leqq x \leqq b$. For convenience we assume

$$p_0(x) \equiv \begin{vmatrix} y_2(x) & y_3(x) & \cdots & y_n(x) \\ y_2'(x) & y_3'(x) & \cdots & y_n'(x) \\ \cdots & \cdots & \cdots & \cdots \\ y_2^{(n-2)}(x) & y_3^{(n-2)}(x) & \cdots & y_n^{(n-2)}(x) \end{vmatrix} \neq 0$$

for $a \leqq x \leqq b$. We now show that under these conditions the y_i are linearly dependent over the range $a \leqq x \leqq b$. First expand (5.45) about the first column to obtain

$$p_0(x)y_1^{(n-1)}(x) + p_1(x)y_1^{(n-2)}(x) + \cdots + p_{n-1}(x)y_1(x) = 0$$

Hence $y_1(x)$ is a solution of the linear differential equation

$$\frac{d^{n-1}y}{dx^{n-1}} + \frac{p_1}{p_0}\frac{d^{n-2}y}{dx^{n-1}} + \cdots + \frac{p_{n-1}}{p_0}y = 0 \tag{5.46}$$

Moreover, if we replace $y_1(x)$ by $y_2(x)$, $y_1'(x)$ by $y_2'(x)$, etc., in the first column of (5.46), the determinant vanishes since two columns are the same. Hence (5.46) is also satisfied by $y_2(x)$. With the same reasoning we find that $y_1(x), y_2(x), y_3(x), \ldots, y_n(x)$ are solutions of (5.46). The reader can verify easily that any linear combination of $y_1, y_2, \ldots, y_n$ is also a solution of (5.46). Now we fix our attention at a point x_0, $a \leqq x_0 \leqq b$. We have

$$\begin{vmatrix} y_1(x_0) & y_2(x_0) & \cdots & y_n(x_0) \\ y_1'(x_0) & y_2'(x_0) & \cdots & y_n'(x_0) \\ \cdots & \cdots & \cdots & \cdots \\ y_1^{(n-1)}(x_0) & y_2^{(n-1)}(x_0) & \cdots & y_n^{(n-1)}(x_0) \end{vmatrix} = 0$$

so that the system of homogeneous equations

$$\sum_{i=1}^{n} c_i y_i^{(j)}(x_0) = 0 \qquad j = 0, 1, 2, \ldots, n-1 \tag{5.47}$$

has a nontrivial solution in the c_i. Consider a set of c_i not all zero which satisfies (5.47), and form

$$y(x) = \sum_{i=1}^{n} c_i y_i(x) \tag{5.48}$$

Equation (5.48) is a solution of (5.46). Moreover $y(x_0) = 0$, $y'(x_0) = 0$, $\ldots$, $y^{(n-2)}(x_0) = 0$ as well as $y^{(n-1)}(x_0) = 0$. It will be shown in Sec. 5.7 that Eq. (5.46) has a unique solution when $n - 1$ initial conditions are imposed. Now $y(x) \equiv 0$ certainly satisfies (5.46) and the initial conditions $y(x_0) = y'(x_0) = y''(x_0) = \cdots = y^{(n-2)}(x_0) = 0$, so that $y(x)$ of (5.48) is identically zero for $a \leqq x \leqq b$. Hence

$$\sum_{i=1}^{n} c_i y_i(x) \equiv 0$$

for $a \leqq x \leqq b$. Q.E.D.

Problems

1. Show that $\sin x$ and $\cos x$ are linearly independent.

2. Show that $\sin x$, $\cos x$, e^{ix} are linearly dependent. Assuming $e^{\pi i/2} = i$, show that $e^{ix} = \cos x + i \sin x$.

3. Consider $y_1 = x^2$, $y_2 = x|x|$, $-1 \leqq x \leqq 1$. Show that $W(y_1, y_2) = 0$ for $-1 \leqq x \leqq 1$. Does this imply that y_1 and y_2 are linearly dependent? Show that y_1 and y_2 are not linearly dependent on the range $-1 \leqq x \leqq 1$. Does this contradict the theorem derived above?

4. Let $y_1(x)$ and $y_2(x)$ be solutions of

$$y'' + p(x)y' + q(x)y = 0$$

for $a \leqq x \leqq b$. Show that $\dfrac{dW(y_1, y_2)}{dx} + p(x)W(y_1, y_2) = 0$ and hence that

$$W(y_1, y_2) = A \exp \left[- \int_a^x p(x)\, dx\right]$$

Show that if $W = 0$ for $x = x_0$ then $W \equiv 0$ for all x on $a \leqq x \leqq b$. How does one determine the constant A? Show that if $W \neq 0$ for $x = x_0$ then $W \neq 0$ for all x on $a \leqq x \leqq b$.

5. Let $y_1(x)$, $y_2(x)$ be linearly independent solutions of

$$y'' + p(x)y = 0 \tag{5.49}$$

for $a \leqq x \leqq b$. Let $y_3(x)$ be any solution of (5.49). Show that

$$y_3(x) = c_1 y_1(x) + c_2 y_2(x)$$

Hint: Show that $W(y_1, y_2, y_3) = 0$ for $a \leqq x \leqq b$.

6. Let $y_1(x)$, $y_2(x)$, $\ldots$, $y_n(x)$ be linearly independent solutions (on the range $a \leqq x \leqq b$) of

$$\frac{d^n y}{dx^n} + p_1(x)\frac{d^{n-1}y}{dx^{n-1}} + \cdots + p_n(x)y = 0 \tag{5.50}$$

If $y(x)$ is any solution of (5.50), show that

$$y(x) = \sum_{k=1}^{n} c_k y_k(x)$$

for $a \leqq x \leqq b$.

5.7. Linear Differential Equations. The differential equation

$$\frac{d^n y}{dx^n} + p_1(x)\frac{d^{n-1}y}{dx^{n-1}} + p_2(x)\frac{d^{n-2}y}{dx^{n-2}} + \cdots + p_n(x)y = p(x) \quad (5.51)$$

is called an nth-order linear differential equation. What can we say about a solution of (5.51) subject to the initial conditions $y(x_0) = y_0$, $y'(x_0) = y_0'$, $y''(x_0) = y_0''$, . . . , $y^{(n-1)}(x_0) = y_0^{(n-1)}$? Assume $y_1(x)$ a solution of (5.51). We note that (5.51) can be written as a system of n first-order equations by the simple device of introducing $n - 1$ new variables. Define $y_2, y_3, \ldots, y_n$ as follows:

$$\begin{aligned}
\frac{dy_1}{dx} &= y_2 \\
\frac{dy_2}{dx} &= y_3 \\
&\cdots\cdots \\
\frac{dy_{n-1}}{dx} &= y_n \\
\frac{dy_n}{dx} &= -p_1(x)y_n - p_2(x)y_{n-1} - \cdots - p_n(x)y_1 + p(x)
\end{aligned} \quad (5.52)$$

The last equation of the system (5.52) is (5.51) in terms of $y_1, y_2, \ldots, y_n$. Conversely, if $[y_1(x), y_2(x), \ldots, y_n(x)]$ is a solution of (5.52), then $y_1(x)$ is easily seen to be a solution of (5.51). Now (5.52) is a special case of the very general system[1]

$$\frac{dy^i}{dx} = f^i(x, y^1, y^2, \ldots, y^n) \qquad i = 1, 2, \ldots, n \quad (5.53)$$

As in Sec. 5.5 it can be shown that, if the f^i satisfy the suitable Lipschitz conditions, there exists a unique solution $y^1(x), y^2(x), \ldots, y^n(x)$ of (5.53) satisfying the initial conditions $y^i(x_0) = y_0^i$, $i = 1, 2, \ldots, n$. The Lipschitz condition for the f^i is

$$|f^i(x, y^1, y^2, \ldots, y^n) - f^i(x, z^1, z^2, \ldots, z^n)| < M \sum_{i=1}^{n} |y^i - z^i| \quad (5.54)$$

for all x in a neighborhood of $x = x_0$, M fixed.

[1] The exponents are superscripts, not powers.

If $p(x)$, $p_i(x)$, $i = 1, 2, \ldots, n$ of (5.52) are continuous in a neighborhood of $x = x_0$, then (5.54) is easily seen to hold. Hence (5.51) has a unique solution subject to the n initial conditions stated above.

The reader can show easily that if $y_1(x), y_2(x), \ldots, y_n(x)$ are linearly independent solutions of the homogeneous equation

$$\frac{d^n y}{dx^n} + p_1(x)\frac{d^{n-1}y}{dx^{n-1}} + \cdots + p_n(x)y = 0 \tag{5.55}$$

and if $\bar{y}(x)$ is any particular solution of (5.51), then

$$y(x) = \sum_{i=1}^{n} c_i y_i(x) + \bar{y}(x) \tag{5.56}$$

is the most general solution of (5.51). Let the reader show that the c_i are uniquely determined from the initial conditions on y and its first $n - 1$ derivatives at $x = x_0$.

In general it is very difficult to find the solution to (5.51), and it is necessary at times to use infinite series in the attempt. This method will be discussed in a later paragraph. There is one case, however, which is the simplest by far. Naturally, one expects the difficulties in solving (5.51) to be alleviated to some extent if the $p_i(x)$ of (5.51) are constants. We study now this case. The homogeneous equation

$$\frac{d^n y}{dx^n} + p_1\frac{d^{n-1}y}{dx^{n-1}} + p_2\frac{d^{n-2}y}{dx^{n-2}} + \cdots + p_n y = 0$$
$$p_i = \text{constant}: i = 1, 2, \ldots, n \tag{5.57}$$

can be solved as follows: Assume a solution of the form $y = e^{mx}$. Substituting into (5.57) yields

$$e^{mx}(m^n + p_1 m^{n-1} + p_2 m^{n-2} + \cdots + p_n) = 0$$

Hence if m is a root of the polynomial equation

$$f(z) \equiv z^n + p_1 z^{n-1} + p_2 z^{n-2} + \cdots + p_n = 0 \tag{5.58}$$

then $y = e^{mx}$ is a solution of (5.57). Equation (5.58) is easily obtained from (5.57). One replaces $\dfrac{d^k y}{dx^k}$ by z^k, $k = 0, 1, 2, \ldots, n$. If the n roots of (5.58) are distinct, call them $m_1, m_2, \ldots, m_n$, then the general solution of (5.57) is

$$y = \sum_{k=1}^{n} c_k e^{m_k x} \tag{5.59}$$

The reader can show that if the m_k are distinct, then

$$W(e^{m_1 x}, e^{m_2 x}, \ldots, e^{m_n x}) \neq 0$$

First let the reader show that

$$W = \exp\left(x \sum_{k=1}^{n} m_k\right) \begin{vmatrix} 1 & 1 & \cdots & 1 \\ m_1 & m_2 & \cdots & m_n \\ m_1^2 & m_2^2 & \cdots & m_n^2 \\ \cdots & \cdots & \cdots & \cdots \\ m_1^{n-1} & m_2^{n-1} & \cdots & m_n^{n-1} \end{vmatrix} \tag{5.60}$$

Next let the reader notice that

$$F(u) \equiv \begin{vmatrix} 1 & 1 & 1 & \cdots & 1 \\ u & m_2 & m_3 & \cdots & m_n \\ u^2 & m_2^2 & m_3^2 & \cdots & m_n^2 \\ \cdots & \cdots & \cdots & \cdots & \cdots \\ u^{n-1} & m_2^{n-1} & m_3^{n-1} & \cdots & m_n^{n-1} \end{vmatrix}$$

is a polynomial of degree $n - 1$ in u. The polynomial equation $F(u) = 0$ certainly vanishes for the $n - 1$ distinct roots $u = m_2, m_3, \ldots, m_n$. If $F(m_1)$ were zero, then

$$\begin{vmatrix} 1 & 1 & 1 & \cdots & 1 \\ m_2 & m_3 & m_4 & \cdots & m_n \\ m_2^2 & m_3^2 & m_4^2 & \cdots & m_n^2 \\ \cdots & \cdots & \cdots & \cdots & \cdots \\ m_2^{n-2} & m_3^{n-2} & \cdot & \cdots & m_n^{n-2} \end{vmatrix} = 0 \qquad \text{Why?}$$

Continuing with this line of reasoning, the reader can show that as a consequence of assuming $W = 0$ one finally obtains

$$\begin{vmatrix} 1 & 1 \\ m_{n-1} & m_n \end{vmatrix} = 0$$

a contradiction, since $m_n \neq m_{n-1}$.

The case for which some of the roots of (5.58) are equal must be treated separately. Before attacking this problem we shall find it beneficial to introduce the operator

$$s \equiv \frac{d}{dx} \tag{5.61}$$

The notation $sf(x)$ means $\dfrac{d}{dx} f(x) = f'(x)$. Similarly,

$$s^2 f(x) \equiv s[sf(x)] = s[f'(x)] = f''(x)$$

and $s^n \equiv \dfrac{d^n}{dx^n}$. We define $(s + a)f(x) = sf(x) + af(x) = f'(x) + af(x)$ for any scalar a. Let the reader show that

$$(s + a)[(s + b)f(x)] = (s + b)[(s + a)f(x)] \tag{5.62}$$

if a and b are constants. Equation (5.57) can be written

$$(s^n + p_1 s^{n-1} + p_2 s^{n-2} + \cdots + p_{n-1}s + p_n)y = 0 \tag{5.63}$$

or

$$(s - m_1)(s - m_2) \cdots (s - m_n)y = 0 \tag{5.64}$$

using the result of (5.62). If a is constant, we have

$$s(e^{ax}y) = e^{ax}sy + ae^{ax}y = e^{ax}(s + a)y$$

Let the reader show by mathematical induction that

$$s^n(e^{ax}y) = e^{ax}(s + a)^n y \tag{5.65}$$

Now let us assume that (5.58) has the distinct roots $m_1, m_2, m_3, \ldots, m_k$ of multiplicity $\alpha_1, \alpha_2, \ldots, \alpha_k$ so that (5.64) can be written

$$f(s)y = (s - m_1)^{\alpha_1}(s - m_2)^{\alpha_2} \cdots (s - m_k)^{\alpha_k}y = 0 \tag{5.66}$$

We note that, if $y(x)$ satisfies $(s - m_k)^{\alpha_k}y(x) = 0$, then $y(x)$ also satisfies (5.66). Now

$$e^{-m_k x}(s - m_k)^{\alpha_k}y(x) \equiv s^{\alpha_k}[e^{-m_k x}y(x)]$$

from (5.65), so that any $y(x)$ which satisfies $(s - m_k)^{\alpha_k}y(x) = 0$ also satisfies $s^{\alpha_k}(e^{-m_k x}y) = 0$, and conversely. Hence

$$\frac{d^{\alpha_k}}{dx^{\alpha_k}}(e^{-m_k x}y) = 0$$

so that

$$y(x) = e^{m_k x}(C_1 + C_2 x + \cdots + C_{\alpha_k - 1}x^{\alpha_k - 1}) \tag{5.67}$$

It is easy to verify that (5.67) is a solution of (5.66). If $\alpha_k = 1$, then $y(x) = C_1 e^{m_k x}$. Equation (5.67) contains α_k constants of integration. Applying the same reasoning to the remaining roots yields n constants of integration. We omit the proof that the solutions thus obtained, namely,

$$e^{m_1 x}, xe^{m_1 x}, \ldots, x^{\alpha_1 - 1}e^{m_1 x}, e^{m_2 x}, xe^{m_2 x}, \ldots, x^{\alpha_2 - 1}e^{m_2 x}, \ldots, e^{m_n x}, xe^{m_n x}, \ldots, x^{\alpha_k - 1}e^{m_k x}$$

are linearly independent.

Example 5.8. The differential equation $y'' - y = 0$ admits the solution $y = e^{mx}$ if $m^2 - 1 = 0$ or $m = \pm 1$. The most general solution is

$$y = C_1 e^x + C_2 e^{-x}$$

If $C_1 = \frac{1}{2}$, $C_2 = -\frac{1}{2}$, then $y = \sinh x$. If $C_1 = C_2 = \frac{1}{2}$, then $y = \cosh x$. It is easy to verify that $\sinh x$ and $\cosh x$ are linearly independent. The most general solution can also be written in the form

$$y = A_1 \sinh x + A_2 \cosh x$$

Example 5.9. We wish to solve $y'' + n^2 y = 0$ subject to the initial condition $y(0) = 0$, $y'(0) = n$, n real. e^{mx} is a solution of $y'' + n^2 y = 0$ if $m^2 + n^2 = 0$ or

$m = \pm in$. Hence the solution is

$$y = Ae^{nix} + Be^{-nix}$$

From $(x = 0, y = 0)$ we have $0 = A + B$, and from $(x = 0, y' = n)$ we have

$$n = (An - Bn)i$$

or $A - B = -i$. Hence $A = \frac{1}{2i}$, $B = -\frac{1}{2i}$, and $y(x) = \frac{1}{2i}(e^{nix} - e^{-nix}) = \sin nx$. Let the reader show that $y = A \sin nx + B \cos nx$ is also a general solution of $y'' + n^2 y = 0$.

Example 5.10. Consider

$$\frac{d^6y}{dx^6} + \frac{d^4y}{dx^4} - 6\frac{d^3y}{dx^3} + 4\frac{d^2y}{dx^2} = 0 \tag{5.68}$$

In operational form (5.68) becomes

$$(s^6 + s^4 - 6s^3 + 4s^2)y = 0$$

or $s^2(s - 1)^2(s^2 + 2s + 4)y = 0$. The zeros of $f(m) = m^2(m - 1)^2(m^2 + 2m + 4)$ are $m = 0, 0, 1, 1, -1 \pm \sqrt{3}\, i$. Hence the general solution of (5.68) is

$$y = e^{0x}(A + Bx) + e^x(C + Dx) + E \exp[(-1 + \sqrt{3}i)x] + F \exp[(-1 - \sqrt{3}i)x]$$

or $\quad y = A + Bx + (C + Dx)e^x + e^{-x}(G \cos \sqrt{3}\, x + H \sin \sqrt{3}\, x)$

It has been explained before that the most general solution of an inhomogeneous equation can be written provided:

(i) A particular solution is known.

(ii) The general solution of the homogeneous equation is known.

The second part has just been considered. It remains to study methods by which a particular solution can be found. Here any guess can be made, and, if successful, no further justification is needed. Quite often a particular solution can be determined by inspection. The work involves finding certain undetermined coefficients. For example:

(*a*) If $p(x)$ of (5.57) is a polynomial, try a polynomial of the same degree.

(*b*) If $p(x) = ae^{rx}$, try Ae^{rx}.

(*c*) If $p(x) = a \sin \omega x + b \cos \omega x$, try $A \sin \omega x + B \cos \omega x$.

A few examples should make clear what we mean.

(*a*) $y'' + y = 2x^2$. We assume a particular solution of the form $y = a + bx + cx^2$. Then $y' = b + 2cx$, $y'' = 2c$ so that of necessity $2c + a + bx + cx^2 \equiv 2x^2$ and $c = 2$, $2c + a = 0$, $b = 0$. Hence $y = A \sin x + B \cos x + 2x^2 - 4$ is the general solution of $y'' + y = 2x^2$.

(*b*) $y'' - y = x - 2e^{2x}$. Assume a particular solution of the form

$$y = a + bx + ce^{2x}$$

Then $y' = b + 2ce^{2x}$, $y'' = 4ce^{2x}$, and of necessity $4ce^{2x} - a - bx - ce^{2x} \equiv x - 2e^{2x}$. This yields $3c = -2$, $a = 0$, $b = -1$. The general solution is

$$y = Ae^x + Be^{-x} - x - \tfrac{2}{3}e^{2x}$$

(c) $y'' + 4y' - 2y = 2 \sin x - 3e^x + 1 - 2x$. We try a particular solution in the form $y = a \sin x + b \cos x + ce^x + d + fx$. Of necessity

$$-a \sin x - b \cos x + ce^x + 4a \cos x - 4b \sin x + 4ce^x + 4f - 2a \sin x - 2b \cos x - 2ce^x - 2d - 2fx \equiv 2 \sin x - 3e^x + 1 - 2x$$

Equating coefficients of $\sin x$, $\cos x$, e^x, etc., yields $-3a - 4b = 2$, $-3b + 4a = 0$, $3c = -3$, $4f - 2d = 1$, $-2f = -2$, so that $a = -\frac{6}{25}$, $b = -\frac{8}{25}$, $c = -1$, $d = \frac{3}{2}$, $f = 1$. The reader can verify that $y = -\frac{6}{25} \sin x - \frac{8}{25} \cos x - e^x + \frac{3}{2} + x$ is a particular solution of (c).

Although in many cases a particular solution is easily obtained by the method of undetermined coefficients, it is important to have a formula valid in all cases. The problem is to find a particular solution of the equation

$$\frac{d^ny}{dx^n} + a_1 \frac{d^{n-1}y}{dx^{n-1}} + \cdots + a_{n-1} \frac{dy}{dx} + a_n y = p(x) \tag{5.69}$$

Let $g(x)$ be a solution of the homogeneous equation

$$\frac{d^ny}{dx^n} + a_1 \frac{d^{n-1}y}{dx^{n-1}} + \cdots + a_{n-1} \frac{dy}{dx} + a_n y = 0 \tag{5.70}$$

which satisfies the initial conditions

$$g(0) = g'(0) = g''(0) = \cdots = g^{(n-2)}(0) = 0 \qquad g^{(n-1)}(0) = 1$$

We prove that

$$y(x) = \int_0^x g(x - t)p(t)\, dt \tag{5.71}$$

is a particular solution of (5.69). Differentiating (5.71) twice yields

$$\begin{aligned} y'(x) &= g(x - x)p(x) + \int_0^x \frac{\partial g(x - t)}{\partial x} p(t)\, dt \\ &= \int_0^x \frac{\partial g(x - t)}{\partial x} p(t)\, dt \\ y''(x) &= \left.\frac{\partial g(x - t)}{\partial x}\right|_{t=x} p(x) + \int_0^x \frac{\partial^2 g(x - t)}{\partial x^2} p(t)\, dt \\ &= \int_0^x \frac{\partial^2 g(x - t)}{\partial x^2} p(t)\, dt \end{aligned}$$

since $g'(0) = 0$. Further differentiation yields

$$\begin{aligned} y^{(n-1)}(x) &= \int_0^x \frac{\partial^{n-1} g(x - t)}{\partial x^{n-1}} p(t)\, dt \\ y^{(n)}(x) &= \left.\frac{\partial^{n-1} g(x - t)}{\partial x^{n-1}}\right|_{t=x} p(x) + \int_0^x \frac{\partial^n g(x - t)}{\partial x^n} p(t)\, dt \\ &= p(x) + \int_0^x \frac{\partial^n g(x - t)}{\partial x^n} p(t)\, dt \end{aligned}$$

since $g^{(n-1)}(0) = 1$. If these values are substituted into (5.69), one obtains

$$\begin{aligned} &y^{(n)}(x) + a_1 y^{(n-1)}(x) + \cdots + a_n y(x) \\ &\quad = p(x) + \int_0^x \left[\frac{\partial^n g(x-t)}{\partial x^n} + a_1 \frac{\partial^{n-1} g(x-t)}{\partial x^{n-1}} + \cdots + a_n g(x-t) \right] p(t)\, dt \\ &\quad = p(x) \end{aligned}$$

The expression

$$\frac{\partial^n g(x-t)}{\partial x^n} + a_1 \frac{\partial^{n-1} g(x-t)}{\partial x^{n-1}} + \cdots + a_n g(x-t) \tag{5.72}$$

vanishes since $g(x)$ satisfies (5.70). The derivatives of $g(x)$ in (5.72) are evaluated at $x - t$, but since $g(x)$ satisfies (5.70) for all x we certainly know that $g(x)$ satisfies (5.70) at $x - t$. We have shown that (5.71) is a particular solution of (5.69); the a_i, $i = 1, 2, \ldots, n$, of (5.69) are constants.

Example 5.11. Consider $y'' - y = \sin x$. We first solve $y'' - y = 0$, which yields the solution $g(x) = Ae^x + Be^{-x}$. We now impose the initial conditions $g(0) = 0$, $g'(0) = 1$. This yields $A + B = 0$, $A - B = 1$, so that $A = \frac{1}{2}$, $B = -\frac{1}{2}$. Thus $g(x) = \frac{1}{2}(e^x - e^{-x}) = \sinh x$. A particular solution of $y'' - y = \sin x$ is

$$\begin{aligned} y_p &= \int_0^x \tfrac{1}{2}(e^{x-t} - e^{-(x-t)}) \sin t\, dt \\ &= \frac{e^x}{2} \int_0^x e^{-t} \sin t\, dt - \frac{e^{-x}}{2} \int_0^x e^t \sin t\, dt \\ &= -\tfrac{1}{2} \sin x + \tfrac{1}{4} e^x - \tfrac{1}{4} e^{-x} \end{aligned}$$

The complete solution is

$$\begin{aligned} y &= Ae^x + Be^{-x} - \tfrac{1}{2} \sin x + \tfrac{1}{4} e^x - \tfrac{1}{4} e^{-x} \\ &= A_1 e^x + B_1 e^{-x} - \tfrac{1}{2} \sin x \end{aligned}$$

Example 5.12. $y''' - 3y'' + 4y' - 2y = e^x \sec x$. The solution of

$$y''' - 3y'' + 4y' - 2y = 0$$

is

$$y = g(x) = Ae^x + e^x(B \sin x + C \cos x)$$

If $g(0) = 0$, $g'(0) = 0$, $g''(0) = 1$, we must have

$$\begin{aligned} A + C &= 0 \\ A + C + B &= 0 \\ A + 2B &= 1 \end{aligned}$$

which yields $B = 0$, $A = 1$, $C = -1$. Thus

$$\begin{aligned} y_p(x) &= \int_0^x [e^{x-t} - e^{x-t} \cos (x - t)] e^t \sec t\, dt \\ &= e^x \int_0^x \sec t\, dt - e^x \int_0^x (\cos x + \sin x \tan t)\, dt \\ &= e^x \ln (\sec x + \tan x) - xe^x \cos x + e^x \sin x \ln \cos x \end{aligned}$$

The general solution is

$$y = e^x[A + B \sin x + C \cos x + \ln (\sec x + \tan x) - x \cos x + \sin x \ln \cos x]$$

Example 5.13. *Method of Variation of Parameters.* Equation (5.71) is a valid particular solution of (5.69) for constant coefficients. We look for a particular solution of

$$\frac{d^n y}{dx^n} + p_1(x) \frac{d^{n-1}y}{dx^{n-1}} + p_2(x) \frac{d^{n-2}y}{dx^{n-2}} + \cdots + p_n(x)y = p(x)$$

To simplify things, we shall discuss the general third-order linear differential equation

$$\frac{d^3y}{dx^3} + p_1(x) \frac{d^2y}{dx^2} + p_2(x) \frac{dy}{dx} + p_3(x)y = p(x) \tag{5.73}$$

Let $y_1(x)$, $y_2(x)$, $y_3(x)$ be linearly independent solutions of the homogeneous equation derived from (5.73).

We know that $y = \sum_{i=1}^{3} A_i y_i$ is the most general solution of the homogeneous equation. The method of variation of parameters consists in attempting to find a particular solution of the inhomogeneous equation by varying the A_i, $i = 1, 2, 3$, that is, by assuming that the A_i, $i = 1, 2, 3$, are not constant. Assume

$$y = u_1y_1 + u_2y_2 + u_3y_3 \tag{5.74}$$

We shall impose three conditions on $\frac{du_i}{dx}$, $i = 1, 2, 3$. Two of these conditions will be

$$\sum_{i=1}^{3} \frac{du_i}{dx} y_i = 0 \qquad \sum_{i=1}^{3} \frac{du_i}{dx} \frac{dy_i}{dx} = 0 \tag{5.75}$$

Differentiating (5.74) and making use of (5.75) yields

$$\frac{dy}{dx} = \sum_{i=1}^{3} u_i \frac{dy_i}{dx} \tag{5.76}$$

Differentiating again and making use of (5.75) yields

$$\frac{d^2y}{dx^2} = \sum_{i=1}^{3} u_i \frac{d^2y_i}{dx^2} \tag{5.77}$$

Finally we have

$$\frac{d^3y}{dx^3} = \sum_{i=1}^{3} u_i \frac{d^3y_i}{dx^3} + \sum \frac{du_i}{dx} \frac{d^2y_i}{dx^2} \tag{5.78}$$

We multiply (5.77) by $p_1(x)$, (5.76) by $p_2(x)$, (5.74) by $p_3(x)$ and add these results to (5.78). If $y(x)$ is to be a solution of (5.73), of necessity

$$\sum_{i=1}^{3} \frac{du_i}{dx} \frac{d^2y_i}{dx^2} = p(x) \tag{5.79}$$

Equations (5.75) and (5.79) are three linear equations in the unknowns $\frac{du_i}{dx}$, $i = 1, 2, 3$. Solving for $\frac{du_1}{dx}$ yields

$$\frac{du_1}{dx} = \frac{\begin{vmatrix} 0 & y_2 & y_3 \\ 0 & y_2' & y_3' \\ p(x) & y_2'' & y_3'' \end{vmatrix}}{\begin{vmatrix} y_1 & y_2 & y_3 \\ y_1' & y_2' & y_3' \\ y_1'' & y_2'' & y_3'' \end{vmatrix}} = \frac{p(x)W(y_2, y_3)}{W(y_1, y_2, y_3)} \tag{5.80}$$

The Wronskian $W(y_1, y_2, y_3)$ does not vanish since y_1, y_2, y_3 are linearly independent. Hence

$$u_1(x) = \int_a^x \frac{p(x)W(y_2, y_3)}{W(y_1, y_2, y_3)}\, dx$$

and in general

$$u_i(x) = \int_a^x \frac{p(x)W(y_{i+1}, y_{i+2})}{W(y_1, y_2, y_3)}\, dx \tag{5.81}$$

where $y_4(x) = y_1(x)$, $y_5(x) = y_2(x)$. The reader can verify that

$$y(x) = \int_a^x \frac{p(t) \sum_{i=1}^{3} y_i(x) W(y_{i+1}(t), y_{i+2}(t))}{W(y_1(t), y_2(t), y_3(t))}\, dt \tag{5.82}$$

is a particular solution of (5.73).

Example 5.14. We consider $y'' + (1/x)y' - (1/x^2)y = x$. It is easy to verify that $y_1(x) = x$, $y_2(x) = 1/x$ are linearly independent solutions of $y'' + (1/x)y' - (1/x^2)y = 0$. We have

$$u_1(x) = \int_0^x \frac{\begin{vmatrix} 0 & y_2(x) \\ p(x) & y_2'(x) \end{vmatrix}}{\begin{vmatrix} y_1(x) & y_2(x) \\ y_1'(x) & y_2'(x) \end{vmatrix}}\, dx = \int_0^x \frac{-x(1/x)\, dx}{\begin{vmatrix} x & \frac{1}{x} \\ 1 & -\frac{1}{x^2} \end{vmatrix}} = \frac{x^2}{4}$$

$$u_2(x) = \int_0^x \frac{\begin{vmatrix} y_1(x) & 0 \\ y_1'(x) & p(x) \end{vmatrix}}{-2/x}\, dx = -\frac{x^4}{8}$$

A particular solution is $y = (x^2/4)(x) + (-x^4/8)(1/x) = x^3/8$. The complete solution is $y = Ax + \frac{B}{x} + \frac{x^3}{8}$.

Problems

1. $Ly'' + Ry' + (1/C)y = 0$, L, R, C constants. Discuss the cases $R^2 - \frac{4L}{C} \gtreqless 0$.
2. $y'' - y = x$
3. $y'''' - 2y'' + y = 5 \sin 2x$
4. $y''' - y' = x + 1$

5. $y''' + y' = 2 \sin x$

6. $y'' - 4y' + 5y = e^{2x} + 4 \sin x$

7. If y_1 is a particular solution of $y'' + p(x)y' + q(x)y = r_1(x)$ and y_2 is a particular solution of $y'' + p(x)y' + q(x)y = r_2(x)$, show that $y_1 + y_2$ is a particular solution of $y'' + p(x)y' + q(x)y = r_1(x) + r_2(x)$.

8. $y'' + y = \tan x$

9. $y'' - 2y' + y = \cos^2 x$

10. $y'' - 4y' + 4y = 2xe^x$

11. Solve $y''' - y'' = 1 + x + \sin x$ by the variation-of-parameter method.

5.8. Properties of Second-order Linear Differential Equations. We consider

$$\frac{d^2y}{dx^2} + p(x)\frac{dy}{dx} + q(x)y = 0 \tag{5.83}$$

(*a*) If $p(x)$ and $q(x)$ are continuous on $a \leqq x \leqq b$, there exists a unique solution $y(x)$ of (5.83) subject to the initial conditions $y(x_0) = y_0$, $y'(x_0) = y_0'$, $a \leqq x_0 \leqq b$. Moreover $\frac{dy}{dx}$ is continuous since $\frac{d^2y}{dx^2}$ exists. We now show that if $y(x)$ is a solution of (5.83) then $y(x)$ cannot have an infinite number of zeros on the interval $a \leqq x \leqq b$ unless $y(x) \equiv 0$. The proof is as follows: Assume $y(x)$ vanishes infinitely often on the interval $a \leqq x \leqq b$. From the Weierstrass-Bolzano theorem there exists a limit point c, $a \leqq c \leqq b$. We can pick out a subsequence $x_1, x_2, \ldots, x_n, \ldots$ which converges to c such that $y(x_n) = 0$, $n = 1, 2, \ldots$. From the theorem of the mean $y(x_n) - y(x_{n-1}) = (x_n - x_{n-1})y'(\xi_n)$ so that $y'(\xi_n) = 0, x_{n-1} \leqq \xi_n \leqq x_n, n = 1, 2, \ldots$. Hence $y'(c) = \lim_{n\to\infty} y'(\xi_n) = 0$ since the ξ_n also approach $x = c$ as a limit and since $y'(x)$ is continuous at $x = c$. Moreover $y(c) = \lim_{n\to\infty} f(x_n) = 0$. But $y(x) \equiv 0$ satisfies (5.83), and $y(c) = 0$, $y'(c) = 0$. Since $y(x)$ is unique, we must have $y(x) \equiv 0$. Q.E.D.

(*b*) Let $y_1(x)$ and $y_2(x)$ be linearly independent solutions of (5.83). Let $y_1(c_1) = y_1(c_2) = 0$ and $y_1(x) \neq 0$ for $a \leqq c_1 < x < c_2 \leqq b$. We show that $y_2(c) = 0$, $c_1 < c < c_2$, that is, the zeros of $y_1(x)$ and $y_2(x)$ separate each other. Assume $y_2(x) \neq 0$ for $c_1 < x < c_2$. Then

$$\varphi(x) \equiv \frac{y_1(x)}{y_2(x)}$$

has no singularities for $c_1 \leqq x \leqq c_2$. Why is it true that $y_2(c_1) \neq 0$, $y_2(c_2) \neq 0$? Moreover $\varphi(c_1) = \varphi(c_2) = 0$. From the theorem of the mean $\varphi'(\xi) = 0$ for $c_1 \leqq \xi \leqq c_2$. But

$$\varphi'(\xi) = \frac{y_2(\xi)y_1'(\xi) - y_1(\xi)y'(\xi)}{y_2^2(\xi)} = \frac{W(y_1, y_2)\,|_{x=\xi}}{y_2^2(\xi)}$$

Since $y_1(x)$ and $y_2(x)$ are linearly independent, we know that $W(y_1, y_2) \neq 0$ for $a \leqq x \leqq b$. Hence $\varphi'(\xi)$ cannot be zero, a contradiction. $y_2(x)$ must be zero for some x, $c_1 < x < c_2$. Q.E.D.

(c) We can write (5.83) in the form

$$\frac{d}{dx}\left[K(x)\frac{dy}{dx}\right] + Q(x)y = 0 \tag{5.84}$$

We multiply (5.83) by $\exp\left(\int_a^x p(t)\,dt\right)$ and write

$$\frac{d}{dx}\left[\exp\left(\int_a^x p(t)\,dt\right)\frac{dy}{dx}\right] + q(x)\exp\left(\int_a^x p(t)\,dt\right)y = 0$$

so that $K(x) = \exp \int_a^x p(t)\,dt$, $Q(x) = q(x) \exp \int_a^x p(t)\,dt$.

An important differential equation is the Sturm-Liouville equation

$$\frac{d}{dx}\left[K(x)\frac{dy}{dx}\right] + \lambda q(x)y = 0 \tag{5.85}$$

where λ is a parameter. Assume $K(a) = K(b) = 0$, and let $y_1(x, \lambda_1)$ be a solution of (5.85) for $\lambda = \lambda_1$, $y_2(x, \lambda_2)$ a solution of (5.85) for $\lambda = \lambda_2$. We show that

$$\int_a^b q(x)y_1(x, \lambda_1)y_2(x, \lambda_2)\,dx = 0 \qquad \lambda_1 \neq \lambda_2 \tag{5.86}$$

We have

$$\begin{aligned}\frac{d}{dx}\left[K(x)\frac{dy_1}{dx}\right] + \lambda_1 q(x)y_1 = 0\\ \frac{d}{dx}\left[K(x)\frac{dy_2}{dx}\right] + \lambda_2 q(x)y_2 = 0\end{aligned} \tag{5.87}$$

Multiply the first equation of (5.87) by y_2 and the second by y_1, and subtract. This yields

$$\frac{d}{dx}\left[K(x)\left(y_2\frac{dy_1}{dx} - y_1\frac{dy_2}{dx}\right)\right] + (\lambda_1 - \lambda_2)q(x)y_1y_2 = 0 \tag{5.88}$$

Equation (5.86) follows if we integrate (5.88) over the range $a \leqq x \leqq b$. The orthogonality property (5.86) will be discussed in greater detail in Chap. 6 dealing with orthogonal polynomials.

(d) Let $y_1(x)$ be a solution of $y'' + G_1(x)y = 0$, and let $y_2(x)$ be a solution of $y'' + G_2(x)y = 0$. Assume further that $y_1(a) = y_2(a) = \alpha > 0$, $y_1'(a) = y_2'(a) = \beta$, $G_1 < G_2$. We show that $y_2(x)$ vanishes before $y_1(x)$ vanishes, $x > a$. The proof is as follows: We have $y_1'' + G_1(x)y_1 = 0$,

$y_2'' + G_2(x)y_2 = 0$ so that

$$y_2y_1'' - y_1y_2'' + (G_1 - G_2)y_1y_2 = 0$$

$$\frac{d}{dx}(y_2y_1' - y_1y_2') + (G_1 - G_2)y_1y_2 = 0$$

$$\int_a^c \frac{d}{dx}(y_2y_1' - y_1y_2')\,dx = \int_a^c (G_2 - G_1)y_1(x)y_2(x)\,dx$$

$$y_2y_1' - y_1y_2'\Big|_a^c = \int_a^c (G_2 - G_1)y_1y_2\,dx$$

$$y_2(c)y_1'(c) - y_1(c)y_2'(c) - \alpha\beta + \alpha\beta = \int_a^c (G_2 - G_1)y_1y_2\,dx \quad (5.89)$$

Let c be the first zero of $y_1(x)$, $c > a$, and assume $y_2(x) \neq 0$ for $a \leqq x \leqq c$. From (5.89) and $y_1(c) = 0$ we have

$$y_2(c)y_1'(c) = \int_a^c (G_2 - G_1)y_1y_2\,dx \quad (5.90)$$

Now $G_2(x) - G_1(x) > 0$ for $a \leqq x \leqq c$, $y_1(x) > 0$ for $a \leqq x < c$, $y_2(x) > 0$ for $a \leqq x \leqq c$. Hence $y_2(c)y_1'(c) > 0$ from (5.90), so that $y_1'(c) > 0$. But

$$y_1'(c) = \lim_{\substack{x \to c \\ x < c}} \frac{y_1(x) - y_1(c)}{x - c} = \lim_{\substack{x \to c \\ x < c}} \frac{y_1(x)}{x - c} \leqq 0$$

since $y_1(x) > 0$ and $x < c$. This is a contradiction. Q.E.D. The same result would have been obtained if $y_1(a) = y_2(a) = \alpha < 0$.

Let the reader show that, if $y_1(c_1) = 0$, $y_1(c_2) = 0$, $y_1(x) \neq 0$ for $c_1 < x < c_2$, then $y_2(x) = 0$ for some x on $c_1 \leqq x \leqq c_2$.

Problems

1. Consider $y'' + p(x)y' + q(x)y = 0$. Let $y = uv$, and determine $u(x)$ so that the resulting second-order differential equation in $v(x)$ does not contain $\frac{dv}{dx}$.

2. Consider Legendre's equation

$$(1 - x^2)\frac{d^2y}{dx^2} - 2x\frac{dy}{dx} + n(n + 1)y = 0 \quad (5.91)$$

Let $y_1(x, \alpha)$, $y_2(x, \beta)$ be solutions of (5.91) for $n = \alpha$ and $n = \beta$, $\alpha \neq \beta$. Show that

$$\int_{-1}^{1} y_1(x, \alpha)y_2(x, \beta)\,dx = 0$$

3. Prove the second statement of (*d*).

4. Give an example of (*b*).

5.9. Differential Equations in the Complex Domain. Up to the present we have discussed differential equations from a real-variable view. Greater insight into the solutions of differential equations can be obtained if we study the differential equations from a complex-variable point of view. The reader may recall that the Taylor-series expansion of

$$f(x) = \frac{1}{1 + x^2}$$

about $x = 0$ converges only for $|x| < 1$. In a way this is puzzling since $1/(1 + x^2)$ has no singularities on the real axis. However, the analytic continuation of $f(x)$, namely, $f(z) = 1/(1 + z^2)$, is known to have poles at $z = \pm i$. The distance of $z = i$ and $z = -i$ from $z = 0$ is 1, so that the Taylor-series expansion of $1/(1 + z^2)$ about $z = 0$ has a radius of convergence equal to unity.

The simplest first-order linear differential equation may be considered without difficulty. If $p(z)$ is analytic at $z = z_0$, the solution of

$$\frac{dw}{dz} + p(z)w = 0 \tag{5.92}$$

is $w(z) = w_0 \exp\left[-\int_{z_0}^{z} p(t)\,dt\right]$ and $w(z)$ is analytic at $z = z_0$.

The first non trivial case is the linear-second order differential equation

$$\frac{d^2w}{dz^2} + p(z)\frac{dw}{dz} + q(z)w = 0 \tag{5.93}$$

This equation is most important to the physicist and engineer. Moreover, the methods used for solving it apply equally well to higher-order equations. In attempting to solve (5.93) we must obviously consider the coefficients $p(z)$ and $q(z)$. Let $z = z_0$ be a point such that $p(z)$ and $q(z)$ are analytic at $z = z_0$. Remember this means that $p(z)$ and $q(z)$ are differentiable in some neighborhood of $z = z_0$. A point z_0 of this type is said to be an ordinary point of (5.93). Let R be the smaller of the two radii of convergence of the series expansions of $p(z)$ and $q(z)$ about $z = z_0$. We shall now prove Theorem 5.2.

THEOREM 5.2. For any two complex numbers w_0, w_0' there exists a unique function $w(z)$ satisfying (5.93) such that $w(z_0) = w_0$, $w'(z_0) = w_0'$. Moreover $w(z)$ is analytic for $|z - z_0| < R$.

We begin the proof by removing the first-order derivative in (5.93). Let $w = uv$, u and v undefined as yet. Then $\frac{dw}{dz} = u\frac{dv}{dz} + v\frac{du}{dz}$,

$$\frac{d^2w}{dz^2} = u\frac{d^2v}{dz^2} + 2\frac{du}{dz}\frac{dv}{dz} + v\frac{d^2u}{dz^2}$$

Substituting into (5.93) yields

$$u\frac{d^2v}{dz^2} + \left(pu + 2\frac{du}{dz}\right)\frac{dv}{dz} + \left(\frac{d^2u}{dz^2} + p\frac{du}{dz} + qu\right)v = 0$$

If we set $pu + 2\frac{du}{dz} = 0$, we see that

$$u(z) = \exp\left[-\tfrac{1}{2}\int_{z_0}^{z} p(t)\,dt\right]$$

Thus the substitution $w = v\exp\left[-\tfrac{1}{2}\int_{z_0}^{z} p(t)\,dt\right]$ reduces (5.93) to

$$\frac{d^2v}{dz^2} + J(z)v = 0 \tag{5.94}$$

Let the reader show that $J(z)$ is analytic for $|z - z_0| < R$. Moreover the reader can also show that if $v(z)$ is a solution of (5.94) then

$$w = v\exp\left[-\tfrac{1}{2}\int_{z_0}^{z} p(t)\,dt\right]$$

is a solution of (5.93). Next we attempt to reduce (5.94) to an integral equation. Assume $v(z)$ satisfies (5.94) such that $v(z_0) = v_0$, $v'(z_0) = v_0'$. Then

$$\int_{z_0}^{z} \frac{d^2v}{dz^2}\,dz = -\int_{z_0}^{z} J(z)v(z)\,dz$$

or

$$\frac{dv}{dz} - v_0' = -\int_{z_0}^{z} J(\tau)v(\tau)\,d\tau$$

Integrating again yields

$$\begin{aligned} v(z) - v_0 - v_0'(z - z_0) &= -\int_{z_0}^{z}\int_{z_0}^{\zeta} J(\tau)v(\tau)\,d\tau\,d\zeta \\ &= -\int_{z_0}^{z} \varphi(\zeta)\,d\zeta \end{aligned}$$

where $\varphi(\zeta) = \int_{z_0}^{\zeta} J(\tau)v(\tau)\,d\tau$. Note that $\varphi(z_0) = 0$. We now integrate by parts and obtain

$$v(z) - v_0 - v_0'(z - z_0) = -\zeta\varphi(\zeta)\Big|_{z_0}^{z} + \int_{z_0}^{z} \zeta J(\zeta)\,v(\zeta)\,d\zeta$$

and

$$v(z) = v_0 + v_0'(z - z_0) + \int_{z_0}^{z} (\zeta - z)J(\zeta)v(\zeta)\,d\zeta \tag{5.95}$$

The reader can show that if $v(z)$ satisfies (5.95) then $v(z)$ also satisfies (5.94). Equation (5.95) is a Volterra integral equation of the first kind. It is a special case of the integral equation

$$v(z) = A(z) + \int_{z_0}^{z} k(z, \zeta)v(\zeta)\,d\zeta \tag{5.96}$$

To find a solution of (5.95), we apply the method of successive approximations due to Picard (see Sec. 5.5). We define the sequence $v_0(z)$,

$v_1(z), v_2(z), \ldots, v_n(z), \ldots$ as follows:

$$
\begin{aligned}
v_0(z) &= v_0 + v_0'(z - z_0) \\
v_1(z) &= v_0 + v_0'(z - z_0) + \int_{z_0}^{z} (\zeta - z)J(\zeta)v_0(\zeta)\,d\zeta \\
v_2(z) &= v_0 + v_0'(z - z_0) + \int_{z_0}^{z} (\zeta - z)J(\zeta)v_1(\zeta)\,d\zeta \\
&\cdots\cdots\cdots\cdots\cdots\cdots \\
v_{n+1}(z) &= v_0 + v_0'(z - z_0) + \int_{z_0}^{z} (\zeta - z)J(\zeta)v_n(\zeta)\,d\zeta \\
&\cdots\cdots\cdots\cdots\cdots\cdots
\end{aligned}
\tag{5.97}
$$

Now

$$
v_{n+1}(z) = v_0(z) + [v_1(z) - v_0(z)] + [v_2(z) - v_1(z)] + \cdots + [v_{n+1}(z) - v_n(z)]
$$

so that the convergence of the sequence $\{v_n(z)\}$ hinges on the convergence of the series $v_0(z) + \sum_{k=0}^{\infty} [v_{k+1}(z) - v_k(z)]$. We can write

$$
v_{k+1}(z) - v_k(z) = \int_{z_0}^{z} (\zeta - z)J(\zeta)[v_k(\zeta) - v_{k-1}(\zeta)]\,d\zeta \tag{5.98}
$$

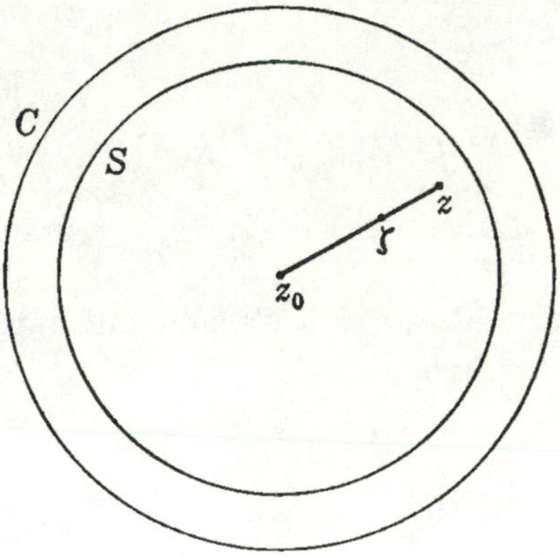

FIG. 5.1

We choose as our path of integration the straight line joining z_0 to z so that $\zeta = z_0 + t(z - z_0)$, $0 \leq t \leq 1$, with $\zeta - z_0 = t(z - z_0)$ and

$$
d\zeta = (z - z_0)\,dt
$$

Let C be the circle of analyticity of $J(z)$ with center at z_0, and let z be an interior point. We construct a circle S with center at z_0 interior to C and containing the given point z in its interior (see Fig. 5.1). Since $J(\zeta)$ and $\zeta - z$ are analytic inside and on S, $J(\zeta)(\zeta - z)$ is bounded by some constant M, that is, $|J(\zeta)(\zeta - z)| < M$ for all z, ζ in and on S. Hence

$$
\begin{aligned}
|v_1(z) - v_0(z)| &= \left| \int_{z_0}^{z} (\zeta - z)J(\zeta)v_0(\zeta)\,d\zeta \right| \\
&< \mu M \int_{z_0}^{z} |d\zeta| = \mu M|z - z_0|
\end{aligned}
$$

where μ is any upper bound of $v_0(z)$ in and on S. Similarly

$$\begin{aligned}|v_2(z) - v_1(z)| &= \left| \int_{z_0}^{z} (\zeta - z)J(\zeta)[v_1(\zeta) - v_0(\zeta)]\, d\zeta \right| \\ &< \mu M^2 \int_{z_0}^{z} |\zeta - z_0|\, |d\zeta| \\ &< \mu M^2 |z - z_0|^2 \int_0^1 t\, dt = \mu M^2 \frac{|z - z_0|^2}{2!}\end{aligned}$$

By mathematical induction the reader can show that

$$|v_{k+1}(z) - v_k(z)| < \mu M^{n+1} \frac{|z - z_0|^{n+1}}{(n+1)!} \tag{5.99}$$

Thus the series representing $v_{n+1}(z)$ is bounded by the series

$$\mu \sum_{k=0}^{n+1} \frac{M^k |z - z_0|^k}{k!}$$

which in turn is bounded by the series of constant terms $\mu \sum_{k=0}^{\infty} \frac{M^k R^k}{k!}$ since $R > |z - z_0|$. We know that the latter series converges to μe^{MR} so that from the Weierstrass M test the series representing $v_{n+1}(z)$ converges uniformly. Since each term of the series representing $v_{n+1}(z)$ is analytic, $v_{n+1}(z)$ converges to an analytic function $v(z)$ (see Prob. 7, Sec. 4.9). We now show that the limiting function, $v(z)$, satisfies (5.95). From (5.97)

$$\begin{aligned}\lim_{n\to\infty} v_{n+1}(z) &= v_0 + v_0'(z - z_0) + \lim_{n\to\infty} \int_{z_0}^{z} (\zeta - z)J(\zeta)v_n(\zeta)\, d\zeta \\ v(z) &= v_0 + v_0'(z - z_0) + \int_{z_0}^{z} \lim_{n\to\infty} (\zeta - z)J(\zeta)v_n(\zeta)\, d\zeta \\ &= v_0 + v_0'(z - z_0) + \int_{z_0}^{z} (\zeta - z)J(\zeta)v(\zeta)\, d\zeta\end{aligned}$$

It is possible to take the limit process inside the integral because of the uniform convergence of $v_n(z)$ to $v(z)$.

Next we prove that $v(z)$ is unique. Let $u(z)$ also satisfy (5.95). It is easy to verify that $r(z) \equiv v(z) - u(z)$ satisfies

$$r(z) = \int_{z_0}^{z} (\zeta - z)J(\zeta)r(\zeta)\, d\zeta \tag{5.100}$$

Since $u(z)$ and $v(z)$ are analytic in and on S, the function $r(z)$ is bounded by some constant K, $|r(\zeta)| < K$. From (5.100) we have

$$|r(z)| < MK \int_{z_0}^{z} |d\zeta| = MK|z - z_0|$$

Applying this inequality to (5.100) yields

$$|r(z)| < M \int_{z_0}^{z} |r(\zeta)|\,|d\zeta| < M^2K \int_{z_0}^{z} |\zeta - z_0|\,|d\zeta|$$
$$< M^2K \frac{|z - z_0|^2}{2!}$$

Continuing this process yields

$$|r(z)| < KM^n \frac{|z - z_0|^n}{n!}$$

for all integers n. Since $\lim_{n\to\infty} \dfrac{KM^n|z - z_0|^n}{n!} = 0$, we can make $|r(z)|$ as small as we please. This is possible only if $r(z) \equiv 0$ so that $u(z) \equiv v(z)$. Q.E.D.

Example 5.15. We consider $w'' - zw' - w = 0$. Certainly $z = 0$ is an ordinary point. An analytic solution is known to exist. The easiest way to find this solution is to let $w = \sum_{n=0}^{\infty} c_n z^n$. The c_n are to be determined by the condition that w satisfy $w'' - zw' - w = 0$. We have

$$w' = \sum_{n=1}^{\infty} nc_n z^{n-1} \qquad w'' = \sum_{n=2}^{\infty} n(n-1)c_n z^{n-2}$$

so that

$$\sum_{n=2}^{\infty} n(n-1)c_n z^{n-2} - \sum_{n=1}^{\infty} nc_n z^n - \sum_{n=0}^{\infty} c_n z^n = 0$$

A power series in z can be identically zero only if the coefficients of z^n, $n = 0, 1, 2, \ldots$, vanish. Hence

$$(n+2)(n+1)c_{n+2} - (n+1)c_n = 0 \qquad n = 0, 1, 2, \ldots$$

We obtain the recursion formula

$$c_{n+2} = \frac{c_n}{n+2} \qquad n = 0, 1, 2, \ldots$$

We note that $c_2 = c_0/2$, $c_4 = c_2/4 = c_0/(2 \cdot 4) = c_0/2^2 2!$,

$$c_6 = \frac{c_4}{6} = \frac{c_0}{(2 \cdot 4 \cdot 6)} = \frac{c_0}{2^3 3!}$$

$\ldots$, $c_{2n} = c_0/2^n n!$. Also $c_3 = c_1/3$, $c_5 = c_3/5 = c_1/(1 \cdot 3 \cdot 5)$,

$$c_7 = \frac{c_5}{7} = \frac{c}{(1 \cdot 3 \cdot 5 \cdot 7)}$$

$\ldots$, $c_{2n+1} = c_1/(2n+1)!!$, where

$$(2n+1)!! = (2n+1)(2n-1)(2n-3) \cdots 5 \cdot 3 \cdot 1$$

The constants c_0 and c_1 are arbitrary constants of integration. The general solution of $w'' - zw' - w = 0$ is

$$w(z) = A \sum_{n=0}^{\infty} \frac{z^{2n}}{2^n n!} + B \sum_{n=0}^{\infty} \frac{z^{2n+1}}{(2n+1)!!}$$

Problems

1. Show that, if $v(z)$ satisfies (5.95), then $v(z_0) = v_0$, $v'(z_0) = v_0'$. Also show that $v(z)$ satisfies (5.94) if $v(z)$ satisfies (5.95).

2. Prove (5.99) by mathematical induction.

3. Solve $w'' - z^2w' + zw = 0$ for $w(z)$ in a Taylor series about $z = 0$.

4. Solve $w'' + 1/(1 - z)w' + w = 0$ in a Taylor series about $z = 0$.

5.10. Singular Points. Let $z = a$ be an isolated singularity of either $p(z)$ or $q(z)$ in (5.93), and surround $z = a$ by a circle, C, of radius ρ, such that $p(z)$ and $q(z)$ are analytic for $0 < |z - a| < \rho$. Now let z_0 be an ordinary point of (5.93), $0 < |z_0 - a| < \rho$ (see Fig. 5.2). From Sec. 5.9 there exists a solution of (5.93) which is analytic in a neighborhood of z_0. Call the solution $w_1(z, z_0)$. The Taylor-series expansion of $w_1(z) \equiv w_1(z, z_0)$ about $z = z_0$ converges up to the nearest singularity of $p(z)$ and $q(z)$, which may be the point $z = a$ or a point on the circle C. In Fig. 5.2 C' is the circle of convergence. Now let Γ be any simple closed curve through z_0 lying inside C and surrounding $z = a$ (see Fig. 5.2). We choose a point z_1 on Γ, z_1 inside C'. Since $w_1(z, z_0)$ is defined at z_1, we can compute $w_1(z_1, z_0)$, $w_1'(z_1, z_0)$. Since z_1 is an ordinary point of (5.93), there exists a unique function $w_{11}(z)$ such that $w_{11}(z)$ satisfies (5.93) and $w_{11}(z_1) = w_1(z_1, z_0)$, $w_{11}'(z_1) = w_1'(z_1, z_0)$. $w_{11}(z)$ is an analytic continuation of $w_1(z, z_0)$. Its domain of definition is the interior of the circle C'' (see Fig. 5.2). This process of analytic continuation can be continued, and the reader can show that in a finite number of such continuations we can reach z_0. Let $w_1^*(z)$ be the analytic function which is the analytic continuation of $w_1(z)$, both $w_1(z)$ and $w_1^*(z)$ analytic in a neighborhood of $z = z_0$. We write $w_1^*(z) = \Lambda w_1(z)$ and look upon Λ as an operator associated with analytic continuation. We leave it as an exercise for the reader to show that, if $w_1(z)$ and $w_2(z)$ are linearly independent solutions of (5.93) for some neighborhood of $z = z_0$, then $\Lambda w_1(z)$ and $\Lambda w_2(z)$ are also linearly independent.

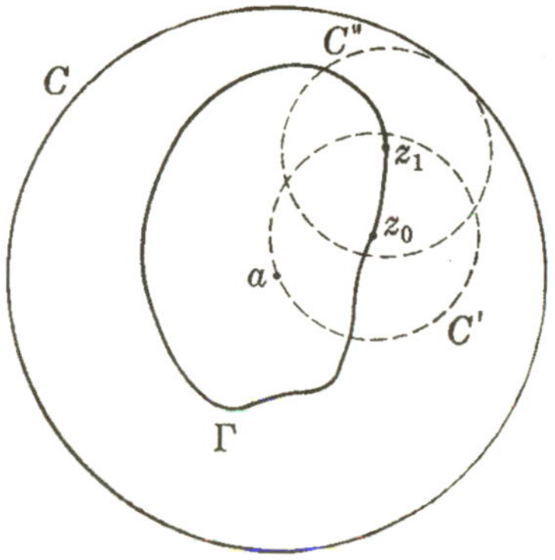

FIG. 5.2

Since $z = a$ is a singular point, we cannot expect necessarily that

$\Lambda w_1(z) \equiv w_1(z)$. Let us now attempt to find a solution $w(z)$ of (5.93) such that

$$\Lambda w(z) = \lambda w(z) \tag{5.101}$$

If $w_1(z)$ and $w_2(z)$ are linearly independent solutions of (5.93) for a neighborhood of $z = z_0$, we have

$$\begin{aligned} w(z) &= c_1 w_1(z) + c_2 w_2(z) \\ \Lambda w_1(z) &= a_{11} w_1(z) + a_{12} w_2(z) \\ \Lambda w_2(z) &= a_{21} w_1(z) + a_{22} w_2(z) \\ \Lambda w(z) &= c_1 \Lambda w_1(z) + c_2 \Lambda w_2(z) \end{aligned} \tag{5.102}$$

The last equation of (5.102) implies that $\Lambda c_1 w_1 = c_1 \Lambda w_1$ and that

$$\Lambda(w_1 + w_2) = \Lambda w_1 + \Lambda w_2$$

In other words, Λ is a linear operator. Let the reader deduce this fact. From (5.101) and (5.102)

$$c_1(a_{11}w_1 + a_{12}w_2) + c_2(a_{21}w_1 + a_{22}w_2) = \lambda(c_1 w_1 + c_2 w_2)$$

which in turn implies

$$\begin{aligned} c_1(a_{11} - \lambda) + c_2 a_{21} &= 0 \\ c_1 a_{12} + c_2(a_{22} - \lambda) &= 0 \end{aligned} \tag{5.103}$$

A nontrivial solution in c_1, c_2 exists if and only if

$$\begin{vmatrix} a_{11} - \lambda & a_{21} \\ a_{12} & a_{22} - \lambda \end{vmatrix} = 0 \tag{5.104}$$

Equation (5.104) is a quadratic equation in λ, so possesses two distinct roots or two equal roots.

Case 1. Let $\varphi_1(z)$, $\varphi_2(z)$ be the functions of (5.101) corresponding to the distinct roots λ_1, λ_2, that is,

$$\Lambda\varphi_i = \lambda_i \varphi_i \qquad i = 1, 2 \tag{5.105}$$

It is an easy task to show that φ_1 and φ_2 are linearly independent.

Let us note the following: Let $z - a = |z - a|e^{i\theta}$ and

$$(z - a)^\alpha = |z - a|^\alpha e^{i\alpha\theta}$$

The analytic continuation of $(z - a)^\alpha$ around Γ increases θ by 2π so that

$$\Lambda(z - a)^\alpha = e^{2\pi i\alpha}(z - a)^\alpha$$

We choose α_j, $j = 1, 2$, so that $e^{2\pi i\alpha_j} = \lambda_j$, $j = 1, 2$. Hence

$$\Lambda(z - a)^{\alpha_j} = \lambda_j(z - a)^{\alpha_j} \qquad j = 1, 2 \tag{5.106}$$

Combining (5.105) and (5.106) yields

$$\begin{aligned}\Lambda[(z-a)^{-\alpha_j}\varphi_j(z)] &= \Lambda(z-a)^{-\alpha_j}\Lambda\varphi_j(z) \qquad \text{Why?}\\ &= \frac{1}{\lambda_j}(z-a)^{-\alpha_j}\lambda_j\varphi_j(z)\\ &= (z-a)^{-\alpha_j}\varphi_j(z) \qquad j = 1, 2 \end{aligned} \tag{5.107}$$

Thus the functions $F_j(z) \equiv (z-a)^{-\alpha_j}\varphi_j(z)$, $j = 1, 2$, are single-valued inside C with a possible singularity at $z = a$. From Laurent's expansion theorem we have

$$(z-a)^{-\alpha_1}\varphi_1(z) = \sum_{n=-\infty}^{\infty} b_n(z-a)^n$$

$$(z-a)^{-\alpha_2}\varphi_2(z) = \sum_{n=-\infty}^{\infty} c_n(z-a)^n$$

or

$$\begin{aligned}\varphi_1(z) &= (z-a)^{\alpha_1}\sum_{n=-\infty}^{\infty} b_n(z-a)^n\\ \varphi_2(z) &= (z-a)^{\alpha_2}\sum_{n=-\infty}^{\infty} c_n(z-a)^n\end{aligned} \tag{5.108}$$

Thus both $\varphi_1(z)$ and $\varphi_2(z)$ have, in general, a branch point and an essential singularity at $z = a$.

Case 2. The reader is referred to MacRobert, "Functions of a Complex Variable," if $\lambda_1 = \lambda_2$. In this case

$$\varphi(z) = (z-a)^{\alpha}\left[w_1(z) + \frac{w_2(z)}{2\pi i}\ln(z-a)\right] \tag{5.109}$$

where $w_1(z)$ and $w_2(z)$ are analytic at $z = a$.

An important case occurs when the functions $\varphi_1(z)$, $\varphi_2(z)$ of (5.108) have no essential singularities. In this case we can write

$$\varphi_j(z) = (z-a)^{\beta_j}\psi_j(z) \qquad j = 1, 2 \tag{5.110}$$

where $\psi_j(z)$ is analytic at $z = a$; $\psi_j(a) \neq 0$. Now

$$\begin{aligned}\varphi_1'' + p(z)\varphi_1' + q(z)\varphi_1 &= 0\\ \varphi_2'' + p(z)\varphi_2' + q(z)\varphi_2 &= 0\end{aligned} \tag{5.111}$$

so that

$$\varphi_2\varphi_1'' - \varphi_1\varphi_2'' + p(z)(\varphi_2\varphi_1' - \varphi_1\varphi_2') = 0$$

Thus

$$\frac{d}{dz}(\varphi_2\varphi_1' - \varphi_1\varphi_2') = -p(z)(\varphi_2\varphi_1' - \varphi_1\varphi_2')$$

and

$$p(z) = -\frac{1}{W(\varphi_1, \varphi_2)}\frac{dW(\varphi_1, \varphi_2)}{dz} \tag{5.112}$$

Making use of (5.110) to (5.112), let the reader show that

$$\begin{aligned} p(z) &= \frac{c_{-1}}{z-a} + p^*(z) \\ q(z) &= \frac{d_{-2}}{(z-a)^2} + \frac{d_{-1}}{z-a} + q^*(z) \end{aligned} \tag{5.113}$$

with $p^*(z)$, $q^*(z)$ analytic at $z = a$. The same result occurs for $\varphi(z)$ of (5.109). We notice that $p(z)$ of (5.113) has at most a simple pole at $z = a$ and $q(z)$ has at most a double pole at $z = a$. This suggests the following definition:

Definition 5.1. $z = z_0$ is said to be a regular singular point of

$$w'' + p(z)w' + q(z)w = 0 \tag{5.114}$$

if:

(i) z_0 is not an ordinary point.
(ii) $(z - z_0)p(z)$ is analytic at $z = z_0$.
(iii) $(z - z_0)^2 q(z)$ is analytic at $z = z_0$.

We now attempt to find a solution of (5.114) in the neighborhood of a regular singular point. To simplify matters, we assume that $z_0 = 0$. The transformation $z' = z - z_0$ transfers the singularity to the origin. Let $w(z) = z^\alpha u(z)$, α an unknown constant, $u(z)$ undefined as yet. For $w = z^\alpha u$ we have $w' = z^\alpha u'(z) + \alpha z^{\alpha-1}u$,

$$w'' = z^\alpha u'' + 2\alpha z^{\alpha-1}u' + \alpha(\alpha - 1)z^{\alpha-2}u$$

If $w(z)$ satisfies (5.114), then $u(z)$ must satisfy

$$z^2u'' + [2\alpha z + p(z)z^2]u' + [\alpha(\alpha - 1) + \alpha z p(z) + z^2 q(z)]u = 0 \tag{5.115}$$

Now

$$\begin{aligned} zp(z) &= p_0 + p_1 z + p_2 z^2 + \cdots \\ z^2 q(z) &= q_0 + q_1 z + q_2 z^2 + \cdots \end{aligned}$$

Equation (5.115) becomes

$$zu''(z) + P(z)u'(z) + Q(z)u(z) = 0 \tag{5.116}$$

provided α is chosen as a root of the indicial equation

$$\alpha(\alpha - 1) + \alpha p_0 + q_0 = 0 \tag{5.117}$$

with

$$\begin{aligned} P(z) &= 2\alpha + zp(z) = 2\alpha + p_0 + p_1 z + p_2 z^2 + \cdots \\ Q(z) &= \alpha(p_1 + p_2 z + \cdots) + q_1 + q_2 z + \cdots \end{aligned}$$

Conversely, if $u(z)$ satisfies (5.116), then $w(z) = z^\alpha u(z)$ will satisfy (5.114) provided α satisfies (5.117). The existence of a solution of (5.114) hinges on the existence of a solution of (5.116). If $u(z)$ satisfies (5.116),

$$u(0) = u_0$$

then a double integration of (5.116) yields

$$zu(z) = [P(0) - 1]u_0z + \int_0^z [2 - P(\tau) + (\tau - z)(Q(\tau) - P'(\tau))]u(\tau)\,d\tau$$

and
$$u(z) = [P(0) - 1]u_0 + \frac{1}{z}\int_0^z B(z, \tau)u(\tau)\,d\tau \qquad z \neq 0 \tag{5.118}$$

with
$$B(z, \tau) = 2 - P(\tau) + (\tau - z)[Q(\tau) - P'(\tau)]$$

We leave it to the reader to show that, if $u(z)$ satisfies (5.118), then $u(z)$ satisfies (5.116). The reader can also show that if we define $u(0) = u_0$ then $u(z)$ of (5.118) is analytic at $z = 0$ so that (5.118) holds for $z = 0$. To do this, the reader must show that $\lim_{z\to 0} u(z) = u_0$.

Now $P(0) = 2\alpha + p_0$ and $\alpha_1 + \alpha_2 = 1 - p_0$ if α_1 and α_2 are the roots of (5.117). If $\alpha_1 = \alpha_2$, then $P(0) \equiv 1$ so that (5.118) becomes

$$u(z) = \frac{1}{z}\int_0^z B(z, \tau)u(\tau)\,d\tau \tag{5.119}$$

If we attempt to prove the existence of a solution of (5.119) by Picard's method of successive approximations, we might expect trouble at $z = 0$. Let our first approximation be $u_0(z) \equiv u_0$, and define

$$u_1(z) = \frac{1}{z}\int_0^z B(z, \tau)u_0\,d\tau \qquad z \neq 0$$

Let the reader show that $\lim u_1(z) = u_0$, so that, if we define $u_1(0) = u_0$, $u_1(z)$ becomes continuous and, indeed, analytic at $z = 0$. Continuing this process yields

$$u_n(z) = \frac{1}{z}\int_0^z B(z, \tau)u_{n-1}(\tau)\,d\tau \tag{5.120}$$

with $u_n(0) = u_0$, $n = 0, 1, 2, \ldots$. We define $u_n(0) = u_0$, and the reader can show that $u_n(z)$ is analytic at $z = 0$. Since $P(0) = 1$, we can write $P(\tau) = 1 + \tau f(\tau)$ and $B(z, \tau) = 1 - \tau f(\tau) + (\tau - z)[Q(\tau) - P'(\tau)]$. We can certainly pick a small circle, C, with center at $z = 0$ such that $|B(z, \tau)| < R < 2$ for all z, τ inside and on the circle. Moreover

$$\begin{aligned}|u_1(z) - u_0(z)| &= \left|\frac{1}{z}\int_0^z \{1 - P(\tau) + (\tau - z)[Q(\tau) - P'(\tau)]\}\,d\tau\right| \\ &= \left|\frac{1}{z}\int_0^z \{-\tau f(\tau) + (\tau - z)[Q(\tau) - P'(\tau)]\}\,d\tau\right|\end{aligned}$$

If we let $\tau = tz$, $0 \leqq t \leqq 1$, t the variable of integration, it is easy to see that

$$|u_1(z) - u_0(z)| < A|z| \qquad A = \text{constant}$$

for all z inside C.

Since
$$|u_2(z) - u_1(z)| = \left|\frac{1}{z}\int_0^z B(z, \tau)[u_1(\tau) - u_0(\tau)]\,d\tau\right|$$

we have

$$|u_2(z) - u_1(z)| < \left|\frac{1}{z}\right| AR \int_0^z |\tau|\,|d\tau| = AR\frac{|z|}{2}$$

Continuing, we obtain

$$|u_n(z) - u_{n-1}(z)| < A\left(\frac{R}{2}\right)^{n-1}|z|$$

Since the series $A|z|\sum_{n=1}^{\infty}(R/2)^{n-1}$ converges uniformly inside and on C, remember $R < 2$, the sequence $\{u_n(z)\}$ converges uniformly to an analytic function $u(z)$. As was done in Sec. 5.9 for $z = z_0$ an ordinary point, it can be shown that $u(z)$ satisfies (5.116) and (5.115). It is then trivial to show that $w(z) = z^\alpha u(z)$ satisfies (5.114).

Example 5.16. $zw'' + w' - w = 0$ or $w'' + (1/z)w' - (1/z)w = 0$. In this case $p(z) = 1/z$, $q(z) = -1/z$ and $zp(z) = 1$, $z^2q(z) = -z$ so that $z = 0$ is a regular singular point. We know that a solution exists in the form $w = z^\alpha u(z)$, where α satisfies a quadratic equation and $u(z)$ is analytic at $z = 0$. One of the simplest ways of determining $u(z)$ is to assume $u(z) = \sum_{n=0}^{\infty} c_n z^n$, $w(z) = \sum_{n=0}^{\infty} c_n z^{n+\alpha}$. The series for $w(z)$ is substituted into $zw'' + w' - w = 0$, and the coefficient of each power of z is equated to zero. More specifically, we have

$$w'(z) = \sum_{n=0}^{\infty} c_n(n+\alpha)z^{n+\alpha-1} \qquad w''(z) = \sum_{n=0}^{\infty} c_n(n+\alpha)(n+\alpha-1)z^{n+\alpha-2}$$

and hence $w(z) = \sum_{n=0}^{\infty} c_n z^{n+\alpha}$ satisfies $zw'' + w' - w = 0$ if and only if

$$\sum_{n=0}^{\infty} c_n(n+\alpha)(n+\alpha-1)z^{n+\alpha-1} + \sum_{n=0}^{\infty} c_n(n+\alpha)z^{n+\alpha-1} - \sum_{n=0}^{\infty} c_n z^{n+\alpha} = 0 \quad (5.121)$$

The lowest power of z which occurs in (5.121) is $z^{\alpha-1}$. The coefficient of $z^{\alpha-1}$ is $c_0\alpha(\alpha - 1) + c_0\alpha \equiv c_0[\alpha(\alpha - 1) + \alpha]$. If (5.121) is to be satisfied, we must have

$$c_0[\alpha(\alpha - 1) + \alpha] = 0$$

c_0 is to be arbitrary so that α must satisfy the quadratic (indicial) equation

$$\alpha(\alpha - 1) + \alpha = 0$$

or $\alpha^2 = 0$. The roots of the indicial equation are $\alpha = 0, 0$. The coefficient of $z^{n+\alpha-1}$ in (5.121) is

$$c_n(n + \alpha)(n + \alpha - 1) + c_n(n + \alpha) - c_{n-1} \qquad n = 1, 2, 3, \ldots$$

If (5.121) is to be satisfied, we must have

$$c_n(n + \alpha)(n + \alpha - 1) + c_n(n + \alpha) - c_{n-1} = 0$$

or

$$c_n = \frac{c_{n-1}}{(n + \alpha)^2} \qquad n = 1, 2, 3, \ldots \tag{5.122}$$

Equation (5.122) is the recursion formula for the c_n, $n = 1, 2, \ldots$. Since $\alpha = 0$, we have

$$c_n = \frac{c_{n-1}}{n^2} \qquad n = 1, 2, \ldots$$

so that $\quad c_1 = \frac{c_0}{1^2} \qquad c_2 = \frac{c_1}{2^2} = \frac{c_0}{1^2 \cdot 2^2} \qquad c_3 = \frac{c^2}{3^2} = \frac{c_0}{1^2 2^2 3^2} = \frac{c_0}{(3!)^2}$

and by induction $c_n = c_0/(n!)^2$. Hence a solution of $zw'' + w' - w = 0$ is

$$w(z) = A \sum_{n=0}^{\infty} \frac{z^n}{(n!)^2}, \; c_0 = A$$

We postpone the discussion for finding a second independent solution of

$$zw'' + w' - w = 0$$

The solution $w(z) = A \sum_{n=0}^{\infty} \frac{z^n}{(n!)^2}$ converges for $0 \leq |z| < \infty$. This is exactly the region of analyticity of $zp(z) = 1$, $z^2q(z) = -z$.

The reader can find a proof in Copson, "Theory of Functions of a Complex Variable," that, if $zp(z)$ and $z^2q(z)$ are analytic for $|z| < R$, then $w'' + p(z)w' + q(z)w = 0$ has a solution $w(z) = \sum_{n=0}^{\infty} c_n z^{n+\alpha}$ and the series $\sum_{n=0}^{\infty} c_n z^n$ converges at least for $|z| < R$. The roots of the indicial equation need not be equal.

Example 5.17. $zw'' + (z - 1)w' + w = 0$. The reader can easily note that $z = 0$ is a regular singular point. Let $w(z) = \sum_{n=0}^{\infty} c_n z^{n+\alpha}$, so that

$$w'(z) = \sum_{n=0}^{\infty} c_n(n + \alpha)z^{n+\alpha-1}$$

$w''(z) = \sum_{n=0}^{\infty} c_n(n + \alpha)(n + \alpha - 1)z^{n+\alpha-2}$. Substituting into

$$zw'' + (z - 1)w' + w = 0$$

yields

$$\sum_{n=0}^{\infty} c_n(n + \alpha)(n + \alpha - 2)z^{n+\alpha-1} + \sum_{n=0}^{\infty} c_n(n + \alpha + 1)z^{n+\alpha} = 0 \qquad (5.123)$$

The smallest power of z occurring in (5.123) is $z^{\alpha-1}$, whose coefficient must be equated to zero. Thus

$$c_0\alpha(\alpha - 2) = 0$$

and α satisfies the indicial equation $\alpha(\alpha - 2) = 0$, so that $\alpha = 0, 2$. We shall see that the larger of these two roots produces no difficulty; the smaller of these two roots does produce a difficulty.

The coefficient of $z^{n+\alpha-1}$, $n = 1, 2, \ldots$, in (5.123) must be set equal to zero. This leads to the recursion formula

$$c_n(n + \alpha)(n + \alpha - 2) + c_{n-1}(n + \alpha) = 0$$

or

$$c_n = -\frac{c_{n-1}}{n + \alpha - 2} \qquad n = 1, 2, 3, \ldots$$

If we try $\alpha = 0$, we have $c_n = -c_{n-1}/(n - 2)$ and c_2 becomes meaningless. We try $\alpha = 2$ and obtain $c_n = -c_{n-1}/n$, so that $c_1 = -c_0$, $c_2 = -c_1/2 = c_0/2$,

$$c_3 = -\frac{c_2}{3} = -\frac{c_0}{3!}$$

$c_4 = -c_3/4 = c_0/4!$, and by induction $c_n = (-1)^n(c_0/n!)$. Hence

$$w(z) = Az^2 \sum_{n=0}^{\infty} (-1)^n \frac{z^n}{n!} = Az^2e^{-z}$$

is a solution of $zw'' + (z - 1)w' + w = 0$.

It is not very difficult to see that a second solution cannot be obtained by the method of series solution used in Examples 5.16 and 5.17 when the roots of the indicial equation differ by an integer. When the roots do differ by an integer the larger of the two roots presents no difficulty.

Let $w(z) = \sum_{n=0}^{\infty} c_n z^{n+\alpha}$ be a solution of (5.114). Then

$$w'(z) = \sum_{n=0}^{\infty} c_n(n + \alpha)z^{n+\alpha-1} \qquad w''(z) = \sum_{n=0}^{\infty} c_n(n + \alpha)(n + \alpha - 1)z^{n+\alpha-2}$$

and substituting into (5.114) yields

$$\sum_{n=0}^{\infty} c_n(n + \alpha)(n + \alpha - 1)z^{n+\alpha-2} + p(z) \sum_{n=0}^{\infty} c_n(n + \alpha)z^{n+\alpha-1} + q(z) \sum_{n=0}^{\infty} c_n z^{n+\alpha} = 0 \qquad (5.124)$$

Since $zp(z) = p_0 + p_1 z + p_2 z^2 + \cdots$, $z^2 q(z) = q_0 + q_1 z + q_2 z^2 + \cdots$, we have

$$\sum_{n=0}^{\infty} [c_n(n+\alpha)(n+\alpha-1) + c_n(n+\alpha)(p_0 + p_1 z + \cdots) + c_n(q_0 + q_1 z + \cdots)]z^{n+\alpha-2} = 0$$

Of necessity

$$\alpha(\alpha - 1) + \alpha p_0 + q_0 = 0 \qquad \text{indicial equation}$$

The coefficient of $z^{n+\alpha-2}$ must be zero, which implies

$$c_n[(n+\alpha)(n+\alpha-1) + (n+\alpha)p_0 + q_0] = -\sum_{s=0}^{n-1} c_s[(s+\alpha)p_{n-s} + q_{n-s}] \qquad n \geqq 1 \tag{5.125}$$

Equation (5.125), the recursion formula for the c_n, will determine c_n, $n = 1, 2, \ldots$, in terms of c_0 unless

$$\begin{aligned} F(\alpha, n) &\equiv (n+\alpha)(n+\alpha-1) + (n+\alpha)p_0 + q_0 = 0 \\ &= (n+\alpha)(n+\alpha-1+p_0) + q_0 = 0 \end{aligned}$$

for some integer n. If α, α_1 are the roots of the indicial equation, $\alpha \geqq \alpha_1$, we have $\alpha + \alpha_1 = 1 - p_0$, $\alpha\alpha_1 = q_0$, so that

$$\begin{aligned} F(\alpha, n) &= (n+\alpha)(n-\alpha_1) + \alpha\alpha_1 \\ &= n(n+\alpha-\alpha_1) \end{aligned}$$

Since $\alpha \geqq \alpha_1$, $F(\alpha, n) \neq 0$ for $n \geqq 1$. The only difficulty would occur if $\alpha - \alpha_1 =$ negative integer. To obtain a second solution when the roots of the indicial equation differ by an integer, we turn to the method of Frobenius.

Method of Frobenius. We use (5.125) to determine $c_1, c_2, \ldots, c_n, \ldots$ in terms of c_0 and α. The specific value of α is not inserted. The series

$$w(z, \alpha) = \sum_{n=0}^{\infty} c_n(\alpha) z^{n+\alpha} \tag{5.126}$$

determined from (5.125) does not satisfy (5.114) but satisfies

$$\frac{\partial^2 w(z,\alpha)}{\partial z^2} + p(z)\frac{\partial w(z,\alpha)}{\partial z} + q(z)w(z,\alpha) = c_0(\alpha-\alpha_1)(\alpha-\alpha_2)z^{\alpha-2} \tag{5.127}$$

where α_1 and α_2 are the roots of the indicial equation. The $c_n(\alpha)$ of (5.126) are well defined for the larger of the two roots, say, α_1. Now let

$c_0 = A(\alpha - \alpha_2)$ so that (5.127) becomes

$$\frac{\partial^2 w(z, \alpha)}{\partial z^2} + p(z)\frac{\partial w(z, \alpha)}{\partial z} + q(z)w(z, \alpha) = A(\alpha - \alpha_1)(\alpha - \alpha_2)^2 z^{\alpha-2} \quad (5.128)$$

The change of c_0 to $A(\alpha - \alpha_2)$ applies also to the coefficients $c_1, c_2, \ldots$ since these coefficients depend on c_0. If $\alpha_1 = \alpha_2$, it is not necessary to replace c_0 by $A(\alpha - \alpha_2)$. We differentiate (5.128) with respect to α. Since α and z are independent variables, we have

$$\frac{\partial^2}{\partial z^2}\left(\frac{\partial w}{\partial \alpha}\right) + p(z)\frac{\partial}{\partial z}\left(\frac{\partial w}{\partial \alpha}\right) + q\left(\frac{\partial w}{\partial \alpha}\right) = 2A(\alpha - \alpha_1)(\alpha - \alpha_2)z^{\alpha-2} + A(\alpha - \alpha_2)^2\frac{\partial}{\partial \alpha}[(\alpha - \alpha_1)z^{\alpha-2}]$$

Evaluating at $\alpha = \alpha_2$ yields

$$\frac{\partial^2}{\partial z^2}\left[\left(\frac{\partial w}{\partial \alpha}\right)_{\alpha=\alpha_2}\right] + p(z)\frac{\partial}{\partial z}\left[\left(\frac{\partial w}{\partial \alpha}\right)_{\alpha=\alpha_2}\right] + q(z)\left(\frac{\partial w}{\partial \alpha}\right)_{\alpha=\alpha_2} = 0 \quad (5.129)$$

Equation (5.129) shows that $\left(\frac{\partial w}{\partial \alpha}\right)_{\alpha=\alpha_2}$ is a solution of (5.114).

Example 5.18. Returning to Example 5.16, we have

$$c_1 = \frac{c_0}{(1+\alpha)^2}, \; c_2 = \frac{c_1}{(2+\alpha)^2} = \frac{c_0}{(1+\alpha)^2(2+\alpha)^2}, \; \ldots, \quad c_n = \frac{c_0}{(1+\alpha)^2(2+\alpha)^2 \cdots (n+\alpha)^2},$$

so that

$$w(z, \alpha) = z^\alpha c_0\left[1 + \frac{z}{(1+\alpha)^2} + \frac{z^2}{(1+\alpha)^2(2+\alpha)^2} + \cdots + \frac{z^n}{(1+\alpha)^2(2+\alpha)^2 \cdots (n+\alpha)^2} + \cdots\right]$$

To compute $\left(\frac{\partial w}{\partial \alpha}\right)_{\alpha=0}$, we need to compute

$$\frac{\partial}{\partial \alpha}\left[\frac{1}{(1+\alpha)^2(2+\alpha)^2 \cdots (n+\alpha)^2}\right]$$

Let $y = (1+\alpha)^{-2}(2+\alpha)^{-2} \cdots (n+\alpha)^{-2}$ so that

$$\ln y = -2\sum_{k=1}^{n} \ln (k + \alpha) \qquad \text{and} \qquad \frac{1}{y}\frac{\partial y}{\partial \alpha} = -2\sum_{k=1}^{n}\frac{1}{k+\alpha}$$

Hence $$\left(\frac{\partial y}{\partial \alpha}\right)_{\alpha=0} = -2y_{\alpha=0}\sum_{k=1}^{n}\frac{1}{k} = -\frac{2}{(n!)^2}\sum_{k=1}^{n}\frac{1}{k} = -\frac{2}{(n!)^2}F(n)$$

where $F(n) = \sum_{k=1}^{n} \frac{1}{k}$. Thus

$$\left(\frac{\partial w}{\partial z}\right)_{\alpha=0} = c_0 \left[\ln z \sum_{n=0}^{\infty} \frac{z^n}{(n!)^2} - 2 \sum_{n=1}^{\infty} \frac{F(n)}{(n!)^2} z^n\right]$$

The second solution of $zw'' + w' - w = 0$ is

$$w(z) = B \left[\ln z \sum_{n=0}^{\infty} \frac{z^n}{(n!)^2} - 2 \sum_{n=1}^{\infty} \frac{F(n)}{(n!)^2} z^n\right]$$

Example 5.19. Referring to Example 5.17, we have

$$c_1 = -\frac{c_0}{\alpha - 1}, \; c_2 = -\frac{c_1}{\alpha} = \frac{c_0}{(\alpha - 1)\alpha}, \; c_3 = -\frac{c_2}{\alpha + 1} = -\frac{c_0}{(\alpha - 1)\alpha(\alpha + 1)}, \; \ldots,$$
$$c_n = \frac{(-1)^n c_0}{(\alpha - 1)\alpha(\alpha + 1) \cdots (\alpha + n - 2)}, \; \ldots$$

We replace c_0 by $A\alpha$ and obtain

$$w(z, \alpha) = Az^\alpha \left[\alpha - \frac{\alpha}{\alpha - 1} z + \frac{z^2}{\alpha - 1} - \frac{z^3}{(\alpha - 1)(\alpha + 1)} + \cdots + (-1)^n \frac{z^n}{(\alpha - 1)(\alpha + 1)(\alpha + 2) \cdots (\alpha + n - 2)} + \cdots\right]$$

To compute $\left(\frac{\partial w}{\partial \alpha}\right)_{\alpha=0}$, we need to compute

$$\frac{\partial}{\partial \alpha}\left[\frac{1}{(\alpha - 1)(\alpha + 1)(\alpha + 2) \cdots (\alpha + n - 2)}\right]_{\alpha=0} \qquad n > 2$$

Let $y = (\alpha - 1)^{-1}(\alpha + 1)^{-1}(\alpha + 2)^{-1} \cdots (\alpha + n - 2)^{-1}$, so that

$$\ln y = -\ln(\alpha - 1) - \sum_{k=1}^{n-2} \ln(\alpha + k)$$

$$\left(\frac{\partial y}{\partial \alpha}\right)_{\alpha=0} = -\frac{1}{(n - 2)!}\left(1 - \sum_{k=1}^{n-2} \frac{1}{k}\right) = \frac{1}{(n - 2)!} \sum_{k=2}^{n-2} \frac{1}{k} \qquad n > 3$$

If $n = 3$, $\left(\frac{\partial y}{\partial \alpha}\right)_{\alpha=0} = 0$. Hence

$$\left(\frac{\partial w}{\partial \alpha}\right)_{\alpha=0} = A \ln z \left(-z^2 + z^3 - \frac{z^4}{2!} + \frac{z^5}{3!} - \cdots\right) + A\left[1 + z - z^2 + \sum_{n=4}^{\infty} \frac{(-1)^n}{(n - 2)!} \sum_{k=2}^{n-2} \frac{1}{k} z^n\right]$$

The second solution of $zw'' + (z - 1) w' + w = 0$ is

$$w(z) = B\left[z^2 e^{-z} \ln z - 1 - z + z^2 - z^2 \sum_{n=2}^{\infty} \frac{F(n)}{n!} z^n\right]$$

where $F(n) = (-1)^n \sum_{k=2}^{n} \frac{1}{k}$.

Problems

1. Derive (5.113).
2. Show that if $u(z)$ satisfies (5.118) then $u(z)$ satisfies (5.116).
3. Derive (5.125).
4. Solve $zw'' - w = 0$.

$$\textit{Ans.} \quad w_1(z) = A \sum_{n=0}^{\infty} \frac{z^{n+1}}{n!(n+1)!}$$

$$w_2(z) = B\left[\ln z \sum_{n=0}^{\infty} \frac{z^{n+1}}{n!(n+1)!} + 1 - z - \frac{z^2}{1!2!}\left(\frac{2}{1} + \frac{1}{2}\right) - \frac{z^3}{2!3!}\left(\frac{2}{1} + \frac{2}{2} + \frac{1}{3}\right) + \cdots\right]$$

5. Solve $z^2(1 - z)w'' + z(1 - 3z)w' - w = 0$ for $w(z)$ in the neighborhood of $z = 0$.
6. Solve $z(1 - z)w'' - (1 + z)w' + w = 0$ for $w(z)$ in the neighborhood of $z = 0$.
7. Solve $4z^2w'' + 4zw' - (z^2 + 1)w = 0$ for $w(z)$ in the neighborhood of $z = 0$.

5.11. The Point at Infinity and the Hypergeometric Function. To determine whether the point at infinity is an ordinary or regular singular point, we let $z = 1/t$ and investigate the point $t = 0$. We have

$$\frac{dw}{dz} = \frac{dw}{dt}\frac{dt}{dz} = -t^2 \frac{dw}{dt}$$

$$\frac{d^2w}{dz^2} = t^4 \frac{d^2w}{dt^2} + 2t^3 \frac{dw}{dt}$$

so that $w'' + p(z)w' + q(z)w = 0$ becomes

$$\frac{d^2w}{dt^2} + \frac{2t - p(1/t)}{t^2}\frac{dw}{dt} + \frac{q(1/t)}{t^4} w = 0 \tag{5.130}$$

Hence $z = \infty$ is an ordinary point if

$$\begin{aligned} &\text{(i)} \quad \frac{2t - p(1/t)}{t^2} \text{ is analytic} \quad \text{at } t = 0 \\ &\text{(ii)} \quad \frac{1}{t^4} q\left(\frac{1}{t}\right) \quad \text{is analytic} \quad \text{at } t = 0 \end{aligned} \tag{5.131}$$

For this case a solution can be found in the form

$$w(z) = a_0 + \frac{a_1}{z} + \frac{a_2}{z^2} + \cdots$$

Again $z = \infty$ is a regular singular point if $z = \infty$ is not an ordinary point and if

$$\text{(iii)}\quad \frac{2t - p(1/t)}{t} \text{ is analytic} \qquad \text{at } t = 0 \tag{5.132}$$

$$\text{(iv)}\quad \frac{1}{t^2}\, q\left(\frac{1}{t}\right) \qquad \text{is analytic} \qquad \text{at } t = 0$$

This implies

$$\frac{1}{t}\, p\left(\frac{1}{t}\right) = p_0 + p_1 t + p_2 t^2 + \cdots$$

$$\frac{1}{t^2}\, q\left(\frac{1}{t}\right) = q_0 + q_1 t + q_2 t^2 + \cdots$$

so that

$$p(z) = \frac{p_0}{z} + \frac{p_1}{z^2} + \cdots + \frac{p_n}{z^{n+1}} + \cdots$$

$$q(z) = \frac{q_0}{z^2} + \frac{q_1}{z^3} + \cdots + \frac{q_n}{z^{n+2}} + \cdots$$

If, moreover, we desire that the origin also be a regular singular point we must have $p(z) = p_0/z$, $q(z) = q_0/z^2$.

Example 5.20. We look for the differential equation (of second order) which has but one regular singular point at $z = 0$. In this case $z = \infty$ is an ordinary point so that

$$\frac{2t - p(1/t)}{t^2} = p_0 + p_1 t + p_2 t^2 + \cdots$$

$$\frac{1}{t^4}\, q\left(\frac{1}{t}\right) = q_0 + q_1 t + q_2 t^2 + \cdots$$

and

$$zp(z) = 2 - \frac{p_0}{z} - \frac{p_1}{z^2} + \cdots$$

$$z^2 q(z) = \frac{q_0}{z^2} + \frac{q_1}{z^3} + \cdots$$

In order that $z = 0$ be a regular singular point, we must have

$$p_0 = p_1 = \cdots = p_n = \cdots = q_0 = q_1 = \cdots = q_n = \cdots = 0$$

Hence $q(z) \equiv 0$, and $p(z) = 2/z$, so that

$$w'' + \frac{2}{z}\, w' = 0$$

We consider now the following problem: We look for the differential equation which has exactly three regular singular points at $z = 0, 1, \infty$; all other points are to be ordinary points. Since $p(z)$ has at most simple poles at $z = 0$, $z = 1$, we know that $z(1 - z)p(z)$ is an entire function. Hence

$$z(1 - z)p(z) = \sum_{n=0}^{\infty} a_n z^n \tag{5.133}$$

for all z. Since $z = \infty$ is to be a regular singular point, condition (iii) must be upheld. From (5.133)

$$p\left(\frac{1}{t}\right) = \frac{t^2 \sum_{n=0}^{\infty} a_n(1/t)^n}{t-1}$$

and
$$\frac{1}{t}\, p\left(\frac{1}{t}\right) = \frac{t \sum_{n=0}^{\infty} a_n(1/t)^n}{t-1}$$

must be analytic at $t = 0$. This is possible if and only if $a_n = 0$, $n > 1$. Thus

$$z(1 - z)p(z) = a_0 + a_1 z$$

and
$$p(z) = \frac{c}{z} + \frac{A}{1-z} \tag{5.134}$$

With the same type of reasoning the reader can show that

$$q(z) = \frac{B(z-\mu)(z-\nu)}{z^2(z-1)^2} \tag{5.135}$$

Let the roots of the indicial equation at $z = \infty$ be $\alpha = a$, $\alpha = b$. We also impose the condition that the indicial equations at $z = 0$ and $z = 1$ have at least one root equal to zero. Now at $z = 0$ we have $p_0 = \lim_{z\to 0} zp(z) = c$ from (5.134). Also $q_0 = \lim_{z\to 0} z^2 q(z) = \mu\nu B$. The indicial equation at $z = 0$ is

$$\alpha(\alpha - 1) + c\alpha + \mu\nu B = 0$$

If $\alpha = 0$ is to be a root of this quadratic, we must have $\mu\nu B = 0$ so that we pick $\mu = 0$ if we desire $q(z) \not\equiv 0$. The other root is $\alpha = 1 - c$. Let the reader show that, if one of the roots of the indicial equation at $z = 1$ is zero, the other is $\alpha = 1 + A$ and

$$q(z) = \frac{B}{z(z-1)} \tag{5.136}$$

At $z = \infty$ we have

$$\frac{2t - p(1/t)}{t^2} = \frac{2t - ct - tA/(t-1)}{t^2} = \frac{2 - c - A/(t-1)}{t}$$

and
$$p_0 = \lim_{t\to 0} t\,\frac{2t - p(1/t)}{t^2} = 2 - c + A$$

Similarly $q_0 = B$, and the indicial equation at $z = \infty$ is

$$\alpha(\alpha - 1) + (2 - c + A)\alpha + B = 0$$

or
$$\alpha^2 + (1 - c + A)\alpha + B = 0$$

If the roots of this equation are a and b, we must have

$$a + b = -1 + c - A$$

$ab = B$, so that $-A = a + b + 1 - c$. Our differential equation is

$$\frac{d^2w}{dz^2} + \left(\frac{c}{z} + \frac{1 - c + a + b}{z - 1}\right)\frac{dw}{dz} + \frac{ab}{z(z-1)}w = 0 \qquad (5.137)$$

Equation (5.137) is the famous hypergeometric differential equation. Its solution is written in the form (after Paperitz)

$$w(z) = P\left\{\begin{matrix} 0 & \infty & 1 & \\ 0 & a & 0 & z \\ 1-c & b & c-a-b & \end{matrix}\right\} \qquad (5.138)$$

The top row denotes the regular singular points, and the other rows contain the roots of the indicial equations at each singular point. Note that the sum of the roots is unity.

The operator $\Theta = z\frac{d}{dz}$ due to Boole is very useful in solving (5.137) by series methods. The reader can show that Θ has the following properties,

$$\begin{aligned} \Theta(c_1\varphi_1 + c_2\varphi_2) &= c_1\Theta\varphi_1 + c_2\Theta\varphi_2 \\ \Theta z^\alpha &= \alpha z^\alpha \\ \Theta(\ln z)^k &= k(\ln z)^{k-1} \\ \Theta^n(z^\alpha\varphi(z)) &= z^\alpha(\Theta + \alpha)^n\varphi \end{aligned} \qquad (5.139)$$

where $\Theta^2\varphi = \Theta(\Theta\varphi)$, etc., and $(\Theta + \alpha)^n\varphi = \sum_{r=0}^{n}\binom{n}{r}\alpha^{n-r}\Theta^r\varphi$.

The reader can show that (5.137) can be written in the form

$$\Theta(\Theta + c - 1)w = z(\Theta + a)(\Theta + b)w \qquad (5.140)$$

We solve (5.137) or (5.140) by series. Let

$$w = \sum_{k=0}^{\infty} c_k z^{k+\alpha}$$

Then

$$\Theta w = \sum_{k=0}^{\infty} c_k\Theta z^{k+\alpha} = \sum_{k=0}^{\infty} c_k(k + \alpha)z^{k+\alpha}$$

$$(\Theta + b)w = \sum_{k=0}^{\infty} c_k(k + \alpha + b)z^{k+\alpha}$$

$$\Theta(\Theta + b)w = \sum_{k=0}^{\infty} c_k(k + \alpha + b)(k + \alpha)z^{k+\alpha}$$

$$z(\Theta + a)(\Theta + b)w = \sum_{k=0}^{\infty} c_k(k + \alpha + b)(k + \alpha + a)z^{k+\alpha+1} \qquad (5.141)$$

Also $$\Theta(\Theta + c - 1)w = \sum_{k=0}^{\infty} c_k(k + \alpha)(k + \alpha + c - 1)z^{k+\alpha} \tag{5.142}$$

Equating (5.141) and (5.142), we obtain the recursion formula

$$c_{k+1}(k + \alpha + 1)(k + \alpha + c) = c_k(k + \alpha + a)(k + \alpha + b)$$

along with the indicial equation

$$\alpha(\alpha + c - 1) = 0$$

For $\alpha = 0$ we have

$$c_{k+1} = \frac{(k + a)(k + b)}{(k + 1)(k + c)} c_k \qquad k = 0, 1, 2, \ldots$$

so that $$c_1 = \frac{ab}{1 \cdot c} c_0$$

$$c_2 = \frac{(a + 1)(b + 1)}{2(c + 1)} = \frac{(a + 1)a(b + 1)b}{(c + 1)c} \frac{c_0}{2!}$$

and, in general,

$$\begin{aligned} c_n &= \frac{(a + n - 1) \cdots a(b + n - 1) \cdots b}{(c + n - 1) \cdots c} \frac{c_0}{n!} \\ &= \frac{c_0\Gamma(c)}{\Gamma(a)\Gamma(b)} \frac{\Gamma(a + n)\Gamma(b + n)}{\Gamma(c + n)} \frac{1}{n!} \\ &= A \frac{\Gamma(a + n)\Gamma(b + n)}{\Gamma(c + n)} \frac{1}{n!} \end{aligned}$$

where $\Gamma(a)$ is the gamma, or factorial, function of Chap. 4. A formal solution of (5.137) is

$$w(z) = A \sum_{n=0}^{\infty} \frac{\Gamma(a + n)\Gamma(b + n)}{\Gamma(c + n)} \frac{z^n}{n!} \tag{5.143}$$

a, b, c cannot be negative integers. Why?

We define the hypergeometric function $F(a, b; c; z)$ by the equation

$$F(a, b; c; z) \equiv \frac{\Gamma(c)}{\Gamma(a)\Gamma(b)} \sum_{n=0}^{\infty} \frac{\Gamma(a + n)\Gamma(b + n)}{\Gamma(c + n)} \frac{z^n}{n!} \tag{5.144}$$

Problems

1. Solve $zw'' + 2w' = 0$ for $w(z)$ in the neighborhood of $z = 0$.
2. Derive (5.135).
3. Let $w(z)$ be a solution of Legendre's differential equation

$$(1 - z^2) \frac{d^2w}{dz^2} - 2z \frac{dw}{dz} + n(n + 1)w = 0$$

Show that

$$w(z) = P\begin{Bmatrix} 1 & -1 & \infty & \\ 0 & 0 & n+1 & z \\ 0 & 0 & -n & \end{Bmatrix}$$

4. Verify (5.139).

5. If c is not an integer, show that a second solution of (5.137) is

$$w(z) = z^{1-c} \sum_{n=0}^{\infty} \frac{\Gamma(n+1+a-c)\Gamma(n+1+b-c)}{\Gamma(n+2-c)} \frac{z^n}{n!}$$

6. What is the interval of convergence of (5.143)?

7. Show that

$$(1-z)^{-\alpha} = F(\alpha, \beta; \beta; z)$$
$$\ln \frac{1}{1-z} = zF(1, 1; 2; z)$$

by expanding the left-hand sides in Taylor-series expansions.

8. Show that when $c = 1$ the second independent solution of (5.137) is

$$\Gamma(a)\Gamma(b)F(a, b; 1; z) \ln z + \sum_{r=1}^{\infty} s_r \frac{\Gamma(a+r)\Gamma(b+r)}{(r!)^2} z^r$$

where $s_r = \sum_{n=1}^{r} \left(\frac{1}{a+n-1} + \frac{1}{b+n-1} - \frac{2}{n}\right)$.

5.12. The Confluence of Singularities. Laguerre Polynomials. In the discussion of the hypergeometric function and its associated differential equation the regular singular points 0, 1, ∞ were distinct. It turns out to be useful to consider what happens when two or more regular singular points approach coincidence, a process usually referred to as the "confluence" of singularities. The reader is referred to Chap. XX of Ince's text on differential equations for more details and for a very useful classification of linear second-order differential equations according to the number, nature, and genesis of their singularities by confluence. We consider Example 5.21 by way of illustration.

Example 5.21. *Kummer's Confluent Hypergeometric Function.* Consider the hypergeometric function

$$w(z) = {}_2F_1(a, b; c; z)$$
$$= \sum_{n=0}^{\infty} \frac{a(a+1)\cdots(a+n-1)b(b+1)\cdots(b+n-1)}{c(c+1)\cdots(c+n-1)} \frac{z^n}{n!} \tag{5.145}$$

for $|z| < 1$, which satisfies

$$z(1-z)w'' + [c - (a+b+1)z]w' - abw = 0 \tag{5.146}$$

Now let $z' = bz$, and then replace z' by z. We obtain

$$
{}_2F_1\left(a, b; c; \frac{z}{b}\right) = \sum_{n=0}^{\infty} \frac{a(a+1)\cdots(a+n-1)b(b+1)\cdots(b+n-1)}{c(c+1)\cdots(c+n-1)b^n} \frac{z^n}{n!} \tag{5.147}
$$

which converges for $|z| < b$. ${}_2F_1$ satisfies

$$
z\left(1 - \frac{z}{b}\right) w'' + \left(c - \frac{a+b+1}{b} z\right) w' - aw = 0 \tag{5.148}
$$

If we let $b \to \infty$, we obtain (formally) the series

$$
{}_1F_1(a; c; z) = \sum_{n=0}^{\infty} \frac{a(a+1)\cdots(a+n-1)}{c(c+1)\cdots(c+n-1)} \frac{z^n}{n!} \tag{5.149}
$$

which certainly converges for $|z| < 1$. Moreover it can be shown that ${}_1F_1(a; c; z)$ satisfies

$$
zw'' + (c - z)w' - aw = 0 \tag{5.150}
$$

obtained formally by letting $b \to \infty$. Equation (5.150) is Kummer's confluent hypergeometric equation. It has a regular singular point at $z = 0$, but $z = \infty$ is an irregular point, that is, $z = \infty$ is neither an ordinary point nor a regular singular point. If c is not an integer, the other solution of (5.150) is

$$
z^{1-c}\, {}_1F_1(a - c + 1; 2 - c; z) \tag{5.151}
$$

If c is a negative integer or zero, (5.149) becomes meaningless, while if c is a positive integer greater than 1, (5.151) becomes meaningless. If $c = 1$, both solutions coincide. The second solution can then be obtained by the method of Frobenius.

We now consider the function ${}_1F_1(a; c + 1; z)$. The series representation of ${}_1F_1(a; c + 1; z)$ will terminate if a is a negative integer, $-m$, for the coefficient of z^{m+1} is

$$
\frac{(-m)(-m+1)\cdots(-m+m+1-1)}{(c+1)(c+2)\cdots(c+n)n!} \equiv 0
$$

Similarly higher powers of z have zero coefficients. Hence ${}_1F_1(-n, c + 1; z)$ is a polynomial of degree n if n is a positive integer. We define

$$
L_n^{(c)}(z) \equiv \frac{(c+1)(c+2)\cdots(c+n)}{n!}\, {}_1F_1(-n; c+1; z) \tag{5.152}
$$

as the associated Laguerre polynomial, n a positive integer. The reader can show that

$$
z\frac{d^2L_n^{(c)}(z)}{dz^2} + (c + 1 - z)\frac{dL_n^{(c)}(z)}{dz} + nL_n^{(c)}(z) = 0 \tag{5.153}
$$

We show now that we can write

$$L_n^{(c)}(z) = \frac{e^z z^{-c}}{n!} \frac{d^n}{dz^n} (e^{-z} z^{n+c}) \tag{5.154}$$

Let $v(z) = e^{-z} z^{n+c}$ so that $\frac{dv}{dz} = e^{-z}[(n + c)z^{n+c-1} - z^{n+c}]$ and

$$z \frac{dv}{dz} = (n + c - z)v$$

Differentiating $n + 1$ times with respect to z yields

$$z \frac{d^{n+2}v}{dz^{n+2}} + (n + 1) \frac{d^{n+1}v}{dz^{n+1}} = (n + c - z) \frac{d^{n+1}v}{dz^{n+1}} - (n + 1) \frac{d^n v}{dz^n}$$

If we let $u = \frac{d^n v}{dz^n}$, we have

$$z \frac{d^2u}{dz^2} + (z + 1 - c) \frac{du}{dz} + (n + 1)u = 0 \tag{5.155}$$

Now let $w(z) = e^z z^{-c} u(z) = e^z z^{-c} \frac{d^n}{dz^n} (e^{-z} z^{n+c})$. If n is a positive integer, it is easy to see that $w(z)$ is a polynomial of degree n. From

$$u(z) = w(z) e^{-z} z^c$$

and the fact that $u(z)$ satisfies (5.155) let the reader show that $w(z)$ satisfies

$$z \frac{d^2w}{dz^2} + (c + 1 - z) \frac{dw}{dz} + nw = 0 \tag{5.156}$$

Since (5.156) is exactly the same as (5.153), we must have

$$L_n^{(c)}(z) = Kw(z) = Ke^z z^{-c} \frac{d^n}{dz^n} (e^{-z} z^{n+c})$$

Let $z = 0$. From (5.152) the reader can easily show that

$$L_n^{(c)}(0) = \frac{(c + 1)(c + 2) \cdots (c + n)}{n!}$$

Let the reader show also that

$$e^z z^{-c} \frac{d^n}{dz^n} (e^{-z} z^{n+c}) \Big|_{z=0} = (c + 1)(c + 2) \cdots (c + n)$$

so that $K = \frac{1}{n!}$. This proves (5.154).

The associated Laguerre polynomials have the interesting and important property

$$\int_0^\infty e^{-x}x^c L_m^{(c)}(x)L_n^{(c)}(x)\,dx = 0 \qquad m \neq n \tag{5.157}$$

To prove (5.157), we need only show that

$$\int_0^\infty e^{-x}x^c x^m L_n^{(c)}(x)\,dx = 0 \qquad \text{for } m < n \tag{5.158}$$

If (5.158) holds, then (5.157) is valid, for $L_m^{(c)}(x)$ is a polynomial of degree $m < n$ and (5.157) is simply a linear sum of integrals of the type (5.158). Equation (5.158) can be obtained by integrating by parts m times after replacing $L_n^{(c)}(x)$ in (5.158) by (5.154).

Problems

1. Show that ${}_1F_1(a; c; z)$ satisfies (5.150).
2. Show that $L_n^{(c)}(z)$ satisfies (5.153).
3. Show that $L_n^{(c)}(0) = \dfrac{(c+1)(c+2)\cdots(c+n)}{n!}$.
4. Prove (5.158).
5. Show that

$$\int_0^\infty e^{-x}x^c L_n^{(c)}(x)L_n^{(c)}(x)\,dx = \frac{\Gamma(c+n+1)}{\Gamma(n+1)}$$

Hint: Evaluate $\int_0^\infty e^{-x}x^c x^n L_n^{(c)}(x)\,dx$.
6. Show that

$$nL_n^{(c)}(z) = (2n + c - 1 - z)L_{n-1}^{(c)}(z) - (n + c - 1)L_{n-2}^{(c)}(z)$$

$$z\frac{dL_n^{(c)}(z)}{dz} = nL_n^{(c)}(z) - (n + c)L_{n-1}^{(c)}(z) \qquad n \geqq 2$$

$$\int_0^t L_n^{(0)}(x)\,dx = L_n^{(0)}(t) - L_{n+1}^{(0)}(t)$$

$L_n^{(0)}$ is called the ordinary Laguerre polynomial.

7. Show that $L_n^{(0)}(z) = \sum_{r=0}^{n} \binom{n}{r} \dfrac{(-z)^r}{r!}$.

5.13. Laplace's Equation. We discuss now a method for solving Laplace's equation. This equation is of some importance in mathematical physics. We consider $\nabla^2 V = 0$ in rectangular coordinates,

$$\frac{\partial^2 V}{\partial x^2} + \frac{\partial^2 V}{\partial y^2} + \frac{\partial^2 V}{\partial z^2} = 0 \tag{5.159}$$

Let us look for a solution in the form

$$V(x, y, z) = X(x)Y(y)Z(z) \tag{5.160}$$

This attempt at finding a solution of (5.159) is called the method of separation of variables. Applying (5.160) to (5.159) yields

$$YZ\frac{d^2X}{dx^2} + ZX\frac{d^2Y}{dy^2} + XY\frac{d^2Z}{dz^2} = 0$$

For $V \neq 0$ we have upon division by V

$$\frac{1}{X}\frac{d^2X}{dx^2} + \frac{1}{Y}\frac{d^2Y}{dy^2} + \frac{1}{Z}\frac{d^2Z}{dz^2} = 0$$

Thus

$$-\frac{1}{X}\frac{d^2X}{dx^2} = \frac{1}{Y}\frac{d^2Y}{dy^2} + \frac{1}{Z}\frac{d^2Z}{dz^2} \tag{5.161}$$

The equality in (5.161) cannot hold unless both sides of (5.161) are constant, for if $u(x) = v(y, z)$, then $\frac{du}{dx} = 0$ and $u \equiv$ constant. Hence

$$-\frac{1}{X}\frac{d^2X}{dx^2} = \text{constant} = \pm k^2$$

or

$$\frac{d^2X}{dx^2} \pm k^2X = 0 \tag{5.162}$$

The solutions of (5.162) are

$$X = A_k \cos kx + B_k \sin kx$$

or

$$X = A_k e^{kx} + B_k e^{-kx} \tag{5.163}$$

or

$$X = Ax + B \qquad \text{if } k = 0$$

Similarly

$$\frac{d^2Y}{dy^2} \pm l^2Y = 0$$

and

$$Y = C_l \cos ly + D_l \sin ly$$

or

$$Y = C_l e^{ly} + D_l e^{-ly} \tag{5.164}$$

or

$$Y = Cy + D \qquad \text{if } l = 0$$

Finally

$$\frac{1}{Z}\frac{d^2Z}{dz^2} = (\pm k^2 \pm l^2)$$

or

$$\frac{d^2Z}{dz^2} + (\mp k^2 \mp l^2)Z = 0 \tag{5.165}$$

The solution of (5.165) depends on the magnitudes of k^2 and l^2 and the signs preceding k^2 and l^2.

The choice of the sign preceding the constants depends on the nature of the physical problem. We illustrate with Example 5.22.

Example 5.22. For steady-state heat flow we have $\nabla^2T = 0$. We consider a two-dimensional semi-infinite slab of width a. The edge given by $x = 0$, $0 \leqq y < \infty$ is kept at constant temperature $T = 0$, as is the edge given by $x = a$, $0 \leqq y < \infty$.

The base of the slab given by $y = 0$, $0 \leqq x \leqq a$ is kept at a steady temperature given by $T = f(x)$. For the steady-state case we have

$$\frac{\partial^2 T}{\partial x^2} + \frac{\partial^2 T}{\partial y^2} = 0$$

If we let $T = X(x)Y(y)$, we obtain as above

$$-\frac{1}{X}\frac{d^2X}{dx^2} = \frac{1}{Y}\frac{d^2Y}{dy^2} = \text{constant}$$

If we choose our constant to be negative, we have $X = A_k e^{kx} + B_k e^{-kx}$. Our boundary condition necessitates, however, that $T = 0$ when $x = 0$ and when $x = a$ for all y. This cannot be achieved for $X = A_k e^{kx} + B_k e^{-kx}$ unless $A_k = B_k = 0$. However, if we choose

$$-\frac{1}{X}\frac{d^2X}{dx^2} = k^2$$

then $X = A_k \cos kx + B_k \sin kx$. Since $\sin (n\pi x/a)$ vanishes for $x = 0$ and $x = a$ provided n is an integer, it seems proper to consider

$$X = B_n \sin \frac{n\pi x}{a}$$

where $k = n\pi/a$. It follows that

$$\frac{d^2Y}{dy^2} - \frac{n^2\pi^2}{a^2} Y = 0$$

so that $Y = C_n e^{(n\pi/a)y} + D_n e^{-(n\pi/a)y}$. We do not expect the temperature to become infinite as y becomes infinite; so we choose $C_n = 0$.

Hence
$$\begin{aligned} T(x, y) &= B_n D_n e^{-(n\pi/a)y} \sin \frac{n\pi x}{a} \\ &= A_n e^{-(n\pi/a)y} \sin \frac{n\pi x}{a} \end{aligned} \tag{5.166}$$

where n is an integer. The final boundary condition is $T(x, 0) = f(x)$. Equation (5.166) cannot, in general, satisfy this boundary condition, unless, by choice, we pick $(x) = A \sin (n\pi x/a)$. Now $\nabla^2 T = 0$ is a linear partial differential equation, and it is an easy thing to show that

$$T(x, y) = \sum_{n=0}^{\infty} A_n e^{-(n\pi/a)y} \sin \frac{n\pi x}{a} \tag{5.167}$$

is also a solution of $\nabla^2 T = 0$ provided the series converges and can be differentiated twice term by term. To satisfy the boundary condition $T(x, 0) = f(x)$, we need

$$f(x) = \sum_{n=0}^{\infty} A_n \sin \frac{n\pi x}{a} \tag{5.168}$$

Can the infinite set of constants A_n, $n = 0, 1, 2, \ldots$, be found so that (5.168) holds? This is the subject of Fourier series and will be discussed in Chap. 6.

Problem 1. Find a solution of $\nabla^2 V = 0$ such that $V = 0$ for $x = 0$, $x = a$, $V = 0$ for $y = 0$, $y = b$, $V = 0$ for $z = +\infty$.

Problem 2. By the method of separation of variables find a solution of

$$\frac{\partial^2 V}{\partial x^2} + \frac{\partial^2 V}{\partial y^2} = k \frac{\partial V}{\partial t}$$

Assume $V(x, y, t) = X(x)Y(y)T(t)$.

Problem 3. Do the same for $\dfrac{\partial^2 V}{\partial x^2} + \dfrac{\partial^2 V}{\partial y^2} = \dfrac{1}{c^2}\dfrac{\partial^2 V}{\partial t^2}$.

We find now a solution of $\nabla^2 V = 0$ in spherical coordinates by use of the method of separation of variables. In spherical coordinates Laplace's equation is

$$\sin\theta \frac{\partial}{\partial r}\left(r^2 \frac{\partial V}{\partial r}\right) + \frac{\partial}{\partial \theta}\left(\sin\theta \frac{\partial V}{\partial \theta}\right) + \frac{1}{\sin\theta}\frac{\partial^2 V}{\partial \varphi^2} = 0$$

Assuming $V(r, \theta, \varphi) = R(r)\Theta(\theta)\Phi(\varphi)$, we obtain upon division by V

$$\frac{\sin^2\theta}{R}\frac{d}{dr}\left(r^2 \frac{dR}{dr}\right) + \frac{\sin\theta}{\Theta}\frac{d}{d\theta}\left(\sin\theta \frac{d\Theta}{d\theta}\right) = -\frac{1}{\Phi}\frac{d^2\Phi}{d\varphi^2} \tag{5.169}$$

Let the reader show that of necessity

$$-\frac{1}{\Phi}\frac{d^2\Phi}{d\varphi^2} = \text{constant}$$

Now physically we expect and desire that

$$V(r, \theta, \varphi) = V(r, \theta, \varphi + 2\pi)$$

Let the reader show that this implies that the constant be chosen as the square of an integer, that is,

$$\frac{d^2\Phi}{d\varphi^2} + n^2\Phi = 0$$

Thus $\Phi = A_n \cos n\varphi + B_n \sin n\varphi$. Equation (5.169) now becomes

$$\frac{1}{R}\frac{d}{dr}\left(r^2 \frac{dR}{dr}\right) = -\frac{1}{\Theta \sin\theta}\frac{d}{d\theta}\left(\sin\theta \frac{d\Theta}{d\theta}\right) + \frac{n^2}{\sin^2\theta} \tag{5.170}$$

and this implies that

$$\frac{1}{R}\frac{d}{dr}\left(r^2 \frac{dR}{dr}\right) = \text{constant} = k$$

or

$$r^2 \frac{d^2R}{dr^2} + 2r\frac{dR}{dr} - kR = 0 \tag{5.171}$$

To solve (5.171), we assume a solution in the form $R(r) = r^m$. This yields $r^m[m(m - 1) + 2m - k] = 0$. Hence r^m is a solution of (5.171) provided $m(m+1) = k$. Similarly $r^{-(m+1)}$ is a solution of (5.171) for $k = m(m+1)$. The most general solution of (5.171) for $k = m(m+1)$ is

$$R(r) = C_m r^m + D_m r^{-(m+1)} \tag{5.172}$$

We need not have guessed at a solution of (5.171), for it is immediately obvious that $r = 0$ is a regular singular point of (5.171). The series method of Sec. 5.10 could be used to obtain (5.172). Equation (5.170) now becomes

$$\sin\theta \frac{d}{d\theta}\left(\sin\theta \frac{d\Theta}{d\theta}\right) + [m(m+1)\sin^2\theta - n^2]\Theta = 0 \tag{5.173}$$

If we let $\mu = \cos\theta$, $d\mu = -\sin\theta\, d\theta$, (5.173) can be written

$$(1-\mu^2)\frac{d}{d\mu}\left[(1-\mu^2)\frac{d\Theta}{d\mu}\right] + [m(m+1)(1-\mu^2) - n^2]\Theta = 0$$

or

$$(1-\mu^2)\frac{d^2\Theta}{d\mu^2} - 2\mu\frac{d\Theta}{d\mu} + \left[m(m+1) - \frac{n^2}{1-\mu^2}\right]\Theta = 0 \tag{5.174}$$

Equation (5.174) is called the associated Legendre differential equation. Its solution in Paperitz's notation is

$$\Theta(\mu) = P\begin{Bmatrix} -1 & \infty & 1 & \\ \frac{1}{2}n & m+1 & \frac{1}{2}n & \mu \\ -\frac{1}{2}n & -m & -\frac{1}{2}n & \end{Bmatrix} \tag{5.175}$$

Problem 4. Deduce (5.175).

Problem 5. If V is independent of φ, that is, $n = 0$, (5.174) becomes

$$(1-\mu^2)\frac{d^2P}{d\mu^2} - 2\mu\frac{dP}{d\mu} + m(m+1)P = 0 \tag{5.176}$$

with Θ replaced by P. For m an integer show that the solution of (5.176) in the neighborhood of $\mu = 0$, written $P_m(\mu)$, is a polynomial of degree m. Also show that

$$\int_{-1}^{1} P_m(\mu)P_n(\mu)\, d\mu = 0 \qquad \text{for } m \neq n$$

The $P_m(\mu)$, $m = 0, 1, 2, \ldots$, are called Legendre polynomials. We shall have a great deal more to say about such polynomials in Chap. 6. A solution of $\nabla^2 V = 0$ for $V = V(r, \theta)$ is

$$V(r, \theta) = \sum_{m=0}^{\infty} (A_m r^m + B_m r^{-(m+1)}) P_m(\cos\theta) \tag{5.177}$$

We have seen that the solution of Laplace's equation in spherical coordinates led to Legendre polynomials. The solution of Laplace's equation in cylindrical coordinates yields the Bessel function. In cylindrical coordinates $\nabla^2 V = 0$ becomes

$$\frac{\partial}{\partial r}\left(r\frac{\partial V}{\partial r}\right) + \frac{1}{r}\frac{\partial^2 V}{\partial \theta^2} + r\frac{\partial^2 V}{\partial z^2} = 0 \tag{5.178}$$

Assuming $V(r, \theta, z) = R(r)\Theta(\theta)Z(z)$ yields

$$\frac{r}{R}\frac{d}{dr}\left(r\frac{dR}{dr}\right) + \frac{r^2}{Z}\frac{d^2Z}{dz^2} = -\frac{1}{\Theta}\frac{d^2\Theta}{d\theta^2} \tag{5.179}$$

If we desire $V(r, \theta, z) = V(r, \theta + 2\pi, z)$, we need

$$-\frac{1}{\Theta}\frac{d^2\Theta}{d\theta^2} = \nu^2 \qquad \nu \text{ an integer}$$

so that

$$\Theta(\theta) = A_\nu \cos \nu\theta + B_\nu \sin \nu\theta \tag{5.180}$$

Similarly

$$\frac{1}{Z}\frac{d^2Z}{dz^2} = k^2 \qquad k \text{ real or a pure imaginary}$$

so that

$$Z(z) = C_k e^{kz} + D_k e^{-kz} \tag{5.181}$$

If k is real, $Z(z)$ is exponential, and if k is a pure imaginary, $Z(z)$ is trigonometric. Equation (5.179) now becomes

$$\frac{d^2R}{dr^2} + \frac{1}{r}\frac{dR}{dr} + \left(k^2 - \frac{\nu^2}{r^2}\right)R = 0 \tag{5.182}$$

If we let $z = kr$, $R = w$, (5.182) becomes

$$\frac{d^2w}{dz^2} + \frac{1}{z}\frac{dw}{dz} + \left(1 - \frac{\nu^2}{z^2}\right)w = 0 \tag{5.183}$$

This is Bessel's differential equation. We see that $z = 0$ is a regular singular point.

Problem 6. Solve (5.182) for $k = 0$ and ν an integer.

Problem 7. Solve (5.183) for the cases ν = integer, $\nu \neq$ integer. *Hint:* Let $z^2 = 4x$, $\Theta = x\dfrac{d}{dx}$, and write (5.183) as

$$(\Theta^2 - \tfrac{1}{4}\nu^2)\, w + xw = 0$$

Let $w = x^\alpha \sum_{r=0}^{\infty} c_r x^r$, and show that

$$w_1(x) = Ax^{\nu/2} \sum_{r=0}^{\infty} \frac{(-x)^r}{r!\Gamma(\nu + r + 1)}$$

If ν is not an integer, a second solution is

$$w_2(x) = Bx^{-\nu/2} \sum_{r=0}^{\infty} \frac{(-x)^r}{r!\Gamma(-\nu + r + 1)}$$

We define the Bessel function of the first kind of order ν to be

$$J_\nu(z) = \left(\frac{z}{2}\right)^\nu \sum_{r=0}^{\infty} \frac{(-1)^r (z/2)^{2r}}{r!\Gamma(\nu + r + 1)} \tag{5.184}$$

An important recurrence formula for the $J_\nu(z)$ is

$$J_{\nu-1}(z) + J_{\nu+1}(z) = \frac{2\nu}{z} J_\nu(z) \tag{5.185}$$

To prove (5.185), we note that

$$\begin{aligned}
J_{\nu-1}(z) + J_{\nu+1}(z) &= \left(\frac{z}{2}\right)^{\nu-1} \sum_{r=0}^{\infty} \frac{(-1)^r (z/2)^{2r}}{r!\Gamma(\nu + r)} \\
&\qquad + \left(\frac{z}{2}\right)^{\nu+1} \sum_{r=0}^{\infty} \frac{(-1)^r (z/2)^{2r}}{r!\Gamma(\nu + r + 2)} \\
&= \left(\frac{z}{2}\right)^{\nu-1} \left[\sum_{r=0}^{\infty} \frac{(-z^2/4)^{r+1}}{r!\Gamma(\nu + r)} - \sum_{r=0}^{\infty} \frac{(-z^2/4)^{r+1}}{r!\Gamma(\nu + r + 2)}\right] \\
&= \left(\frac{z}{2}\right)^{\nu-1} \left\{\frac{1}{\Gamma(\nu)}\right. \\
&\qquad \left. + \sum_{r=1}^{\infty} \left[\frac{1}{r!\Gamma(\nu + r)} - \frac{1}{(r-1)!\Gamma(\nu + r + 1)}\right] \left(-\frac{z^2}{4}\right)^r\right\} \\
&= \left(\frac{z}{2}\right)^{\nu-1} \left[\frac{1}{\Gamma(\nu)} + \sum_{r=1}^{\infty} \frac{\nu + r - r}{r!\Gamma(\nu + r + 1)} \left(-\frac{z^2}{4}\right)^r\right] \\
&= \left(\frac{z}{2}\right)^{\nu-1} \left[\frac{1}{\Gamma(\nu)} + \sum_{r=1}^{\infty} \frac{\nu}{r!\Gamma(\nu + r + 1)} \left(-\frac{z^2}{4}\right)^r\right]
\end{aligned}$$

However,

$$\begin{aligned}
\frac{2\nu}{z} J_\nu(z) &= \frac{2\nu}{z} \left(\frac{z}{2}\right)^\nu \left[\frac{1}{\Gamma(\nu + 1)} + \sum_{r=1}^{\infty} \frac{(-z^2/4)^r}{r!\Gamma(v + r + 1)}\right] \\
&= \left(\frac{z}{2}\right)^{\nu-1} \left[\frac{\nu}{\Gamma(\nu + 1)} + \sum_{r=1}^{\infty} \frac{\nu(-z^2/4)^r}{r!\Gamma(\nu + r + 1)}\right]
\end{aligned}$$

Equation (5.185) is seen to hold since $\nu/\Gamma(\nu + 1) = 1/\Gamma(\nu)$.

Problem 8. Show that

$$J_{\nu-1}(z) - J_{\nu+1}(z) = 2J'_\nu(z) \tag{5.186}$$

Problem 9. From (5.185) and (5.186) show that

$$\frac{\nu}{z} J_\nu(z) + J'_\nu(z) = J_{\nu-1}(z)$$

$$\frac{\nu}{z} J_\nu(z) - J'_\nu(z) = J_{\nu+1}(z)$$

Problem 10. Show that

$$J_{\frac{1}{2}}(z) = \left(\frac{2}{\pi z}\right)^{\frac{1}{2}} \sin z$$

Problem 11. Show that, if $\alpha \neq \beta$, $\nu > -1$,

$$(\alpha^2 - \beta^2) \int_0^x t J_\nu(\alpha t) J_\nu(\beta t)\, dt = x \left[J_\nu(\alpha x) \frac{dJ_\nu(\beta x)}{dx} - J_\nu(\beta x) \frac{dJ_\nu(\alpha x)}{dx} \right]$$

$$2\alpha^2 \int_0^x t[J_\nu(\alpha t)]^2\, dt = (\alpha^2 x^2 - \nu^2)[J_\nu(\alpha x)]^2 + \left[x \frac{dJ_\nu(\alpha x)}{dx} \right]^2$$

Problem 12. From the results of Prob. 11 show that

$$\int_0^1 t J_\nu(\alpha t) J_\nu(\beta t)\, dt = 0$$

provided $J(\alpha) = J(\beta) = 0$, $\alpha \neq \beta$, $\nu > -1$. Also show that for this case

$$\int_0^1 t[J_\nu(\alpha t)]^2\, dt = \tfrac{1}{2}[J_{\nu+1}(\alpha)]^2$$

5.14. Hermite Polynomials. The differential equation

$$\frac{d^2 w}{dt^2} - t\frac{dw}{dt} + nw = 0 \tag{5.187}$$

arises in quantum mechanics in connection with the one-dimensional harmonic oscillator. We shall see that the Hermite polynomials defined by

$$H_n(t) = (-1)^n e^{t^2/2} \frac{d^n e^{-t^2/2}}{dt^n} \tag{5.188}$$

satisfy (5.187) for any nonnegative integer n. The reader can easily verify that $H_n(t)$ is a polynomial of degree n. From (5.188) we have

$$\begin{aligned}
\frac{dH_n(t)}{dt} &= (-1)^n t e^{t^2/2} \frac{d^n e^{-t^2/2}}{dt^n} + (-1)^n e^{-t^2/2} \frac{d^n}{dt^n}(-te^{-t^2/2}) \\
&= (-1)^n e^{t^2/2} \left(t\frac{d^n e^{-t^2/2}}{dt^n} - t\frac{d^n e^{-t^2/2}}{dt^n} - n\frac{d^{n-1} e^{-t^2/2}}{dt^{n-1}} \right) \\
&= (-1)^{n-1} e^{t^2/2} n \frac{d^{n-1} e^{-t^2/2}}{dt^{n-1}}
\end{aligned}$$

so that

$$\frac{dH_n(t)}{dt} = nH_{n-1}(t) \tag{5.189}$$

Problem 1. Show that

$$H_{n+2}(t) = tH_{n+1}(t) - (n+1)H_n(t) \tag{5.190}$$

Problem 2. From (5.189), (5.190) show that $\frac{d^2H_n}{dt^2} - t\frac{dH_n}{dt} + nH_n = 0$.

From (5.189) we have

$$\frac{d^2H_n(t)}{dt^2} = n\,\frac{dH_{n-1}(t)}{dt} = n(n-1)H_{n-2}(t)$$

and, in general,

$$\frac{d^rH_n(t)}{dt^r} = n(n-1)\cdots(n-r+1)H_{n-r}(t)$$

and

$$\frac{1}{r!}\frac{d^rH_n(t)}{dt^r} = \binom{n}{r}H_{n-r}(t) \tag{5.191}$$

From Taylor's expansion theorem and using (5.191) we obtain

$$H_n(t+s) = \sum_{r=0}^{\infty} \frac{d^rH_n(t)}{dt^r}\frac{s^r}{r!} = \sum_{r=0}^{\infty}\binom{n}{r}H_{n-r}(t)s^r \tag{5.192}$$

We now look for a generating function $\varphi(z, t)$ in the sense that

$$\varphi(z, t) = \sum_{n=0}^{\infty} H_n(t)\,\frac{z^n}{n!} \tag{5.193}$$

If $\varphi(z, t)$ can be found, then the Taylor-series expansion of $\varphi(z, t)$ in powers of z will yield the Hermite polynomials. If $\varphi(z, t)$ does exist, then formally we have

$$\frac{\partial\varphi}{\partial t} = \sum_{n=0}^{\infty} H_n'(t)\,\frac{z^n}{n!}$$

assuming that one can differentiate term by term. From (5.189) we have

$$\begin{aligned}\frac{\partial\varphi}{\partial t} &= \sum_{n=1}^{\infty} H_{n-1}(t)\,\frac{z^n}{(n-1)!} = z\sum_{n=1}^{\infty} H_{n-1}(t)\,\frac{z^{n-1}}{(n-1)!}\\ &= z\varphi(z, t)\end{aligned}$$

Integration yields

$$\varphi(z, t) = e^{zt+\mu(z)}$$

where $\mu(z)$ is an arbitrary function of integration. Thus

$$\begin{aligned}\frac{\partial\varphi}{\partial z} &= e^{zt+\mu(z)}[t + \mu'(z)]\\ \frac{\partial^2\varphi}{\partial z^2} &= e^{zt+\mu(z)}\{\mu''(z) + [t+\mu'(z)]^2\}\end{aligned} \tag{5.194}$$

From (5.193) we have formally

$$\frac{\partial \varphi}{\partial z} = \sum_{n=1}^{\infty} H_n(t) \frac{z^{n-1}}{(n-1)!} = \sum_{n=0}^{\infty} H_{n+1}(t) \frac{z^n}{n!}$$

$$\frac{\partial^2 \varphi}{\partial z^2} = \sum_{n=0}^{\infty} H_{n+2}(t) \frac{z^n}{n!}$$

so that

$$\begin{aligned} \frac{\partial^2 \varphi}{\partial z^2} - t\frac{\partial \varphi}{\partial z} &= \sum_{n=0}^{\infty} [H_{n+2}(t) - tH_{n+1}(t)] \frac{z^n}{n!} \\ &= -\sum_{n=0}^{\infty} (n+1)H_n(t) \frac{z^n}{n!} \\ &= -z\frac{\partial \varphi}{\partial z} - \varphi \end{aligned}$$

Making use of (5.194) yields

$$\mu''(z) + [\mu'(z)]^2 + (z+t)\mu'(z) + zt = -1 \tag{5.195}$$

Since z and t are independent, we find that $\mu(z)$ must satisfy $\mu' + z = 0$. Now $\mu(z) = -z^2/2$ satisfies (5.195). The function

$$\varphi(z, t) \equiv e^{-z^2/2+zt} \tag{5.196}$$

can be shown to be the generating function for the $H_n(t)$. There is no trouble in differentiating the series expansion of $\varphi(z, t)$ term by term since $\varphi(z, t)$ is analytic for all z, t in the complex domain.

Finally, we show that

$$\frac{1}{\sqrt{2\pi}} \int_{-\infty}^{\infty} e^{-t^2/2} H_m(t)H_n(t)\,dt = \left.\begin{matrix} 0 & \text{if } m \neq n \\ n! & \text{if } m = n \end{matrix}\right\} \tag{5.197}$$

We consider

$$\begin{aligned} I &= \int_{-\infty}^{\infty} e^{-t^2/2} H_m(t)H_n(t)\,dt \qquad m \geqq n \\ &= \int_{-\infty}^{\infty} (-1)^m H_n(t) \frac{d^m e^{-t^2/2}}{dt^m}\,dt \end{aligned}$$

Integration by parts yields

$$\begin{aligned} I &= (-1)^m H_n(t) \frac{d^{m-1}e^{-t^2/2}}{dt^{m-1}}\bigg|_{-\infty}^{\infty} + (-1)^{m-1} \int_{-\infty}^{\infty} \frac{dH_n}{dt}\frac{d^{m-1}e^{-t^2/2}}{dt^{m-1}}\,dt \\ &= (-1)^{m-1} n \int_{-\infty}^{\infty} H_{n-1}(t) \frac{d^{m-1}e^{-t^2/2}}{dt^{m-1}}\,dt \qquad \text{Why?} \end{aligned}$$

Further integration by parts yields

$$I = (-1)^{m-n} n(n-1) \cdots (n-m+1) \int_{-\infty}^{\infty} \frac{d^{m-n} e^{-t^2/2}}{dt^{m-n}} \, dt$$

If $m > n$, let the reader deduce that $I = 0$. If $m = n$,

$$I = n! \int_{-\infty}^{\infty} e^{-t^2/2} \, dt = \sqrt{2\pi} \, n!$$

REFERENCES

Agnew, R. P.: "Differential Equations," McGraw-Hill Book Company, Inc., New York, 1942.

Bateman, H.: "Partial Differential Equations of Mathematical Physics," Dover Publications, New York, 1944.

Cohen, A.: "Differential Equations," D. C. Heath and Company, Boston, 1933.

Goursat, E.: "Differential Equations," Ginn & Company, Boston, 1917.

Ince, E. L.: "Ordinary Differential Equations," Dover Publications, New York, 1944.

Kells, L. M.: "Elementary Differential Equations," McGraw-Hill Book Company, Inc., New York, 1947.

Miller, F. H.: "Partial Differential Equations," John Wiley & Sons, Inc., New York, 1941.

Petrovsky, I. G.: "Lectures on Partial Differential Equations," Interscience Publishers, Inc., New York, 1954.

CHAPTER 6

ORTHOGONAL POLYNOMIALS, FOURIER SERIES, AND FOURIER INTEGRALS

6.1. Orthogonality of Functions. A function $f(x)$ defined over the interval (a, b) may be thought of as an infinite dimensional vector. $f(x_0)$ is the component of $f(x)$ at $x = x_0$. Let $g(x)$ be defined also over the interval (a, b). In vector analysis one obtained the scalar, dot, or inner product of two vectors in a cartesian coordinate system by multiplying corresponding components and summing. Obviously we cannot sum $f(x)g(x)$ for all x, $a \leqq x \leqq b$. The next best thing is to define

$$S(f, g) = \int_a^b f(x)g(x)\,dx \tag{6.1}$$

as the scalar product of f and g. If $S(f, g) = 0$, we say that f and g are orthogonal on the range (a, b). Generalizing, if

$$\int_a^b \rho(x)f(x)g(x)\,dx = 0 \tag{6.2}$$

we say that f and g are orthogonal relative to the weight, or density, function, $\rho(x)$, on the interval (a, b).

6.2. Generating Orthogonal Polynomials. In Chap. 5 we noticed that important classes of polynomials such as the Legendre, Laguerre, and Hermite polynomials arose in connection with the solutions of various differential equations. In this section we shall show how to generate orthogonal polynomials. The various theorems proved here will apply to every type of polynomial generated. We proceed as follows:

Let (a, b) be any closed interval, $-\infty \leqq a \leqq x \leqq b \leqq \infty$, and let $\rho(x)$ be any real-valued function satisfying the following conditions:

$$\begin{aligned} &\text{(i) } \rho(x) \geqq 0 && \text{for } a < x < b \\ &\text{(ii) } \int_\alpha^\beta \rho(x)\,dx > 0 && \text{for } a \leqq \alpha < \beta \leqq b \end{aligned} \tag{6.3}$$

(iii) The moments of $\rho(x)$, say, μ_n, defined by

$$\mu_n = \int_a^b x^n \rho(x)\,dx \qquad n = 0, 1, 2, \ldots$$

exist and are finite for $n = 0, 1, 2, \ldots$. We call $\rho(x)$ a weight, or density, function.

We now proceed to construct a sequence, or family, of polynomials,

$$P_0(x) \equiv 1, P_1(x), P_2(x), \ldots, P_n(x), \ldots$$

such that

$$P_n(x) = x^n + \sigma_n x^{n-1} + \cdots \tag{6.4}$$

and such that

$$\int_a^b \rho(x)P_m(x)P_n(x)\,dx = 0 \qquad m \neq n \tag{6.5}$$

Condition (6.5) states that $P_m(x)$ and $P_n(x)$ are orthogonal to each other relative to the weight factor $\rho(x)$. Note that $P_n(x)$ has its leading coefficient equal to unity. We apply mathematical induction to generate the family $\{P_n(x)\}$. We have already defined $P_0(x) \equiv 1$. Let

$$P_1(x) = x + \sigma_1$$

To fulfill condition (6.5), we need

$$\int \rho(x)(x + \sigma_1)\,dx = 0 \tag{6.6}$$

The limits of integration are omitted with the understanding that they remain fixed throughout the discussion. Equation (6.6) yields

$$\sigma_1 = -\frac{\int x\rho(x)\,dx}{\int \rho(x)\,dx}$$

[see (ii) and (iii) of (6.3)]. This choice of σ_1 yields $P_1(x)$ orthogonal to $P_0(x)$. Now assume that we have constructed the sequence of orthogonal polynomials $P_0(x), P_1(x), P_2(x), \ldots, P_k(x)$ satisfying (6.5). We know that k is at least 1. We now show that we can extend our constructed set of polynomials to include the polynomial $P_{k+1}(x)$ while preserving the important orthogonality condition (6.5). Let

$$P_{k+1}(x) = x^{k+1} + \sum_{r=0}^{k} c_r P_r(x) \tag{6.7}$$

where c_r, $r = 0, 1, 2, \ldots, k$, are $k + 1$ undetermined constants. We desire

$$\int \rho(x)P_s(x)P_{k+1}(x)\,dx = 0 \qquad s = 0, 1, 2, \ldots, k \tag{6.8}$$

With the aid of (6.7), (6.8) becomes

$$\int x^{k+1}\rho(x)\,dx + \sum_{r=0}^{k} c_r \int \rho(x)P_s(x)P_r(x)\,dx = 0 \tag{6.9}$$

However, for $s \neq r$ we have $\int \rho(x)P_s(x)P_r(x)\,dx = 0$ provided $s \leq k$,

$r \leq k$, by our assumption on the set $P_0, P_1, P_2, \ldots, P_k(x)$. Hence of necessity

$$c_s = -\frac{\int x^{k+1}\rho(x)\,dx}{\int \rho(x)P_s^2(x)\,dx} \qquad s = 0, 1, 2, \ldots, k \tag{6.10}$$

The numerator of c_s exists [see (iii) of (6.3)], and the denominator of c_s is different from zero and is a linear combination of various moments of $\rho(x)$, each of which exists. We leave this as an exercise for the reader. It is a trivial procedure to reverse the steps taken to derive the c_s, $s = 0, 1, 2, \ldots, k$; that is, if the c_s of (6.10) are substituted into (6.7), one easily shows that $P_{k+1}(x)$ is orthogonal to $P_r(x)$, $r = 0, 1, 2, \ldots, k$, relative to $\rho(x)$. The principle of finite mathematical induction states that a sequence of polynomials $P_0(x), P_1(x), P_2(x), \ldots, P_n(x), \ldots$ exists satisfying (6.5).

Problem 1. Prove by mathematical induction that the sequence of polynomials $\{P_n(x)\}$ is unique relative to the interval (a, b) and the density function $\rho(x)$.

For the interval $(0, 1)$ and for $\rho(x) = 1$ the orthogonal polynomials are the well-known Legendre polynomials. We generate now $P_1(x)$, $P_2(x)$. For $P_1(x) = x + \sigma_1$ we have

$$\sigma_1 = -\frac{\int_0^1 x\,dx}{\int_0^1 dx} = -\frac{1}{2}$$

so that $P_1(x) = x - \frac{1}{2}$. Let $P_2(x) = x^2 + a_1P_1 + A_0P_0$. From

$$\int_0^1 P_0P_2\,dx = 0$$

we have $\int_0^1 x^2\,dx + a_0 \int_0^1 dx = 0$ so that $a_0 = -\frac{1}{3}$. From

$$\int_0^1 P_1P_2\,dx = 0$$

we have $\int_0^1 (x - \frac{1}{2})x^2\,dx + a_1 \int_0^1 (x - \frac{1}{2})^2\,dx = 0$, so that $a_1 = -1$. Thus $P_2(x) = x^2 - (x - \frac{1}{2}) - \frac{1}{3} = x^2 - x + \frac{1}{6}$.

Problem 2. For the interval $(0, \infty)$ and for $\rho(x) = e^{-x}$ generate $L_1(x)$ and $L_2(x)$. These are the Laguerre polynomials.

Problem 3. Construct the Hermite polynomials $H_1(x)$, $H_2(x)$ for the interval $(-\infty, \infty)$, $\rho(x) = e^{-x^2/2}$.

An interesting result concerning orthogonal polynomials is the following: Let $Q(x)$ be any polynomial of degree n. There exists a unique set of constants $c_0, c_1, \ldots, c_n$ such that

$$Q(x) = \sum_{r=0}^{n} c_r P_r(x) \qquad c_n \neq 0$$

The constants depend, of course, on the family of polynomials under discussion. The proof is by induction. If $Q(x)$ is of degree zero, say $Q(x) = a$, then $Q(x) = aP_0(x)$. Assume the theorem true for all polynomials of degree $\leqq k$. Now let $Q(x)$ be any polynomial of degree $k + 1$, $Q(x) = a_0x^{k+1} + a_1x^k + \cdots + a_{k+1}$, $a_0 \neq 0$. Then $Q(x) - a_0P_{k+1}(x)$ is a polynomial of degree $\leqq k$. By our assumption

$$Q(x) - a_0P_{k+1}(x) = \sum_{r=0}^{k} c_r P_r(x)$$

so that

$$Q(x) = \sum_{r=0}^{k+1} c_r P_r(x) \qquad c_{k+1} = a_0$$

Problem 4. Show that c_r, $r = 0, 1, 2, 3, \ldots, k+1$, in the preceding line are unique.

Problem 5. Show that $\int \rho(x)x^n P_m(x)\,dx = 0$ for $m > n$.

Problem 6. If $Q(x)$ is a polynomial of degree $m < n$, show that

$$\int \rho(x)Q(x)P_n(x)\,dx = 0$$

Problem 7. Show that the conditions

$$\int \rho(x)x^k P_n(x)\,dx = 0 \qquad k = 0, 1, 2, \ldots, n-1 \tag{6.11}$$

yield a unique polynomial $P_n(x) = x^n + \sigma_n x^{n-1} + \cdots$ and that

$$\int \rho(x)P_m(x)P_n(x)\,dx = 0$$

for $m \neq n$ if $P_m(x)$ is generated in the same manner. Equation (6.11) could have been taken as the starting point for developing orthogonal polynomials.

Problem 8. For $a \neq -\infty$, $b \neq +\infty$ find a linear transformation $x = r\bar{x} + s$ such that $\bar{x} = 0$ when $x = a$, $\bar{x} = 1$ when $x = b$. For $a \neq \pm\infty$, $b = +\infty$ find a linear transformation which maps $x = a$ into $\bar{x} = 0$, $x = +\infty$ into $\bar{x} = +\infty$.

Problem 9. Show that $\int xP_{n-1}P_n\,\rho\,dx = \int P_n^2\rho\,dx$.

6.3. Normalizing Factors. Examples of Orthogonal Polynomials. The constants

$$\gamma_n^2 = \int P_n^2(x)\rho(x)\,dx \qquad n = 0, 1, 2, \ldots \tag{6.12}$$

are called the normalizing factors of the $P_n(x)$. It follows that

$$\int \frac{\rho(x)P_n^2(x)}{\gamma_n^2}\,dx = \int \left[\frac{\sqrt{\rho(x)}\,P_n(x)}{\gamma_n}\right]^2 dx = 1$$

We define the orthonormal set of functions $\{\varphi_n(x)\}$ by

$$\varphi_n(x) = \frac{1}{\gamma_n} \sqrt{\rho(x)}\, P_n(x) \qquad n = 0, 1, 2, \ldots \tag{6.13}$$

The $\varphi_n(x)$ are not, in general, polynomials. They possess the attractive property

$$\int \varphi_i(x)\varphi_j(x)\, dx = \delta_{ij} \begin{array}{ll} = 0 & \text{if } i \neq j \\ = 1 & \text{if } i = j \end{array}$$

The $\{\varphi_n(x)\}$ form what is known as a normalized orthogonal system of functions associated with the family of orthogonal polynomials $\{P_n(x)\}$.

We now list a few families of orthogonal polynomials.

	Interval	Weight function	Polynomial
1	$0 \leqq x \leqq 1$	1	Legendre
2	$-1 \leqq x \leqq 1$	$(1 - x^2)^{-\frac{1}{2}}$ or $(1 - x^2)^{\frac{1}{2}}$	Tchebysheff of 1st and 2d kinds
3	$0 \leqq x \leqq \infty$	e^{-x}	Laguerre
4	$0 \leqq x \leqq \infty$	$x^c e^{-x}$, $c > 0$	Generalized Laguerre
5	$-\infty \leqq x \leqq \infty$	$e^{-x^2/2}$	Hermite

Example 6.1. Let $a = 0$, $b = 1$ with $\rho(x) = 1$. We choose

$$P_n(x) = 1 + a_1x + a_2x^2 + \cdots + a_nx^n$$

From $\int_0^1 x^kP_n(x)\,dx = 0$ for $k < n$ we have

$$\frac{1}{k+1} + \frac{a_1}{k+2} + \cdots + \frac{a_n}{k+n+1} = 0 \qquad \text{for } k = 0, 1, 2, \ldots, n-1$$

We can solve these n linear equations for the a_i, $i = 1, 2, \ldots, n$, as follows: We have

$$\frac{1}{k+1} + \frac{a_1}{k+2} + \cdots + \frac{a_n}{k+n+1} = \frac{Q(k)}{(k+1)(k+2)\cdots(k+n+1)}$$

where $Q(k)$ is a polynomial in k of degree at most n. Since $Q(k)$ vanishes for $k = 0, 1, 2, \ldots, n-1$, of necessity

$$Q(k) = Ck(k-1)(k-2)\cdots(k-n+1)$$

(see Lemma 1, Sec. 6.4). Thus

$$1 + a_1\frac{k+1}{k+2} + a_2\frac{k+1}{k+3} + \cdots + a_n\frac{k+1}{k+n+1} = C\frac{k(k-1)\cdots(k-n+1)}{(k+2)\cdots(k+n+1)}$$

If we set $k = -1$, we obtain

$$1 = \frac{C(-1)(-2)\cdots(-n)}{1 \cdot 2 \cdot 3 \cdots n}$$

so that $C = (-1)^n$. Thus

$$(\alpha)\quad \frac{1}{k+1} + \frac{a_1}{k+2} + \cdots + \frac{a_r}{k+r+1} + \cdots + \frac{a_n}{k+n+1} = \frac{(-1)^n k(k-1)\cdots(k-n+1)}{(k+1)(k+2)\cdots(k+n+1)}$$

To solve for a_r, we multiply the above by $k + r + 1$ and then set $k = -(r + 1)$. This yields

$$\begin{aligned} a_r &= \frac{(-1)^n(r+1)(r+2)\cdots(r+n)(-1)^n}{(-r)(-r+1)(-r+2)\cdots(-1)1\cdot 2\cdots(n-r)} \\ &= (-1)^r \frac{(r+n)!}{r!r!(n-r)!} = (-1)^r \binom{n+r}{r}\binom{n}{r} \end{aligned}$$

Hence
$$P_n(x) = \sum_{r=0}^{n} a_r x^r = \sum_{r=0}^{n} (-1)^r \binom{n}{r}\binom{n+r}{r} x^r$$

The Legendre polynomial with leading coefficient unity is

$$P_n^*(x) = \frac{(-1)^n}{\binom{2n}{n}} \sum_{r=0}^{n} (-1)^r \binom{n}{r}\binom{n+r}{r} x^r$$

We now determine $\int_0^1 P_n^2(x)\rho(x)\,dx$. We have

$$\begin{aligned} \int_0^1 P_n^2(x)\,dx &= \int_0^1 a_n x^n P_n(x)\,dx \\ &= a_n \sum_{r=0}^{n} a_r \int_0^1 x^n x^r\,dx \\ &= a_n \sum_{r=0}^{n} \frac{a_r}{n+r+1} = a_n \frac{(-1)^n n!n!}{(2n+1)!} \end{aligned}$$

from (α) above. Thus

$$\int_0^1 P_n^2(x)\,dx = \binom{2n}{n}\binom{n}{n}\frac{n!n!}{(2n+1)!} = \frac{(2n)!}{(2n+1)!} = \frac{1}{2n+1}$$

6.4. The Zeros of the Orthogonal Polynomials. We wish to show that the zeros of $P_n(x)$ of Sec. 6.2 are real, distinct, and lie in the fundamental interval. We first state and prove two lemmas.

LEMMA 1. If $x = r$ is a root of $P(x) = 0$, then $x - r$ is a factor of $P(x)$, $P(x)$ a polynomial. The proof is as follows: We have upon division

$$\frac{P(x)}{x-r} = Q(x) + \frac{R}{x-r} \qquad R = a = \text{constant}$$

or
$$P(x) = Q(x)(x-r) + R$$

If $P(r) = 0$, then $R = 0$ so that $P(x) = Q(x)(x - r)$. Q.E.D.

LEMMA 2. If $x = a + i \cdot b$ is a zero of a real polynomial, $P(x)$, a, b real, then $a - i \cdot b$ is also a zero of $P(x)$. It is easy to see that

$$\begin{aligned} P(a + ib) &= U(a, b) + iV(a, b) \\ P(a - ib) &= U(a, b) - iV(a, b) \end{aligned}$$

Since $P(a + ib) = 0$, we have $U = V = 0$ so that $P(a - ib) = 0$.

We now prove the general theorem stated above. Let $P_n(x)$ be a real orthogonal polynomial. If $x = a + ib$, $b \neq 0$, is a zero of $P_n(x)$, then $[x - (a + ib)][x - (a - ib)] \equiv (x - a)^2 + b^2$ is a factor of $P_n(x)$. This factor is positive for all x in our fundamental interval (a, b). This is true for all complex zeros of $P_n(x)$. Now let $x = x_1$, $x = x_2$, . . . , $x = x_N$ be the real zeros of $P_n(x)$ of odd multiplicity lying on the interval (a, b). Certainly $N \leqq n$. Assume $N < n$, and let

$$Q(x) = (x - x_1)(x - x_2) \cdots (x - x_N)P_n(x)$$

Let the reader deduce that $Q(x) > 0$ for all x on (a, b) except for $x = x_1$, x_2, . . . , x_N or $Q(x) < 0$ for all x on (a, b) except for $x = x_1, x_2, \ldots, x_N$. We have, however,

$$\int_a^b \rho(x)Q(x)\,dx = \int_a^b \rho(x)(x - x_1) \cdots (x - x_N)P_n(x)\,dx = 0 \quad (6.14)$$

since $(x - x_1)(x - x_2) \cdots (x - x_N)$ is polynomial of degree $N < n$ (see Prob. 6, Sec. 6.2). But $\int_a^b \rho(x)Q(x)\,dx > 0$ or $\int_a^b \rho(x)Q(x)\,dx < 0$ from the nature of $Q(x)$, a contradiction to (6.14). Hence $N = n$, and the theorem is proved.

6.5. The Difference Equation for Orthogonal Polynomials. We obtain now an equation involving $P_{n-1}(x)$, $P_n(x)$, and $P_{n+1}(x)$. We have

$$\begin{aligned} P_{n+1}(x) &= x^{n+1} + \sigma_{n+1}x^n + \cdots \\ P_n(x) &= x^n + \sigma_n x^{n-1} + \cdots \end{aligned}$$

so that $\quad P_{n+1}(x) - xP_n(x) = (\sigma_{n+1} - \sigma_n)x^n + \cdots$

and $\quad P_{n+1}(x) - xP_n(x) = (\sigma_{n+1} - \sigma_n)P_n(x) + \cdots$

since $x^n = P_n(x) - \sigma_n x^{n-1} - \cdots$. Hence

$$P_{n+1}(x) - xP_n(x) - (\sigma_{n+1} - \sigma_n)P_n(x) \equiv P_{n+1}(x) - (x + \sigma_{n+1} - \sigma_n)P_n(x)$$

is a polynomial of degree at most $n - 1$. From a previous theorem

$$\begin{aligned} P_{n+1}(x) - (x + \sigma_{n+1} - \sigma_n)P_n(x) &= \sum_{r=0}^{n-1} c_r P_r(x) \\ &= c_{n-1}P_{n-1} + \sum_{r=0}^{n-2} c_r P_r(x) \quad (6.15) \end{aligned}$$

We multiply (6.15) by $\rho(x)P_{n-1}(x)$ and integrate over the interval (a, b). Using the orthogonality property and the fact that

$$\int xP_{n-1}(x)P_n(x)\rho(x)\,dx = \int P_n^2(x)\rho(x)\,dx = \gamma_n^2$$

(see Prob. 9, Sec. 6.2), we obtain $-\gamma_n^2 = c_{n-1}\gamma_{n-1}^2$. Thus

$$P_{n+1}(x) - (x + \sigma_{n+1} - \sigma_n)P_n(x) + \frac{\gamma_n^2}{\gamma_{n-1}^2} P_{n-1}(x) = \sum_{r=0}^{n-2} c_r P_r(x) \tag{6.16}$$

is a polynomial of degree at most $n - 2$. To show that $c_r = 0$ for $r = 0, 1, 2, \ldots, n - 2$, one need only multiply (6.16) by $\rho(x)P_s(x)$, $s \leqq n - 2$, and perform an integration over (a, b). Hence

$$P_{n+1}(x) - (x + \sigma_{n+1} - \sigma_n)P_n(x) + \frac{\gamma_n^2}{\gamma_{n-1}^2} P_{n-1}(x) \equiv 0 \qquad n \geqq 1 \tag{6.17}$$

Equation (6.17) is the difference equation satisfied by the orthogonal polynomials. Using (6.17), it is possible to show that the zeros of $P_n(x)$, $P_{n+1}(x)$ interlace in the sense that, if $x_1 < x_2 < \cdots < x_n$ are the zeros of $P_n(x)$ and if $y_1 < y_2 < \cdots < y_n < y_{n+1}$ are the zeros of $P_{n+1}(x)$, then $a < y_1 < x_1 < y_2 < x_2 < \cdots < y_n < x_n < y_{n+1} < b$. We omit this proof.

6.6. The Christoffel-Darboux Identity. From (6.17) we have

$$P_{n+1}(t) - (t + \sigma_{n+1} - \sigma_n)P_n(t) + \frac{\gamma_n^2}{\gamma_{n-1}^2} P_{n-1}(t) = 0 \tag{6.18}$$

so that

$$\begin{aligned} &P_n(t)P_{n+1}(x) - (x + \sigma_{n+1} - \sigma_n)P_n(x)P_n(t) \\ &\qquad\qquad + \frac{\gamma_n^2}{\gamma_{n-1}^2} P_n(t)P_{n-1}(x) = 0 \\ &P_n(x)P_{n+1}(t) - (t + \sigma_{n+1} - \sigma_n)P_n(x)P_n(t) \\ &\qquad\qquad + \frac{\gamma_n^2}{\gamma_{n-1}^2} P_n(x)P_{n-1}(t) = 0 \end{aligned} \tag{6.19}$$

upon multiplying (6.17) by $P_n(t)$ and (6.18) by $P_n(x)$. Subtracting and dividing by γ_n^2 yields

$$\frac{P_{n+1}(x)P_n(t) - P_{n+1}(t)P_n(x)}{\gamma_n^2} - \frac{P_n(x)P_{n-1}(t) - P_n(t)P_{n-1}(x)}{\gamma_{n-1}^2} = \frac{(x - t)P_n(x)P_n(t)}{\gamma_n^2} \qquad n \geqq 1 \tag{6.20}$$

We have also

$$\frac{P_1(x)P_0(t) - P_1(t)P_0(x)}{\gamma_0^2} = \frac{(x - \sigma_1) - (t - \sigma_1)}{\gamma_0^2} = \frac{(x - t)P_0(x)P_0(t)}{\gamma_0^2} \tag{6.21}$$

If we let $n = 1, 2, 3, \ldots, k$ in (6.20), add, and make use of (6.21), we obtain

$$\frac{P_{k+1}(x)P_k(t) - P_{k+1}(t)P_k(x)}{\gamma_k^2(x - t)} = \sum_{n=0}^{k} \frac{P_n(x)P_n(t)}{\gamma_n^2} \tag{6.22}$$

Equation (6.22) is the Christoffel-Darboux identity. We define the kernel, $K_k(x; t)$, by the equation

$$K_k(x; t) = \sum_{n=0}^{k} \frac{P_n(x)P_n(t)}{\gamma_n^2} \tag{6.23}$$

Problem 10. Evaluate $K_k(x; x)$ by L'Hospital's rule.

Problem 11. If $x = c$, $x = d$ are distinct zeros of $P_k(x)$, show that $K_k(c; d) = 0$. What if $c = d$?

Problem 12. Prove that

$$\int_a^b K_k(x; t)\rho(t)\,dt = 1 \tag{6.24}$$

6.7. Fourier Coefficients and Partial Sums. We may redefine the orthogonal polynomials so that

$$\int_a^b \rho(x)p_m(x)p_n(x)\,dx = \delta_{mn} \begin{matrix} = 0 & \text{if } m \neq n \\ = 1 & \text{if } m = n \end{matrix} \tag{6.25}$$

by simply letting $p_n(x) = (1/\gamma_n)P_n(x)$, $n = 0, 1, 2, \ldots$ (see Sec. 6.3). Now let us consider a real-valued function of x whose Taylor-series expansion about $x = 0$ exists. We have

$$f(x) = \sum_{n=0}^{\infty} \frac{f^{(n)}(0)}{n!} x^n \tag{6.26}$$

We see that $f(x)$ can be expanded as a linear combination of terms from the sequence of polynomials

$$1, x, x^2, x^3, \ldots, x^n, \ldots$$

$$f(x) = \sum_{n=0}^{\infty} c_n x^n \qquad c_n = \frac{f^{(n)}(0)}{n!} \tag{6.27}$$

It seems logical to ask whether a given function $f(x)$ can be expanded in terms of the infinite sequence of polynomials

$$p_0(x), p_1(x), p_2(x), \ldots, p_n(x), \ldots \tag{6.28}$$

If this is the case, we may write

$$f(x) = \sum_{n=0}^{\infty} c_n p_n(x) \tag{6.29}$$

To determine the coefficients c_n, $n = 0, 1, 2, \ldots$, we multiply (6.29) by $\rho(x)p_k(x)$ and integrate. This yields

$$\int_a^b \rho(x)p_k(x)f(x)\,dx = \int_a^b \sum_{n=0}^{\infty} c_n\rho(x)p_k(x)p_n(x)\,dx \tag{6.30}$$

Assuming further that the series expansion converges uniformly on (a, b) yields

$$\int_a^b \rho(x)p_k(x)f(x)\,dx = \sum_{n=0}^{\infty} c_n \int_a^b \rho(x)p_k(x)p_n(x)\,dx = c_k \tag{6.31}$$

The c_k, $k = 0, 1, 2, \ldots$, of (6.31) are defined as the "Fourier" coefficients of $f(x)$. If $\int_a^b \rho(x)f^2(x)\,dx$ exists, then from the Schwarz-Cauchy inequality for integrals we have

$$|c_k|^2 \leqq \int_a^b \rho(x)f^2(x)\,dx \int_a^b \rho(x)p_k^2(x)\,dx \tag{6.32}$$

Thus c_k, given by

$$c_k = \int_a^b \rho(x)p_k(x)f(x)\,dx \qquad k = 0, 1, 2, \ldots \tag{6.33}$$

exists provided $\int_a^b \rho(x)f^2(x)\,dx$ exists. Thus one can speak of the Fourier coefficients of $f(x)$ regardless of the existence or nonexistence of the expansion of $f(x)$ in the form given by (6.29).

If the c_k, $k = 0, 1, 2, \ldots, n$, of (6.33) exist, we can form the series

$$s_n(x) = \sum_{i=1}^{n} c_i p_i(x) \tag{6.34}$$

$s_n(x)$ is called the nth partial sum of the Fourier series associated with $f(x)$. With the aid of (6.23) and (6.33) we have

$$\begin{aligned} s_n(x) &= \int_a^b \sum_{i=0}^{n} \rho(t)f(t)p_i(t)p_i(x)\,dt \\ &= \int_a^b K_n(x; t)f(t)\rho(t)\,dt \end{aligned} \tag{6.35}$$

The difference between $f(x)$ and $s_n(x)$ is called the nth remainder in the Fourier series of $f(x)$. We have

$$\begin{aligned} R_n(x) &= f(x) - s_n(x) \\ &= \int_a^b f(x)K_n(x; t)\rho(t)\, dt - \int_a^b K_n(x; t)f(t)\rho(t)\, dt \\ &= \int_a^b K_n(x; t)\rho(t)[f(x) - f(t)]\, dt \end{aligned} \tag{6.36}$$

since $\int_a^b K_n(x; t)\rho(t)\, dt = 1$ [see (6.24)]. Now

$$K_n(x; t) = \frac{\gamma_{n+1}}{\gamma_n} \frac{[p_{n+1}(x)p_n(t) - p_{n+1}(t)p_n(x)]}{x - t}$$

so that

$$R_n(x) = \frac{\gamma_{n+1}}{\gamma_n} \int_a^b [p_{n+1}(x)p_n(t) - p_{n+1}(t)p_n(x)] \frac{f(x) - f(t)}{x - t} \rho(t)\, dt \tag{6.37}$$

The reader should explain the apparent difficulty at $t = x$. If

$$\lim_{n\to\infty} R_n(x) = 0$$

the Fourier series, $\sum_{n=0}^{\infty} c_n p_n(x)$, converges to $f(x)$.

6.8. Bessel's Inequality. We show first that $R_n(x) = f(x) - s_n(x)$ is orthogonal to $p_k(x)$, $k = 0, 1, 2, \ldots, n$. We have

$$\begin{aligned} \int_a^b [f(x) - s_n(x)]p_k(x)\rho(x)\, dx &= \int_a^b f(x)p_k(x)\rho(x)\, dx \\ &\qquad - \sum_{i=0}^{n} c_i \int_a^b p_i(x)p_k(x)\rho(x)\, dx \\ &= c_k - \sum_{i=0}^{n} c_i\delta_{ik} \\ &= c_k - c_k = 0 \end{aligned} \tag{6.38}$$

We show next that

$$\int_a^b f^2(x)\rho(x)\, dx \geqq \sum_{k=0}^{n} c_k^2 \tag{6.39}$$

The proof is as follows: Clearly

$$\int_a^b [f(x) - s_n(x)]^2\rho(x)\, dx \geqq 0$$

so that

$$\int_a^b f^2(x)\rho(x)\, dx - 2\int_a^b f(x)s_n(x)\rho(x)\, dx + \int_a^b s_n^2(x)\rho(x)\, dx \geqq 0 \tag{6.40}$$

However,

$$\int_a^b f(x)s_n(x)\rho(x)\,dx = \sum_{k=0}^{n} c_k \int_a^b f(x)p_k(x)\rho(x)\,dx$$
$$= \sum_{k=0}^{n} c_k^2$$

and
$$\int_a^b s_n^2(x)\rho(x)\,dx = \sum_{j=0}^{n}\sum_{k=0}^{n} c_j c_k \int_a^b p_j(x)p_k(x)\rho(x)\,dx$$
$$= \sum_{j=0}^{n}\sum_{k=0}^{n} c_j c_k \delta_{jk}$$
$$= \sum_{k=0}^{n} c_k^2$$

so that (6.39) follows as a consequence. Since (6.39) holds for all n, the sequence $t_n = \sum_{k=0}^{n} c_k^2$, $n = 0, 1, 2, \ldots$, is bounded. Moreover, this sequence is monotonic nondecreasing. This implies that $\lim_{n\to\infty} t_n$ exists. Let the reader deduce that

$$\int_a^b f^2(x)\rho(x)\,dx \geqq \sum_{k=0}^{\infty} c_k^2 \tag{6.41}$$

Formula (6.41) is Bessel's inequality.

Problem 13. Why is it true that $\lim_{k\to\infty} c_k = 0$?

Problem 14. If $f(x)$ is a polynomial of degree n, show that

$$\int_a^b f^2(x)\rho(x)\,dx = \sum_{k=0}^{n} c_k^2$$

Problem 15. For (α, β) such that $a \leqq \alpha \leqq \beta \leqq b$ assume that

$$d_n = \int_\alpha^\beta g(x)p_n(x)\rho(x)\,dx$$

exists. Let $f(x) = g(x)$ for $\alpha \leqq x \leqq \beta$, $f(x) = 0$ otherwise. Show that $\lim_{n\to\infty} d_n = 0$.

6.9. A Minimal Property of the Partial Fourier Sums. We look for a polynomial $Q(x)$ of degree $\leqq n$ such that

$$I = \int_a^b [f(x) - Q(x)]^2\rho(x)\,dx$$

shall be a minimum. Let $s_n(x)$ be the nth partial Fourier sum of $f(x)$. Then

$$\begin{aligned} I &= \int_a^b [f(x) - s_n(x) + s_n(x) - Q(x)]^2\rho(x)\,dx \\ &= \int_a^b [f(x) - s_n(x)]^2\rho(x)\,dx + 2\int_a^b [f(x) - s_n(x)][s_n(x) - Q(x)]\rho(x)\,dx \\ &\qquad + \int_a^b [s_n(x) - Q(x)]^2\rho(x)\,dx \end{aligned}$$

The second integral vanishes from the results of Sec. 6.8 [see (6.38)] since $s_n(x) - Q(x)$ is a polynomial of degree $\leqq n$. The first integral has a fixed value. Hence I will be a minimum when $\int_a^b [s_n(x) - Q(x)]^2\rho(x)\,dx$ is a minimum. This obviously occurs if we choose $Q(x) \equiv s_n(x)$. This is a highly important property of $s_n(x)$.

Problem 16. Let $Q_n(x)$ be a polynomial of degree n with leading coefficient unity. If $\int_a^b Q_n^2(x)\rho(x)\,dx$ is to be a minimum, show that $Q_n(x) \equiv P_n(x)$.

Problem 17. Let $\{P_n(x)\}$ be a family of polynomials with leading coefficients equal to unity, $P_n(x)$ of degree n such that $\int_a^b P_n^2(x)\rho(x)\,dx$ is a minimum. Without using the result of Prob. 16 show that

$$\int_a^b P_i(x)P_j(x)\rho(x)\,dx = 0 \qquad \text{if } i \neq j$$

Problem 18. Show that $\gamma_{n+1}^2 < \gamma_n^2$ for the fundamental interval $(-1, 1)$. *Hint:*

$$\int_{-1}^1 P_{n+1}^2(x)\rho(x)\,dx \leqq \int_{-1}^1 x^2P_n^2(x)\rho(x)\,dx < \int_{-1}^1 P_n^2(x)\rho(x)\,dx$$

6.10. Complete and Closed Sets of Orthogonal Polynomials. A set of orthogonal polynomials is said to be complete if for every continuous function $f(x)$ we have

$$\lim_{n\to\infty} \int_a^b R_n^2(x)\rho(x)\,dx = 0 \tag{6.42}$$

where $R_n(x)$ is defined by (6.36). From Sec. 6.8 we have

$$\int_a^b R_n^2(x)\rho(x)\,dx = \int_a^b f^2(x)\rho(x)\,dx - \sum_{k=0}^n c_k^2$$

so that (6.42) is equivalent to the statement

$$\int_a^b f^2(x)\rho(x)\,dx = \sum_{k=0}^\infty c_k^2 \tag{6.43}$$

We shall prove in Sec. 6.11 that the orthogonal polynomials generated above are complete for any finite range (a, b).

DEFINITION 6.1. A set of orthogonal polynomials is said to be closed if the vanishing of all the Fourier coefficients of a continuous function $f(x)$ implies $f(x) \equiv 0$. In other words, if $f(x)$ is continuous and if

$$\int_a^b f(x)p_n(x)\rho(x)\,dx = 0 \qquad n = 0, 1, 2, \ldots \tag{6.44}$$

then $f(x) \equiv 0$.

Problem 19. Show that a complete set of orthogonal polynomials is a closed set.

Problem 20. Show that (6.44) implies $\lambda_n = \int_a^b x^n f(x)\rho(x)\,dx = 0$ for $n = 0, 1, 2, \ldots$, and conversely.

6.11. Completeness of the Orthogonal Polynomials for a Finite Range. By a linear transformation the fundamental interval (a, b) can be reduced to the interval $-1 \leqq x \leqq 1$ if a and b are finite. The famous Weierstrass approximation theorem states that a continuous function on a closed interval can be approximated arbitrarily closely by a polynomial (see Courant-Hilbert, "Mathematische Physik," vol. 1). In our case we wish to approximate a continuous function $f(x)$ on $-1 \leqq x \leqq 1$ to within $\sqrt{\epsilon/\mu_0}$, where ϵ is an assigned positive number and μ_0 is the first moment, $\mu_0 = \int_{-1}^1 \rho(x)\,dx$. The Weierstrass approximation theorem states that there is a polynomial $Q(x)$ such that

$$|f(x) - Q(x)| < \sqrt{\frac{\epsilon}{\mu_0}} \qquad -1 \leqq x \leqq 1 \tag{6.45}$$

We denote the degree of Q by m so that, from Sec. 6.9 for $n \geqq m$,

$$\int_{-1}^1 [f(x) - s_n(x)]^2\rho(x)\,dx \leqq \int_{-1}^1 [f(x) - Q(x)]^2\rho(x)\,dx$$

and

$$\int_{-1}^1 [f(x) - s_n(x)]^2\rho(x)\,dx < \frac{\epsilon}{\mu_0}\int_{-1}^1 \rho(x)\,dx = \epsilon$$

Expanding $\int_{-1}^1 [f(x) - s_n(x)]^2\rho(x)\,dx$ as in Sec. 6.8, we have

$$\int_{-1}^1 f^2(x)\rho(x)\,dx - \sum_{k=0}^n c_k^2 < \epsilon$$

But $\sum_{k=0}^\infty c_k^2 - \sum_{k=0}^n c_k^2 < \epsilon$ for n sufficiently large so that

$$\left| \int_{-1}^1 f^2(x)\rho(x)\,dx - \sum_{k=0}^\infty c_k^2 \right| < 2\epsilon$$

Since $\int_{-1}^{1} f^2(x)\rho(x)\,dx$ and $\sum_{k=0}^{\infty} c_k^2$ are fixed constants whose difference can be made as small as we please, we must have

$$\int_{-1}^{1} f^2(x)\rho(x)\,dx = \sum_{k=0}^{\infty} c_k^2$$

This completes the proof [see (6.43)]. As an immediate corollary it follows that for $f(x)$ continuous the relations $\int_{-1}^{1} x^n f(x)\rho(x)\,dx = 0$, $n = 0, 1, 2, \ldots$, imply $f(x) \equiv 0$.

6.12. The Local Character of Convergence of the Fourier Series of $f(x)$. We are now in a position to examine the Fourier remainder $R_n(x)$ [see (6.36)] for the range $-1 \leqq x \leqq 1$. Let us assume that the $p_n(x)$ are uniformly bounded on this range. This means that $|p_n(x)| < M < \infty$, $-1 \leqq x \leqq 1$, for all n, M a constant. Let δ be any small positive number. Then for $-1 \leqq x_0 \leqq 1$

$$\begin{aligned} R_n(x_0) &= f(x_0) - s_n(x_0) \\ &= \frac{\gamma_{n+1}}{\gamma_n}\left[\int_{-1}^{x_0-\delta} \mu_n(x, t)\,dt + \int_{x_0-\delta}^{x_0+\delta} \mu_n(x, t)\,dt + \int_{x_0+\delta}^{1} \mu_n(x, t)\,dt\right] \end{aligned} \tag{6.46}$$

where $\quad \mu_n(x, t) = [p_{n+1}(x)p_n(t) - p_{n+1}(t)p_n(x)]\rho(t)\,\dfrac{f(x_0) - f(t)}{x_0 - t}$

[see (6.37)]. For the intervals $-1 \leqq t \leqq x_0 - \delta$, $x_0 + \delta \leqq t \leqq 1$ the function $\dfrac{f(x_0) - f(t)}{x_0 - t}$ is well behaved. Hence

$$\lim_{n\to\infty} \int_{-1}^{x_0-\delta} \mu_n(x, t)\,dt = 0$$

$$\lim_{n\to\infty} \int_{x_0+\delta}^{1} \mu_n(x, t)\,dt = 0$$

(see Prob. 15, Sec. 6.8). Hence

$$\lim_{n\to\infty} R_n(x_0) = \lim_{n\to\infty} \frac{\gamma_{n+1}}{\gamma_n} \int_{x_0-\delta}^{x_0+\delta} [p_{n+1}(x_0)p_n(t) - p_n(x_0)p_{n+1}(t)]\rho(t)\,\frac{f(x_0) - f(t)}{x_0 - t}\,dt$$

We need not worry about $\lim_{n\to\infty} \gamma_{n+1}/\gamma_n$ since $0 < \gamma_{n+1}/\gamma_n < 1$ (see Prob. 18, Sec. 6.9). Whether $\lim_{n\to\infty} R_n(x_0)$ is zero or not depends entirely on the behavior of $f(t)$ in the neighborhood of $t = x_0$. If, for example,

$$\left|\frac{f(x_0) - f(t)}{x_0 - t}\right| < A < \infty$$

A = constant, for all t near x_0, then

$$|R_n(x_0)| \leqq 6M^2 A \delta \mu_0$$

for n sufficiently large. Since δ can be chosen arbitrarily small, it follows that $\lim_{n\to\infty} R_n(x_0) = 0$.

For a complete discussion of orthogonal polynomials the reader is urged to read G. Szego, "Orthogonal Polynomials," *American Mathematical Society Colloquium Publications*, vol. 23.

6.13. Fourier Series of Trigonometric Functions. In the early part of the nineteenth century the French mathematician Fourier made a tremendous contribution to the field of mathematical physics. A study of heat motion (see Example 22, Chap. 5) led Fourier to the idea of expanding a real valued function $f(x)$ in a trigonometric series.

$$\begin{aligned} f(x) &= \tfrac{1}{2}a_0 + a_1 \cos x + a_2 \cos 2x + \cdots \\ &\qquad + b_1 \sin x + b_2 \sin 2x + \cdots \\ &= \tfrac{1}{2}a_0 + \sum_{n=1}^{\infty} a_n \cos nx + \sum_{n=1}^{\infty} b_n \sin nx \end{aligned} \tag{6.47}$$

At the moment let us not concern ourselves with the validity of (6.47). We wish to determine a_n, $n = 0, 1, 2, \ldots$, and b_n, $n = 1, 2, \ldots$, if (6.47) were valid. From trigonometry we have

$$\begin{aligned} \sin nx \sin mx &= \tfrac{1}{2}[\cos (n - m)x - \cos (n + m)x] \\ \sin nx \cos mx &= \tfrac{1}{2}[\sin (n - m)x + \sin (n + m)x] \\ \cos nx \cos mx &= \tfrac{1}{2}[\cos (n - m)x + \cos (n + m)x] \end{aligned}$$

so that

$$\left.\begin{aligned} \int_{-\pi}^{\pi} \sin nx \sin mx\, dx &= \begin{matrix} 0 & \text{if } n \neq m \\ \pi & \text{if } n = m \end{matrix} \\ \int_{-\pi}^{\pi} \sin nx \cos mx\, dx &= 0 \\ \int_{-\pi}^{\pi} \cos nx \cos mx\, dx &= \begin{matrix} 0 & \text{if } n \neq m \\ \pi & \text{if } n = m \end{matrix} \end{aligned}\right\} \tag{6.48}$$

provided m and n are integers.

If we multiply (6.47) by $\cos mx$ and integrate term by term (we are not concerned at present with the validity of term-by-term integration), we obtain by use of (6.48)

$$a_m = \frac{1}{\pi}\int_{-\pi}^{\pi} \cos mx\, f(x)\, dx \qquad m = 0, 1, 2, \ldots$$

Similarly (6.49)

$$b_m = \frac{1}{\pi}\int_{-\pi}^{\pi} \sin mx\, f(x)\, dx \qquad m = 1, 2, 3, \ldots$$

The a's and b's of (6.49) are called the Fourier "coefficients" of $f(x)$. These coefficients can exist [if the integrals of (6.49) exist] regardless of the possibility of expanding $f(x)$ in the series given by (6.47). We can replace the interval $(-\pi, \pi)$ by the interval $(0, 2\pi)$ if we so desire. The results of (6.48) and (6.49) hold for the interval $(0, 2\pi)$.

Example 6.2. We shall see later that $f(x) = x$, for $-\pi \leqq x \leqq \pi$, has a Fourier-series development. We calculate the Fourier coefficients given by (6.49).

$$
\begin{aligned}
a_m &= \frac{1}{\pi}\int_{-\pi}^{\pi} x \cos mx\, dx = \frac{1}{\pi}\left(\frac{x \sin mx}{m}\bigg|_{-\pi}^{\pi} - \frac{1}{m}\int_{-\pi}^{\pi} \sin mx\, dx\right) \\
&= \frac{1}{\pi}\frac{1}{m^2}[\cos m\pi - \cos(-m\pi)] = 0 \qquad m \neq 0 \\
a_0 &= \frac{1}{\pi}\int_{-\pi}^{\pi} x\, dx = 0 \\
b_m &= \frac{1}{\pi}\int_{-\pi}^{\pi} x \sin mx = -\frac{2}{m}\cos m\pi = -\frac{2}{m} \qquad \text{for } m \text{ even} \\
&\phantom{= \frac{1}{\pi}\int_{-\pi}^{\pi} x \sin mx = -\frac{2}{m}\cos m\pi =\ } \frac{2}{m} \qquad \text{for } m \text{ odd}
\end{aligned}
$$

Hence
$$f(x) = 2(\sin x - \tfrac{1}{2}\sin 2x + \tfrac{1}{3}\sin 3x - \tfrac{1}{4}\sin 4x + \cdots) \tag{6.50}$$

Since the right-hand side of (6.50) is periodic, its graph is given by Fig. 6.1. We note

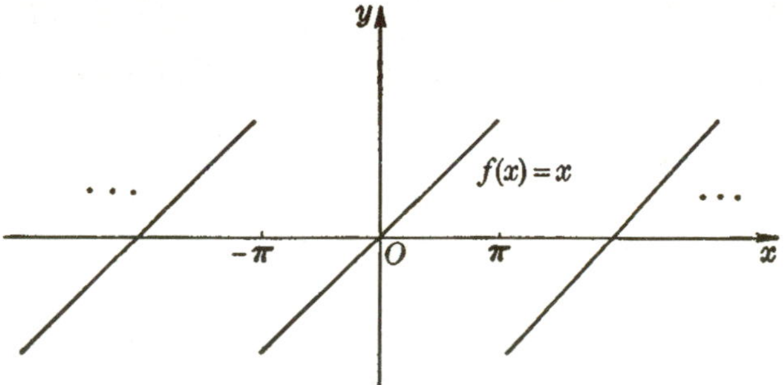

FIG. 6.1

that $f(\pi/2) = \pi/2 = 2(1 - \frac{1}{3} + \frac{1}{5} - \frac{1}{7} + \cdots)$. This checks the result obtained from

$$
\begin{aligned}
\frac{\pi}{4} = \tan^{-1} 1 &= \int_0^1 \frac{dx}{1 + x^2} = \int_0^1 (1 - x^2 + x^4 + x^6 + \cdots)\, dx \\
&= 1 - \tfrac{1}{3} + \tfrac{1}{5} - \tfrac{1}{7} + \cdots
\end{aligned}
$$

provided the series can be integrated term by term.

Example 6.3. The function $f(x)$ given by

$$
\begin{aligned}
f(x) &= 0 \qquad -\pi \leqq x \leqq 0 \\
f(x) &= x \qquad 0 \leqq x \leqq \pi
\end{aligned}
$$

will be seen to have a Fourier-series expansion. The Fourier coefficients of $f(x)$ are

$$a_n = \frac{1}{\pi}\int_{-\pi}^{0} 0\cdot\cos nx\,dx + \frac{1}{\pi}\int_0^{\pi} x\cos nx\,dx = \frac{1}{n^2\pi}[(-1)^n - 1]$$

$$b_n = \frac{1}{\pi}\int_0^{\pi} x\sin nx\,dx = \frac{(-1)^{n+1}}{n}$$

$$a_0 = \frac{1}{\pi}\int_0^{\pi} x\,dx = \frac{\pi}{2}$$

Thus

$$f(x) = \frac{\pi}{4} + \sum_{n=1}^{\infty}\left[\frac{(-1)^n - 1}{\pi n^2}\cos nx + \frac{(-1)^{n+1}}{n}\sin nx\right]$$

$$= \frac{\pi}{4} - \frac{2}{\pi}\sum_{n=1}^{\infty}\frac{\cos(2n-1)x}{(2n-1)^2} - \sum_{n=1}^{\infty}\frac{(-1)^n\sin nx}{n} \tag{6.51}$$

For $x = 0$ the right-hand side of (6.51) becomes

$$\frac{\pi}{4} - \frac{2}{\pi}\sum_{n=1}^{\infty}\frac{1}{(2n-1)^2} \overset{?}{=} f(0) = 0$$

Is $\sum_{n=1}^{\infty}\frac{1}{(2n-1)^2} = \frac{\pi^2}{8}$? In complex-variable theory it can be shown that

$$\frac{\tan z}{z} = 2\sum_{n=0}^{\infty}\frac{1}{(n+\frac{1}{2})^2\pi^2 - z^2}$$

If we allow z to tend to zero, we obtain $\sum_{n=0}^{\infty}\frac{1}{(2n+1)^2} = \frac{\pi^2}{8}$, the desired result. The limit process can be justified.

For $x = -\pi$ and $x = \pi$ the right-hand side of (6.51) becomes

$$\frac{\pi}{4} - \frac{2}{\pi}\sum_{n=1}^{\infty}\frac{\cos(2n-1)\pi}{(2n-1)^2} = \frac{\pi}{4} + \frac{2}{\pi}\sum_{n=1}^{\infty}\frac{1}{(2n-1)^2}$$

$$= \frac{\pi}{4} + \frac{2}{\pi}\frac{\pi^2}{8} = \frac{\pi}{2}$$

Notice that $\frac{f(-\pi) + f(\pi)}{2} = \frac{0+\pi}{2} = \frac{\pi}{2}$. The right-hand side of (6.51) when evaluated at either end point, $x = \pm\pi$, yields the mean of $f(x)$ at these end points. The graph of the right-hand side of (6.51) is given by Fig. 6.2.

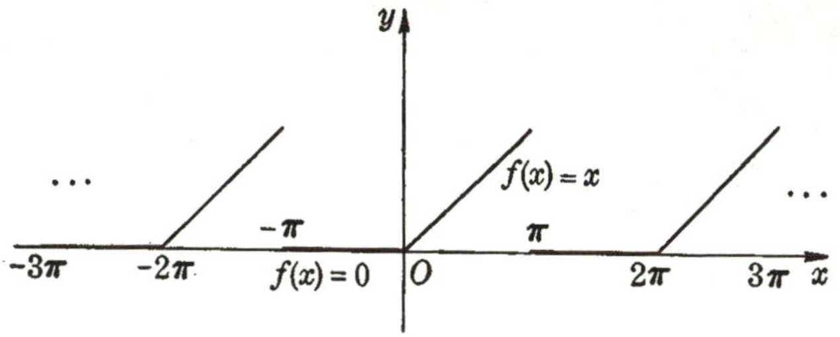

Fig. 6.2

Example 6.4. Let $f(z)$ be analytic for $|z| > 0$. Then

$$f(z) = \sum_{n=-\infty}^{\infty} a_n z^n$$
$$a_n = \frac{1}{2\pi i} \oint_C \frac{f(\zeta)\, d\zeta}{\zeta^{n+1}} \tag{6.52}$$

where C can be chosen as the circle $|z| = 1$. On C we have $\zeta = e^{i\varphi}$, $d\zeta = ie^{i\varphi}\, d\varphi$, so that

$$f(z) = \sum_{n=-\infty}^{\infty} a_n z^n$$
$$a_n = \frac{1}{2\pi} \int_0^{2\pi} f(e^{i\varphi})e^{-ni\varphi}\, d\varphi$$

For z on the unit circle $|z| = 1$ we have $z = e^{i\theta}$ and

$$f(e^{i\theta}) = \sum_{n=-\infty}^{\infty} a_n e^{ni\theta}$$
$$a_n = \frac{1}{2\pi} \int_0^{2\pi} f(e^{i\varphi})e^{-ni\varphi}\, d\varphi \tag{6.53}$$

If we define $g(\theta) \equiv f(e^{i\theta})$, (6.53) becomes

$$g(\theta) = \sum_{n=-\infty}^{\infty} a_n e^{ni\theta}$$
$$a_n = \frac{1}{2\pi} \int_0^{2\pi} g(\varphi)e^{-in\varphi}\, d\varphi \tag{6.54}$$

Example 6.5. We can employ the results of Sec. 5.7 to evaluate the Fourier coefficients of $f(x)$. A particular solution of

$$y'' + n^2 y = f(x) \tag{6.55}$$

is given by

$$y(x) = \int_0^x f(t)g(x - t)\, dt$$

where $g(x)$ is a solution of $y'' + n^2 y = 0$, $g(0) = 0$, $g'(0) = 1$. Thus

$$y(x) = \frac{1}{n} \int_0^x f(t) \sin n(x - t)\, dt \tag{6.56}$$

is a solution of (6.55) with $y(0) = 0$, $y'(0) = 0$. Evaluating (6.56) at $x = 2\pi$ yields

$$y(2\pi) = -\frac{1}{n} \int_0^{2\pi} f(t) \sin nt\, dt$$

so that

$$b_n = \frac{1}{\pi} \int_0^{2\pi} f(t) \sin nt\, dt = -\frac{ny(2\pi)}{\pi} \tag{6.57}$$

An analogue computer can be used to solve (6.55) subject to the initial conditions $y(0) = 0$, $y'(0) = 0$.

A graphical solution can be obtained for $y(x)$. The height of the ordinate at $x = 2\pi$ yields $y(2\pi)$, which in turn yields b_n of (6.57). Differentiating (6.56) and evaluating at $x = 2\pi$ yields

$$a_n = \frac{1}{\pi} \int_0^{2\pi} f(t) \cos nt \, dt = \frac{1}{\pi} y'(2\pi) \tag{6.58}$$

Problems

Find the Fourier series for the following functions defined in the interval $-\pi < x < \pi$:

1. $f(x) = 0$ for $-\pi < x \leqq 0$, $f(x) = 1$ for $0 < x < \pi$
2. $f(x) = |x|$ for $-\pi < x < \pi$
3. $f(x) = e^x$ for $-\pi < x < \pi$
4. $f(x) = \cos^2 x$ for $-\pi < x < \pi$
5. $f(x) = \cos a$ for $|x| \leqq a \neq 0$, $f(x) = \cos x$ otherwise
6. Let $f(x)$ be an even function, that is, $f(x) = f(-x)$. Show that

$$b_n = \frac{1}{\pi} \int_{-\pi}^{\pi} f(x) \sin nx \, dx = 0$$

7. Let $f(x)$ be an odd function, that is, $f(x) = -f(-x)$. Show that

$$a_n = \frac{1}{\pi} \int_{-\pi}^{\pi} f(x) \cos nx \, dx = 0$$

8. Show that any function can be written as the sum of an odd function and an even function.
9. Find the Fourier series of $f(x) = |\sin x|$, $-\pi \leqq x \leqq \pi$.

6.14. Convergence of the Trigonometric Fourier Series. We wish now to investigate the convergence to $f(x)$ of the Fourier series

$$\tfrac{1}{2}a_0 + \sum_{n=1}^{\infty} a_n \cos nx + \sum_{n=1}^{\infty} b_n \sin nx \tag{6.59}$$

$$a_n = \frac{1}{\pi} \int_{-\pi}^{\pi} f(t) \cos nt \, dt \qquad n = 0, 1, 2, \ldots$$

$$b_n = \frac{1}{\pi} \int_{-\pi}^{\pi} f(t) \sin nt \, dt$$

Some preliminary discussions are necessary. First let us consider

$$\lim_{k \to \infty} \int_{-\pi}^{\pi} F(t) \sin kt \, dt \tag{6.60}$$

Intuitively we feel that for very large k the function $F(x) \sin kx$ will be positive about as often as it is negative (with the same absolute value) since $\sin kx$ will oscillate very rapidly for large k. Since the integral $\int_{-\pi}^{\pi} F(t) \sin kt \, dt$ represents an area, we shall not be surprised if

$$\lim_{k \to \infty} \int_{-\pi}^{\pi} F(t) \sin kt \, dt = 0 \tag{6.61}$$

Certainly much will depend on the nature of $F(x)$. It is easy to see that (6.61) holds true if $F(x)$ has a continuous first derivative for $-\pi \leqq x \leqq \pi$. Integration by parts yields

$$\int_{-\pi}^{\pi} F(x) \sin kx \, dx = \frac{-F(x) \cos kx}{k} \Big|_{-\pi}^{\pi} + \frac{1}{k} \int_{-\pi}^{\pi} F'(x) \cos kx \, dx$$

From the fact that $F'(x)$ is bounded on the closed interval $-\pi \leqq x \leqq \pi$ it follows immediately that (6.61) holds true.

Actually we can weaken or make less stringent the conditions on $F(x)$ in order that (6.61) hold true. Nothing need be said about $F'(x)$. Let $F(x)$ and $[F(x)]^2$ be integrable on the interval $-\pi \leqq x \leqq \pi$. We now define $s_n(x)$ by

$$s_n(x) = \tfrac{1}{2}a_0 + \sum_{k=1}^{n} a_k \cos kx + \sum_{k=1}^{n} b_k \sin kx \tag{6.62}$$

$$a_k = \frac{1}{\pi} \int_{-\pi}^{\pi} F(t) \cos kt \, dt$$

$$b_k = \frac{1}{\pi} \int_{-\pi}^{\pi} F(t) \sin kt \, dt$$

$s_n(x)$ is called the nth partial Fourier sum of $F(x)$. Note that $s_n(x)$ is a finite trigonometric series. Moreover $s_n(x)$ is continuous. From

$$[F(x) - s_n(x)]^2 = [F(x)]^2 - 2F(x)s_n(x) + [s_n(x)]^2$$

we have

$$\int_{-\pi}^{\pi} [F(t) - s_n(t)]^2 \, dt = \int_{-\pi}^{\pi} [F(t)]^2 \, dt - 2 \int_{-\pi}^{\pi} F(t)s_n(t) \, dt + \int_{-\pi}^{\pi} [s_n(t)]^2 \, dt$$

One can easily show (see Sec. 6.8) that

$$\int_{-\pi}^{\pi} F(t)s_n(t) \, dt = \frac{\pi a_0^2}{2} + \pi \sum_{k=1}^{n} (a_k^2 + b_k^2)$$

$$\int_{-\pi}^{\pi} [s_n(t)]^2 \, dt = \frac{\pi a_0^2}{2} + \pi \sum_{k=1}^{n} (a_k^2 + b_k^2)$$

From the fact that $\int_{-\pi}^{\pi} [F(t) - s_n(t)]^2 \, dt \geqq 0$ we deduce that

$$\frac{a_0^2}{2} + \sum_{k=1}^{n} (a_k^2 + b_k^2) \leqq \frac{1}{\pi} \int_{-\pi}^{\pi} [F(t)]^2 \, dt \tag{6.63}$$

The left-hand side is monotonic increasing with n and is bounded by the constant $\frac{1}{\pi}\int_{-\pi}^{\pi} [F(t)]^2\,dt$. Hence

$$\frac{a_0^2}{2} + \sum_{k=1}^{\infty} (a_k^2 + b_k^2) \leqq \frac{1}{\pi}\int_{-\pi}^{\pi} [F(t)]^2\,dt \tag{6.64}$$

This is Bessel's inequality [see (6.41)]. Since the series of (6.64) converges, of necessity,

$$\lim_{k\to\infty} a_k = 0 \qquad \lim_{k\to\infty} b_k = 0$$

$\lim_{k\to\infty} b_k = 0$ is precisely the statement (6.61).

A simple class of functions which are both integrable and integrable square is the set of functions which have a finite number of bounded discontinuities in the sense that if $x = c$ is a point of discontinuity of such a function then $\lim_{\substack{x\to c\\ x<c}} f(x)$ and $\lim_{\substack{x\to c\\ x>c}} f(x)$ both exist but are not equal. We write

$$\lim_{\substack{x\to c\\ x<c}} f(x) = f(c - 0)$$
$$\lim_{\substack{x\to c\\ x>c}} f(x) = f(c + 0)$$

A function which has a finite number of bounded discontinuities of the type described above is said to be sectionally, or piecewise, continuous. Now let $f(x)$ be sectionally continuous for $-\pi \leqq x \leqq \pi$, and assume that $f(x)$ has the period 2π, that is, $f(x + 2\pi) = f(x)$. We make use of the periodicity of $f(x)$ in (6.67) below. It is not necessary, however, that $f(\pi) = f(-\pi)$, since the value of $f(x)$ at one point does not affect the value of the integral [see (6.67)]. In most cases $f(x)$ is defined only for the interval $-\pi \leqq x \leqq \pi$. We easily make $f(x)$ periodic by defining $f(x)$ at other values of x by the condition $f(x + 2\pi) = f(x)$.

The nth Fourier partial sum of $f(x)$ can be written as

$$\begin{aligned} s_n(x) &= \frac{1}{2\pi}\int_{-\pi}^{\pi} f(t)\,dt + \frac{1}{\pi}\sum_{k=1}^{n}\int_{-\pi}^{\pi} f(t)\cos kt\cos kx\,dt \\ &\qquad + \frac{1}{\pi}\sum_{k=1}^{n}\int_{-\pi}^{\pi} f(t)\sin kt\sin kx\,dt \\ &= \frac{1}{\pi}\int_{-\pi}^{\pi} f(t)\left[\frac{1}{2} + \sum_{k=1}^{n}\cos k(t - x)\right]dt \end{aligned}$$

It is important to note that $s_n(x)$ is a finite trigonometric series and that $s_n(x)$ is continuous. As long as the Fourier coefficients exist, it is always possible to construct $s_n(x)$. Thus $s_n(x)$ exists independent of the possible development of $f(x)$ as a Fourier series.

We desire to show that, under suitable restrictions concerning the first derivative of $f(x)$,

$$\lim_{n\to\infty} s_n(x) = \tfrac{1}{2}[f(x+0) + f(x-0)] \tag{6.65}$$

From

$$e^{ki(t-x)} = \cos k(t-x) + i \sin k(t-x)$$
$$e^{-ki(t-x)} = \cos k(t-x) - i \sin k(t-x)$$

the reader can show that

$$\sum_{k=1}^{n} \cos k(t-x) = -\frac{1}{2} + \frac{\sin (n+\frac{1}{2})(t-x)}{2 \sin \frac{1}{2}(t-x)} \tag{6.66}$$

Thus

$$\begin{aligned} s_n(x) &= \frac{1}{\pi}\int_{-\pi}^{\pi} f(t) \frac{\sin (n+\frac{1}{2})(t-x)}{2\sin \frac{1}{2}(t-x)}\, dt \\ &= \frac{1}{\pi}\int_{-\pi-x}^{\pi-x} f(x+u) \frac{\sin (n+\frac{1}{2})u}{2\sin (u/2)}\, du \\ &= \frac{1}{\pi}\int_{-\pi}^{\pi} f(x+u) \frac{\sin (n+\frac{1}{2})u}{2\sin (u/2)}\, du \end{aligned} \tag{6.67}$$

The last integral results from the fact that the integrand has period 2π. From (6.66) the reader can deduce that

$$1 = \frac{1}{\pi}\int_{-\pi}^{\pi} \frac{\sin (n+\frac{1}{2})u}{2\sin (u/2)}\, du$$

so that

$$f(x) = \frac{1}{\pi}\int_{-\pi}^{\pi} f(x) \frac{\sin (n+\frac{1}{2})u}{2\sin (u/2)}\, du$$

and

$$\begin{aligned} s_n(x) - f(x) &= \frac{1}{\pi}\int_{-\pi}^{\pi} [f(x+u) - f(x)] \frac{\sin (n+\frac{1}{2})u}{2\sin (u/2)}\, du \\ &= \frac{1}{\pi}\int_{-\pi}^{\pi} \frac{f(x+u)-f(x)}{u} \frac{u}{2\sin (u/2)} \sin (n+\tfrac{1}{2})u\, du \end{aligned}$$

Now consider x fixed, and define

$$F(u) = \frac{f(x+u)-f(x)}{u} \frac{u}{2\sin (u/2)}$$

$F(u)$ will be sectionally continuous if $f'(x)$ exists, for in this case

$$\lim_{u\to 0} F(u) = f'(x)$$

Hence $\lim_{n\to\infty} \int_{-\pi}^{\pi} F(u) \sin (n+\frac{1}{2})u\, du = 0$ (see Prob. 2 of this section).

We have shown that, if $f(x)$ is sectionally continuous for $-\pi \leqq x \leqq \pi$ and if $f(x)$ has a period 2π, then, at any point x such that $f'(x)$ exists, the Fourier series of $f(x)$ given by (6.59) converges to $f(x)$.

It may be that $f'(x)$ does not exist at the point x but that

$$\lim_{\substack{u \to 0 \\ u > 0}} \frac{f(x+u) - f(x+0)}{u} = f'(x+0)$$

$$\lim_{\substack{u \to 0 \\ u < 0}} \frac{f(x+u) - f(x-0)}{u} = f'(x-0)$$

do exist separately. We call $f'(x+0)$ and $f'(x-0)$ the right- and left-hand derivatives of $f(x)$, respectively. Now let the reader show that

$$\begin{aligned} s_n(x) - \tfrac{1}{2}[f(x+0) + f(x-0)] &= \frac{1}{\pi}\int_0^\pi \frac{f(x+u) - f(x+0)}{u} \frac{u}{2\sin(u/2)} \sin(n+\tfrac{1}{2})u\,du \\ &\quad + \frac{1}{\pi}\int_{-\pi}^0 \frac{f(x+u) - f(x-0)}{u} \frac{u}{2\sin(u/2)} \sin(n+\tfrac{1}{2})u\,du \end{aligned}$$

It follows that if $f'(x+0)$ and $f'(x-0)$ exist then

$$\lim_{n\to\infty} s_n(x) = \tfrac{1}{2}[f(x+0) + f(x-0)] \tag{6.68}$$

We formulate the above results in the following statement:

THEOREM 6.1. Let $f(x)$ be sectionally continuous for $-\pi \leqq x \leqq \pi$, and let $f(x)$ be periodic with period 2π. At any point x such that $f(x)$ has both a right- and left-hand derivative the Fourier series of $f(x)$ will converge to the mean value of $f(x)$ at x defined by $\frac{1}{2}[f(x+0) + f(x-0)]$.

It should be emphasized that a sufficient condition for $f(x)$ to be written as a Fourier series has been given. There are no known necessary and sufficient conditions for $f(x)$ to be developed in terms of a Fourier series. Study of the Lebesgue integral yields greater insight into the development of Fourier series. References to this line of study are given at the end of this chapter.

Problems

1. From the identity

$$2\sin\frac{u}{2}\cos ku = \sin(k+\tfrac{1}{2})u - \sin(k-\tfrac{1}{2})u$$

deduce that

$$\sum_{k=1}^{n} \cos ku = -\frac{1}{2} + \frac{\sin(n+\frac{1}{2})u}{2\sin(u/2)}$$

Hint: Let $k = 1, 2, \ldots, n$, and add.

2. Under the restriction that $F(x)$ and $[F(x)]^2$ be integrable we have shown that

$$\lim_{n\to\infty} \int_{-\pi}^{\pi} F(t) \sin nt = 0$$

$$\lim_{n\to\infty} \int_{-\pi}^{\pi} F(t) \cos nt = 0$$

For the same $F(x)$ why is it true that $\lim_{n\to\infty} \int_{-\pi}^{\pi} F(t) \cos (t/2) \sin nt \, dt = 0$? Show that $\lim_{n\to\infty} \int_{-\pi}^{\pi} F(t) \sin (n + \frac{1}{2})t \, dt = 0$.

3. Let $f(x)$ have a continuous derivative for $-\pi \leqq x \leqq \pi$, so that $|f'(x)| < M$ for $-\pi \leqq x \leqq \pi$. Show that the Fourier coefficients of $f(x)$ satisfy

$$|a_k| < \frac{2M}{k} \qquad k = 1, 2, 3, \ldots$$

What further restriction can be imposed upon $f(x)$ so that $|b_k| < 2M/k$?

4. Let $f(x)$ and $[f(x)]^2$ be integrable for the range $-\pi \leqq x \leqq \pi$. Consider the finite trigonometric series

$$S_n(x) = \tfrac{1}{2}a_0 + \sum_{k=1}^{n} a_k \cos kx + \sum_{k=1}^{n} b_k \sin kx$$

Show that $J = \int_{-\pi}^{\pi} [f(x) - S_n(x)]^2 \, dx$ is a minimum for

$$a_k = \frac{1}{\pi} \int_{-\pi}^{\pi} f(x) \cos kx \, dx \qquad k = 0, 1, 2, \ldots$$

$$b_k = \frac{1}{\pi} \int_{-\pi}^{\pi} f(x) \sin kx \, dx \qquad k = 1, 2, 3, \ldots$$

5. Let $f(x) = \sin (1/x)$, $x \neq 0$, $f(0) = 1$. Is $f(x)$ sectionally continuous?

6. If $f'(x)$ is sectionally continuous for $-\pi \leqq x \leqq \pi$, show that $f(x)$ is continuous for $-\pi < x < \pi$.

7. If $f(x)$ is a continuous function such that $f(x + \pi/3) = f(x)$, show that its Fourier series has the form

$$\frac{a_0}{2} + \sum_{n=1}^{\infty} a_n \cos 6nx + \sum_{n=1}^{\infty} b_n \sin 6nx$$

8. Let (a_n, b_n) be the Fourier coefficients of $f(x)$, $\{\alpha_n, \beta_n\}$ the Fourier coefficients of $g(x)$. Find the Fourier coefficients of

$$h(x) = \int_{-\pi}^{\pi} f(x - t) g(t) \, dt$$

9. Let $f(x)$ be continuous on the closed interval $a \leqq x \leqq b$. Subdivide this interval into $a = x_0, x_1, x_2, \ldots, x_n = b$. Now write

$$\int_a^b f(x) \sin \alpha x \, dx = \sum_{i=1}^{n} \int_{x_{i-1}}^{x_i} f(x) \sin \alpha x \, dx$$

$$= \sum_{i=1}^{n} \int_{x_{i-1}}^{x_i} [f(x) - f(x_{i-1})] \sin \alpha x \, dx + \sum_{i=1}^{n} \int_{x_{i-1}}^{x_i} f(x_{i-1}) \sin \alpha x \, dx$$

From the fact that $f(x)$ is uniformly continuous deduce that

$$\lim_{\alpha \to \infty} \int_a^b f(x) \sin \alpha x \, dx = 0$$

Extend this result for $f(x)$ sectionally continuous on $a \leqq x \leqq b$.

6.15. Differentiation and Integration of Trigonometric Fourier Series. We have seen that under suitable restrictions the trigonometric Fourier series of $f(x)$ will converge to $f(x)$. Now we are concerned with term-by-term differentiation of a Fourier series. Let us return to the Taylor-series expansion of $f(z)$. If $\sum_{n=0}^{\infty} a_n z^n$ converges to $f(z)$ for $0 \leqq |z| < r$, we know that $\sum_{n=1}^{\infty} a_n n z^{n-1}$ will converge to $f'(z)$ for $0 \leqq |z| < r$. In the case of Fourier series, however, a simple example illustrates that the new Fourier series obtained by term-by-term differentiation may not converge to $f'(x)$ even though $f'(x)$ exists. The Fourier series of $f(x) = x$, $-\pi < x \leqq \pi$, is given by

$$2(\sin x - \tfrac{1}{2} \sin 2x + \tfrac{1}{3} \sin 3x - \tfrac{1}{4} \sin 4x + \cdots)$$

Term-by-term differentiation yields

$$2(\cos x - \cos 2x + \cos 3x - \cos 4x + \cdots)$$

This series cannot converge since $\cos nx$ does not tend to zero as n becomes infinite.

Certainly term-by-term differentiation of a Fourier series leads to a new Fourier series. If this new series converges to $f'(x)$, we shall wish to make certain that $f'(x)$ has a convergent Fourier series. From Sec. 6.14 we have seen that a sufficient condition for this is that $f'(x)$ be sectionally continuous and that $f'(x)$ have periodicity 2π. It is not necessary that $f'(-\pi) = f'(\pi)$. Let the reader show that, if $f'(x)$ is sectionally continuous, then $f(x)$ is continuous. Furthermore if $f'(x)$ has periodicity 2π, the reader can show that $f(-\pi) = f(\pi)$ is sufficient to guarantee that $f(x)$ has periodicity 2π. We state and prove now the following theorem:

THEOREM 6.2. Let $f(x)$ be continuous in the interval $-\pi \leqq x \leqq \pi$, $f(-\pi) = f(\pi)$, and let $f'(x)$ be sectionally continuous with periodicity 2π. At each point x for which $f''(x)$ exists the Fourier series of $f(x)$ can be differentiated term by term, and the resulting Fourier series will converge to $f'(x)$.

The proof is as follows: From Sec. 6.14, $f'(x)$ can be written

$$f'(x) = \tfrac{1}{2}a_0' + \sum_{n=1}^{\infty} a_n' \cos nx + \sum_{n=1}^{\infty} b_n' \sin nx$$

with
$$a'_n = \frac{1}{\pi}\int_{-\pi}^{\pi} f'(x)\cos nx\,dx \qquad n = 0, 1, 2, \ldots$$
$$b'_n = \frac{1}{\pi}\int_{-\pi}^{\pi} f'(x)\sin nx\,dx \qquad n = 1, 2, 3, \ldots$$

Integration by parts yields

$$\begin{aligned} a'_n &= \frac{1}{\pi} f(x)\cos nx\Big|_{-\pi}^{\pi} + \frac{n}{\pi}\int_{-\pi}^{\pi} f(x)\sin nx\,dx \\ &= \frac{n}{\pi}\int_{-\pi}^{\pi} f(x)\sin nx\,dx = nb_n \end{aligned}$$

since $f(-\pi) = f(\pi)$ by assumption. b_n is the nth Fourier sine coefficient of $f(x)$. Similarly

$$b'_n = -na_n$$

so that
$$f'(x) = \sum_{n=1}^{\infty} -na_n \sin nx + \sum_{n=1}^{\infty} nb_n \cos nx \tag{6.69}$$

The Fourier series of (6.69), however, is the Fourier series obtained by term-by-term differentiation of

$$f(x) = \tfrac{1}{2}a_0 + \sum_{n=1}^{\infty} a_n \cos nx + \sum_{n=1}^{\infty} b_n \sin nx$$

This proves the theorem stated above.

We turn now to the question of term-by-term integration of a Fourier series. The reader knows from real-variable theory that integration tends to smooth out discontinuities, whereas differentiation tends to introduce discontinuities. As an example, consider the function $f(x)$ defined as follows:

$$\begin{aligned} f(x) &= 0 \qquad \text{for } -\infty < x \leqq 0 \\ f(x) &= x \qquad \text{for } 0 \leqq x \leqq 1 \\ f(x) &= 2 \qquad \text{for } 1 < x < \infty \end{aligned}$$

$f'(x)$ does not exist at $x = 0$ and at $x = 1$. Moreover $f(x)$ is discontinuous at $x = 1$. However, the Riemann integral of $f(x)$ defined by

$$F(x) = \int_{-\infty}^{x} f(t)\,dt$$

is continuous for $-\infty < x < \infty$. At any point $x \neq 1$ [$x = 1$ is the only discontinuity of $f(x)$] we have $F'(x) = f(x)$. Thus $F'(0)$ exists, whereas $f'(0)$ does not exist. Let the reader show that $F'(1)$ does not exist.

We find that very little is required to integrate a Fourier series term by term. We state and prove the following theorem:

THEOREM 6.3. Let $f(x)$ be sectionally continuous for $-\pi \leqq x \leqq \pi$. If $\frac{1}{2}a_0 + \sum_{n=1}^{\infty} a_n \cos nx + \sum_{n=1}^{\infty} b_n \sin nx$ is the Fourier series of $f(x)$, then

$$\int_{-\pi}^{x} f(x)\,dx = \frac{a_0}{2}(x+\pi) + \sum_{n=1}^{\infty} \frac{a_n}{n} \sin nx - \sum_{n=1}^{\infty} \frac{b_n}{n}(\cos nx - \cos n\pi) \tag{6.70}$$

Equation (6.70) holds whether the Fourier series corresponding to $f(x)$ converges or not. Equation (6.70) is not a Fourier series unless $a_0 = 0$.

The proof is as follows: Since $f(x)$ is sectionally continuous and hence Riemann-integrable, we note that $F(x)$ defined by

$$F(x) = \int_{-\pi}^{x} f(x)\,dx - \tfrac{1}{2}a_0 x \tag{6.71}$$

is continuous. Moreover $F'(x) = f(x) - \frac{1}{2}a_0$ except at points of discontinuity of $f(x)$. If $x = c$ is a point of discontinuity of $f(x)$, the reader can easily verify that $F'(c-0)$ and $F'(c+0)$ exist. From (6.71) it follows that $F(-\pi) = F(\pi) = \frac{1}{2}a_0\pi$. $F(x)$ can be made periodic by defining $F(x)$ for $|x| > \pi$ by the equation $F(x + 2\pi) = F(x)$. From Sec. 6.14, $F(x)$ has a Fourier-series development

$$\tfrac{1}{2}[F(x+0) + F(x-0)] = \tfrac{1}{2}A_0 + \sum_{n=1}^{\infty} A_n \cos nx + \sum_{n=1}^{\infty} B_n \sin nx$$

or

$$F(x) = \tfrac{1}{2}A_0 + \sum_{n=1}^{\infty} A_n \cos nx + \sum_{n=1}^{\infty} B_n \sin nx \tag{6.72}$$

since $F(x+0) = F(x-0)$ from the continuity of $F(x)$. Integration by parts yields

$$\begin{aligned} A_n &= \frac{1}{\pi}\int_{-\pi}^{\pi} F(x) \cos nx\,dx \\ &= \frac{F(x)}{n\pi} \sin nx \Big|_{-\pi}^{\pi} - \frac{1}{n\pi}\int_{-\pi}^{\pi} F'(x) \sin nx\,dx \\ &= -\frac{1}{n\pi}\int_{-\pi}^{\pi} [f(x) - \tfrac{1}{2}a_0] \sin nx\,dx \\ &= -\frac{1}{n\pi}\int_{-\pi}^{\pi} f(x) \sin nx\,dx = -\frac{1}{n} b_n \qquad n \neq 0 \end{aligned}$$

Similarly $B_n = (1/n)a_n$. We now write

$$F(x) = \tfrac{1}{2}A_0 - \sum_{n=1}^{\infty} \frac{1}{n} b_n \cos nx + \sum_{n=1}^{\infty} \frac{1}{n} a_n \sin nx$$

From $F(\pi) = \frac{1}{2}a_0\pi$ we deduce that $A_0 = a_0\pi + \sum_{n=1}^{\infty} \frac{1}{n} b_n \cos n\pi$. Using these results along with (6.71) yields (6.70). Q.E.D.

Problems

1. Differentiate and integrate the Fourier series for those functions of Probs. 1, 2, 3, 4, 5, 9, Sec. 6.13, which admit these processes.

2. Sum the series

$$1 - \frac{1}{3^3} + \frac{1}{5^3} - \frac{1}{7^3} + \cdots + \frac{(-1)^n}{(2n+1)^3} + \cdots$$

3. It can be shown that if $f(x)$ is continuous for $-\pi \leqq x \leqq \pi$, $f(\pi) = f(-\pi)$, and if $f'(x)$ is sectionally continuous for $-\pi \leqq x \leqq \pi$, then the Fourier series of $f(x)$ converges uniformly and absolutely for $-\pi \leqq x \leqq \pi$. Multiply

$$f(x) = \tfrac{1}{2}a_0 + \sum_{n=1}^{\infty} a_n \cos nx + \sum_{n=1}^{\infty} b_n \sin nx$$

by $f(x)$, integrate term by term (this is permissible since the new series converges uniformly), and show that

$$\int_{-\pi}^{\pi} [f(x)]^2 \, dx = \pi \left[\tfrac{1}{2}a_0^2 + \sum_{n=1}^{\infty} (a_n^2 + b_n^2) \right]$$

This identity is due to Parseval.

6.16. The Fourier Integral. If $f(x)$ is defined for x on the range $-\infty < x < \infty$ and if $f(x)$ is not periodic, it is obvious that we cannot represent $f(x)$ by a trigonometric Fourier series since such series, of necessity, must be periodic. We shall show, however, that under certain conditions it will be possible to obtain a Fourier integral representation of $f(x)$.

THEOREM 6.4. Let $f(x)$ be sectionally continuous in every finite interval (a, b), and let $f(x)$ be absolutely integrable for the infinite range $-\infty < x < \infty$, that is, $\int_{-\infty}^{\infty} |f(x)| \, dx = A < \infty$. At every point x, $-\infty < x < \infty$, such that $f(x)$ has a right- and left-hand derivative, $f(x)$ is represented by its Fourier integral as follows:

$$\tfrac{1}{2}[f(x+0) + f(x-0)] = \frac{1}{\pi} \int_0^{\infty} d\alpha \int_{-\infty}^{\infty} f(t) \cos \alpha(t - x) \, dt \quad (6.73)$$

We proceed to the proof in the following manner: The result of Prob. 9, Sec. 6.14, as well as the methods of that section, enables us to state that

$$\tfrac{1}{2}[f(x+0) + f(x-0)] = \lim_{\alpha\to\infty} \frac{1}{\pi} \int_a^b f(t) \frac{\sin \alpha(t-x)}{t-x}\, dt \tag{6.74}$$

at any point x for which $f(x)$ has a right- and left-hand derivative, $a < x < b$. Now we investigate the existence of

$$\int_{-\infty}^{\infty} f(t) \frac{\sin \alpha(t-x)}{t-x}\, dt \tag{6.75}$$

For $a < x$ we observe that

$$\int_{-\infty}^{a} \left| \frac{f(t) \sin \alpha(t-x)}{t-x} \right| dt \leqq \frac{1}{|a-x|} \int_{-\infty}^{a} |f(t)|\, dt$$

The convergence of $\int_{-\infty}^{\infty} |f(t)|\, dt$ shows that for any $\epsilon/2 > 0$ an A exists such that for $a \leqq A$

$$\left| \int_{-\infty}^{a} \frac{f(t) \sin \alpha(t-x)}{t-x}\, dt \right| < \frac{\epsilon}{2}$$

independent of α. Similarly one shows that for any $\epsilon/2 > 0$ a number B exists such that for $b \geqq B$

$$\left| \int_{b}^{\infty} \frac{f(t) \sin \alpha(t-x)}{t-x}\, dt \right| < \frac{\epsilon}{2}$$

independent of α. Thus the right-hand side of (6.74) can be made to differ from

$$\lim_{\alpha\to\infty} \frac{1}{\pi} \int_{-\infty}^{\infty} f(t) \frac{\sin \alpha(t-x)}{t-x}\, dt$$

by any arbitrary $\epsilon > 0$ provided $-a$ and b are chosen sufficiently large. This implies immediately that

$$\tfrac{1}{2}[f(x+0) + f(x-0)] = \lim_{\alpha\to\infty} \frac{1}{\pi} \int_{-\infty}^{\infty} f(t) \frac{\sin \alpha(t-x)}{t-x}\, dt \tag{6.76}$$

Now $\int_0^\alpha \cos \mu(t-x)\, d\mu = \dfrac{\sin \alpha(t-x)}{t-x}$, so that (6.76) may be written

$$\tfrac{1}{2}[f(x+0) + f(x-0)] = \lim_{\alpha\to\infty} \frac{1}{\pi} \int_{-\infty}^{\infty} f(t)\, dt \int_0^\alpha \cos \mu(t-x)\, d\mu \tag{6.77}$$

Before we pass to the limit as $\alpha \to \infty$, we wonder if we can interchange the order of the double integration. Let us first consider

$$G(\mu) = \int_{-\infty}^{\infty} f(t) \cos \mu(t - x)\, dt$$

x a parameter and $0 \leqq \mu \leqq \alpha$, α fixed. We show that $G(\mu)$ is continuous. For $\mu = \mu_0$ we have

$$\begin{aligned} G(\mu) - G(\mu_0) = &\int_{-\infty}^{-T} f(t)[\cos \mu(t - x) - \cos \mu_0(t - x)]\, dt \\ &+ \int_{-T}^{T} f(t)[\cos \mu(t - x) - \cos \mu_0(t - x)]\, dt \\ &+ \int_{T}^{\infty} f(t)[\cos \mu(t - x) - \cos \mu_0(t - x)]\, dt \end{aligned}$$

The first integral is less than $2 \int_{-\infty}^{-T} |f(t)|\, dt$, and the last integral is less than $2 \int_{T}^{\infty} |f(t)|\, dt$. Since we are given that $\int_{-\infty}^{\infty} |f(t)|\, dt$ exists, a T can be found such that these integrals can each be made less than $\frac{1}{3}\epsilon$ for any $\epsilon > 0$. T is obtained after ϵ is chosen. From

$$\cos \mu(t - x) - \cos \mu_0(t - x) = (\mu - \mu_0) \sin \bar{\mu}(t - x)$$

$\bar{\mu}$ between μ_0 and μ, we note that the middle integral is less than

$$|\mu - \mu_0| \int_{-\infty}^{\infty} |f(t)|\, dt$$

Since $\int_{-\infty}^{\infty} |f(t)|\, dt = A$ is a finite number, we can find a $\delta > 0$ so that the middle integral is also less than $\epsilon/3$ for $|\mu - \mu_0| < \delta < \epsilon/3A$. Hence for any $\epsilon > 0$ there exists a $\delta > 0$ such that

$$|G(\mu) - G(\mu_0)| < \epsilon$$

for $|\mu - \mu_0| < \delta$. This is the definition of continuity, so that $G(\mu)$ is continuous at $\mu = \mu_0$ and, in fact, for all μ. Hence

$$H(\alpha) = \int_0^{\alpha} d\mu \int_{-\infty}^{\infty} f(t) \cos \mu(t - x)\, dt$$

exists, and $\dfrac{dH}{d\alpha} = \displaystyle\int_{-\infty}^{\infty} f(t) \cos \mu(t - x)\, dt$. By considering

$$K(\alpha) = \int_{-\infty}^{\infty} dt \int_0^{\alpha} f(t) \cos \mu(t - x)\, d\mu$$

let the reader show that $\dfrac{dK}{d\alpha} = \dfrac{dH}{d\alpha}$. Since $K(0) = H(0)$, of necessity

$$\int_{-\infty}^{\infty} dt \int_0^{\alpha} f(t) \cos \mu(t - x)\, d\mu = \int_0^{\alpha} d\mu \int_{-\infty}^{\infty} f(t) \cos \mu(t - x)\, dt$$

Allowing α to tend to infinity enables us to rewrite (6.77) in the form given by (6.73).

If $f(x)$ is continuous at x, (6.73) may be written

$$f(x) = \frac{1}{\pi}\int_0^\infty d\alpha \int_{-\infty}^\infty f(t)\cos\alpha(t - x)\,dt \tag{6.78}$$

From

$$\cos\alpha(t - x) = \tfrac{1}{2}(e^{i\alpha(x-t)} + e^{-i\alpha(x-t)})$$

one readily obtains

$$\begin{aligned} f(x) &= \frac{1}{2\pi}\int_{-\infty}^\infty d\alpha \int_{-\infty}^\infty f(t)e^{i\alpha(x-t)}\,dt \\ &= \frac{1}{2\pi}\int_{-\infty}^\infty e^{i\alpha x}\,d\alpha \int_{-\infty}^\infty f(t)e^{-i\alpha t}\,dt \end{aligned} \tag{6.79}$$

Equation (6.79) can be written in the symmetric form

$$\begin{aligned} f(x) &= \frac{1}{\sqrt{2\pi}}\int_{-\infty}^\infty e^{i\alpha x}g(\alpha)\,d\alpha \\ g(\alpha) &= \frac{1}{\sqrt{2\pi}}\int_{-\infty}^\infty e^{-i\alpha t}f(t)\,dt \end{aligned} \tag{6.80}$$

We say that $g(\alpha)$ is the Fourier transform of the real function $f(x)$, and $f(x)$ is called the inverse Fourier transform of $g(\alpha)$.

A more rigorous treatment of the Fourier integral requires the use of the Lebesgue integral.

Example 6.6. We consider the function $f(x)$ defined as follows: $f(x) = 0$ for $x < 0$ and $f(x) = e^{-\beta x}$ for $x \geqq 0$, $\beta > 0$. Fourier's integral certainly applies to this function, for $f(x)$ has a simple discontinuity at $x = 0$, and $\int_{-\infty}^\infty |f(x)|\,dx$ exists. From (6.79)

$$\begin{aligned} g(\alpha) &= \frac{1}{\sqrt{2\pi}}\int_0^\infty e^{-i\alpha t}e^{-\beta t}\,dt = \frac{1}{\sqrt{2\pi}}\frac{1}{\beta + i\alpha} \\ f(x) &= \frac{1}{(\sqrt{2\pi})^2}\int_{-\infty}^\infty \frac{e^{i\alpha x}}{\beta + i\alpha}\,d\alpha = \frac{1}{2\pi}\int_{-\infty}^\infty \frac{e^{i\alpha x}}{\beta + i\alpha}\,d\alpha \\ &= \frac{1}{2\pi}\int_{-\infty}^\infty \frac{\beta\cos\alpha x + \alpha\sin\alpha x}{\beta^2 + \alpha^2}\,d\alpha \\ &= \frac{1}{\pi}\int_0^\infty \frac{\beta\cos\alpha x + \alpha\sin\alpha x}{\beta^2 + \alpha^2}\,d\alpha \end{aligned}$$

From the definition of $f(x)$ one has for $x > 0$

$$\begin{aligned} e^{-\beta x} &= \frac{1}{\pi}\int_0^\infty \frac{\beta\cos\alpha x + \alpha\sin\alpha x}{\beta^2 + \alpha^2}\,d\alpha \\ 0 &= \frac{1}{\pi}\int_0^\infty \frac{\beta\cos\alpha x - \alpha\sin\alpha x}{\beta^2 + \alpha^2}\,d\alpha \end{aligned}$$

so that

$$\begin{aligned} \frac{\pi}{2\beta} e^{-\beta x} &= \int_0^\infty \frac{\cos \alpha x}{\beta^2 + \alpha^2}\, d\alpha \\ \frac{\pi}{2} e^{-\beta x} &= \int_0^\infty \frac{\alpha \sin \alpha x}{\beta^2 + \alpha^2}\, d\alpha \end{aligned} \tag{6.81}$$

Let the reader obtain (6.81) directly by contour integration.

Example 6.7. Suppose $f(x)$ is an even function. Equation (6.78) can be written

$$\begin{aligned} f(x) &= \frac{2}{\pi} \int_0^\infty d\alpha \int_0^\infty f(t) \cos \alpha t \cos \alpha x\, dt \\ &= \frac{2}{\pi} \int_0^\infty \cos \alpha x\, dx \int_0^\infty f(t) \cos \alpha t\, dt \end{aligned} \tag{6.82}$$

If we wish to solve the integral equation

$$e^{-\alpha} = \int_0^\infty \cos \alpha t\, f(t)\, dt$$

for $f(x)$, we apply (6.82) and obtain

$$\begin{aligned} f(x) &= \frac{2}{\pi} \int_0^\infty e^{-\alpha} \cos \alpha x\, dx \\ &= \frac{2}{\pi} \frac{1}{1 + x^2} \end{aligned}$$

Notice that $f(x)$ is an even function and that this function has the necessary properties for the application of the Fourier integral formula.

Example 6.8. Let $p = \frac{d}{dt}$, and assume $Z(p)$ such that $Z(p)e^{i\omega t} = Z(i\omega)e^{i\omega t}$. A simple example is $Z(p) = a_0p^n + a_1p^{n-1} + \cdots + a_{n-1}p + a_n$, where $p^2 = \frac{d^2}{dt^2}$, etc. We consider the equation

$$Z(p)\Theta_o(t) = \Theta_i(t) \tag{6.83}$$

We assume that the input $\Theta_i(t)$ is given, and we desire to find the output $\Theta_o(t)$ such that (6.83) will hold. Let $H_i(\omega)$ and $H_o(\omega)$ be the Fourier transforms of $\Theta_i(t)$ and $\Theta_o(t)$, respectively. Then

$$\begin{aligned} H_i(\omega) &= \frac{1}{\sqrt{2\pi}} \int_{-\infty}^\infty e^{-it\omega}\Theta_i(t)\, dt \\ H_0(\omega) &= \frac{1}{\sqrt{2\pi}} \int_{-\infty}^\infty e^{-it\omega}\Theta_0(t)\, dt \end{aligned} \tag{6.84}$$

and

$$\begin{aligned} \Theta_i(t) &= \frac{1}{\sqrt{2\pi}} \int_{-\infty}^\infty e^{it\omega}H_i(\omega)\, d\omega \\ \Theta_0(t) &= \frac{1}{\sqrt{2\pi}} \int_{-\infty}^\infty e^{it\omega}H_0(\omega)\, d\omega \end{aligned}$$

Substituting into (6.83) yields

$$Z(p) \int_{-\infty}^\infty e^{it\omega}H_o(\omega)\, d\omega = \int_{-\infty}^\infty e^{it\omega}H_i(\omega)\, d\omega$$

Assuming that $Z(p)$ may be placed inside the integral yields

$$\int_{-\infty}^\infty Z(i\omega)e^{it\omega}H_o(\omega)\, d\omega = \int_{-\infty}^\infty e^{it\omega}H_i(\omega)\, d\omega$$

Since the two integrals are equal for the full range of t, it appears logical that

$$Z(i\omega)H_o(\omega) = H_i(\omega)$$

Hence
$$\Theta_o(t) = \frac{1}{\sqrt{2\pi}} \int_{-\infty}^{\infty} \frac{e^{it\omega}H_i(\omega)}{Z(i\omega)}\, d\omega$$

$Z(p)$ is called the operational impedance, and its reciprocal, $Y(p)$, is called the transfer function. We now have

$$\Theta_o(t) = \frac{1}{\sqrt{2\pi}} \int_{-\infty}^{\infty} e^{it\omega}H_i(\omega)Y(i\omega)\, d\omega$$

Making use of (6.84) yields

$$\begin{aligned}\Theta_o(t) &= \frac{1}{2\pi} \int_{-\infty}^{\infty} e^{it\omega}Y(i\omega) \int_{-\infty}^{\infty} \Theta_i(\tau)e^{-i\tau\omega}\, d\tau\, d\omega \\ &= \frac{1}{2\pi} \int_{-\infty}^{\infty} \int_{-\infty}^{\infty} \Theta_i(\tau)Y(i\omega)e^{i\omega(t-\tau)}\, d\tau\, d\omega \\ &= \int_{-\infty}^{\infty} \Theta_i(\tau)y(t-\tau)\, d\tau \qquad (6.85)\end{aligned}$$

where
$$y(t-\tau) = \frac{1}{2\pi} \int_{-\infty}^{\infty} Y(i\omega)e^{i\omega(t-\tau)}\, d\omega$$

We have assumed that the order of integration could be interchanged. Equation (6.85) can be written

$$\Theta_o(t) = \int_{-\infty}^{\infty} \Theta_i(t-\tau)y(\tau)\, d\tau \qquad (6.86)$$

$y(\tau)$ is appropriately called the memory function because of the term $\Theta_i(t-\tau)$. Let the reader compare (6.86) with the particular solution of a linear differential equation with constant coefficients given by (5.71).

No attempt at rigor has been made in obtaining the results of this example.

Problems

1. Suppose that $f(x)$ of (6.74) is an even function. Show that

$$\tfrac{1}{2}[f(x+0) + f(x-0)] = \frac{2}{\pi} \int_0^{\infty} d\alpha \int_0^{\infty} f(t) \cos \alpha t \cos \alpha x\, dt \qquad x \geqq 0$$

If $f(x)$ is odd, show that

$$\tfrac{1}{2}[f(x+0) + f(x-0)] = \frac{2}{\pi} \int_0^{\infty} d\alpha \int_0^{\infty} f(t) \sin \alpha t \sin \alpha x\, dt \qquad x \geqq 0$$

2. Find the Fourier transform $g(\alpha)$ of the function $f(x)$ defined as follows: $f(x) = 0$ for $x < 0$, $f(x) = 1/a$ for $0 \leqq x \leqq a$, $f(x) = 0$ for $x > a$. Find $\lim_{a \to 0} g(\alpha)$.

3. Find the Fourier transform for $f(x)$ defined as follows: $f(x) = 0$ for $x < 0$, $f(x) = e^{-x}$ for $x \geqq 0$. Show that

$$\int_0^{\infty} \frac{\cos \alpha x + \alpha \sin \alpha x}{1 + \alpha^2}\, d\alpha = \begin{cases} 0 & \text{for } x < 0 \\ \frac{\pi}{2} & \text{for } x = 0 \\ \pi e^{-x} & \text{for } x > 0 \end{cases}$$

4. Consider $g(\alpha) = \int_{-\infty}^{\infty} e^{-i\alpha t} f'(t)\,dt$ such that $\lim_{x \to \pm\infty} f(x) = 0$. Show that

$$g(\alpha) = i\alpha \int_{-\infty}^{\infty} f(t) e^{-i\alpha t}\,dt$$

6.17. Nonlinear Differential Equations. We discuss the nonlinear differential equation

$$\frac{d^2x}{dt^2} + x + \mu f\left(x, \frac{dx}{dt}\right) = 0 \qquad \mu \ll 1 \tag{6.87}$$

which can be written as the system

$$\begin{aligned} \frac{dx}{dt} &= y \\ \frac{dy}{dt} &= -x - \mu f(x, y) \end{aligned} \tag{6.88}$$

The solution of (6.87) for $\mu = 0$ is $x = A \cos (t + B)$,

$$y = \frac{dx}{dt} = -A \sin (t + B)$$

The method of Kryloff-Bogoliuboff is to vary A and B in the hope that an approximate solution to (6.88) may be found. This is essentially the variation-of-parameter method discussed in Chap. 5.

We write

$$\begin{aligned} x &= r \cos \theta \\ y &= -r \sin \theta \end{aligned}$$

and obtain

$$\begin{aligned} \frac{dx}{dt} &= \frac{dr}{dt} \cos \theta - r \sin \theta \frac{d\theta}{dt} \\ \frac{dy}{dt} &= -\frac{dr}{dt} \sin \theta - r \cos \theta \frac{d\theta}{dt} \end{aligned}$$

Equations (6.88) can now be written

$$\begin{aligned} \frac{dr}{dt} \cos \theta - r \sin \theta \frac{d\theta}{dt} &= -r \sin \theta \\ -\frac{dr}{dt} \sin \theta - r \cos \theta \frac{d\theta}{dt} &= -r \cos \theta - \mu f(r \cos \theta, -r \sin \theta) \end{aligned}$$

so that

$$\begin{aligned} \frac{dr}{dt} &= \mu \sin \theta\, f(r \cos \theta, -r \sin \theta) \\ -r \frac{d\theta}{dt} &= -r - \mu \cos \theta\, f(r \cos \theta, -r \sin \theta) \qquad \mu \ll 1 \end{aligned} \tag{6.89}$$

For the range of values (r, θ) such that

$$\mu f(r \cos \theta, r \sin \theta) \ll 1$$

one has to a first approximation

$$\begin{aligned} \frac{dr}{dt} &= 0 \\ r\frac{d\theta}{dt} &= r \end{aligned} \tag{6.90}$$

Integration yields $\theta = \theta_0 + t$. We replace θ of (6.89) by $\theta_0 + t$ so that

$$\frac{dr}{dt} = \mu \sin(\theta_0 + t) f(r\cos(\theta_0 + t), -r\sin(\theta_0 + t)) \tag{6.91}$$

Since $\frac{dr}{dt} \ll 1$, the value of r will not change appreciably over the interval $(t, t + 2\pi)$. Integrating (6.91) on this basis [r is assumed constant on the right-hand side of (6.91)] yields

$$\begin{aligned} r(t+2\pi) - r(t) &= \mu \int_t^{t+2\pi} \sin(\theta_0+t) f(r\cos(\theta_0+t), -r\sin(\theta_0+t))\,dt \\ &= \mu \int_0^{2\pi} \sin\psi\, f(r\cos\psi, -r\sin\psi)\,d\psi \\ &= \mu F(r) \end{aligned} \tag{6.92}$$

where
$$F(r) = \int_0^{2\pi} \sin\psi\, f(r\cos\psi, -r\sin\psi)\,d\psi \tag{6.93}$$

The periodicity of $\sin\psi\, f(r\cos\psi, -r\sin\psi)$ introduced the new limits of integration. Thus

$$\frac{r(t + 2\pi) - r(t)}{2\pi} = \frac{\mu}{2\pi} F(r)$$

Now $\frac{r(t + 2\pi) - r(t)}{2\pi}$ represents the average change of $r(t)$ for t ranging from t to $t + 2\pi$. Since $\frac{dr}{dt} \ll 1$, the average change of $r(t)$ can be replaced by $\frac{dr}{dt}$ itself, at least to a first approximation. This yields

$$\frac{dr}{dt} = \frac{\mu}{2\pi} F(r) \tag{6.94}$$

The integration of (6.94) yields the first approximation for $r(t)$. The same method applied to the second equation of (6.89) yields

$$\begin{aligned} \frac{d\theta}{dt} &= 1 + \frac{\mu}{2\pi r} \int_0^{2\pi} \cos\psi\, f(r\cos\psi, -r\sin\psi)\,d\psi \\ &= 1 + \frac{\mu}{2\pi r} G(r) \end{aligned}$$

and
$$\theta = \theta_0 + \left[1 + \frac{\mu}{2\pi r} G(r)\right] t \tag{6.95}$$

with
$$G(r) = \int_0^{2\pi} \cos\psi\, f(r\cos\psi, -r\sin\psi)\,d\psi \tag{6.96}$$

Before applying Fourier-series methods to obtain improved approximations, we discuss an example.

Example 6.9. Van der Pol's equation is

$$\frac{d^2x}{dt^2} + x - \mu(1 - x^2)\frac{dx}{dt} = 0 \qquad \mu \ll 1$$

Here $f(x, y) = -(1 - x^2)y$, and

$$f(r \cos \psi, -r \sin \psi) = (1 - r^2 \cos^2 \psi)r \sin \psi$$

so that
$$F(r) = r\int_0^{2\pi} \sin^2 \psi(1 - r^2 \cos^2 \psi)\, d\psi = \pi r\left(1 - \frac{r^2}{4}\right)$$

Equation (6.94) becomes

$$\frac{dr}{dt} = \frac{\mu r}{2}\left(1 - \frac{r^2}{4}\right)$$

whose solution is

$$r(t) = \frac{r_0 e^{-\mu t/2}}{\sqrt{1 + \frac{1}{4}r_0^2(e^{\mu t} - 1)}} \tag{6.97}$$

At $t = 0$, $r(0) = r_0$. For $r_0 \neq 0$ we have

$$\lim_{t \to \infty} r(t) = 2$$

Since $r^2 = x^2 + y^2$, we note that to a first approximation the motion in the phase plane $\left(x, y = \dfrac{dx}{dt}\right)$ resembles a spiral motion into the limit cycle circle $x^2 + y^2 = r^2 = 4$. From (6.95) we obtain $G(r) = 0$, so that $\theta = \theta_0 + t$, which yields nothing new.

To obtain an improved approximation to (6.87) we proceed as follows: Let

$$x(t) = r(t) \cos \theta(t) + \mu\alpha(r) + \mu \sum_{n=2}^{\infty} \alpha_n(r) \cos n\theta(t) + \mu \sum_{n=2}^{\infty} \beta_n(r) \sin n\theta(t) \tag{6.98}$$

and let us attempt to make (6.98) a solution of (6.87). $r(t)$ and $\theta(t)$ will be taken as the first approximation given by the solution of (6.94) along with (6.95). We shall attempt to find $\alpha(r)$, $\alpha_n(r)$, $\beta_n(r)$, $n = 2, 3, 4, \ldots$. Terms involving μ^2 will be ignored. From (6.98) we obtain

$$\frac{dx}{dt} = \frac{dr}{dt} \cos \theta - r \sin \theta \frac{d\theta}{dt} - \mu \sum_{n=2}^{\infty} n\alpha_n \sin n\theta \frac{d\theta}{dt} + \mu \sum_{n=2}^{\infty} n\beta_n \cos n\theta \frac{d\theta}{dt} \tag{6.99}$$

The term $\mu \frac{d\alpha}{dt} = \mu \frac{d\alpha}{dr}\frac{dr}{dt} = \mu^2 \frac{F(r)}{2\pi}\frac{d\alpha}{dr}$ is of the order of μ^2 and so is neglected. The same reasoning applies to the terms involving

$$\mu \frac{d\alpha_n}{dt} = \mu \frac{d\alpha_n}{dr}\frac{dr}{dt} = \frac{\mu^2}{2\pi}\frac{d\alpha_n}{dr} F(r)$$
$$\mu \frac{d\beta_n}{dt} = \mu \frac{d\beta_n}{dr}\frac{dr}{dt} = \frac{\mu^2}{2\pi}\frac{d\beta_n}{dr} F(r)$$

Differentiating again yields

$$\frac{d^2x}{dt^2} = -2\frac{dr}{dt}\frac{d\theta}{dt}\sin\theta - r\cos\theta\left(\frac{d\theta}{dt}\right)^2 - \mu\sum_{n=2}^{\infty} n^2\alpha_n \cos n\theta\left(\frac{d\theta}{dt}\right)^2 - \mu\sum_{n=2}^{\infty} n^2\beta_n \sin n\theta\left(\frac{d\theta}{dt}\right)^2 \quad (6.100)$$

Terms involving $\frac{d^2r}{dt^2}, \frac{d^2\theta}{dt^2}$ have been omitted since

$$\frac{d^2r}{dt^2} = \frac{\mu}{2\pi}F'(r)\frac{dr}{dt} = \frac{\mu^2}{4\pi^2}F(r)F'(r)$$
$$\frac{d^2\theta}{dt^2} = \frac{\mu^2}{4\pi^2}\left[\frac{rG'(r) - G(r)}{r^2}\right]F(r) \qquad r \neq 0$$

Moreover
$$\left(\frac{d\theta}{dt}\right)^2 = 1 + \frac{\pi}{\pi r}G(r)$$
$$\frac{dr}{dt}\frac{d\theta}{dt} = \frac{\mu}{2\pi}F(r)$$

when we neglect terms of the order of μ^2. Substituting $x(t)$ of (6.98) and $\frac{d^2x}{dt^2}$ of (6.100) into (6.87) yields

$$\mu\alpha(r) - \frac{\mu}{\pi}F(r)\sin\theta - \frac{\mu}{\pi}G(r)\cos\theta - \mu\sum_{n=2}^{\infty}(n^2-1)\alpha_n\cos n\theta - \mu\sum_{n=2}^{\infty}(n^2-1)\beta_n\sin n\theta = -\mu f(r\cos\theta, -r\sin\theta) \quad (6.101)$$

In the term $\mu f(x, \dot{x})$ we have replaced x by $r\cos\theta$, $\dot{x}$ by $-r\sin\theta$, again neglecting terms of the order of μ^2.

Since the left-hand side of (6.101) is a Fourier series, we expand the right-hand side of (6.101) in a Fourier series,

$$f(r\cos\theta, -r\sin\theta) = a_0(r) + \sum_{n=1}^{\infty} a_n(r)\cos n\theta + \sum_{n=1}^{\infty} b_n(r)\sin n\theta$$

$$\text{with} \quad a_0(r) = \frac{1}{2\pi}\int_0^{2\pi} f(r\cos\theta, -r\sin\theta)\,d\theta$$

$$a_n(r) = \frac{1}{\pi}\int_0^{2\pi} \cos n\theta\, f(r\cos\theta, -r\sin\theta)\,d\theta \qquad n = 1, 2, 3, \ldots$$

$$b_n(r) = \frac{1}{\pi}\int_0^{2\pi} \sin n\theta\, f(r\cos\theta, -r\sin\theta)\,d\theta$$

Equating coefficients of $\cos n\theta$ and $\sin n\theta$ in (6.101) yields

$$F(r) = \int_0^{2\pi} \sin\theta\, f(r\cos\theta, -r\sin\theta)\,d\theta$$

$$G(r) = \int_0^{2\pi} \cos\theta\, f(r\cos\theta, -r\sin\theta)\,d\theta$$

$$\alpha(r) = -\frac{1}{2\pi}\int_0^{2\pi} f(r\cos\theta, -r\sin\theta)\,d\theta \tag{6.102}$$

$$\alpha_n(r) = \frac{1}{\pi(n^2-1)}\int_0^{2\pi} \cos n\theta\, f(r\cos\theta, -r\sin\theta)\,d\theta$$

$$\beta_n(r) = \frac{1}{\pi(n^2-1)}\int_0^{2\pi} \sin n\theta\, f(r\cos\theta, -r\sin\theta)\,d\theta \qquad n = 2, 3, 4, \ldots$$

The values of $F(r)$, $G(r)$ given by (6.102) agree with their previous definitions. It can be shown that $x(t)$ given by (6.98) satisfies (6.87) with accuracy of order μ^2 for $0 \leqq t < \infty$.

Example 6.10. In Example 6.9, $f(x, \dot{x}) = (x^2 - 1)\dot{x}$, so that

$$f(r\cos\theta, -r\sin\theta) = (1 - r^2\cos^2\theta)r\sin\theta = \left(r - \frac{r^3}{4}\right)\sin\theta - \frac{r^3}{4}\sin 3\theta$$

Applying (6.102) yields

$$\begin{aligned} \alpha(r) &= 0 \\ \alpha_n(r) &= 0 \qquad n = 2, 3, 4, \ldots \\ \beta_3(r) &= -\frac{r^3}{32} \\ \beta_n(r) &= 0 \qquad n = 2, 4, 5, \ldots \end{aligned}$$

Thus the improved first approximation is

$$x(t) = r(t)\cos(\theta_0 + t) - \mu\frac{r^3}{32}\sin 3(\theta_0 + t)$$

with $r(t)$ given by (6.97).

Problems

1. The differential equation of a simple pendulum is $\ddot{\theta} + (g/l)\sin\theta = 0$. Take $\sin\theta \approx \theta - \theta^3/6$, and show that the period of oscillation depends on the square of the amplitude.

2. Solve $\ddot{x} + x + \mu\dot{x}|\dot{x}| = 0$, $\mu \ll 1$, for the improved first approximation.

3. Solve $\ddot{x} + x + \mu(\text{sign } \dot{x})x = 0$, $\mu \ll 1$, for the improved first approximation.

4. Solve $\ddot{x} + x + \mu(-\alpha + \beta\dot{x}^2)x = 0$, $\mu \ll 1$, for the improved first approximation. Show that a limit cycle occurs and that the radius of this limit cycle is $r_0 = \sqrt{4\alpha/3\beta}$.

5. Consider the differential equation

$$\dddot{x} + \dot{x} + \mu x^2 = 0 \qquad \mu \ll 1 \tag{6.103}$$

written as the system

$$\begin{aligned} \frac{dx}{dt} &= y \\ \frac{dy}{dt} &= z \\ \frac{dz}{dt} &= -y - \mu x^2 \end{aligned} \tag{6.104}$$

such that, at $t = t_0$, $x = x_0$, $y = y_0$, $z = z_0$. The solution of $\dddot{x} + \dot{x} = 0$ is

$$x(t) = A + B\cos(t + C)$$

$y = \dfrac{dx}{dt} = -B\sin(t + C)$, $z = \dfrac{dy}{dt} = -B\cos(t + C)$. This suggests the transformation

$$\begin{aligned} x &= w - r\cos\theta \\ y &= r\sin\theta \\ z &= r\cos\theta \end{aligned}$$

in an attempt to find an approximate solution of (6.104). Let the reader show that to a first approximation

$$\begin{aligned} \theta &= t + \theta_0 \\ \frac{dr}{dt} &= \mu w r \\ \frac{dw}{dt} &= -\mu\left(w^2 + \frac{r^2}{2}\right) \end{aligned}$$

and $w(t) = \pm\dfrac{1}{2}\dfrac{\sqrt{a^4 - r^4}}{r}$ with $(x_0 + z_0)^2 = \dfrac{1}{4}\dfrac{a^4 - r_0^4}{r_0^2}$, $r_0^2 = y_0^2 + z_0^2$

REFERENCES

Carslaw, H. S.: "Introduction to the Fourier's Series and Integrals," Dover Publications, Inc., New York, 1930.

Churchill, R. V.: "Fourier Series and Boundary Value Problems," McGraw-Hill Book Company, Inc., New York, 1941.

Courant, R., and D. Hilbert: "Methoden der Mathematischen Physik," Springer-Verlag OHG, Berlin, 1931.

Franklin, P.: "Fourier Methods," McGraw-Hill Book Company, Inc., New York, 1949.

Minorski, N.: "Introduction to Nonlinear Mechanics," J. W. Edwards, Publisher, Inc., Ann Arbor, Mich., 1947.

Sneddon, I. N.: "Fourier Transforms," McGraw-Hill Book Company, Inc., New York, 1951.

Stoker, J. J.: "Nonlinear Vibrations," Interscience Publishers, Inc., New York, 1950.

Szego, G.: Orthogonal Polynomials, *American Mathematical Society Colloquium*, vol. 23, 1939.

CHAPTER 7

THE STIELTJES INTEGRAL, LAPLACE TRANSFORM, AND CALCULUS OF VARIATIONS

7.1. Functions of Bounded Variation. In an attempt to define arc length of a curve one is led to consider functions of bounded variation. Let us consider a simple curve, Γ, given parametrically by $x = f(t)$, $y = \varphi(t)$, $\alpha \leqq t \leqq \beta$. Two distinct values of t are assumed to yield two distinct points on Γ, so that as t varies from α to β, the point P with coordinates $x = f(t)$, $y = \varphi(t)$ moves continuously from one end of the curve to the other without retracing its motion. We now subdivide the interval (α, β) into

$$\alpha = t_0 < t_1 < t_2 < \cdots < t_k < t_{k+1} < \cdots < t_{n-1} < t_n = \beta$$

For $t = t_k$ we have the point P_k with coordinates $x_k = f(t_k)$, $y_k = \varphi(t_k)$, P_k on Γ, $k = 0, 1, 2, \ldots, n$. The straight-line distance from P_{k-1} to P_k is given by

$$\{[f(t_k) - f(t_{k-1})]^2 + [\varphi(t_k) - \varphi(t_{k-1})]^2\}^{\frac{1}{2}}$$

The sum total of these straight-line arcs is

$$S_n = \sum_{k=1}^{n} \{[f(t_k) - f(t_{k-1})]^2 + [\varphi(t_k) - \varphi(t_{k-1})]^2\}^{\frac{1}{2}} \tag{7.1}$$

Now if for all manner of subdivisions of $\alpha \leqq t \leqq \beta$ a constant A exists such that

$$\begin{aligned} \sum_{k=1}^{n} |f(t_k) - f(t_{k-1})| &< A \\ \sum_{k=1}^{n} |\varphi(t_k) - \varphi(t_{k-1})| &< A \end{aligned} \tag{7.2}$$

then $S_n < \sqrt{2}\, A$. Since the set of numbers $\{S_n\}$ is bounded, of necessity, a least upper bound (supremum), L, will exist for this set. We define L as the length of arc of Γ. For a discussion of the supremum see Chap. 10. Conversely, one easily shows that if the $\{S_n\}$ are bounded for all subdivisions then a constant A exists satisfying (7.2) for all subdivisions. The inequality of (7.2) leads to the following definition:

Let $f(x)$ be defined on the bounded interval $a \leqq x \leqq b$. If a constant A exists such that for all possible finite subdivisions of (a, b) into

$$x_0 = a < x_1 < x_2 < \cdots x_k < \cdots < x_{n-1} < x_n = b$$

we have

$$\sum_{k=1}^{n} |f(x_k) - f(x_{k-1})| < A \tag{7.3}$$

we say that $f(x)$ is of *bounded variation* on $a \leqq x \leqq b$.

A bounded monotonic nondecreasing or nonincreasing function is always of bounded variation. If $f(x)$ is a monotonic nondecreasing function, then

$$\sum_{k=1}^{n} |f(x_k) - f(x_{k-1})| = \sum_{k=1}^{n} [f(x_k) - f(x_{k-1})] = f(b) - f(a)$$

Thus (7.3) is satisfied for $A > f(b) - f(a)$. Another example of a function of bounded variation is the following: Assume $f(x)$ has a continuous derivative for $a \leqq x \leqq b$. Then

$$|f(x_k) - f(x_{k-1})| = |(x_k - x_{k-1})f'(\xi_k)| < M(x_k - x_{k-1})$$

since $|f'(x)| < M$ for $a \leqq x \leqq b$. Thus

$$\sum_{k=1}^{n} |f(x_k) - f(x_{k-1})| < M \sum_{k=1}^{n} |x_k - x_{k-1}| = M(b - a)$$

which proves our statement.

Continuity of $f(x)$ is not sufficient to guarantee that $f(x)$ is of bounded variation. For example, consider $f(x) = x \sin (1/x)$, $x \neq 0$, $f(0) = 0$, $0 \leqq x \leqq 2/\pi$. Since

$$\lim_{x\to 0} f(x) = \lim_{x\to 0} x \sin \frac{1}{x} \leqq \lim_{x\to 0} |x| = 0 = f(0)$$

$f(x)$ is continuous at $x = 0$. Moreover $f(x)$ is easily seen to be continuous for $x \neq 0$. Let us subdivide $\left(0, \frac{2}{\pi}\right)$ into $\left(0, \frac{2}{(2n+1)\pi}, \frac{2}{2n\pi}, \ldots, \frac{2}{\pi}\right)$. Then

$$\sum_{k=1}^{n} |f(x_k) - f(x_{k-1})| = \frac{4}{\pi}\left[1 + \tfrac{1}{3} + \tfrac{1}{5} + \cdots + \frac{1}{2n+1}\right]$$

Since $\sum_{k=0}^{\infty} \frac{1}{2k+1}$ diverges, no constant A exists satisfying (7.3) for all modes of subdivision.

A fundamental result concerning functions of bounded variation is the following theorem:

A necessary and sufficient condition that $f(x)$ be of bounded variation on $a \leqq x \leqq b$ is that $f(x)$ be written as the difference of two positive monotonic nondecreasing functions. That the condition is sufficient follows almost immediately from previous considerations concerning monotonic nondecreasing functions. Now let us assume that $f(x)$ is of bounded variation on (a, b). Let x be any number on the interval (a, b). Let us subdivide (a, x) into

$$a = x_0 < x_1 < x_2 < \cdots < x_{k-1} < x_k < \cdots < x_n = x$$

Then

$$S_n = \sum_{k=1}^{n} |f(x_k) - f(x_{k-1})| < A \tag{7.4}$$

Some of the terms of S_n are such that $f(x_r) \geqq f(x_{r-1})$, whereas other terms are such that $f(x_s) < f(x_{s-1})$. We write

$$S_n = P_n + N_n$$

where P_n is the sum of terms of S_n for which $f(x_r) \geqq f(x_{r-1})$ and N_n is the sum of terms of S_n for which $f(x_s) < f(x_{s-1})$. One easily shows that $P_n - N_n = f(x) - f(a)$ so that

$$\begin{aligned} S_n &= f(x) - f(a) + 2N_n \\ S_n &= -f(x) + f(a) + 2P_n \end{aligned} \tag{7.5}$$

Since $S_n < A$ for all methods of subdividing the interval (a, x), we know from real-variable theory (see Chap. 10) that a least upper bound exists for the set $\{S_n\}$. We call this least upper bound, $V(a, x)$, the total variation of $f(x)$ on the interval (a, x). The suprema (least upper bounds) of $\{N_n\}$ and $\{P_n\}$ are written as $N(a, x)$, $P(a, x)$, respectively. From (7.5) we have

$$\begin{aligned} V(a, x) &= f(x) - f(a) + 2N(a, x) \\ V(a, x) &= -f(x) + f(a) + 2P(a, x) \end{aligned} \tag{7.6}$$

so that

$$f(x) = f(a) + P(a, x) - N(a, x) \tag{7.7}$$

From the very definitions of $P(a, x)$, $N(a, x)$ we note that they are monotonic nondecreasing functions of x [see (7.6)]. $f(a) + P(a, x)$ is monotonic nondecreasing, which proves the theorem.

Problem 1. Let $f(x) = \sin x$ for $0 \leqq x \leqq 2\pi$. Write $f(x)$ as the difference of two monotonic nondecreasing functions of x for this interval.

Problem 2. If $f(x)$ is of bounded variation and continuous for $a \leqq x \leqq b$, show that $P(a, x)$ and $N(a, x)$ are continuous for $a \leqq x \leqq b$.

7.2. The Stieltjes Integral. An important generalization of the Riemann integral is the Stieltjes integral. The Stieltjes integral is defined as follows: Let $f(x)$ and $g(x)$ be real-valued functions of the real variable x for $a \leqq x \leqq b$. Subdivide the interval (a, b) into

$$a = x_0 < x_1 < x_2 < \cdots < x_{k-1} < x_k < \cdots < x_n = b$$

and let δ be largest of the numbers $x_k - x_{k-1}$, $k = 1, 2, \ldots, n$. Now form the sum

$$S_n = \sum_{k=1}^{n} f(\xi_k)[g(x_k) - g(x_{k-1})] \tag{7.8}$$

where ξ_k is any number such that $x_{k-1} \leqq \xi_k \leqq x_k$. If $\lim\limits_{n\to\infty} S_n$ exists independent of the choice of the ξ_k and the method of subdivision, provided $\delta \to 0$, we call this limit the Stieltjes integral of $f(x)$ relative to $g(x)$ on (a, b), written

$$\int_a^b f(x)\, dg(x) \tag{7.9}$$

In the special case $g(x) \equiv x$, (7.9) reduces to the Riemann integral.

If $f(x)$ is bounded and if $g(x)$ is a bounded monotonic nondecreasing function of x, then S_n of (7.8) satisfies the following inequality,

$$\begin{aligned} m[g(b) - g(a)] &\leqq \sum_{k=1}^{n} m_k[g(x_k) - g(x_{k-1})] \\ &\leqq S_n \leqq \sum_{k=1}^{n} M_k[g(x_k) - g(x_{k-1})] \leqq M[g(b) - g(a)] \end{aligned}$$

where m_k is the infemum (greatest lower bound) of $f(x)$ for $x_{k-1} \leqq x \leqq x_{k-1}$, M_k is the supremum (least upper bound) of $f(x)$ for $x_{k-1} \leqq x \leqq x_k$, m and M are the infemum and supremum of $f(x)$, respectively, for $a \leqq x \leqq b$. The supremum of the sums $\sum_{k=1}^{n} m_k[g(x_k) - g(x_{k-1})]$ can be called the lower Darboux-Stieltjes integral, L, and the infemum of the sums $\sum_{k=1}^{n} M_k[g(x_k) - g(x_{k-1})]$ can be called the upper Darboux-Stieltjes integral, U. If these two integrals are equal, we say that the Riemann-Stieltjes integral exists and write

$$L = U = \int_a^b f(x)\, dg(x)$$

This definition is easily seen to be equivalent to the one given above.

If $f(x)$ is continuous in (a, b), it is a simple matter to prove that $L = U$. Now if $g(x)$ is a function of bounded variation, it can be written as the difference of two monotonic nondecreasing functions. It follows immediately that (7.9) exists if $f(x)$ is continuous and $g(x)$ is of bounded variation.

Example 7.1. Let $f(x)$ be continuous for $0 \leqq x \leqq 1$, and let $g(x) = 0$ for $0 \leqq x < \frac{1}{2}$, $g(x) = 1$ for $\frac{1}{2} \leqq x \leqq 1$. For any subdivision not containing $x = \frac{1}{2}$ we have $dg(x) = 0$. The subdivision covering $x = \frac{1}{2}$ yields $dg(x) = 1$. Thus $S_n = f(\xi)$, where ξ is any number near $x = \frac{1}{2}$. Since $\lim_{n\to\infty} f(\xi) = f(\frac{1}{2})$, we have $\int_0^1 f(x)\, dg(x) = f(\frac{1}{2})$.

Problem 3. Let $f(x) = x + \frac{1}{2}$, $g(x) = x^2$ for $0 \leqq x \leqq 1$. Show that

$$\int_0^1 f(x)\, dg(x) = \tfrac{7}{6}$$

Problem 4. If $f(x)$ and $g'(x)$ are continuous for $a \leqq x \leqq b$, show that

$$\int_a^b f(x)\, dg(x) = \int_a^b f(x)g'(x)\, dx$$

The last integral is a Riemann integral.

Problem 5. If $f(x)$ and $g(x)$ have a common point of discontinuity, show that the Stieltjes integral of $f(x)$ relative to $g(x)$ does not exist provided the range of integration covers the point of discontinuity.

Problem 6. If $h(x)$ is nondecreasing, $f(x)$ and $g(x)$ continuous with $f(x) \geqq g(x)$, show that $\int_a^b f(x)\, dh(x) \geqq \int_a^b g(x)\, dh(x)$.

Problem 7. Let $S_n = \sum_{k=1}^{n} g(\xi_k)[f(x_k) - f(x_{k-1})]$, $x_{k-1} \leqq \xi_k \leqq x_k$. Show that

$$S_n = g(\xi_n)f(b) - g(\xi_1)f(a) - \sum_{k=1}^{n-1} f(x_k)[g(\xi_{k+1}) - g(\xi_k)]$$

with $x_0 = a$, $x_n = b$. Assume $g(x)$ of bounded variation, $f(x)$ continuous, and $g(x)$ continuous at $x = a$ and $x = b$. Show that

$$\int_a^b g(x)\, df(x) = g(b)f(b) - g(a)f(a) - \int_a^b f(x)\, dg(x)$$

Problem 8. Let $f(x) = g(x) + ih(x)$ be continuous for $a \leqq x \leqq b$, and assume $\alpha(x) = \beta(x) + i\gamma(x)$ to be of bounded variation. Show that

$$\int_a^b f(x)\, d\alpha(x) = \int_a^b g(x)\, d\beta(x) - \int_a^b h(x)\, d\gamma(x) + i\int_a^b g(x)\, d\gamma(x) + i\int_a^b h(x)\, d\beta(x)$$

Problem 9. Consider the sequence of continuous functions, $f_n(x)$, $n = 1, 2, \ldots$. Assume $\sum_{n=1}^{\infty} f_n(x)$ converges uniformly for $a \leqq x \leqq b$, and let $g(x)$ be of bounded variation for this interval. Show that

$$\int_a^b \sum_{n=1}^{\infty} f_n(x)\,dg(x) = \sum_{n=1}^{\infty} \int_a^b f_n(x)\,dg(x)$$

Problem 10. How would you define $\int_a^\infty f(x)\,dg(x)$, and under what conditions would this integral exist?

7.3. The Laplace Transform. Let $g(t)$ be a complex function of the real variable t defined for $t \geqq 0$. Let $g(t)$ be of bounded variation on the finite interval $0 \leqq t \leqq R$, R arbitrary. The function e^{-zt} with $z = x + iy$ is continuous for all t, so that the integral

$$\int_0^R e^{-zt}\,dg(t) \tag{7.10}$$

exists for all complex z. It may be that, for a given value of z,

$$\lim_{R\to\infty} \int_0^R e^{-zt}\,dg(t) \tag{7.11}$$

exists. If this is the case, we write

$$\int_0^\infty e^{zt}\,dg(t) = \lim_{R\to\infty} \int_0^R e^{-zt}\,dg(t) \tag{7.12}$$

Equation (7.12) is called an improper integral, and the right-hand side of (7.12) is called the Cauchy value of the improper integral.

Those values of z for which (7.12) exists define a function of z, written

$$f(z) = \int_0^\infty e^{-zt}\,dg(t) \tag{7.13}$$

$f(z)$ is called the Laplace-Stieltjes transform of $g(t)$.

We consider now the region of z for which $f(z)$ of (7.13) exists. First we investigate three special cases. Let $g(t) = \int_0^t e^u\,du$, so that $g(t)$ is monotonic increasing, and hence is of bounded variation for $0 \leqq t \leqq R$, R arbitrary. Since $dg(t) = e^{e^t}\,dt$, we have

$$\int_0^R e^{-(x+iy)t} e^{e^t}\,dt = \int_0^R e^{e^t - xt} \cos yt\,dt - i \int_0^R e^{e^t - xt} \sin yt\,dt$$

Since $e^t - xt$ increases beyond bound for any fixed x, we leave it to the reader to show that $\lim_{R\to\infty} \int_0^R e^{-zt} e^{e^t}\,dt$ fails to exist for all z. As a second example, let $g(t) = \int_0^t e^{-u}\,du$. Let the reader show that $\lim_{R\to\infty} \int_0^R e^{-zt}\,dg(t)$ exists for all z. Finally, let $g(t) = t$, so that

$$\lim_{R\to\infty} \int_0^R e^{-zt}\,dg(t) = \lim_{R\to\infty} \int_0^R e^{-zt}\,dt = \lim_{R\to\infty} \frac{1 - e^{-zR}}{z} = \frac{1}{z}$$

provided $x = \text{Rl } z > 0$. Hence $\int_0^\infty e^{-zt}\,dt = f(z)$ exists for $\text{Rl } z > 0$.

Let us consider now a general case. We assume that $\int_0^\infty e^{-z_0t}\,dg(t)$ exists, with $z_0 = x_0 + iy_0$. Further, let us assume that a constant A exists such that

$$\left| \int_0^u e^{-z_0t}\,dg(t) \right| < A \tag{7.14}$$

for $u \geqq 0$. Define $h(u)$ by the equation

$$h(u) = \int_0^u e^{-z_0t}\,dg(t) \tag{7.15}$$

so that $dh(u) = e^{-z_0u}\,dg(u)$ and $dg(t) = e^{z_0t}\,dh(t)$. Then

$$\int_0^R e^{-zt}\,dg(t) = \int_0^R e^{-(z-z_0)t}\,dh(t) \tag{7.16}$$

Integration by parts (see Prob. 7, Sec. 7.2) yields

$$\int_0^R e^{-zt}\,dg(t) = e^{-(z-z_0)R} \int_0^R e^{-z_0t}\,dg(t) + (z - z_0) \int_0^R h(t)e^{-(z-z_0)t}\,dt$$

If $\text{Rl } z > \text{Rl } z_0$, then $\lim_{R\to\infty} e^{-(z-z_0)R} \int_0^R e^{-z_0t}\,dg(t) = 0$, since

$$\lim_{R\to\infty} e^{-(z-z_0)R} = 0$$

and $\left| \int_0^R e^{-z_0t}\,dg(t) \right| < A$ for all R. Moreover $\left| \int_0^\infty h(t)e^{-(z-z_0)t}\,dt \right| \leqq \int_0^\infty |h(t)|e^{-(x-x_0)t}\,dt < A/(x - x_0)$, for $x > x_0$. Allowing R to become infinite in (7.16) yields

$$\int_0^\infty e^{-zt}\,dg(t) = (z - z_0) \int_0^\infty h(t)e^{-(z-z_0)t}\,dt$$

for $\text{Rl } z > \text{Rl } z_0$. Thus

$$f(z) = \int_0^\infty e^{-zt}\,dg(t) \tag{7.17}$$

exists for $\text{Rl } z > \text{Rl } z_0$. We have shown that, if (7.17) converges for $z_0 = x_0 + iy_0$, then $f(z)$ of (7.17) is well defined for $z = x + iy$ provided $x > x_0$. There may be singularities of $f(z)$ on the line $z = x_0 + iy$, $-\infty < y < \infty$, and in the half plane $\text{Rl } z < x_0$.

The ordinary Laplace transform of $f(t)$ is defined as

$$F(z) = L[f(t)] = \int_0^\infty e^{-zt}f(t)\,dt \tag{7.18}$$

provided the improper integral exists. If the Laplace transform of $f'(t)$ also exists, one can integrate by parts to obtain

$$\begin{aligned} L[f'(t)] &= \int_0^\infty e^{-zt}f'(t)\,dt = e^{-zt}f(t)\Big|_0^\infty + z\int_0^\infty e^{-zt}f(t)\,dt \\ L[f'(t)] &= zL[f(t)] - f(0) \end{aligned} \tag{7.19}$$

provided

$$\lim_{t\to\infty} e^{-zt}f(t) = 0 \tag{7.20}$$

Equation (7.20) will certainly hold if $|f(t)|$ is bounded for $t \geqq 0$ and if $\mathrm{Rl}\, z > 0$. Further application of (7.19) yields

$$\begin{aligned} L[f''(t)] &= zL[f'(t)] - f'(0) \\ &= z^2L[f(t)] - zf(0) - f'(0) \end{aligned} \tag{7.21}$$

If $F(z)$ is the Laplace transform of $f(t)$, we say that $f(t)$ is the inverse Laplace transform of $F(z)$, written $f(t) = L^{-1}[F(z)]$.

Example 7.2. We find the Laplace transform of sin at. We have

$$F(z) = \int_0^\infty e^{-zt} \sin at\,dt$$

Integration by parts yields

$$\int_0^R e^{-zt}\sin at\,dt = \frac{e^{-zt}}{a^2+z^2}(-z\sin at - a\cos at)\Big|_{t=0}^{t=R}$$

If $\mathrm{Rl}\, z > 0$, we have

$$\int_0^\infty e^{-zt}\sin at\,dt = \frac{a}{a^2+z^2}$$

Table 7.1 lists Laplace transforms for some simple cases of $f(t)$.

Problem 11. Derive the results of Table 7.1.

Problem 12. Show that $L[af(t) + bg(t)] = aL[f(t)] + bL[g(t)]$.

Example 7.3. Let us consider the differential equation

$$\frac{d^2y}{dt^2} + 3\frac{dy}{dt} - 4y = 0 \tag{7.22}$$

subject to the initial conditions $y(0) = y_0$, $y'(0) = y_0'$. Assuming that the solution of (7.22) and its derivatives are such that their Laplace transforms exist, we can apply (7.19) and (7.21) and the result of Prob. 12 to obtain

$$z^2L[y(t)] - zy_0 - y_0' + 3zL[y(t)] - 3y_0 - 4L[y(t)] = 0$$

so that $$L[y(t)] = \frac{(z+3)y_0 + y_0'}{(z+4)(z-1)} = \frac{y_0 - y_0'}{5}\frac{1}{z+4} + \frac{4y_0 + y_0'}{5}\frac{1}{z-1}$$

$$= \frac{y_0 - y_0'}{5}L[\varphi_1(t)] + \frac{4y_0+y_0'}{5}L[\varphi_2(t)]$$

From Table 7.1 we see that $\varphi_1(t) = e^{-4t}$, $\varphi_2(t) = e^t$, so that the suggested solution of (7.22) is

$$y(t) = \frac{y_0 - y_0'}{5} e^{-4t} + \frac{4y_0 + y_0'}{5} e^t \tag{7.23}$$

One easily checks that $y(t)$ of (7.23) is the required solution. It is to be noted that the Laplace-transform method for solving (7.22) introduces the initial conditions in a natural manner.

TABLE 7.1

	$f(t)$	$F(z) = \int_0^\infty e^{-zt} f(t)\, dt$
1	1	$\frac{1}{z}$, Rl $z > 0$
2	e^{at}	$\frac{1}{z-a}$, Rl $z >$ Rl a
3	$\sin at$	$\frac{a}{z^2+a^2}$, Rl $z > 0$
4	$\cos at$	$\frac{z}{z^2+a^2}$, Rl $z > 0$
5	$\sinh at$	$\frac{a}{z^2-a^2}$, Rl $z > \|a\|$
6	$\cosh at$	$\frac{z}{z^2-a^2}$, Rl $z > \|a\|$
7	$\frac{t}{2a} \sin at$	$\frac{z}{(z^2+a^2)^2}$, Rl $z > 0$
8	$\frac{1}{2a^3} (\sin at - at \cos at)$	$\frac{1}{(z^2+a^2)^2}$, Rl $z > 0$
9	$J_n(t)$	$\frac{(\sqrt{z^2+1} - z)^n}{\sqrt{z^2+1}}$
10	$\frac{t^{n-1}}{(n-1)!}$	$\frac{1}{z^n}$, Rl $z > 0$

Problem 13. Solve $\frac{d^2y}{dx^2} + 2\frac{dy}{dx} = 1 - 3x$ by the Laplace-transform method, $y(0) = 0$, $y'(0) = 1$.

Problem 14. Let $f(t) = 0$ for $t < 0$. Show that

$$\int_0^\infty e^{-zt} f(t - x)\, dt = \int_x^\infty e^{-zt} f(t - x)\, dt = e^{-zx} \int_0^\infty e^{-zt} f(t)\, dt$$

provided the integrals exist.

Example 7.4. Let us attempt to solve the wave equation $\frac{\partial^2 y(x, t)}{\partial x^2} = \frac{1}{c^2}\frac{\partial^2 y(x, t)}{\partial t^2}$ subject to the boundary conditions $y(x, 0) = \frac{\partial y(x, 0)}{\partial t} = 0$ for all x, and $y(0, t) = f(t)$ for $t \geqq 0$. We assume $y(\infty, t) = \lim_{x\to\infty} y(x, t) = 0$ for $t \geqq 0$. Physically, we have an elastic string from $x = 0$ to $x = \infty$, initially at rest. At the origin the string is constrained to move in such a manner that $y(0, t) = f(t)$, $f(t)$ a given function of time. If we multiply the wave equation by e^{-zt} and integrate from $t = 0$ to $t = \infty$, we obtain

$$\frac{\partial^2}{\partial x^2}\int_0^\infty y(x, t)e^{-zt}\,dt = \frac{1}{c^2} z^2 L[y(x, t)]$$

provided we assume $\frac{\partial^2}{\partial x^2}\int_0^\infty y(x, t)e^{-zt}\,dt = \int_0^\infty \frac{\partial^2 y(x, t)}{\partial x^2} e^{-zt}\,dt$. Thus $L[y]$ satisfies

$$\frac{\partial^2}{\partial x^2} L[y] = \frac{z^2}{c^2} L[y] \tag{7.24}$$

A solution of (7.24) is

$$L[y] \equiv \int_0^\infty y(x, t)e^{-zt}\,dt = Ae^{(z/c)x} + Be^{-(z/c)x} \tag{7.25}$$

where A and B can be functions of z. Since $y(x, t)$ tends to zero as x becomes infinite we choose $A = 0$. At $x = 0$ we need

$$\int_0^\infty f(t)e^{-zt}\,dt = B(z)$$

From Prob. 14 we have

$$\int_0^\infty y(x, t)e^{-zt}\,dt = e^{-zx/c}\int_0^\infty f(t)e^{-zt}\,dt = \int_0^\infty f\left(t - \frac{x}{c}\right)e^{-zt}\,dt \tag{7.26}$$

Equation (7.26) suggests that $y(x, t) = f(t - x/c)$ for $t \geqq x/c$ provided $f(t - x/c) = 0$ for $t < x/c$ (see Prob. 14). It is a simple matter to show that $f(t - x/c)$ satisfies the wave equation and the boundary conditions. Of course one needs the fact that $f(t)$ be twice differentiable.

Example 7.5. We wish to find the function $f(t)$ such that

$$\int_0^\infty f(t)e^{-zt}\,dt = (z^2 - 1)^{-\frac{1}{2}} \tag{7.27}$$

Let us assume that $f(t)$ has the Taylor-series expansion, $\sum_{n=0}^\infty a_n t^n$. Without justification let us assume that term-by-term integration is permissible. Hence

$$\int_0^\infty f(t)e^{-zt}\,dt = \sum_{n=0}^\infty a_n \int_0^\infty t^n e^{-zt}\,dt = \sum_{n=0}^\infty \frac{n!}{z^{n+1}} a_n$$

Now for $|z| > 1$ we know that

$$\frac{1}{\sqrt{z^2 - 1}} = \frac{1}{z\sqrt{1 - (1/z)^2}} = \sum_{m=0}^\infty \frac{(2m)!}{m!m!2^{2m}} \frac{1}{z^{2m+1}}$$

Comparing the two Laurent series, we see that $a_n = 0$ if n is odd and, if $n = 2m$,

$$a_{2m} = \frac{1}{m!m!2^{2m}}, \text{ so that}$$

$$f(t) = \sum_{m=0}^{\infty} \frac{1}{m!m!} \left(\frac{t}{2}\right)^{2m} \tag{7.28}$$

We note that $f(it) = \sum_{m=0}^{\infty} \frac{(-1)^m}{m!m!} \left(\frac{t}{2}\right)^{2m} = J_0(t)$, where $J_0(t)$ is the Bessel function of order zero of the first kind. One can start with $f(t)$ of (7.28), justify the interchange of integration and summation, and show that (7.27) results.

Problems

1. Solve $\frac{d^2y}{dx^2} + y = \sin x$ with $y(0) = 0$, $y'(0) = 0$.

2. Solve $\frac{d^2y}{dx^2} - 4y = 3e^{2x}$ with $y(0) = 0$, $y'(0) = 1$.

3. Solve $2x \frac{\partial y}{\partial t} + \frac{\partial y}{\partial x} = 2x$, $y(x, 0) = 1$, $y(0, t) = 1$, for $y(x, t)$.

Ans. $y(x, t) = 1 + t$ for $0 < t < x^2$, $y(x, t) = 1 + x^2$ for $t > x^2$.

4. By the inversion formula (see Sec. 7.4) solve for $f(t)$ if

$$\frac{2az}{(z^2 + a^2)^2} = \int_0^\infty e^{-zt} f(t)\, dt$$

Ans. $f(t) = t \sin at$ (see Table 7.1).

5. From Prob. 4 show that

$$\int_0^\infty e^{-zt}\, t \cos at\, dt = \frac{z^2 - a^2}{(z^2 + a^2)^2}$$

6. Find $f(t)$ such that

$$\int_0^\infty e^{-zt} f(t)\, dt = \frac{z}{\sqrt{z^2 - 1}}$$

7.4. The Inversion Theorem. Let $g(t)$ satisfy the requirements which enable one to write

$$g(w) = \frac{1}{2\pi} \int_{-\infty}^{\infty} dv \int_{-\infty}^{\infty} g(t) \cos v(w - t)\, dt \tag{7.29}$$

(see Sec. 6.16). We assume that $\int_0^\infty f(w)\, dw$ converges absolutely, and choose $g(t) = e^{-xt}f(t)$ for $t \geqq 0$, $g(0) = 0$ for $t < 0$. Since

$$0 = \frac{1}{2\pi} \int_{-\infty}^{\infty} dv \int_{-\infty}^{\infty} g(t) \sin v(w - t)\, dt$$

we have

$$\begin{aligned} g(w) &= \frac{1}{2\pi} \int_{-\infty}^{\infty} e^{ivw}\, dv \int_{-\infty}^{\infty} g(t) e^{-ivt}\, dt \\ e^{-xw} f(w) &= \frac{1}{2\pi} \int_{-\infty}^{\infty} e^{ivw}\, dv \int_0^{\infty} e^{-xt} f(t) e^{-ivt}\, dt \end{aligned} \tag{7.30}$$

Now assume

$$F(z) = \int_0^\infty e^{-zt} f(t)\, dt$$

exists for Rl $z \geqq x > 0$. Then

$$\int_{x-iy}^{x+iy} F(z)e^{wz}\, dz = \int_{x-iy}^{x+iy} e^{wz}\, dz \int_0^\infty e^{-zt} f(t)\, dt$$
$$= ie^{wx} \int_{-y}^{y} e^{iwv}\, dv \int_0^\infty e^{-xt} f(t) e^{-ivt}\, dt \qquad (7.31)$$

on letting $z = x + iv$, v a new variable of integration. On letting y become infinite and comparing (7.31) with (7.30) we see that

$$f(w) = \frac{1}{2\pi i} \int_{x-i\infty}^{x+i\infty} F(z)e^{wz}\, dz \qquad (7.32)$$

If $F(z)$ is the Laplace transform of $f(w)$, (7.32) enables one to find $f(w)$ in terms of $F(z)$. This is the Laplace-transform inversion theorem. Equation (7.32) can often be evaluated by the calculus of residues.

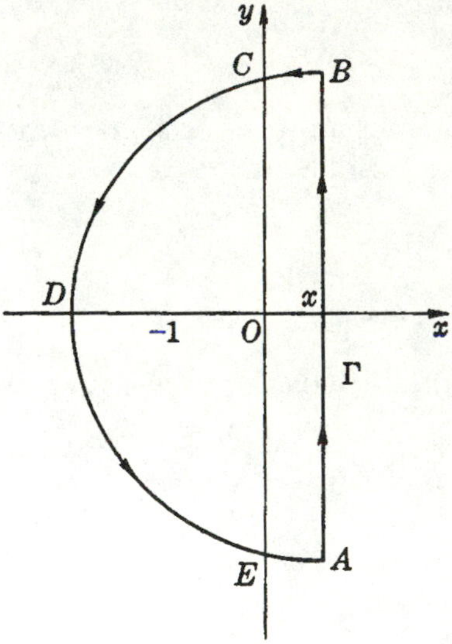

FIG. 7.1

Example 7.6. Let $F(z) = 1/(z+1)$. Then from (7.32)

$$f(w) = \frac{1}{2\pi i} \int_{x-i\infty}^{x+i\infty} \frac{e^{wz}}{z+1}\, dz$$

with $x > 0$. We consider $\oint \frac{e^{wz}}{z+1}\, dz$ around the path given in Fig. 7.1. Let the reader show that, as the radius of the semicircle becomes infinite, the integrals of $e^{wz}/(z+1)$ tend to zero along BC, CDE, EA. The residue of $e^{wz}/(z+1)$ at $z = -1$ is e^{-w}. Thus

$$f(w) = \frac{1}{2\pi i} [2\pi i e^{-w}] = e^{-w}$$

7.5. The Calculus of Variations. The calculus of variations owes its beginning to a problem proposed by Johann Bernoulli near the completion of the seventeenth century. Suppose two points to be fixed in a vertical plane. What curve joining these two points will be such that a particle sliding (without friction) down this curve under the influence of gravity will go from the upper to the lower point in a minimum of time? This is the problem of the brachistochrone (shortest time). A problem of a similar nature is the following: What curve joining two fixed points is such that its rotation about a fixed line will yield a minimum surface of revolution? This is the soap-film problem. A third problem asks the following question: What curve lying on a sphere and joining two fixed points of the sphere is such that the distance along the curve from one point to the other is a minimum? This is the problem of geodesics.

Let us obtain a mathematical formulation of these problems.

1. *Brachistochrone.* Let the particle P move along any curve given by $y = y(x)$ (see Fig. 7.2). The speed of the particle is given by $v^2 = 2gy$ or $\frac{ds}{dt} = \sqrt{2g}\, y^{\frac{1}{2}}$. Hence

$$dt = \frac{1}{\sqrt{2g}} \frac{ds}{y^{\frac{1}{2}}} = \frac{1}{\sqrt{2g}} \frac{\sqrt{dx^2 + dy^2}}{y^{\frac{1}{2}}}$$
$$= \frac{1}{\sqrt{2g}} \sqrt{\frac{1 + (y')^2}{y}}\, dx$$

The total time of descent is given by

$$T = \frac{1}{\sqrt{2g}} \int_0^{x_0} \sqrt{\frac{1 + (y')^2}{y}}\, dx \tag{7.33}$$

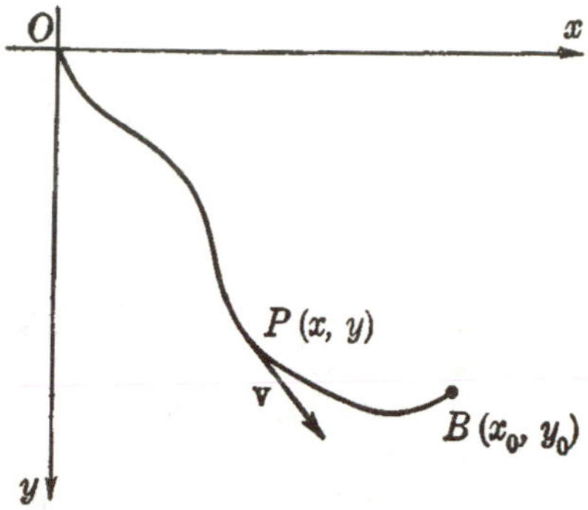

FIG. 7.2

To solve the brachistochrone problem, one must find the function $y(x)$ which makes (7.33) a minimum.

2. *Minimum Surface of Revolution.* If the curve $y(x)$ lying above the x axis is rotated about the x axis, the surface of revolution generated in this manner is

$$S = 2\pi \int y\, ds = 2\pi \int_{x_1}^{x_2} y \sqrt{1 + (y')^2}\, dx \tag{7.34}$$

To solve the soap-film problem, one must find $y(x)$ such that S of (7.34) is a minimum.

3. *Geodesics of a Sphere.* In a Euclidean space we have

$$ds^2 = dx^2 + dy^2 + dz^2$$

For a sphere $x = r \sin\theta \cos\varphi$, $y = r \sin\theta \sin\varphi$, $z = r\cos\theta$, so that

$$ds^2 = r^2(d\theta^2 + \sin^2\theta\, d\varphi^2) = r^2 \left[1 + \sin^2\theta \left(\frac{d\varphi}{d\theta}\right)^2\right] d\theta^2.$$

If $\varphi = \varphi(\theta)$ is any curve on the sphere, the distance between two points of the sphere joined by $\varphi = \varphi(\theta)$ is given by

$$L = r \int_{\theta_1}^{\theta_2} \sqrt{1 + \sin^2\theta \left(\frac{d\varphi}{d\theta}\right)^2}\, d\theta \tag{7.35}$$

To find the geodesics of a sphere, one attempts to find $\varphi = \varphi(\theta)$ such that L of (7.35) is a minimum.

Formulas (7.33) to (7.35) are special cases of a more general case. Let $f(x, y, y')$ be a function of the three variables x, y, y'. We wish to determine $y = y(x)$, and hence $y' = y'(x)$, such that

$$I = \int_{x_1}^{x_2} f(x, y, y')\, dx \tag{7.36}$$

will be an extremal (minimum or maximum) subject to the restriction that $y(x_1) = y_1$, $y(x_2) = y_2$. For (7.33), (7.34), (7.35) we have

$$f(x, y, y') = \frac{1}{\sqrt{2g}}\left[\frac{1 + (y')^2}{y}\right]^{\frac{1}{2}}$$

$f(x, y, y') = 2\pi y[1 + (y')^2]^{\frac{1}{2}}$, $f(x, y, y') = r\sqrt{1 + \sin^2 x (y')^2}\, dx$ with $x = \theta$, $y = \varphi$, respectively.

Let us see how a problem in the calculus of variations differs from an extremal problem of the ordinary calculus. In the latter case we are given the function $y = y(x)$. For each real x there corresponds a unique real number y. Thus $y = y(x)$ maps a set of real numbers into another set of real numbers. The relation $y = 1/x$, $0 < x \leqq 1$, maps the interval $0 < x \leqq 1$ into the interval $y \geqq 1$. A simple problem in the ordinary calculus is to find a number x which yields the minimum or maximum value of $y = y(x)$. Now (7.36) may also be looked upon as a mapping. For any function $y(x)$, (7.36) defines a real number I. Thus (7.36) is a mapping of a function space [the space of $y(x)$] into the real-number space. Our problem is to select a member of the function space which yields a minimum or maximum value for I of (7.36). To resolve this question, we reduce the problem to one of the ordinary calculus.

7.6. The Euler-Lagrange Equation. Let us assume that there exists a function $y(x)$ which makes I of (7.36) an extremal. We now consider the class of functions

$$Y(x, \lambda) = y(x) + \lambda\eta(x) \tag{7.37}$$

where $\eta(x)$ is an arbitrary differentiable function such that $\eta(x_1) = 0$, $\eta(x_2) = 0$ and λ is a real parameter. We have $Y(x_1, \lambda) = y(x_1) = y_1$, $Y(x_2, \lambda) = y(x_2) = y_2$. For $\lambda = 0$ we have $Y(x, 0) = y(x)$. As we vary λ and $\eta(x)$, we obtain a family of curves passing through the two given points $P_1(x_1, y_1)$, $P_2(x_2, y_2)$. Moreover $\lambda = 0$ yields the desired curve, $y = y(x)$. The value of I for any member of (7.37) is

$$I(\lambda) = \int_{x_1}^{x_2} f(x, y + \lambda\eta, y' + \lambda\eta')\, dx \tag{7.38}$$

For any fixed $\eta(x)$ we know that $I(\lambda)$ is an extremal for $\lambda = 0$. From the calculus, of necessity, $\left.\dfrac{dI}{d\lambda}\right|_{\lambda=0} = 0$. Assuming continuity of $\dfrac{\partial f}{\partial y}$ and $\dfrac{\partial f}{\partial y'}$ we have

$$\begin{aligned}\left.\frac{dI}{d\lambda}\right|_{\lambda=0} &= \int_{x_1}^{x_2}\left(\frac{\partial f}{\partial y}\eta + \frac{\partial f}{\partial y'}\eta'\right)dx \\ &= \int_{x_1}^{x_2}\frac{\partial f}{\partial y}\eta\,dx + \frac{\partial f}{\partial y'}\eta(x)\Big|_{x_1}^{x_2} - \int_{x_1}^{x_2}\frac{d}{dx}\left(\frac{\partial f}{\partial y'}\right)\eta\,dx \qquad (7.39)\end{aligned}$$

upon integration by parts. Since $\eta(x_1) = \eta(x_2) = 0$, we have

$$0 = \left.\frac{dI}{d\lambda}\right|_{\lambda=0} = \int_{x_1}^{x_2}\left[\frac{\partial f}{\partial y} - \frac{d}{dx}\left(\frac{\partial f}{\partial y'}\right)\right]\eta(x)\,dx \qquad (7.40)$$

It is a simple matter to show that, if $M(x)$ is continuous on $x_1 \leqq x \leqq x_2$, then the vanishing of $\int_{x_1}^{x_2} M(x)\eta(x)\,dx$ for arbitrary $\eta(x)$ implies $M(x) \equiv 0$. We leave this as an exercise for the reader. Hence if $y(x)$ is the required solution, of necessity, $y(x)$ must satisfy the differential equation

$$\frac{d}{dx}\left(\frac{\partial f}{\partial y'}\right) - \frac{\partial f}{\partial y} = 0 \qquad (7.41)$$

Equation (7.41) is the important Euler-Lagrange equation. It can be written as a second-order differential equation in the form

$$\frac{\partial^2 f}{(\partial y')^2}\frac{d^2 y}{dx^2} + \frac{\partial^2 f}{\partial y\,\partial y'}\frac{dy}{dx} + \frac{\partial^2 f}{\partial x\,\partial y'} - \frac{\partial f}{\partial y} = 0 \qquad (7.42)$$

If we differentiate $f - y'\dfrac{\partial f}{\partial y'}$ with respect to x along the curve $y(x)$ satisfying (7.41), we obtain

$$\frac{\partial f}{\partial x} + \frac{\partial f}{\partial y}y' + \frac{\partial f}{\partial y'}y'' - y''\frac{\partial f}{\partial y'} - y'\frac{d}{dx}\left(\frac{\partial f}{\partial y'}\right) \equiv \frac{\partial f}{\partial x}$$

by making use of (7.41). Thus

$$\frac{d}{dx}\left(f - y'\frac{\partial f}{\partial y'}\right) = \frac{\partial f}{\partial x} \qquad (7.43)$$

for an extremal. If f is explicitly independent of y, one has a first integral of (7.41) given by $\dfrac{\partial f}{\partial y'} = \text{constant}$. If f is explicitly independent of x, $\dfrac{\partial f}{\partial x} = 0$ and (7.43) yields the first integral,

$$f - y'\frac{\partial f}{\partial y'} = \text{constant} \qquad (7.44)$$

One may obtain the Euler-Lagrange equation from the following point of view: First let us consider the equation $y(x) = x^3 - 2x$, so that

$$y(x + \lambda\,dx) = (x + \lambda\,dx)^3 - 2(x + \lambda\,dx)$$

Differentiating with respect to λ yields $\frac{\partial y}{\partial \lambda} = 3(x + \lambda\,dx)^2\,dx - 2\,dx$, so that $\left.\frac{\partial y(x + \lambda\,dx)}{\partial y}\right|_{\lambda=0} = (3x^2 - 2)\,dx = dy$, the differential of y. Now we have seen that (7.36) can be looked upon as a mapping of a function space into a real-number space. We call this mapping a functional of y, written $I = I[y]$. The first variation, or differential, of I may be defined as

$$\begin{aligned}\delta I &= \lim_{\lambda\to 0} \left.\frac{I[y + \lambda\,\delta y] - I[y]}{\lambda}\right|_{\lambda=0} \\ &= \left.\frac{\partial I[y + \lambda\,\delta y]}{\partial \lambda}\right|_{\lambda=0}\end{aligned} \tag{7.45}$$

where $\delta y(x)$ is any variation in $y(x)$. Applying (7.45) to (7.36) yields

$$\begin{aligned}I[y + \lambda\,\delta y] &= \int_{x_1}^{x_2} f(x, y + \lambda\,\delta y, y' + \lambda\,\delta y')\,dx \\ \left.\frac{\partial I}{\partial \lambda}\right|_{\lambda=0} &= \int_{x_1}^{x_2} \left(\frac{\partial f}{\partial y}\,\delta y + \frac{\partial f}{\partial y'}\,\delta y'\right) dx\end{aligned} \tag{7.46}$$

Integrating $\int_{x_1}^{x_2} \frac{\partial f}{\partial y'}\,\delta y'\,dx$ by parts yields

$$\delta I = \int_{x_1}^{x_2} \left[\frac{\partial f}{\partial y} - \frac{d}{dx}\left(\frac{\partial f}{\partial y'}\right)\right] \delta y\,dx$$

provided $\delta y(x_1) = \delta y(x_2) = 0$. Equation (7.41) holds, of necessity, if $\delta I = 0$ for arbitrary δy.

In the calculus it is necessary to examine the second derivative of $y(x)$, and, at times, higher derivatives, in order to determine the type of extremal (maximum, minimum, point of inflection) encountered at the point $x = x_1$ for which $\frac{dy}{dx} = 0$. Similarly, in the calculus of variations it is necessary to examine the second variation in I to determine what type of extremal is obtained from the solution of the Euler-Lagrange equation. Let the reader show that the second variation of I can be written as

$$\delta^2 I = \int_{x_1}^{x_2} \left[\frac{\partial^2 f}{\partial y^2}(\delta y)^2 + 2\,\frac{\partial^2 f}{\partial y\,\partial y'}\,\delta y\,\delta y' + \frac{\partial^2 f}{(\partial y')^2}(\delta y')^2\right] dx \tag{7.47}$$

Example 7.7. Equations (7.33) and (7.34) are special cases of

$$I = \int_{x_1}^{x_2} g(y)\,\sqrt{1 + (y')^2}\,dx$$

We have $f = g(y)\,[1 + (y')^2]^{\frac{1}{2}}$, $\frac{\partial f}{\partial x} = 0$, so that a first integral is obtained from (7.44). Let the reader show that (7.44) yields

$$\frac{g(y)}{\sqrt{1 + (y')^2}} = \text{constant} = a$$

From $\tan\theta = \frac{dy}{dx}$ we have $\cos\theta = 1/\sqrt{1 + (y')^2}$ so that

$$g(y) = a\sec\theta \tag{7.48}$$

Moreover $g'(y)\,dy = a\sec\theta\tan\theta\,d\theta$ from (7.48), and $dx = \cot\theta\,dy$ so that

$$dx = \frac{a\sec\theta\,d\theta}{g'(y)}$$

yielding

$$x = b + a\int_0^\theta \frac{\sec\theta\,d\theta}{g'(y)} \tag{7.49}$$

Given $g(y)$, one solves (7.48) for y as a function of θ. Then integration of (7.49) yields x as a function of θ. This parametric representation of x and y as functions of θ yields the required curve which extremalizes I.

In the brachistochrone problem $g(y) = y^{-\frac{1}{2}}$, so that $y = c\cos^2\theta = (c/2)(1 + \cos 2\theta)$, $c = 1/a^2$, from (7.48). Thus $g'(y) = -\frac{1}{2}y^{-\frac{3}{2}} = (-2c^{\frac{3}{2}}\cos^3\theta)^{-1}$, and (7.49) yields

$$\begin{aligned} x(\theta) &= b - 2c\int_0^\theta \cos^2\theta\,d\theta = b - c\int_0^\theta (1 + \cos 2\theta)\,d\theta \\ x(\theta) &= b - \frac{c}{2}(2\theta + \sin 2\theta) \\ y(\theta) &= \frac{c}{2}(1 + \cos 2\theta) \end{aligned} \tag{7.50}$$

Equation (7.50) is the parametric equation of a cycloid.

Example 7.8. *Variable-end-point Problem.* We are given the fixed curve Γ, $y = \varphi(x)$, and the functional

$$I = \int_{x_1}^{x_2 = b} F(x, y, y')\,dx$$

We wish to find the curve $y = y(x)$ joining the fixed point $A(x_1, y_1)$ and $B(b, \varphi(b))$, where B is a point on Γ such that I is an extremal. The coordinate $x = b$ is unknown. If we consider the curve $y(x) + \delta y(x)$, we have

$$I[y + \delta y] = \int_{x_1}^{b+\delta x} F(x, y + \delta y, y' + \delta y')\,dx$$

The upper limit has changed since the end point of $y(x) + \delta y(x)$ is constrained to lie on $y = \varphi(x)$. Let the reader show that

$$I[y + \delta y] - I[y] = \int_{x_1}^{b} [F(x, y + \delta y, y' + \delta y') - F(x, y, y')]\,dx + \int_b^{b+\delta x} F(x, y + \delta y, y' + \delta y')\,dx \tag{7.51}$$

From (7.51) it is logical to define δI by

$$\delta I = \int_{x_1}^{b} \left(\frac{\partial F}{\partial y} \delta y + \frac{\partial F}{\partial y'} \delta y' \right) dx + F(x, y, y') \Big|_{x=b} \delta x \tag{7.52}$$

Now $\delta y = 0$ at $x = a$. We must compute δy at $x = b$. We have

$$y(b + \delta x) + \delta y(b + \delta x) = \varphi(b + \delta x)$$

$y(b) = \varphi(b)$. Thus $y(b + \delta x) - y(b) + \delta y(b + \delta x) = \varphi(b + \delta x) - \varphi(b)$. Applying the law of the mean one easily shows that

$$\delta y \Big|_{x=b} = [\varphi'(b) - y'(b)]\, \delta x$$

Integrating the second part of the integral of (7.52) by parts yields

$$\delta I = \int_{x_1}^{b} \left[\frac{\partial F}{\partial y} - \frac{d}{dx}\left(\frac{\partial F}{\partial y'} \right) \right] \delta y\, dx + \left[F + \frac{\partial F}{\partial y'} (\varphi' - y') \right]_{x=b} \delta x$$

If $\delta I = 0$ for arbitrary δy, of necessity

$$\left[F + (\varphi' - y') \frac{\partial F}{\partial y'} \right]_{x=b} = 0 \tag{7.53}$$

along with the Euler-Lagrange equation.

Let us apply (7.53) to the problem of the minimum surface of revolution with $\varphi(x)$ arbitrary. We have $F(x, y, y') = y[1 + (y')^2]^{\frac{1}{2}}$, $\dfrac{\partial F}{\partial y'} = yy'[1 + (y')^2]^{-\frac{1}{2}}$, so that (7.53) becomes

$$\{y[1 + (y')^2]^{\frac{1}{2}} + yy'[1 + (y')^2]^{-\frac{1}{2}}(\varphi' - y')\}_{x=b} = 0$$

or $y'\varphi' = -1$ at $x = b$. Hence, at their point of intersection, $y(x)$ and $\varphi(x)$ intersect at right angles.

Problems

1. Show that the solution of the soap-film problem is $y = a \cosh \dfrac{x - b}{a}$.

2. Show that the geodesics on a sphere are arcs of great circles.

3. Derive (7.47).

4. Show that the Euler-Lagrange equation for the functional

$$I[y] = \int_{x_1}^{x_2} f(x, y, y', y'')\, dx$$

is

$$\frac{d^2}{dx^2}\left(\frac{\partial f}{\partial y''} \right) - \frac{d}{dx}\left(\frac{\partial f}{\partial y'} \right) + \frac{\partial f}{\partial y} = 0$$

5. Show that there are no extremals or stationary values for the functional

$$I[y] = \int_{x_1}^{x_2} y\, dx$$

6. Consider the functional

$$I[x_1, x_2, \ldots, x_n] = \int_{t_1}^{t_2} f(x_1, x_2, \ldots, x_n, \dot{x}_1, \dot{x}_2, \ldots, \dot{x}_n, t)\, dt$$

Show that the Euler-Lagrange equations are

$$\frac{d}{dt}\left(\frac{\partial f}{\partial \dot{x}_i}\right) - \frac{\partial f}{\partial x_i} = 0 \qquad i = 1, 2, \ldots, n \tag{7.54}$$

Explain why it is necessary that

$$f(x_1, x_2, \ldots, x_n, \lambda\dot{x}_1, \lambda\dot{x}_2, \ldots, \lambda\dot{x}_n, t) = \lambda f(x_1, x_2, \ldots, x_n, \dot{x}_1, \dot{x}_2, \ldots, \dot{x}_n, t)$$

7. Apply (7.54) of Prob. 6 to

$$I[x, y] = \int_{t_1}^{t_2} \left[\frac{m}{2}(\dot{x}^2 + \dot{y}^2) - mgy\right] dt$$

8. In Chap. 3 we saw that Lagrange's equations of motion could be written as

$$\frac{d}{dt}\left(\frac{\partial L}{\partial \dot{q}i}\right) - \frac{\partial L}{\partial qi} = 0$$

Show that this implies that the functional $I = \int_{t_1}^{t_2} L\,dt$ is an extremal for Newtonian motion. For a conservative system, $T + V = \text{constant} = h$. Show that Newtonian motion for a conservative system is such that $\int_{t_1}^{t_2} 2T\,dt$ is an extremal subject to the condition $T + V = \text{constant}$.

9. Consider the functional $I[z] = \iint\limits_S F\left(x, y, z, \frac{\partial z}{\partial x}, \frac{\partial z}{\partial y}\right) dy\,dx$, the region of integration, S, having the simple closed curve C as its boundary. Show that, for $I[z]$ to be an extremal, $z = z(x, y)$ must satisfy

$$\frac{\partial F}{\partial z} - \frac{\partial^2 F}{\partial x\,\partial p} - \frac{\partial^2 F}{\partial y\,\partial q} = 0$$

with

$$p = \frac{\partial z}{\partial x}, q = \frac{\partial z}{\partial y} \text{ in } F$$

7.7. The Problem of Constraints. In the ordinary calculus one solved problems of the following type: Given the function of two variables, $z = f(x, y)$, at what point $P(x, y)$ is z an extremal subject to the constraint $\varphi(x, y) = \text{constant}$? We know that if $f(x, y)$ is continuous on the closed and bounded curve C, $\varphi(x, y) = \text{constant}$, there will exist points on C at which z takes on minimum and maximum values. One way to solve this problem is to solve for y from $\varphi(x, y) = \text{constant}$, obtaining $y = \psi(x)$, and then to extremalize $z = f(x, \psi(x))$, a one-dimensional problem. Another method is due to Lagrange. For an extremal, $\frac{dz}{dx} = 0$, so that $\frac{\partial f}{\partial x} + \frac{\partial f}{\partial y}\frac{d\psi}{dx} = 0$. This same equation can be obtained as follows: Consider $w = f(x, y) + \lambda\varphi(x, y)$, λ a parameter. Compute $\frac{\partial w}{\partial x}$ and $\frac{\partial w}{\partial y}$ as if x and y were independent variables. Setting these two partial derivatives equal to zero yields

$$\frac{\partial w}{\partial x} = \frac{\partial f}{\partial x} + \lambda \frac{\partial \varphi}{\partial x} = 0$$

$$\frac{\partial w}{\partial y} = \frac{\partial f}{\partial y} + \lambda \frac{\partial \varphi}{\partial y} = 0$$

Eliminating λ yields

$$\frac{\partial f}{\partial x} + \frac{\partial f}{\partial y}\left(\frac{-\partial \varphi/\partial x}{\partial \varphi/\partial y}\right) = 0$$

Along the curve $\varphi(x, y) =$ constant, we have $\frac{\partial \varphi}{\partial x} + \frac{\partial \varphi}{\partial y}\frac{dy}{dx} = 0$, so that $\frac{dy}{dx} = \frac{d\psi}{dx} = -\frac{\partial \varphi/\partial x}{\partial \varphi/\partial y}$, and *Lagrange's method of λ multipliers* yields $\frac{\partial f}{\partial x} + \frac{\partial f}{\partial y}\frac{d\psi}{dx} = 0$, the required equation for an extremal.

The simplest constraint problem in the calculus of variations is the following: We wish to extremalize the functional

$$I[y] = \int_a^b f(x, y, y')\, dx \tag{7.55}$$

subject to the constraint

$$\int_a^b \varphi(x, y, y')\, dx = \text{constant} \tag{7.56}$$

As in Sec 7.6 we let $Y(x, \lambda) = y(x) + \lambda_1\eta_1(x) + \lambda_2\eta_2(x)$,

$$\eta_1(a) = \eta_1(b) = \eta_2(a) = \eta_2(b) = 0$$

and we desire to extremalize

$$I(\lambda_1, \lambda_2) = \int_a^b f(x, y + \lambda_1\eta_1 + \lambda_2\eta_2, y' + \lambda_1\eta_1' + \lambda_2\eta_2')\, dx$$

subject to the condition

$$J(\lambda_1, \lambda_2) = \int_a^b \varphi(x, y + \lambda_1\eta_1 + \lambda_2\eta_2, y' + \lambda_1\eta_1' + \lambda_2\eta_2')\, dx = \text{constant}$$

$I(\lambda_1, \lambda_2)$ is to be an extremal at $\lambda_1 = \lambda_2 = 0$.

Let the reader show that the Euler-Lagrange equation becomes

$$\frac{d}{dx}\left[\frac{\partial}{\partial y'}(f + \lambda\varphi)\right] - \frac{\partial}{\partial y}(f + \lambda\varphi) = 0 \tag{7.57}$$

The solution $y(x, \lambda)$ of (7.57) is substituted into (7.56) to eliminate λ.

Example 7.9. Let us find $y(x)$ such that the functional $I[y] = \int_0^a y\, dx$ is an extremal subject to the constraint $\int_0^a \sqrt{1 + (y')^2}\, dx = \text{constant} = b$. Since the constraint states that the length of arc of the curve $y(x)$ be a constant, this type of problem is called an isoperimetric problem. We have $f + \lambda\varphi = y + \lambda\sqrt{1 + (y')^2}$,

$$\frac{\partial}{\partial y'}(f + \lambda\varphi) = \frac{\lambda y'}{\sqrt{1 + (y')^2}}, \qquad \frac{\partial}{\partial y}(f + \lambda\varphi) = 1$$

so that (7.57) yields

$$\frac{d}{dx}\left[\frac{y'}{\sqrt{1 + (y')^2}}\right] = \frac{1}{\lambda}$$

or $y''/[1 + (y')^2]^{\frac{3}{2}} = 1/\lambda$. We recall that $y''/[1 + (y')^2]^{\frac{3}{2}}$ is the curvature, so that $y(x)$ has constant curvature, and hence must be an arc of a circle. The radius of the circle is λ, which can be obtained from $\int_0^a \sqrt{1 + (y')^2}\, dx = b$.

Example 7.10. If the Riemannian metric (see Sec. 3.6) is extremalized subject to the constraint $\int_{s_0}^{s_1} \varphi_\alpha(x^1, x^2, \ldots, x^n) \frac{dx^\alpha}{ds}\, ds$, where φ_α, $\alpha = 1, 2, 3, 4$, is the electromagnetic vector potential, one obtains the motion of a charged particle in a gravitational and electromagnetic field,

$$\frac{d^2x^i}{ds^2} + \Gamma^i_{jk}\frac{dx^j}{ds}\frac{dx^k}{ds} + \frac{e}{m}F^i_\alpha \frac{dx^\alpha}{ds} = 0, \qquad \lambda = \frac{e}{m}$$

Problems

1. Find $y(x)$ which extremalizes the functional $I[y] = \int_a^b (y')^2\, dx$ subject to the constraint $\int_a^b xy\, dx =$ constant.

2. Find the curve of constant length joining two fixed points with the lowest center of mass.

3. The area underneath the curve $y = f(x)$ from $x = a$ to $x = b$ is rotated about the x axis. Find the maximum volume obtained for $\int_a^b y\, dx =$ constant $= c$.

4. Derive (7.57).

REFERENCES

Bliss, G. A.: "Calculus of Variations and Multiple Integrals," University of Chicago Press, Chicago, 1938.

Carslaw, H. S.: "Conduction of Heat in Solids," Oxford University Press, New York, 1947.

Churchill, R. V.: "Modern Operational Mathematics in Engineering," McGraw-Hill Book Company, Inc., New York, 1944.

Doetsche, G.: "Handbuch der Laplace Transformation," Birkhauser, Basel, 1950.

Jaeger, J. C.: "An Introduction to the Laplace Transformation," Methuen & Co., Ltd., London, 1949.

Weinstock, R.: "Calculus of Variations," McGraw-Hill Book Company, Inc., New York, 1952.

CHAPTER 8

GROUP THEORY AND ALGEBRAIC EQUATIONS

8.1. Introduction. The study of groups owes its beginning in an attempt to solve algebraic equations of degree higher than 4. The linear equation $ax + b = 0$, $a \neq 0$, has for its solution $x = -b/a$. The solution of the quadratic equation $ax^2 + bx + c = 0$, $a \neq 0$, is known to be $x_1 = (1/2a)(-b + \sqrt{b^2 - 4ac})$, $x_2 = (1/2a)(-b - \sqrt{b^2 - 4ac})$. We note that x_1 and x_2 are written in terms of a finite number of operations involving addition, subtraction, multiplication, division, and root extractions. The operations are performed on the coefficients of the quadratic equation. We say that the quadratic equation is solvable by radicals. The cubic and quartic equations are solvable also by radicals. Lagrange attempted to extend this result to algebraic equations of degree higher than 4. He was unsuccessful, but his work laid the foundation which enabled Galois and Abel in the early part of the nineteenth century to grapple successfully with this problem. In general, an algebraic equation of degree higher than 4 cannot be solved by radicals. The equation $x^5 - 1 = 0$ can be solved by radicals, however. It remained for Cauchy to systematically begin the study of group theory proper. The theory of groups plays an outstanding role in the unification of mathematics. Its applications in mathematics are widespread, and, moreover, it has served an important role in the development of the modern quantum theory of physics.

8.2. Definition of a Group. We consider first some elementary examples of groups. Let us consider the set of all rationals, excluding zero, subject to the rule of multiplication. We note that the product of two rationals is again a rational. If a, b, c are rationals, then the associative law, $(ab)c = a(bc)$, holds. The unique rational 1 has the property that $1 \cdot a = a \cdot 1 = a$, for all rationals a. Finally, for any rational a, there exists a unique rational, $1/a = a^{-1}$, such that $aa^{-1} = a^{-1}a = 1$. Let the reader show that the four elements $(1, -1, i, -i)$ possess these same properties under the operation of multiplication. Let us consider the 90°, 180°, 270°, and 360° = 0° rotations in a plane about a fixed point. Let us denote these rotations by A_1, A_2, A_3, and $A_4 = E$, respectively. By A_2A_1 we mean a 90° rotation followed by a 180° rotation, etc. We note that $A_iA_j = A_k$. A 270° rotation followed by a 180° rotation is equivalent to a 450° = 90° rotation. Moreover $A_iE = EA_i = A_i$ for $i = 1, 2, 3, 4$. E is the identity element in the sense that the rotation E leaves a body invariant. We also note that $A_1A_3 = A_3A_1 = E$, $A_2A_2 = E$, $EE = E$, so that every element has a unique inverse. From the fact that

$$e^{i\theta_1}e^{i\theta_2} = e^{i(\theta_1+\theta_2)},\ e^{0i} = 1,\ e^{(\pi/2)i} = i,\ e^{\pi i} = -1,\ e^{(3\pi/2)i} = -i$$

let the reader deduce a correspondence between the rotations discussed above and the four elements $(1, -1, i, -i)$ under multiplication. The examples above lead us to the formal definition of a group. A discussion of sets can be found in Sec. 10.7.

Let G consist of a set of objects $A, B, C, \ldots$. An operator, $\otimes$, is associated with every pair of elements of G, $\otimes(A, B) = A \otimes B$. For convenience we call the operator $\otimes$ multiplication and write $A \otimes B = AB$. The set G is said to be a group relative to the operator $\otimes$ if:

I. For every A and B of G, $AB = C$ implies C is a member of G. This is the closure property under $\otimes$.

II. For all A, B, and C of G,

$$(AB)C = A(BC)$$

This is the associative law.

III. There exists a unique element, E, of G, such that

$$AE = EA = A$$

for all A in G. E is called the identity, or unit, element.

IV. For every A of G there exists a unique element, written A^{-1}, such that

$$AA^{-1} = A^{-1}A = E$$

We call A^{-1} the inverse of A. It follows that A is the inverse of A^{-1}.

One can replace III and IV by:

III′. For every A and B of G there exist unique X and Y of G such that $AX = B$, $YA = B$.

By choosing $B = A$ we see that every element A has a unique right and left identity, from (III′). Thus $AE_1 = A$, $E_2A = A$. We show now that $E_1 = E_2$. Now $A(E_2A) = AA = (AE_2)A$, so that $A = AE_2$ from (III′). Hence $E_2 = E_1$ from (III′). We show next that the identity element for A is the same as that for B for all A and B. We have $AE_A = E_AA = A$, $BE_B = E_BB = B$. Now $B(E_BA) = (BE_B)A = BA$, so that, from (III′), $E_BA = A$, which implies $E_B = E_A$. Let the reader show that (III′) implies (IV).

Example 8.1. Let x and y be elements of a set such that $x^2 = e$, $y^3 = e$, $yxy = x$. We consider the set of elements $(x, y, y^2 = y \cdot y, xy, xy^2, e)$, e the unit element. Let us construct a multiplication table for these elements. We write $x \cdot x = x^2$, $y \cdot y = y^2$, $y \cdot y \cdot y = y^3$, etc. From $x^2 = e$ we note that x is its own inverse, and from $y^3 = e$ we note that y^2 is the inverse of y. If we desire to compute $(xy^2)x$, we note that

$$(xy^2)x = (xy^2)(yxy) = xy^3xy = xexy = xxy = x^2y = ey = y$$

(see Table 8.1).

We note that each row and column of Table 8.1 contains the six elements of our set with no repetitions. If $x \neq e$, $y \neq e$, let the reader show that the six elements are distinct, and hence form a group.

TABLE 8.1

	e	x	y	y^2	xy	xy^2
e	e	x	y	y^2	xy	xy^2
x	x	e	xy	xy^2	y	y^2
y	y	xy^2	y^2	e	x	xy
y^2	y^2	xy	e	y	xy^2	x
xy	xy	y^2	xy^2	x	e	y
xy^2	xy^2	y	x	xy	y^2	e

Problems

1. Verify Table 8.1.

2. Consider the set of square matrices, $\|a_{ij}\|$, $i, j = 1, 2, \ldots, n$, such that $|a_{ij}| \neq 0$. Show that this set of matrices is a group under multiplication.

3. An Abelian group is one for which $AB = BA$ for all A and B of G. Is the group of Example 8.1 an Abelian group? Show that the group of Prob. 2 is non-Abelian.

4. A finite group is one containing a finite number of distinct elements. Show that we can replace (III) and (IV) by (III″): $AB = AC$ implies $B = C$, and $BA = CA$ implies $B = C$, for finite groups.

5. We define $a \cdot a = a^2$, $a \cdot a \cdot a = a^3$, etc. A group is said to be cyclic if a single element generates every element of the group, that is, an element a exists such that, if x is any element of the group, then $x = a^n$ for some positive integer n. Give an example of a cyclic group.

6. Show that $A = (A^{-1})^{-1}$.

7. If A and B are elements of a group G, show that $(AB)^{-1} = B^{-1}A^{-1}$. Generalize this result.

8. Show that the set of rational integers (positive and negative integers including zero) form a group relative to the operation of addition. Do the set of rational integers form a group relative to the operation of multiplication?

8.3. Finite Groups. A finite group is a group consisting of a finite number of distinct elements. We deduce now some theorems concerning finite groups and illustrate each theorem with an example.

THEOREM 8.1. *The order of a subgroup is a divisor of the order of the complete group.* The order of a group is the number of distinct elements of the group. H is a subgroup of G if H is a group and if, furthermore, every element of H belongs to G. If at least one member of G is not a member of H, we say that H is a proper subgroup of G. The proof of Theorem 8.1 is as follows: Let H consist of the elements $E, A, B, \ldots, F$. If $H \equiv G$, there is no problem. Assume H a proper subgroup of G, and let X be any element of G not in H. Construct the set H_1 of elements $XE, XA, XB, \ldots, XF$. Let the reader show that these elements are distinct. Moreover, if $XA = B$, then $X = BA^{-1}$ is a member of H since H is a group. But X is not a member of H so that $XA \neq B$. Hence every element of H_1 is distinct from every element of H. If the members of H and H_1 exhaust G, then $g = 2h$, where g is the order of G and h is the order of H. If this is not the case, let Y be a member of G not in H or H_1. We now construct the set H_2 consisting of YE, $YA, \ldots, YF$. One easily shows that the members of H_2 are distinct from each other and are distinct from the elements of H and H_1. If H, H_1, H_2 exhaust G, then $g = 3h$. If not, we continue in the same manner. Eventually we must exhaust G since G has a finite number of elements. Thus $g = nh$, and h divides g.

Example 8.2. The group of Table 8.1 consists of six elements. A subgroup of this group is $H(x, x^2 = e)$. The order of H is 2, and $6 = 2 \cdot 3$. Another proper subgroup of G is $K(y, y^2, y^3 = e)$, $6 = 3 \cdot 2$. Is it possible for G to have a subgroup of order 4?

THEOREM 8.2. *Every subgroup of a cyclic group is a cyclic group.* The definition of a cyclic group is given in Prob. 5, Sec. 8.2. Let G consist of $A, A^2, \ldots, A^g = E$. Let H be a proper subgroup of G with elements

$$A^b, A^{b_1}, \ldots, A^{2b}, \ldots, A^r = E \qquad b < b_1 < \cdots < r$$

Since $b_1 > b$, we have $b_1 = qb + s$, $0 \leqq s < b$. Then

$$A^{b_1} = A^{qb+s} = A^{qb}A^s$$

and $A^{b_1 - qb} = A^s$. If $s \neq 0$, then A^s is a member of H, a contradiction, since b was assumed to be the smallest exponent of A such that A^b is in H. Hence $s = 0$, and $b_1 = qb = 2b$, since $A^bA^b = A^{2b}$ is in H. The only elements of H are of the form A^{mb}, so that H is cyclic.

Example 8.3. Let G be a cyclic group of order 8, so that

$$(a, a^2, a^3, a^4, a^5, a^6, a^7, a^8 = e)$$

are the elements of G. Consider the subgroup $H(a^2, a^4, a^6, a^8 = e)$. We note that H is cyclic since $a^4 = (a^2)^2$, $a^6 = (a^2)^3$, $a^8 = (a^2)^4$.

THEOREM 8.3. A criterion for a subgroup is the following: Let G be a finite group, and let S be a subset of G such that the product of any two elements of S is again an element of S. Then S is a subgroup of G.

Certainly the closure and associative properties hold for S. Now let A be any element of S. Then $A, A^2, \ldots, A^r, \ldots$ belong to S. There can exist only a finite number of distinct elements of the type A^k. Thus $A^n = A^m$ for $n > m$, and $A^{n-m} = E$ belongs to S. Moreover $AA^{n-m-1} = E$, so that $A^{n-m-1} = A^{-1}$ belongs to S. Q.E.D.

Example 8.4. Let us consider the permutations of the integers (1, 2, 3). We obtain the six permutations (1, 2, 3), (1, 3, 2), (2, 1, 3), (2, 3, 1), (3, 1, 2), (3, 2, 1). We can consider the particular permutation (2, 3, 1) as being obtained from a substitution of the integers 1, 2, 3, in the sense that $1 \to 2$, $2 \to 3$, $3 \to 1$, written $\begin{pmatrix}123\\231\end{pmatrix}$. In this way we obtain the six elements

$$S_1 = \begin{pmatrix}123\\123\end{pmatrix} \quad S_2 = \begin{pmatrix}123\\132\end{pmatrix} \quad S_3 = \begin{pmatrix}123\\213\end{pmatrix} \quad S_4 = \begin{pmatrix}123\\231\end{pmatrix} \quad S_5 = \begin{pmatrix}123\\312\end{pmatrix} \quad S_6 = \begin{pmatrix}123\\321\end{pmatrix}$$

If we consider a triangle with vertices labeled 1, 2, 3, respectively, then S_6 states that we interchange the labels 1 and 3 and leave label 2 invariant. Let us consider any function of three variables, $f(x_1, x_2, x_3)$. The operation of S_2 on f yields

$$S_2 f(x_1, x_2, x_3) = f(x_1, x_3, x_2)$$

If we follow this by the operation S_5, we obtain

$$S_5 S_2 f(x_1, x_2, x_3) = S_5 f(x_1, x_3, x_2) = f(x_3, x_2, x_1)$$

since S_5 permutes 1 into 3, 2 into 1, and 3 into 2. Thus

$$S_5 S_2 f(x_1, x_2, x_3) = S_6 f(x_1, x_2, x_3)$$

and it is natural to define $S_5 S_2 = S_6$, written $\begin{pmatrix}123\\312\end{pmatrix}\begin{pmatrix}123\\132\end{pmatrix} = \begin{pmatrix}123\\321\end{pmatrix}$. We can look upon $\begin{pmatrix}123\\312\end{pmatrix}\begin{pmatrix}123\\132\end{pmatrix}$ as follows: Starting with the right-hand side, we see that 1 goes into 1; then, moving to the left, we see that 1 goes into 3. The final result is the permutation of 1 into 3. Again, on the right-hand side, 2 goes into 3, and, moving to the left, 3 goes into 2, so that the end result is to leave 2 invariant. $3 \to 2$ followed by $2 \to 1$ yields $3 \to 1$. The product yields $\begin{pmatrix}123\\321\end{pmatrix} = S_6$. It follows that S_1 plays the role of the identity element of the group, and we leave it as an exercise to show that the elements S_i, $i = 1, 2, \ldots, 6$, do, indeed, form a group relative to multiplication defined above. The order of the group is $3! = 6$. Let the reader obtain a generalization for the permutation group of order $n!$, often called the *symmetric* group.

We consider now the function

$$f(x_1, x_2, x_3) = (x_1 - x_2)(x_2 - x_3)(x_3 - x_1)$$

We note that $S_1 f = f$, $S_2 f = -f$, $S_3 f = -f$, $S_4 f = f$, $S_5 f = f$, $S_6 f = -f$, so that S_1, S_4, and S_5 leave f invariant. These elements are called the even permutations. It is a simple matter to show that the product of two even permutations is again an even permutation. Hence, from Theorem 8.3, the set (S_1, S_4, S_5) is a subgroup of the symmetric group of order 3!. Do (S_2, S_3, S_6) form a group? These are the odd permutations.

Any subgroup of a symmetric group is called a *regular permutation* group.

Problems

1. Show that the elements of Example 8.4 form a group. Construct the multiplication table for this group. Is this group Abelian? Construct all the proper subgroups.

2. Show that the product of two even permutations is an even permutation. Consider the cases of the product of an even with an odd permutation and the product of two odd permutations.

3. We can write $S_4 = \begin{pmatrix}123\\231\end{pmatrix} \equiv (123)$ in the sense that $1 \to 2$, $2 \to 3$, $3 \to 1$. S_6 can be written $S_6 = (13)(2)$. Do the same for S_1, S_2, S_3, S_5.

4. Show that the permutation group of order $n!$ contains a subgroup of order $n!/2$.

5. C is called the *commutator* of two elements A and B of a group if $C = (AB)^{-1}BA$. For any element X of G show that $X^{-1}CX$ is the commutator of $X^{-1}AX$ and $X^{-1}BX$.

8.4. Isomorphisms. Let us consider two groups G_1 and G_2. If a one-to-one correspondence can be established such that

$$A \leftrightarrow A'$$
$$B \leftrightarrow B'$$

implies

$$AB \leftrightarrow A'B'$$

for all A and B in G, we say that the two groups G_1 and G_2 are isomorphic to each other and write $G_1 \cong G_2$. A', B', . . . belong to G_2.

An isomorphism of two groups implies that the two groups are equivalent in the sense that we are using two different languages to describe the elements of the groups. It is apparent that any theorem obtained from the fact that G_1 is a group will also hold for G_2.

Example 8.5. We consider a cyclic group of order 4 with elements A, A^2, A^3, $A^4 = E$ along with a subgroup of the symmetric group of order 4! Let the reader show that the elements

$$S_1 = \begin{pmatrix}1\,2\,3\,4\\2\,3\,4\,1\end{pmatrix} \qquad S_2 = \begin{pmatrix}1\,2\,3\,4\\3\,4\,1\,2\end{pmatrix} \qquad S_3 = \begin{pmatrix}1\,2\,3\,4\\4\,1\,2\,3\end{pmatrix} \qquad S_4 = \begin{pmatrix}1\,2\,3\,4\\1\,2\,3\,4\end{pmatrix}$$

form a cyclic group with $S_2 = S_1^2$, $S_3 = S_1^3$, $S_4 = E = S_1^4$. It is a simple matter to show that the correspondence

$$S_i \leftrightarrow A^i \qquad i = 1, 2, 3, 4$$

is an isomorphism. We leave this as an exercise for the reader.

An important theorem due to Cayley is stated as follows:

THEOREM 8.4. *Every finite group G is isomorphic to a regular permutation group.* Let the elements of the group be written as $E = A_1$, A_2,

$\ldots, A_n$. Let A_k be any member of G, and consider the elements

$$A_kA_1, A_kA_2, \ldots, A_kA_j, \ldots, A_kA_n$$

Since A_kA_1 belongs to G, it must be equal to an A_r, and by A_{k1} we mean A_r. Similarly $A_kA_2 = A_{k2}$, etc. Let A_k correspond to

$$A_k \leftrightarrow \begin{pmatrix} 1 & 2 & \cdots & n \\ k1 & k2 & \cdots & kn \end{pmatrix} \qquad k1 = r, \text{ etc.} \tag{8.1}$$

for $k = 1, 2, \ldots, n$. We wish to show that (8.1) represents an isomorphism. We have from (8.1)

$$A_j \leftrightarrow \begin{pmatrix} 1 & 2 & \cdots & n \\ j1 & j2 & \cdots & jn \end{pmatrix}$$

$$A_jA_k = A_{jk} \leftrightarrow \begin{pmatrix} 1 & 2 & \cdots & n \\ jk1 & jk2 & \cdots & jkn \end{pmatrix}$$

But $\begin{pmatrix} 1 & 2 & \cdots & n \\ j1 & j2 & \cdots & jn \end{pmatrix}\begin{pmatrix} 1 & 2 & \cdots & n \\ k1 & k2 & \cdots & kn \end{pmatrix} = \begin{pmatrix} 1 & 2 & \cdots & n \\ jk1 & jk2 & \cdots & jkn \end{pmatrix}$, which establishes the isomorphism. The isomorphism of Example 8.5 was established in this fashion.

An *automorphism* is an isomorphism of a group with itself. A simple example of an automorphism is as follows: Let A_i, $i = 1, 2, \ldots, n$ be the elements of a group G, and let X be any element of G. For convenience we choose $X \neq E$. Let us construct the elements $X^{-1}A_iX$, $i = 1, 2, \ldots, n$. If G is Abelian, we have $X^{-1}A_iX = A_iX^{-1}X = A_iE = A_i$. Let us assume that G is non-Abelian. Generally, then, $X^{-1}A_iX \neq A_i$. It is a simple matter to prove, in any case, that $X^{-1}A_iX \neq X^{-1}A_jX$, $i \neq j$, for if $X^{-1}A_iX = X^{-1}A_jX$, then

$$A_i = XX^{-1}A_iXX^{-1} = XX^{-1}A_jXX^{-1} = A_j$$

a contradiction if $i \neq j$. Since the set S of elements $X^{-1}A_iX$, $i = 1, 2, \ldots, n$, contains n distant elements and since every member of S belongs to G, $S \equiv G$. We show now that the correspondence

$$A_i \leftrightarrow X^{-1}A_iX \qquad i = 1, 2, \ldots, n \tag{8.2}$$

is an automorphism. We have $A_j \leftrightarrow X^{-1}A_jX$,

$$A_iA_j \leftrightarrow X^{-1}A_iA_jX = X^{-1}A_iXX^{-1}A_jX = (X^{-1}A_iX)(X^{-1}A_jX)$$

which proves the automorphism of the correspondence (8.2).

Problems

1. If G_1 and G_2 are isomorphic and if G_1 is cyclic, show that G_2 is cyclic.
2. Find the regular permutation group which is isomorphic to the group of Table 8.1.

3. Show that all cyclic groups of order n are isomorphic (see Prob. 1).

4. An automorphism of the type (8.2) of the last paragraph is called an inner automorphism. Show that the only inner automorphism of a cyclic group is the trivial automorphism, $A_i \leftrightarrow A_i$, $i = 1, 2, \ldots, n$. This is called the identity automorphism.

5. Prove that the product of two automorphisms is an automorphism. *Hint:* If $A_i \leftrightarrow A_i'$ is an automorphism and $A_i \leftrightarrow A_i''$ is an automorphism, then under the second automorphism we have $A_i' \leftrightarrow (A_i')''$, so that $A_i \leftrightarrow (A_i')''$ is another correspondence, called the product of the two automorphisms. Show that $A_i \leftrightarrow (A_i')''$ represents an automorphism.

6. Show that the inner automorphisms of a group form a group.

7. Show that the automorphisms of a group form a group.

8. Find the inner automorphisms of the group given in Table 8.1.

8.5. Cosets. Conjugate Subgroups. Normal Subgroups. Let G be a group and H a proper subgroup of G. Let the elements of H be H_1, H_2, $\ldots$, H_m, and let A be an element of G not in H. We consider the elements H_1A, H_2A, $\ldots$, H_nA. The reader can quickly verify that $H_iA \neq H_jA$ for $i \neq j$, and $H_iA \neq H_j$. If $H_iA = H_j$, then $A = H_i^{-1}H_j$ is in H, a contradiction. Thus the set of elements H_iA, $i = 1, 2, \ldots, n$, are distinct and do not belong to H if A does not belong to H. This set is not a subgroup of G since it does not contain the identity element E. We call this set a *coset*, written as HA. If the elements of H and HA exhaust G, we can write $G = H + HA$. If not, we consider an element B of G, B not in H and HA. We construct the coset HB and omit the proof that the elements of HB are distinct from the elements of H and HA. We can continue this process until we exhaust G, if G is a finite group. Thus

$$G = H + HA + HB + \cdots + HC \tag{8.3}$$

Example 8.6. We consider the symmetric group of order 4!. Let H consist of the elements S_1, S_2, S_3, S_4 of Example 8.5. Let A be the element $\begin{pmatrix}1&2&3&4\\2&1&4&3\end{pmatrix}$. Now

$$S_1A = \begin{pmatrix}1&2&3&4\\2&3&4&1\end{pmatrix}\begin{pmatrix}1&2&3&4\\2&1&4&3\end{pmatrix} = \begin{pmatrix}1&2&3&4\\3&2&1&4\end{pmatrix}$$

$$S_2A = \begin{pmatrix}1&2&3&4\\3&4&1&2\end{pmatrix}\begin{pmatrix}1&2&3&4\\2&1&4&3\end{pmatrix} = \begin{pmatrix}1&2&3&4\\4&3&2&1\end{pmatrix}$$

$$S_3A = \begin{pmatrix}1&2&3&4\\4&1&2&3\end{pmatrix}\begin{pmatrix}1&2&3&4\\2&1&4&3\end{pmatrix} = \begin{pmatrix}1&2&3&4\\1&4&2&3\end{pmatrix}$$

$$S_4A = \begin{pmatrix}1&2&3&4\\1&2&3&4\end{pmatrix}\begin{pmatrix}1&2&3&4\\2&1&4&3\end{pmatrix} = \begin{pmatrix}1&2&3&4\\2&1&4&3\end{pmatrix}$$

Note that $S_iA \neq S_jA$, $i \neq j$, and that $S_iA \neq S_j$ for all i, j. The elements S_iA, $i = 1, 2, 3, 4$, are members of the coset HA. The element $B = \begin{pmatrix}1&2&3&4\\2&4&1&3\end{pmatrix}$ is not a member of H or HA. We consider

$$S_1B = \begin{pmatrix} 1 & 2 & 3 & 4 \\ 2 & 3 & 4 & 1 \end{pmatrix} \begin{pmatrix} 1 & 2 & 3 & 4 \\ 2 & 4 & 1 & 3 \end{pmatrix} = \begin{pmatrix} 1 & 2 & 3 & 4 \\ 3 & 1 & 2 & 4 \end{pmatrix}$$

$$S_2B = \begin{pmatrix} 1 & 2 & 3 & 4 \\ 3 & 4 & 1 & 2 \end{pmatrix} \begin{pmatrix} 1 & 2 & 3 & 4 \\ 2 & 4 & 1 & 3 \end{pmatrix} = \begin{pmatrix} 1 & 2 & 3 & 4 \\ 4 & 2 & 3 & 1 \end{pmatrix}$$

$$S_3B = \begin{pmatrix} 1 & 2 & 3 & 4 \\ 4 & 1 & 2 & 3 \end{pmatrix} \begin{pmatrix} 1 & 2 & 3 & 4 \\ 2 & 4 & 1 & 3 \end{pmatrix} = \begin{pmatrix} 1 & 2 & 3 & 4 \\ 1 & 3 & 4 & 2 \end{pmatrix}$$

$$S_4B = \begin{pmatrix} 1 & 2 & 3 & 4 \\ 1 & 2 & 3 & 4 \end{pmatrix} \begin{pmatrix} 1 & 2 & 3 & 4 \\ 2 & 4 & 1 & 3 \end{pmatrix} = \begin{pmatrix} 1 & 2 & 3 & 4 \\ 2 & 4 & 1 & 3 \end{pmatrix}$$

The elements S_iB, $i = 1, 2, 3, 4$, are members of the coset HB. Note that $S_iB \neq S_j$ for all i, j, $S_iB \neq S_jA$ for all i, j, $S_iB \neq S_jB$ for $i \neq j$. Continuing, we can obtain $G = H + HA + HB + HC + HD + HF$.

THEOREM 8.5. *The elements X and Y belong to the same coset if and only if XY^{-1} is in H.* Assume XY^{-1} a member of H, so that $XY^{-1} = H_1$, H_1 in H. Then $X = H_1Y$, and $H_iX = H_iH_1Y = H_jY$ for all i, with $H_j = H_iH_1$. Thus every member of HX is a member of HY. From $Y = H_1^{-1}X$, $H_iY = H_iH_1^{-1}X = H_kX$ for all i, we have that every member of HY belongs to HX. Thus $HX = HY$. Conversely, if $HX = HY$, then H_1 and H_2 exist such that $H_1X = H_2Y$ so that $XY^{-1} = H_1^{-1}H_2$, a member of H. Q.E.D.

DEFINITION 8.1. If $B = X^{-1}AX$, we say that B is conjugate to A, with A, B, X in G. We note the following:

1. Every element is conjugate to itself, $A = E^{-1}AE$.

2. If B is conjugate to A, then A is conjugate to B, since $B = X^{-1}AX$ implies $A = XBX^{-1} = (X^{-1})^{-1}BX^{-1}$.

3. If A and B are conjugate to C, then A and B are conjugate. We have

$$A = X^{-1}CX \qquad B = Y^{-1}CY = Y^{-1}(XAX^{-1})Y = (X^{-1}Y)^{-1}A(X^{-1}Y)$$

Let us consider a subgroup H of G consisting of the elements E, A, B, ... Let X be a member of G not in H. We form the elements $X^{-1}EX = E$, $X^{-1}AX$, $X^{-1}BX$, . . . and designate this set as $X^{-1}HX$. Since

$$(X^{-1}AX)(X^{-1}BX) = X^{-1}(AB)X$$

we have from Theorem 8.3 that the set $X^{-1}HX$ is a subgroup of G. We have used the fact that H is a group, which implies that AB is in H if A and B are in H. We call H and $X^{-1}HX$ conjugate subgroups. By considering $Y^{-1}HY$, etc., one can construct the complete set of conjugate subgroups of H.

DEFINITION 8.2. If H is identical with all its conjugate subgroups, then H is called a *normal*, or *invariant*, subgroup of G.

Example 8.7. The even permutations of a symmetric group form a subgroup, H. If X is any odd permutation, $X^{-1}AX$ is an even permutation if A is an even permutation. Thus $X^{-1}HX \equiv H$ for all X so that H is a normal subgroup of G.

THEOREM 8.6. *The intersection of two normal subgroups is a normal subgroup.* By the intersection of two subgroups H_1, H_2 we mean the set of elements belonging to both H_1 and H_2, written $H_1 \cap H_2$. It is obvious that the identity element E belongs to $H_1 \cap H_2$. If A and B belong to $H_1 \cap H_2$, then AB belongs to H_1 and to H_2 so that AB belongs to $H_1 \cap H_2$. From Theorem 8.3, $H_1 \cap H_2$ is a subgroup of the finite group G. We must show now that $H_1 \cap H_2$ is a normal subgroup if H_1 and H_2 are normal subgroups. The set of elements $X^{-1}(H_1 \cap H_2)X$ belong to H_1 since any element of $H_1 \cap H_2$ belongs to H_1 and, moreover, H_1 is normal. The same statement applies to H_2, so that every element of $X^{-1}(H_1 \cap H_2)X$ belongs to $H_1 \cap H_2$. Thus $H_1 \cap H_2$ is identical with all its conjugate groups so that $H_1 \cap H_2$ is normal.

Problems

1. The elements $\begin{pmatrix}1&2&3\\1&2&3\end{pmatrix}$, $\begin{pmatrix}1&2&3\\2&1&3\end{pmatrix}$ form a subgroup H of the symmetric group of order 3!. Find the right-hand cosets HX of G. Find the left-hand cosets XH of G. Show that H is not a normal subgroup of G.

2. Show that the subgroup H, consisting of the elements S_i, $i = 1, 2, 3, 4$ of Example 8.5, is a normal subgroup of the symmetric group of order 4!.

3. If H is a subgroup of a cyclic group G, show that H is normal.

4. A group G is said to be *simple* if it contains no proper normal subgroups. Show that all groups of prime order are simple. Show that all simple Abelian groups are of prime order.

5. Let M and N be normal subgroups of G containing only the identity E in common. Show that if M_1 and N_1 are any two elements of M and N, respectively, then $M_1N_1 = N_1M_1$. *Hint:* Consider $C = M_1^{-1}N_1^{-1}M_1N_1$, and show that $C = E$.

6. Consider the set of elements $E, A_1, A_2, \ldots$ of G such that $A_i^{-1}XA_i = X$ for all X of G. Show that this set is an Abelian subgroup of G, called the *central* of G. Show that the central of G is normal. Find the central of the group given by Table 8.1. Find the central of the symmetric group of order 3!.

8.6. Factor, or Quotient, Groups. The set of rational integers form a group relative to addition (see Prob. 8, Sec. 8.2). The zero element plays the role of the identity element. A proper subgroup of this group is the set of even integers including zero. The odd integers yield a coset of the group of even integers, so that $I = I_0 + I_1$, where I is the group of rational integers, I_0 is the group of even integers, and I_1 is the set of odd integers. I_1 is obviously not a group relative to addition since I_1 does not contain the identity element. Let the reader show that I_0 is a normal subgroup of I. If we add any element of I_0 to any other element of I_0, we obtain another element of I_0. We may thus write $I_0 + I_0 = I_0$. If we add any element of I_0 to any element of I_1, we obtain an element of I_1, so that $I_0 + I_1 = I_1 + I_0 = I_1$. Finally $I_1 + I_1 = I_0$. Thus if we abstract in the sense that we denote all even integers by the single element I_0 and denote all odd integers by the single element I_1, we note that the two elements I_0, I_1 form a group relative to addition, with I_0 serving as the identity element. We generalize this result as follows:

Let N be any normal subgroup of the group G. We consider N along with its cosets,

$$N, NA_1, NA_2 \ . \ . \ . \ , NA_r \tag{8.4}$$

Since N is normal, we have $X^{-1}NX \approx N$, that is, every element of the set $X^{-1}NX$ belongs to N, and, conversely, every element of N belongs to $X^{-1}NX$. Thus $NX \approx XN$, and $NA_i \approx A_iN$, $i = 1, 2, \ . \ . \ . \ , r$, so that the right cosets of N are equivalent to the left cosets of N. Let us consider the product of two cosets of N. We have

$$(NA_i)(NA_j) = N(A_iN)A_j \approx N(NA_i)A_j = NN(A_iA_j) = N(A_iA_j)$$

with $N(A_iA_j)$ another coset of N. Conversely,

$$N(A_iA_j) = NN(A_iA_j) = N(NA_i)A_j \approx (NA_i)(NA_j)$$

If we look upon a coset as a single element, we note that the set (8.4) forms a group. The element N is the identity element for this group abstracted from the group G and the normal subgroup N.

DEFINITION 8.3. The group whose elements are constructed from the normal subgroup N of G and the cosets of N is called the *factor group*, or *quotient group*, of N, written G/N. The order, or number of elements, of G/N is called the *index* of the factor group.

THEOREM 8.7. Let H be a proper subgroup of G, written $H \subset G$. If N is a normal subgroup of G such that $N \subset H \subset G$, then $H/N \subset G/N$. Let the reader show that, if N is a normal subgroup of G, then N is a normal subgroup of the subgroup H, provided $N \subset H \subset G$. Thus

$$H = N + NA_1 + NA_2 + \cdots + NA_s$$

and $G/H = (N, NA_1, \ . \ . \ . \ , NA_s)$. Now

$$G = N + NA_1 + NA_2 + \cdots + NA_s + NX + \cdots + NY$$

with $G/N = (N, NA_1, \ . \ . \ . \ , NA_s, NX, \ . \ . \ . \ , NY)$. It is obvious that $G/H \subset G/N$.

DEFINITION 8.4. The product HK of two subgroups H and K of G is the set of elements h_ik_j, where h_i ranges over H and k_j ranges over K.

THEOREM 8.8. If H and K are normal subgroups of G, their product $L = HK$ is a normal subgroup of G. First we show that HK is a subgroup of G and that $HK \approx KH$. Certainly the associative law holds since H and K are in G. Moreover $k_1h_2k_1^{-1} = h_3$ since H is normal, so that $k_1h_2 = h_3k_1$ and $(h_1k_1)(h_2k_2) = (h_1h_3)(k_1k_2) = hk \subset HK$. The identity element of HK is e, the identity element of G. We leave it to the reader to show that $HK \approx KH$ and that the inverse of every element of HK is again an element of HK. Finally, for any element X of G,

$$X^{-1}(HK)X = X^{-1}(HXX^{-1}K)X = (X^{-1}HX)(X^{-1}KX) \approx HK$$

since H and K are normal. Thus HK is identical with all its conjugate subgroups so that HK is normal.

DEFINITION 8.5. By a *maximal normal group* N of G one understands that N is not contained properly in any normal subgroup of G other than G itself.

THEOREM 8.9. Let N_1 and N_2 be normal maximal subgroups of G, D their intersection, written $D = N_1 \cap N_2$. Then

$$G/N_1 \cong N_1/D \qquad G/N_2 \cong N_2/D \tag{8.5}$$

The intersection of two subgroups of G is the set of elements common to both N_1 and N_2. D is nonvacuous since the identity element obviously belongs to D. The reader should refer to Sec. 8.4 for the definition of an isomorphism.

From Theorem 8.8 the product N_1N_2 is a normal group. Obviously N_1N_2 contains both N_1 and N_2. Since N_1 and N_2 are maximal, it is necessary that $N_1N_2 = G$. Since any subgroup of a normal group is normal, it makes sense to speak of N_1/D. Now

$$N_1 = D + DA_1 + DA_2 + \cdots + DA_r \tag{8.6}$$

with $A_i \neq A_j$ for $i \neq j$, and the cosets DA_i are distinct. Let L be the set of elements belonging to N_2, N_2A_i, $i = 1, 2, \ldots, r$, so that

$$L = N_2 + N_2A_1 + N_2A_2 + \cdots + N_2A_r$$

We show first that $L \equiv G$. We have

$$\begin{aligned} G = N_1N_2 \approx N_2N_1 &= N_2(D + DA_1 + \cdots + DA_r) \\ &\approx N_2 + N_2A_1 + \cdots + N_2A_r \end{aligned}$$

since $N_2D \approx N_2$. The cosets N_2A_i, $i = 1, 2, \ldots, r$, are distinct, for $N_2A_i \approx N_2A_j$ implies $N_2A_iA_j^{-1} \approx N_2$, which further implies that $A_iA_j^{-1}$ is a member of N_2. Moreover $A_iA_j^{-1}$ is a member of N_1 since A_i and A_j belong to N_1 [see (8.6)]. Thus $A_iA_j^{-1}$ belongs to both N_1 and N_2 and hence to D. This implies $DA_iA_j^{-1} \approx D$, $DA_i \approx DA_j$, a contradiction to (8.6). The correspondence of cosets, $DA_i \leftrightarrow N_2A_i$, $i = 1, 2, \ldots, r$, with $D \leftrightarrow N_2$ yields the isomorphism $N_1/D \cong G/N_2$. Similarly $N_2/D \cong G/N_1$. Q.E.D.

Example 8.8. We consider the cyclic group G of order 6 with elements $a, a^2, a^3, a^4, a^5, a^6 = e$. The three proper subgroups of G are $N_1(a^2, a^4, a^6 = e)$, $N_2(a^3, a^6 = e)$, $N_3(e)$. Let the reader show that N_1 and N_2 are maximal normal subgroups of G. We also have that $D = N_3 = N_1 \cap N_2$. We have

$$\begin{aligned} G &= (a^2, a^4, e) + (a^2, a^4, e)a = (a^2, a^4, e) + (a, a^3, a^5) \\ G &= (a^3, e) + (a^3, e)a + (a^3, e)a^2 = (a^3, e) + (a^4, a) + (a^5, a^2) \\ G/N_1 &= [(a^2, a^4, e), (a, a^3, a^5)] \\ G/N_2 &= [(a^3, e), (a^4, a), (a^5, a^2)] \\ N_1/D &= [(a^2), (a^4), (e)] \\ N_2/D &= [(a^3), (e)] \end{aligned}$$

The correspondence $(e) \leftrightarrow (a^2, a^4, e)$, $(a^3) \leftrightarrow (a, a^3, a^5)$ is an isomorphism, $N_2/D \cong G/N_1$. Similarly $N_1/D \cong G/N_2$. By $(a^2, a^4, e) \cdot (a, a^3, a^5)$ we mean all elements obtained by multiplying elements of (a^2, a^4, e) with elements of (a, a^3, a^5). Note that $(a^2, a^4, e) \cdot (a, a^3, a^5) = (a, a^3, a^5)$ so that the element $[(a^2), (a^4), (e)]$ is the unit element of G/N_1.

Problems

1. Find the maximal normal subgroups of the cyclic group of order 12. Consider any two such maximal normal subgroups, and show the isomorphism (8.5) by constructing their cosets and finding their intersection.
2. Find the normal subgroups of the group of Table 8.1.
3. A group is said to be simple if it has no normal subgroup other than itself and the identity element. Show that any group of prime order is simple.
4. Show that G/H is simple if and only if H is maximal.
5. Let N be a normal subgroup of G, H any subgroup of G. Let $D = H \cap N$, $L = HN$. Show that HN is a group, D a normal subgroup of H, $L/N \cong H/D$.
6. Find the maximal normal subgroups of the symmetric group of order 24.

8.7. Series of Composition. The Jordan-Hölder Theorem. Let N be a normal subgroup of G. One then obtains the factor, or quotient, group

$$G/N = [N, NA_1, NA_2, \ldots, NA_r]$$

where NA_i, $i = 1, 2, \ldots, r$, is a coset consisting of the elements N_jA_i with N_j ranging over the group N. Let $T = [N, NB_1, \ldots, NB_s]$ be a normal subgroup of G/N. We prove the following theorem:

THEOREM 8.10. Every normal group of a factor group G/N yields a normal group of G; each normal group of G which contains N corresponds to a normal group of the factor group. The first part of Theorem 8.10 states that the set

$$H = N + NB_1 + \cdots + NB_s$$

is a normal subgroup of G. Now

$$\begin{aligned} X^{-1}HX &= X^{-1}(N + NB_1 + \cdots + NB_s)X \\ &= X^{-1}NX + X^{-1}NB_1X + \cdots + X^{-1}NB_sX \end{aligned}$$

Since N is normal, we have $X^{-1}NX \approx N$ for all X in G. Since T is normal, $(NX)^{-1}(NB_i)NX$ is in T. Thus $X^{-1}(B_iN)X$ is in T and hence is an element of H. Thus $X^{-1}HX$ is an element of H for all X of G, which proves that H is normal. It is obvious that $G \supset H \supset N$. Now assume that $G \supset H \supset N$ with both H and N normal. From

$$\begin{aligned} H &= N + NA_1 + NA_2 + \cdots + NA_s \\ G &= N + NA_1 + NA_2 + \cdots + NA_s + NA_{s+1} + \cdots + NA_r \end{aligned}$$

it follows that $H/N \subset G/N$. We wish to show that H/N is a normal subgroup of G/N. From the fact that H is normal and that A_i belongs to H we have that $A_j^{-1}A_iA_j$ belongs to H. Thus

$$\begin{aligned} (NA_j)^{-1}(NA_i)(NA_j) &= A_j^{-1}A_iNA_j \\ &\approx (A_j^{-1}A_iA_j)N \qquad \text{since } N \text{ is normal} \\ &\approx N(A_j^{-1}A_iA_j) \qquad \text{since } N \text{ is normal} \\ &= N(NA_k) \approx NA_k \end{aligned}$$

Thus the conjugates of H/N are members of H/N so that H/N is normal. It may be that $A_j^{-1}A_iA_j = E$, the identity element, so that $N(A_j^{-1}A_iA_j) = NE = N$. Q.E.D.

DEFINITION 8.6. Let N_1 be a maximal normal subgroup of G, N_2 a maximal normal subgroup of N_1, etc. We obtain a series $G, N_1, N_2, \ldots, N_k = E$, called a *series of composition*.

The factor groups $G/N_1, N_1/N_2, \ldots, N_{k-1}/N_k$ are all simple groups (see Prob. 3, Sec. 8.6) for the definition of a simple group. If N_i/N_{i+1} were not simple, Theorem 8.10 states that a normal group M would exist such that $N_i \supset M \supset N_{i+1}$, a contradiction, since N_{i+1} is assumed maximal relative to N_i. These simple groups are called the *prime factors of composition* of G. The orders of $G/N_1, N_1/N_2, \ldots$ are called the factors of composition of G.

THEOREM 8.11. (*Jordan-Hölder*). *For two series of composition of a finite group G the prime-factor groups are isomorphic.*

This means that if we consider two arbitrary series of composition of G, say

$$\begin{aligned} G &= H_0, H_1, H_2, \ldots, H_r = E \\ G &= K_0, K_1, K_2, \ldots, K_s = E \end{aligned} \tag{8.7}$$

with prime-factor groups

$$\begin{aligned} &G/H_1, H_1/H_2, \ldots, H_{r-1}/H_r \\ &G/K_1, K_1/K_2, \ldots, K_{s-1}/K_s \end{aligned} \tag{8.8}$$

then $r = s$ and a one-to-one correspondence, $i \leftrightarrow j$, can be set up such that

$$H_i/H_{i+1} \cong K_j/K_{j+1} \tag{8.9}$$

with i and j ranging from 0 to r.

The theorem is evidently true if the order of G is prime, for in this case there exists only one normal subgroup of G, the identity element, and $G/E \cong G/E$. The proof of Theorem 8.11 is by induction. Assume the theorem true for any group G whose order, g, can be written as the product of n prime factors or less. We know that the theorem is true for $n = 1$. Now let G be any group whose order can be written as the product of $n + 1$ prime factors, and let us consider two arbitrary series of composition [see (8.7) and (8.8)]. If $H_1 = K_1$, then $G/H_1 \cong G/K_1$ and the set of elements of H_1 is a group whose order contains at most n prime factors. By hypothesis $H_i/H_{i+1} \cong K_j/K_{j+1}$ for $i \geqq 1$. We assume, therefore, that H_1 and K_1 are distinct maximal normal subgroups of G. From Theorem 8.9

$$G/H_1 \cong K_1/D_1 \qquad G/K_1 \cong H_1/D_1 \tag{8.10}$$

where $D_1 = H_1 \cap K_1$. Since G/H_1 and G/K_1 are simple, of necessity, K_1/D_1 and H_1/D_1 are simple. Hence D_1 is maximal (see Prob. 4, Sec. 8.6). Now form the series of composition

$$\begin{aligned} G, H_1, D_1, D_2, \ldots, D_m = E \\ G, K_1, D_1, D_2, \ldots, D_m = E \end{aligned} \tag{8.11}$$

From (8.10) and the fact that $D_1/D_2 \cong D_1/D_2$, $D_2/D_3 \cong D_2/D_3$, etc., we see that the two series of (8.11) satisfy Theorem 8.11. But the series $(H_1, H_2, \ldots, H_r)$, $(H_1, D_1, \ldots, D_m)$ satisfy Theorem 8.11 by assumption, as do the series $(K_1, K_2, \ldots, K_r)$, $(K_1, D_1, \ldots, D_m)$. Thus Theorem 8.11 holds for the two series $(H_1, H_2, \ldots, H_r)$, $(K_1, K_2, \ldots, K_r)$, so that with (8.10) it is seen that the Jordan-Hölder theorem holds for any group G whose order can be written as a product of $n + 1$ prime factors. This concludes the proof by induction. Example 8.8 is an illustration of the Jordan-Hölder theorem.

Problem. Construct the multiplication table for the octic group whose elements are u^iv^j ($i = 0, 1, 2, 3$; $j = 0, 1$) with $u^4 = 1$, $v^2 = 1$, $vu = u^3v$. Find the different series of composition and prime-factor groups for the octic group, and show that the Jordan-Hölder theorem applies.

8.8. Group Characters. Representation of Groups. If to each element s of a group G we can assign a nonvanishing number, real or complex, written $X(s)$, such that

$$X(s)X(t) = X(st) \tag{8.12}$$

we say that we have a one-dimensional representation of the group and we call X a group character. From (8.12) we have

$$X(s)X(e) = X(se) = X(s)$$

so that $X(e) = 1$, where e is the unit element of the group G. A trivial group character is $X(s) = 1$ for all s of G.

Example 8.9. Let G be a cyclic group of order n with generator a, $a^n = e$. Let us define $X(a^r)$ by the equation

$$X(a^r) = e^{(2\pi i/n)r} \qquad r = 1, 2, \ldots, n \tag{8.13}$$

We have $X(a^r)X(a^s) = e^{(2\pi i/n)r}e^{(2\pi i/n)s} = e^{(2\pi i/n)(r+s)} = X(a^{r+s})$, so that (8.13) defines a character of the group. A set of characters $X_0, X_1, \ldots, X_{n-1}$ can be defined by

$$X_\mu(a^r) = e^{(2\pi i/n)\mu r} \qquad \mu = 0, 1, 2, \ldots, n-1 \tag{8.14}$$

Let us note that

$$\sum_{r=1}^{n} \bar{X}_\mu(a^r)X_\nu(a^r) = \sum_{r=1}^{n} e^{(2\pi i/n)r(\nu-\mu)} = \begin{matrix} n \\ 0 \end{matrix} \quad \left.\begin{matrix} \text{if } \mu = \nu \\ \text{if } \mu \neq \nu \end{matrix}\right\} \tag{8.15}$$

The group characters of (8.14) are orthogonal in the sense that (8.15) holds. Moreover

$$\sum_{\mu=0}^{n-1} \bar{X}_\mu(a^s)X_\mu(a^r) = \sum_{\mu=0}^{n-1} e^{(2\pi i/n)(r-s)\mu} = \begin{matrix} n \\ 0 \end{matrix} \quad \left.\begin{matrix} \text{if } r = s \\ \text{if } r \neq s \end{matrix}\right\} \tag{8.16}$$

Let us now consider a set of $n \times n$ matrices $\mathbf{A}_1, \mathbf{A}_2, \ldots, \mathbf{A}_n$ and a group G with elements $g_1, g_2, \ldots, g_n$. We assume that a one-to-one correspondence exists such that

$$\mathbf{A}_i \leftrightarrow g_i$$
$$\mathbf{A}_j \leftrightarrow g_j$$

implies

$$\mathbf{A}_i\mathbf{A}_j \leftrightarrow g_ig_j$$

If such is the case, it is easy to show that the set of matrices is a group, A, isomorphic to the group G. We say that the group of matrices, A, is a representation of the group G. Any matrix $\mathbf{A} = \|a^i_j\|$ can be thought of as representing an affine (linear) transformation $y^i = a^i_jx^j$ (see Chap. 1), so that a representation of a group implies that a group of affine transformations is isomorphic to the group G. Let the reader show that, if $g_i \leftrightarrow \mathbf{A}_i$ is a representation of G, then $g_i \leftrightarrow \mathbf{B}^{-1}\mathbf{A}_i\mathbf{B}$ is also a representation of G, $|\mathbf{B}| \neq 0$.

Example 8.10. A representation of the symmetric group of order 6 is easy to construct. The identity permutation $\begin{pmatrix} 1\,2\,3 \\ 1\,2\,3 \end{pmatrix}$ can be looked upon as the transformation of coordinates $\bar{x} = x$, $\bar{y} = y$, $\bar{z} = z$, or

$$\begin{aligned} \bar{x} &= 1x + 0y + 0z \\ \bar{y} &= 0x + 1y + 0z \\ \bar{z} &= 0x + 0y + 1z \end{aligned}$$

yielding the unit matrix

$$\mathbf{A}_1 = \begin{Vmatrix} 1 & 0 & 0 \\ 0 & 1 & 0 \\ 0 & 0 & 1 \end{Vmatrix}$$

The permutation $\begin{pmatrix} 1\,2\,3 \\ 2\,1\,3 \end{pmatrix}$ can be looked upon as the transformation of coordinates

$$\begin{aligned} \bar{x} &= y = 0x + 1y + 0z \\ \bar{y} &= x = 1x + 0y + 0z \\ \bar{z} &= z = 0x + 0y + 1z \end{aligned}$$

yielding the matrix

$$\mathbf{A}_2 = \begin{Vmatrix} 0 & 1 & 0 \\ 1 & 0 & 0 \\ 0 & 0 & 1 \end{Vmatrix}$$

Let the reader obtain the matrices corresponding to the permutations

$$\begin{pmatrix} 1\,2\,3 \\ 1\,3\,2 \end{pmatrix} \quad \begin{pmatrix} 1\,2\,3 \\ 2\,3\,1 \end{pmatrix} \quad \begin{pmatrix} 1\,2\,3 \\ 3\,1\,2 \end{pmatrix} \quad \begin{pmatrix} 1\,2\,3 \\ 3\,2\,1 \end{pmatrix}$$

Since every finite group is isomorphic to a regular permutation group (see Theorem 8.5), it follows very simply from Example 8.10 that one can always obtain a representation of a finite group.

It may happen that a matrix **B** exists such that $\mathbf{B}^{-1}\mathbf{A}_i\mathbf{B}$ has the form

$$\begin{Vmatrix} \mathbf{B}_1(\mathbf{A}_i) & 0 & \cdots & 0 \\ 0 & \mathbf{B}_2(\mathbf{A}_i) & \cdots & 0 \\ \cdots & \cdots & \cdots & \cdots \\ 0 & 0 & \cdots & \mathbf{B}_k(\mathbf{A}_i) \end{Vmatrix} \qquad i = 1, 2, \ldots, n \tag{8.17}$$

where $\mathbf{B}_r(\mathbf{A}_i)$, $r = 1, 2, \ldots, k$, are matrices. If this is so, the original representation is said to be reducible. One easily shows that the set of matrices $\{\mathbf{B}_r(\mathbf{A}_1), \mathbf{B}_r(\mathbf{A}_2), \ldots, \mathbf{B}_r(\mathbf{A}_n)\}$ is a representation of G. Thus (8.17) yields k new representations of G. It may be that further reductions are possible. If not, the representation given by (8.17) is said to be irreducible. In this case the correspondence is written

$$g_i \leftrightarrow \Gamma(\mathbf{A}_i) = \mathbf{B}_1(\mathbf{A}_i) + \mathbf{B}_2(\mathbf{A}_i) + \cdots + \mathbf{B}_k(\mathbf{A}_i) \tag{8.18}$$

where the sum in (8.18) denotes the matrix (8.17). Methods for determining the irreducible representations of a group are laborious. The reader may consult the references cited at the end of this chapter for detailed information on irreducible representations.

Problems

1. Derive (8.16).

2. Find a representation for the group of Table 8.1.

3. If the correspondence $g_i \leftrightarrow \mathbf{A}_i$ is a representation, show that $X(g_i) = |\mathbf{A}_i|$ is a character of the group G.

4. Find a character for the group given by Table 8.1 other than the trivial character $X(g_i) = 1$ for all g_i of G.

8.9. Continuous Transformation Groups. Let us consider the transformation of coordinates given by

$$\begin{aligned} x_1 &= f(x, y; t) \\ y_1 &= \varphi(x, y; t) \end{aligned} \tag{8.19}$$

where t is a parameter ranging continuously over a given range of values. The transformation (8.19) will be said to be a one-parameter transformation group if the following is true:

1. There exists a t_0 such that

$$x = f(x, y; t_0)$$
$$y = \varphi(x, y; t_0)$$

This value of t, $t = t_0$, leaves the point (x, y) invariant. This corresponds to the identity transformation.

2. The result of applying two successive transformations is identical to a third transformation of the family given by (8.19). This implies that if $x_2 = f(x_1, y_1; t_1)$, $y_2 = \varphi(x_1, y_1; t_1)$ then

$$x_2 = f(f(x, y; t), \varphi(x, y; t); t_1) = f(x, y; t_2)$$
$$y_2 = \varphi(f(x, y; t), \varphi(x, y; t); t_1) = \varphi(x, y; t_2)$$

where t_2 is a function of t and t_1.

3. The associative law must hold.

4. Each transformation has a unique inverse. This implies that we can solve (8.19) uniquely for x and y such that

$$x = f(x_1, y_1; t_1)$$
$$y = \varphi(x_1, y_1; t_1)$$

with t_1 some function of t.

Example 8.11. The translations $x_1 = x + b$, $y_1 = y + 2b$ satisfy the requirements for a one-parameter group. The identity transformation occurs for $b = 0$, and the inverse transformation is obtained by replacing b by $-b$.

The rotations

$$x_1 = x \cos \theta - y \sin \theta$$
$$y_1 = x \sin \theta + y \cos \theta$$

form a one-parameter group. $\theta = 0$ yields the identity transformation. If

$$x_2 = x_1 \cos \theta_1 - y_1 \sin \theta_1$$

$y_2 = x_1 \sin \theta_1 + y_1 \cos \theta_1$, then

$$x_2 = x \cos (\theta + \theta_1) - y \sin (\theta + \theta_1) \qquad y_2 = x \sin (\theta_1 + \theta_2) + y \cos (\theta_1 + \theta_2)$$

The parameter θ is additive for two successive transformations.

A third example is the group $x_1 = ax$, $y_1 = a^2y$. The identity transformation occurs for $a = 1$. The parameter is multiplicative for two successive transformations.

Infinitesimal transformations can be found as follows: Let t_0 be that value of the parameter t which yields the identity transformation. Then

$$x = f(x, y; t_0)$$
$$y = \varphi(x, y; t_0) \tag{8.20}$$

If (8.19) is continuous in t, a small change in t will yield a transformation differing from (8.20) by a small amount. The transformation

$$\begin{aligned} x_1 &= f(x, y; t_0 + \epsilon) \\ y_1 &= \varphi(x, y; t_0 + \epsilon) \end{aligned} \tag{8.21}$$

is said to be an infinitesimal in the sense that x_1 and y_1 differ from x and y, respectively, by small amounts when ϵ is sufficiently small. Subtracting (8.20) from (8.21) yields

$$\begin{aligned} \delta x &= x_1 - x = f(x, y; t_0 + \epsilon) - f(x, y; t_0) \\ \delta y &= y_1 - y = \varphi(x, y; t_0 + \epsilon) - \varphi(x, y; t_0) \end{aligned}$$

If f and φ are differentiable in a neighborhood of $t = t_0$, one has

$$\begin{aligned} \delta x &= \epsilon \left(\frac{\partial f}{\partial t}\right)_{t_0} = \xi(x, y)\, \delta t \\ \delta y &= \epsilon \left(\frac{\partial \varphi}{\partial t}\right)_{t_0} = \eta(x, y)\, \delta t \end{aligned} \tag{8.22}$$

except for infinitesimals of higher order, $\epsilon = \delta t$.

Example 8.12. For the rotation group of Example 8.11 one has $\left(\frac{\partial f}{\partial \theta}\right)_{\theta=0} = -y$, $\left(\frac{\partial \varphi}{\partial \theta}\right)_{\theta=0} = x$, so that (8.22) becomes

$$\delta x = -y\, \delta t \qquad \delta y = x\, \delta t$$

Let us consider the change in the value of a function $F(x, y)$ when the point (x, y) undergoes an infinitesimal transformation. One has

$$\delta F = F(x_1, y_1) - F(x, y) = \frac{\partial F}{\partial x}\, \delta x + \frac{\partial F}{\partial y}\, \delta y$$

except for infinitesimals of higher order. From (8.22) one obtains

$$\delta F = \left(\xi \frac{\partial F}{\partial x} + \eta \frac{\partial F}{\partial y}\right) \delta t \tag{8.23}$$

The operator U defined by

$$U = \xi \frac{\partial}{\partial x} + \eta \frac{\partial}{\partial y} \tag{8.24}$$

is very useful. Equation (8.23) can be written as $\delta F = (UF)\, \delta t$. If $t = 0$ corresponds to the identity transformation, we can replace δt by t, remembering that t is small.

It is interesting to note that if one is given the system of differential equations

$$\begin{aligned} \frac{dx_1}{dt} &= \xi(x_1, y_1) \qquad x_1(0) = x \\ \frac{dy_1}{dt} &= \eta(x_1, y_1) \qquad y_1(0) = y \end{aligned} \tag{8.25}$$

then the solution of (8.25), written

$$\begin{aligned} x_1 &= f(x, y; t) \\ y_1 &= \varphi(x, y; t) \end{aligned} \tag{8.26}$$

is a one-parameter group additive in t. To show this, we first note that (8.26) yields a curve starting at the fixed point (x, y). As t varies, x_1 and y_1 vary (see Fig. 8.1).

Now let us consider the system

$$\left.\begin{aligned} \frac{dx_2}{dt_1} &= \xi(x_2, y_2) \qquad & x_2 = x_1 \text{ at } t_1 = t \\ \frac{dy_2}{dt_1} &= \eta(x_2, y_2) \qquad & y_2 = y_1 \text{ at } t_1 = t \end{aligned}\right\} \tag{8.27}$$

If we let $t_2 = t_1 - t$, t fixed, (8.27) becomes

$$\left.\begin{aligned} \frac{dx_2}{dt_2} &= \xi(x_2, y_2) \qquad & x_2 = x_1 \text{ at } t_2 = 0 \\ \frac{dy_2}{dt_2} &= \eta(x_2, y_2) \qquad & y_2 = y_1 \text{ at } t_2 = 0 \end{aligned}\right\} \tag{8.28}$$

Since (8.28) has exactly the same form and initial conditions as (8.25), of necessity,

$$\begin{aligned} x_2 &= f(x_1, y_1; t_2) = f(x_1, y_1; t_1 - t) \\ y_2 &= \varphi(x_1, y_1; t_2) = \varphi(x_1, y_1; t_1 - t) \end{aligned} \tag{8.29}$$

At $t = t_1$ we have $x_2 = x_1$, $y_2 = y_1$. It is obvious that (8.29) yields the same curve as given in Fig. 8.1, and for $t_1 > t$ we obtain an extension of Γ to the point (x_2, y_2). Thus the product of two transformations belongs to the group, and the parameter t is additive. The establishment of the inverse transformation is left to the reader.

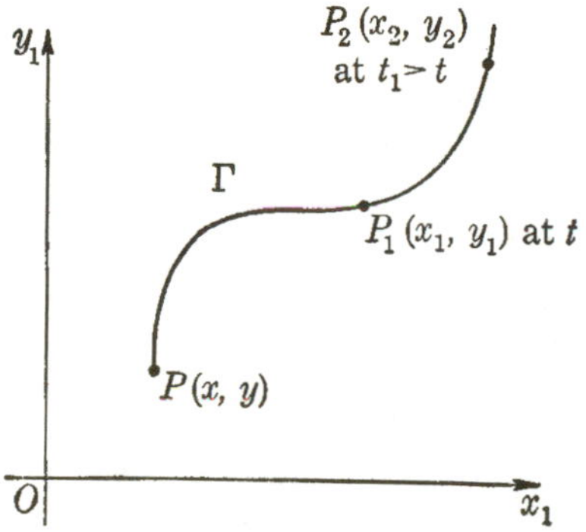

Fig. 8.1

Let us now begin with a group transformation given by (8.19) and attempt to establish a system of differential equations involving x_1, y_1, and t. From the group property we have

$$\begin{aligned} x_2 &= f(x, y; \beta(t, t_1)) \equiv f(x_1, y_1; t_1) \\ y_2 &= \varphi(x, y; \beta(t, t_1)) \equiv \varphi(x_1, y_1; t_1) \end{aligned} \tag{8.30}$$

when x_1 and y_1 are replaced by the values as given by (8.19). We choose $t = 0$ as that value of the parameter which yields the identity element.

Of necessity, $\beta(t, 0) = t$, since $f(x_1, y_1; 0) = x_1 = f(x, y; t)$. We differentiate (8.30) with respect to t_1, keeping x, y, and t fixed. Thus

$$\frac{\partial f(x, y; \beta)}{\partial \beta} \frac{\partial \beta}{\partial t_1} = \frac{\partial f(x_1, y_1; t_1)}{\partial t_1}$$

Evaluating at $t_1 = 0$ yields

$$\frac{\partial f(x, y; t)}{\partial t} \left(\frac{\partial \beta}{\partial t_1}\right)_{t_1=0} = \left.\frac{\partial f(x_1, y_1; t_1)}{\partial t_1}\right|_{t_1=0} = \xi(x_1, y_1)$$
$$\frac{\partial \varphi(x, y; t)}{\partial t} \left(\frac{\partial \beta}{\partial t_1}\right)_{t_1=0} = \left.\frac{\partial \varphi(x, y_1; t_1)}{\partial t_1}\right|_{t_1=0} = \eta(x_1, y_1)$$

From (8.19), $\dfrac{dx_1}{dt} = \dfrac{\partial f(x, y; t)}{\partial t}$, $\dfrac{dy_1}{dt} = \dfrac{\partial \varphi(x, y; t)}{\partial t}$, so that

$$\begin{aligned} \frac{dx_1}{dt} &= \lambda(t)\xi(x_1, y_1) \\ \frac{dy_1}{dt} &= \lambda(t)\eta(x_1, y_1) \end{aligned} \tag{8.31}$$

where $\lambda(t) = \left[\left(\dfrac{\partial \beta}{\partial t_1}\right)_{t_1=0}\right]^{-1}$. If t is additive, we have $\beta(t, t_1) = t + t_1$, $\lambda(t) = 1$ and (8.31) becomes (8.25). In any case

$$\frac{dx_1}{\xi(x_1, y_1)} = \frac{dy_1}{\eta(x_1, y_1)} = \lambda(t)\, dt \tag{8.32}$$

so that the change of variable, $u = \int \lambda(t)\, dt$, reduces (8.32) to (8.25). Thus one can always find a new parameter such that the group transformation is additive relative to the parameter.

Example 8.13. We consider $\dfrac{dx_1}{dt} = x_1^2$, $\dfrac{dy_1}{dt} = 1$, with $x_1 = x$, $y_1 = y$ at $t = 0$. The solution is $x_1 = x/(1 - xt)$, $y_1 = t + y$. Let the reader show that an additive group has been obtained.

Let us now consider an arbitrary function of x and y, say, $F(x, y)$. The value of $F(x_1, y_1)$ for t small can be obtained as follows: Since $F(x_1, y_1) = F(f(x, y; t), \varphi(x, y; t))$, we expand F in a Taylor series about $t = 0$, assuming that this is possible. Thus

$$F(x_1, y_1) = F(x_1, y_1)\Big|_{t=0} + \left.\frac{dF(x_1, y_1)}{dt}\right|_{t=0} t + \left.\frac{d^2F}{dt^2}\right|_{t=0} \frac{t^2}{2!} + \cdots \tag{8.33}$$

Now $F(x_1, y_1)\Big|_{t=0} = F(x, y)$. Also

$$\left.\frac{dF(x_1, y_1)}{dt}\right|_{t=0} = \left(\frac{\partial F}{\partial x_1}\frac{dx_1}{dt} + \frac{\partial F}{\partial y_1}\frac{dy_1}{dt}\right)_{t=0}$$
$$= \left[\frac{\partial F}{\partial x_1}\xi(x_1, y_1) + \frac{\partial F}{\partial y_1}\eta(x_1, y_1)\right]_{t=0}$$
$$= \frac{\partial F(x, y)}{\partial x}\xi(x, y) + \frac{\partial F(x, y)}{\partial y}\eta(x, y) = UF \qquad \text{see (8.24)}$$

provided the group transformation is additive. Moreover

$$\left.\frac{d^2F(x_1, y_1)}{dt^2}\right|_{t=0} = \left[\frac{\partial^2 F(x_1, y_1)}{\partial x_1^2}\xi^2(x_1, y_1) + 2\frac{\partial^2 F(x_1, y_1)}{\partial y_1\,\partial x_1}\xi(x_1, y_1)\eta(x_1, y_1)\right.$$
$$\left. + \frac{\partial^2 F(x_1, y_1)}{\partial y_1^2}\eta^2(x_1, y_1) + \frac{\partial F}{\partial x_1}\frac{d\xi(x_1, y_1)}{dt} + \frac{\partial F}{\partial y_1}\frac{d\eta(x_1, y_1)}{dt}\right]_{t=0}$$

From

$$U^2F(x, y) = \left(\xi\frac{\partial}{\partial x} + \eta\frac{\partial}{\partial y}\right)\left[\xi\frac{\partial F(x, y)}{\partial x} + \eta\frac{\partial F(x, y)}{\partial y}\right]$$

one easily shows that $\left.\frac{d^2F(x_1, y_1)}{dt^2}\right|_{t=0} = U^2F(x, y)$. By mathematical induction it can be shown that

$$\left.\frac{d^nF(x_1, y_1)}{dt^n}\right|_{t=0} = U^nF(x, y)$$

Equation (8.33) can be written in the form

$$F(x_1, y_1) = F(x, y) + (UF)t + (U^2F)\frac{t^2}{2!} + \cdots \tag{8.34}$$

For the special case $F(x_1, y_1) \equiv x_1$ we have

$$x_1(x, y, t) = x + (Ux)t + (U^2x)\frac{t^2}{2!} + \cdots$$

Example 8.14. For $\xi(x, y) = -y$, $\eta(x, y) = x$ we have

$$U = -y\frac{\partial}{\partial x} + x\frac{\partial}{\partial y} \qquad Ux = -y \qquad U^2x = -x \qquad U^3x = y \qquad U^4x = x$$

so that
$$x_1 = x - yt - x\frac{t^2}{2!} + y\frac{t^3}{3!} + x\frac{t^4}{4!} - y\frac{t^5}{5!} - \cdots$$
$$= x\left(1 - \frac{t^2}{2!} + \frac{t^4}{4!} - \cdots\right) - y\left(t - \frac{t^3}{3!} + \frac{t^5}{5!} - \cdots\right)$$
$$= x\cos t - y\sin t$$

Similarly $y_1 = x\sin t + y\cos t$, and the rotation group is obtained.

Problems

1. Show that the Einstein-Lorentz transformations form a one-parameter group

$$\bar{x} = \frac{x - V}{\sqrt{1 - V^2/c^2}}$$
$$\bar{t} = \frac{t - V/c^2 x}{\sqrt{1 - V^2/c^2}}$$

Note that the parameter V is not additive.

2. Find the transformation group obtained by integrating

$$\frac{dx_1}{dt} = y_1 \qquad \frac{dy_1}{dt} = x_1 \qquad \begin{matrix} x_1 = x \\ y_1 = y \end{matrix} \qquad \text{at } t = 0$$

3. If $\Omega(x, y)$ is an invariant under a group transformation $\Omega(x, y) = \Omega(x_1, y_1)$, show that $U\Omega(x, y) = 0$.

4. Solve $U\Omega(x, y) = 0$ for the rotation group, and show that $x^2 + y^2$ is an invariant for this group.

5. Given $\xi(x, y) = -y$, $\eta(x, y) = -x$, find the transformation group.

6. The differential equation $\frac{dy}{dx} = f\left(\frac{y}{x}\right)$ remains invariant under the group transformation $x = tx_1$, $y = ty_1$, for all t, since $\frac{y}{x} = \frac{y_1}{x_1}$ and $\frac{dy}{dx} = \frac{t\,dy_1}{t\,dx_1} = \frac{dy_1}{dx_1}$. Let

$$u = \frac{y}{x} = \frac{y_1}{x_1} = u_1$$

$v = \ln y = \ln y_1 + \ln t = v_1 + a$, $\ln t = a$. The equation $\frac{dy}{dx} = f\left(\frac{y}{x}\right)$ becomes $F\left(u, v, \frac{du}{dv}\right) = 0$, and $F\left(u, v, \frac{du}{dv}\right)$ is invariant under the translation

$$u = u_1, \; v = v_1 + a$$

Thus F must be independent of v, so that $F_1\left(u, \frac{du}{dv}\right) = 0$ can be integrated by a separation of variables. Integrate $\frac{dy}{dx} = \frac{x^2 + y^2}{x^2 - y^2}$.

8.10. Symmetric Functions. A function $F(x_1, x_2, \ldots, x_n)$ is a symmetric function in $x_1, x_2, \ldots, x_n$ if any permutation on the subscripts $1, 2, \ldots, n$ leaves f invariant. The function

$$f(x_1, x_2, x_3) = x_1x_2x_3 + x_1^2 + x_2^2 + x_3^2$$

is invariant under the symmetric group of order 6. If we expand the product $(x - x_1)(x - x_2) \cdots (x - x_n)$, we obtain

$$(x - x_1)(x - x_2) \cdots (x - x_n) = x^n - \sigma_1 x^{n-1} + \sigma_2 x^{n-2} - \sigma_3 x^{n-3} + \cdots (-1)^n \sigma_n$$

with
$$\begin{aligned}
\sigma_1 &= x_1 + x_2 + x_3 + \cdots + x_n \\
\sigma_2 &= x_1x_2 + x_1x_3 + \cdots + x_1x_n + x_2x_3 + \cdots + x_{n-1}x_n \\
\sigma_3 &= \Sigma x_i x_j x_k \qquad i \neq j \neq k \neq i \\
&\cdots\cdots\cdots\cdots\cdots \\
\sigma_n &= x_1x_2x_3 \cdots x_n
\end{aligned} \tag{8.35}$$

The σ_i, $i = 1, 2, \ldots, n$, of (8.35) are called the fundamental symmetric functions. It is apparent that any function of the σ's is a symmetric function. A few examples lead us to suspect that the converse is also true. We note that $f(x_1, x_2) = x_1^3 + x_2^3$ can be written as

$$(x_1 + x_2)^3 - 3x_1x_2(x_1 + x_2) = \sigma_1^3 - 3\sigma_1\sigma_2$$

with $\sigma_1 = x_1 + x_2$, $\sigma_2 = x_1x_2$. The symmetric function

$$f(x_1, x_2, x_3) = x_1^2x_2x_3 + x_2^2x_1x_3 + x_3^2x_1x_2$$

can be written as $f(x_1, x_2, x_3) = (x_1x_2x_3)(x_1 + x_2 + x_3) = \sigma_3\sigma_1$.

THEOREM 8.12. *The Fundamental Theorem of Symmetric Functions.* If $f(x_1, x_2, \ldots, x_n)$ is a symmetric function (multinomial) in $x_1, x_2, \ldots, x_n$, then $f(x_1, x_2, \ldots, x_n) \equiv g(\sigma_1, \sigma_2, \ldots, \sigma_n)$, that is, f can be written as a multinomial in the fundamental symmetric functions $\sigma_1, \sigma_2, \ldots, \sigma_n$.

The proof of the theorem is by mathematical induction. The theorem is certainly true for a function of one variable since $f(x_1) = f(\sigma_1)$. Let us assume the theorem true for a function of $n - 1$ variables. Thus if $f(x_1, x_2, \ldots, x_{n-1})$ is symmetric, then $f \equiv g(\sigma_1, \sigma_2, \ldots, \sigma_{n-1})$. Now let $f(x_1, x_2, \ldots, x_n)$ be any symmetric multinomial. Then $f(x_1, x_2, \ldots, x_{n-1}, 0)$ is a symmetric function in $x_1, x_2, \ldots, x_{n-1}$; so by assumption we can write $f(x_1, x_2, \ldots, x_{n-1}, 0) = g((\sigma_1)_0, (\sigma_2)_0, \ldots, (\sigma_{n-1})_0)$, where $\sigma_1 = x_1 + x_2 + \cdots + x_n$, $(\sigma_1)_0 = x_1 + x_2 + \cdots + x_{n-1} + 0$, etc., that is, $(\sigma_j)_0$ is the value of σ_j evaluated at $x_n = 0$, for $j = 1, 2, \ldots, n$. Now if $f(x_1, x_2, \ldots, x_n)$ is a multinomial of degree 1, that is, $f \equiv a(x_1 + x_2 + \cdots + x_n)$, then the theorem is certainly true. We assume the theorem true for any symmetric multinomial of degree $< k$. This is a double mathematical induction type of proof. Now we consider $h(x_1, x_2, \ldots, x_n) \equiv f(x_1, x_2, \ldots, x_n) - g(\sigma_1, \sigma_2, \ldots, \sigma_{n-1})$. It is obvious that h is also symmetric. Now

$$\begin{aligned}
h(x_1, x_2, \ldots, x_{n-1}, 0) = f(x_1, x_2, \ldots, x_{n-1}, 0) \\
- g((\sigma_1)_0, (\sigma_2)_0, \ldots, (\sigma_{n-1})_0) = 0
\end{aligned}$$

so that $x_n = 0$ is a zero of $h(x_1, x_2, \ldots, x_n)$, which implies that x_n is a factor of h. Since h is symmetric, of necessity $x_1, x_2, \ldots, x_{n-1}$ are also factors of h. Thus

$$h(x_1, x_2, \ldots, x_n) = x_1x_2 \cdots x_n s(x_1, x_2, \ldots, x_n) = \sigma_n s$$

where s is a symmetric function, and

$$f(x_1, x_2, \ldots, x_n) = g(\sigma_1, \sigma_2, \ldots, \sigma_{n-1}) + \sigma_n s(x_1, x_2, \ldots, x_n)$$

The degree of s is $k - n < k$, so that, by the induction hypothesis, $s(x_1, x_2, \ldots, x_n)$ can be expressed in terms of the fundamental symmetric functions, yielding

$$f(x_1, x_2, \ldots, x_n) = g(\sigma_1, \sigma_2, \ldots, \sigma_{n-1}) + \sigma_n t(\sigma_1, \sigma_2, \ldots, \sigma_n)$$

This concludes the proof of the fundamental theorem of symmetric functions.

Example 8.15. We consider

$$f(x_1, x_2, x_3) = x_1^2 + x_2^2 + x_3^2 + 2x_1x_2 + 2x_1x_3 + 2x_2x_3 - 3x_1^2x_2^2x_3 - 3x_1^2x_3^2x_2 - 3x_2^2x_3^2x_1$$

With patience one can show that f is symmetric. Let us consider

$$f(x_1, x_2, 0) = x_1^2 + x_2^2 + 2x_1x_2 = (x_1 + x_2)^2 = [(\sigma_1)_0]^2$$

The function $f(x_1, x_2, x_3) - \sigma_1^2$ should have the factor $x_1x_2x_3$. We note that

$$f - \sigma_1^2 = -3x_1x_2x_3(x_1x_2 + x_1x_3 + x_2x_3) = -3\sigma_2\sigma_3$$

so that $f = \sigma_1^2 - 3\sigma_2\sigma_3$.

Problems

1. Show that

$$(1 + x_1^2)(1 + x_2^2)(1 + x_3^2) = 1 + \sigma_1^2 - 2\sigma_2 + \sigma_2^2 - 2\sigma_1\sigma_3 + \sigma_3^2$$

2. Let α, β, γ be the zeros of $f(x) = x^3 + px^2 + qx + r$. Show that

$$\frac{\alpha^2 + \beta^2}{\alpha + \beta} + \frac{\alpha^2 + \gamma^2}{\alpha + \gamma} + \frac{\beta^2 + \gamma^2}{\beta + \gamma} = \frac{2q^2 - 2p^2q + 4pr}{pq - r}$$

3. Referring to Prob. 2, show that

$$\alpha^2\beta^2 + \alpha^2\gamma^2 + \beta^2\gamma^2 = q^2 - 2pr$$

4. Consider

$$\begin{aligned} s_1 &= x_1 + x_2 + \cdots + x_n \\ s_2 &= x_1^2 + x_2^2 + \cdots + x_n^2 \\ s_3 &= x_1^3 + x_2^3 + \cdots + x_n^3 \\ &\cdots\cdots\cdots\cdots\cdots \\ s_n &= x_1^n + x_2^n + \cdots + x_n^n \end{aligned}$$

Express σ_3 in terms of s_1, s_2, s_3.

8.11. Polynomials. We shall be interested in polynomials of the type

$$p(x) = \sum_{k=0}^{n} c_k x^k = c_0 + c_1 x + c_2 x^2 + \cdots + c_n x^n \tag{8.36}$$

The coefficients c_i, $i = 0, 1, 2, \ldots, n$, are assumed to belong to a field F. The reader is referred to Sec. 4.1 for the properties of a field. If a set of elements belong to a field, we can perform the simple operations of addition, multiplication, and their inverses on these elements. The zero element and unit element belong to a field and the distributive laws, $a(b + c) = ab + ac$, $(b + c)a = ba + ca$, hold. The complex numbers form a field. A subfield of the field of complex numbers is the field of real numbers. The rational numbers form a subfield of the field of real numbers. Let us consider the set of real numbers of the form $a + b\sqrt{2}$, where a and b are rational numbers. The sum and product of two such numbers is evidently a number of this set. The unit element is $1 = 1 + 0\sqrt{2}$, and the zero element is $0 = 0 + 0\sqrt{2}$. We must show that every nonzero number has a unique inverse. From the fact that $\sqrt{2}$ is irrational we have $a + b\sqrt{2} = 0$ if and only if $a = b = 0$. Assume $a + b\sqrt{2} \neq 0$. Then it is easy to show that $[a/a^2 - 2b^2] + [(-b)/a^2 - 2b^2]\sqrt{2}$ is the unique inverse of $a + b\sqrt{2}$. Let the reader solve

$$(a + b\sqrt{2})(x + y\sqrt{2}) = 1 + 0\sqrt{2}$$

for x and y, $a^2 - 2b^2 \neq 0$. We call this field an algebraic extension of the field of rationals, written $R(\sqrt{2})$. We shall discuss algebraic extensions of a field in the next section.

If $f(x)$ and $g(x)$ are polynomials with coefficients in a field F, we say that $g(x)$ divides $f(x)$ if a polynomial $h(x)$ with coefficients in F exists such that $f(x) = g(x)h(x)$. An important theorem concerning polynomials is as follows: Given two polynomials with coefficients in F, say, $f(x)$ of degree m, $g(x)$ of degree n, there exist two polynomials with coefficients in F, $r(x)$ of degree $< n$, $s(x)$ of degree $< m$, such that

$$r(x)f(x) + s(x)g(x) = d(x) \tag{8.37}$$

where $d(x)$ is the greatest common divisor of $f(x)$ and $g(x)$. All other divisors of both $f(x)$ and $g(x)$ are divisors of $d(x)$. The coefficients of $d(x)$ are in F.

Proof. We consider the set of all polynomials of the form $a(x)f(x) + b(x)g(x)$, with $a(x)$ and $b(x)$ arbitrary. These polynomials have degrees (exponent of highest power of x) greater than or equal to zero. A constant is a polynomial of degree zero. Let $a(x) = r(x)$, $b(x) = s(x)$ be those polynomials for which the degree of $a(x)f(x) + b(x)g(x)$ is a minimum, but not identically zero, and let

$$d(x) = r(x)f(x) + s(x)g(x)$$

Any divisor of both $f(x)$ and $g(x)$ is obviously a divisor of $d(x)$. We show now that $d(x)$ divides both $f(x)$ and $g(x)$. If $d(x)$ does not divide $f(x)$, then by division (always possible since the coefficients are in a field)

$$f(x) = q(x)d(x) + t(x) \qquad 0 \leqq \text{degree of } t(x) < \text{degree of } d(x)$$

$d(x) \not\equiv 0$. Hence $qd = qrf + qsg = f - t$, and

$$(1 - qr)f + (-qs)g = t(x) \tag{8.38}$$

From the definition of $d(x)$, (8.38) is impossible unless $t(x) \equiv 0$, so that $d(x)$ divides $f(x)$, written $d(x)/f(x)$. Similarly $d(x)/g(x)$. Let the reader show that degree of $r(x) <$ degree of $g(x)$, degree of $s(x) <$ degree of $f(x)$.

Example 8.16. Let $f(x) = x^2 + 1$, $g(x) = x$. Then $1 \cdot (x^2 + 1) + (-x)x \equiv 1$, so that 1 is the greatest common divisor of $f(x)$ and $g(x)$. We say that $f(x)$ and $g(x)$ are relatively prime.

Problems

1. Show that the integers do not form a field.

2. Show that the set of numbers $x + iy\sqrt{3}$, x and y ranging over all rational numbers, forms a field.

3. Show that $x + 1$ is the greatest common divisor of $f(x) = x^3 + 1$, $g(x) = (x + 1)^2$. Find $r(x)$ and $s(x)$ such that $r(x)f(x) + s(x)g(x) = x + 1$. Are $r(x)$ and $s(x)$ unique?

4. Let $f(x)$ and $g(x)$ be polynomials with coefficients in a field F_1, with F_1 a subfield of the field F. If $f(x)$ and $g(x)$ are relatively prime with respect to the field F_1, show that $f(x)$ and $g(x)$ are relatively prime relative to the field F. How does this apply to Prob. 3?

5. Do the set of numbers $a + b\sqrt{2} + c\sqrt{3}$, a, b, c rational, form a field? What of $a + b\sqrt{2} + c\sqrt{3} + d\sqrt{6}$?

6. Let F be any field. If we label the zero and unit elements 0 and 1, respectively, and define $1 + 1 = 2$, etc., show that every field contains the rationals as a subfield.

8.12. The Algebraic Extension of a Field F. Let $p(x) = \sum_{k=0}^{n} c_k x^k$ be a polynomial with coefficients in a field F. We say that $p(x)$ is *irreducible* in F if $p(x)$ cannot be written as a product (nontrivial) of two polynomials with coefficients in F. We say that ρ is a zero of $p(x)$ or a root of $p(x) = 0$ if $p(\rho) = 0$. Kronecker has shown that one can always extend the field F in such a fashion that a new number ρ is introduced, yielding $p(\rho) = 0$. The solution of $x^2 + 1 = 0$ involves the invention of the number $i = \sqrt{-1}$.

THEOREM 8.13. Let $p(x)$ be an irreducible polynomial in F such that $p(\rho) = 0$. If $q(x)$ has coefficients in F, then $q(\rho) = 0$ implies that $p(x)/q(x)$.

Proof. Since $p(x)$ is irreducible, $p(x)/q(x)$, or $p(x)$ and $q(x)$ are relatively prime. From Sec. 8.11

$$A(x)p(x) + B(x)q(x) \equiv 1$$

for some $A(x)$, $B(x)$. Thus $A(\rho)p(\rho) + B(\rho)q(\rho) = 0 = 1$, a contradiction. Thus $p(x)$ divides $q(x)$.

Two corollaries follow immediately from Theorem 8.13.

COROLLARY 1. $p(x)$ is that irreducible polynomial of smallest degree such that $p(\rho) = 0$.

COROLLARY 2. If we make the leading coefficient of $p(x)$ unity, $p(x) = x^n + c_{n-1}x^{n-1} + \cdots + c_0$, then $p(x)$ is unique.

DEFINITION 8.7. The n zeros of $p(x)$, say, $\rho_1, \rho_2, \ldots, \rho_n$, are called *conjugates* of each other. The zeros of $x^2 - 2$, namely, $\sqrt{2}$, $-\sqrt{2}$, are conjugates of each other. It follows from Theorem 8.13 that any one of the n conjugates of $p(x)$ determines $p(x)$ uniquely, assuming that the leading coefficient of $p(x)$ is unity.

THEOREM 8.14. Let ρ be a zero of the irreducible polynomial $p(x)$ with coefficients in a field F. The numbers $1, \rho, \rho^2, \ldots, \rho^{n-1}$ are linearly independent relative to F, since if constants c_i, $i = 0, 1, 2, \ldots, n - 1$, exist in F such that $\sum_{k=0}^{n-1} c_k\rho^k = 0$, then $t(x) = \sum_{k=0}^{n-1} c_kx^k$ is a polynomial in F of degree $< n$ with $t(\rho) = 0$, a contradiction to Corollary 1.

An important result is the following: Given the irreducible polynomial $p(x)$ with coefficients in F such that $p(\rho) = 0$, one can obtain a field F_1 containing ρ such that F is a subfield of F_1. We call F_1 an *algebraic extension of* F and write $F_1 = F(\rho)$. The elements of $F_1(\rho)$ are of the form

$$\sigma = \frac{\sum_{i=0}^{r} a_i\rho^i}{\sum_{j=0}^{s} b_j\rho^j} \tag{8.39}$$

with $\sum_{j=0}^{s} b_j\rho^j \neq 0$, a_i in F, b_j in F.

Let the reader show that the elements of the type given by (8.39) form a field. By using the fact that $\rho^n + c_{n-1}\rho^{n-1} + \cdots + c_0 = 0$ one has

$$\begin{aligned}
\rho^n &= -(c_0 + c_1\rho + \cdots + c_{n-1}\rho^{n-1}) \\
\rho^{n+1} &= -(c_0\rho + c_1\rho^2 + \cdots + c_{n-1}\rho^n) \\
&= -(c_0\rho + c_1\rho^2 + \cdots + c_{n-2}\rho^{n-1}) \\
&\qquad + c_{n-1}(c_0 + c_1\rho + \cdots + c_{n-1}\rho^{n-1})
\end{aligned}$$

so that powers of ρ higher than $n - 1$ can always be reduced to lower powers. For example, if $\rho^3 + \rho + 1 = 0$, then

$$\rho^3 = -(\rho + 1) \qquad \rho^4 = -\rho - \rho^2$$
$$\rho^5 = -\rho^2 - \rho^3 = -\rho^2 + \rho + 1 \qquad \text{etc.}$$

We show next that the elements σ of (8.39) can actually be written as polynomials in ρ. The denominator of (8.39) is $\sum_{j=0}^{s} b_j\rho^j$. Let

$$h(x) = \sum_{j=0}^{s} b_j x^j$$

Since $h(\rho) \neq 0$, of necessity $h(x)$ and $p(x)$ are relatively prime. From Sec. 8.11 we have

$$\begin{aligned} A(x)h(x) + B(x)p(x) &= 1 \\ A(\rho)h(\rho) + B(\rho)p(\rho) &= 1 = A(\rho)h(\rho) \end{aligned}$$

so that the inverse of $h(\rho)$ is $A(\rho)$. Thus

$$\sigma(\rho) = A(\rho) \sum_{i=0}^{r} a_i\rho^i = \sum_{i=0}^{n-1} d_i\rho^i$$

by using the fact that $p(\rho) = 0$ to eliminate higher powers of ρ.

Example 8.17. The field $R(\sqrt{2})$ is obtained by considering $x^2 - 2 = 0$, $\rho^2 - 2 = 0$, $\rho = \sqrt{2}$. The same field is obtained if we consider $\rho = -\sqrt{2}$. This is not true for all cases. The coefficients of $x^2 - 2$ belong to the field of rationals. The elements of $R(\sqrt{2})$ are of the type $a + b\rho = a + b\sqrt{2}$, a and b rational.

Next we consider the polynomial equation $p(x) \equiv x^2 - 3 = 0$. It is easy to show that $p(x)$ is irreducible in the field $R(\sqrt{2})$ since $\sqrt{3} = a + b\sqrt{2}$, a and b rational, is impossible. Let the reader show this. Now the coefficients of $x^2 - 3$, namely, 1 and -3, belong to $R(\sqrt{2})$. Hence we obtain an algebraic extension of $R(\sqrt{2})$ by considering the set of numbers of the type $(a + b\sqrt{2}) + (c + d\sqrt{2})\sqrt{3}$, a, b, c, d rational. This is the field $R(\sqrt{2}, \sqrt{3})$, and its elements are of the form $a + b\sqrt{2} + c\sqrt{3} + d\sqrt{6}$, a, b, c, d rational. In a similar manner we could construct $R(\sqrt{3}, \sqrt{2})$. It is evident that $R(\sqrt{2}, \sqrt{3}) = R(\sqrt{3}, \sqrt{2})$.

THEOREM 8.15. Let α be any number of the algebraic extension $F(\rho)$ obtained from $p(\rho) = 0$. Then α is the zero of a polynomial $f(x)$ with coefficients in F. We say that $f(x) = 0$ is the *principal equation* of $x = \alpha$.

Proof. Let the conjugates of $p(x) = 0$ be $\rho = \rho_1, \rho_2, \rho_3, \ldots, \rho_n$. Since α is in $F(\rho)$, of necessity,

$$\alpha = \alpha_1 = a_0 + a_1\rho_1 + a_2\rho_1^2 + \cdots + a_{n-1}\rho_1^{n-1}$$

We form the conjugates of α_1 defined by

$$\begin{aligned} \alpha_2 &= a_0 + a_1\rho_2 + a_2\rho_2^2 + \cdots + a_{n-1}\rho_2^{n-1} \\ \alpha_3 &= a_0 + a_1\rho_3 + a_2\rho_3^2 + \cdots + a_{n-1}\rho_3^{n-1} \\ &\cdots\cdots\cdots\cdots\cdots\cdots\cdots\cdots \\ \alpha_n &= a_0 + a_1\rho_n + a_2\rho_n^2 + \cdots + a_{n-1}\rho_n^{n-1} \end{aligned} \tag{8.40}$$

Certainly $f(x) \equiv (x - \alpha_1)(x - x_2) \cdots (x - \alpha_n)$ has α_1 as a zero. We need only show that the coefficients of $f(x)$ are in F. Now

$$f(x) = x^n - \left(\sum_{j=1}^{n} \alpha_j\right) x^{n-1} + \cdots + (-1)^n \alpha_1\alpha_2 \cdots \alpha_n$$

Moreover $\Sigma\alpha_j$, $\Sigma\alpha_j\alpha_k$, etc., are certainly symmetric in the ρ_i, $i = 1, 2, \ldots, n$, and so can be written as polynomials involving the fundamental symmetric functions,

$$\begin{gathered} \rho_1 + \rho_2 + \cdots + \rho_n \\ \rho_1\rho_2 + \rho_1\rho_3 + \cdots + \rho_{n-1}\rho_n \\ \cdots\cdots\cdots\cdots\cdots \\ \rho_1\rho_2 \cdots \rho_n \end{gathered} \tag{8.41}$$

The elements of (8.41) are simply the coefficients of $p(x)$, except for some negative factors, as seen by expanding $(x - \rho_1)(x - \rho_2) \cdots (x - \rho_n)$. Thus the coefficients of $f(x)$ are in F.

Example 8.18. Referring to Example 8.17, we consider $\alpha = \alpha_1 = 3 - 4\sqrt{2}$. Then $\alpha_2 = 3 + 4\sqrt{2}$, and

$$\begin{aligned} f(x) &= (x - \alpha_1)(x - \alpha_2) = (x - 3 + 4\sqrt{2})(x - 3 - 4\sqrt{2}) \\ &= (x - 3)^2 - 32 \\ &= x^2 - 6x - 23 \end{aligned}$$

The coefficients of $f(x)$ are in the field of rationals.

Example 8.19. Let us find the principal equation of $\alpha = 1 + \rho^2$, where ρ is a zero of $p(x) = x^3 + 3x + 2$. Let the zeros of $p(x)$ be $\rho = \rho_1, \rho_2, \rho_3$. Then

$$\begin{aligned} f(x) &= [x - (1 + \rho_1^2)][x - (1 + \rho_2^2)][x - (1 + \rho_3^2)] \\ &= x^3 - [3 + \rho_1^2 + \rho_2^2 + \rho_3^2]x^2 + [(1 + \rho_1^2)(1 + \rho_2^2) + (1 + \rho_1^2)(1 + \rho_3^2) \\ &\qquad + (1 + \rho_2^2)(1 + \rho_3^2)] - (1 + \rho_1^2)(1 + \rho_2^2)(1 + \rho_3^2) \end{aligned}$$

From

$$\begin{aligned} \sigma_1 &= \rho_1 + \rho_2 + \rho_3 = 0 \\ \sigma_2 &= \rho_1\rho_2 + \rho_2\rho_3 + \rho_3\rho_1 = 3 \\ \sigma_3 &= \rho_1\rho_2\rho_3 = -2 \end{aligned}$$

we have

$$\begin{aligned} \rho_1^2 + \rho_2^2 + \rho_3^2 &= (\rho_1 + \rho_2 + \rho_3)^2 - 2(\rho_1\rho_2 + \rho_2\rho_3 + \rho_3\rho_1) \\ &= 0 - 2 \cdot 3 = -6 \end{aligned}$$

Also $(1 + \rho_1^2)(1 + \rho_2^2)(1 + \rho_3^2) = 1 + \sigma_1^2 - 2\sigma_2 + \sigma_2^2 - 2\sigma_1\sigma_3 + \sigma_3^2 = 8$ (see Prob. 1, Sec. 8.10). Moreover

$$\begin{aligned} &(1 + \rho_1^2)(1 + \rho_2^2) + (1 + \rho_1^2)(1 + \rho_3^2) + (1 + \rho_2^2)(1 + \rho_3^2) \\ &\qquad = 3 + 2(\rho_1^2 + \rho_2^2 + \rho_3^2) + (\rho_1\rho_2 + \rho_1\rho_3 + \rho_3\rho_2)^2 - 2\rho_1\rho_2\rho_3(\rho_1 + \rho_2 + \rho_3) \\ &\qquad = 3 + 2(-6) + 9 = 0 \end{aligned}$$

Hence $f(x) = x^3 + 3x^2 - 8$. We check that $1 + \rho^2$ satisfies

$$(1 + \rho^2)^3 + 3(1 + \rho^2)^2 - 8 = 0$$

We have

$$\begin{aligned} f(1 + \rho^2) &= \rho^6 + 6\rho^4 + 9\rho^2 - 4 \\ &= (-2 - 3\rho)^2 + 6\rho(-2 - 3\rho) + 9\rho^2 - 4 = 0 \end{aligned}$$

since $\rho^3 + 3\rho + 2 = 0$, or $\rho^3 = -2 - 3\rho$.

Problems

1. Show that $x^3 + x + 1 = 0$ has no rational roots. *Hint:* Assume $x = m/n$ is a zero of $p(x) = x^3 + x + 1$, m and n integers in lowest form. Show that of necessity $m = \pm 1$, $n = \pm 1$. By testing $x = \pm 1$ show that $p(x)$ has no rational zeros. Why is $p(x)$ irreducible in the field of rationals?

2. The polynomial $p(x) = x^4 + x^3 + 2x^2 + x + 1$ has no rational zeros. However, $p(x)$ is reducible in the field of rationals since

$$x^4 + x^3 + 2x^2 + x + 1 = (x^2 + x + 1)(x^2 + 1)$$

Is this a contradiction?

3. Let ρ be a zero of $p(x) = x^3 + x + 1$ (see Prob. 1). Express $1 + \rho + \rho^2/1 - \rho + \rho^2$ as a second-degree polynomial in ρ.

4. Let $f(x) = a_0 + a_1x + \cdots + a_nx^n$, a_i, $i = 0, 1, 2, \ldots, n$, integers. If a prime number p exists such that p does not divide a_n, p divides a_i for $i < n$, p^2 does not divide a_0, then $f(x)$ is irreducible in the field of rationals. This criterion is due to *Eisenstein.* Show that $x^n - p$ is irreducible in the field of rationals, p a prime. Show that $f(x + 1) = x^2 + 2x + 2 = (x + 1)^2 + 1$ is irreducible in the field of rationals. Why can one immediately make the same statement for $f(x) = x^2 + 1$?

5. Consider $p(x) = x^4 + 3x^2 + 9$. Let the zeros of $p(x)$ be $\rho, -\rho, \sigma, -\sigma$. Show that the principal equation of $\alpha = 1 + 6\rho + \rho^3$ is $f(x) = [(x - 1)^2 - 81]^2$.

6. Consider $x^2 - 2 = 0$, $x^3 - 2 = 0$, and obtain the field $R(\sqrt{2}, \sqrt[3]{2})$.

7. A polynomial is said to be separable if it has no multiple zeros. Show that if $p(x)$ is irreducible it is separable.

8. Consider $f(x) = (x - \alpha_1)(x - \alpha_2) \cdots (x - \alpha_n)$, the α_i defined by (8.40). If $f(x)$ is irreducible, then $f(x) = [f(x)]^1$. If $f(x)$ is reducible in F, then

$$f(x) = f_1(x)f_2(x) \cdots f_r(x)$$

Assume $f_1(x)$ is a function for which $f_1(\alpha_1) = 0$, so that $f_1(\alpha_1(x))$ is zero for $x = \rho_1$. Why does $p(x)/f_1(\alpha_1(x))$? Hence explain why $f(\alpha_1(x))$ is zero for $x = \rho_1, \rho_2, \ldots, \rho_n$, so that $f(\alpha_1) = f(\alpha_2) = \cdots = f(\alpha_n) = 0$. If the α_i are not distinct, show that $f(x) = [f_1(x)]^k$, otherwise $f(x) = f_1(x)$.

8.13. The Galois Resolvent. Let $f(x)$ be a principal equation of $x = \alpha$. We say that α is a *primitive* number of $F(\rho)$ if $f(x)$ is irreducible in F.

THEOREM 8.16. If $\alpha = \alpha_1$ is a primitive number of $F(\rho)$, then

$$F(\alpha) = F(\rho)$$

that is, every number of $F(\alpha)$ belongs to $F(\rho)$, and conversely.

Proof. Let $f(x) = (x - \alpha_1)(x - \alpha_2) \cdots (x - \alpha_n)$ be the principal equation of $\alpha = \alpha_1$. Define $\varphi(x)$ by

$$\varphi(x) = f(x)\left(\frac{\rho_1}{x - \alpha_1} + \frac{\rho_2}{x - \alpha_2} + \cdots + \frac{\rho_n}{x - \alpha_n}\right)$$

The ρ_i, $i = 1, 2, \ldots, n$, are the conjugate zeros of $p(x)$. Then

$$\varphi(\alpha_1) = \rho_1(\alpha_1 - \alpha_2)(\alpha_1 - \alpha_3) \cdots (\alpha_1 - \alpha_n) = \rho_1 f'(\alpha_1)$$

Since $f(x)$ is assumed irreducible, $f'(\alpha_1) \neq 0$ (see Prob. 6, Sec. 8.12).

Thus $\rho_1 = \varphi(\alpha_1)/f'(\alpha_1)$, so that ρ_1 is a number in $F(\alpha_1)$. Any number of $F(\rho_1)$ thus belongs to $F(\alpha_1)$. Why? Since α_1 is a polynomial in ρ_1, it follows that any number in $F(\alpha_1)$ is a number in $F(\rho_1)$. Q.E.D.

THEOREM 8.17. Let ρ_1 be a zero of $p_1(x)$, σ_1 a zero of $p_2(x)$, $p_1(x)$ and $p_2(x)$ irreducible in F. We can form the fields $F(\rho_1)$, $F(\sigma_1)$. If $p_2(x)$ is reducible in $F(\rho_1)$, we consider the irreducible factor of $p_2(x)$ having σ_1 as a zero. In this way we can form $F(\rho_1, \sigma_1)$ (see Example 8.17). Let the reader show that $F(\rho_1, \sigma_1) = F(\sigma_1, \rho_1)$. We show that an irreducible polynomial in F exists yielding a field $F(\tau)$ such that $F(\tau) = F(\rho_1, \sigma_1)$.

Proof. Let

$$\begin{aligned}\tau_1 &= \alpha\rho_1 + \beta\sigma_1\\ \tau_2 &= \alpha\rho_1 + \beta\sigma_2\\ &\cdots\cdots\cdots\\ \tau_n &= \alpha\rho_1 + \beta\sigma_n\\ \tau_{n+1} &= \alpha\rho_2 + \beta\sigma_1\\ &\cdots\cdots\cdots\\ \tau_{nm} &= \alpha\rho_m + \beta\sigma_n\end{aligned}$$

with α and β in F. The ρ_i, $i = 1, 2, \ldots, m$, are the conjugates of $p_1(x) = 0$, and the σ_i, $i = 1, 2, \ldots, n$, are the conjugates of $p_2(x) = 0$. We choose α and β so that the τ_j, $j = 1, 2, \ldots, mn$, are all distinct. This can be done since, if $\alpha\rho_i + \beta\sigma_j = \alpha\rho_k + \beta\sigma_l$, then

$$\frac{\alpha}{\beta} = \frac{\sigma_l - \sigma_k}{\rho_i - \rho_k} \qquad i \neq k \tag{8.42}$$

There are only a finite number of ratios in (8.42), so that α and β can be chosen such that (8.42) fails to hold for all $i \neq k$, $j \neq l$. For $i = k$ we have $\beta\sigma_j \neq \beta\sigma_l$ if $\sigma_j \neq \sigma_l$. For $j = l$ we have $\alpha\rho_i \neq \alpha\rho_k$ for $i \neq k$. Next we form

$$\begin{aligned}g(x) &= (x - \tau_1)(x - \tau_2) \cdots (x - \tau_{mn})\\ &= x^{mn} - (\Sigma\tau_i)x^{mn-1} + \cdots + (-1)^{mn}\tau_1\tau_2 \cdots \tau_{mn}\end{aligned} \tag{8.43}$$

Any interchange of two ρ's leaves the coefficients of $g(x)$ invariant, as does any interchange of two σ's. Thus the coefficients of $g(x)$ are in F. That factor of $g(x)$ irreducible in F having τ_1 as a zero generates $F(\tau_1)$ with $\tau_1 = \alpha\rho_1 + \beta\sigma_1$. Since $\tau_1 = \alpha\rho_1 + \beta\sigma_1$, $F(\tau_1)$ is a subfield of $F(\rho_1, \sigma_1)$. We now show that any number η of $F(\rho_1, \sigma_1)$ belongs to $F(\tau_1)$ so that $F(\tau_1) = F(\rho_1, \sigma_1)$. If η is in $F(\rho_1, \sigma_1)$, η is of the form

$$\begin{aligned}\eta = \eta_1 &= (a_0 + a_1\rho_1 + \cdots + a_{m-1}\rho_1^{m-1})\\ &\qquad + (b_0 + b_1\rho_1 + \cdots + b_{m-1}\rho_1^{m-1})\sigma_1\\ &\qquad\qquad + \cdots + (c_0 + c_1\rho_1 + \cdots + c_{m-1}\rho_1^{m-1})\sigma_1^{n-1}\\ &= \sum_{j=0}^{n-1}\sum_{i=0}^{m-1} d_{ij}\rho_1^i\sigma_1^j\end{aligned}$$

with the d_{ij} in F. The $mn - 1$ other conjugates of ρ_1 can be formed by replacing ρ_1 by $\rho_2, \rho_3, \ldots, \rho_m$ and σ_1 by $\sigma_2, \sigma_3, \ldots, \sigma_n$ in all possible ways. We define $\varphi(x)$ by

$$\varphi(x) = g(x)\left(\frac{\eta_1}{x - \tau_1} + \frac{\eta_2}{x - \tau_2} + \cdots + \frac{\eta_{mn}}{x - \tau_{mn}}\right) \tag{8.44}$$

Let the reader show that $\eta_1 = \varphi(\tau_1)/g'(\tau_1)$, $g'(\tau_1) \neq 0$, since $g(x) = 0$ has no multiple roots. Hence η_1 belongs to $F(\tau_1)$. Q.E.D.

Example 8.20. We consider $x^2 - 2 = 0$, $x^2 - 3 = 0$ with $\rho_1 = \sqrt{2}$, $\sigma_1 = \sqrt{3}$. We form $\tau_1 = \sqrt{2} + \sqrt{3}$, $\tau_2 = \sqrt{2} - \sqrt{3}$, $\tau_3 = -\sqrt{2} + \sqrt{3}$, $\tau_4 = -\sqrt{2} - \sqrt{3}$. No two of the τ's are equal. Then

$$g(x) = (x - \tau_1)(x - \tau_2)(x - \tau_3)(x - \tau_4) = x^4 - 10x^2 + 1$$

The coefficients of $g(x)$ are in the field of rationals. It can be easily shown that $g(x)$ is irreducible in F. Thus $g(x) = 0$ generates $F(\sqrt{2} + \sqrt{3})$, which has elements of the form

$$a + b(\sqrt{2} + \sqrt{3}) + c(\sqrt{2} + \sqrt{3})^2 + d(\sqrt{2} + \sqrt{3})^3 = x + y\sqrt{2} + u\sqrt{3} + v\sqrt{6}$$

x, y, u, v rational. Thus $F(\sqrt{2} + \sqrt{3}) = F(\sqrt{2}, \sqrt{3})$ (see Example 8.16).

A special case of Theorem 8.17 occurs if $p_1(x) \equiv p_2(x)$ and $\rho_1 \neq \sigma_1 = \rho_2$. The field $F(\rho_1, \rho_2)$ can be generated. Continuing, we can adjoin to F all the roots of $p(x) = 0$, obtaining $F(\rho_1, \rho_2, \ldots, \rho_n)$. A single number τ exists such that $F(\tau) = F(\rho_1, \rho_2, \ldots, \rho_n)$. The irreducible polynomial having τ as a zero is called the *Galois resolvent* of $p(x) = 0$. It is a polynomial of degree $\leq n!$, since we need only find constants $\{\alpha_k\}$ such that $\tau = \sum_{k=1}^{n} \alpha_k \rho^k$ and the other τ's obtained from the $n! - 1$ permutations of the ρ's are different from each other. The rest of the proof proceeds as in Theorem 8.17.

DEFINITION 8.8. If $\rho_1, \rho_2, \ldots, \rho_n$ are the zeros of the irreducible polynomial $p(x)$ such that

$$F(\rho_1) = F(\rho_2) = \cdots = F(\rho_n)$$

then $p(x)$ or $p(x) = 0$ is said to be *normal* and $F(\rho)$ is called a *normal extension* of F. In this case $F(\rho_1) = F(\rho_1, \rho_2, \ldots, \rho_n)$ and $p(x)$ is its own Galois resolvent.

Example 8.21. Consider $p(x) = x^3 - 3x + 1 = 0$. Let ρ be a zero of $p(x)$. By division $x^3 - 3x + 1 = (x - \rho)(x^2 + \rho x + \rho^2 - 3)$. The other two roots of $p(x) = 0$ are $\rho_{1,2} = (-1 \pm \sqrt{12 - 3\rho^2})/2$. We show that $12 - 3\rho^2$ is a perfect square in $F(\rho)$. Assume $12 - 3\rho^2 = (a + b\rho + c\rho^2)^2$, a, b, c rational. Using the fact that $\rho^3 = 3\rho - 1$, $\rho^4 = 3\rho^2 - \rho$, we have $12 = a^2 - 2bc$, $0 = -c^2 + 2ab + 6bc$, $-3 = b^2 + 3c^2 + 2ac$. It is easy to check that $a = 4$, $b = -1$, $c = -2$ is a rational solution so that $\rho_1 = 2 - \rho - \rho^2$, $\rho_2 = \rho^2 - 2$, which implies $F(\rho) = F(\rho_1) = F(\rho_2)$, and $p(x)$ is normal.

Problems

1. Consider $p_1(x) = x^2 - 2 = 0$, $p_2(x) = x^4 - 2 = 0$, with $p_1(x)$, $p_2(x)$ irreducible in the field R of rationals. Find $R(\sqrt{2}, \sqrt[4]{2})$.

2. Find a Galois resolvent for $p(x) = x^3 + 2x + 1$.

3. Show that $p(x) = x^3 - 3(r^2 + 3s^2)x + 2r(r^2 + 3s^2)$ is normal for r and s integers provided $p(x)$ is irreducible in the field of rationals.

4. The discriminant of $p(x) = x^3 + px + q$ is $\Delta = -4p^3 - 27q^2$. Show that the discriminant of $p(x)$ of Prob. 3 is a perfect square.

5. Show that $\eta_1 = \varphi(\tau_1)/g'(\tau_1)$, $\varphi(x)$ defined by (8.44).

8.14. Automorphisms. The Galois Group. Let us consider the field $R(\sqrt{2})$ composed of all elements of the form $a + b\sqrt{2}$, a and b rational. We consider the one-to-one correspondence between $a + b\sqrt{2}$ and its conjugate, $a - b\sqrt{2}$, written $a + b\sqrt{2} \leftrightarrow a - b\sqrt{2}$. Let us look more closely at this correspondence. If

$$\begin{aligned} x + y\sqrt{2} &\leftrightarrow x - y\sqrt{2} \\ u + v\sqrt{2} &\leftrightarrow u - v\sqrt{2} \end{aligned}$$

then

$$\begin{aligned} (x + y\sqrt{2}) + (u + v\sqrt{2}) &= (x + u) + (y + v)\sqrt{2} \\ &\leftrightarrow (x + u) - (y + v)\sqrt{2} \\ &= (x - y\sqrt{2}) + (u - v\sqrt{2}) \\ (x + y\sqrt{2})(u + v\sqrt{2}) &= (xu + 2yv) + (xv + yu)\sqrt{2} \\ &\leftrightarrow (xu + 2yv) - (xv + yu)\sqrt{2} \\ &= (x - y\sqrt{2})(u - v\sqrt{2}) \end{aligned}$$

A one-to-one mapping of a field into itself, written $\alpha \leftrightarrow \alpha'$, or $\alpha' = f(\alpha)$, $\alpha = f^{-1}(\alpha')$, such that $\alpha \leftrightarrow \alpha'$, $\beta \leftrightarrow \beta'$, imply $\alpha + \beta \leftrightarrow \alpha' + \beta'$, $\alpha\beta \leftrightarrow \alpha'\beta'$, for all α and β of the field, is called an automorphism of the field. From above we see that the mapping $a + b\sqrt{2} \leftrightarrow a - b\sqrt{2}$ is an automorphism of $R(\sqrt{2})$. Every field has at least one automorphism attached to it, the identity mapping, $\alpha \leftrightarrow \alpha$, for all α of the field. Not all one-to-one mappings of a field into itself yield an automorphism. Witness the mapping $x \leftrightarrow 2x$, with x ranging over the field of real numbers. For $y \leftrightarrow 2y$ we have $xy \leftrightarrow 2(xy) \neq (2x)(2y)$. Under an automorphism the zero element maps into itself, as does the unit element, for $\alpha \leftrightarrow \alpha'$, $0 \leftrightarrow x$, imply $\alpha + 0 = \alpha \leftrightarrow \alpha' + x = \alpha'$, so that $x = 0$, and $\alpha \leftrightarrow \alpha'$, $1 \leftrightarrow y$, imply $\alpha \cdot 1 = \alpha \leftrightarrow \alpha' y = \alpha'$, so that $y = 1$. We say that 0 and 1 are left invariant under any automorphism. Let the reader show that the rationals of a field F remain invariant for all automorphisms of F. It is also a simple matter to show that the automorphisms of a field form a group. If $\alpha = f_1(\beta)$, $\beta = f_2(\gamma)$ are automorphisms A_1 and A_2, we define A_1A_2 as the automorphism $\alpha = f_1(f_2(\gamma)) = f_3(\gamma)$. Let the reader show that under this definition of multiplication the set of automorphisms of F form a group. The unit element of the group is the identity automorphism.

Let us return to the irreducible polynomial $p(x)$ with coefficients in F. We shall be interested in the extension field $N = F(\rho_1, \rho_2, \ldots, \rho_n)$ with ρ_i, $i = 1, 2, \ldots, n$, such that $p(\rho_i) = 0$. We consider only those automorphisms of $F(\rho_1, \rho_2, \ldots, \rho_n)$ which leave the elements of F invariant. The identity automorphism is one of this type. Let the reader show that the set of all automorphisms of $F(\rho_1, \rho_2, \ldots, \rho_n)$ leaving the elements of F invariant form a group.

DEFINITION 8.9. The group G of automorphisms of $F(\rho_1, \rho_2, \ldots, \rho_n)$ leaving F invariant is called the *Galois group of* $p(x) = 0$, or the *Galois group of* $F(\rho_1, \rho_2, \ldots, \rho_n)$ *relative to* F.

THEOREM 8.18. If $p(x)$ is normal (see Sec. 8.13), the Galois group of $p(x)$ contains exactly n automorphisms.

Proof. First of all we have

$$F(\rho_1, \rho_2, \ldots, \rho_n) = F(\rho_1) = F(\rho_2) = \cdots = F(\rho_n)$$

since $p(x)$ is assumed normal. If $p(x) = \sum_{k=0}^{n} a_k x^k$, then $a_k \leftrightarrow a_k$ for any automorphism of the Galois group G. Let ρ_1 correspond to σ for any automorphism of G. Then $0 = \sum_{k=0}^{n} a_k \rho_1^k \leftrightarrow \sum_{k=0}^{n} a_k \sigma^k = 0$, so that σ must be a zero of $q(x)$. Thus $(\rho_1 \leftrightarrow \rho_1)$, $(\rho_1 \leftrightarrow \rho_2)$, $\ldots$, $(\rho_1 \leftrightarrow \rho_n)$ are the only possible elements of G. We must remember that $(\rho_1 \leftrightarrow \rho_i)$ completely determines an automorphism of G since any element of $F(\rho_1, \rho_2, \ldots, \rho_n)$ is of the form $\sum_{k=0}^{n-1} b_k \rho_1^k$, with b_k, $k = 0, 1, 2, \ldots, n-1$ in F. If $p(x)$ were not normal, we could not make this statement. If $p(x)$ is not normal, the order of G is less than or equal to $n!$ Why? Finally, we shall wish to use the fact that the Galois group G can be made isomorphic to a regular permutation group (see Theorem 8.4).

THEOREM 8.19. Let $p(x)$ be normal. If α is any element of $F(\rho = \rho_1)$ which remains invariant under all automorphisms of G, then α is an element of F.

Proof. We know that all elements of F remain invariant under G. We wish to prove the converse if $p(x)$ is normal. Let $\alpha = \alpha_1$ be an element of $F(\rho_1)$, and consider its conjugates. Then

$$\begin{aligned}
\alpha_1 &= b_0 + b_1\rho_1 + b_2\rho_1^2 + \cdots + b_{n-1}\rho_1^{n-1} \\
\alpha_2 &= b_0 + b_1\rho_2 + b_2\rho_2^2 + \cdots + b_{n-1}\rho_2^{n-1} \\
&\cdots\cdots\cdots\cdots\cdots\cdots\cdots\cdots \\
\alpha_n &= b_0 + b_1\rho_n + b_2\rho_n^2 + \cdots + b_{n-1}\rho_n^{n-1}
\end{aligned}$$

and $f(x) = (x - \alpha_1)(x - \alpha_2) \cdots (x - \alpha_n)$ is the principal equation of $\alpha = \alpha_1$ (see Theorem 8.15), and the coefficients of $f(x)$ are in F. But $\alpha_1 = \alpha_2 = \cdots = \alpha_n$ since α_1 is left invariant under G. Thus

$$f(x) = (x - \alpha)^n = x^n - n\alpha x^{n-1} + \cdots + (-1)^n \alpha_n$$

so that $-n\alpha$ is in F, which implies that α is in F. Q.E.D.

DEFINITION 8.10. If the Galois group is cyclic, the equation $p(x) = 0$ is said to be cyclic.

THEOREM 8.20. If $p(x) = \sum_{k=0}^{n} a_k x^k = 0$ is cyclic and normal, we can solve for the roots of $p(x) = 0$ by a finite number of rational operations (addition, multiplication, etc.) and root extractions on elements of $F(w)$, where $w^n = 1$, $w \neq 1$, $arg\ w = 2\pi/n$. The degree of $p(x)$ is n, and the coefficients of $p(x)$ are in F. It can also be shown that $F(w)$ can be generated by a finite number of rational operations and root extractions on the elements of F. We omit proof of this latter statement.

Proof. Since $p(x)$ is normal, G is composed of exactly n elements. Moreover G is assumed to be cyclic so that a single element generates G. We represent this element by the permutation $P = \begin{pmatrix} 1\,2\,3 & \cdots & n \\ 2\,3\,4 & \cdots & 1 \end{pmatrix}$. Let us consider the set of elements

$$\begin{aligned}
\xi_1 &= \rho_1 + \rho_2 + \rho_3 + \cdots + \rho_n = -\frac{a_{n-1}}{a_n} \\
\xi_2 &= \rho_1 + w\rho_2 + w^2\rho_3 + \cdots + w^{n-1}\rho_n \\
\xi_3 &= \rho_1 + w^2\rho_3 + w^4\rho_3 + \cdots + w^{2(n-1)}\rho_n \\
&\cdots\cdots\cdots\cdots\cdots\cdots\cdots\cdots \\
\xi_n &= \rho_1 + w^{n-1}\rho_2 + w^{2(n-1)}\rho_3 + \cdots + w^{(n-1)^2}\rho_n
\end{aligned} \tag{8.45}$$

The ξ_i, $i = 1, 2, \ldots, n$, certainly belong to $F(\rho, w)$, $\rho = \rho_1$. Now $P(\xi_2) = \rho_2 + w\rho_3 + \cdots + w^{n-1}\rho_1 = \xi_2/w$, considering w as a parameter in ξ_2. Thus $[P(\xi_2)]^n = \xi_2^n$ since $w^n = 1$. But $[P(\xi_2)]^n = P(\xi_2^n)$ since raising ξ_2 to the nth power and then permuting the indices $1, 2, \ldots, n$ is equivalent to first permuting the ρ's and then raising the new entity to the nth power. Thus $P(\xi_2^n) = \xi_2^n$ so that ξ_2^n is invariant under P of G. From $P^2(\xi_2^n) = P[P(\xi_2^n)] = P(\xi_2^n) = \xi_2^n$, etc., we see that ξ_2^n is invariant for all elements of G. The same result applies to $\xi_1^n, \xi_3^n, \xi_4^n, \ldots, \xi_n^n$ Now

$$\xi_2^n = a_0(\rho_1, \rho_2, \ldots, \rho_n) + a_1(\rho_1, \rho_2, \ldots, \rho_n)w + \cdots + a_{n-1}(\rho_1, \rho_2, \ldots, \rho_n)w^{n-1}$$

by direct expansion of ξ_2, since $w^n = 1$. We wish to show that $a_0, a_1, \ldots, a_{n-1}$ are in F so that ξ_2^n is in $F(w)$. Under any permutation of G we have $a_0 \leftrightarrow a_0'$, $a_1 \leftrightarrow a_1'$, $a_2 \leftrightarrow a_2'$, $\ldots$, $a_{n-1} \leftrightarrow a_{n-1}'$. Since ξ_2^n is invariant under G, we have

$$a_0 + a_1w + a_2w^2 + \cdots + a_{n-1}w^{n-1} = a'_0 + a'_1w + a'_2w^2 + \cdots + a'_{n-1}w^{n-1} \qquad (8.46)$$

for all w such that $w^n = 1$. Now the equation

$$q(x) = b_0 + b_1x + \cdots + b_{n-1}x^{n-1} = 0$$

has exactly $n - 1$ roots. Equation (8.46) is of this type yet has n distinct zeros, namely, $1, w, w^2, \ldots, w^{n-1}$. Thus (8.46) can only be true for all w satisfying $w^n = 1$ if $a_0 = a'_0, a_1 = a'_1, \ldots, a_{n-1} = a'_{n-1}$. Hence $a_0, a_1, \ldots, a_{n-1}$ are invariant under G, so belong to F by Theorem 8.19. Hence ξ_2^n is in $F(w)$. The same result applies to $\xi_1^n, \xi_3^n, \ldots, \xi_n^n$. The $\xi_i, i = 1, 2, \ldots, n$, of (8.45) are nth roots of elements in $F(w)$, $w = e^{2\pi i/n}$.

We now look upon (8.45) as a linear system of equations in the unknowns $\rho_1, \rho_2, \ldots, \rho_n$. This system has a unique solution provided the Vandermonde determinant

$$D(w) = \begin{vmatrix} 1 & 1 & 1 & \cdots & 1 \\ 1 & w & w^2 & \cdots & w^{n-1} \\ 1 & w^2 & w^4 & \cdots & w^{2(n-1)} \\ \cdots & \cdots & \cdots & \cdots & \cdots \\ 1 & w^{n-1} & w^{2(n-1)} & \cdots & w^{(n-1)^2} \end{vmatrix} \qquad (8.47)$$

is different from zero. Let the reader show that

$$D(w) = \prod_{\substack{i=0 \\ j=0}}^{n-2} (w^i - w^j) \neq 0, j > i$$

Hence the $\rho_i, i = 1, 2, \ldots, n$, can be expressed as linear functions of the $\sqrt[n]{\xi_i}$ with coefficients in $F(w)$. Q.E.D.

Example 8.22. In Example 8.21 we saw that $p(x) = x^3 - 3x + 1 = 0$ is normal. It is also cyclic since its Galois group is of order 3. Let the roots of $p(x) = 0$ be ρ_1, ρ_2, ρ_3. Then

$$\begin{aligned} \xi_1 &= \rho_1 + \rho_2 + \rho_3 = 0 \\ \xi_2 &= \rho_1 + w\rho_2 + w^2\rho_3 \\ \xi_3 &= \rho_1 + w^2\rho_2 + w\rho_3 \end{aligned} \qquad (8.48)$$

with $w^3 = 1, w \neq 1, w^2 + w + 1 = 0$. We must find ξ_2^3, ξ_3^3 using the additional fact that $\rho_1\rho_2 + \rho_2\rho_3 + \rho_3\rho_1 = -3, \rho_1\rho_2\rho_3 = -1$. Now

$$\begin{aligned} \xi_2\xi_3 &= \rho_1^2 + \rho_2^2 + \rho_3^2 + (w + w^2)(\rho_1\rho_2 + \rho_2\rho_3 + \rho_3\rho_1) \\ &= (\rho_1 + \rho_2 + \rho_3)^2 + (\rho_1\rho_2 + \rho_2\rho_3 + \rho_3\rho_1)(w + w^2 - 2) \\ &= 9 \\ \xi_2^3 + \xi_3^3 &= 2(\rho_1^3 + \rho_2^3 + \rho_3^3) - 3(\rho_1^2\rho_2 + \rho_2^2\rho_3 + \rho_3^2\rho_1 + \rho_1^2\rho_3 + \rho_2^2\rho_1 + \rho_3^2\rho_2) + 12\rho_1\rho_2\rho_3 \\ &= 2(\rho_1 + \rho_2 + \rho_3)^3 - 9(\rho_1 + \rho_2 + \rho_3)(\rho_1\rho_2 + \rho_2\rho_3 + \rho_3\rho_1) + 27\rho_1\rho_2\rho_3 \\ &= -27 \\ (\xi_2^3 - \xi_3^3)^2 &= (\xi_2^3 + \xi_3^3)^2 - 4\xi_2^3\xi_3^3 = (-27)^2 - 4(9)^3 = -2{,}187 \\ \xi_2^3 - \xi_3^3 &= -27\sqrt{3}\,i \end{aligned}$$

Hence $\xi_2^3 = -27\left(\frac{1+\sqrt{3}\,i}{2}\right)$, $\xi_3^3 = -27\left(\frac{1-\sqrt{3}\,i}{2}\right)$. Note that $\xi_2^3\xi_3^3 = 9^3$. The cube roots of ξ_2^3, ξ_3^3 must be chosen so that $\xi_2\xi_3 = 9$. Note that ξ_2^3 and ξ_3^3 belong to $R(w) = R(\sqrt{3}\,i)$. It is now a simple matter to solve (8.48) for ρ_1, ρ_2, ρ_3, which we omit.

Before concluding this chapter let us consider the following: We consider $p(x) = x^4 + 9$, irreducible in the field of rationals. Let the roots of $x^4 + 9 = 0$ be $\rho, -\rho, \sigma, -\sigma$. From $\rho(-\rho)\sigma(-\sigma) = 9$ we see that $\sigma = \pm 3/\rho = \pm 3\rho^3/\rho^4 = \mp\frac{1}{3}\rho^3$. Hence $p(x)$ is normal. The automorphisms comprising the Galois group of $p(x)$ are $(\rho \leftrightarrow \rho)$, $(\rho \leftrightarrow -\rho)$, $(\rho \leftrightarrow \sigma)$, $(\rho \leftrightarrow -\sigma)$, and we can represent G by the permutation group

$$P_1 = \begin{pmatrix} 1\,2\,3\,4 \\ 1\,2\,3\,4 \end{pmatrix} \qquad P_2 = \begin{pmatrix} 1\,2\,3\,4 \\ 2\,1\,4\,3 \end{pmatrix} \qquad P_3 = \begin{pmatrix} 1\,2\,3\,4 \\ 3\,4\,1\,2 \end{pmatrix} \qquad P_4 = \begin{pmatrix} 1\,2\,3\,4 \\ 4\,3\,2\,1 \end{pmatrix}$$

with $\rho_1 = \rho$, $\rho_2 = -\rho$, $\rho_3 = \sigma$, $\rho_4 = -\sigma$. This group is not cyclic. The subgroup $G_1 = (P_1, P_2)$ leaves invariant a certain number of elements of $F(\rho)$. Let us see which of these elements are invariant under G_1. P_1 leaves all elements invariant. If $\alpha = a + b\rho + c\rho^2 + d\rho^3$ is left invariant under P_2, of necessity

$$a + b\rho + c\rho^2 + d\rho^3 = a + b(-\rho) + c(-\rho)^2 + d(-\rho)^3$$

which implies that $b = d = 0$. Hence the elements $a + b\rho^2$ are left invariant under G_1. It is easy to show that these elements form a subfield $F(\rho^2)$ of $F(\rho)$. G_1 is called the Galois group of $F(\rho)$ relative to $F(\rho^2)$. Note that $F(\rho) \supset F(\rho^2) \supset F$.

The principal equation of $a + b\rho^2$ is

$$\begin{aligned} f(x) &= [x - (a + b\rho^2)][x - (a + b\rho^2)][x - (a + b\sigma^2)][x - (a + b\sigma^2)] \\ &= [(x - a)^2 - b(\rho^2 + \sigma^2)(x - a) + b^2\rho^2\sigma^2]^2 \\ &= [(x - a)^2 + 9b^2]^2 \end{aligned}$$

so that $a + b\rho^2$ satisfies a quadratic equation [of lower degree than $p(x)$] with coefficients in F. Thus we can solve a quadratic equation for ρ^2, obtaining ρ^2 in terms of rational operations and extractions of roots of elements in F. Another square root yields ρ. Of course we know all this in advance since it is trivial to solve for the roots of $x^4 + 9 = 0$. Let us note, however, that G_1 in the above example is a maximal normal subgroup of G and that E is a maximal normal subgroup of G_1. The factor groups G/G_1 and G_1/E are of order 2 (a prime number). The fact that 2 is a prime number and that a group whose order is prime is cyclic (see Theorem 8.20 for the importance of this fact) leads to the all-important theorem, which we state without proof.

THEOREM 8.21. If G is the Galois group of an equation $p(x) = 0$ relative to its coefficient field F, a necessary and sufficient condition that

$p(x) = 0$ be solvable by radicals relative to F is that the factors of composition of G be entirely primes.

It can be shown that in general a polynomial of degree greater than 4 cannot be solved by radicals. The equation $x^5 - 1 = 0$, however, can be solved by radicals. Factoring, we have

$$(x - 1)(x^4 + x^3 + x^2 + x + 1) = 0$$

Considering $x^4 + x^3 + x^2 + x + 1 = 0$, let

$$x + \frac{1}{x} = y \qquad x^2 + \frac{1}{x^2} + 2 = y^2$$

so that upon division by x^2 one has $y^2 + y - 1 = 0$. One solves for y and then for x from $x^2 - yx + 1 = 0$.

Problems

1. Solve the quadratic $x^2 + bx + c = 0$ by the method of this section.
2. Show that the automorphisms of a field form a group.
3. Show that $D(w)$ of (8.47) is given by $D(w) = \Pi(w^i - w^j)$, $j > i$.
4. Solve for the roots of $x^5 = 1$.
5. Find the polynomial with coefficients in the field of rationals having

$$\rho = \sqrt{\sqrt[3]{2} + 1}$$

as a zero.

REFERENCES

Albert, A. A.: "Modern Higher Algebra," University of Chicago Press, Chicago, 1937.
———: "Introduction to Algebraic Theories," University of Chicago Press, Chicago, 1941.
Artin, E.: "Galois Theory," Edward Brothers, Inc., Ann Arbor, Mich., 1942.
Birkhoff, G., and S. MacLane: "A Survey of Modern Algebra," The Macmillan Company, New York, 1941.
Dickson, L. E.: "First Course in the Theory of Equations," John Wiley & Sons, Inc., New York, 1922.
———: "Modern Algebraic Theories," Benj. H. Sanborn & Co., Chicago, 1930.
MacDuffee, C. C.: "An Introduction to Abstract Algebra," John Wiley & Sons, Inc., New York, 1940.
Murnaghan, F. C.: "The Theory of Group Representations," Johns Hopkins Press, Baltimore, 1938.
Pontrjagin, L.: "Topological Groups," Princeton University Press, Princeton, N.J., 1946.
Van der Waerden, B. L.: "Modern Algebra," Frederick Ungar Publishing Company, New York, 1940.
Weisner, L.: "Introduction to the Theory of Equations," The Macmillan Company, New York, 1938.
Weyl, H.: "The Classical Groups," Princeton University Press, Princeton, N.J., 1946.

CHAPTER 9

PROBABILITY THEORY AND STATISTICS

9.1. Introduction. It appears that the mathematical theory of probability owes its formation to the inquisitiveness of a professional gambler. In the seventeenth century Antoine Gombaud, Chevalier de Mere, proposed some simple problems involving games of chance to the famous French philosopher, writer, and mathematician, Blaise Pascal. It is to mathematicians Pascal and Fermat that probability theory owes its origin, though since their early works a great number of mathematicians have contributed to its development. Its applications range from statistics to quantum theory. An understanding of probability theory is necessary to undertake studies of the modern theory of games and information theory. It will be useful, however, to consider some elementary combinatorial analysis before we attempt a definition of probability.

9.2. Permutations and Combinations. Let us assume that we have a collection of white balls numbered 1 to m and a collection of red balls numbered 1 to n. In how many ways can we choose exactly one red and one white ball? To each white ball we can associate n red balls. Since there are m white balls, it appears that there are $m \cdot n$ ways of choosing exactly one white and one red ball. In this example we note that the choice of a white ball does not affect the choice of a red ball, and conversely. The events are said to be independent. We state without formal proof the following theorem:

THEOREM 9.1. If there are m ways of performing a first event and n ways of performing an independent second event, there are $m \cdot n$ ways of performing both events.

The use of Theorem 9.1 enables us to solve the following problem: Given a group of objects numbered 1 to n, in how many ways can we order these objects? We have n choices for the object which is to be placed first in our order. After a choice has been made there are $n - 1$ choices for the object that is to be placed second in the order. Continuing, we see that there are $n(n - 1)(n - 2) \cdots 2 \cdot 1 = n!$ ways of arranging or permuting the objects. If we wish to consider the number of arrangements or permutations of n objects taken r at a time, we arrive at the answer $n(n - 1) \cdots (n - r + 1)$. We write

$$
\begin{aligned}
{}_nP_r = {}^nP_r = P_r^n &= n(n-1)(n-2)\cdots(n-r+1) \\
&= \frac{n(n-1)\cdots(n-r+1)(n-r)\cdots 1}{(n-r)(n-r-1)\cdots 1} \\
&= \frac{n!}{(n-r)!} \qquad (9.1)
\end{aligned}
$$

Example 9.1. How many four-digit numbers can be formed from the integers 1, 2, . . . , 9, using any integer at most once? We have $P_4^9 = 9!/5! = 3{,}024$ as our answer. If we were to use the integers as many times as we pleased, our answer would be $9^4 = 6{,}561$.

Example 9.2. Given 3 red flags, 4 white flags, and 6 black flags, how many signals can we send using the 13 flags? Let us assume for the moment that we can distinguish between the red flags by numbering them 1, 2, 3, with the same statement concerning the white and black flags. It is then obvious that 13! different signals could be sent. Now the red flags can be permuted in $3! = 6$ ways. Each permutation, however, yields no new signal if the red flags are indistinguishable. We must divide 13! by 3! to account for this characteristic of the red flags. By considering the white and black flags we arrive at the correct answer, $13!/3!4!6!$.

The permutations of the integers 1, 2, 3, 4 taken two at a time are listed below:

1 2	1 3	1 4	2 3	2 4	3 4
2 1	3 1	4 1	3 2	4 2	4 3

A particular arrangement of objects considered independent of their order or permutation is called a combination. We see that there are 6 combinations of the integers 1, 2, 3, 4 taken two at a time. Although 12 and 21 are different permutations, they yield a single combination. If we wish to hire 3 secretaries from a group of 12, we are usually not interested in the order of hiring the secretaries. Now let us see in how many ways we can choose r objects from among n, the order of the choices being immaterial. Call this number ${}_nC_r = {}^nC_r = C_r^n = \binom{n}{r}$. Every combination will yield $r!$ permutations. Thus ${}^nC_r\, r! = {}^nP_r$ so that

$$
{}^nC_r = \frac{n!}{r!(n-r)!} \qquad (9.2)
$$

Note that $1 = \binom{n}{n} = n!/n!0!$ so that 0! is chosen to be 1. What does ${}^nC_0$ signify?

Example 9.3. In obtaining 5 cards from among 52 cards in draw poker we are not interested in the order in which the cards are dealt. Thus there are ${}^{52}C_5$ combinations of cards which can be dealt. Let the reader show that ${}^{52}C_5 = 2{,}598{,}960$. Let us determine how many full houses (three of a kind and a pair) can be dealt. If we consider the particular full house consisting of a pair of fives and three aces, we note that there are ${}^4C_3 = 4$ ways of obtaining three aces. Thus there are $6 \cdot 4 = 24$ different combinations yielding three aces and a pair of fives. But there are 13 choices for the rank of the card yielding three of a kind, and then 12 choices remain for the choice of the rank of the pair. Thus there are $24 \cdot 13 \cdot 12 = 3{,}744$ different full houses that can be dealt.

Example 9.4. Let us consider a rectangular $m \times n$ grid. If we start at the lower left-hand corner and are allowed to move only to the right and up, how many different paths can we traverse to reach the upper right-hand corner? All told, we must move a total of $m + n$ blocks. Once we choose any m blocks from among the $m + n$ blocks for our horizontal motions, of necessity, the remaining n blocks will be vertical motions. Thus there are ${}^{m+n}C_m = (m + n)!/m!n!$ different paths. Note that the answer is symmetric in m and n.

Example 9.5. Let us consider $(x + y)^n$, n a positive integer. In the product $(x + y)(x + y) \cdots (x + y)$ we note that terms of the type $x^r y^s$ will occur. Since a term from each factor must be used exactly once, of necessity, $r + s = n$. The term $x^r y^s$ can occur in exactly ${}^nC_r$ ways since we can choose x from any r of the n factors. Thus $(x + y)^n = \sum_{r=0}^{n} \binom{n}{r} x^r y^{n-r}$. If we choose $x = y = 1$, we have $2^n = \sum_{r=0}^{n} \binom{n}{r}$.

From $(1 + x)^n = \sum_{r=0}^{n} \binom{n}{r} x^r$, $(1 + x)^n = \sum_{s=0}^{n} \binom{n}{s} x^s$, we have

$$(1 + x)^{2n} = \sum_{k=0}^{2n} \binom{2n}{k} x^k = \sum_{s=0}^{n} \sum_{r=0}^{n} \binom{n}{r}\binom{n}{s} x^{r+s} = \sum_{k=0}^{2n} \sum_{r=0}^{k} \binom{n}{r}\binom{n}{k-r} x^k$$

Equating coefficients of x^k yields

$$\binom{2n}{k} = \sum_{r=0}^{k} \binom{n}{r}\binom{n}{k-r}$$

Problems

1. In how many ways can an ordered pair of dice be rolled? *Ans.* 36
2. Which number will occur most often when a pair of dice is rolled? *Ans.* 7
3. How many handshakes can 10 people perform two at a time? *Ans.* 45
4. How many five-card poker hands can be dealt consisting of exactly one pair (other three cards different)? *Ans.* 1,098,240
5. Considering the grid of Example 9.4, show that the grid contains

$$\frac{(m + 1)(n + 1)mn}{4}$$

rectangles.

6. Consider a function of p variables. How many nth distinct derivatives can be formed? *Hint:* The grid of Example 9.4 is useful if we designate the horizontal lines by $x_1, x_2, \ldots, x_p$. *Ans.* ${}^{n+p-1}C_{p-1}$
7. By differentiating $(1 + x)^n$ show that $\binom{n}{r} = \binom{n-1}{r-1} + \binom{n-1}{r}$.
8. Show that $\sum_{r=0}^{n} (-1)^r \binom{n}{r} = 0$.

9. By considering $(1 + x)^n(1 - x)^n = (1 - x^2)^n$ show that

$$\sum_{r=0}^{2k} (-1)^r \binom{n}{r} \binom{n}{2k - r} = (-1)^k \binom{n}{k}$$

10. In how many ways can the integers 1, 2, 3, 4, 5 be ordered such that no integer corresponds to its order in the sequence? For example, 21453 is such an ordering, while 21354 is not, since 3 occupies the third position in the sequence.

Ans. $5!\left(\frac{1}{2!} - \frac{1}{3!} + \frac{1}{4!} - \frac{1}{5!}\right)$

9.3. The Meaning and Postulates of Probability Theory. Whereas the psychologist, economist, political scientist, etc., can, with some measure of success, communicate their ideas to the layman, the mathematician, unfortunately, realizes that he has no hope of describing his chief mathematical fields of interest with a reasonable prospect of being understood. A modicum of hope prevails when we consider probability theory from its most elementary viewpoint. It is rare indeed to find a man who has not evinced interest at one time or another in breaking the bank at Monte Carlo. However, one need only consider the vast number of persons who believe that someday they will discover an invincible system of gambling, to realize how little the layman actually understands the basic ideas of probability theory. Probability theory has been called the science of common sense. If this is the case, it is strange indeed that two persons using common sense can differ so greatly from each other in solving a problem involving probability theory.

Let us now investigate the meaning of the word "probability." Webster's New International Dictionary states that probability is "the likelihood of the occurrence of any particular form of an event, estimated as the ratio of the number of ways in which that form might occur to the whole number of ways in which the event might occur in any form."[1] The same dictionary defines "likelihood" as "Probability; as, it will rain in all *likelihood*."[1] Let us consider the question, What is the probability that it will rain on Sept. 9, 1957, in Los Angeles? To some mathematicians this question is meaningless. There is only one Sept. 9, 1957. We cannot compute the ratio of the number of times it has rained on Sept. 9, 1957, to the total number of days comprising Sept. 9, 1957, rain or shine, at least not before Sept. 9, 1957. After Sept. 9, 1957, the ratio would be zero or 1 depending on the weather. It is up to the meteorologist to give us a precise answer. Either it will or it will not rain on Sept. 9, 1957, in Los Angeles. If there is no front within 1,000 miles of Los Angeles on Sept. 8, 1957, the meteorologist would predict no rain with a high degree

[1] By permission. From "Webster's New International Dictionary," Second Edition, copyright, 1934, 1939, 1945, 1950, 1953, 1954, by G. & C. Merriam Company.

of "probability." But the word "probability" would not be used as a mathematician defines probability. Let us now ask, What is the probability that a five occur if a die be rolled? This question is, in a sense, very much like the question asked above. Theoretically, if we knew the initial position of the die and if we knew the stresses and strains of the die along with the external forces, we could predict exactly which of the six numbers of the die would occur face up. The same statement can be made about the weather. Unfortunately for the meteorologist there are too many variables involved in attempting to predict the exact state of the weather. The hope of the meteorologist is to reduce the number of relevant factors to a minimum in an attempt to predict the weather. The die problem differs from the weather problem in the following way: If we have the patience and time, we can continue to roll the die as often as we please. If after n throws we note that the number five has occurred r_n times, we can form the ratio r_n/n. One would be naïve in calling r_n/n the probability of rolling a five, even if n is large. There are some who would define the probability of rolling a five as

$$p = \lim_{n \to \infty} \frac{r_n}{n} \tag{9.3}$$

The limit of a sequence cannot be found unless one knows every term of the sequence (the nth term of the sequence must be given for all n). To compute (9.3), one would have to perform an infinite number of experiments, an obvious impossibility.

Let us turn, for the moment, to the science of physics. Newton's second law of motion states that the force acting on a particle is proportional to the time rate of change of momentum of the particle. The particle of Newton's second law of motion is an idealized point mass. No such mass occurs in nature. This does not act as a deterrent to the physicist. The motion of a gyroscope is computed on the basis that ideal rigid bodies exist. The close correlation between experiment and theory gives the physicist confidence in his so-called laws of nature. The mathematician working with probability theory encounters the same difficulty. He realizes that a die or a roulette wheel is not perfect. Man, however, has the ability to make abstractions. He visualizes a perfect die, and, furthermore, he postulates that if a die be rolled all the six numbers on the die are "equally likely" to occur. It is, of course, impossible to prove that the occurrence of each number of a die is equally likely. With these idealized assumptions and definitions the mathematician can predict the probability of winning at dice. The success of the gambling houses in Las Vegas is sufficient evidence to the professional gambler that the idealized science of probability theory is on firm ground. The pure mathematician is not interested in games of chance, per se. Probability theory, to the pure mathematician, is simply a set of axioms and definitions from which he derives certain consequences or theorems. We now consider the axioms of probability theory. The reader is urged to read Sec. 10.7 concerning the union, intersection, complement, etc., of sets. We shall find it advantageous to consider a simple example before treating the general case.

If a pair of dice are rolled, the following events can happen as listed below:

$$E: \quad \begin{matrix} (1,1) & (2,1) & (3,1) & (4,1) & (5,1) & (6,1) \\ (1,2) & (2,2) & (3,2) & (4,2) & (5,2) & (6,2) \\ (1,3) & (2,3) & (3,3) & (4,3) & (5,3) & (6,3) \\ (1,4) & (2,4) & (3,4) & (4,4) & (5,4) & (6,4) \\ (1,5) & (2,5) & (3,5) & (4,5) & (5,5) & (6,5) \\ (1,6) & (2,6) & (3,6) & (4,6) & (5,6) & (6,6) \end{matrix}$$

E is the collection of all possible events. The elements of E are called elementary events. A subset of E, say, F_1, is the set of events guaranteeing the occurrence of the number 1 on at least one of the two dice.

$$F_1: \quad (1,1), (1,2), (1,3), (1,4), (1,5), \\ (1,6), (2,1), (3,1), (4,1), (5,1), (6,1)$$

Another subset of E is the set F_2 consisting of the events (1, 2), (2, 1), and (6, 6). The union, or sum, of F_1 and F_2 is the set of all events or elements belonging to either F_1 or F_2 or both, written $F_1 \cup F_2 = F_1 + F_2$. In this example $F_1 + F_2$ consists of all the elements of F_1 plus the event (6, 6). The intersection, or product, of F_1 and F_2, written $F_1 \cap F_2 = F_1F_2$, is the set of elements belonging to both F_1 and F_2. In this example F_1F_2 is composed of the elements (1, 2) and (2, 1). If two sets F_1 and F_2 have no elements in common, we say that their intersection is the null set, written $F_1F_2 = 0$. The complement of a set F is the set of elements of E not in F. What is the complement of the set F_1 defined above? The complement of the full set E is the null set 0. Now let G be the set of all subsets of E including E and the null set 0. The elements of G are called random events. G is said to be a *field of sets* in the sense that the sum, product, and complement are again elements of G. Let us now attach to each element of E (these elementary individual events are also elements of G) the nonnegative number $\frac{1}{36}$. There is no mystery as to the choice of the number $\frac{1}{36}$. First we note that E is composed of 36 elements. The assumption that all 36 events are equally likely implies that any single event has a probability of $\frac{1}{36}$ of occurring. If A is any subset of E consisting of r events, we define the probability of an element of A occurring as $p(A) = r/36$. Thus $P(F_1) = \frac{11}{36}$, $P(F_2) = \frac{3}{36}$, $P(E) = 1$. We define $P(0) = \frac{0}{36} = 0$. Thus to each set A of G we have defined a real nonnegative number. If A and B have no element in common, then $P(A + B) = P(A) + P(B)$.

Formally, a *field of probability* is defined as follows: Let E be any collection of elements $x, y, z, \ldots$, which are called elementary events, and let G be a collection of subsets of E. Assume that the following postulates or axioms are satisfied:

I. G is a field of sets.

II. G contains E and the null set.

III. To each set A in G there corresponds a nonnegative real number, written $P(A)$. The number $P(A)$ is called the probability that an element of A will occur.

IV. $P(E) = 1$, that is, one of the events of E is sure to happen.

V. If A and B of G have no element in common, then

$$P(A + B) = P(A) + P(B)$$

The single toss of a coin is a simple example of a field of probability. E is composed of the elements H (for heads) and T (for tails). G contains the sets $H, T, E = (H, T), 0$. We define $p(H) = \frac{1}{2}$, $p(T) = \frac{1}{2}$, $p(H, T) = 1$, $p(0) = 0$. $p(E) = 1$ implies that a head or tail will certainly occur.

Events, such as tossing a coin, rolling dice, spinning a roulette wheel, yield finite probability fields since there are only a finite number of different events which can occur. If 5 cards are dealt from a pack of 52 cards, there are exactly ${}^{52}C_5$ different events which can occur. It is logical to postulate that the probability of any single event occurring from among the ${}^{52}C_5$ different events be given by $p = 1/{}^{52}C_5$. This is what is meant by an "honest" deal.

It is important that the reader understand Postulate V. In rolling a die the event $A = (1, 3)$ means that A occurs if a one or a three is face up on the die. The event $B = (4)$ means that B occurs if a four is face up. Now A and B have no elements in common, that is, if A occurs, B cannot occur, and conversely. We say that A and B are mutually exclusive events. For a perfect die, $p(1, 3) = \frac{2}{6}$, $p(4) = \frac{1}{6}$, so that Postulate V states that $p(A + B) = \frac{2}{6} + \frac{1}{6} = \frac{1}{2}$. The probability that a one, three, or four occurs is $\frac{1}{2}$, the expected answer. A simple working rule for the reader is the following:

RULE 1. If A and B are mutually exclusive events with probabilities $p(A)$ and $p(B)$, respectively, then the probability that either A or B occurs is

$$p(A + B) = p(A) + p(B)$$

The reader should give an example for which $p(A + B) < p(A) + p(B)$.

Problems

1. What is the probability of throwing a seven with two dice? *Ans.* $\frac{1}{6}$

2. What is the probability of throwing a four, eight, or twelve with two dice? *Ans.* $\frac{1}{4}$

3. Five cards are dealt from a deck of 52 cards. Find the probability of obtaining the following poker hands: three of a kind, a flush, a straight.

$$Ans. \ \frac{88}{4{,}165}, \frac{33}{16{,}660}, \frac{10 \cdot 4^5}{{}^{52}C_5}$$

4. A set of balls are numbered from 1 to 20. Two balls are drawn simultaneously at random. What is the probability that their sum is 14? *Hint:* There are ${}^{20}C_2 = 190$ ways of drawing two balls. *Ans.* $\frac{3}{95}$

5. In Prob. 4 what is the probability that the sum of the two numbers is 14 if a ball is drawn and replaced and then a second ball is drawn? *Ans.* $\frac{13}{400}$

6. Five coins are tossed. What is the probability that exactly three heads occur? What is the probability that more than two heads occur? *Ans.* $\frac{5}{16}, \frac{1}{2}$

7. If n coins are tossed, show that the probability that exactly r heads occur is ${}^nC_r/2^n$.

8. If a coin is tossed and a pair of dice are rolled, what is the probability that a head occurs and a total of 5 is rolled? *Ans.* $\frac{1}{18}$

9. Show that Postulates IV and V imply $p(0) = 0$.

10. Let the complement of A be denoted by $\bar{A}$ so that $A + \bar{A} = E$. Show that $p(\bar{A}) = 1 - p(A)$.

11. Let A and B be elements of a field G. If $p(A) \neq 0$, define $p_A(B)$ by

$$p_A(B) = \frac{p(AB)}{p(A)}$$

$p_A(B)$ is called the conditional probability of the event B under the condition A. Show that

$$p_B(A)p(B) = p_A(B)p(A)$$

Give a geometric interpretation of $p_A(B)$, assuming A and B are point sets in a plane.

9.4. Theorem of Bayes. Let us consider the following simple example: Assume that one throws a dart at a target whose area is taken to be 1. We assume further that the dart is sure to strike the target and that the dart is equally likely to strike anywhere on the target. This means that if A is any area of the target then the probability that the dart strikes a point of A is simply the area of A, written $p(A)$. Now let A and B be overlapping areas (see Fig. 9.1).

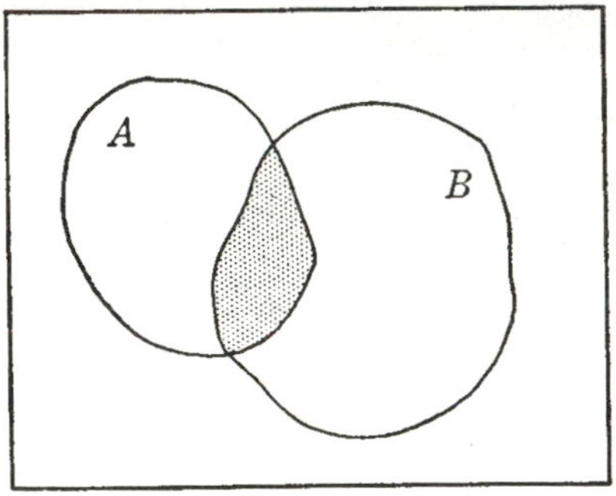

FIG. 9.1

Let us consider a second person located in another room who knows that a dart will be thrown at the target. The probability that the dart falls in A will be given by $p(A)$. Let us now assume that after the dart is thrown our second person asks the following question: Has the dart fallen in the area B? He receives the truthful answer, "yes." How does this affect this person's opinion as to the probability that the dart has also fallen in A? We can answer this question in the following manner: It is fairly obvious that if it is known that the dart lies in B then the only way it can also lie in A is for the dart to have struck the region $A \cap B$, and the probability of this happening before it is known that the dart has fallen in B is simply the area of $A \cap B$, written $p(A \cap B)$. Since the dart is known to be somewhere in B, it appears that the ratio of the area of $A \cap B$ to the area of B will yield the probability that the dart also lies in $A \cap B$. For example, if $A \cap B$ is $\frac{1}{3}$ the area of B, then the chance that the dart is in $A \cap B$ if it is known that it is in B is simply $\frac{1}{3}$. We write

$$p_B(A) = \frac{p(A \cap B)}{p(B)} \qquad p(B) \neq 0 \tag{9.4}$$

and define $p_B(A)$ as the probability that the event A has happened if it is known that the event B has happened. We may also say that $p_B(A)$ is the *conditional probability* of the event A under the condition B.

It is to be noted that (9.4) is a definition and requires no proof. Experimentally, however, one might attempt to verify (9.4) to some extent. Every time an observer learns that the dart is in B, he records whether the dart is in A or is not in A. He is not concerned with those experiments for which the dart lies outside of B. The ratio of the number of successes to the total number of tries (the dart must be in B) yields a fraction lying between zero and 1. For a large number of experiments it is reasonable to hope that this number is close to the number $p_B(A)$.

Since the roles of A and B can be interchanged, we have from (9.4)

$$p_A(B) = \frac{p(B \cap A)}{p(A)} \qquad p(A) \neq 0 \tag{9.5}$$

Combining (9.4) and (9.5) yields one form of Bayes's theorem,

$$p_B(A) = \frac{p(A)p_A(B)}{p(B)} \tag{9.6}$$

since $p(B \cap A) = p(A \cap B)$. $p(A)$ is called the *a priori probability* that A occurs, whereas $p_B(A)$ is called the *a posteriori probability* that A occurs under the hypothesis that B occurs.

Example 9.6. A group of 100 girls contains 30 blondes and 70 brunettes. Twenty-five of the blondes are blue-eyed, and the rest are brown-eyed, whereas 55 of the brunettes are brown-eyed, and the rest are blue-eyed. A girl is picked at random. The a priori probability that she is a blonde and blue-eyed is $\frac{25}{100} = \frac{1}{4}$. If we pick a girl at random and find out that she is blue-eyed, what is the probability that she is blonde? Let x denote the quality of being blonde and y denote the quality of being blue-eyed. We are interested in determining $p_y(x)$. Now $p(y) = \frac{40}{100} = 0.4$, since 40 of the 100 girls are blue-eyed. Moreover, $p(x \cap y) = \frac{1}{4}$. Applying (9.4) yields

$$p_y(x) = \frac{p(x \cap y)}{p(y)} = \frac{5}{8}$$

Note that there are 25 blue-eyed blondes and 15 blue-eyed brunettes, so that the appearance of a blue-eyed girl means that the probability of the girl being a blonde is $\frac{25}{40} = \frac{5}{8}$. In information theory one determines the ratio of the conditional probability to the a priori probability and defines the amount of "bits" of information as the logarithm (base 2) of this ratio. In this example, $I = \log_2 \frac{\frac{5}{8}}{0.3} = \log_2 \frac{25}{12} \approx 1$.

We can extend the result of (9.5) as follows: Let E be a collection of elementary probability events $A_1, A_2, \ldots, A_n$ with the A_i, $i = 1, 2, \ldots, n$, mutually exclusive, so that

$$E = A_1 + A_2 + \cdots + A_n = \bigcup_{i=1}^{n} A_i$$

Now let B be any subset of E. Then

$$B = B \cap E = B \cap A_1 + B \cap A_2 + \cdots + B \cap A_n \tag{9.7}$$

Let the reader deduce that $B \cap A_i$, $i = 1, 2, \ldots, n$, are mutually exclusive. Thus

$$\begin{aligned} p(B) &= p(B \cap A_1) + p(B \cap A_2) + \cdots + p(B \cap A_n) \\ &= \sum_{i=1}^{n} p(B \cap A_i) \end{aligned} \tag{9.8}$$

Applying (9.5) yields

$$p(B) = \sum_{i=1}^{n} p(A_i)p_{A_i}(B) \tag{9.9}$$

Substituting (9.9) into (9.6) yields Bayes's formula,

$$p_B(A_j) = \frac{p(A_j)p_{A_j}(B)}{\sum_{i=1}^{n} p(A_i)p_{A_i}(B)} \tag{9.10}$$

It is important to realize that the event B need not be a subset of E in the ordinary sense. For example, the events $A_1, A_2, \ldots, A_n$ might be different urns containing various assortments of red and white balls. The event B could be the successive drawing of three red balls from any particular urn, each ball being returned to its urn before the next drawing. Formulas (9.7) to (9.10) would still hold. By $B \cap A_1 = A_1 \cap B$

we mean the composite event of choosing A_1 and then drawing three successive red balls as described above.

Example 9.7. Let us consider two urns. Urn I contains three red and five white balls, and urn II contains two red and five white balls. An urn is chosen at random ($p(\mathrm{I}_1) = \frac{1}{2}$, $p(\mathrm{I}_2) = \frac{1}{2}$), and a ball chosen at random from this urn turns out to be red ($p_{\mathrm{I}}(R) = \frac{3}{8}$, $p_{\mathrm{II}}(R) = \frac{2}{7}$). What is the a posteriori probability that the red ball came from urn I? Applying (9.10) yields

$$p_R(\mathrm{I}) = \frac{p(\mathrm{I})p_{\mathrm{I}}(R)}{p(\mathrm{I})p_{\mathrm{I}}(R) + p(\mathrm{II})p_{\mathrm{II}}(R)} = \frac{\frac{1}{2}\cdot\frac{3}{8}}{\frac{1}{2}\cdot\frac{3}{8}+\frac{1}{2}\cdot\frac{2}{7}} = \frac{21}{37}$$

Example 9.8. Two balls are placed in an urn as follows: A coin is tossed twice, and a white ball is placed in the urn if a head occurs, with a red ball placed in the urn if a tail occurs. Let A_0, A_1, A_2 represent the events of the urn containing none, one, and two red balls, respectively. Balls are drawn from the urn three times in succession (always returned before the next drawing), and it is found that on all three occasions a red ball was drawn. What is the probability that both balls in the urn are red? Let B be the event of drawing three successive red balls. We are interested in computing $p_B(A_2)$. Applying (9.10) yields

$$p_B(A_2) = \frac{p(A_2)p_{A_2}(B)}{p(A_0)p_{A_0}(B) + p(A_1)p_{A_1}(B) + p(A_2)p_{A_2}(B)}$$

From $p(A_0) = \frac{1}{4}$, $p(A_1) = \frac{1}{2}$, $p(A_2) = \frac{1}{4}$, $p_{A_0}(B) = 0$, $p_{A_1}(B) = (\frac{1}{2})^3 = \frac{1}{8}$, $p_{A_2}(B) = 1$, we obtain $p_B(A_2) = \frac{4}{5}$. Let the reader show that, if B_n represents the successive drawing of n red balls, then $p_{B_n}(A_2)$ tends to 1 as $n \to \infty$. Is this reasonable to expect? We note that $p_{A_1}(B)$ represents the probability of drawing three successive red balls from an urn containing one red and one white ball. We liken this problem to that of finding the probability of tossing three successive heads. The event space for this case is (H, H, H), (H, H, T), (H, T, H), (H, T, T), (T, H, H), (T, H, T), (T, T, H), (T, T, T), involving eight equally likely events. The a priori probability of the event (H, H, H) is $\frac{1}{8}$. We note that, for a single toss of the coin, $p(H) = \frac{1}{2}$. Is it reasonable to expect that $p(H, H, H) = p(H)p(H)p(H)$? This question is answered in the next section.

Problems

1. Three urns contain, respectively, 2 red and 3 black balls, 1 red and 4 black balls, 3 red and 1 black ball. An urn is chosen at random and a ball drawn from it. If the ball is red, what is the probability that it came from the first or second urn? *Ans.* $\frac{4}{9}$

2. Urn I contains two red and three black balls. Urn II contains three red and two black balls. A ball is chosen at random from urn I and placed in urn II. A ball is then chosen at random from urn II. If this ball is red, what is the probability that a red ball was transferred from urn I to urn II? *Ans.* $\frac{8}{17}$

9.5. Independent Events. Events Not Mutually Exclusive. If we rewrite formula (9.5), we have

$$p(A \cap B) = p(A)p_A(B) \tag{9.11}$$

Formula (9.11) states that the probability that both A and B happen is the probability that A happens times the probability that B will happen if A happens. What can we surmise if $p(A \cap B) = p(A)p(B)$? Of necessity, $P_A(B) = p(B)$, so that the a priori probability of B happening, $p(B)$, does not depend on the event A. We say that A and B are independent events. Two events are independent, by definition, if

$$p(A \cap B) = p(A)p(B) \tag{9.12}$$

A set of probability events, $A_1, A_2, \ldots, A_n$, are said to be mutually independent if

$$p(A_i \cap A_j) = p(A_i)p(A_j) \qquad i \neq j \tag{9.13}$$

for $i, j = 1, 2, \ldots, n$.

If we consider, for example, the tossing of a coin and the rolling of a die, we obtain the event space of 12 elements,

$$\begin{matrix} (H, 1) & (H, 2) & (H, 3) & (H, 4) & (H, 5) & (H, 6) \\ (T, 1) & (T, 2) & (T, 3) & (T, 4) & (T, 5) & (T, 6) \end{matrix}$$

We can assign the a priori probability of $\frac{1}{12}$ to each event. The probability that a head occurs and a six occurs is taken to be $\frac{1}{12}$. If we look upon the rolling of a die as completely independent of the tossing of a coin, we note that $p(H) = \frac{1}{2}$, $p(6) = \frac{1}{6}$, and $p(H \cap 6) = \frac{1}{2} \cdot \frac{1}{6} = \frac{1}{12}$. Assigning the a priori probability of $\frac{1}{12}$ is equivalent to assuming that the two events are independent.

Example 9.9. A die is rolled n times. We compute the probability that a six occur at least once. The a priori probability of failing to throw a six on any given toss of the die is $\frac{5}{6}$. If we assume that each successive roll of the die is independent of the previous throws, the probability of not obtaining a six in n throws is $(\frac{5}{6})^n$. Hence the probability of rolling at least 1 six in n tosses is $1 - (\frac{5}{6})^n$. For $n = 3$, $p < \frac{1}{2}$, while for $n = 4$, $p > \frac{1}{2}$.

Example 9.10. The game of craps is played as follows: A pair of dice is rolled, and the player wins immediately if a seven or eleven occurs. He loses if a two, three, or twelve arises on the first throw. If a four, five, six, eight, nine, or ten is thrown, the player continues to roll the dice until he duplicates his first toss or until a seven occurs. The numbers two, three, eleven, twelve are disregarded after the first toss. If the player duplicates his first toss before rolling a seven, he wins; otherwise he loses. We compute the probability that the man rolling the dice will win. At gambling establishments an opponent of the roller of the dice (except the establishment) does not win if the player rolls a twelve. The probability of winning on the first toss is

$$p(7) + p(11) = \frac{8}{36}$$

since there are 6 ways of rolling a seven, 2 ways of rolling an eleven, and 36 total possible throws of two dice. The player can also win by rolling a four and then rolling a four before a seven. The probability of rolling a four is $\frac{3}{36}$, and the probability of rolling a four before a seven is $\frac{3}{9}$. Thus the probability of rolling a four and then rolling another four before a seven is $\frac{3}{36} \cdot \frac{3}{9} = \frac{1}{36}$. The same reasoning for the numbers five, six, eight, nine, ten yields

$$p = \frac{8}{36} + 2(\frac{1}{36} + \frac{2}{45} + \frac{25}{396}) = \frac{244}{495} = 0.49292 \cdots$$

Example 9.11. *Bernoulli Trials.* Let p be the probability of success of a certain experiment, $q = 1 - p$ the probability of its failure. Assume that the experiment is performed n times, the probability of success remaining constant, while the result of each experiment is independent of all the others. Such a sequence is called a Bernoulli sequence. The simplest example is illustrated by the repeated toss of a coin. We determine now the probability of attaining exactly r successes in n trials. In the case of the coin problem a particular successful sequence would be the sequence $H_1H_2 \cdots H_rT_1T_2 \cdots T_{n-r}$ if we were interested in obtaining exactly r heads. There are obviously ${}^nC_r$ different sequences containing exactly r heads and $n - r$ tails. The probability of obtaining any particular sequence is $(\frac{1}{2})^n$. Thus the probability of obtaining exactly r heads in n tosses of a coin is given by ${}^nC_r/2^n$. Let the reader show that

$$P_r = {}^nC_r\, p^r q^{n-r} \tag{9.14}$$

is the probability of obtaining exactly r successes in n trials for the general Bernoulli sequence. It can be shown that the number r which makes P_r a maximum for a given n and p is the greatest integer less than or equal to $(n + 1)p$.

We now consider events which may not be mutually exclusive. Let E be an event space with subsets $A_1, A_2, \ldots, A_n$. The A_i, $i = 1, 2, \ldots, n$, need not be mutually exclusive (see Fig. 9.2).

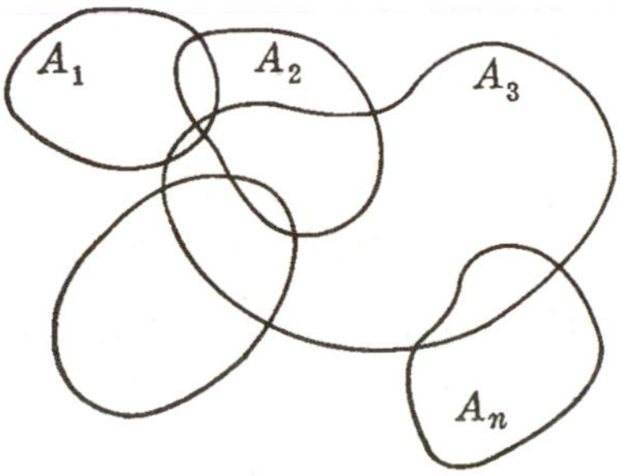

FIG. 9.2

Certainly

$$p(A_1) + p(A_2) + p(A_3) \geqq p(A_1 \cup A_2 \cup A_3)$$

since there may be overlapping regions of A_1, A_2, A_3. Geometrically, it is easy to verify that

$$A_1 \cup A_2 \cup A_3 = A_1 + A_2 + A_3 - A_1 \cap A_2 - A_2 \cap A_3 - A_3 \cap A_1 + A_1 \cap A_2 \cap A_3 \tag{9.15}$$

In $A_1 + A_2 + A_3$ we have counted $A_1 \cap A_2$ twice; so we substract $A_1 \cap A_2$, with a similar statement for $A_1 \cap A_3$, $A_2 \cap A_3$. In $A_1 + A_2 + A_3$ we have counted $A_1 \cap A_2 \cap A_3$ three times, and then we have subtracted $A_1 \cap A_2 \cap A_3$ three times in $A_1 \cap A_2$, etc. Thus we add $A_1 \cap A_2 \cap A_3$ to obtain (9.15). In general

$$p\left(\bigcup_{i=1}^{n} A_i\right) = \sum_{i=1}^{n} p(A_i) - \sum_{j \neq i = 1}^{n} p(A_i \cap A_j) + \sum_{i,j,k} p(A_i \cap A_j \cap A_k) + \cdots + (-1)^{n+1} p\left(\bigcap_{i=1}^{n} A_i\right) \tag{9.16}$$

Example 9.12. Balls numbered 1 to n are placed at random in n urns numbered 1 to n, one ball to each urn. What is the probability that the number of at least one ball corresponds to the number of the urn in which it is placed? Let A_i be the event of having the ith ball in the ith urn regardless of the distribution of the rest of the balls. Then $p(A_i) = 1/n$ for $i = 1, 2, \ldots, n$. $p(A_i \cap A_j)$ represents the probability that both the ith and jth ball be in their proper urns. From (9.5),

$$p(A_i \cap A_j) = p(A_j)p_{A_j}(A_i)$$

Now $p_{A_j}(A_i) = 1/(n-1)$, since, if it is known that the jth ball is in the proper urn, there is one chance in $n - 1$ that the ith ball will be in its proper urn. Thus $p(A_i \cap A_j) = 1/n(n-1)$. We note also that there are ${}^nC_2$ different $A_i \cap A_j$. Continuing, it is easy to see that

$$\begin{aligned} P_n &= n\binom{1}{n} - \binom{n}{2}\frac{1}{n(n-1)} + \binom{n}{3}\frac{1}{n(n-1)(n-2)} + \cdots \\ &= 1 - \frac{1}{2!} + \frac{1}{3!} - \frac{1}{4!} + \cdots + (-1)^{n+1}\frac{1}{n!} \end{aligned}$$

Let the reader show that $\lim_{n\to\infty} P_n = (e-1)/e$.

We conclude this section by considering the simple one-dimensional random-walk problem. Assume that a person starts at the origin. A coin is tossed, with the result that if a head occurs the person will move one unit to the right and if a tail occurs the person will move one unit to the left. The process is repeated n times. What is the probability that the final position is located at $x = r$? Let u be the number of moves to the right and v the number of moves to the left that can occur so that the final position will be at $x = r$. Then $u - v = r$, $u + v = n$, so that $u = (r+n)/2$, $v = (n-r)/2$. There are ${}^nC_u$ different ways in which one can move to the right, the rest of the moves being to the left. The probability of ending at $x = r$ is thus ${}^nC_{(r+n)/2}/2^n$. This problem is similar to that of Example 9.4.

Problems

1. What is the probability that a seven occurs at least once in n tosses of a pair of dice? *Ans.* $1 - (\frac{5}{6})^n$

2. Derive (9.14).

3. Show that

$$p[A_1 \cap (A_2 \cup A_3)] = p(A_1 \cap A_2) + p(A_1 \cap A_3) - p(A_1 \cap A_2 \cap A_3)$$

What does this formula reduce to if A_1, A_2, and A_3 are mutually independent?

4. Show that the probability of having at least r successes in a sequence of n Bernoulli trials is $P = \sum_{i=r}^{n} \binom{n}{i} p^i q^{n-i}$.

5. Show that $\sum_{i=0}^{n} \binom{n}{i} p^i(1-p)^{n-i} = 1$.

6. A coin is tossed in a Bernoulli sequence. What is the probability of obtaining m heads before n tails appear? *Hint:* If there are at least m heads in the first $m + n - 1$ tosses of the coin, the game is won.

$$Ans.\ \frac{1}{2^{m+n-1}} \sum_{r=m}^{m+n-1} \binom{m+n-1}{r}$$

7. Generalize the random-walk problem to the case of two dimensions with $p = \frac{1}{4}$ for each of the possible motions.

8. A and B have, respectively, $n + 1$ and n coins. If they toss their coins simultaneously, what is the probability that A will have more heads than B? *Ans.* $p = \frac{1}{2}$

9. What is the probability of getting exactly 2 sixes in three throws of a die?

$$Ans.\ \frac{3 \cdot 5^4}{6^6}$$

10. Coin 1 has a probability of p of getting a head, and coin 2 has a probability of q of getting a tail. We start with coin 1 and keep tossing it until a tail occurs, whereupon we switch to coin 2. Whenever a tail occurs on coin 2, we switch to coin 1, etc. What is the probability that the nth toss will be performed on coin 1? *Hint:* Let P_n be the desired probability so that $P_{n+1} = P_n p + (1 - P_n)q$, $P_1 = 1$. Show that

$$P_{n+1} - P_n = (p - q)(P_n - P_{n-1}) = (p - 1)(p - q)^{n+1}$$

$$Ans.\ P_n = \frac{q - (p - 1)(p - q)^{n+1}}{1 + q - p}$$

9.6. Continuous Probability and Distribution Functions. In previous discussions the probability event space consisted of a finite number of elementary events. A simple example illustrates the extension of this idea. We consider an idealized spinner whose pointer can assume any one of the directions $0 \leqq x < 2\pi$, x measured in radians. We say that there is zero probability that the direction of the pointer be less than zero or greater than 2π since we are concerned only with angles between zero and 2π radians. The direction of the pointer has been put into a correspondence with the real-number system. The set of all possible directions is called a one-dimensional random, or stochastic, variable, usually denoted by ξ. In this example it makes very little sense to ask the following question: What is the probability that a random spin yields the direction θ_0? θ_0 is just one possible event of an infinite number of different events which may occur. It does make sense to ask the following question: What is the probability that the random event ξ be less than or equal to x? We define the *a priori distribution function* $F(x)$ as

$$\begin{aligned} F(x) &= 0 \qquad \text{for } x < 0 \\ F(x) &= P(\xi \leqq x) = \frac{x}{2\pi} \qquad \text{for } 0 \leqq x < 2\pi \qquad (9.17) \\ F(x) &= 1 \qquad \text{for } x \geqq 2\pi \end{aligned}$$

Figure 9.3 shows a graphic representation of $F(x)$.

For any x and y we note that (9.17) yields

$$P(x \leqq \xi \leqq y) = F(y) - F(x)$$

For $y = x + \Delta x$ we have $P(x \leqq \xi \leqq x + \Delta x) = F(x + \Delta x) - F(x) = \Delta F$. ΔF is the probability that ξ lies between x and $x + \Delta x$. We also note that $F(x)$ is a monotonic nondecreasing function of x with $F(-\infty) = 0$, $F(+\infty) = 1$.

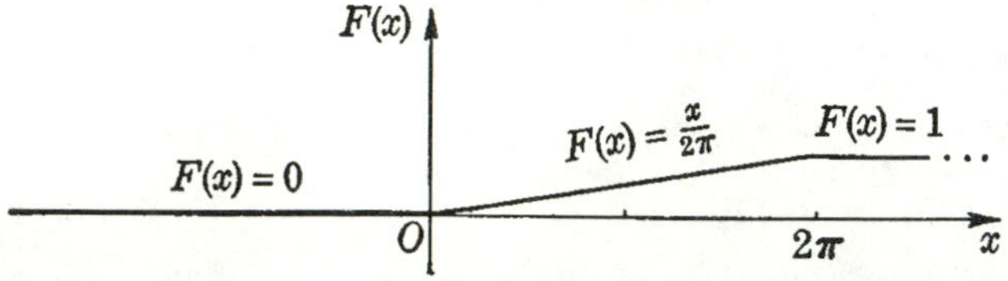

FIG. 9.3

For the more general case we consider a nonnegative function $p(x)$ such that $p(x)\,dx$ represents the probability that the random event ξ lie between x and $x + dx$, except for infinitesimals of higher order. We further desire

$$\int_{-\infty}^{\infty} p(x)\,dx = 1 \tag{9.18}$$

$p(x)$ is called the *probability function* of the random variable ξ. The distribution function is simply

$$F(x) = P(\xi \leqq x) = \int_{-\infty}^{x} p(x)\,dx \tag{9.19}$$

Note that $F(x)$ is nondecreasing since $p(x)$ is nonnegative, with

$$F(-\infty) = 0 \qquad F(\infty) = 1$$

$dF(x) = p(x)\,dx$ is the probability that ξ lies between x and $x + dx$. A simple example of a probability function is $p(x) = \dfrac{1}{\pi}\dfrac{1}{1 + x^2}$, with $F(x) = \dfrac{1}{\pi}\displaystyle\int_{-\infty}^{x} \frac{dx}{1 + x^2} = \frac{1}{\pi}\left(\tan^{-1} x + \frac{\pi}{2}\right)$.

The nth moment of a distribution is defined by

$$a_n = \int_{-\infty}^{\infty} x^n p(x)\,dx \qquad n = 0, 1, 2, \ldots \tag{9.20}$$

provided the integrals exist. For $p(x) = \dfrac{1}{\pi}\dfrac{1}{1 + x^2}$ only a_0 exists. From $dF(x) = p(x)\,dx$ we can write

$$a_n = \int_{-\infty}^{\infty} x^n\,dF(x) \qquad n = 0, 1, 2, \ldots \tag{9.21}$$

Equation (9.21) is the Stieltjes integral of Chap. 7. The discrete case can be handled by use of this integral. Let $x = 1$ represent the occurrence of a head and $x = 2$ the occurrence of a tail when a coin is tossed with $p(1) = \frac{1}{2}$, $p(2) = \frac{1}{2}$, $p(x) = 0$ otherwise. Then $F(x) = 0$ for $x < 1$, $F(x) = \frac{1}{2}$ for $1 \leqq x < 2$, $F(x) = 1$ for $x \geqq 2$. We can consider that two point masses contribute to the probability function. From $dF(x) = 0$ for $x < 1$, $dF = \frac{1}{2}$ at $x = 1$, $dF = 0$ for $1 < x < 2$, $dF = \frac{1}{2}$ at $x = 2$, $dF = 0$ for $x > 2$, we note that

$$F(x) = \int_{-\infty}^{x} dF = \begin{array}{ll} 0 & \text{for } x < 1 \\ \frac{1}{2} & \text{for } x = 1 \\ \frac{1}{2} & \text{for } 1 < x < 2 \\ = 1 & \text{for } x \geqq 2 \end{array}$$

Geometrically (see Fig. 9.4), we note that $F(x)$ is monotonic nondecreasing. Discontinuities in $F(x)$ occur at $x = 1$, $x = 2$, because point masses

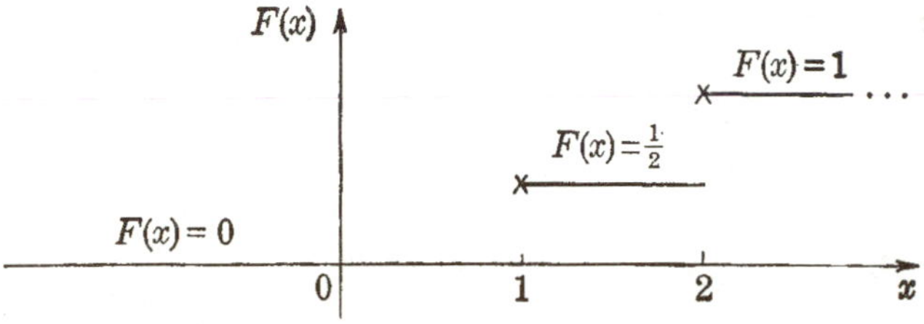

FIG. 9.4

are situated there. To compute a_1, we note that

$$a_1 = \int_{-\infty}^{\infty} x\, dF(x) = 1 \cdot \tfrac{1}{2} + 2 \cdot \tfrac{1}{2} = \tfrac{3}{2}$$

Had we chosen $x_1 = -1$, $x_2 = 1$, we would have obtained

$$a_1 = (-1)\tfrac{1}{2} + (1)\tfrac{1}{2} = 0$$

Note that $\frac{3}{2}$ is the mean of $x = 1$ and $x = 2$.

Example 9.13. *The Gaussian Distribution.* Let us assume that a dart is thrown at the xy plane under the following assumptions:

I. If $p(x, y)\, dy\, dx$ is the probability that the dart fall in the area bounded by x and $x + dx$, y and $y + dy$, then $p(x, y)$ depends only on the distance of (x, y) from the origin and is independent of $\theta = \tan^{-1}(y/x)$. Thus

$$p(x, y)\, dy\, dx = q(r^2) r\, dr\, d\theta = q(x^2 + y^2)\, dy\, dx$$
$$p(x, y) \equiv q(x^2 + y^2)$$

We assume further that $p(x, y)$ is differentiable.

II. $p(x, y) \to 0$ as x or y becomes infinite.

III. $\int_{-\infty}^{\infty} \int_{-\infty}^{\infty} p(x, y)\, dy\, dx = 1$.

We attempt to determine $p(x, y)$. First we note that

$$\left[\int_{-\infty}^{\infty} p(x, y)\, dy\right] dx \left[\int_{-\infty}^{\infty} p(x, y)\, dx\right] dy = p(x, y)\, dy\, dx$$

since $P_1(x \leqq \xi \leqq x + dx, -\infty < y < \infty) = \left[\int_{-\infty}^{\infty} p(x, y)\, dy\right] dx$,

$$P_2(-\infty < x < \infty, y \leqq \eta \leqq y + dy) = \left[\int_{-\infty}^{\infty} p(x, y)\, dx\right] dy$$

and $$P(x \leqq \xi \leqq x + dx, y \leqq \eta \leqq y + dy) = P_1P_2 = p(x, y)\, dy\, dx$$

Define $R(x) = \int_{-\infty}^{\infty} p(x, y)\, dy$, $S(y) = \int_{-\infty}^{\infty} p(x, y)\, dx$, so that

$$R(x)S(y) = q(x^2 + y^2) \tag{9.22}$$

Differentiating (9.22) yields

$$R(x)\frac{dS(y)}{dy} = 2yq'(x^2 + y^2)$$

$$S(y)\frac{dR(x)}{dx} = 2xq'(x^2 + y^2)$$

and $$\frac{1}{yS(y)}\frac{dS(y)}{dy} = \frac{1}{xR(x)}\frac{dR(x)}{dx} = \text{constant} = a \tag{9.23}$$

From (9.23), $R(x) = Ae^{ax^2}$, $S(y) = Be^{ay^2}$, and

$$p(x, y) = q(x^2 + y^2) = Ce^{a(x^2+y^2)} \tag{9.24}$$

Condition II implies that a is negative. If we choose $a = -1/2\sigma^2$, of necessity, $C = 1/2\pi\sigma^2$ from condition III. Thus

$$p(x, y) = \frac{1}{2\pi\sigma^2} e^{-(1/2\sigma^2)(x^2+y^2)} \tag{9.25}$$

Equation (9.25) is the normal distribution of Gauss. For the one-dimensional case

$$\varphi(x) = \frac{1}{\sigma\sqrt{2\pi}} e^{-(1/2\sigma^2)(x-m)^2} \tag{9.26}$$

where m represents a displacement from the origin. Let the reader show that

$$\sigma^2 = \int_{-\infty}^{\infty} (x - m)^2\varphi(x)\, dx$$

Thus σ^2 is the second moment of $\varphi(x)$ relative to the center m.

Example 9.14. *Tchebyscheff's Theorem.* Let ξ be a random variable with a probability function $p(x)$. Let $g(\xi)$ be a nonnegative function of ξ, and let S be the set of points such that $g(\xi) \geqq K > 0$. $\bar{S}$ will denote the rest of the real axis. The expected or mean value of $g(\xi)$ is defined as

$$E[g(\xi)] = \int_{-\infty}^{\infty} g(x)p(x)\, dx$$

The probability that $g(\xi)$ be greater than or equal to K when ξ is chosen at random is simply the probability that ξ be found in S, so that $P[g(\xi) \geqq K] = \int_S p(x)\, dx$. Now

$$E[g(\xi)] = \int_S g(x)p(x)\, dx + \int_{\bar{S}} g(x)p(x)\, dx$$

$$\geqq \int_S g(x)p(x)\, dx \geqq K\int_S p(x)\, dx$$

since $g(x)$ and $p(x)$ are nonnegative. We obtain Tchebyscheff's theorem,

$$P[g(\xi) \geqq K] \leqq \frac{E[g(\xi)]}{K} \tag{9.27}$$

What difficulties arise if $g(x) = 1$ for x irrational, $g(x) = 0$ for x rational, $K = \frac{1}{2}$?

If m is the first moment or mean of ξ and σ^2 is the variance of ξ defined by

$$\sigma^2 = \int_{-\infty}^{\infty} (x - m)^2 p(x)\, dx$$

let the reader deduce the Bienaymé-Tchebyscheff inequality,

$$P(|\xi - m| \geqq k\sigma) \leqq \frac{1}{k^2} \tag{9.28}$$

if we choose $g(\xi) = (\xi - m)^2$, $K = k^2\sigma^2$.

Example 9.15. *The Buffon Needle Problem.* A board is ruled by equidistant parallel lines, two consecutive lines being d units apart. A needle of length $a < d$ is thrown at random on the board. What is the probability that the needle will intersect one of the lines? Let x and y describe the position of the needle (see Fig. 9.5).

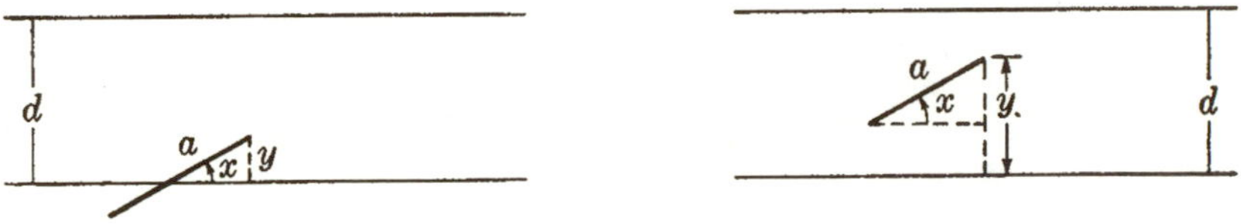

FIG. 9.5

If $y \leqq a \sin x$, our situation is favorable, while if $d > y > a \sin x$, an unfavorable case occurs. We assume that x and y are independent random variables. Thus

$$p_1(x \leqq \xi \leqq x + dx) = \frac{dx}{\pi} \qquad p_2(y \leqq \eta \leqq y + dy) = \frac{dy}{d}$$

and $p(x, y)\, dy\, dx = (1/\pi d)\, dy\, dx$. We are interested in $P(0 \leqq x \leqq \pi,\ 0 \leqq y \leqq a \sin x)$, which is defined by

$$\int_0^{\pi} \int_0^{a \sin x} p(x, y)\, dy\, dx = \frac{1}{\pi d} \int_0^{\pi} \int_0^{a \sin x} dy\, dx = \frac{2a}{\pi d}$$

Example 9.16. Let ξ be a random variable with a probability function $p(x)$. The probability that ξ lies between x_1 and $x_1 + dx_1$ as the result of a single experiment is $p(x_1)\, dx_1$. If we repeat the experiment, the probability that ξ lies between x_2 and $x_2 + dx_2$ is $p(x_2)\, dx_2$. The joint probability that both results happen as the result of two experiments (assumed independent) is

$$p(x_1)p(x_2)\, dx_1\, dx_2 \tag{9.29}$$

Equation (9.29) is a special case of the single random variable (ξ, η) with probability function $p(x, y)$. We note that

$$\int_{-\infty}^{\infty} \int_{-\infty}^{\infty} p(x_1)p(x_2)\, dx_1\, dx_2 = 1$$

We now ask the following question: What is the probability that, as the result of two experiments, $x_1 + x_2 \geqq K$, K a constant? In Fig. 9.6 we note that this is just the

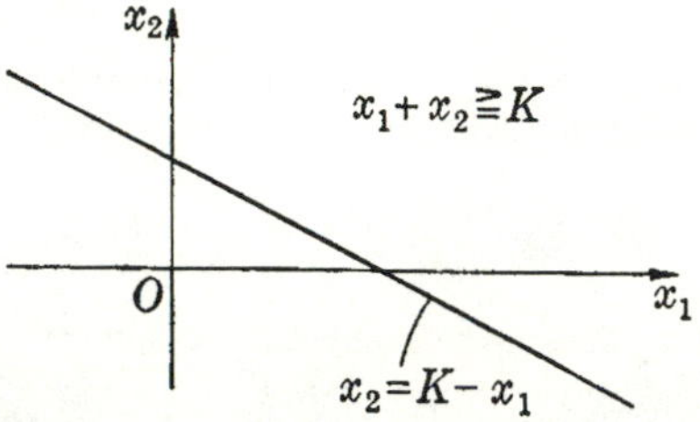

FIG. 9.6

probability that the point (x_1, x_2) lie above the line $x_1 + x_2 = K$, so that

$$P(x_1 + x_2 \geqq K) = \int_{-\infty}^{\infty} \int_{K-x_1}^{\infty} p(x_1)p(x_2)\, dx_2\, dx_1 \tag{9.30}$$

From (9.30) we have

$$P(x_1 + x_2 \geqq K + dK) = \int_{-\infty}^{\infty} \int_{K+dK-x_1}^{\infty} p(x_1)p(x_2)\, dx_2\, dx_1 \tag{9.31}$$

so that
$$P(K \leqq x_1 + x_2 \leqq K + dK) = \int_{-\infty}^{\infty} \int_{K-x_1}^{K+dK-x_1} p(x_1)p(x_2)\, dx_2\, dx_1 \tag{9.32}$$

by subtracting (9.31) from (9.30). Since $\int_{K-x_1}^{K+dK-x_1} p(x_2)\, dx_2 \approx p(K - x_1)\, dK$, we note that the probability function for $x_1 + x_2 = u$ is

$$p(u) = \int_{-\infty}^{\infty} p(x_1)p(u - x_1)\, dx_1 \tag{9.33}$$

Problems

1. Show that $\varphi(x)$ of (9.26) satisfies $\int_{-\infty}^{\infty} \varphi(x) = 1$.

2. Derive (9.28).

3. Find the four-dimensional joint probability function for the two-needle Buffon problem.

4. Show that $\sigma^2 \equiv \int_{-\infty}^{\infty} (x - m)^2 p(x)\, dx = a_2 - m^2$, a_2 defined by (9.20).

5. Let ξ and η be random variables with Gaussian probability functions ($m = 0$). Show that $u = \xi + \eta$ has a Gaussian distribution.

6. Find a_1, a_2, a_3 for $\varphi(x)$ of (9.26). See (9.20) for the definition of a_i, $i = 1, 2, 3, \ldots$

7. If ξ is a random variable with a Gaussian distribution ($m = 0$), find the probability function for the random variable ξ^2. *Hint:* Let $u = \xi^2$ so that

$$\begin{aligned} P(u \geqq t) &= P(\xi^2 \geqq t) = P(|\xi| \geqq \sqrt{t}) \qquad t \geqq 0 \\ &= \frac{1}{\sqrt{2\pi}\,\sigma} \int_{\sqrt{t}}^{\infty} e^{-x^2/2\sigma^2}\, dx + \frac{1}{\sqrt{2\pi}\,\sigma} \int_{-\infty}^{-\sqrt{t}} e^{-x^2/2\sigma^2}\, dx \end{aligned}$$

$$P(u \geqq t + dt) = P(\xi^2 \geqq t + dt) = P(|\xi| \geqq \sqrt{t + dt}) \approx P\left(|\xi| \geqq \sqrt{t} + \frac{dt}{2\sqrt{t}}\right)$$

Then show that $P(t \leqq u \leqq t + dt) = (1/\sigma \sqrt{2\pi t})\, e^{-t/2\sigma^2}\, dt$, so that

$$p(t) = \frac{1}{\sigma \sqrt{2\pi t}} e^{-t/2\sigma^2} \text{ for } t > 0 \qquad p(t) = 0 \text{ for } t \leqq 0$$

8. Let ξ be a random variable with probability function $p(x)$. Let $u = u(\xi)$ be a strictly increasing (or decreasing) function of ξ. Show that the probability function for the random variable u is $q(u) = p(x(u)) \left| \frac{dx(u)}{du} \right|$, where $x(u)$ is the inverse function of $u(x)$. Apply this result to Prob. 7.

9. Consider a distribution of points in the plane with a density function given by (9.25). At $t = 0$ each point moves in a direction φ with speed V and with a probability function $p(\varphi) = \varphi/2\pi$, $0 \leqq \varphi < 2\pi$. Show that at time t the probability function in space and direction of motion is given by

$$p(r, \theta, \varphi, t) = \frac{1}{4\pi^2\sigma^2} e^{-(1/2\sigma^2)[r^2+V^2t^2-2rVt\cos(\theta-\varphi)]}$$

Show that $p(r, \theta, \varphi, t)$ satisfies $\frac{\partial p}{\partial t} + \nabla \cdot (p\mathbf{v}) = 0$ with

$$\mathbf{v} = V \cos \varphi\, \mathbf{i} + V \sin \varphi\, \mathbf{j} = V \cos (\varphi - \theta)\mathbf{e}_r + V \sin (\varphi - \theta)\mathbf{e}_\theta$$

9.7. The Characteristic Function. Bernoulli's Theorem. The Central-limit Theorem. Let ξ be a random variable with a probability function $p(x)$. Let us assume that the Fourier integral of $p(x)$ exists. Then

$$\varphi(t) = \int_{-\infty}^{\infty} e^{itx} p(x)\, dx \tag{9.34}$$

is called the characteristic function of $p(x)$, with t real. We note that $\varphi(0) = \int_{-\infty}^{\infty} p(x)\, dx = 1$. For t real, n a positive integer, we have

$$\left| \int_{-\infty}^{\infty} x^{2n} e^{itx} p(x)\, dx \right| \leqq \int_{-\infty}^{\infty} x^{2n} p(x)\, dx = a_{2n}$$

If we can differentiate inside the integral, then

$$\left.\frac{d^n\varphi}{dt^n}\right|_{t=0} = i^n \int_{-\infty}^{\infty} x^n p(x)\, dx = i^n a_n \tag{9.35}$$

Hence, if we can find $\varphi(t)$, it will be possible to find the nth moment from (9.35).

Example 9.17. Let $p(x) = e^{-x}$ for $x \geqq 0$, $p(x) = 0$ otherwise. Then

$$\varphi(t) = \int_0^\infty e^{x(it-1)}\, dx = \frac{1}{1 - it}$$

$$\left.\frac{d\varphi}{dt}\right|_{t=0} = \left.\frac{i}{(1 - it)^2}\right|_{t=0} = i$$

$$\left.\frac{d^2\varphi}{dt^2}\right|_{t=0} = \left.\frac{-2}{(1 - it)^3}\right|_{t=0} = -2$$

so that $a_1 = 1$, $a_2 = 2$. Since

$$\frac{1}{1-it} = \sum_{n=0}^{\infty} \varphi^{(n)}(0)\frac{t^n}{n!} = 1 + it + i^2t^2 + \cdots + i^nt^n + \cdots$$

we have $\varphi^{(n)}(0) = i^n n!$ and $a_n = n!$. This result is also obvious since

$$\Gamma(n+1) = \int_0^\infty x^n e^{-x}\,dx = n!$$

Example 9.18. *Bernoulli's Theorem.* For a Bernoulli sequence of n events we have

$$P_r = \binom{n}{r} p^r(1-p)^{n-r} \tag{9.36}$$

for the probability of obtaining exactly r successes [see (9.14)].

The analogue of (9.34) for the discrete case is

$$\begin{aligned}\varphi(t) &= \sum_{r=0}^{n} e^{irt}P_r \\ &= \int_{-\infty}^{\infty} e^{itx}\,dF(x)\end{aligned} \tag{9.37}$$

Applying (9.36) yields

$$\begin{aligned}\varphi(t) &= \sum_{r=0}^{n} \binom{n}{r} (pe^{it})^r(1-p)^{n-r} \\ &= [pe^{it} + (1-p)]^n\end{aligned}$$

Thus $\varphi(0) = 1$, $\varphi'(0) = ipn$, $\varphi''(0) = -n(n-1)p^2 - np$. Hence $a_1 = m = pn$, $a_2 = n(n-1)p^2 + np$, so that $\sigma^2 = a_2 - m^2 = np(1-p)$ (see Prob. 4, Sec. 9.6).

Applying (9.28) with $k = \epsilon\sqrt{\dfrac{n}{p(1-p)}}$, $\epsilon > 0$, yields

$$P(|\xi - np| \geqq \epsilon n) \leqq \frac{p(1-p)}{\epsilon^2 n} \leqq \frac{1}{4\epsilon^2 n} \tag{9.38}$$

since $p(1-p) \leqq \frac{1}{4}$ for $0 \leqq p \leqq 1$. Formula (9.38) is equivalent to

$$P\left(\left|\frac{\xi}{n} - p\right| \geqq \epsilon\right) \leqq \frac{1}{4\epsilon^2 n} \tag{9.39}$$

Formula (9.39) is a form of Bernoulli's theorem. In other words, the probability that the frequence of occurrence, ξ/n, differs from its mean value p by a quantity of absolute value at least equal to ϵ tends to zero as $n \to \infty$ however small ϵ is chosen. This essentially means that the greater n is the more certain we are that ξ/n differs little from p, where ξ is the total number of successes in n trials.

Let us return to the characteristic function. From (9.34) we note that, if $\varphi(t)$ is known, then

$$p(x) = \frac{1}{2\pi}\int_{-\infty}^{\infty} e^{-itx}\varphi(t)\,dt \tag{9.40}$$

Hence $p(x)$ is uniquely determined if the characteristic function is known. Equation (9.40) follows from the Fourier integral theorem. If ξ and η are independent stochastic variables with probability functions $p_1(x)$, $p_2(y)$, respectively, it follows from the method of Example 9.16 that

$$p(u) = \int_{-\infty}^{\infty} p_1(x)p_2(u - x)\,dx \tag{9.41}$$

is the probability function for the stochastic variable $\xi + \eta$. The characteristic function of $p(u)$ is

$$\begin{aligned} \varphi(t) &= \int_{-\infty}^{\infty} e^{itu}p(u)\,du \\ &= \int_{-\infty}^{\infty}\int_{-\infty}^{\infty} e^{itu}p_1(x)p_2(u - x)\,dx\,du \end{aligned} \tag{9.42}$$

Now

$$\varphi_1(t) = \int_{-\infty}^{\infty} e^{itx}p_1(x)\,dx$$

$$\varphi_2(t) = \int_{-\infty}^{\infty} e^{ity}p_2(y)\,dy$$

so that

$$\begin{aligned} \varphi_1(t)\varphi_2(t) &= \int_{-\infty}^{\infty}\int_{-\infty}^{\infty} e^{it(x+y)}p_1(x)p_2(y)\,dy\,dx \\ &= \int_{-\infty}^{\infty}\int_{-\infty}^{\infty} e^{itu}p_1(x)p_2(u - x)\,du\,dx \end{aligned} \tag{9.43}$$

by letting $u = x + y$. A comparison of (9.42) with (9.43) yields

$$\varphi(t) = \varphi_1(t)\varphi_2(t) \tag{9.44}$$

provided the order of integration can be interchanged.

Another interesting result is the following: If $\varphi_1(t)$, $\varphi_2(t)$, . . . , $\varphi_n(t)$, . . . is a sequence of characteristic functions which converges to $\varphi(t)$, obtained from a sequence of probability functions $p_1(x)$, $p_2(x)$, . . . , $p_n(x)$, . . . , it may be possible that

$$p(x) = \frac{1}{2\pi}\int_{-\infty}^{\infty} e^{-itx}\varphi(t)\,dt \tag{9.45}$$

where $\varphi(t)$ and $p(x)$ are limits of their respective sequences. In order that (9.45) hold, we must have

$$p(x) \equiv \lim_{n\to\infty} p_n(x) = \lim_{n\to\infty} \frac{1}{2\pi}\int_{-\infty}^{\infty} e^{-itx}\varphi_n(t)\,dt = \frac{1}{2\pi}\int_{-\infty}^{\infty} e^{-itx}\varphi(t)\,dt \tag{9.45$'$}$$

The reader is referred to Sec. 10.22 for a discussion of this type of problem.

Let us consider now the following example, which will illustrate one aspect of the central-limit theorem involving probabilities. Let ξ be a random variable, and, to be momentarily specific, we shall consider the toss of a coin. The random variable ξ will have the value $x = 1$ if a head occurs and the value $x = -1$ if a tail occurs. If we toss the coin five times, we can record the numbers $(x_1, x_2, x_3, x_4, x_5)$ with x_i equal to 1 or -1, $i = 1, 2, 3, 4, 5$. The set $(x_1, x_2, x_3, x_4, x_5)$ represents a point in a five-dimensional space. We can repeat the experiment of tossing five coins as often as we please and obtain a set of 5-tuples, $(x_1^i, x_2^i, x_3^i, x_4^i, x_5^i)$, $i = 1, 2, \ldots$. We may look upon the aggregate $x_1^1, x_1^2, x_1^3, \ldots$ as a set of results yielding information about the random variable ξ. The same statement can be made concerning the set $x_2^1, x_2^2, x_2^3, \ldots, x_2^n, \ldots$, etc. The set of 5-tuples can also be looked upon as defining a new random variable. For the more general case we let $p(x)$ be the probability function for the random variable ξ, and we consider the n-tuple of independent events $(x_1, x_2, \ldots, x_n)$. The probability that a point lie in the volume bounded by x_1 and $x_1 + dx_1$, x_2 and $x_2 + dx_2$, x_n and $x_n + dx_n$ is given by

$$p(x_1)p(x_2) \cdots p(x_n)\, dx_1\, dx_2 \cdots dx_n$$

If S_n is the random variable $\sum_{i=1}^{n} x_i$, an extension of Example 9.16 shows that the probability function for S_n is given by

$$P_n\left(u = \sum_{i=1}^{n} x_i\right)$$
$$= \int_{-\infty}^{\infty} \cdots \int_{-\infty}^{\infty} p(x_1)p(x_2) \cdots p\left(u - \sum_{i=1}^{n-1} x_i\right) dx_{n-1} \cdots dx_1 \quad (9.46)$$

The characteristic function is

$$\begin{aligned}\Phi_n(t) &= \int_{-\infty}^{\infty} P_n(u)e^{iut}\, du \\ &= \int_{-\infty}^{\infty} \cdots \int_{-\infty}^{\infty} e^{i(x_1+x_2+\cdots+x_n)t} p(x_1) \cdots p(x_n)\, dx_1 \cdots dx_n \quad (9.47)\end{aligned}$$

by letting $u = x_1 + x_2 + \cdots + x_n$, $du = dx_n$, and reversing the order of integration. Equation (9.47) becomes

$$\Phi_n(t) = [\varphi(t)]^n \quad (9.48)$$

with $\varphi(t) = \int_{-\infty}^{\infty} p(x)e^{itx}\, dx$. Let us assume that the first moment of $p(x)$ is zero so that σ^2 is its second moment. Thus $\varphi(0) = 1$, $\varphi'(0) = 0$, $\varphi''(0) = -\sigma^2$. If $\varphi(t)$ has a continuous third derivative in a neighborhood of $t = 0$, then

$$\varphi(t) = 1 - \sigma^2 \frac{t^2}{2!} + \alpha(t)$$

with $|\alpha(t)| < At^3$ (see Sec. 10.23).

If T_n is the random variable $S_n/\sigma\sqrt{n}$, let the reader show that the characteristic function for T_n is $\psi_n(t) = \Phi_n(t/\sigma\sqrt{n})$. Hence

$$\psi_n(t) = \left[\varphi\left(\frac{t}{\sigma\sqrt{n}}\right)\right]^n$$
$$= \left[1 - \frac{t^2}{2n} + \alpha\left(\frac{t}{\sigma\sqrt{n}}\right)\right]^n$$

with $|\alpha(t/\sigma\sqrt{n})| < At^3/\sigma^3 n^{\frac{3}{2}}$. From $\ln(1+z) = z + \beta(z)$ with $|\beta(z)| < Bz^2$ for z small, one has that

$$\lim_{n\to\infty} \ln \psi_n(t) = \lim_{n\to\infty} n \ln\left[1 - \frac{t^2}{2n} + \alpha\left(\frac{t}{\sigma\sqrt{n}}\right)\right] = -\frac{t^2}{2} \tag{9.49}$$

From (9.49) it follows that

$$\lim_{n\to\infty} \psi_n(t) = e^{-t^2/2}$$

The probability function $P(x)$ associated with the characteristic function $e^{-t^2/2}$ is

$$P(x) = \frac{1}{2\pi}\int_{-\infty}^{\infty} e^{-t^2/2}e^{-itx}\,dt = \frac{1}{\sqrt{2\pi}}e^{-x^2/2} \tag{9.50}$$

If we accept (9.45′), we have shown that if $\xi_1, \xi_2, \ldots, \xi_n, \ldots$ are random variables with the same probability function $p(x)$, with a mean equal to zero and second moment equal to σ^2, then $T_n = \sum_{i=1}^{n} \xi_i/\sigma\sqrt{n}$ is a random variable whose probability function approaches the normal distribution $(1/\sqrt{2\pi})e^{-x^2/2}$ as n becomes infinite. We say that the sequence of random variables obeys the central-limit law.

Problems

1. Derive (9.49).

2. Derive (9.50).

3. Show that the characteristic function for Cauchy's distribution function,

$$p(x) = \frac{1}{\pi}\frac{1}{1+x^2}, \text{ is } \varphi(t) = e^{-|t|}$$

4. Let $p(x) = 1$ for $-\frac{1}{2} \leqq x \leqq \frac{1}{2}$, $p(x) = 0$ otherwise. Show that

$$\varphi(t) = \frac{2}{t}\sin\frac{t}{2}$$

5. If ξ is a random variable with characteristic function $\varphi(t)$, show that $e^{-ita}\varphi(t)$ is the characteristic function for the random variable $\xi - a$, $a =$ constant.

6. Consider a sequence of 1,000 Bernoulli trials with $p = \frac{1}{2}$. Show that the probability that the experimental ratio r/n will differ from $\frac{1}{2}$ by less than 0.01 is greater than or equal to $\frac{3}{4}$.

9.8. The χ^2 Distribution. Application to Statistics. From the definition of the gamma function (see Sec. 4.16) we have

$$\Gamma(z) = \int_0^\infty e^{-x}x^{z-1}\,dx$$
$$= a^z \int_0^\infty e^{-ay}y^{z-1}\,dy \qquad a > 0$$

so that

$$\int_0^\infty \frac{a^z}{\Gamma(z)} e^{-ay}y^{z-1}\,dy = 1 \qquad a > 0 \tag{9.51}$$

The function

$$f(y; a, z) = \frac{a^z}{\Gamma(z)} e^{-ay}y^{z-1} \qquad \text{for } y > 0$$
$$= 0 \qquad \text{for } y \leqq 0 \tag{9.52}$$

is nonnegative for $a > 0$, $z > 0$, and $\int_{-\infty}^{\infty} f(y; a, z)\,dy = 1$. We can look upon $f(y; a, z)$ as a probability function. Its characteristic function is

$$\varphi(t) = \int_{-\infty}^{\infty} e^{ity}f(y; a, z)\,dy$$
$$= \frac{a^z}{\Gamma(z)} \int_0^\infty e^{-(a-it)y}y^{z-1}\,dy$$
$$= \frac{a^z}{(a - it)^z\Gamma(z)} \int_0^\infty e^{-v}v^{z-1}\,dv = \frac{a^z}{(a - it)^z} = \frac{1}{(1 - it/a)^z} \tag{9.53}$$

Now let ξ be a random variable with probability function

$$p(x) = \left(\frac{1}{\sqrt{2\pi}}\right) e^{-x^2/2}$$

The probability function for ξ^2 is $P(x) = (1/\sqrt{2\pi x})\, e^{-x/2}$ for $x > 0$, $P(x) = 0$ otherwise (see Prob. 7, Sec. 9.6). If $\xi_1, \xi_2, \ldots, \xi_n$ are n independent random variables with the same probability function $p(x)$ given above, the characteristic function for the random variable

$$\chi^2 = \sum_{i=1}^{n} \xi_i^2 \tag{9.54}$$

is the product of the characteristic functions for each random variable ξ_i^2, $i = 1, 2, \ldots, n$ [see (9.44)]. Now the characteristic function for $P(x)$ is

$$\varphi(t) = \int_0^\infty \frac{e^{-x/2}}{\sqrt{2\pi x}} e^{itx}\,dx = \int_0^\infty \frac{e^{-(\frac{1}{2}-it)x}}{\sqrt{2\pi x}}\,dx$$
$$= \frac{1}{\sqrt{2\pi}}\left(\frac{1}{2} - it\right)^{-\frac{1}{2}} \int_0^\infty e^{-v^2}\,dv = (1 - 2it)^{-\frac{1}{2}}$$

The characteristic function for χ^2 is

$$\Phi(t) = (1 - 2it)^{-n/2} \tag{9.55}$$

Comparing (9.55) with (9.53), we note that $z = n/2$, $a = \frac{1}{2}$, so that the probability function associated with the random variable χ^2 is $f(y; \frac{1}{2}, n/2)$,

$$\begin{aligned} k_n(y) \equiv f\left(y; \frac{1}{2}, \frac{n}{2}\right) &= \frac{1}{2^{n/2}\Gamma(n/2)} y^{n/2-1}e^{-y/2} \qquad &\text{for } y > 0 \\ &= 0 &\text{for } y \leqq 0 \end{aligned} \tag{9.56}$$

The distribution function for $k_n(y)$ is

$$K_n(x) = P(\chi^2 \leqq x) = \frac{1}{2^{n/2}\Gamma(n/2)} \int_0^x y^{n/2-1}e^{-y/2}\,dy \tag{9.57}$$

The distribution defined by $K_n(x)$ is called the χ^2 distribution, principally associated with K. Pearson.

The χ^2 test of significance arises in the following manner: Let ξ be a random variable with a known probability function $p(x)$. Let us divide the interval $-\infty < x < \infty$ into m parts, say, $-\infty < x < x_1$, $x_1 \leqq x \leqq x_2$, . . . , $x_{m-1} \leqq x < \infty$. We have

$$P_i(x_{i-1} \leqq x \leqq x_i) = \int_{x_{i-1}}^{x_i} p(x)\,dx$$

with $\sum_{i=1}^{m} P_i = 1$. Let N be the number of times we sample ξ, the result of any sample being independent of the previous samples. The expected number of samples which fall in the ith interval is given by NP_i. If r_i is the actual number of samples falling in the ith interval, then

$$\sum_{i=1}^{m} (r_i - NP_i)^2$$

certainly constitutes some measure of discrepancy between the theoretical and experimental results. K. Pearson found that

$$\chi^2 = \sum_{i=1}^{m} \frac{(r_i - NP_i)^2}{NP_i} \tag{9.58}$$

yielded a practical means for measuring the reliability of the experimental results. If we let $y_i = (r_i - NP_i)/\sqrt{NP_i}$, then $\chi^2 = \sum_{i=1}^{m} y_i^2$, and

$\sum_{i=1}^{m} y_i \sqrt{P_i} = \frac{1}{\sqrt{N}} \sum_{i=1}^{m} r_i - \sqrt{N} \sum_{i=1}^{m} P_i = \sqrt{N} - \sqrt{N} = 0$. Hence the random variables y_i are not independent. The y_i lie in the hyperplane $\sum_{i=1}^{m} y_i \sqrt{P_i} = 0$, a subspace of the m-dimensional Euclidean space. It can be shown that, as N becomes infinite (N is the total number of samples or trials), then the distribution of χ^2 approaches $K_{m-1}(x)$ [see (9.57)]. The reader is referred to Cramér, "Mathematical Methods of Statistics," Chap. 30, for proof of this statement. Thus

$$\lim_{N\to\infty} P(\chi^2 \geqq \chi_0^2) = \int_{x_0}^{\infty} \frac{1}{2^{(m-1)/2}\,\Gamma\left(\frac{m-1}{2}\right)} y^{(m-3)/2} e^{-y/2}\, dy \qquad (9.59)$$

One can determine the value of the integral of (9.59) from a table of the χ^2 distribution. It is important to note that (9.59) is independent of the original probability function $p(x)$.

Example 9.19. A coin was tossed 5,000 times and heads appeared 2,512 times. Under the assumption that the coin is "true," we have $p(H) = \frac{1}{2}$, $p(T) = \frac{1}{2}$, $m = 2$, $N = 5{,}000$, $P_1 = P_2 = \frac{1}{2}$; $r_1 = 2{,}512$, $r_2 = 2{,}488$. Hence

$$\chi^2 = \frac{(2{,}512 - 2{,}500)^2}{2{,}500} + \frac{(2{,}488 - 2{,}500)^2}{2{,}500}$$
$$= 0.115$$

In a table of the χ^2 distribution it is found that the probability that χ^2 exceed 3.841 is 0.05. Since $0.115 < 3.841$, we feel that there is no inconsistency in the assumption that the coin is true. The 5 per cent level is taken as a fairly significant level. In the tables one also finds that $P(\chi^2 \geqq 0.115)$ is about 0.73, which means that we have a probability of about 73 per cent of obtaining a deviation from the expected result at least as great as that actually observed. We are therefore not too worried about the fact that experimentally we obtained $\chi^2 = 0.115$.

Problems

1. A coin is tossed 5,000 times, with heads occurring 3,000 times. Evaluate χ^2 for this experiment. Would you be suspicious that the coin is "true"?

2. A die is rolled 6,000 times with the following occurrences: the number of times 1, 2, 3, 4, 5, 6 occurred, respectively, was 1,020; 1,032; 981; 977; 1,011; 979. Show that $\chi^2 = 2.876$. What is your opinion as to the "truthness" of the die?

3. A random variable ξ has the probability function $p(x) = (1/\sqrt{2\pi})e^{-x^2/2}$. If in 1,000 experiments one obtained 534 values of $x < 0$ and 466 values of $x \geqq 0$, would you consider that the experiment was biased?

9.9. Monte Carlo Methods and the Theory of Games. The method of Monte Carlo is essentially a device for making use of probability theory to approximate the solution of a mathematical or physical problem. A few examples will illustrate the method.

Let $p(x)$ be the probability function of a random variable, ξ, defined on the range $a \leqq x \leqq b$, with $\int_a^b p(x)\,dx = 1$, $p(x) = 0$ for x outside the interval (a, b). The expected, or mean, value of $f(x)$ has been defined as

$$\bar{f} = \int_a^b f(x)p(x)\,dx$$

A sequence of values $x_1, x_2, \ldots, x_n$, is chosen at random subject to the condition that the probability that the random variable ξ lie between x and $x + dx$ be given by $p(x)\,dx$. For n large one hopes that

$$\int_a^b f(x)p(x)\,dx \approx \frac{1}{n}\sum_{i=1}^{n} f(x_i) \tag{9.60}$$

If we wish to find an approximate value of $\int_a^b F(x)\,dx$, we note that

$$\int_a^b F(x)\,dx = \int_a^b \frac{F(x)}{p(x)}\,p(x)\,dx \tag{9.61}$$

so that $f(x) = F(x)/p(x)$. A simple choice for $p(x)$ is $p(x) = 1/(b - a)$.

The choice $p(x) = 1/(b - a)$ implies that one can construct an experiment such that all numbers on the interval (a, b) have equal likelihood of occurrence. It is extremely doubtful that such an experiment can be devised. The toss of a single coin can be used to generate the number zero if a head occurs, the number 1 if a tail occurs. An infinite number of ordered tosses of a single coin, or an infinite number of ordered coins tossed simultaneously, would yield a sequence $a_0, a_1, a_2, \ldots, a_n, \ldots,$ with $a_n = 0$ or 1 for all $n \geqq 0$. This, in turn, yields the number

$$x = \sum_{n=0}^{\infty} \frac{a_n}{2^n} \tag{9.62}$$

with $0 \leqq x \leqq 1$. Mathematicians have constructed tables of random numbers to avoid the cumbersome process of obtaining a random variable by experiment. Let the reader deduce a method for obtaining $\xi = \sum_{n=0}^{N} \frac{a_n}{2^n}$ by use of a bowl, containing an equal number of red and white balls, and a scoop with $N + 1$ ordered holes. The author chose 25 numbers from a table of random numbers and obtained $\frac{1}{25}\sum_{i=1}^{25} x_i = 0.5030$, reasonably close to $\int_0^1 x\,dx$.

We now consider a less trivial example. The game to be played concerns three spinners (see Fig. 9.7). If a pointer comes to rest in the area

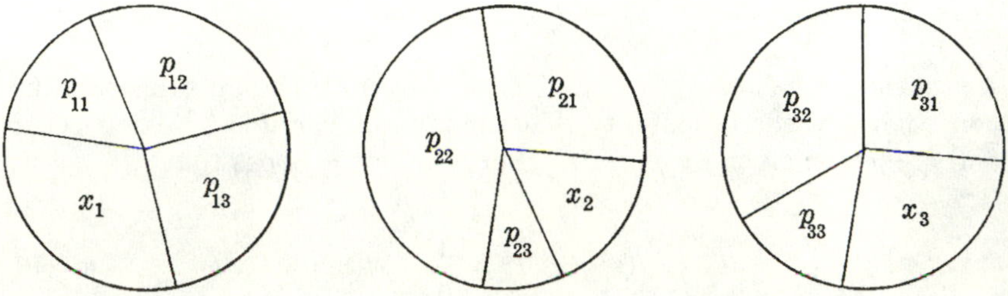

FIG. 9.7

designated by p_{ij} we move from the ith spinner to the jth spinner. If the pointer comes to rest in the area x_i, $i = 1, 2, 3$, the game is over. The areas are to be unity so that $x_i = 1 - \sum_{j=1}^{3} p_{ij}$, $i = 1, 2, 3$, and the p's and x's represent probabilities. We ask the following question: If we begin the game at spinner I, what is the probability, f_{11}, that the game will end at spinner I? We can be successful as follows:

1. x_1 occurs on the first spin.
2. p_{11} occurs n times in succession, and then x_1 occurs, $n = 1, 2, 3, \ldots$.
3. p_{12} occurs followed by the probability, f_{21}, that if we are at spinner II the game will end at spinner I.
4. p_{11} occurs n times in succession, $n \geqq 1$, and then (3) occurs.
5. Replace p_{12} and f_{21} of (3) and (4) by p_{13} and f_{31}.

Let the reader show that

$$\begin{aligned} f_{11} &= (x_1 + p_{11}x_1 + p_{11}^2x_1 + \cdots) + (p_{12}f_{21} + p_{11}p_{12}f_{21} \\ &\qquad + p_{11}^2p_{12}f_{21} + \cdots) + (p_{13}f_{31} + p_{11}p_{13}f_{31} + p_{11}^2p_{13}f_{31} + \cdots) \\ &= \frac{x_1 + p_{12}f_{21} + p_{13}f_{31}}{1 - p_{11}} \end{aligned}$$

By the reasoning used above one can readily show that

$$\begin{aligned} (1 - p_{ii})f_{ii} &= x_i + \sum_{k \neq i} p_{ik}f_{ki} \qquad & i &= 1, 2, 3 \\ (1 - p_{ii})f_{ij} &= \sum_{k \neq i} p_{ik}f_{kj} \qquad & j \neq i &= 1, 2, 3 \end{aligned} \tag{9.63}$$

From (9.63) one obtains

$$\begin{aligned}(p_{11}-1)f_{12}+p_{12}f_{22}+p_{13}f_{32}&=0\\ p_{21}f_{12}+(p_{22}-1)f_{22}+p_{23}f_{32}&=-x_2\\ p_{31}f_{12}+p_{32}f_{22}+(p_{33}-1)f_{32}&=0\end{aligned}$$

If we solve for f_{12}, we have

$$f_{12}=\frac{\begin{vmatrix}0 & p_{12} & p_{13}\\ -x_2 & p_{22}-1 & p_{23}\\ 0 & p_{32} & p_{33}-1\end{vmatrix}}{\begin{vmatrix}p_{11}-1 & p_{12} & p_{13}\\ p_{21} & p_{22}-1 & p_{23}\\ p_{31} & p_{32} & p_{33}-1\end{vmatrix}}=\frac{x_2\begin{vmatrix}p_{12} & p_{13}\\ p_{32} & p_{33}-1\end{vmatrix}}{|\Delta|}$$

so that $-f_{12}/x_2$ is the inverse element of p_{21} in the matrix

$$\|\Delta\|=\left\|\begin{matrix}p_{11}-1 & p_{12} & p_{13}\\ p_{21} & p_{22}-1 & p_{23}\\ p_{31} & p_{32} & p_{33}-1\end{matrix}\right\|$$

(see Sec. 1.3). Let the reader show that $-f_{22}/x_2$ is the inverse element of p_{22} for $\|\Delta\|$, etc. Thus probability theory can be applied to approximate the elements of the inverse matrix of a given matrix. The quantities f_{ij}, $i, j = 1, 2, 3$, are obtained by experiment.

An analogy exists between the diffusion process of heat motion and the two-dimensional random-walk problem. Suppose that a particle located at (x, y) moves to one of the four positions $(x + 1, y)$, $(x - 1, y)$, $(x, y + 1)$, $(x, y - 1)$, with equal probability, $p = \frac{1}{4}$. What is the probability $P(x, y; t)$ that a particle will arrive at (x, y) after t steps if it starts from the origin $(0, 0)$? For a particle to arrive at (x, y) in t steps, it is obvious that it will have had to arrive at one of the four positions $(x, y - 1)$, $(x, y + 1)$, $(x - 1, y)$, $(x + 1, y)$ in $t - 1$ steps. Thus

$$\begin{aligned}P(x, y; t) = \tfrac{1}{4}[P(x, y-1; t-1) &+ P(x, y+1; t-1)\\ &+ P(x-1, y; t-1) + P(x+1, y; t-1)]\end{aligned} \tag{9.64}$$

Let us subtract $P(x, y; t - 1)$ from both sides of (9.64). Then

$$\begin{aligned}&P(x, y; t) - P(x, y; t-1)\\ &\quad= \tfrac{1}{4}[P(x+1, y; t-1) - 2P(x, y; t-1) + P(x-1, y; t-1)]\\ &+ \tfrac{1}{4}[P(x, y+1; t-1) - 2P(x, y; t-1) + P(x, y-1; t-1)]\end{aligned} \tag{9.65}$$

In Sec. 10.13 one notes that, if $f''(x)$ is continuous for $a \leqq x \leqq b$, then

$$f''(x) = \lim_{h\to 0}\frac{f(x+2h) - 2f(x+h) + f(x)}{h^2} \tag{9.66}$$

[see (10.22)]. In the calculus of finite differences the expression $[f(x+2) - 2f(x+1) + f(x)]$ represents the second difference of $f(x)$. Thus

$$\begin{aligned} \Delta f(x) &= f(x+1) - f(x) \\ \Delta f(x+1) &= f(x+2) - f(x+1) \qquad (9.67) \\ \Delta^2 f(x) &= \Delta f(x+1) - \Delta f(x) = f(x+2) - 2f(x+1) + f(x) \end{aligned}$$

. . . .

Thus (9.65) may be written

$$\Delta_t P = \tfrac{1}{4}(\Delta_x^2 P + \Delta_y^2 P) \qquad (9.68)$$

which may be compared with the diffusion equation

$$\frac{\partial P}{\partial t} = k\left(\frac{\partial^2 P}{\partial x^2} + \frac{\partial^2 P}{\partial y^2}\right) \qquad (9.69)$$

A very fine subdivision of space and time must be used, however, in order to freely interchange (9.68) and (9.69). One traces the histories of a large number of particles with initial lattice distributions to obtain the distributions after a given number of steps. These histories yield an approximate solution to (9.69).

We conclude this section with a brief discussion of the theory of games. A simple example illustrates the basic factors encountered in the theory of games. We consider a two-person zero-sum game as exemplified by the square array given by (9.70).

X \ Y	1	2
1	−6	2
2	5	4

(9.70)

X and Y choose a row and column, respectively, and the choice of each is made without any express knowledge of the other's choice. The number in the ith row and jth column (the choices of X and Y, respectively) designates the number of units that Y must pay X. If $a_{ij} < 0$, then X pays Y an amount $|a_{ij}|$. The total sum earned by X and Y is zero, and it is in this sense that we have a two-person zero-sum game. We assume that X will attempt to earn as much as possible and that Y will attempt to hold his losses to a minimum. We see that in this particular game the possible courses of action of X and Y, the consequences of these actions, and the objectives of X and Y are fully known. The theory of games seeks to analyze objectively the conflict of X and Y and, moreover, attempts to determine an optimal course of action for each player.

The game given by (9.70) is very easily analyzed. If X chooses row 1, the least he can gain is -6. If he chooses row 2, the least he can gain is 4 units. Since $4 > -6$, it is obvious that X will always gain at least 4 units by choosing row 2, no matter what the choice of Y. Since Y knows that X will always choose row 2, of necessity, Y will always choose column 2 to hold his losses to a minimum. Since X and Y will always choose row 2 and column 2, respectively, with probability 1, we say that X and Y use pure strategies.

Let us note the following: Let $f(x, y)$ denote the quantity in row x and column y. The smallest number in row x is designated by $\min_y f(x, y)$. The largest of these numbers is written

$$U = \max_x \min_y f(x, y) \tag{9.71}$$

Thus $U = \min_y f(x_0, y) \geqq \min_y f(x, y)$ for all x. Let the reader deduce that the choice of $x = x_0$ by X guarantees that X will always earn at least U units. Let us determine the maximum amount Y can possibly lose if he plays wisely whatever the action of X. The largest number in row y is designated by $\max_x f(x, y)$. The smallest of all such numbers obtained by varying y is

$$V = \min_y \max_x f(x, y) \tag{9.72}$$

Thus $V = \max_x f(x, y_0) \leqq \max_x f(x, y)$ for all y. Let the reader deduce that the choice of $y = y_0$ will enable Y to hold his loss to an amount at most V. For the square array given by (9.70) we have

$$U = V = f(2, 2) = 4$$

and it is because of this that X always chooses the second row and Y always chooses the second column. Let us note that the element $f(2, 2) = 4$ is the minimum element of the second row and the maximum element of the second column. It is for this reason that $f(2, 2)$ is called a saddle element. Let the reader deduce that the existence of a saddle element implies that $U = V$. In any case one always has $U \leqq V$. The proof proceeds as follows: Let $g(x) = \min_y f(x, y)$, $U = g(x_0) = \max_x g(x)$, $h(y) = \max_x f(x, y)$, $V = h(y_0) = \min_y h(y)$. Then

$$\begin{aligned} U = g(x_0) = \max_x g(x) &= \max_x [\min_y f(x, y)] = \min_y f(x_0, y) \\ &\leqq f(x_0, y_0) \leqq \max_x f(x, y_0) = h(y_0) = V \end{aligned} \tag{9.73}$$

The reader is urged to verify every step of (9.73).

We now consider a case for which $U < V$, given by (9.74).

X \ Y	1	2
1	3	2
2	-1	4

(9.74)

Let the reader note the nonexistence of a saddle element with

$$U = 2 < V = 3$$

Thus X is sure to win 2 units per game if he always chooses $x = 1$, while Y can never lose more than 3 units if he always chooses $y = 1$. If Y were to notice, however, that X always chose $x = 1$, then it would be expedient for Y to choose $y = 2$. Can it pay for X to change his strategy so that he can realize more than 2 units per game on the average? The answer is "yes"! Let us suppose that X and Y use mixed strategies in the following sense: X chooses $x = 1$ with probability p and chooses $x = 2$ with probability $1 - p$; Y chooses $y = 1$ with probability q and chooses $y = 2$ with probability $1 - q$. The total expectation of X is

$$\begin{aligned} E &= 3pq + 2p(1 - q) + (-1)(1 - p)q + 4(1 - p)(1 - q) \\ &= q(6p - 5) + 4 - 2p \end{aligned} \tag{9.75}$$

X now reasons as follows: What mixed strategy should I use in order to guarantee a given expectation no matter what the mixed strategy employed by Y? If $6p - 5 > 0$ or $p > \frac{5}{6}$, then E is a minimum for $q = 0$, so that $E = 4 - 2p < \frac{7}{3}$. On the other hand, if $p < \frac{5}{6}$, E is a minimum for $q = 1$ so that $E = 4p - 1 < \frac{7}{3}$. For $6p - 5 = 0$ or $p = \frac{5}{6}$ we have $E = \frac{7}{3}$ independent of q. Thus, no matter what the strategy of Y, X can be assured of earning $\frac{7}{3}$ of a unit per game by mixing his strategy. X chooses row 2 if a six occurs on a die and otherwise chooses row 1. Let the reader show that Y can choose a mixed strategy such that he will never lose more than $\frac{7}{3}$ of a unit per game no matter what the strategy of X.

Let us now consider the following game exemplified by (9.76), with $\alpha > 0$, $\delta > 0$:

X \ Y	1	2	3
1	1	-1	0
2	-1	1	0
3	α	α	$-\delta$

(9.76)

Let $p_1, p_2, 1 - (p_1 + p_2)$ be the probabilities associated with the choices of $x = 1, 2, 3$, respectively, and let $q_1, q_2, 1 - (q_1 + q_2)$ be the probabilities associated with the choices of $y = 1, 2, 3$, respectively. The expectation of X is

$$\begin{aligned} E &= p_1(q_1 - q_2) + p_2(-q_1 + q_2) \\ &\qquad + [1 - (p_1 + p_2)][\alpha(q_1 + q_2) - \delta(1 - (q_1 + q_2))] \\ &= q_1[p_1 - p_2 + (\alpha + \delta)(1 - p_1 - p_2)] \\ &\quad + q_2[-p_1 + p_2 + (\alpha + \delta)(1 - p_1 - p_2)] - \delta(1 - p_1 - p_2) \end{aligned} \tag{9.77}$$

Let us examine E of (9.77). For $p_1 = p_2 = \frac{1}{2}$ we have $E = 0$ independent of q_1 and q_2. Thus a mixed strategy occurs for X such that X cannot lose or gain. Y has a good strategy (pure) with $q_1 = q_2 = 0$, $q_3 = 1 - q_1 - q_2 = 1$. Why is it obvious that the only good strategy for X is the mixed strategy discussed above? However, we can show that good strategies exist for Y other than $q_1 = q_2 = 0$, $q_3 = 1$. We can write

$$\begin{aligned} E &= p_1[(1 - \alpha - \delta)q_1 - (1 + \alpha + \delta)q_2 + \delta] \\ &\quad + p_2[-(1 + \alpha + \delta)q_1 + (1 - \alpha - \delta)q_2 + \delta] + (\alpha + \delta)(q_1 + q_2) - \delta \end{aligned}$$

It is obvious that E will be independent of p_1 and p_2 if the coefficients of p_1 and p_2 are zero. The solutions of these two equations yield $q_1 = q_2 = \delta/2(\alpha + \delta)$, which in turn yields $E = 0$ so that Y can suffer no loss for this choice of q_1 and q_2. Thus a good strategy which is mixed exists for Y. Let the reader show that, if $q_1 = q_2 < \delta/2(\alpha + \delta)$, then $E = (1 - p_1 - p_2)[2(\alpha + \delta)q_1 - \delta] \leqq 0$ for all choices of p_1 and p_2, so that an infinite number of good mixed strategies exist for Y.

Problems

1. Generalize the results of (9.63) for the case of n spinners.

2. Let $f(x)$ be a continuous monotonic increasing function defined for $0 \leqq x \leqq 1$. If a random variable ξ be chosen from $(0, 1)$, show that the probability that $f(\xi) \leqq y_0$, $f(0) \leqq y_0 \leqq f(1)$ is given by $p = f^{-1}(y_0)$, so that $f(p) = y_0$. Use this method to find an approximate value of $\sqrt{2}$ by use of a table of random numbers.

3. Let us consider the plane of a two-dimensional random-walk problem with boundary points at (x_i, y_i), $i = 1, 2, \ldots, r$. To each point (x_i, y_i) we associate a value $U(x_i, y_i)$. Let $V(x, y)$ be the expected value of a particle whose motion starts at (x, y) and eventually ends at one of the boundary points. Show that

$$V(x, y) = \sum_{i=1}^{r} U(x_i, y_i) \cdot P_i$$

with P_i the probability that the walk terminates at (x_i, y_i) if it begins at (x, y). Also show that

$$V(x, y) = \tfrac{1}{4}[V(x + 1, y) + V(x - 1, y) + V(x, y + 1) + V(x, y - 1)]$$

and compare this result with $\dfrac{\partial^2 V}{\partial x^2} + \dfrac{\partial^2 V}{\partial y^2} = 0$.

4. Show that the game of "paper, rock, and scissor" is exemplified by the matrix below. What strategies should be used by X and Y?

$X \backslash Y$	P 1	R 2	S 3
P 1	0	1	−1
R 2	−1	0	1
S 3	1	−1	0

REFERENCES

Cramer, H.: "Mathematical Methods of Statistics," Princeton University Press, Princeton, N.J., 1946.

Doob, J. L.: "Stochastic Processes," John Wiley & Sons, Inc., New York, 1953.

Feller, W.: "Introduction to Probability Theory," John Wiley & Sons, Inc., New York, 1950.

Mood, A. M.: "Introduction to the Theory of Statistics," McGraw-Hill Book Company, Inc., New York, 1950.

Morgenstern, O., and J. von Neumann: "Theory of Games and Economic Behavior," Princeton University Press, Princeton, N.J., 1947.

Munroe, M. E.: "Theory of Probability," McGraw-Hill Book Company, Inc., New York, 1951.

Uspensky, J. V.: "Introduction to Mathematical Probability," McGraw-Hill Book Company, Inc., New York, 1937.

CHAPTER 10

REAL-VARIABLE THEORY

10.1. Introduction. We desire to preface the study of the real-number system by first introducing a brief view of our philosophy of mathematics and science. Historians are generally in agreement that in many ways Greek mathematics was the precursor to modern mathematics. One begins the study of geometry with a set of postulates or axioms along with some definitions of the fundamental entities in which one is interested. From this beginning one deduces through the use of Aristotelian logic the many theorems concerning triangles, circles, etc. To the novice the axioms appear to be self-evident truths. We do not hold to this view: a postulate is nothing more than a man-invented rule of the game, the game in this case being mathematics. Actually the mathematician is not interested in the truth, per se, of his postulates. The postulates of a mathematical system can be chosen arbitrarily subject to the condition that they be consistent with each other. The proof of the self-consistency of the postulates is no easy task, if, indeed, such proofs exist. A discussion of such matters lies beyond the scope of this text.

Finally, we should like to add that any theorems deducible from the postulates are a consequence of the postulates themselves. We admit this is a trivial and obvious remark. Too often, however, the scientist forgets this fact. He is prone to believe that the laws of nature are discovered. It is our personal belief that this is not the case. We lean toward the more esoteric point of view that the so-called laws of nature are invented by man. Newton described the motion of the planet Mercury relative to the sun in a fairly accurate manner by use of $\mathbf{f} = m\mathbf{a}$ and $\mathbf{f} = -(GmM/r^3)\mathbf{r}$. Because of this one feels that these two equations are laws of nature discovered by Newton. On the other hand, by postulating that a gravitational point mass yields a four-dimensional Riemannian space and that a particle moves along a geodesic of this space, Einstein also obtained the motion of Mercury relative to the sun. It is well known that Einstein's predicted motion of Mercury is more accurate than Newton's predicted motion. Are we to assume that from the time of Newton to Einstein we had a "true" law of nature and that now we discard this law and accept Einstein's theory as the truth? It is just a matter of time before a physicist will invent a new set of postulates which will explain more fully the physical results obtained by experiment. Moreover it is not our belief that we continually approach the true laws of nature.

10.2. The Positive Integers. We begin a study of the real-number system by first discussing the positive integers from a postulational point of view. The positive integers will be undefined in the sense that they can be anything the reader desires subject to the condition that they satisfy the postulates set forth below. We shall attempt to be rigorous, but shall not be as rigorous as might be possible. First we do not feel qualified to attempt this task, and second it would require a treatise in itself to elaborate on the logical and philosophical aspects and implications of the material upon which we shall discourse.

The positive integers have been characterized by Peano through the following postulates:

P_a: There exists a positive integer, called one, and written 1.

P_b: Every positive integer x has a unique successor x'.

P_c: There is no positive integer x such that $x' = 1$.

P_d: If $x' = y'$, then $x = y$.

P_e: Let S be any collection of positive integers. If S contains the integer 1, and if, moreover, any integer x of S implies that x' is in S, then S is the complete collection of integers, I.

The postulate P_a guarantees that there is at least one member of our set of positive integers so that we may all agree that there is something we can talk about. The equal sign ($=$) involved in the relation $x = y$ of P_d means that x and y are identical elements in the sense that we may replace x by y or y by x in any operation concerning these integers. The postulate P_e differs in some respects from the other postulates. P_e is a rule for determining when we have the complete set of integers at hand, and is used chiefly in deducing new laws which the integers obey.

As postulated above, the integers could be any sequence $(a_1, a_2, \ldots, a_n, \ldots)$ which the reader has encountered in calculus courses. It is in this sense that the integers are undefined. It is obvious that more will have to be said before we can identify the collection of integers defined above with the ordinary integers of our everyday working world. We shall accept as intuitive the notion and idea of a set, or collection, of objects. The Peano postulates will make themselves clearer to the student as he proceeds to read the text.

The operation of addition ($+$) relative to the positive integers is defined by the postulates

$$A_a: \qquad a' = a + 1$$
$$A_b: \qquad a + b' = (a + b)'$$

The postulate A_b implies that $a + b$ is an integer (closure property).

We are now in a position to deduce some new results concerning the positive integers.

THEOREM 10.1. *Addition is associative* $(a + b) + c = a + (b + c)$. We let S be the collection of integers $x, y, z, \ldots$ which satisfy Theorem 10.1 for all a and b. Thus x is in S if $(a + b) + x = a + (b + x)$ for all integers a and b. We first show that 1 is an element of S. Now

$$\begin{aligned} (a + b) + 1 &= (a + b)' && \text{from } A_a \\ &= a + b' && \text{from } A_b \\ &= a + (b + 1) && \text{from } A_a \end{aligned}$$

so that 1 belongs to S. Now let x be any element of S. This means that $(a + b) + x = a + (b + x)$ for all a and b. Now

$$\begin{aligned} (a + b) + x' &= [(a + b) + x]' && \text{from } A_b \\ &= [a + (b + x)]' && \text{since } x \text{ is in } S \\ &= a + (b + x)' && \text{from } A_a \\ &= a + (b + x') && \text{from } A_b \end{aligned}$$

so that x' belongs to S whenever x is in S. From P_e, S is the complete set of integers, $S \equiv I$. Q.E.D.

THEOREM 10.2. *Addition is commutative* $a + b = b + a$ (see Prob. 1).

THEOREM 10.3. *Law of Cancellation.* If $a + b = a + c$, then $b = c$. We proceed by induction on a. First, if

$$1 + b = 1 + c$$

then

$$\begin{aligned} b + 1 &= c + 1 && \text{from Theorem 10.2} \\ b' &= c' && \text{from } A_a \\ b &= c && \text{from } P_d \end{aligned}$$

Hence 1 belongs to S. Now let x be any element of S. This means that whenever $x + b = x + c$ then $b = c$. Now assume $x' + b = x' + c$. Then

$$\begin{aligned} b + x' &= c + x' && \text{from Theorem 10.2} \\ (b + x)' &= (c + x)' && \text{from } A_b \\ b + x &= c + x && \text{from } P_d \\ x + b &= x + c && \text{from Theorem 10.2} \\ b &= c && \text{since } x \text{ is in } S \end{aligned}$$

Thus x' belongs to S whenever x belongs to S. From P_e, $S \equiv I$. Q.E.D.

THEOREM 10.4. If $b + a = c + a$, then $b = c$ (See Prob. 2).

In what follows we shall use the symbol $(\Rightarrow)$ to mean "implies." Thus Theorem 10.4 may be written $b + a = c + a \Rightarrow b = c$. Implication in both directions will be represented by $(\Leftrightarrow)$. Thus

$$b + a = c + a \Leftrightarrow b = c$$

The reader should realize that $b = c \Rightarrow b + a = c + a$ does not mean that we have introduced the axiom about adding equals to equals. All we have done to the expression $b + a$ is to replace b by c (permissible from the definition of equality of two integers).

Equality (=) belongs to the class of relations, R, called equivalence relations. We note that

$$a = a$$
$$a = b \Leftrightarrow b = a$$
$$a = b,\ b = c \Rightarrow a = c$$

In general a relation R is called an *equivalence relation* relative to a set of elements $(a, b, c, \ldots)$ if

R_1:	aRa	reflexive
R_2:	$aRb \Leftrightarrow bRa$	symmetric
R_3:	$aRb,\ bRc \Rightarrow aRc$	transitive

An equivalence relation R enables one to distinguish between various classes of elements. For example, we say that aRb holds if and only if a and b yield the same remainder when divided by 2. Thus any integer belongs either to the class of even integers or to the class of odd integers. The reader should check that the elements of the class of even integers satisfy R_1, R_2, R_3. We write $aRb \Leftrightarrow a = b \bmod 2$.

Now we introduce multiplication by the two postulates given below:

$$M_a{:}\quad a \cdot 1 = a$$
$$M_b{:}\quad a \cdot b' = a \cdot b + a$$

M_b assumes that $a \cdot b$ is an integer for all a and b. We shall omit the dot whenever it is convenient to do so. Thus M_b can be written $ab' = ab + a$.

THEOREM 10.5. *Multiplication is left-distributive with respect to addition,* $a(b + c) = ab + ac$. We prove this theorem by induction on c. Let S be the collection of integers $(x, y, z, \ldots)$ such that

$$a(b + x) = ab + ax$$

for x in S and for all a and b. First

$$a(b + 1) = ab' = ab + a = ab + a \cdot 1$$

from A_a, M_b. Thus 1 is in S. Now let x be in S. We have

$$\begin{aligned} a(b + x') &= a(b + x)' && \text{from } A_b \\ &= a(b + x) + a && \text{from } M_b \\ &= (ab + ax) + a && \text{since } x \text{ is in } S \\ &= ab + (ax + a) && \text{from Theorem 10.1} \\ &= ab + ax' && \text{from } M_b \end{aligned}$$

so that x' is in S if x is in S. Thus $S = I$.

THEOREM 10.6. $1 \cdot a = a$ (see Prob. 3).

THEOREM 10.7. *Multiplication is associative,* $(ab)c = a(bc)$ (see Prob. 4).

THEOREM 10.8. *Multiplication is commutative,* $ab = ba$ (see Prob. 5).

THEOREM 10.9. *Multiplication is right-commutative,*

$$(b + c)a = ba + ca$$

(see Prob. 6).

DEFINITION 10.1. If $a = b + c$, we say that a is greater than b or that b is less than a, written $a > b$ or $b < a$, respectively. We can state four important ordering theorems regarding inequalities:

$$
\begin{aligned}
O_a&: \quad a, b \text{ given} \Rightarrow a = b \text{ or } a > b \text{ or } a < b \\
O_b&: \quad a > b,\ b > c \Rightarrow a > c \\
O_c&: \quad a > b \Rightarrow a + c > b + c \\
O_d&: \quad a > b \Rightarrow ac > bc
\end{aligned}
$$

We leave these theorems as exercises for the reader (see Probs. 7 and 12). A set obeying O_a is said to be *totally ordered.*

To place the integers in the realm of our everyday working experiences, we now postulate that the integer 1, which occurred in the first of Peano's postulates, P_a, will be the adjective which reasonable, sane, prudent human beings attach to single entities when they speak intelligently of one book, one day, one world, etc. Of course the number 1 as given in Peano's postulates is a noun, not an adjective. B. Russell associates "one" with the class of all single entities. Trouble occurs, however, when one tries to define a class of elements.

The successor of one, namely, $1' = 1 + 1$, will be called the integer two, written 2, and addition will again mean what we wish it to mean, namely, that one book plus one book implies two books. We now feel free to speak of the class of positive integers $I(1, 2, 3, \ldots, n, \ldots)$.

Problems

1. Prove Theorem 10.2. *Hint:* First prove that $a + 1 = 1 + a$ for all a.
2. Prove Theorem 10.4.
3. Prove Theorem 10.6.
4. Prove Theorem 10.7.
5. Prove Theorem 10.8.
6. Prove Theorem 10.9.
7. Prove ordering theorems O_b, O_c, O_d.
8. Prove that $1 \cdot a = a$.
9. Prove that $b'a = ba + a$.
10. Show that $1 \leqq$ all integers a. *Hint:* Let S be the class of all integers x such that $x \geqq 1$. Then show that $S = I$.
11. If $a' > b$, show that $a > b$ or $a = b$.
12. Using the results of Probs. 10, 11, prove O_a.

13. Prove that every nonempty set of integers contains a smallest integer. First define what is meant by a smallest integer of a collection of integers. Also use the results of Probs. 10, 11.

10.3. The Rational Integers. An extension of the positive integers which includes the zero element and the negative integers is obtained in the following manner: Given any pair of positive integers, we postulate the existence of a difference function d which operates on the positive integers a, b subject to the condition that

$$d(a, b) = d(x, y) \tag{10.1}$$

if and only if $a + y = b + x$. Equality as defined by (10.1) certainly is an equivalence relation. We note that the above definition depends only on addition of positive integers (Sec. 10.2).

Let us see what properties are possessed by d. Obviously

$$d(a, a) = d(b, b)$$

Moreover $d(a + c, b + c) = d(a, b)$. If we desire that $d(a, b)$ behave like $b - a$ of ordinary arithmetic, of necessity we are forced to postulate that addition be defined by

$$d(a, b) + d(x, y) = d(a + x, b + y) \tag{10.2}$$

We now note that

$$d(a, b) + d(c, c) = d(a + c, b + c) = d(a, b) \tag{10.3}$$

from (10.2) and (10.1). Hence $d(c, c)$ has the property of being a zero element for the operation of addition (+).

We define multiplication by the following rule:

$$d(a, b)d(x, y) = d(ay + bx, by + ax) = d(x, y)d(a, b) \tag{10.4}$$

What of the number $d(a, a + 1)$? We see that

$$\begin{aligned} d(x, y)d(a, a + 1) &= d(xa + x + ya, ax + ya + y) \\ d(x, y) &= d(a, a + 1)d(x, y) \end{aligned} \tag{10.5}$$

so that $d(a, a + 1)$ has the unit property as regards multiplication.

It now can easily be shown that the collection of elements $d(1, x + 1)$ when x ranges over the set of positive integers will satisfy all the Peano postulates outlined in Sec. 10.2. The set of numbers $d(a, b)$ contains the positive integers as a subclass and in its totality represents the class of rational integers (positive integers, negative integers, and the zero element). We let the reader show that the zero element $d(a, a)$ has the properties

$$\begin{aligned} d(a, a)d(x, y) &= d(a, a) \\ d(x, y) + d(y, x) &= d(a, a) \end{aligned} \tag{10.6}$$

We say that $d(y, x)$ is the negative of $d(x, y)$, and conversely. We may, if we please, write $d(a, a) = 0$, $d(1, a + 1) = a$, $d(a + 1, 1) = -a$, $d(a, b) = b - a$. The ordering postulates can easily be extended to cover the new collection $d(a, b)$. We say that

$$d(a, b) > d(x, y)$$

if and only if

$$b + x > a + y$$

Let the reader show that

$$\begin{aligned} d(a, b) > d(x, y) &\Rightarrow d(a, b) + d(u, v) > d(x, y) + d(u, v) \\ d(a, b) > d(x, y) &\Rightarrow d(a, b)d(z, u + z) > d(x, y)d(z, u + z) \\ d(a, b) > d(x, y) &\Rightarrow d(x, y)d(u + z, z) > d(a, b)d(u + z, z) \end{aligned} \tag{10.7}$$

What is the ordinary meaning of (10.7)?

Let us note some properties of the rational integers as outlined above. First we have a set of elements called the rational integers. Associated with this set of elements are two operations called addition and multiplication. The elements satisfy the following properties relative to the operations defined above:

1. Closure properties: $a + b$, ab are elements of our set when a, b are elements of the set.
2. Associative laws: $a + (b + c) = (a + b) + c$, $a(bc) = (ab)c$.
3. Unit elements: $a + 0 = a = 0 + a$, $a \cdot 1 = a = 1 \cdot a$.
4. Distributive laws: $a(b + c) = ab + ac$, $(a + b)c = ac + bc$.
5. An inverse element exists for each element relative to addition, that is, $a + (-a) = (-a) + a = 0$ for all a.

We say that the rational integers form a *ring* with unit element 1. We also note that, if $ab = 0$, then $a = 0$ or $b = 0$. Why?

We give an example of a ring such that $ab = 0$ does not imply $a = 0$ or $b = 0$. Such rings are said to have zero divisors. Let us consider the six following classes: Into class one we place all integers which yield a remainder 1 upon division by 6. Class two contains all integers which yield a remainder 2 upon division by 6. Class j has the obvious property that it consists of those integers which yield a remainder j upon division by 6. We represent these various classes by the symbols 1, 2, 3, 4, 5, 0. Addition and multiplication are defined in the ordinary way. Thus $2 \cdot 5 = 10 = 4 \bmod 6$, since 10 yields a remainder of 4 when divided by 6. $3 + 4 = 7 = 1 \bmod 6$, etc. It is obvious that $2 \cdot 3 = 0 \bmod 6$, but neither $2 = 0 \bmod 6$ nor $3 = 0 \bmod 6$. We have constructed the ring of integers modulo 6.

10.4. The Rational Numbers. We can extend the ring of rational integers by postulating the existence of a rational function $R(a, b)$ where a, b are any two rational integers, $a \neq 0$. We postulate that R is to have the following properties:

$$
\begin{aligned}
&R(a, b) = R(x, y) \Leftrightarrow bx = ay \\
&R(a, b) + R(x, y) = R(ax, bx + ay) \\
&R(a, b)R(x, y) = R(ax, by) \\
&R(a, b) > R(x, y) \Leftrightarrow bx > ay \qquad \text{if } ax > 0 \\
&R(a, b) > R(x, y) \Leftrightarrow bx < ay \qquad \text{if } ax < 0
\end{aligned} \tag{10.8}
$$

We leave it to the reader to show that the set of elements $R(1, b)$ has all the properties of the set of rational integers b. For all practical considerations we may write $R(1, b) \equiv b$, and we shall obviously write $R(a, b) = b/a$. The set of rationals has a further property not possessed by the ring of rational integers. For every $R(a, b)$, $b \neq 0$, $a \neq 0$, there exists a rational $R(b, a)$ such that

$$R(a, b)R(b, a) = R(ab, ba) = R(1, 1) = 1$$

Thus every nonzero element has an inverse with respect to multiplication. We say that the rationals form a field.

The reader should have no trouble showing that the integers modulo 7 form a field. The inverse of 2 is 4 since $2 \cdot 4 = 8 = 1 \bmod 7$.

The rationals form a totally ordered field. We feel, however, that the rationals are not complete in the sense that we should like to deal with numbers like $\sqrt{2}$, e, π, etc. For the irrationality of $\sqrt{2}$, see Sec. 10.5. We shall have to extend the field of rationals. This will be done in Sec. 10.6.

10.5. Some Theorems Concerning the Integers

THEOREM 10.10. Let f, g be positive integers. There exist integers h, r such that $f = gh + r$, $0 \leqq r < g$. Our proof is by induction on f. Let $f = 1$. Then

$$
\begin{aligned}
1 &= 1 \cdot 1 + 0 \qquad \text{if } g = 1 \text{ so that } h = 1, r = 0 \\
1 &= g \cdot 0 + 1 \qquad \text{if } g > 1
\end{aligned}
$$

Now assume Theorem 10.10 holds for f. Then

$$f = gh + r \qquad 0 \leqq r < g$$

and

$$f + 1 = gh + (r + 1)$$

Since $r < g$, we have $r + 1 < g$ or $r + 1 = g$. If $r + 1 < g$, the theorem holds. If $r + 1 = g$, then $f + 1 = g(h + 1) + 0$ and Theorem 10.10 holds. From P_e, the theorem is true for all integers f. We leave it as an exercise to show that the h, r of Theorem 10.10 are unique.

THEOREM 10.11. We now derive a theorem regarding the greatest common divisor of two integers a, b. Consider the collection of integers (>0) of the form $ax + by > 0$. Since every collection of positive integers has a least integer, there exist integers x_0, y_0 such that

$$ax_0 + by_0 = d > 0$$

is the least integer of our collection. We now show that d is the greatest common divisor. First assume that d does not divide a. Then $a = ds + r$, $0 < r < d$ from Theorem 10.10. Thus

$$ax_0s + by_0s = ds = a - r$$

and $(1 - x_0s)a + (-y_0s)b = r$, a contradiction since $r < d$, unless $r = 0$, in which case d divides a, d/a. Similarly d/b. Now any other divisor, d_1, of a and b must divide $ax_0 + by_0$ (why?), and hence d_1/d. Therefore d is the greatest common divisor of a and b, written $d = (a, b)$. As an immediate corollary, if a and b are relatively prime, $d = 1$ and there exist integers x, y such that $ax + by = 1$.

THEOREM 10.12. If $(a, b) = 1$ and a/bc, then a/c.

Proof. $ax + by = 1$ since $(a, b) = 1$, so that $axc + byc = c$. But $bc = ad$ since a/bc. Hence $a(xc + yd) = c$, which in turn implies that a/c.

THEOREM 10.13. *The Fundamental Theorem of Arithmetic.* An integer N can be written uniquely as a product of primes.

Proof. Assume

$$N = p_1p_2 \cdots p_r = q_1q_2 \cdots q_s$$

It is obvious that we can consider the p_i, q_j as primes since a nonprime factor can be broken into further factors. Continuing this process eventually reduces all factors of N into primes. Furthermore this can be done only a finite number of times since all primes are greater than or equal to 2. From above $p_1/q_1q_2 \cdots q_s$. If $(p_1, q_1) \neq 1$, then $p_1 = q_1$. Why? If $(p_1, q_1) = 1$, then $p_1/q_2q_3 \cdots q_s$ from Theorem 10.12. We continue this process until eventually p_1/q_i for an $i = 1, 2, \ldots, s$. Since q_i is assumed prime, we must have $p_1 = q_i$. We now cancel these equal factors and continue our reasoning in the same manner. In this way we obviously exhaust all the p_i and q_j. Thus, except for the order, the number N is factored uniquely as a product of primes.

THEOREM 10.14. The $\sqrt{2}$ is irrational.

Proof. Assume $\sqrt{2}$ rational so that $\sqrt{2} = p/q$. From Theorem 10.13 we have $p = p_1p_2 \cdots p_r$, $q = q_1q_2 \cdots q_s$, with the p_i, q_j unique. Thus upon squaring we have

$$N = 2q_1q_1q_2q_2 \cdots q_sq_s = p_1p_1p_2p_2 \cdots p_rp_r$$

Since the prime factor 2 must occur an odd number of times on the left and can occur only an even number of times on the right, a contradiction occurs. Q.E.D.

Problems

1. Show that the collection of elements $d(a, b)$ satisfies the Peano postulates as the elements a, b range over the Peano integers.

2. Prove (10.5).

3. Prove (10.6).

4. Show that the set of elements $R(1, b)$ has all the properties of the set of rational integers b.

5. Show that the integers modulo 7 form a field.

6. Show that $\sqrt{p}$ is irrational for p prime.

7. Show that $\sqrt{2} + \sqrt{3}$ is irrational.

8. Consider the set of polynomials with rational coefficients. Deduce theorems for these polynomials analogous to the theorems of Sec. 10.5.

10.6. An Extension of the Rationals. The Real-number System. The rationals obey the two following rules:

1. They form a field.
2. They are totally ordered.

On the other hand, the rationals are not complete. Pythagoras realized that there existed so-called numbers, $\sqrt{2}$, etc., which are not rational numbers. If we wish to include the $\sqrt{2}$ into our number system, we must enlarge the field of rationals. R. Dedekind gave a fairly satisfactory treatment of irrationals. We discuss an extension of the rationals which differs little from Dedekind's approach and which is due to B Russell.

Definition 10.2. A *Russell* number is a set, R, of rationals having the following properties.

1. R contains only rationals and is nonempty.
2. R does not contain all the rationals.
3. If α is a rational in R and if β is a rational less than α, then β is in R.

Two examples of Russell numbers are given as follows:

1. The set of all rationals $\leqq 1$ is a Russell number. It will turn out to be the unit of our constructed field of real numbers.

2. The negative rationals and all positive rationals whose squares are less than 2 form a Russell set or number. We shall designate it by the symbol $\sqrt{2}$.

Definition 10.3. Two Russell numbers are said to be equal or equivalent if and only if every rational of R_1 is contained in R_2, and conversely, with one possible exception.

The set of all rationals less than 2, which defines the Russell number, two, is equivalent to the Russell number consisting of all rationals less than or equal to 2. The only member of the second Russell number not found in the first Russell number is the rational two.

Definition 10.4. The Russell number R_1 is said to be greater than the Russell number R_2 if, and only if, at least two rationals of R_1 are not members of R_2.

We note that, given R_1 and R_2, only one of $R_1 = R_2$, $R_1 > R_2$, $R_1 < R_2$ can occur.

We must now define addition of two R numbers in such a way that the result will yield an R number. Let R_1 and R_2 be two R numbers. Construct the set of rationals obtained by adding in all possible ways the rationals of R_1 with those of R_2. We leave it to the reader to show that this new set of rationals defines an R number. What property will the set of all rationals less than zero have? Is this Russell number needed for a field? Does every R number have a negative? The student should readily answer these questions.

It is a bit more difficult to define multiplication. We shall define multiplication for two Russell numbers R_1, R_2 which are greater than the zero Russell number. We first delete the negatives of R_1 and R_2. Then we multiply the positive rationals of R_1 with the positive rationals of R_2 in all possible ways. To this acquired set we adjoin the negative rationals. We leave it to the reader to show that the newly acquired set is a Russell number. An equivalent definition would be the following: Let $C(R_1)$, the complement of R_1, be the set of rationals not in R_1. Show that

$$C[C(R_1) \cdot C(R_2)]$$

is a Russell number. By $C(R_1) \cdot C(R_2)$ we mean the complete set of rationals obtained by the product of rationals of $C(R_1)$ with rationals of $C(R_2)$ in all possible ways. We leave it to the reader to extend the definition for the other cases of R_1 and R_2.

It is now an easy problem to show that the complete set of Russell numbers form a totally ordered field which contains the rationals as a subfield. Everything hinges essentially on the fact that we reduce our computations to the field of rationals. Let the reader show that the set of Russell numbers forms a totally ordered field.

The field of Russell numbers has an additional property not possessed by the rationals. To exhibit this property, we first define the supremum (least upper bound) of a set of numbers (Russell).

DEFINITION 10.5. The number s is said to be the *supremum* of a set of numbers S: $(x, y, z, \ldots)$ if and only if:

1. $s \geqq x$ for all x in S.
2. If $t < s$, there exists an element y of S such that $y > t$. We leave it to the reader to show that 1 is the supremum of the set of all rationals less than 1. Also, $\sqrt{2}$ is the supremum of the set of all rationals whose squares are less than 2. Let the reader show that a set of numbers cannot have two distinct suprema. Let the reader also define the *infemum* (greatest lower bound) of a set.

DEFINITION 10.6. A set, S, of Russell numbers is bounded above if a Russell number, N, exists such that $N > x$ for all x in S.

THEOREM 10.15. A supremum s exists for every set of numbers which has an upper bound.

Our proof is as follows: Let S be a set of Russell numbers bounded above by N. We construct a new Russell number as follows: Let s consist of all rationals which belong to any Russell number of S. We first show that s is a Russell number.

1. s obviously contains only rationals and is not the empty set.

2. Since N contains rationals not in any of the Russell numbers of S, there are rationals not in s.

3. Let α be a rational of s and β a rational $<\alpha$. Since α belongs to a Russell number, say, S_1 of S, then β also is in S_1 (why?), so that β belongs to s from the definition of s.

Thus s is a Russell number (see Definition 10.5). We next demonstrate that s is the supremum of the set S.

1. $s \geqq$ all numbers of S, for otherwise there is at least one element, S_1, of S such that $S_1 > s$. This implies that S_1 contains a rational not in s, which is impossible from the construction of s.

2. Let r be a Russell number less than s, and let β, γ be elements of s not in r. This is possible since $s > r$. Now β, γ belong to at least one set, S_1, of S from the construction of s. It is easy to see that $S_1 > r$.

From Definition 10.5, s is the supremum of the set S.

The set of Russell numbers thus

1. Forms a field.

(A) 2. Is totally ordered.

3. Satisfies the property that a supremum exists for every set of numbers bounded above.

Any set of elements satisfying the three properties of (A) is unique in the sense that they differ from the Russell numbers in name only. Thus the real-number system as set forth above is unique as regards its construction.

To show this uniqueness, we let S and $\bar{S}$ be two totally ordered fields with the supremum property. We first match the zero elements and the unit elements.

$$0 \leftrightarrow \bar{0}$$
$$1 \leftrightarrow \bar{1}$$

Then we match the rational integers,

$$2 \leftrightarrow \bar{2}$$
$$3 \leftrightarrow \bar{3}$$
$$\cdots\cdots$$
$$n \leftrightarrow \bar{n}$$
$$\cdots\cdots$$

Then we match the rationals,

$$r_\alpha \leftrightarrow \bar{r}_\alpha$$

Now consider any element, s, of S which has not as yet been put into one-to-one correspondence with an element of $\bar{S}$. We consider the set of all rationals of S less than s. It is obvious that s is the supremum of this set. The rationals of $\bar{S}$ which correspond to the above-mentioned set of rationals of S will be bounded above. Why? A supremum $\bar{s}$ will exist for this set. We match $s \leftrightarrow \bar{s}$. Thus every member of S can be mapped into a number of $\bar{S}$. We leave it to the reader to show that every number of $\bar{S}$ is exhausted (mapped completely) by this method.

Under the above mapping we realize that

$$a \leftrightarrow \bar{a}$$
$$b \leftrightarrow \bar{b}$$

implies

$$a + b \leftrightarrow \overline{(a + b)} = \bar{a} + \bar{b}$$
$$ab \leftrightarrow \overline{(ab)} = \bar{a}\bar{b}$$

Such a correspondence, or mapping, is called an isomorphism, and it is in this sense that we say the sets S and $\bar{S}$ are equivalent.

We can now prove the *Archimidean ordering postulate* by use of the supremum. Let $r > 0$, and consider the sequence of numbers

$$r, 2r, 3r, \ldots, nr, \ldots$$

We maintain that there is an integer m such that $mr > 1$. If this were not so, the sequence constructed above would be bounded above and so a supremum s would exist for the sequence. From the properties of s we have $s \geqq nr$ for all n. Now consider $t = s - r < s$. An element pr exists such that $pr > t = s - r$. Thus $pr + r > s$ or $(p + 1)r > s$, a contradiction.

We list the Archimidean ordering postulates. Any one of the postulates implies the existence of the others.

AO_1: $r > 0 \Rightarrow \exists$ integer $n \ni nr > 1$. ($\exists \equiv$ there exists, $\ni \equiv$ such that.)

AO_2: $\alpha > \beta > 0 \Rightarrow \exists$ rational $r \ni \alpha > r > \beta$.

AO_3: $\alpha > 0 \Rightarrow \exists$ rational $r \ni \alpha > r > 0$.

AO_4: (Eudoxus) If for every rational $r > \beta$ it follows that $r > \alpha$, then $\beta \geqq \alpha$.

We now prove AO_2 from AO_1. If $\alpha > \beta > 0$, then $\alpha - \beta > 0$. From AO_1 an integer n exists such that $n(\alpha - \beta) > 1$ or $n\alpha > n\beta + 1$. Now $\beta > 0$ so that an integer m exists such that $m > \beta n$. Let m be the smallest of all such integers. Thus $m > \beta n \geqq m - 1$. Hence

$$n\alpha > n\beta + 1 \geqq m > n\beta$$

so that

$$\alpha > \frac{m}{n} > \beta$$

Problems

1. If b is a Russell number $\neq 0$, how does one find the number b^{-1} such that $bb^{-1} = b^{-1}b = 1$?

2. Extend the definition of multiplication of two Russell numbers for the cases for which both numbers are not positive.

3. Show that the Russell numbers form a field.

4. Show from the definition of the supremum that it is unique for a set of elements.

5. Define the infemum of a set.

6. Prove AO_3 from AO_4, AO_4 from AO_3, AO_1 from AO_4.

10.7. Point-set Theory. For convenience, we shall consider only points or elements of the real-number system in what follows. Any set of real numbers will be called a linear set. The field of real numbers can be put into one-to-one correspondence with the points of a straight line in the usual fashion encountered in analytic geometry and the calculus. All the definitions and theorems here proved for linear sets can easily be extended to finite-dimensional spaces.

DEFINITION 10.7. The set of points $\{x\}$ satisfying $a \leqq x \leqq b$, a, b finite, will be called a closed interval. If we omit the end points, that is, consider those x which satisfy $a < x < b$, we say that the interval is open (open at both ends). For example, $0 \leqq x \leqq 1$ is a closed interval, while $0 < x < 1$ is an open interval.

DEFINITION 10.8. A linear set of points will be said to be bounded if there exists an open interval containing the set. It must be emphasized that the ends of the interval are to be finite numbers, thus excluding $-\infty$ and $+\infty$.

An alternative definition would be the following: A set S of real numbers is bounded if there exists a finite number N such that $-N < x < N$ for all x in S.

The set of numbers whose squares are less than 3 is certainly bounded, for if $x^2 < 3$, then obviously $-2 < x < 2$. However, the set of numbers whose cubes are less than 3 is unbounded, for $x^3 < 3$ is at least satisfied by all the negative numbers. This set, however, is bounded above. By this we mean that there exists a finite number N such that $x < N$ if $x^3 < 3$. Certainly $N = 2$ does the trick. Generally speaking, a set S of elements is bounded above if a finite number N exists such that $x < N$ for all x in S. Let the reader frame a definition for sets bounded below.

We shall, in the main, be concerned with sets which contain an infinite number of distinct points. The rational numbers in the interval $0 < x < 1$ form such a collection.

DEFINITION 10.9. *Limit Point.* A point p will be called a limit point of a set S if every open interval containing p contains an infinite number of distinct elements of S.

For example, let S be the sequence of numbers

$$\left(\frac{1}{2}, \frac{1}{3}, \frac{1}{4}, \ldots, \frac{1}{n}, \ldots\right)$$

It is easy to verify that any open interval containing the origin O contains an infinite number of elements of S. Thus O is a limit point of the set S. Note that in this case the limit point O does not belong to S. It is also at once apparent that a set S containing only a finite number of points cannot have a limit point.

Let the reader show that a point q is not a limit point of a set S if at least one open interval containing q exists such that no points of S (except possibly q itself) are in this open interval.

DEFINITION 10.10. *Neighborhood.* A neighborhood of a point is any open interval containing that point. A deleted neighborhood N_p of p is a set of points belonging to a neighborhood of p with the point p removed. The set of points x such that $\frac{1}{20} < x < \frac{1}{19}$, $x \neq \frac{39}{100}$, is a deleted neighborhood of $x = \frac{39}{100}$.

The definition of a limit point can be reframed to read: p is a limit point of a set S if every deleted neighborhood of p contains at least one point of S.

DEFINITION 10.11. *Interior Point.* A point p is said to be an interior point of a set S if a neighborhood N_p of p exists such that every element of N_p belongs to S. Does p belong to S? If S is the set of points $0 \leqq x \leqq 1$, then 0 and 1 are not interior points since every neighborhood of 0 or 1 contains points that are not in S. All other points of this set, however, are interior points.

DEFINITION 10.12. *Boundary Point.* A point p is a boundary point of a set S if every neighborhood of p contains points in S and points not in S. If S is the set $0 \leqq x \leqq 1$, then 0 and 1 are the only boundary points. A boundary point need not belong to the set. 1 is the only boundary point of the set S which consists of points x such that $x > 1$.

DEFINITION 10.13. *Exterior Point.* A point p is an exterior point of a set S if it is not an interior or boundary point of S.

DEFINITION 10.14. *Complement of a Set.* The complement of a set S is the set of points not in S. The complement $C(S)$ has a relative meaning, for it depends on the set T in which S is embedded. If S, for example, is the set of real numbers $-1 \leqq x \leqq 1$, then the complement of S relative to the real-number system is the set of points $|x| > 1$. The complement of $-1 \leqq x \leqq 1$ relative to the set $-1 \leqq x \leqq 1$ is the null set (no elements). The complement of the set of rationals relative to the reals is the set of irrationals, and conversely.

DEFINITION 10.15. *Open Set.* A set S is said to be open (not to be confused with open interval) if every point of S is an interior point of S.

For example, the set of points $\{x\}$ which satisfies either $0 < x < 1$ or $6 < x < 8$ is an open set.

DEFINITION 10.16. *Closed Set.* A set which contains all its limit points is called a closed set. For example, the set

$$\left(0, \frac{1}{2}, \frac{1}{3}, \frac{1}{4}, \ldots, \frac{1}{n}, \ldots\right)$$

is closed since its only limit point is 0, which it contains.

DEFINITION 10.17. The set of all points (or elements) which belong to S_1 or S_2 is called the *union* $(S_1 \cup S_2)$ or sum $(S_1 + S_2)$ of the two sets S_1, S_2. The shaded area below is $S_1 + S_2$ (see Fig. 10.1). The union, $S = \bigcup_{(\alpha)} S_\alpha$, of any number of sets, $\{S_\alpha\}$, is the set of all points $\{x\}$, x in at least one of the S_α.

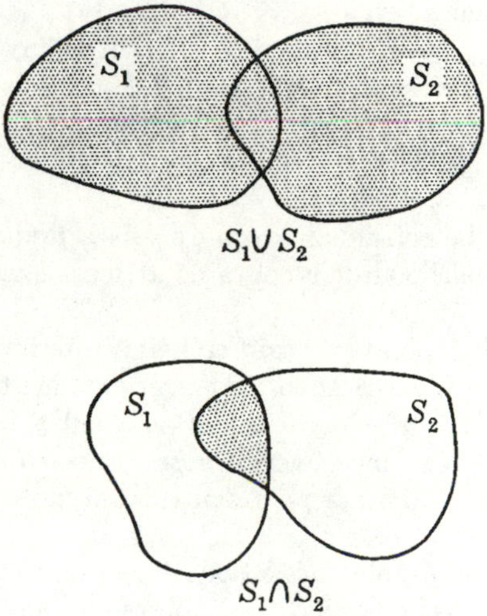

FIG. 10.1

DEFINITION 10.18. The set of all points which belong to both S_1 and S_2 simultaneously is called the *intersection*, or product, written $S_1 \cap S_2$ or $S_1 \cdot S_2$, of the two sets S_1, S_2. Graphically, the shaded area is $S_1 \cap S_2$.

The theory of sets and its application to logic and mathematics was given great impetus by the English mathematician, George Boole (1815–1864).

Two sets S_1, S_2 are said to be equal, $S_1 = S_2$, if every element of S_1 belongs to S_2, and conversely. If S_1 contains every element of S_2, we say that S_2 is a subset of S_1, written $S_1 \supseteqq S_2$. If S_2 is a subset of S_1 and if, furthermore, at least one element of S_1 is not in S_2, we say that S_2 is a proper subset of S_1, written

$$S_1 \supset S_2 \qquad \text{or} \qquad S_2 \subset S_1$$

Some obvious facts of set theory are:
(1) $A + B = B + A$
(2) $(A + B) + C = A + (B + C)$
(3) $AB = BA$
(4) $(AB)C = A(BC)$
(5) $A + A = A$, $AA = A$
(6) $A \subseteqq A + B$, $AB \subseteqq A$, $AB \subseteqq B$

(7) $A \subseteqq C, B \subseteqq C \Rightarrow A + B \subseteqq C$
(8) $A \supseteqq C, B \supseteqq C \Rightarrow AB \supseteqq C$
(9) $AB \subseteqq A(B + C)$
(10) $A(B + C) = AB + AC$

We proceed to prove (10). Let x be any element of $A(B + C)$. Then x belongs to A and to either B or C. Thus x belongs to either AB or AC so that x is an element of $AB + AC$. Thus $A(B + C) \subseteqq AB + AC$. Conversely, $AB \subseteqq A(B + C)$, $AC \subseteqq A(B + C)$ from (9). From (7), $AB + AC \subseteqq A(B + C)$, so that $A(B + C) = AB + AC$.

Problems

1. What are the limit points of the set $0 \leqq x \leqq 1$? Is the set closed, open? What are the boundary points?

2. The same as Prob. 1 with the point $x = \frac{1}{2}$ removed.

3. Show that the set of all boundary points (the boundary of a set) of a set S is a closed set.

4. Prove that the set of all limit points of a set S is a closed set. The set $\bar{S}$ consisting of S and its limit points is called the closure of S. Is $\bar{S}$ a closed set?

5. Prove that the complement of a closed set is an open set, and conversely.

6. Why is a set S which contains only a finite number of elements a closed set?

7. Prove that the intersection of any number of closed sets is a closed set.

8. Prove that the union of any number of open sets is an open set.

9. A union of an infinite number of closed sets is not necessarily closed. Give an example which verifies this.

10. An intersection of an infinite number of open sets is not necessarily open. Give an example which verifies this.

11. Show that $(A + B) \cdot (A + C) = A + BC$.

12. The set of elements of A which are not in B is represented by $A - B$. Show that

$$A(B - C) = AB - AC$$
$$(A - B) + (B - A) = (A + B) - AB$$
$$A + (B - A) = A + B$$
$$A(B - A) = 0 \qquad \text{the null set}$$

13. Show that, if $A + X = S$, $AX = 0$, then $X = C(A)$. The null set (no elements) is represented by 0.

14. If $A \subseteqq S$, we call $S - A = C(A)$ the complement of A relative to S. Show that

$$A \subseteqq B \Leftrightarrow C(B) \subseteqq C(A)$$
$$C(A + B) = C(A) \cdot C(B)$$
$$C(AB) = C(A) + C(B)$$

10.8. The Weierstrass-Bolzano Theorem. We are now in a position to determine a sufficient condition for the existence of a limit point.

THEOREM 10.16. A limit point p exists for every infinite bounded linear set of points, S.

The proof proceeds as follows: We construct a new set T. Into T we place all points which are less than an infinite number of points of S.

T is not empty since S is bounded below. Moreover T is bounded above, for S is bounded above. From Theorem 10.15 a supremum p exists for the set T. We now show that p is a limit point of the set S. Consider any neighborhood of p, say,

$$p - \epsilon < x < p + \delta \qquad \epsilon > 0 \qquad \delta > 0$$

Since $p - \epsilon < p$, there is a member t of T such that $t > p - \epsilon$. Since t is less than an infinite number of elements of S (why?), $p - \epsilon$ is less than an infinite number of S. Thus an infinite number of points of S lie to the right of the point $p - \epsilon$. On the other hand, the point $p + \delta$ has only a finite number of points of S greater than it, for otherwise $p + \delta$ would be a member of T, a contradiction, since p is the supremum of the set T. Thus the open interval

$$p - \epsilon < x < p + \delta$$

contains an infinite number of elements of S. Since ϵ, δ are arbitrary, every open interval containing p contains an infinite number of elements of S. This proves the existence of a limit point p.

Problems

1. Show that the limit point p of Theorem 10.16 is the supremum of all limit points of S.

2. Prove the existence of a least or smallest limit point for an infinite bounded linear set S.

3. Prove that a bounded monotonic increasing set of points has a unique limit. Show that the sequence

$$s_n = \left(1 + \frac{1}{n}\right)^n \qquad n = 1, 2, 3, \ldots$$

has a unique limit. We call it e.

4. Let S_1 and S_2 be two closed and bounded linear sets. Consider the totality of distances obtained by finding the distance from any point s_1 in S_1 to any point s_2 in S_2. Show that there exist two points $\bar{s}_1$, $\bar{s}_2$ of S_1 and S_2, respectively, such that $|\bar{s}_2 - \bar{s}_1|$ is the maximum distance between the two sets.

5. Let S be a closed and bounded set. Define the diameter of the set, and show that two points s_1, s_2 of S exist such that $|s_2 - s_1|$ is the maximum distance between any two points of S.

6. Consider the infinite dimension linear vector space whose elements are of the form $\mathbf{a} = (a_1, a_2, \ldots, a_n, \ldots)$, the a_i real, such that $\sum_{i=1}^{\infty} a_i^2$ converges, written

$$\sum_{i=1}^{\infty} a_i^2 < A \neq \infty$$

If $\mathbf{b} = (b_1, b_2, \ldots, b_n, \ldots)$ with $\sum_{i=1}^{\infty} b_i^2 < B \neq \infty$, then by definition

$$\mathbf{a} + \mathbf{b} = (a_1 + b_1, a_2 + b_2, \ldots, a_n + b_n, \ldots)$$

and $x\mathbf{a} = (xa_1, xa_2, \ldots, xa_n, \ldots)$. We can define the distance $\rho(\mathbf{a}, \mathbf{b})$ between two elements of this space (*Hilbert*) by the formula

$$\rho(\mathbf{a}, \mathbf{b}) = \left[\sum_{n=1}^{\infty} (b_n - a_n)^2\right]^{\frac{1}{2}}$$

Show that $\rho(\mathbf{a}, \mathbf{b})$ converges. *Hint:* Consider the sum

$$\sum_{n=1}^{\infty} (\lambda a_n + b_n)^2 \geqq 0$$

or

$$\left(\sum_{n=1}^{\infty} a_n^2\right)\lambda^2 + 2\left(\sum_{n=1}^{\infty} a_n b_n\right)\lambda + \sum_{n=1}^{\infty} b_n^2 \geqq 0$$

and prove the *Schwarz-Cauchy inequality,*

$$\left(\sum_{n=1}^{\infty} a_n b_n\right)^2 \leqq \left(\sum_{n=1}^{\infty} a_n^2\right)\left(\sum_{n=1}^{\infty} b_n^2\right)$$

Also show that

(1) $\rho(\mathbf{a}, \mathbf{b}) = 0 \Leftrightarrow \mathbf{a} = \mathbf{b}$

(2) $\rho(\mathbf{a}, \mathbf{b}) = \rho(\mathbf{b}, \mathbf{a})$

(3) $\rho(\mathbf{a}, \mathbf{b}) + \rho(\mathbf{b}, \mathbf{c}) \geqq \rho(\mathbf{a}, \mathbf{c})$

Show that the infinite bounded set of points

$$\begin{aligned} \mathbf{e}_1 &= (1, 0, 0, \ldots, 0, \ldots) \\ \mathbf{e}_2 &= (0, 1, 0, \ldots, 0, \ldots) \\ &\cdots\cdots\cdots\cdots \\ \mathbf{e}_n &= (0, 0, 0, \ldots, 1, 0, \ldots) \\ &\cdots\cdots\cdots\cdots \end{aligned}$$

cannot have a limit point. By a spherical neighborhood of a point

$$\mathbf{p} = (p_1, p_2, \ldots, p_n, \ldots)$$

we mean the set of points

$$\mathbf{x} = (x_1, x_2, \ldots, x_n, \ldots)$$

which satisfy

$$\sum_{n=1}^{\infty} (x_n - p_n)^2 < R^2$$

7. Let L be a limit point of a set S. Show that one can pick out a sequence of elements from S, say, $s_1, s_2, s_3, \ldots, s_n, \ldots$, such that

$$\lim_{n\to\infty} s_n = L$$

in the sense that every neighborhood of L contains an infinite number of the s_i. Furthermore s_{n+1} is closer to L than s_n. We call such an approach a sequential approach to L.

10.9. Theorem of Nested Sets. Let $M_1, M_2, \ldots, M_n, \ldots$ be a sequence of closed and bounded sets such that $M_1 \supset M_2 \supset M_3 \supset \cdots \supset M_n \supset \cdots$. We show that there is at least one point p which belongs to all the sets $M_1, M_2, \ldots, M_n, \ldots$. First we choose an element p_1 from M_1, then p_2 from $M_2, \ldots, p_n$ from M_n, etc. The set of points $(p_1, p_2, \ldots, p_n, \ldots)$ belongs to M_1. This set is bounded since M_1 is bounded. From the Weierstrass-Bolzano theorem the set has at least one limit point p which belongs to M_1, since M_1 is closed. Similarly p is a limit point of the set $(p_2, p_3, \ldots, p_n, \ldots)$. These points belong to M_2 so that $p \epsilon M_2$. Finally p is a limit point of the set $(p_n, p_{n+1}, \ldots)$ so that $p \epsilon M_n$. Thus p belongs to each M_i, $i = 1, 2, \ldots, n, \ldots$. If the diameters of the sets M_n tend to zero, then p is unique (see Prob. 5, Sec. 10.8) for the definition of the diameter of a set.

Some mathematicians believe it necessary to postulate the existence of the set $(p_1, p_2, \ldots, p_n, \ldots)$. More generally they desire the following postulate: Let $S = \{S_\alpha\}$ be any collection of sets. Then there exists a set P of elements $\{p_\alpha\}$ such that a point p_i of P exists for each set S_i of S, p_i belonging to the set S_i. This is essentially the *axiom of choice* (*Zermelo*).

10.10. The Heine-Borel Theorem. Let S be any closed and bounded set, and let T be any collection of open intervals having the property that, if x is an element of S, then there exists an open interval T_x of the collection T such that x is an element of T_x. The Heine-Borel theorem states that there exists a subcollection T' of T which contains a finite number of open intervals and such that every element x of S is contained in one of the finite collection of open intervals that comprise T'.

Before proceeding to the proof we point out the following:

1. Both the set S and the collection of open sets T are given beforehand, since it is a simple matter to choose a single open interval which completely covers a bounded set S.

2. S is to be closed, for consider the set $S(1, \frac{1}{2}, \frac{1}{3}, \ldots, 1/n, \ldots)$, and let T consist of the following open intervals:

$$T_1: \qquad x \ni |x - 1| < \frac{1}{2^2}$$

$$T_2: \qquad x \ni \left|x - \frac{1}{2}\right| < \frac{1}{3^2}$$

$$\cdots\cdots\cdots\cdots\cdots\cdots$$

$$T_n: \qquad x \ni \left|x - \frac{1}{n}\right| < \frac{1}{(n+1)^2}$$

$$\cdots\cdots\cdots\cdots\cdots\cdots$$

It is easy to see that we cannot reduce the covering of S by eliminating any of the given T_n, for there is no overlapping of these open intervals. Each T_j is required to cover the point $1/j$ of S.

Proof of the Theorem. Let S be contained in the interval $-N \leqq x \leqq N$. This is possible since S is assumed bounded. Now divide this interval into the two intervals (1) $-N \leqq x \leqq 0$, (2) $0 \leqq x \leqq N$. Any element x of S lies in either (1) or (2). If the origin O is in S, then O lies in both (1) and (2). The points of S in (1) form a closed set, as do the points of S in (2). Why? Now if the theorem is false, it will not be possible to cover the points of S in both (1) and (2) by a finite number of open intervals which form a subset of T. Thus the points of S in either (1) or (2), or possibly both, require an infinite number of covering sets of T. Assume the elements of S in (1) still require an infinite covering; call these points the set S_1. Do the same for S_1 by subdividing the interval $-N \leqq x \leqq 0$ into two parts. Continue this process, repeating the argument used above. In this way we construct a sequence of sets

$$S \supset S_1 \supset S_2 \supset \cdots \supset S_n \supset \cdots$$

such that each S_i is a closed set (proof left to reader) and such that diameters of the $S_n \to 0$ as $n \to \infty$. From the theorem of nested sets (Sec. 10.9) there exists a unique point p which is contained in every S_i, $i = 1, 2, \ldots$, and $p \epsilon S$. Since p is in S, an open interval T_p exists which covers p. This T_p has a finite nonzero diameter, so that eventually one of the S_i, say, S_m, will be contained in T_p. Why? But by assumption all the elements of S_m require an infinite number of the $\{T\}$ to cover them. This is a direct contradiction to the fact that a single T_p covers them. Hence our original assumption was wrong, and the theorem is proved.

Note that our proof was by contradiction. Certain mathematicians from the intuitional school of thought object to this type of proof. They wish to have the finite subset of T, say, $T_1, T_2, \ldots, T_r$, exhibited, such that every point p of S belongs to at least one of the T_j, $j = 1, 2, \ldots, r$.

Problem. Assuming the Heine-Borel theorem, show that the Weierstrass-Bolzano theorem holds. *Hint:* If S has no limit points, it is a closed set. If p is a point of S and is not a limit point, a neighborhood N_p of p exists such that N_p contains no points of S other than p itself. Continue the proof.

We now show that, if we assume the Heine-Borel theorem true for any linear point set, then the existence of the supremum can be established. Let S be an infinite bounded set of real numbers. Why can we dispense with the case for which S contains only a finite number of elements? Since S is bounded above by an integer N, we can speak of the set T consisting of all numbers x such that $x \geqq$ every s of S, $x \leq N$. The set T is bounded. Is T closed? Yes, for if t is a limit point of T, and if

$t < s_0$, $s_0 \epsilon S$, then we can find a neighborhood of t which excludes s_0, so that the points of T in this neighborhood are less than s_0, a contradiction. Thus $t \geqq s_0 \epsilon S$, so that $t \epsilon T$, and T is closed. Every member of T satisfies the first criterion for the existence of a supremum. Hence, if no member of T is the actual supremum of the set S, the second criterion for the existence of a supremum must be violated for every member of T. Thus every point $t \epsilon T$ is contained in a neighborhood N_t, such that no member of S is in this neighborhood. These neighborhoods form a covering of T. Assuming the Heine-Borel theorem, we can remove all but a finite number of these neighborhoods and still cover the set T. Let the finite collection of neighborhoods be designated by

$$N: \qquad a_i < x < b_i \qquad i = 1, 2, 3, \ldots, r$$

Let a_u be the smallest of the a_i, $i = 1, 2, \ldots, r$. We assert that a_u belongs to T. Let the reader verify this. Then the neighborhood to which a_u originally belonged has the property that no point of S is in this neighborhood. This is a contradiction since those points to the left of $a_u (< a_u)$ are not covered by the neighborhoods N. Hence there must be a supremum of the set S. Q.E.D.

10.11. Cardinal Numbers. We conclude the study of point sets with a brief discussion of the Cantor theory of countable and uncountable sets.

DEFINITION 10.19. If two sets A and B are such that a correspondence exists between their elements in such a way that for each x in A there corresponds a unique y in B, and conversely, we say that the two sets are in one-to-one correspondence and that they have the same cardinal number.

Thus, without counting, a savage having 7 goats can make a fair trade with one having 7 wives. They need only pair each goat with each wife.

DEFINITION 10.20. A set in one-to-one correspondence with the set of integers $(1, 2, 3, \ldots, n)$ is said to have *cardinal* number n.

DEFINITION 10.21. A set which can be put into one-to-one correspondence with the Peano set of positive integers is said to be *countable*, or *denumerable*, or *countably infinite*, or *denumerably infinite*.

We note that the set of integers $1^2, 2^2, 3^2, \ldots, n^2, \ldots$ is countably infinite and at the same time is a subset of all the positive integers. The cardinal number of the positive integers is called aleph zero ($\aleph_0$).

DEFINITION 10.22. An infinite collection which is not denumerable is said to be *uncountable*, or *nondenumerable*.

THEOREM 10.17. A countable collection of countable sets is a countable set. Let the sets be $S_1, S_2, \ldots, S_n, \ldots$. This can be arranged since we have a countable collection of sets $\{S_\alpha\}$. Since the elements of S_1 are assumed countable, we may arrange them as follows:

Similarly

$$
\begin{array}{ll}
S_1: & a_{11}, a_{12}, a_{13}, \ldots, a_{1n}, \ldots \\
S_2: & a_{21}, a_{22}, a_{23}, \ldots, a_{2n}, \ldots \\
S_3: & a_{31}, a_{32}, a_{33}, \ldots, a_{3n}, \ldots \\
\cdots & \cdots\cdots\cdots\cdots\cdots\cdots \\
S_n: & a_{n1}, a_{n2}, a_{n3}, \ldots, a_{nn}, \ldots \\
\cdots & \cdots\cdots\cdots\cdots\cdots\cdots
\end{array}
$$

Now consider the collection of elements

$$a_{11}, a_{12}, a_{21}, a_{13}, a_{31}, a_{22}, \ldots, a_{1n}, a_{2(n-1)}, \ldots, a_{n1}, \ldots$$

This collection is countable since we have a first element, a second element, etc. But this sequence exhausts every element of $\{S_i\}$. This is done by a diagonalization process. Thus the proof of the theorem is demonstrated.

We can now easily prove that the set of rationals on the interval $0 \leqq x \leqq 1$ is countable. Into the set S_i we place all fractions m/i,

$$0 \leqq m \leqq i \qquad i = 1, 2, 3, \ldots$$

There are a countable number of S_i each containing a finite number of fractions. The complete set of all such fractions yields the set of rationals of $(0 \leqq x \leqq 1)$, this set being countable. Now let the reader show that the set of rationals, $\{r\}$, for which $-\infty < r < \infty$ is a countable set.

Cantor showed that the set of real numbers in the interval $0 \leqq x \leqq 1$ is not a countable set. Assume the set countable. Then the elements of $0 \leqq x \leqq 1$ can be arranged in a sequence

$$
\begin{array}{l}
s_1 = 0.a_{11}\, a_{12}\, a_{13} \cdots a_{1n} \cdots \\
s_2 = 0.a_{21}\, a_{22}\, a_{23} \cdots a_{2n} \cdots \\
\cdots\cdots\cdots\cdots\cdots\cdots \\
s_n = 0.a_{n1}\, a_{n2}\, a_{n3} \cdots a_{nn} \cdots \\
\cdots\cdots\cdots\cdots\cdots\cdots
\end{array}
$$

where the a_{mn} are the integers zero through 9.

The s_j, $j = 1, 2, 3, \ldots, n, \ldots$ are written in decimal notation. Now consider the number $s = 0.b_1\, b_2 \cdots b_n \cdots$, where

$$b_1 = a_{11} \pm 2 \qquad b_2 = a_{22} \pm 2 \qquad \ldots \qquad b_n = a_{nn} \pm 2 \qquad \ldots$$

the -2 occurring if $a_{ii} = 8$ or 9. The number b is certainly not among any of the s_j, $j = 1, 2, 3, \ldots, n, \ldots$ since it differs from every one of these s_j. But the collection $\{s_j\}$ was supposed to exhaust the real numbers on the interval $0 \leqq x \leqq 1$. Thus a contradiction occurs, and the real numbers are not countable.

Problems

1. Show that the set of *algebraic numbers*, those numbers which satisfy polynomial equations with integral coefficients, is a countable set.

2. Why are the irrationals uncountable?

3. Consider the Cantor set obtained as follows: From the interval $0 \leqq x \leqq 1$ delete the middle third, leaving the end points $\frac{1}{3}$, $\frac{2}{3}$. Repeat this process for the remaining intervals, always keeping the end points. The set that remains after a countable number of such operations is the Cantor set. Since each removed set is open, show that the Cantor set is a closed set. If the numbers $0 \leqq x \leqq 1$ are written as decimal expansions to the base 3, show that the Cantor set can be put into one-to-one correspondence with the set of reals on $0 \leqq x \leqq 1$ when these numbers are written as decimal expansions to the base 2. Show that every point of the Cantor set is a limit point of the set.

4. Show that a limit point exists for a nondenumerable collection S of linear points. The set need not be bounded. *Hint:* Consider the intervals $N \leqq x \leqq N + 1$ for the rational integers N. At least one of these intervals must contain a nondenumerable number of S. Why?

5. Show that a limit point exists and belongs to a nondenumerable collection S of linear points.

10.12. Functions of a Real Variable. Limits and Continuity. Let G be any linear set of real numbers. G may be an open or closed set, a finite set of points, the continuum, a closed interval, etc. We write $G = \{x\}$, where x is any element of G. Assume that to each x of G there corresponds a unique real number y, $x \to y$, and let $H = \{y\}$ be the totality of real numbers obtained through the correspondence. This mapping of the set G into the set H defines a real-valued function, usually written

$$y = f(x) \qquad x \text{ in } G$$

We say that y is a function of x in the sense that for each x there exists a rule for determining the corresponding y. The notation $f(x)$ simply means that f operates on x according to some predetermined rule. For example, if $y = f(x) = x^2$, $-3 \leqq x < 5$, the rule for determining y given any x of the set $-3 \leqq x < 5$ simply is to square the given number x. Consider Table 10.1. The set G consists of the numbers $x = -1, 2, 5, 14$,

Table 10.1

x	-1	2	5	14
y	3	8	-2	0

and the set H consists of the corresponding values $y = 3, 8, -2, 0$, respectively. The table defines $y = f(x)$, x in G. The reader is familiar with the use of cartesian coordinates for obtaining a visual picture of $y = f(x)$.

An important concept in the study of functions is the limit process. One reason why the student of elementary calculus encounters difficulty in understanding the theory of limits is simply that the concept of a limit rarely occurs in physical reality. The statement $\lim_{x \to x_0} x^2 = x_0^2$ is quite confusing to beginners. To the novice this statement appears to be trivially true, for is not $x^2 = x_0^2$ when $x = x_0$? It must be understood that the statement $x \to x_0$ is an abstract entity invented by man. We visualize a variable x which is orced to approach the constant x_0 in such a manner that the difference between x and x_0 tends to zero. This leads us to the following definition:

DEFINITION 10.23. An *infinitesimal* is a variable which approaches zero as a limit. η is said to be an infinitesimal if $|\eta|$ becomes and remains less than any preassigned $\epsilon > 0$.

If η_1 and η_2 are infinitesimals, then eventually $|\eta_1| < \epsilon/2$, $|\eta_2| < \epsilon/2$, for arbitrary $\epsilon > 0$, so that $|\eta_1 + \eta_2| \leqq |\eta_1| + |\eta_2| < \epsilon$, and $\eta_1 + \eta_2$ is also an infinitesimal. Let the reader show that the difference and product of two infinitesimals are again infinitesimals and that the product of a bounded function with an infinitesimal is again an infinitesimal. Let x be a variable which in succession takes on the values $1, \frac{1}{2}, \frac{1}{3}, \ldots, 1/n, \ldots.$ Certainly x is an infinitesimal since $1/n \to 0$ as n becomes infinite. For any value of x the sum $S = x + x + x + \cdots$ is infinite. On the other hand each term of S tends to zero. Hence an infinite sum of infinitesimals need not be an infinitesimal. For another example, let x be an infinitesimal, so that $x^3/(1 + x^2)^n$ is also an infinitesimal for $n = 0, 1, 2, \ldots.$ Now

$$S = \sum_{n=0}^{\infty} \frac{x^3}{(1 + x^2)^n} = x(1 + x^2)$$

so that S is also an infinitesimal. An infinite sum of infinitesimals may or may not be an infinitesimal. Let the reader show by specific examples that the quotient of two infinitesimals may or may not be an infinitesimal.

Let us return to the statement $\lim_{x \to x_0} x^2 = x_0^2$. To show that this statement holds, we must prove that $x^2 - x_0^2$ is an infinitesimal whenever $x - x_0$ is an infinitesimal. Now $|x^2 - x_0^2| = |x - x_0| \cdot |x + x_0|$, and $|x + x_0|$ is bounded for x near x_0. Since $|x - x_0|$ is an infinitesimal (by our choice), of necessity $x^2 - x_0^2$ is an infinitesimal, so that $x^2 - x_0^2 \to 0$, or $x^2 \to x_0^2$, whenever $x \to x_0$.

Let the reader prove the following results: If $\lim_{x \to a} f(x) = M$,

$$\lim_{x \to a} g(x) = N$$

then

(1) $\lim_{x \to a} [f(x) \pm g(x)] = M \pm N$

(2) $\lim_{x \to a} [f(x)g(x)] = MN$

(3) $\lim_{x \to a} \left[\frac{f(x)}{g(x)}\right] = \frac{M}{N} \qquad N \neq 0$

A knowledge of limits enables us to introduce the concept of continuity of a function. We consider the functional relationship $y = f(x)$, and we let $x = x_0$ be a point of G with $y_0 = f(x_0)$. What shall we mean when we say that $f(x)$ is continuous at $x = x_0$? Intuitively we feel that, if x is any point of G which is close to x_0, then $f(x)$ will be continuous at $x = x_0$ if the corresponding value y will be close to y_0. Actually we desire $y - y_0$ to be an infinitesimal if $x - x_0$ is an infinitesimal.

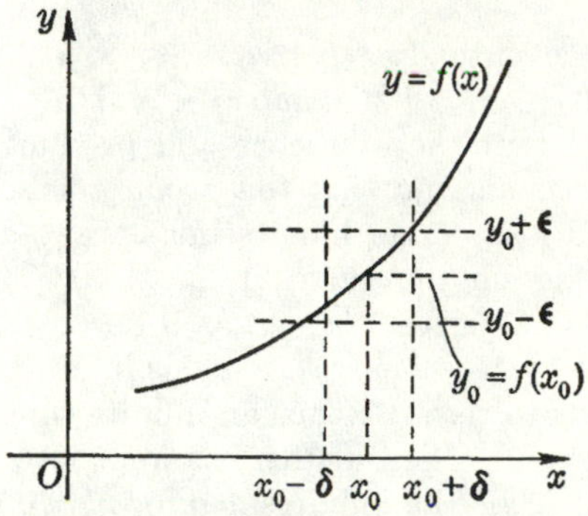

FIG. 10.2

DEFINITION 10.24. $f(x)$ is said to be continuous at $x = x_0$ if for any $\epsilon > 0$ there exists a $\delta(\epsilon) > 0$ such that

$$|f(x) - f(x_0)| < \epsilon$$

whenever $|x - x_0| < \delta$, x in G.

Let us look at the graph of Fig. 10.2. For any $\epsilon > 0$ we construct the horizontal lines $y = y_0 + \epsilon$, $y = y_0 - \epsilon$. If we can draw two distinct vertical lines $x = x_0 + \delta$, $x = x_0 - \delta$ such that, for every point x of G in the open interval $x_0 - \delta < x < x_0 + \delta$, $f(x)$ has a numerical value lying between $f(x_0) - \epsilon$, $f(x_0) + \epsilon$, we say that $f(x)$ is continuous at $x = x_0$. The distinct vertical lines must exist for every $\epsilon > 0$. Note that the maximum size of δ will depend on ϵ. The smaller ϵ is, the smaller δ will have to be. Moreover the point x_0 will influence the value of δ. The steeper the curve, the smaller δ becomes.

Two equivalent definitions of continuity are as follows: $f(x)$ is continuous at $x = x_0$ if:

(a) $\lim_{x \to x_0} f(x) = f(x_0)$, x in G.

(b) For any $\epsilon > 0$ there exists a δ neighborhood of x_0 such that $|f(x_1) - f(x_2)| < \epsilon$ whenever x_1 and x_2 are in G and belong to the δ neighborhood of x_0.

Functions are discontinuous for two reasons:

1. $\lim_{x \to x_0} f(x)$ does not exist.
2. $\lim_{x \to x_0} f(x)$ exists but is not equal to $f(x_0)$.

Example 10.1. Consider $f(x) = x + 1$, $x \neq 2$, $f(2) = 5$. We have

$$\lim_{x\to 2} (x + 1) = 3$$

However $f(2) = 5 \neq 3$, so that $f(x)$ is discontinuous at $x = 2$. If we were to redefine the value of $f(x)$ at $x = 2$ to be 3 instead of 5, we would remove the discontinuity. In this example $f(x)$ has a removable type of discontinuity.

Example 10.2. Consider $f(x) = \dfrac{2^{1/x}}{(1 + 2^{1/x})}$, $x \neq 0$, $f(0) = 1$. We have

$$\lim_{\substack{x\to 0\\ x>0}} \frac{2^{1/x}}{1 + 2^{1/x}} = 1 \qquad \lim_{\substack{x\to 0\\ x<0}} \frac{2^{1/x}}{1 + 2^{1/x}} = 0$$

Hence $\lim_{x\to 0} f(x)$ does not exist for all manner of approach of x to zero. It is obvious that $f(x)$ does not have a removable type of discontinuity at $x = 0$.

Example 10.3. $f(x) = \sin(1/x)$, $x \neq 0$, $f(0) = 2$. Since $f(x)$ oscillates between -1 and $+1$ as $x \to 0$, $\lim_{x\to 0} f(x)$ does not exist and $f(x)$ is discontinuous at $x = 0$.

Example 10.4. $f(x) = 1/x$, $x \neq 0$, $f(0) = 41$ Some authors consider that $1/x$ is discontinuous at $x = 0$ since division by zero is undefined. In this example $f(x)$ is discontinuous at $x = 0$ since $\lim_{x\to 0} 1/x$ does not exist. On the other hand $f(x) = (x^2 - 1)/(x - 1)$, $x \neq 0$, $f(0) = 2$ is continuous at $x = 1$ even though $(x^2 - 1)/(x - 1)$ is undefined at $x = 1$.

Example 10.5. If $f(x) = x^2$ for $0 \leqq x \leqq 1$, we note that $f(x)$ is continuous everywhere on this interval. Why is $f(x)$ continuous at $x = 0$ and $x = 1$? If $x = c$ is a point of this interval, we have $|x^2 - c^2| = |x - c| \cdot |x + c| < 3|x - c|$ so that $|x^2 - c^2| < \epsilon$ whenever $|x - c| < \epsilon/3 = \delta$. We have found a value of δ independent of the point $x = c$. We say, therefore, that $f(x) = x^2$ is *uniformly* continuous on the interval $0 \leqq x \leqq 1$.

Example 10.6. Let $f(x) = 1/x$ for $\alpha \leqq x \leqq 1$, $\alpha > 0$. It is easily seen that $f(x)$ is continuous for every x of this range. For $x = c$ we have

$$\left|\frac{1}{x} - \frac{1}{c}\right| = \left|\frac{x - c}{xc}\right| \leqq \frac{1}{\alpha^2}|x - c|$$

Thus $\left|\frac{1}{x} - \frac{1}{c}\right| < \epsilon$ if $\frac{1}{\alpha^2}|x - c| < \epsilon$ or $|x - c| < \alpha^2\epsilon = \delta$. Since $\delta = \alpha^2\epsilon$ is independent of $x = c$, we have uniform continuity. On the other hand $f(x)$ is also continuous at every point of the interval $0 < x \leqq 1$. But $f(x)$ is not uniformly continuous on this range since $\delta \to 0$ as $\alpha \to 0$. For any $\epsilon > 0$ we cannot find a $\delta > 0$ such that $\left|\frac{1}{x} - \frac{1}{c}\right| < \epsilon$ whenever $|x - c| < \delta$ for all c on the range $0 < x \leqq 1$. Note that the interval $0 < x \leqq 1$ is not a closed set.

Definition 10.25. Let $f(x)$ be defined over the range G. $f(x)$ is said to be uniformly continuous on G if for every $\epsilon > 0$ there exists a $\delta = \delta(\epsilon)$ such that

$$|f(x_1) - f(x_2)| < \epsilon$$

whenever $|x_1 - x_2| < \delta$, x_1 and x_2 in G.

Theorem 10.18. If G is a closed and bounded set and if $f(x)$ is continuous at every point of G, then $f(x)$ is uniformly continuous on G.

Proof. We base the proof of this theorem on the Weierstrass-Bolzano theorem. First let us assume that $f(x)$ is not uniformly continuous on the closed and bounded set G. This means that there exists at least one $\epsilon = \epsilon_0$ such that for any $\delta > 0$ there will exist x_1, $\bar{x}_1$ such that $|f(x_1) - f(\bar{x}_1)| > \epsilon_0$ with $|x_1 - \bar{x}_1| < \delta$. We take $\delta = 1$ and obtain the pair of numbers x_1, $\bar{x}_1$. Then we take $\delta = \frac{1}{2}$ and obtain the pair $(x_2, \bar{x}_2)$. For $\delta = 1/n$ we have the pair $(x_n, \bar{x}_n)$ such that $|f(x_n) - f(\bar{x}_n)| > \epsilon_0$ with $|x_n - \bar{x}_n| < 1/n = \delta$. By this process we generate the two sequences $(x_1, x_2, \ldots, x_n, \ldots)$, $(\bar{x}_1, \bar{x}_2, \ldots, \bar{x}_n, \ldots)$. Since G is a closed and bounded set, the sequence $(x_1, x_2, \ldots, x_n, \ldots)$ has a limit point $x = c$ belonging to G. We pick out a subsequence which converges sequentially to $x = c$. The corresponding points from the sequence $\{\bar{x}_n\}$ have the same limit, $x = c$, since $|x_n - \bar{x}_n| < 1/n$. Let these two sequences be designated by $\{x_n'\}$, $\{\bar{x}_n'\}$. Since $f(x)$ is continuous at $x = c$, we have $|f(x_n') - f(c)| < \epsilon_0/2$, $|f(\bar{x}_n') - f(c)| < \epsilon_0/2$, for n sufficiently large. Thus $|f(x_n') - f(\bar{x}_n')| < \epsilon_0$, a contradiction, since for all pairs x_n', $\bar{x}_n'$, $|f(x_n') - f(\bar{x}_n')| > \epsilon_0$. Q.E.D.

THEOREM 10.19. If $f(x)$ and $g(x)$ are continuous at $x = a$, then $f(x) \pm g(x)$, $f(x)g(x)$, $f(x)/g(x)$ with $g(a) \neq 0$, are continuous at $x = a$. The proof is left to the reader.

THEOREM 10.20. If $f(x)$ is continuous over a closed and bounded set G, then $f(x)$ is bounded.

Proof. Assume $f(x)$ is not bounded. There exists an x_1 such that $f(x_1) > 1$, an x_2 such that $f(x_2) > 2$, . . . , an x_n such that $f(x_n) > n$, . . . The set $(x_1, x_2, \ldots, x_n, \ldots)$ has a limit point $\bar{x}$ belonging to G since G is a closed and bounded set. There exists a subsequence $(x_1', x_2', \ldots, x_n', \ldots)$ which converges sequentially to $\bar{x}$. Thus from continuity

$$\lim_{x_n' \to \bar{x}} f(x_n') = f(\bar{x})$$

But $f(x_n') \to \infty$ as $x_n' \to \bar{x}$, a contradiction, since $f(x)$ has a specific real value at $x = \bar{x}$.

THEOREM 10.21. If $f(x)$ is continuous on a closed and bounded set G, then a point x_0 of G exists such that $f(x_0) \geqq f(x)$ for all x of G.

Proof. From Theorem 10.20 $f(x)$ is bounded above as x ranges over G. Thus the set of values $\{f(x)\}$ has a supremum, say, S. Thus $S \geqq f(x)$ for all x of G. Choose x_1 such that $f(x_1) > S - 1$, x_2 such that $f(x_2) > S - \frac{1}{2}$, . . . , x_n such that $f(x_n) > S - 1/n$, ... Why is this possible? The set $(x_1, x_2, \ldots, x_n, \ldots)$ has a limit $\bar{x}$ belonging to G. We pick out a subsequence $\{x_n'\}$ which converges sequentially to $\bar{x}$, with $x_n' = x_m$. Now $f(x_m) > S - 1/m$, so that

$$\lim_{m \to \infty} f(x_m) \geqq \lim_{m \to \infty} \left(S - \frac{1}{m}\right) = S$$

From continuity, $\lim_{n\to\infty} f(x'_n) = \lim_{m\to\infty} f(x_m) = f(\bar{x})$, so that $f(\bar{x}) \geqq S$. But $S \geqq f(\bar{x})$, so that $f(\bar{x}) = S$. Q.E.D.

THEOREM 10.22. Let $f(x)$ be continuous on the range $a \leqq x \leqq b$. If $f(a) < 0, f(b) > 0$, there exists a point c such that $f(c) = 0$ with $a < c < b$.

Proof. Let T be the set of points of $a \leqq x \leqq b$ such that $f(x) < 0$, and if $f(x_1) < 0$, $x_2 < x_1$, then $f(x_2) < 0$. We speak only of x on the range $a \leqq x \leqq b$. T is not empty since $x = a$ belongs to T. T is bounded above since $x = b$ does not belong to T. Hence T has a supremum; call it $x = c$. Now $f(c) = 0$, or $f(c) > 0$, or $f(c) < 0$. Assume $f(c) > 0$. Since $f(x)$ is continuous at $x = c$, we can find a δ such that

$$|f(x) - f(c)| < \epsilon = \frac{f(c)}{2}$$

whenever $|x - c| < \delta$, $\delta > 0$. Thus $f(x) > f(c)/2 > 0$ whenever $|x - c| < \delta$. But $f(x) < 0$ for $x < c$, a contradiction, so that $f(c) \leqq 0$. Let the reader show that $f(c) < 0$ cannot occur, so that $f(c) = 0$.

Problems

1. For the following functions find a $\delta = \delta(\epsilon)$ such that $|f(x_1) - f(x_2)| < \epsilon$ whenever $|x_1 - x_2| < \delta$:

(*a*) $f(x) = 2x^3$ for $-1 \leqq x \leqq 1$
(*b*) $f(x) = \sin x$ for $0 \leqq x \leqq 2\pi$
(*c*) $f(x) = \sqrt[3]{1 - x^2}$ for $-1 \leqq x \leqq 1$

2. Show that $g(x) = x \sin (1/x)$, $x \neq 0$, $f(0) = 0$, is continuous at $x = 0$.

3. Show that $y = \sin (1/x)$ is not uniformly continuous on the range $0 < x \leqq 1$.

4. Find an example of a function $f(x)$ which is continuous on a bounded set, but $f(x)$ is not bounded.

5. Find an example of a function $f(x)$ which is continuous on a closed set, but $f(x)$ is not bounded.

6. Prove Theorem 10.19.

7. Discuss the continuity of $f(x) = 0$ when x is rational, $f(x) = x$ when x is irrational, $f(0) = 0$.

8. Let $f(x) = 0$ when x is irrational, $f(x) = 1/q$ when $x = p/q$, $(p, q) = 1$. Show that $f(x)$ is discontinuous at the rational points and is continuous at the irrational points. *Hint:* Show that there are only a finite number of rational numbers p/q with $(p, q) = 1$ such that $1/q > \epsilon$ for any $\epsilon > 0$.

9. Prove Theorem 10.18 by making use of the Heine-Borel theorem.

10. Prove Theorem 10.20 by making use of Theorem 10.18.

11. Let $f(x) = g(x)$ for x rational. If $f(x)$ and $g(x)$ are continuous for all x, show that $f(x) = g(x)$ for all x.

12. Let $f(x)$ and $g(x)$ be continuous on the range $a \leqq x \leqq b$. Let $h(x) = f(x)$ if $f(x) \geqq g(x)$, $h(x) = g(x)$ if $g(x) \geqq f(x)$. Show that

$$h(x) = \tfrac{1}{2}(f(x) + g(x)) + \tfrac{1}{2}|f(x) - g(x)|$$

Also show that $h(x)$ is continuous for $a \leqq x \leqq b$.

13. If $f(x)$ is defined only for a finite set of values of x, show that $f(x)$ is continuous at these points.

14. If $f(x)$ is continuous at $x = c$ with $f(c) > 0$, show that a $\delta > 0$ exists such that, for $|x - c| < \delta$, $f(x) > 0$, x in G.

15. If $f(x)$ is continuous on a closed and bounded set G, show that a point x_0 of G exists such that $f(x_0) \leqq f(x)$ for all x in G.

10.13. The Derivative. Let $f(x)$ be defined on a range G, and let c be a limit point of G. We consider any sequence of G which converges to c, say, $\{x_n\}$. Next we compute

$$\lim_{x_n \to c} \frac{f(x_n) - f(c)}{x_n - c} \tag{10.9}$$

If this limit exists and is independent of the particular sequence which converges to c, we say that $f(x)$ is differentiable at $x = c$. We represent this limit or first derivative by $f'(c)$.

The definition above is equivalent to the statement that for any $\epsilon > 0$ there exists a $\delta > 0$ and a number, $f'(c)$, such that

$$\left| \frac{f(x) - f(c)}{x - c} - f'(c) \right| < \epsilon$$

whenever $0 < |x - c| < \delta$, x in G.

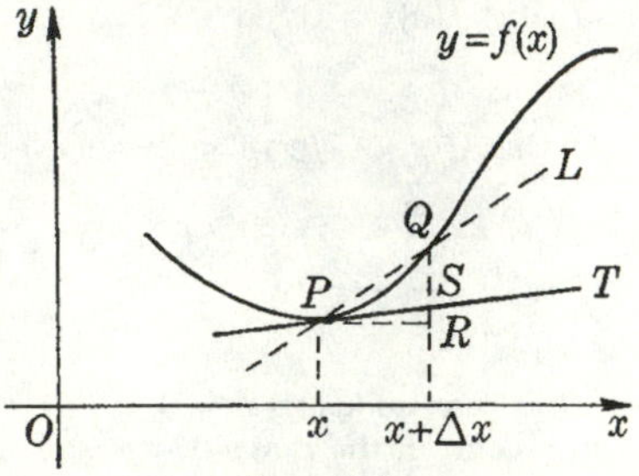

FIG. 10.3

It follows almost immediately that if $f'(c)$ exists then $f(x)$ is continuous at $x = c$, for

$$|f(x) - f(c) - f'(c)(x - c)| \to 0$$

as $x \to c$ implies $f(x) \to f(c)$ as $x \to c$.

The reader is well aware of the geometric interpretation of the first derivative (see Fig. 10.3). In Fig. 10.3,

$$\overline{PR} = \Delta x \qquad \overline{RQ} = f(x + \Delta x) - f(x)$$

the slope of the secant line L joining P to Q is $\dfrac{f(x + \Delta x) - f(x)}{\Delta x}$, the slope of the tangent line T is defined as $f'(x)$, and $\overline{RS} = f'(x)\,\Delta x$ is called the differential of $f(x)$.

Example 10.7. Let $f(x) = x^2 \sin(1/x)$, $x \neq 0$, $f(0) = 0$. To see whether $f(x)$ is differentiable at $x = 0$, we compute

$$\lim_{x \to 0} \frac{f(x) - f(0)}{x - 0} = \lim_{x \to 0} \frac{x^2 \sin(1/x)}{x} = \lim_{x \to 0} x \sin \frac{1}{x} = 0$$

Hence $f'(0) = 0$. Let the reader show that $f'(x)$ is not continuous at $x = 0$ by showing that $\lim_{x \to 0} f'(x)$ does not exist.

The reader can verify that, if $f(x)$ and $g(x)$ are differentiable at $x = c$, then

$$\begin{aligned} \frac{d}{dx}[f(x) \pm g(x)]_{x=c} &= f'(c) + g'(c) \\ \frac{d}{dx}[f(x)g(x)]_{x=c} &= f(c)g'(c) + g(c)f'(c) \\ \frac{d}{dx}\left[\frac{f(x)}{g(x)}\right]_{x=c} &= \frac{g(c)f'(c) - f(c)g'(c)}{[g(c)]^2} \qquad g(c) \neq 0 \end{aligned} \tag{10.10}$$

The following differentiation rules are known to hold:

$$\begin{aligned} \frac{d\,(\text{constant})}{dx} &= 0 & \frac{d\cos x}{dx} &= -\sin x \\ \frac{dx^n}{dx} &= nx^{n-1} & \frac{de^x}{dx} &= e^x \\ \frac{d\sin x}{dx} &= \cos x & \frac{d\ln x}{dx} &= \frac{1}{x} \end{aligned}$$

Example 10.8. *Implicit Differentiation.* If $y = f(u)$ and $u = \varphi(x)$, we say that y depends implicitly on x. Let $u_0 = \varphi(x_0)$, and assume that $\frac{df}{du}$ exists at $u = u_0$, $\frac{d\varphi}{dx}$ exists at $x = x_0$. We show that

$$\frac{dy}{dx} = \frac{df}{du}\frac{d\varphi}{dx} = \frac{dy}{du}\frac{du}{dx}$$

at $x = x_0$, $u = u_0$. From

$$\begin{aligned} \Delta y &= f(u_0 + \Delta u) - f(u_0) \\ \Delta u &= \varphi(x_0 + \Delta x) - \varphi(x_0) \end{aligned}$$

we have

$$\frac{\Delta y}{\Delta x} = \frac{f(u_0 + \Delta u) - f(u_0)}{\Delta u}\,\frac{\varphi(x_0 + \Delta x) - \varphi(x_0)}{\Delta x} \qquad \left.\begin{aligned} \Delta u &\neq 0 \\ \Delta x &\neq 0 \end{aligned}\right\} \tag{10.11}$$

If, as $\Delta x \to 0$, Δu does not vanish infinitely often, then eventually $\Delta u \neq 0$, and

$$\frac{dy}{dx} = \lim_{\Delta x \to 0} \frac{\Delta y}{\Delta x} = \left.\frac{df}{du}\right|_{u=u_0} \cdot \left.\frac{d\varphi}{dx}\right|_{x=x_0}$$

since the limit of a product is the product of the respective limits, provided they exist. On the other hand, we cannot apply (10.11) if $\Delta u = 0$ infinitely often as $\Delta x \to 0$. For this case, however, $\frac{du}{dx} = \lim_{\Delta x \to 0} \frac{\Delta u}{\Delta x} = 0$. But for those values of Δx for which $\Delta u = 0$ we have $\frac{\Delta y}{\Delta x} = \frac{f(u_0) - f(u_0)}{\Delta x} = 0$, so that $\lim_{\Delta x \to 0} \frac{\Delta y}{\Delta x} = 0$. Hence $\frac{dy}{dx} = \frac{dy}{du}\frac{du}{dx}$ still holds since both sides of this equation vanish.

Example 10.9. *The Rule of Leibnitz.* The derivative of a product $u(x)v(x)$ is

$$\frac{d}{dx}(uv) = u\frac{dv}{dx} + v\frac{du}{dx}$$

If we differentiate again, we obtain $\frac{d^2}{dx^2}(uv) = u\frac{d^2v}{dx^2} + 2\frac{du}{dx}\frac{dv}{dx} + v\frac{d^2u}{dx^2}$. Let the reader prove by mathematical induction or otherwise that

$$\frac{d^n}{dx^n}(uv) = u\frac{d^nv}{dx^n} + \binom{n}{1}\frac{du}{dx}\frac{d^{n-1}v}{dx^{n-1}} + \binom{n}{2}\frac{d^2u}{dx^2}\frac{d^{n-2}v}{dx^{n-2}} + \cdots$$
$$= \sum_{k=0}^{n}\binom{n}{k}\frac{d^ku}{dx^k}\frac{d^{n-k}v}{dx^{n-k}} \tag{10.12}$$

with $\dfrac{d^0u}{dx^0} \equiv u$, $\dfrac{d^0v}{dx^0} = v$. As an example of Leibnitz's rule we consider the Legendre polynomials $P_n(x)$, defined as $P_n(x) = \dfrac{1}{2^n n!}\dfrac{d^n(x^2-1)^n}{dx^n}$, $P_0(x) = 1$, $n = 0, 1, 2, \ldots$. We have

$$\begin{aligned}
P_n(x) &= \frac{1}{2^n n!}\frac{d^n}{dx^n}(x^2-1)^n = \frac{1}{2^n n!}\frac{d^{n-1}}{dx^{n-1}}[2nx(x^2-1)^{n-1}] \\
&= \frac{1}{2^n n!}\left[2nx\frac{d^{n-1}}{dx^{n-1}}(x^2-1)^{n-1} + 2n(n-1)\frac{d^{n-2}}{dx^{n-2}}(x^2-1)^{n-1}\right] \\
&= xP_{n-1}(x) + \frac{n-1}{2^{n-1}(n-1)!}\frac{d^{n-2}}{dx^{n-2}}(x^2-1)^{n-1} \\
\frac{dP_n(x)}{dx} &= x\frac{dP_{n-1}(x)}{dx} + P_{n-1}(x) + (n-1)P_{n-1}(x) \\
&= x\frac{dP_{n-1}(x)}{dx} + nP_{n-1}(x)
\end{aligned} \tag{10.13}$$

Moreover

$$\begin{aligned}
P_n(x) &= \frac{1}{2^n n!}\frac{d^n}{dx^n}[(x^2-1)(x^2-1)^{n-1}] \\
&= \frac{1}{2^n n!}\left[(x^2-1)\frac{d^n}{dx^n}(x^2-1)^{n-1} + 2nx\frac{d^{n-1}}{dx^{n-1}}(x^2-1)^{n-1}\right. \\
&\qquad \left. + n(n-1)\frac{d^{n-2}}{dx^{n-2}}(x^2-1)^{n-1}\right] \\
&= \frac{x^2-1}{2n}\frac{dP_{n-1}(x)}{dx} + xP_{n-1}(x) + \frac{n-1}{2^n(n-1)!}\frac{d^{n-2}}{dx^{n-2}}(x^2-1)^{n-1} \\
\frac{dP_n(x)}{dx} &= \frac{1}{2n}\frac{d}{dx}\left[(x^2-1)\frac{dP_{n-1}(x)}{dx}\right] + P_{n-1}(x) + x\frac{dP_{n-1}(x)}{dx} + \frac{n-1}{2}P_{n-1}(x)
\end{aligned} \tag{10.14}$$

Combining (10.13) with (10.14) yields

$$\frac{d}{dx}\left[(1-x^2)\frac{dP_{n-1}(x)}{dx}\right] + n(n-1)P_{n-1}(x) = 0$$

or

$$\frac{d}{dx}\left[(1-x^2)\frac{dP_n(x)}{dx}\right] + n(n+1)P_n(x) = 0 \tag{10.15}$$

Equation (10.15) is Legendre's differential equation (see Sec. 5.13).

From Theorem 10.21 we readily prove Theorem 10.23 due to Rolle.

THEOREM 10.23. *Rolle's Theorem.* Let $f(x)$ be continuous in the interval $a \leqq x \leqq b$, and let $f(x)$ possess a derivative at every point of this interval. If $f(a) = f(b) = 0$, there exists a point $x = c$, $a < c < b$, such that $f'(c) = 0$.

Proof. From Theorem 10.21 there exists at least one point $x = c$ such that $f(c) \geqq f(x)$ for $a \leqq x \leqq b$. Now

$$f'(c) = \lim_{\substack{h \to 0 \\ h > 0}} \frac{f(c + h) - f(c)}{h} \leqq 0$$

since $f(c + h) \leqq f(c)$. Moreover

$$f'(c) = \lim_{\substack{h \to 0 \\ h < 0}} \frac{f(c + h) - f(c)}{h} \geqq 0$$

Of necessity, $f'(c) = 0$. Q.E.D. The condition of the theorem can be weakened by assuming that $f'(c)$ exist at $x = c$, with $f(c)$ a local supremum or infemum of $f(x)$. Let the reader give a geometric interpretation of Rolle's theorem, and let him construct an example of a continuous function, $f(x)$, $f(a) = f(b) = 0$, such that nowhere is $f'(x) = 0$, $a \leqq x \leqq b$.

Law of the Mean. Let $f(x)$ and $\varphi(x)$ be differentiable functions on the interval $a \leqq x \leqq b$. We show that a point $x = c$ exists such that

$$[\varphi(b) - \varphi(a)]f'(c) = [f(b) - f(a)]\varphi'(c) \tag{10.16}$$

The function

$$\psi(x) \equiv \varphi(x)[f(b) - f(a)] - f(x)[\varphi(b) - \varphi(a)] - \varphi(a)f(b) + \varphi(b)f(a)$$

vanishes for $x = a$ and $x = b$. $\psi(x)$ was obtained by finding A, B, C such that $\psi(x) = A\varphi(x) + Bf(x) + C$ vanishes for $x = a$, $x = b$. Applying Rolle's theorem to $\psi(x)$ yields the theorem of the mean. As a special case, choose $\varphi(x) \equiv x$, so that (10.16) becomes

$$f(b) - f(a) = f'(c)(b - a) \qquad a < c < b \tag{10.17}$$

Let the reader give a geometric interpretation of (10.17).

L'Hospital's Rule. From (10.16) we have

$$\frac{f(b) - f(a)}{\varphi(b) - \varphi(a)} = \frac{f'(c)}{\varphi'(c)} \qquad a < c < b \tag{10.18}$$

provided $\varphi(b) - \varphi(a) \neq 0$, $\varphi'(c) \neq 0$. Let us assume that $f(a) = 0$, $\varphi(a) = 0$, so that

$$\frac{f(b)}{\varphi(b)} = \frac{f'(c)}{\varphi'(c)} \qquad a < c < b$$

and

$$\lim_{b \to a} \frac{f(b)}{\varphi(b)} = \lim_{b \to a} \frac{f'(c)}{\varphi'(c)} = \lim_{c \to a} \frac{f'(c)}{\varphi'(c)} \tag{10.19}$$

since $a < c < b$. We can rewrite (10.19) as

$$\lim_{x \to a} \frac{f(x)}{\varphi(x)} = \lim_{x \to a} \frac{f'(x)}{\varphi'(x)} \tag{10.20}$$

which is L'Hospital's rule. We cannot apply (10.20) unless $f(a) = 0$, $\varphi(a) = 0$. What can be done if $f'(a) = 0$, $\varphi'(a) = 0$? If $f'(x)$ and $\varphi'(x)$ are continuous at $x = a$, and if $\varphi'(a) \neq 0$, then (10.20) can be written

$$\lim_{x \to a} \frac{f(x)}{\varphi(x)} = \frac{f'(a)}{\varphi'(a)} \tag{10.21}$$

provided $f(a) = 0$, $\varphi(a) = 0$.

Example 10.10.

$$\lim_{x \to 0} \frac{\sin x - x + x^3/6}{x^5} = \lim_{x \to 0} \frac{\cos x - 1 + x^2/2}{5x^4}$$

Now $g(x) = \cos x - 1 + x^2/2$ and $h(x) = 5x^4$ both vanish at $x = 0$ and are differentiable. Reapplying (10.20) yields

$$\lim_{x \to 0} \frac{\sin x - x + x^3/6}{x^5} = \lim_{x \to 0} \frac{-\sin x + x}{20x^3} = \lim_{x \to 0} \frac{-\cos x + 1}{60x^2}$$
$$= \lim_{x \to 0} \frac{\sin x}{120x} = \lim_{x \to 0} \frac{\cos x}{120} = \frac{1}{120}$$

We complete this section with a discussion of the second derivative. Let $f(x)$ be differentiable in a neighborhood of $x = c$. The second derivative of $f(x)$ at $x = c$ is defined as

$$f''(c) = \lim_{k \to 0} \frac{f'(c + k) - f'(c)}{k}$$

provided this limit exists. Now $f'(x) = \lim_{h \to 0} \frac{f(x + h) - f(x)}{h}$ so that

$$f'(c) = \lim_{h \to 0} \frac{f(c + h) - f(c)}{h} \qquad f'(c + k) = \lim_{h \to 0} \frac{f(c + k + h) - f(c + k)}{h}$$

and $$f''(c) = \lim_{k \to 0} \frac{\lim_{h \to 0} \frac{f(c + k + h) - f(c + k)}{h} - \lim_{h \to 0} \frac{f(c + h) - f(c)}{h}}{k}$$
$$= \lim_{k \to 0} \lim_{h \to 0} \frac{f(c + k + h) - f(c + k) - f(c + h) + f(c)}{hk}$$

The order in which we take these limits is important. It can be shown, however, that under certain conditions we can replace k by h so that only one limit process is necessary. We would thus have

$$f''(c) = \lim_{h \to 0} \frac{f(c + 2h) - 2f(c + h) + f(c)}{h^2} \tag{10.22}$$

We now investigate under what conditions (10.22) yields $f''(c)$. Let $U(c) = f(c + h) - f(c)$, $U(c + h) = f(c + 2h) - f(c + h)$, so that

$$U(c + h) - U(c) = f(c + 2h) - 2f(c + h) + f(c)$$

Applying the law of the mean as given by (10.17) yields

$$hU'(\bar{c}) = h[f'(\bar{c} + h) - f'(\bar{c})] = f(c + 2h) - 2f(c + h) + f(c)$$

Applying (10.17) once more yields

$$h^2 f''(\bar{\bar{c}}) = f(c + 2h) - 2f(c + h) + f(c) \tag{10.23}$$

with $c < \bar{\bar{c}} < c + h$. To obtain (10.23), we assumed that $f''(x)$ exists in a neighborhood of $x = c$. If, moreover, $f''(x)$ is continuous at $x = c$, we note that (10.22) results.

Problems

1. Derive the results of (10.10).

2. Show that if $y = x^{m/n}$, m and n integers, then $\frac{dy}{dx} = \frac{m}{n} x^{m/n-1}$ by assuming

$$\frac{dx^n}{dx} = nx^{n-1}$$

for any positive integer n.

3. Show that $\lim_{x \to 0} \frac{\cos x - 1 + x^2/2}{x^4} = \frac{1}{24}$.

4. Let $u(x) = \tan^{-1} x$. From $(1 + x^2)\frac{d^2u}{dx^2} + 2x\frac{du}{dx} = 0$ find $\frac{d^nu}{dx^n}$ at $x = 0$ for $n = 2, 3, 4, \ldots$, by the use of Leibnitz's rule.

5. Show that $\frac{d^n}{dx^n}(e^{ax}f(x)) = e^{ax}(D + a)^{(n)}f(x)$ with $D = \frac{d}{dx}$, $D^2 = \frac{d^2}{dx^2}$, $\ldots$,

$$(D + a)^{(n)} = \sum_{k=0}^{n} \binom{n}{k} a^{n-k}D^k$$

6. Find an expression for $D^n[f(x) \cos \lambda x]$.

7. Prove Leibnitz's rule by mathematical induction.

8. If $f'(x) = 0$ for $a \leqq x \leqq b$, show that $f(x) =$ constant for $a \leqq x \leqq b$.

9. Show that (10.17) can be written as $f(b) - f(a) = (b - a)f'(a + \theta(b - a))$, $0 < \theta < 1$.

10.14. Functions of Two or More Variables. We say that $f(x, y)$ is continuous at $x = x_0$, $y = y_0$ if for any $\epsilon > 0$ there exists a $\delta > 0$ such that

$$|f(x, y) - f(x_0, y_0)| < \epsilon$$

whenever $|x - x_0|^2 + |y - y_0|^2 < \delta$. In general $f(x_1, x_2, \ldots, x_n)$ is continuous at $(x_1^0, x_2^0, \ldots, x_n^0)$ if for any $\epsilon > 0$ there exists a $\delta > 0$ such that

$$|f(x_1, x_2, \ldots, x_n) - f(x_1^0, x_2^0, \ldots, x_n^0)| < \epsilon$$

whenever $\sum_{i=1}^{n} |x_i - x_i^0|^2 < \delta$.

Theorems 10.20 and 10.21 can be shown to be true for a continuous function of n variables, n finite.

An example of a function which is continuous in x and is continuous in y but is not continuous in both x and y is as follows: Define

$$f(x, y) = \frac{xy}{x^2 + y^2} \qquad x^2 + y^2 \neq 0$$
$$f(0, 0) = 0$$

Then $\lim_{x\to 0} f(x, 0) = \lim_{x\to 0} 0 = 0 = f(0, 0)$, so that $f(x, y)$ is continuous in x at (0, 0). The same statement applies to the variable y. However,

$$\lim_{\substack{x\to 0\\ y\to 0}} f(x, y) \neq f(0, 0)$$

since $\lim_{\substack{x\to 0\\ y\to 0}} f(x, y)$ does not exist. This can be seen by letting $y = mx$, $m \neq 0$, and noting that $\lim_{\substack{x\to 0\\ y\to 0}} f(x, y) = \lim_{x\to 0} \frac{mx^2}{x^2 + m^2x^2} = \frac{m}{1 + m^2}$. Since m is arbitrary, $\lim_{\substack{x\to 0\\ y\to 0}} f(x, y)$ does not exist.

One defines the partial derivatives of $f(x, y)$ at the point (x_0, y_0) as follows:

$$\begin{aligned} f_1(x_0, y_0) &= \left.\frac{\partial f}{\partial x}\right|_{\substack{x=x_0\\ y=y_0}} = \lim_{\Delta x\to 0} \frac{f(x_0 + \Delta x, y_0) - f(x_0, y_0)}{\Delta x} \\ f_2(x_0, y_0) &= \left.\frac{\partial f}{\partial y}\right|_{\substack{x=x_0\\ y=y_0}} = \lim_{\Delta y\to 0} \frac{f(x_0, y_0 + \Delta y) - f(x_0, y_0)}{\Delta y} \end{aligned} \tag{10.24}$$

provided these limits exist. Further derivatives can be computed, and we write

$$f_{11} = \frac{\partial^2 f}{\partial x^2} \qquad f_{12} = \frac{\partial^2 f}{\partial x\, \partial y} \qquad f_{21} = \frac{\partial^2 f}{\partial y\, \partial x} \qquad f_{22} = \frac{\partial^2 f}{\partial y^2}$$

Example 10.11. Consider

$$f(x, y) = \frac{x^3 y}{x^2 + y^2} \qquad x^2 + y^2 \neq 0$$
$$f(0, 0) = 0$$

Then

$$f_1(0, 0) = \left.\frac{\partial f}{\partial x}\right|_{0,0} = \lim_{h\to 0} \frac{f(h, 0) - f(0, 0)}{h} = \lim_{h\to 0} \frac{0}{h} = 0$$

$$f_1(0, y) = \left.\frac{\partial f}{\partial x}\right|_{0,y} = \lim_{h\to 0} \frac{f(h, y) - f(0, y)}{h} = \lim_{h\to 0} \frac{h^2 y}{h^2 + y^2} = 0$$

$$f_{21}(0, 0) = \left.\frac{\partial^2 f}{\partial y\, \partial x}\right|_{0,0} = \lim_{h\to 0} \frac{f_1(0, h) - f_1(0, 0)}{h} = \lim_{h\to 0} \frac{0}{h} = 0$$

$$f_2(0, 0) = \left.\frac{\partial f}{\partial y}\right|_{0,0} = \lim_{h\to 0} \frac{f(0, h) - f(0, 0)}{h} = \lim_{h\to 0} \frac{0}{h} = 0$$

$$f_2(x, 0) = \left.\frac{\partial f}{\partial y}\right|_{x,0} = \lim_{h\to 0} \frac{f(x, h) - f(x, 0)}{h} = \lim_{h\to 0} \frac{x^3}{x^2 + h^2} = x$$

$$f_{12}(0, 0) = \left.\frac{\partial^2 f}{\partial x\, \partial y}\right|_{0,0} = \lim_{h\to 0} \frac{f_2(h, 0) - f_2(0, 0)}{h} = \lim_{h\to 0} \frac{h}{h} = 1$$

In this example, $\left.\frac{\partial^2 f}{\partial y\,\partial x}\right|_{0,0} \neq \left.\frac{\partial^2 f}{\partial x\,\partial y}\right|_{0,0}$. We can show, however, that, if f_{12} and f_{21}, are continuous, then $f_{12} = f_{21}$.

Proof. Define

$$\begin{aligned} U &= f(x + \Delta x, y + \Delta y) - f(x, y + \Delta y) - f(x + \Delta x, y) + f(x, y) \\ \Phi(y) &= f(x + \Delta x, y) - f(x, y) \\ \Phi(y + \Delta y) &= f(x + \Delta x, y + \Delta y) - f(x, y + \Delta y) \end{aligned}$$

so that

$$U = \Phi(y + \Delta y) - \Phi(y)$$

If we apply the law of the mean to U, we have

$$\begin{aligned} U &= \Delta y \frac{\partial \Phi(y + \theta_1 \Delta y)}{\partial y} \qquad 0 < \theta_1 < 1 \\ &= \Delta y \left[\frac{\partial f(x + \Delta x, y + \theta_1 \Delta y)}{\partial y} - \frac{\partial f}{\partial y}(x, y + \theta_1 \Delta y)\right] \end{aligned}$$

Again applying the law of the mean to the variable x ($y + \theta_1 \Delta y$ can be considered as a constant as far as x is concerned) yields

$$U = \Delta y\, \Delta x \frac{\partial^2 f(x + \theta_2 \Delta x, y + \theta_1 \Delta y)}{\partial x\, \partial y} \qquad \begin{matrix} 0 < \theta_1 < 1 \\ 0 < \theta_2 < 1 \end{matrix}$$

By interchanging the role of x and y one easily shows that

$$U = \Delta x\, \Delta y \frac{\partial^2 f(x + \theta_3 \Delta x, y + \theta_4 \Delta y)}{\partial y\, \partial x} \qquad \begin{matrix} 0 < \theta_3 < 1 \\ 0 < \theta_4 < 1 \end{matrix}$$

Equating these two values of U, dividing by $\Delta y \cdot \Delta x$, and finally allowing $\Delta x \to 0$, $\Delta y \to 0$, we obtain

$$\frac{\partial^2 f(x, y)}{\partial y\, \partial x} = \frac{\partial^2 f(x, y)}{\partial x\, \partial y}$$

provided these partial derivatives are continuous.

Example 10.12. *Total Differentials.* Let $z = f(x, y)$ have continuous first partial derivatives. Now

$$\begin{aligned} z + \Delta z &= f(x + \Delta x, y + \Delta y) \\ \Delta z &= f(x + \Delta x, y + \Delta y) - f(x, y) \\ &= [f(x + \Delta x, y + \Delta y) - f(x, y + \Delta y)] + [f(x, y + \Delta y) - f(x, y)] \end{aligned}$$

If we apply the law of the mean to both terms above, we obtain

$$\Delta z = \frac{\partial f(x + \theta_1 \Delta x, y + \Delta y)}{\partial x} \Delta x + \frac{\partial f(x, y + \theta_2 \Delta y)}{\partial y} \Delta y$$

with $0 < \theta_1 < 1$, $0 < \theta_2 < 1$. Under the assumption of continuity of the first partial derivatives we have

$$\begin{aligned} \frac{\partial f(x + \theta_1 \Delta x, y + \Delta y)}{\partial x} &= \frac{\partial f(x, y)}{\partial x} + \epsilon_1 \\ \frac{\partial f(x, y + \theta_2 \Delta y)}{\partial y} &= \frac{\partial f(x, y)}{\partial y} + \epsilon_2 \end{aligned}$$

where $\epsilon_1 \to 0$, $\epsilon_2 \to 0$ as $\Delta x \to 0$, $\Delta y \to 0$. Hence Δz becomes

$$\Delta z = \frac{\partial f(x, y)}{\partial x} \Delta x + \frac{\partial f(x, y)}{\partial y} \Delta y + \epsilon_1 \Delta x + \epsilon_2 \Delta y \tag{10.25}$$

The principal part of Δz is defined to be

$$dz = \frac{\partial f}{\partial x}\Delta x + \frac{\partial f}{\partial y}\Delta y$$

If $z = f(x, y) \equiv x$, $dz = dx$, $\frac{\partial f}{\partial x} = 1$, $\frac{\partial f}{\partial y} = 0$, and $\Delta x = dx$. Similarly $\Delta y = dy$, so that

$$dz = \frac{\partial f}{\partial x} dx + \frac{\partial f}{\partial y} dy = \frac{\partial z}{\partial x} dx + \frac{\partial z}{\partial y} dy \tag{10.26}$$

where dz is called the total differential of z. If x and y are functions of t, then from (10.25)

$$\frac{\Delta z}{\Delta t} = \frac{\partial f(x, y)}{\partial x}\frac{\Delta x}{\Delta t} + \frac{\partial f(x, y)}{\partial y}\frac{\Delta y}{\Delta t} + \epsilon_1 \frac{\Delta x}{\Delta t} + \epsilon_2 \frac{\Delta y}{\Delta t}$$

If $\frac{dx}{dt}$ and $\frac{dy}{dt}$ exist, we obtain

$$\frac{dz}{dt} = \frac{\partial f}{\partial x}\frac{dx}{dt} + \frac{\partial f}{\partial y}\frac{dy}{dt} \tag{10.27}$$

since $\epsilon_1 \to 0$, $\epsilon_2 \to 0$ as $\Delta t \to 0$. Remember that if $\Delta t \to 0$ of necessity $\Delta x \to 0$, $\Delta y \to 0$.

In the general case, if $u = f(x^1, x^2, \ldots, x^n)$, the total differential of u is defined as

$$du = \sum_{\alpha=1}^{n} \frac{\partial f}{\partial x^\alpha} dx^\alpha = \frac{\partial f}{\partial x^\alpha} dx^\alpha \tag{10.28}$$

The operator form of (10.26) is

$$d[\quad] = dx \frac{\partial}{\partial x}[\quad] + dy \frac{\partial}{\partial y}[\quad] \tag{10.29}$$

where the bracket can represent any differentiable function of x and y. In particular,

$$\begin{aligned} d\left(\frac{\partial f}{\partial x}\right) &= \frac{\partial}{\partial x}\left(\frac{\partial f}{\partial x}\right) dx + \frac{\partial}{\partial y}\left(\frac{\partial f}{\partial x}\right) dy \\ &= \frac{\partial^2 f}{\partial x^2} dx + \frac{\partial^2 f}{\partial y\, \partial x} dy \\ d\left(\frac{\partial f}{\partial y}\right) &= \frac{\partial}{\partial x}\left(\frac{\partial f}{\partial y}\right) dx + \frac{\partial}{\partial y}\left(\frac{\partial f}{\partial y}\right) dy \\ &= \frac{\partial^2 f}{\partial x\, \partial y} dx + \frac{d^2 f}{\partial y^2} dy \end{aligned}$$

and

$$d^2 z = d(dz) = \frac{\partial^2 f}{\partial x^2} dx + 2\frac{\partial^2 f}{\partial x\, \partial y} dx\, dy + \frac{\partial^2 f}{\partial y^2} dy^2 \tag{10.30}$$

provided $\frac{\partial^2 f}{\partial x\, \partial y} = \frac{\partial^2 f}{\partial y\, \partial x}$. Symbolically, we can write (10.30) as

$$d^2 z = \left[\frac{\partial z}{\partial x} dx + \frac{\partial z}{\partial y} dy\right]^{(2)}$$

In general

$$d^n z = \left[\frac{\partial z}{\partial x} dx + \frac{\partial z}{\partial y} dy\right]^{(n)} \tag{10.31}$$

Example 10.13. *Change of Coordinates.* Let $z = f(x, y)$, and consider the change of coordinates given by $x = x(u, v)$, $y = y(u, v)$, so that z depends implicitly on u and v. From (10.25)

$$\frac{\Delta z}{\Delta u} = \frac{\partial f(x, y)}{\partial x}\frac{\Delta x}{\Delta u} + \frac{\partial f(x, y)}{\partial y}\frac{\Delta y}{\Delta u} + \epsilon_1 \frac{\Delta x}{\Delta u} + \epsilon_2 \frac{\Delta y}{\Delta u}$$

Allowing Δu to approach zero yields

$$\frac{\partial z}{\partial u} = \frac{\partial f}{\partial x}\frac{\partial x}{\partial u} + \frac{\partial f}{\partial y}\frac{\partial y}{\partial u} \tag{10.32}$$

Similarly,

$$\frac{\partial z}{\partial v} = \frac{\partial f}{\partial x}\frac{\partial x}{\partial v} + \frac{\partial f}{\partial y}\frac{\partial y}{\partial v} \tag{10.33}$$

Symbolically,

$$\frac{\partial[\ \ ]}{\partial r} = \frac{\partial x}{\partial r}\frac{\partial[\ \ ]}{\partial x} + \frac{\partial y}{\partial r}\frac{\partial[\ \ ]}{\partial y} \tag{10.34}$$

where the brackets can represent any function of x and y with x and y depending on any variable r. Of course $\frac{\partial f}{\partial x}$ and $\frac{\partial f}{\partial y}$ must, in general, be continuous. In particular

$$\begin{aligned}\frac{\partial}{\partial u}\left(\frac{\partial z}{\partial x}\right) &= \frac{\partial x}{\partial u}\frac{\partial}{\partial x}\left(\frac{\partial z}{\partial x}\right) + \frac{\partial y}{\partial u}\frac{\partial}{\partial y}\left(\frac{\partial z}{\partial x}\right)\\ &= \frac{\partial^2 z}{\partial x^2}\frac{\partial x}{\partial u} + \frac{\partial^2 z}{\partial y\,\partial x}\frac{\partial y}{\partial u}\\ \frac{\partial}{\partial u}\left(\frac{\partial z}{\partial y}\right) &= \frac{\partial x}{\partial u}\frac{\partial}{\partial x}\left(\frac{\partial z}{\partial y}\right) + \frac{\partial y}{\partial u}\frac{\partial}{\partial y}\left(\frac{\partial z}{\partial y}\right)\\ &= \frac{\partial^2 z}{\partial x\,\partial y}\frac{\partial x}{\partial u} + \frac{\partial^2 z}{\partial y^2}\frac{\partial y}{\partial u}\end{aligned}$$

Example 10.14. Consider $z = z(x, y)$, $x = r\cos\theta$, $y = r\sin\theta$. Then

$$\begin{aligned}\frac{\partial z}{\partial r} &= \frac{\partial z}{\partial x}\frac{\partial x}{\partial r} + \frac{\partial z}{\partial y}\frac{\partial y}{\partial r} = \frac{\partial z}{\partial x}\cos\theta + \frac{\partial z}{\partial y}\sin\theta\\ \frac{\partial^2 z}{\partial r^2} &= \frac{\partial}{\partial r}\left(\frac{\partial z}{\partial r}\right) = \cos\theta\,\frac{\partial}{\partial r}\left(\frac{\partial z}{\partial x}\right) + \sin\theta\,\frac{\partial}{\partial r}\left(\frac{\partial z}{\partial y}\right)\\ &= \cos\theta\left[\frac{\partial^2 z}{\partial x^2}\frac{\partial x}{\partial r} + \frac{\partial^2 z}{\partial y\,\partial x}\frac{\partial y}{\partial r}\right] + \sin\theta\left[\frac{\partial^2 z}{\partial x\,\partial y}\frac{\partial x}{\partial r} + \frac{\partial^2 z}{\partial y^2}\frac{\partial y}{\partial r}\right]\\ &= \cos\theta\left[\frac{\partial^2 z}{\partial x^2}\cos\theta + \frac{\partial^2 z}{\partial y\,\partial x}\sin\theta\right] + \sin\theta\left[\frac{\partial^2 z}{\partial x\,\partial y}\cos\theta + \frac{\partial^2 z}{\partial y^2}\sin\theta\right]\end{aligned}$$

Let the reader show that

$$\begin{aligned}\frac{\partial z}{\partial \theta} &= -r\sin\theta\,\frac{\partial z}{\partial x} + r\cos\theta\,\frac{\partial z}{\partial y}\\ \frac{\partial^2 z}{\partial \theta^2} &= -r\cos\theta\,\frac{\partial z}{\partial x} - r\cos\theta\,\frac{\partial z}{\partial y} - r\sin\theta\left[-\frac{\partial^2 z}{\partial x^2}r\sin\theta + \frac{\partial^2 z}{\partial y\,\partial x}r\cos\theta\right]\\ &\qquad + r\cos\theta\left[-\frac{\partial^2 z}{\partial x\,\partial y}r\sin\theta + \frac{\partial^2 z}{\partial y^2}r\cos\theta\right]\end{aligned}$$

Thus

$$\nabla^2 z \equiv \frac{\partial^2 z}{\partial x^2} + \frac{\partial^2 z}{\partial y^2} = \frac{\partial^2 z}{\partial r^2} + \frac{1}{r^2}\frac{\partial^2 z}{\partial \theta^2} + \frac{1}{r}\frac{\partial z}{\partial r}$$

Problems

1. Let $z = x^2 - y^2$, $x = e^t$, $y = \sin t$. Find $\dfrac{dz}{dt}$ by two methods.

2. Let $u = x^2 - y^2$, $v = x + y$. Find $\dfrac{\partial x}{\partial u}, \dfrac{\partial y}{\partial u}, \dfrac{\partial x}{\partial v}$, and $\dfrac{\partial y}{\partial v}$. *Hint:*

$$\frac{\partial u}{\partial u} = 1 = 2x\frac{\partial x}{\partial u} - 2y\frac{\partial y}{\partial u},$$

$0 = \dfrac{\partial u}{\partial v} = 2x\dfrac{\partial x}{\partial v} - 2y\dfrac{\partial y}{\partial v}$, etc. Why is it true that $\dfrac{\partial u}{\partial v} = 0$?

3. Let $x = r\sin\theta\cos\varphi$, $y = r\sin\theta\sin\varphi$, $z = r\cos\theta$. If $V = V(x, y, z)$, show that

$$\begin{aligned}\nabla^2 V &= \frac{\partial^2 V}{\partial x^2} + \frac{\partial^2 V}{\partial y^2} + \frac{\partial^2 V}{\partial z^2}\\ &= \frac{1}{r^2\sin\theta}\left[\frac{\partial}{\partial r}\left(r^2\sin\theta\frac{\partial V}{\partial r}\right) + \frac{\partial}{\partial\theta}\left(\sin\theta\frac{\partial V}{\partial\theta}\right) + \frac{\partial}{\partial\varphi}\left(\frac{1}{\sin\theta}\frac{\partial V}{\partial\varphi}\right)\right]\end{aligned}$$

4. Let $f(x_1, x_2, \ldots, x_n)$ be homogeneous of degree k, that is,

$$f(tx_1, tx_2, \ldots, tx_n) = t^k f(x_1, x_2, \ldots, x_n)$$

Differentiate with respect to t, set $t = 1$, and prove a result due to Euler,

$$kf = \sum_{\alpha=1}^{n} x_\alpha \frac{\partial f}{\partial x_\alpha}$$

5. Prove (10.31) by mathematical induction.

6. If $f(x_1, x_2, \ldots, x_n)$ is homogeneous of degree k (see Prob. 4), show by mathematical induction that

$$k(k-1)\cdots(k-p+1) = \left(x_1\frac{\partial f}{\partial x_1} + x_2\frac{\partial f}{\partial x_2} + \cdots + x_n\frac{\partial f}{\partial x_n}\right)^{(p)}$$

10.15. Implicit-function Theorems

THEOREM 10.24. Let $x = x_0$, $y = y_0$ satisfy $F(x, y) = 0$, and let us suppose that F, $\dfrac{\partial F}{\partial x}$, and $\dfrac{\partial F}{\partial y}$ are continuous in a neighborhood of $x = x_0$, $y = y_0$, say, for $|x - x_0| < l$, $|y - y_0| < l$. If $\dfrac{\partial F}{\partial y} \neq 0$ at $x = x_0$, $y = y_0$, there exists one and only one continuous function of the independent variable x, say, $y = f(x)$, such that $F(x, f(x)) \equiv 0$ with $y_0 = f(x_0)$. In other words, if x_1 is any number near x_0, there exists a unique y_1 such that $f(x_1, y_1) = 0$. It is in this sense that we speak of y as a function of x.

Proof. Assume $\dfrac{\partial F}{\partial y} > 0$ at $P(x_0, y_0)$. The same reasoning applies to the case $\dfrac{\partial F}{\partial y} < 0$. From continuity of $\dfrac{\partial F(x, y)}{\partial y}$ there exists a neighborhood of $P(x_0, y_0)$ such that $\dfrac{\partial F}{\partial y} > Q > 0$, Q fixed. We can decrease the size of this neighborhood, if necessary, so that it is bounded by $|x - x_0| < l$, $|y - y_0| < l$. In this new neighborhood of $x = x_0$, $y = y_0$, given by $|x - x_0| < l_0$, $|y - y_0| < l_0$, we have F, $\frac{\partial F}{\partial x}$, $\frac{\partial F}{\partial y}$ continuous and $\frac{\partial F}{\partial y} > Q > 0$. Let $R > 0$ be any upper limit of $\left|\frac{\partial F}{\partial x}\right|$ in this neighborhood. For any x and y satisfying $|x - x_0| < l_0$, $|y - y_0| < l_0$ we have

$$\begin{aligned} F(x, y) &= F(x, y) - F(x_0, y_0) = F(x, y) - F(x_0, y) + F(x_0, y) - F(x_0, y_0) \\ &= (x - x_0)\frac{\partial F(x + \theta_1(x - x_0), y)}{\partial x} + (y - y_0)\frac{\partial F(x_0, y_0 + \theta_2(y - y_0))}{\partial y} \end{aligned}$$

Let $y = y_0 + \epsilon$, $0 < \epsilon < l_0$. Then

$$\begin{aligned} F(x, y_0 + \epsilon) &= (x - x_0)\frac{\partial F(x + \theta_1(x - x_0), y_0 + \epsilon)}{\partial x} + \epsilon\frac{\partial F(x_0, y_0 + \epsilon\theta_2)}{\partial y} \\ &> (x - x_0)\frac{\partial F(x + \theta_1(x - x_0), y_0 + \epsilon)}{\partial x} + \epsilon Q \end{aligned}$$

Since $\epsilon Q > 0$, we shall have $F(x, y_0 + \epsilon) > 0$ if $|x - x_0| < \epsilon Q/R$. Similarly $F(x, y_0 - \epsilon) < 0$ for $|x - x_0| < \epsilon Q/R$. From Theorem 10.22 a y exists such that $F(x, y) = 0$. We have shown that, for any x satisfying $|x - x_0| < \epsilon Q/R \leqq l_0$, a number y exists such that $F(x, y) = 0$. We say that y is a function of x and write $y = f(x)$, so that $F(x, f(x)) = 0$.

Next we show that y is a single-valued function of x. Assume that, for any number x_1 satisfying $|x - x_0| < \epsilon Q/R \leqq l_0$, there exist two numbers y_1 and y_2 such that $F(x_1, y_1) = 0$ and $F(x_1, y_2) = 0$. Applying the law of the mean to $F(x_1, y_2) - F(x_1, y_1) \equiv 0$ yields

$$(y_2 - y_1)\frac{\partial F(x_1, y_1 + \theta(y_2 - y_1))}{\partial y} = 0$$

Since $\frac{\partial F}{\partial y} > 0$, of necessity $y_2 = y_1$, so that $y = f(x)$ is single-valued. To show that y is a continuous function of x, we note that if $F(x_1, y_1) = 0$, $F(x_2, y_2) = 0$, then

$$F(x_2, y_2) - F(x_1, y_2) + F(x_1, y_2) - F(x_1, y_1) = 0$$

$$(x_2 - x_1)\frac{\partial F(x_1 + \theta_1(x_2 - x_1), y_2)}{\partial x} + (y_2 - y_1)\frac{\partial F(x_1, y_1 + \theta_2(y_2 - y_1))}{\partial y} = 0$$

Since $\frac{\partial F}{\partial y} \neq 0$ and $\frac{\partial F}{\partial x}$ is bounded, it is obvious that $y_2 \to y_1$ as $x_2 \to x_1$. This is exactly the property of continuity.

Theorem 10.24 is a special case of Theorem 10.25.

THEOREM 10.25. Let $F(x_1, x_2, \ldots, x_n, z)$ satisfy the following conditions:

1. F is continuous at the point $P(a_1, a_2, \ldots, a_n, z_0)$.
2. The first partial derivatives of F exist at P and are continuous.

3. $F(a_1, a_2, \ldots, a_n, z_0) = 0$.

4. $\dfrac{\partial F}{\partial z} \neq 0$ at P.

By the method used in Theorem 10.24 it can be shown that a neighborhood of $P_0(a_1, a_2, \ldots, a_n)$ exists such that $z = f(x_1, x_2, \ldots, x_n)$ and $F(x_1, x_2, \ldots, x_n, f(x_1, x_2, \ldots, x_n)) \equiv 0$ with $z_0 = f(a_1, a_2, \ldots, a_n)$. Moreover z is a continuous function in the variables $x_1, x_2, \ldots, x_n$.

For the case $F(x, y, z) = 0$, $z = f(x, y)$, $F(a, b, c) = 0$ we have

$$F(x, y, z) - F(a, b, c) \equiv 0 \qquad \text{if } z = f(x, y)$$

so that

$$F(x, y, z) - F(a, y, z) + F(a, y, z) - F(a, b, z) + F(a, b, z) - F(a, b, c) = 0$$

Applying the law of the mean to each difference yields

$$(x - a)\,\frac{\partial F(a + \theta_1(x - a), y, z)}{\partial x} + (y - b)\,\frac{\partial F(a, b + \theta_2(y - b), z)}{\partial y} + (z - c)\,\frac{\partial F(a, b, c + \theta_3(z - c))}{\partial z} = 0$$

If $y = b$, we have

$$\lim_{\Delta x \to 0} \frac{\Delta z}{\Delta x} = \frac{\partial z}{\partial x} = \lim_{x \to a} \frac{z - c}{x - a} = -\frac{\partial F/\partial x}{\partial F/\partial z}$$

Similarly

$$\frac{\partial z}{\partial y} = -\frac{\partial F/\partial y}{\partial F/\partial z} \tag{10.35}$$

Example 10.15. Consider $F(x, y, z) = xe^z + yz - y$. At the point $(1, 1, 0)$ we have $F(1, 1, 0) = 0$, and $\dfrac{\partial F}{\partial z} = xe^z + y \neq 0$ at $(1, 1, 0)$. Thus $F(x, y, z) = 0$ defines z as a function of x and y, $z = f(x, y)$, for a neighborhood of $(1, 1)$ such that $0 = f(1, 1)$. To compute $\dfrac{\partial z}{\partial x}$ we use (10.35). Thus $\dfrac{\partial F}{\partial x} = e^z$, $\dfrac{\partial F}{\partial z} = xe^z + y$, so that $\dfrac{\partial z}{\partial x} = -\dfrac{e^z}{xe^z + y}$. Notice that we do not claim to solve for z explicitly in terms of x and y.

THEOREM 10.26. Let $F_1(x, u, v) = 0$, $F_2(x, u, v) = 0$ be two equations involving the variables (x, u, v). Suppose that the following conditions are satisfied:

1. $F_1(x_0, u_0, v_0) = 0$, $F_2(x_0, u_0, v_0) = 0$.

2. F_1 and F_2 have continuous first partial derivatives for a neighborhood of $(x_0; u_0, v_0)$. Of necessity F_1 and F_2 are continuous.

3. The *Jacobian* of F_1 and F_2 relative to u and v, written

$$J\left(\frac{F_1, F_2}{u, v}\right) = \begin{vmatrix} \dfrac{\partial F_1}{\partial u} & \dfrac{\partial F_1}{\partial v} \\ \dfrac{\partial F_2}{\partial u} & \dfrac{\partial F_2}{\partial v} \end{vmatrix}$$

does not vanish at $(x_0; u_0, v_0)$.

Under these conditions there exists one and only one system of continuous functions

$$u = \varphi_1(x) \qquad v = \varphi_2(x)$$

such that $F(x, \varphi_1(x), \varphi_2(x)) \equiv 0, F_2(x, \varphi_1(x), \varphi_2(x)) \equiv 0$ with $u_0 = \varphi_1(x_0)$, $v_0 = \varphi_2(x_0)$.

Proof. If $J \neq 0$, then $\frac{\partial F_2}{\partial u} \neq 0$ or $\frac{\partial F_2}{\partial v} \neq 0$. Without loss of generality we take $\frac{\partial F_2}{\partial v} \neq 0$. Also we note that J is continuous, so that $J \neq 0$ for some neighborhood of $(x_0; u_0, v_0)$, and, moreover, $\frac{\partial F_2}{\partial v} \neq 0$ for a neighborhood of $(x_0; u_0, v_0)$. From Theorem 10.25 we can solve for v as a unique function of x and u, $v = f(x, u)$, such that $F_2(x, u, f(x, u)) \equiv 0$. Hence

$$\frac{\partial F_2}{\partial u} + \frac{\partial F_2}{\partial v}\frac{\partial f}{\partial u} = 0 \qquad \frac{\partial f}{\partial u} = -\frac{\partial F_2/\partial u}{\partial F_2/\partial v}$$

by considering F_2 as a function of x and u. From $F_1(x, u, v)$, $v = f(x, u)$ we note that F_1 is a function of x and u, so that

$$\begin{aligned}
\left.\frac{\partial F_1}{\partial u}\right|_{x=\text{constant}} &= \left.\frac{\partial F_1}{\partial u}\right|_{\substack{x=\text{constant}\\ v=\text{constant}}} + \left.\frac{\partial F_1}{\partial v}\right|_{\substack{x=\text{constant}\\ u=\text{constant}}} \left.\frac{\partial f}{\partial u}\right|_{x=\text{constant}} \\
&= \left.\frac{\partial F_1}{\partial u}\right|_{\substack{x=\text{constant}\\ v=\text{constant}}} + \left.\frac{\partial F_1}{\partial v}\right|_{\substack{x=\text{constant}\\ u=\text{constant}}} \left[-\frac{\partial F_2/\partial u}{\partial F_2/\partial v}\right] \\
&= \frac{J\left(\dfrac{F_1, F_2}{u, v}\right)}{\partial F_2/\partial v}
\end{aligned}$$

Since $J \neq 0$, we have $\frac{\partial F_1(x, u, f(x, u))}{\partial u} \neq 0$ so that from Theorem 10.24 we can solve $F_1(x, u, f(x, u)) = 0$ for $u = \varphi_1(x)$, $\varphi_1(x)$ unique. Hence $v = f(x, u) = f(x, \varphi_1(x)) = \varphi_2(x)$. Q.E.D. Theorem 10.26 holds if we consider $F_1(x_1, x_2, \ldots, x_n, u, v) = 0$, $F_2(x_1, x_2, \ldots, x_n, u, v) = 0$. The proof proceeds in the same manner, making use of Theorem 10.25.

Example 10.16. If $x = f(u, v)$, $y = \varphi(u, v)$, then $F_1(x, y, u, v) \equiv f(u, v) - x = 0$, $F_2(x, y, u, v) \equiv \varphi(u, v) - y = 0$. We can solve for u and v as functions of x and y provided

$$J\left(\frac{F_1, F_2}{u, v}\right) = \begin{vmatrix} \frac{\partial F_1}{\partial u} & \frac{\partial F_1}{\partial v} \\ \frac{\partial F_2}{\partial u} & \frac{\partial F_2}{\partial v} \end{vmatrix} = \begin{vmatrix} \frac{\partial f}{\partial u} & \frac{\partial f}{\partial v} \\ \frac{\partial \varphi}{\partial u} & \frac{\partial \varphi}{\partial v} \end{vmatrix} = \begin{vmatrix} \frac{\partial x}{\partial u} & \frac{\partial x}{\partial v} \\ \frac{\partial y}{\partial u} & \frac{\partial y}{\partial v} \end{vmatrix} \neq 0 \qquad (10.36)$$

along with the condition that the first partials of x and y be continuous. If $x = r \cos \theta$,

$y = r \sin \theta$, then

$$J\left(\frac{x, y}{r, \theta}\right) = \begin{vmatrix} \frac{\partial x}{\partial r} & \frac{\partial x}{\partial \theta} \\ \frac{\partial y}{\partial r} & \frac{\partial y}{\partial \theta} \end{vmatrix} = \begin{vmatrix} \cos \theta & -r \sin \theta \\ \sin \theta & r \cos \theta \end{vmatrix} = r$$

For $r \neq 0$ we can solve for r and θ uniquely in terms of x and y in the neighborhood of any point $x_0 = r_0 \cos \theta_0$, $y_0 = r_0 \sin \theta_0$. Indeed,

$$r = (x^2 + y^2)^{\frac{1}{2}} \qquad \theta = \tan^{-1} \frac{y}{x} \qquad -\pi \leqq \theta < \pi$$

By mathematical induction one can extend the results of Theorem 10.26. We state Theorem 10.27 but give no proof.

THEOREM 10.27. Let $f_i = f_i(x_1, x_2, \ldots, x_m, u_1, u_2, \ldots, u_n)$, $i = 1, 2, \ldots, n$ satisfy the following requirements:

1. Each f_i, $i = 1, 2, \ldots, n$ has continuous first partial derivatives at the point $P(a_1, a_2, \ldots, a_m, b_1, b_2, \ldots, b_n)$.
2. $f_i = 0$ at P, $i = 1, 2, \ldots, n$.
3. $J\left(\frac{f_1, f_2, \ldots, f_n}{u_1, u_2, \ldots, u_n}\right) = \left|\frac{\partial f_i}{\partial u_j}\right| \neq 0$ at P.

There exists one and only one system of continuous functions,

$$u_i = \varphi_i(x_1, x_2, \ldots, x_m) \qquad i = 1, 2, \ldots, n$$

which satisfy (2) such that $\varphi_i(a_1, a_2, \ldots, a_m) = b_i$, $i = 1, 2, \ldots, n$.

Example 10.17. For the coordinate transformation

$$y_i = f_i(x_1, x_2, \ldots, x_n) \qquad i = 1, 2, \ldots, n$$

we note that $F_i \equiv f_i(x_1, x_2, \ldots, x_n) - y_i = 0$, so that we can solve for x_i as a function of $y_1, y_2, \ldots, y_n$ provided

$$\left|\frac{\partial F_i}{\partial x_j}\right| = \left|\frac{\partial f_i}{\partial x_j}\right| = \left|\frac{\partial y_i}{\partial x_j}\right| \neq 0$$

assuming that the first partials are continuous.

For the system of linear equations $y^i = a^i_\alpha x^\alpha$, $i = 1, 2, \ldots, n$, we have

$$\left|\frac{\partial f_i}{\partial x^j}\right| = |a^i_j|$$

so that we can solve for x^i, $i = 1, 2, \ldots, n$, as a function of $y^1, y^2, \ldots, y^n$, provided $|a^i_j| \neq 0$ (see Sec. 1.2).

THEOREM 10.28. Let $u_i = f_i(x_1, x_2, \ldots, x_n)$, $i = 1, 2, \ldots, n$, have continuous first partial derivatives at $P(x_1^0, x_2^0, \ldots, x_n^0)$. A necessary and sufficient condition that a functional relationship of the form $F(u_1, u_2, \ldots, u_n) \equiv 0$ exist is that

$$J\left(\frac{u_1, u_2, \ldots, u_n}{x_1, x_2, \ldots, x_n}\right) \equiv 0 \tag{10.37}$$

Proof. First assume that $F(u_1, u_2, \ldots, u_n) \equiv 0$. Then

$$\frac{\partial F}{\partial x^j} = \sum_{\alpha=1}^{n} \frac{\partial F}{\partial u_\alpha} \frac{\partial u_\alpha}{\partial x_j} \equiv 0 \qquad j = 1, 2, \ldots, n \tag{10.38}$$

Since $\sum_{\alpha=1}^{n} \frac{\partial F}{\partial u_\alpha} \frac{\partial u_\alpha}{\partial x_j} = 0$ can be looked upon as a linear system of n equations in the n quantities, $\frac{\partial F}{\partial u_\alpha}$, $\alpha = 1, 2, \ldots, n$, with $\frac{\partial F}{\partial u_\alpha} \neq 0$ for at least one α of necessity,

$$\left| \frac{\partial u_\alpha}{\partial x_j} \right| = 0$$

(see Sec. 1.4). Hence (10.39) has been shown to be a necessary condition for the existence of a functional relationship involving the u_i, $i = 1, 2, \ldots, n$. Why is $\frac{\partial F}{\partial u_\alpha} \neq 0$ for at least one α?

Conversely, let us assume that (10.39) holds, so that

$$\begin{vmatrix} \frac{\partial f_1}{\partial x_1} & \frac{\partial f_1}{\partial x_2} & \cdots & \frac{\partial f_1}{\partial x_n} \\ \frac{\partial f_2}{\partial x_1} & \frac{\partial f_2}{\partial x_2} & \cdots & \frac{\partial f_2}{\partial x_n} \\ \cdots & \cdots & \cdots & \cdots \\ \frac{\partial f_n}{\partial x_1} & \frac{\partial f_n}{\partial x_2} & \cdots & \frac{\partial f_n}{\partial x_n} \end{vmatrix} = 0 \tag{10.39}$$

For the sake of simplicity we assume that the minor of $\frac{\partial f_n}{\partial x_n}$ does not vanish, so that

$$\begin{vmatrix} \frac{\partial f_1}{\partial x_1} & \frac{\partial f_1}{\partial x_2} & \cdots & \frac{\partial f_1}{\partial x_{n-1}} \\ \frac{\partial f_2}{\partial x_1} & \frac{\partial f_2}{\partial x_2} & \cdots & \frac{\partial f_2}{\partial x_{n-1}} \\ \cdots & \cdots & \cdots & \cdots \\ \frac{\partial f_{n-1}}{\partial x_1} & \frac{\partial f_{n-1}}{\partial x_2} & \cdots & \frac{\partial f_{n-1}}{\partial x_{n-1}} \end{vmatrix} \neq 0 \tag{10.40}$$

Now let

$$\begin{aligned} y_1 &= u_1 = f_1(x_1, x_2, \ldots, x_n) \\ y_2 &= u_2 = f_2(x_1, x_2, \ldots, x_n) \\ &\cdots\cdots\cdots\cdots\cdots\cdots \\ y_{n-1} &= u_{n-1} = f_{n-1}(x_1, x_2, \ldots, x_n) \\ y_n &= x_n \end{aligned} \tag{10.41}$$

From (10.40) and (10.41) we have

$$J\left(\frac{y_1, y_2, \ldots, y_n}{x_1, x_2, \ldots, x_n}\right) = \begin{vmatrix} \frac{\partial f_1}{\partial x_1} & \frac{\partial f_1}{\partial x_2} & \cdots & \frac{\partial f_1}{\partial x_n} \\ \frac{\partial f_2}{\partial x_1} & \frac{\partial f_2}{\partial x_2} & \cdots & \frac{\partial f_2}{\partial x_n} \\ \cdots & \cdots & \cdots & \cdots \\ \frac{\partial f_{n-1}}{\partial x_1} & \frac{\partial f_{n-1}}{\partial x_2} & \cdots & \frac{\partial f_{n-1}}{\partial x_n} \\ 0 & 0 & & 1 \end{vmatrix} \neq 0$$

From Theorem 10.27 we can solve for the x_i, $i = 1, 2, \ldots, n$, as functions of the y_i, $i = 1, 2, \ldots, n$, so that

$$x_i = \varphi_i(y_1, y_2, \ldots, y_n) \qquad i = 1, 2, \ldots, n \tag{10.42}$$

and $$u_n = f_n(x_1, x_2, \ldots, x_n) = F(y_1, y_2, \ldots, y_n)$$

along with $u_1 = y_1$, $u_2 = y_2$, $\ldots$, $u_{n-1} = y_{n-1}$ [see (10.41)]. From

$$J\left(\frac{u_1, u_2, \ldots, u_n}{y_1, y_2, \ldots, y_n}\right) = J\left(\frac{u_1, u_2, \ldots, u_n}{x_1, x_2, \ldots, x_n}\right) \cdot J\left(\frac{x_1, x_2, \ldots, x_n}{y_1, y_2, \ldots, y_n}\right) = 0$$

(see Sec. 1.2) we have

$$\begin{vmatrix} 1 & 0 & 0 & \cdots & 0 \\ 0 & 1 & 0 & \cdots & 0 \\ \cdots & \cdots & \cdots & \cdots & \cdots \\ 0 & 0 & 0 & \cdots & 0 \\ \frac{\partial F}{\partial y_1} & \frac{\partial F}{\partial y_2} & \frac{\partial F}{\partial y_3} & \cdots & \frac{\partial F}{\partial y_n} \end{vmatrix} = 0$$

so that $\dfrac{\partial F}{\partial y_n} = 0$ and $u_n = F(y_1, y_2, \ldots, y_{n-1}) = F(u_1, u_2, \ldots, u_{n-1})$. Thus $G(u_1, u_2, \ldots, u_n) = F(u_1, u_2, \ldots, u_{n-1}) - u_n \equiv 0$. G is the functional relationship predicted by the theorem. Q.E.D.

Example 10.18. Consider $u = x + y + z$, $v = x - y - z$, $w = 2x$. Then

$$J\left(\frac{u, v, w}{x, y, z}\right) = \begin{vmatrix} 1 & 1 & 1 \\ 1 & -1 & -1 \\ 2 & 0 & 0 \end{vmatrix} = 0$$

It is obvious that $G(u, v, w) = u + v - w \equiv 0$.

Problems

1. Let $x = r \sin\theta \cos\varphi$, $y = r \sin\theta \sin\varphi$, $z = r\cos\theta$. Show that

$$J\left(\frac{x, y, z}{r, \theta, \varphi}\right) = r^2 \sin\theta.$$

2. In Prob. 1 solve r, θ, φ in terms of x, y, z.

3. Let $u = xyz$, $v = xy + yz + zx$, $w = x + y + z$. Show that

$$J\left(\frac{u,\ v,\ w}{x,\ y,\ z}\right) = (x - y)(y - z)(z - x)$$

4. For $J \neq 0$ in Prob. 3 solve for $\dfrac{\partial x}{\partial u}$. *Hint:* Assuming that we can solve for x, y, z in terms of u, v, w since $J \neq 0$, we have, upon implicit differentiation with respect to u,

$$\begin{aligned}
\frac{\partial u}{\partial u} &= 1 = yz\frac{\partial x}{\partial u} + xz\frac{\partial y}{\partial u} + xy\frac{\partial z}{\partial u} \\
\frac{\partial v}{\partial u} &= 0 = (y + z)\frac{\partial x}{\partial u} + (x + z)\frac{\partial y}{\partial u} + (y + x)\frac{\partial z}{\partial u} \\
\frac{\partial w}{\partial u} &= 0 = \frac{\partial x}{\partial u} + \frac{\partial y}{\partial u} + \frac{\partial z}{\partial u}
\end{aligned}$$

Solve this linear system for $\dfrac{\partial x}{\partial u}$.

5. Let $u = x + y + z$, $v = xy + yz + zx$, $w = x^2 + y^2 + z^2$. Show that

$$J\left(\frac{u,\ v,\ w}{x,\ y,\ z}\right) = 0$$

Find a functional relationship between u, v, w.

6. Consider a function of $(x^1, x^2, x^3, \ldots, x^n)$, say, $f(x^1, x^2, \ldots, x^n)$. We define the *Hessian* of F by

$$H = \left|\frac{\partial^2 f}{\partial x^i\,\partial x^j}\right|$$

Under the coordinate transformation $x^\alpha = a^\alpha_\beta y^\beta$ the function f becomes $F(y^1, y^2, \ldots, y^n)$. Show that

$$\left|\frac{\partial^2 F}{\partial y^i\,\partial y^j}\right| = \left|\frac{\partial^2 f}{\partial x^i\,\partial x^j}\right|\Delta^2 \qquad \Delta = |a^i_j|$$

10.16. The Riemann Integral. Let us consider any bounded function $f(x)$ defined over the range $a \leqq x \leqq b$. $f(x)$ need not be continuous, but we desire $|f(x)| < A$. We subdivide the range (a, b) into n arbitrary subdivisions,

$$a = x_0 < x_1 < x_2 < \cdots < x_{i-1} < x_i < x_{i+1} < \cdots < x_n = b$$

and form the sum

$$S_n = \sum_{i=1}^{n} M_i(x_i - x_{i-1}) \tag{10.43}$$

where M_i is the supremum of $f(x)$ for the interval $x_{i-1} \leqq x \leqq x_i$. The sum S_n obviously will depend on the manner in which we subdivide the interval (a, b). The infemum of all such sums obtained in this manner is called the *upper Darboux* integral of $f(x)$, written

$$\bar{S} = \overline{\int_a^b} f(x)\,dx \tag{10.44}$$

Since $S_n \geqq -A(b-a)$, the infemum $\bar{S}$ exists for all bounded functions. Similarly, we can form the sum

$$s_n = \sum_{i=1}^{n} m_i(x_i - x_{i-1}) \tag{10.45}$$

with m_i the infemum of $f(x)$ for the interval $x_{i-1} \leqq x \leqq x_{i-1}$. The supremum of all such sums obtained in this manner is called the *lower Darboux* integral of $f(x)$, written

$$\underline{S} = \underline{\int_a^b} f(x)\, dx \tag{10.46}$$

The reader can easily verify that $\underline{S} \leqq \bar{S}$.

If $\bar{S} = \underline{S}$, we say that $f(x)$ is Riemann-integrable (R-integrable) and we write

$$I = \int_a^b f(x)\, dx \tag{10.47}$$

We now investigate some conditions for which $f(x)$ is R-integrable.

1. Let $f(x) \equiv \text{constant} = c$. Then

$$S_n = \sum_{i=1}^{n} c(x_i - x_{i-1}) = c \sum_{i=1}^{n} (x_i - x_{i-1}) = c(b-a)$$
$$s_n = c(b-a)$$

so that $\bar{S} = \underline{S} = c(b-a) = \int_a^b c\, dx$.

2. Let $f(x) = x$ for $0 \leqq x \leqq 1$. We consider the subdivision $0 < 1/n < 2/n < \cdots < i/n < \cdots < 1$. Then

$$S_n = \sum_{i=1}^{n} \frac{i}{n}\left(\frac{i}{n} - \frac{i-1}{n}\right) = \frac{1}{n^2}\sum_{i=1}^{n} i = \frac{n(n+1)}{2n^2} = \frac{1}{2}\left(1 + \frac{1}{n}\right)$$

The infemum of all sums obtained by equally spaced divisions is seen to be $\frac{1}{2}$. Similarly

$$s_n = \sum_{i=1}^{n} \frac{i-1}{n}\left(\frac{i}{n} - \frac{i-1}{n}\right) = \frac{1}{n^2}\sum_{i=1}^{n} (i-1) = \frac{1}{2}\left(1 - \frac{1}{n}\right)$$

The supremum of all such sums is $\frac{1}{2}$. Let the reader prove that $\bar{S} = \underline{S} = \frac{1}{2}$ for all manners of subdivisions (not necessarily equally spaced), so that $\int_0^1 x\, dx = \frac{1}{2}$.

3. *A continuous function is always R-integrable for the finite range* (a, b).

Proof. Since $f(x)$ is continuous, it is uniformly continuous for

$$a \leqq x \leqq b$$

Hence for any $\epsilon > 0$ there exists a finite subdivision of $a \leqq x \leqq b$ such that the difference between the supremum and infemum of $f(x)$ on any interval of the subdivision is less than $\epsilon/(b - a)$. This yields

$$|S_n - s_n| < (b - a)\frac{\epsilon}{b - a} = \epsilon$$

If necessary we subdivide further so that $|\bar{S} - S_n| < \epsilon$, $|\underset{\sim}{S} - s_n| < \epsilon$. Why can this be done? One must recall the definitions and properties of the supremum and infemum. This is left for the reader. Hence $|\bar{S} - \underset{\sim}{S}| < 3\epsilon$. Since ϵ can be chosen arbitrarily small, of necessity $\bar{S} - \underset{\sim}{S} = 0$, $\bar{S} = \underset{\sim}{S}$. Q.E.D. It must be remembered that $\bar{S}$ and $\underset{\sim}{S}$ are fixed numbers.

4. Any bounded monotonic increasing or decreasing function is R-integrable (see Prob. 10).

5. An example of a function which is not R-integrable is as follows: $f(x) = 1$ for x irrational, $f(x) = 0$ for x rational, $0 \leqq x \leqq 1$. Let the reader show that $\bar{S} = 1$, $\underset{\sim}{S} = 0$.

We list some properties of the Riemann integral. It is assumed that the integrals under discussion exist.

(1) $\int_a^b cf(x)\,dx = c\int_a^b f(x)\,dx$

(2) $\int_a^b f(x)\,dx = -\int_b^a f(x)\,dx$

(3) $\int_a^b f(x)\,dx = \int_a^c f(x)\,dx + \int_c^b f(x)\,dx$

(4) If $f(x) \geqq 0$, then $\int_a^b f(x)\,dx \geqq 0$ for $b \geqq a$.

(5) $\int_a^b [f(x) + \varphi(x)]\,dx = \int_a^b f(x)\,dx + \int_a^b \varphi(x)\,dx$

(6) $\int_a^b |f(x)|\,dx \geqq \int_a^b f(x)\,dx,\ b \geqq a$

(7) $\left[\int_a^b \varphi(x)f(x)\,dx\right]^2 \leqq \int_a^b \varphi^2(x)\,dx \int_a^b f^2(x)\,dx$

Proof.

$$\int_a^b [\lambda\varphi(x) + f(x)]^2\,dx = \lambda^2 \int_a^b \varphi^2(x)\,dx + 2\lambda\int_a^b \varphi(x)f(x)\,dx + \int_a^b f^2(x)\,dx \geqq 0$$

for $b > a$. If $y = A\lambda^2 + 2B\lambda + C \geqq 0$ for all real values of λ, then $A\lambda^2 + 2B\lambda + C = 0$ has either two equal real roots or no real roots.

Of necessity $B^2 - 4AC \leqq 0$, which yields

$$\left[\int_a^b \varphi(x)f(x)\,dx\right]^2 \leqq \int_a^b \varphi^2(x)\,dx \int_a^b f^2(x)\,dx \tag{10.48}$$

(8) $\left|\int_a^b f(x)\,dx\right| \leqq M|b - a|$, M = supremum of $f(x)$ for $a \leqq x \leqq b$

(9) *The First Theorem of the Mean for Integrals.* If $f(x)$ is continuous on the range $a \leqq x \leqq b$, then

$$\int_a^b f(x)\,dx = (b - a)f(\xi) \qquad a \leqq \xi \leqq b$$

The proof of (9) is left as an exercise for the reader (see Prob. 8).

(10) $\int_a^a f(x)\,dx = 0$

We should like to emphasize that, if $f(x)$ is R-integrable on the range (a, b), we can write

$$I = \int_a^b f(x)\,dx = \int_a^b f(t)\,dt$$

since the value of the integral does not depend on the variable which is used to describe $f(x)$. Of course I depends on the upper and lower limits, b and a, respectively. Moreover I does depend on the form of $f(x)$. Once $f(x)$ is chosen, however, we need not use x as the variable of integration. If $f(x)$ is R-integrable for the range (a, b), then $\int_a^x f(t)\,dt$ exists for $a \leqq x \leqq b$. Each value of x determines a value of $\int_a^x f(t)\,dt$, so that a single-valued function, $F(x)$, can be defined by

$$F(x) = \int_a^x f(t)\,dt \tag{10.49}$$

We could write $F(x) = \int_a^x f(x)\,dx$, but confusion is avoided if we use (10.49). From (10.49) we have

$$F(x + \Delta x) = \int_a^{x+\Delta x} f(t)\,dt$$

and

$$\begin{aligned} F(x + \Delta x) - F(x) &= \int_a^{x+\Delta x} f(t)\,dt - \int_a^x f(t)\,dt \\ &= \int_x^a f(t)\,dt + \int_a^{x+\Delta x} f(t)\,dt = \int_x^{x+\Delta x} f(t)\,dt \end{aligned} \tag{10.50}$$

Since $f(x)$ is bounded for $a \leqq x \leqq b$, we have

$$|F(x + \Delta x) - F(x)| \leqq A\,\Delta x \tag{10.51}$$

Hence $F(x)$ is continuous since $F(x + \Delta x) \to F(x)$ as $\Delta x \to 0$.

From (10.50) we note that

$$F(x + \Delta x) - F(x) = f(\xi)\,\Delta x \qquad x \leqq \xi \leqq x + \Delta x$$

[see (9) above], provided $f(x)$ is continuous on $(x, x + \Delta x)$. Hence

$$F'(x) = \frac{dF(x)}{dx} = \lim_{\Delta x \to 0} \frac{F(x + \Delta x) - F(x)}{\Delta x} = \lim_{\Delta x \to 0} f(\xi) = f(x)$$

$$\frac{d}{dx} \int_a^x f(t)\,dt = f(x) \tag{10.52}$$

We can obtain (10.52) without recourse to the law of the mean. We have

$$\frac{F(x + \Delta x) - F(x)}{\Delta x} - f(x) = \frac{1}{\Delta x} \int_x^{x+\Delta x} f(t)\,dt - \frac{1}{\Delta x} \int_x^{x+\Delta x} f(x)\,dt$$
$$= \frac{1}{\Delta x} \int_x^{x+\Delta x} [f(t) - f(x)]\,dt$$

If $f(t)$ is continuous at $t = x$, we have $|f(t) - f(x)| < \epsilon$ for $|t - x| < \delta$. Choose $|t - x| < \delta$ so that

$$\left| \frac{F(x + \Delta x) - F(x)}{\Delta x} - f(x) \right| \leqq \frac{1}{|\Delta x|} \left| \int_x^{x+\Delta x} \epsilon\,dt \right| = \epsilon$$

Since ϵ can be chosen arbitrarily small, we note that

$$F'(x) = \lim_{\Delta x \to 0} \frac{F(x + \Delta x) - F(x)}{\Delta x} = f(x)$$

We obtain the fundamental theorem of the integral calculus as follows: Let $G(x)$ be any function whose derivative is $f(x)$, $f(x)$ continuous. Then $G'(x) = F'(x)$ for $a \leqq x \leqq b$. From Prob. 8, Sec. 10.13, we note that $F(x) = G(x) + C$, C = constant. Hence

$$G(x) + C = \int_a^x f(t)\,dt$$

For $x = a$ we have $G(a) + C = 0$ so that

$$G(x) - G(a) = \int_a^x f(t)\,dt$$

For $x = b$ we obtain

$$G(b) - G(a) = \int_a^b f(x)\,dx \tag{10.53}$$

Hence, to evaluate $\int_a^b f(x)\,dx$, we find any function $G(x)$ whose derivative is $f(x)$; the difference between $G(b)$ and $G(a)$ yields $\int_a^b f(x)\,dx$.

Problems

1. From $(k+1)^2 - k^2 = 2k+1$ we have $\sum_{k=1}^{n} [(k+1)^2 - k^2] = n + 2\sum_{k=1}^{n} k$, so that $\sum_{k=1}^{n} k = \frac{1}{2}[(n+1)^2 - 1 - n] = n(n+1)/2$. Consider

$$(k+1)^3 - k^3 = 3k^2 + 3k + 1$$

and show that $\sum_{k=1}^{n} k^2 = \frac{1}{6}n(n+1)(2n+1)$.

2. Without recourse to the fundamental theorem of the integral calculus show that $\int_0^1 x^2\, dx = \frac{1}{3}$ by making use of the result of Prob. 1.

3. If $f(x)$ is continuous for $a \leqq x \leqq b$, $b > a$, and if $\int_a^x f(t)\, dt = 0$ for all x on (a, b), show that $f(x) \equiv 0$.

4. If $f(x)$ is continuous on (a, b), $b > a$, and if, for any continuous function $\varphi(x)$, $\int_a^b f(x)\varphi(x)\, dx = 0$, show that $f(x) \equiv 0$ on (a, b).

5. If $f(x)$ is continuous for $a \leqq x \leqq b$, $b > a$, and if $f(x) \geqq 0$ on (a, b),

$$\int_a^b f(x)\, dx = 0$$

show that $f(x) \equiv 0$ on (a, b).

6. Let $f(x) = 1$ for $0 \leqq x \leqq \frac{1}{2}$, $f(x) = 2$ for $\frac{1}{2} < x \leqq 1$, and consider

$$F(x) = \int_0^x f(t)\, dt,$$

$0 \leqq x \leqq 1$. Obtain a graph of $F(x)$. Why does $F'(x)$ not exist at $x = \frac{1}{2}$? Does this contradict (10.52)?

7. Consider the interval $0 \leqq x \leqq 1$. Define $f(x)$ as follows: $f(x) = 0$ if x is irrational, $f(x) = 1/q$ if $x = p/q$, p and q integers in lowest form. Show that $f(x)$ is R-integrable on $0 \leqq x \leqq 1$.

8. Prove the first theorem of the mean for integrals. If $f(x)$ is continuous on (a, b), $\varphi(x)$ R-integrable on (a, b), and $\varphi(x)$ of the same sign on (a, b), then

$$\int_a^b f(x)\varphi(x)\, dx = f(\xi) \int_a^b \varphi(x)\, dx \qquad a \leqq \xi \leqq b$$

9. If $f'(x)$ and $\varphi'(x)$ are R-integrable, show that

$$\begin{aligned}\int_a^b f(x)\varphi'(x)\, dx &= f(x)\varphi(x)\Big|_a^b - \int_a^b \varphi(x)f'(x)\, dx \\ &= f(b)\varphi(b) - f(a)\varphi(a) - \int_a^b \varphi(x)f'(x)\, dx \qquad (10.54)\end{aligned}$$

Equation (10.54) represents an integration by parts.

10. Prove that any bounded monotonic function is R-integrable.

10.17. Integrals with a Parameter. Let us consider the integral

$$F(t) = \int_a^b f(x, t)\, dx \qquad t_0 \leqq t \leqq t_1 \tag{10.55}$$

After $f(x, t)$ is integrated with respect to x, what remains depends on the parameter t, $t_0 \leqq t \leqq t_1$. Of course $F(t)$ depends also on a and b, but for the present we assume that these numbers are fixed and finite. For example,

$$F(t) = \int_0^1 \frac{dx}{x^2 + t^2} = \frac{1}{t} \tan^{-1} \frac{1}{t} \qquad t > 0$$

is a well-defined function of t for $t > 0$.

We now determine some conditions for which $F(t)$ will be continuous We desire

$$\lim_{t \to \bar{t}} F(t) = F(\bar{t}) = \int_a^b f(x, \bar{t})\, dx \tag{10.56}$$

If $\dfrac{\partial f(x, t)}{\partial t}$ exists and is a bounded R-integrable function with respect to x for $t_0 \leqq t \leqq t_1$, then

$$\begin{aligned} |F(t) - F(\bar{t})| &= \left| \int_a^b [f(x, t) - f(x, \bar{t})]\, dx \right| \\ &= |t - \bar{t}| \left| \int_a^b \frac{\partial f(x, \bar{\bar{t}})}{\partial t}\, dx \right| \leqq M|t - \bar{t}| \end{aligned} \tag{10.57}$$

by applying the law of the mean to $f(x, t) - f(x, \bar{t})$. From (10.57) we see that $\lim_{t \to \bar{t}} F(t) = F(\bar{t})$. A less stringent condition which does not imply the existence of $\dfrac{\partial f(x, t)}{\partial t}$ and which enables us to prove the continuity of $F(t)$ is as follows: Let $f(x, t)$ be uniformly continuous in t with respect to x so that for any $\epsilon > 0$ there exists a $\delta > 0$, δ independent of x, such that $|f(x, t) - f(x, \bar{t})| < \epsilon$ whenever $|t - \bar{t}| < \delta$. One notes immediately that

$$|F(t) - F(\bar{t})| \leqq \int_a^b \epsilon |dx| = \epsilon |b - a|$$

Since ϵ can be chosen arbitrarily small, we see that $F(t) \to F(\bar{t})$ as $t \to \bar{t}$.

Of more particular interest is the possibility of showing that

$$\frac{dF(t)}{dt} \equiv \frac{d}{dt} \int_a^b f(x, t)\, dx = \int_a^b \frac{\partial f(x, t)}{\partial t}\, dx \tag{10.58}$$

Let us assume that $\dfrac{\partial f(x, t)}{\partial t}$ is continuous for $a \leqq x \leqq b$, $t_0 \leqq t \leqq t_1$. We define $G(t)$ by

$$G(t) = \int_a^b \frac{\partial f(x, t)}{\partial t}\, dx$$

Then
$$\int_{t_0}^t G(t)\, dt = \int_{t_0}^t \int_a^b \frac{\partial f(x, t)}{\partial t}\, dx\, dt = \int_a^b \int_{t_0}^t \frac{\partial f(x, t)}{\partial t}\, dt\, dx$$
$$= \int_a^b [f(x, t) - f(x, t_0)]\, dx$$
$$= F(t) - F(t_0) \tag{10.59}$$

The interchange of the order of integration can be justified since the integrand is continuous. From (10.52) we obtain

$$\frac{dF(t)}{dt} = \frac{d}{dt}\int_a^b f(x, t)\, dx = G(t) = \int_a^b \frac{\partial f(x, t)}{\partial t}\, dx$$

Example 10.19. We consider

$$F(t) = \int_0^\pi e^{t \cos x}\, dx \qquad |t| < T \tag{10.60}$$

It is obvious that $\frac{\partial}{\partial t}(e^{t\cos x}) = \cos x\, e^{t \cos x}$ is continuous in x and t for $0 \leqq x \leqq \pi$, $|t| < T$, so that

$$\frac{dF}{dt} = \int_0^\pi \cos x\, e^{t \cos x}\, dx$$

It is easy to justify a further differentiation so that

$$\frac{d^2F}{dt^2} = \int_0^\pi \cos^2 x\, e^{t\cos x}\, dx$$
$$= F(t) - \int_0^\pi \sin^2 x\, e^{t\cos x}\, dx$$
$$= F(t) + \int_0^\pi \sin x\, d\left(\frac{1}{t}\, e^{t\cos x}\right)$$
$$= F(t) + \frac{\sin x}{t}\, e^{t\cos x}\Big|_{x=0}^{x=\pi} - \frac{1}{t}\int_0^\pi \cos x\, e^{t\cos x}\, dx$$
$$= F(t) - \frac{1}{t}\frac{dF(t)}{dt}$$

Bessel's differential equation (see Sec. 5.13) for $n = 0$ is $\frac{d^2w}{dz^2} + \frac{1}{z}\frac{dw}{dz} + w = 0$. If $z = it$, we have $\frac{d^2w}{dt^2} + \frac{1}{t}\frac{dw}{dt} - w = 0$, which is satisfied by $F(t)$ of (10.60).

Problems

1. Show that

$$\int_0^1 \frac{dx}{(x^2 + t)^n} = \frac{d^n}{dt^n}\left[\frac{1}{\sqrt{t}} \tan^{-1} \frac{1}{\sqrt{t}}\right]$$

2. From $\int_a^b \cos tx\, dx = (1/t)(\sin bt - \sin at)$ compute $\int_a^b x^{2p} \cos tx\, dx$, p a positive integer.

3. Show that

$$\int_0^\pi \frac{\sin x\,dx}{\sqrt{1 - 2\alpha\cos x + \alpha^2}} = \begin{cases} 2 & 0 < \alpha \leqq 1 \\ \dfrac{2}{\alpha} & \alpha > 1 \end{cases}$$

4. Given $F(t) = \int_{\varphi(t)}^{\psi(t)} f(x, t)\,dt$, show that

$$\frac{dF(t)}{dt} = \int_{\varphi(t)}^{\psi(t)} \frac{\partial f(x, t)}{\partial t}\,dx + \frac{d\psi(t)}{dt} f(\psi(t), t) - \frac{d\varphi(t)}{dt} f(\varphi(t), t)$$

for certain restrictions on $\dfrac{\partial f(x, t)}{\partial t}$, $\varphi(t)$, $\psi(t)$.

5. By two methods show that

$$\frac{d}{dt}\int_t^{t^2} (2x + t)\,dx = 4t^3 + 3t^2 - 4t$$

10.18. Improper and Infinite Integrals. The function $f(x) = x^{-\frac{1}{2}}$, $x > 0$, $f(0) = 1$, is well defined on the interval $0 \leqq x \leqq 1$. Since $f(x)$ is not bounded on $0 \leqq x \leqq 1$, we are not certain that it is R-integrable on this range. However, for $0 < \epsilon \leqq x \leqq 1$, we have

$$F(\epsilon) = \int_\epsilon^1 \frac{dx}{\sqrt{x}} = 2 - 2\sqrt{\epsilon}$$

As $\epsilon \to 0$, $F(\epsilon) \to 2$ and we define

$$\int_0^1 \frac{dx}{\sqrt{x}} = \lim_{\substack{\epsilon \to 0 \\ \epsilon > 0}} \int_\epsilon^1 \frac{dx}{\sqrt{x}} = 2 \tag{10.61}$$

We notice that $f(0)$ could have been defined in any way we please without affecting the value of the improper integral $\int_0^1 \frac{dx}{\sqrt{x}}$. Let the reader show that the improper integral $\int_0^1 \frac{dx}{x}$ fails to exist.

A Practical Rule. Let $f(x)$ be R-integrable for the range $a + \epsilon \leqq x \leqq b$, for all $\epsilon > 0$. Define $\psi(x)$ by

$$\psi(x) = (x - a)^\mu f(x)$$

1. If for $\mu < 1$ we have $|\psi(x)| < M =$ constant for $a \leqq x \leqq b$, then $\lim_{\substack{\epsilon \to 0 \\ \epsilon > 0}} \int_{a+\epsilon}^b f(x)\,dx$ exists.

2. If for $\mu \geqq 1$ we have $|\psi(x)| > m > 0$ for $a \leqq x \leqq b$, then

$$\lim_{\substack{\epsilon \to 0 \\ \epsilon > 0}} \int_{a+\epsilon}^b f(x)\,dx$$

fails to exist.

These results are left as an exercise for the reader. One easily sees that $\lim\limits_{\substack{\epsilon \to 0 \\ \epsilon > 0}} \int_{a+\epsilon}^{b} \frac{dx}{(x-a)^\mu}$ exists if $\mu < 1$ and fails to exist if $\mu \geqq 1$.

If $f(x)$ is continuous for $a \leqq x \leqq b$ except at $x = c$, $a < c < b$, we define

$$\int_a^b f(x)\,dx = \lim_{\substack{\epsilon \to 0 \\ \epsilon > 0}} \int_a^{c-\epsilon} f(x)\,dx + \lim_{\substack{\delta \to 0 \\ \delta > 0}} \int_{c+\delta}^{b} f(x)\,dx \tag{10.62}$$

provided these limits exist. It is important that ϵ and δ approach zero independently.

Example 10.20. We consider $\int_0^a \frac{dx}{\sqrt{a^2 - x^2}}$. The upper limit may give us trouble. However, $f(x) = (a - x)^{-\frac{1}{2}}(a + x)^{\frac{1}{2}}$, so that $\psi(x) = (a - x)^{\frac{1}{2}}f(x) = (a + x)^{\frac{1}{2}}$ is bounded for $0 \leqq x \leqq a$. Since $\mu = \frac{1}{2}$ the integral exists and, actually,

$$\int_0^a \frac{dx}{\sqrt{a^2 - x^2}} = \frac{\pi}{2}$$

The integral $\int_a^\infty f(x)\,dx$ is called an infinite integral and is defined as follows: Let $f(x)$ be R-integrable for $a \leqq x \leqq X$, X arbitrary. We define

$$\int_a^\infty f(x)\,dx = \lim_{X \to \infty} \int_a^X f(x)\,dx \tag{10.63}$$

provided this limit exists. For example,

$$\int_0^\infty \frac{dx}{1 + x^2} = \lim_{X \to \infty} \int_0^X \frac{dx}{1 + x^2} = \lim_{X \to \infty} \tan^{-1} X = \frac{\pi}{2}$$

Let the reader verify that $\int_0^\infty \cos x\,dx$ does not exist. If $\int_a^\infty |f(x)|\,dx$ exists, we say that $f(x)$ is absolutely integrable.

From (10.63) we note that for any $\epsilon > 0$ an X_0 exists such that

$$\left| \int_a^\infty f(x)\,dx - \int_a^X f(x)\,dx \right| < \epsilon \qquad \text{for } X \geqq X_0$$

or

$$\left| \int_X^\infty f(x)\,dx \right| < \epsilon \qquad \text{for } X \geqq X_0 \tag{10.64}$$

One sees immediately that if $\int_a^\infty |f(x)|\,dx$ exists then $\int_a^\infty f(x)\,dx$ exists, since

$$\left| \int_X^\infty f(x)\,dx \right| \leqq \int_X^\infty |f(x)|\,dx$$

A simple test for the convergence or existence of $\int_a^\infty f(x)\,dx$ is as follows: If a function $\varphi(x)$ exists such that $\int_a^\infty |\varphi(x)|\,dx$ converges, and if $|\varphi(x)| > |f(x)|$ for $x \geqq a$, then $\int_a^\infty f(x)\,dx$ converges. We leave the proof of this statement as a simple exercise for the reader.

Another simple test for the convergence of an infinite integral is based upon the integration-by-parts formula. From

$$\int_a^X u(x)\,dv(x) = u(X)v(X) - u(a)v(a) - \int_a^X v(x)\,du(x)$$

we note that, if $\lim_{X\to\infty} u(X)v(X)$ exists and if $\int_a^\infty v(x)\,du(x)$ exists, then $\int_a^\infty u(x)\,dv(x)$ exists.

Example 10.21. We consider $\int_0^\infty \frac{\sin x}{x}\,dx$. The origin need not concern us since $\lim_{x\to 0} \frac{\sin x}{x} = 1$. We have

$$\int_{\pi/2}^X \frac{\sin x}{x}\,dx = \int_{\pi/2}^X \frac{1}{x}\,d\,(-\cos x) = -\frac{\cos X}{X} - \int_{\pi/2}^X \frac{\cos x}{x^2}\,dx$$

Now $\lim_{X\to\infty} \frac{\cos X}{X} = 0$ and $\lim_{X\to\infty} \int_{\pi/2}^X \frac{\cos x}{x^2}\,dx$ exists since $\int_{\pi/2}^\infty \frac{dx}{x^2}$ exists. Since

$$\int_0^{\pi/2} \frac{\sin x}{x}\,dx$$

exists, we see that $\int_0^\infty \frac{\sin x}{x}\,dx$ exists. Its value is $\frac{\pi}{2}$.

We now consider an infinite integral involving a parameter. Let $\varphi(t)$ be defined by

$$\varphi(t) = \int_a^\infty f(x, t)\,dx \qquad t_0 \leqq t \leqq t_1 \tag{10.65}$$

We assume that the integral of (10.65) exists for each value of t on the range $t_0 \leqq t \leqq t_1$. If we fix our attention on a specific value of t, we have for any $\epsilon > 0$ an X_0 such that

$$\left| \int_X^\infty f(x, t)\,dx \right| < \epsilon \qquad X \geqq X_0 \tag{10.66}$$

The value of X_0 (for a fixed ϵ) may well depend on t. If, for any $\epsilon > 0$, there exists an X_0 independent of t such that (10.66) holds for all t on the range $t_0 \leqq t \leqq t_1$, we say that the infinite integral converges uniformly with respect to t.

Let the reader show that if $g(x) \geqq |f(x, t)|$ for $t_0 \leqq t \leqq t_1$, and if $\int_a^\infty g(x)\,dx$ exists, then $\int_a^\infty f(x, t)\,dx$ converges uniformly.

Example 10.22. We consider $F(t) = \int_0^\infty e^{-x^2} \cos xt\,dx$. Since $e^{-x^2} \geqq |e^{-x^2} \cos xt|$ for all t and since $\int_0^\infty e^{-x^2}\,dx = \sqrt{\pi}/2$ exists, $F(t)$ exists for all t.

We now show that if $f(x, t)$ is uniformly continuous in t for $a \leqq x \leqq X$, X finite but arbitrary, then $\varphi(t)$ of (10.65) is continuous in t provided $\int_a^\infty f(x, t)\, dx$ converges uniformly. We have

$$\begin{aligned}\varphi(t + \Delta t) - \varphi(t) &= \int_a^\infty [f(x, t + \Delta t) - f(x, t)]\, dx \\ &= \int_a^X [f(x, t + \Delta t) - f(x, t)]\, dx \\ &\qquad + \int_X^\infty f(x, t + \Delta t)\, dx - \int_X^\infty f(x, t)\, dx\end{aligned}$$

From uniform convergence we have for any $\epsilon/3 > 0$ an X such that

$$\left| \int_X^\infty f(x, t + \Delta t)\, dx \right| < \frac{\epsilon}{3} \qquad \left| \int_X^\infty f(x, t)\, dx \right| < \frac{\epsilon}{3}$$

From Sec. 10.17 we note that for any $\epsilon/3 > 0$ a $\delta > 0$ exists such that

$$\left| \int_a^X [f(x, t + \Delta t) - f(x, t)]\, dx \right| < \frac{\epsilon}{3}$$

for $|\Delta t| < \delta$. This is the property that $\int_a^X f(x, t)\, dx$ is continuous in t, X finite. Hence

$$|\varphi(t + \Delta t) - \varphi(t)| < \epsilon$$

for $|\Delta t| < \delta$, $\delta > 0$, so that $\varphi(t)$ is continuous. Q.E.D.

It is interesting to find a sufficient condition which enables one to write

$$\frac{d\varphi(t)}{dt} = \frac{d}{dt} \int_a^\infty f(x, t)\, dx = \int_a^\infty \frac{\partial f(x, t)}{\partial t}\, dx \tag{10.67}$$

Let the reader first show that, if $G(t) = \int_a^\infty \frac{\partial f(x, t)}{\partial t}\, dx$ converges uniformly, then

$$\int_{t_0}^t \int_a^\infty \frac{\partial f(x, t)}{\partial t}\, dx\, dt = \int_a^\infty \int_{t_0}^t \frac{\partial f(x, t)}{\partial t}\, dt\, dx$$

$$\int_{t_0}^t G(t)\, dt = \varphi(t) - \varphi(t_0) \tag{10.68}$$

provided $\frac{\partial f(x, t)}{\partial t}$ is continuous for $x \geqq a$, $t_0 \leqq t \leqq t_1$ [see (10.59)]. Differentiation of (10.68) leads to (10.67).

Example 10.23. We consider $F(t)$ given in Example 10.22. From

$$f(x, t) = e^{-x^2} \cos 2xt$$

we have $\frac{\partial f}{\partial t} = -2xe^{-x^2} \sin 2xt$. To show that $\int_0^\infty \frac{\partial f(x, t)}{\partial t}\, dx$ converges uniformly for all t, we note that $\left| \int_0^\infty \frac{\partial f(x, t)}{\partial t} dx \right| \leqq \int_0^\infty 2xe^{-x^2} dx = 1$. Hence

$$\begin{aligned} \frac{dF(t)}{dt} &= -\int_0^\infty 2xe^{-x^2} \sin 2xt \, dx \\ &= \int_0^\infty \sin 2xt \, de^{-x^2} = -2t \int_0^\infty e^{-x^2} \cos 2xt \, dx \\ &= -2tF(t) \end{aligned}$$

Integration yields $F(t) = Ae^{-t^2} = \int_0^\infty e^{-x^2} \cos 2xt \, dx$. At $t = 0$ we have

$$A = \int_0^\infty e^{-x^2} dx = \frac{\sqrt{\pi}}{2},$$

so that

$$\int_0^\infty e^{-x^2} \cos 2xt \, dx = \frac{\sqrt{\pi}}{2} e^{-t^2}$$

Problems

1. Show that $\int_0^1 \sqrt{\frac{1 - a^2t^2}{1 - t^2}} \, dt$ exists for $0 \leqq a < 1$.

2. Show that $\int_0^{1/k} \frac{dx}{\sqrt{(1 - x^2)(1 - k^2x^2)}}$ exists.

3. Show that $\int_0^{\pi/2} \ln \sin x \, dx$ exists.

4. Show that $\pi/2 - \tan^{-1} t = \int_0^\infty (1/x)e^{-tx} \sin x \, dx, \; t \geqq 0$.

5. Show that $\int_0^\infty \frac{\cos tx}{1 + x^2} dx$ converges uniformly for all t. Evaluate this integral by contour integration.

6. Show that $\int_0^\infty x \sin e^x \, dx = \int_0^\infty xe^{-x} d(-\cos e^x)$ exists.

7. Let $f(x, t) = xt$. Show that

$$\int_0^\infty \int_{-1}^1 f(x, t) \, dt \, dx \neq \int_{-1}^1 \int_0^\infty f(x, t) \, dx \, dt$$

8. Consider $F(y) = \int_0^\infty \frac{\sin^2 xy}{x^2} dx$, and show that $F(y) = \frac{\pi}{2} y$.

10.19. Methods of Integration. We first consider the special indefinite integral

$$\int \frac{dx}{(x^2 + 1)(x - 2)} \tag{10.69}$$

In the elementary calculus the reader was told to expand $(x^2 + 1)^{-1}(x - 2)^{-1}$ in partial fractions, obtaining

$$\begin{aligned} \frac{1}{(x^2 + 1)(x - 2)} &= \frac{Ax + B}{x^2 + 1} + \frac{C}{x - 2} \\ &= \frac{(A + C)x^2 + (B - 2A)x + C - 2B}{(x^2 + 1)(x - 2)} \end{aligned} \tag{10.70}$$

Since (10.70) holds for all x, one has $A + C = 0$, $B - 2A = 0$, $C - 2B = 1$, so that $A = -\frac{1}{5}$, $B = -\frac{2}{5}$, $C = \frac{1}{5}$. The integral (10.69) can be written

$$\int \frac{dx}{(x^2+1)(x-2)} = -\frac{1}{5}\int \frac{x\,dx}{x^2+1} - \frac{2}{5}\int \frac{dx}{x^2+1} + \frac{1}{5}\int \frac{dx}{x-2}$$
$$= -\frac{1}{10}\ln(x^2+1) - \frac{2}{5}\tan^{-1} x + \frac{1}{5}\ln(x-2) + \text{constant}$$

The results of Sec. 8.11 justify the expansion (10.70). Since $x^2 + 1$ and $x - 2$ are relatively prime (there is no polynomial of degree $\geqq 1$ which divides both $x^2 + 1$ and $x - 2$), there exist two polynomials $P(x)$, $Q(x)$ such that

$$P(x)(x^2+1) + Q(x)(x-2) = 1 \qquad \begin{aligned} &\deg P(x) < \deg (x-2) = 1 \\ &\deg Q(x) < \deg (x^2+1) = 2 \end{aligned}$$

Thus $P(x) = C$, $Q(x) = Ax + B$, and (10.70) results upon division by $(x^2 + 1)(x - 2)$. To evaluate the integral

$$\int \frac{x\,dx}{(x^2+1)(x-2)}$$

one notes that

$$\frac{x}{(x^2+1)(x-2)} = -\frac{1}{5}\frac{x^2}{x^2+1} - \frac{2}{5}\frac{x}{x^2+1} + \frac{1}{5}\frac{x}{x-2}$$
$$= -\frac{1}{5}\left(1 - \frac{1}{x^2+1}\right) - \frac{2}{5}\frac{x}{x^2+1} + \frac{1}{5}\left(1 + \frac{2}{x-2}\right)$$

so that

$$\int \frac{x\,dx}{(x^2+1)(x-2)} = \tfrac{1}{5}\tan^{-1} x - \tfrac{1}{5}\ln(x^2+1) + \tfrac{2}{5}\ln(x-2) + C$$

If the zeros of a polynomial $P(x)$ are known, one can evaluate

$$\int \frac{Q(x)}{P(x)}\,dx \tag{10.71}$$

provided $Q(x)$ is a polynomial. One writes

$$\frac{1}{P(x)} = \frac{a_1}{x-r_1} + \frac{a_2}{x-r_2} + \cdots + \frac{a_k}{x-r_k} + \cdots + \frac{a_n}{x-r_n}$$

with $a_k = \lim\limits_{x\to r_k} \dfrac{x-r_k}{P(x)}$, $k = 1, 2, \ldots, n$. One then divides $Q(x)$ by $x - r_k$ and performs a series of simple integrations for $k = 1, 2, \ldots, n$.

If $P(x)$ is a polynomial such that $x = r$ is a kth-fold zero of $P(x)$ we have

$$P(x) = (x - r)^k Q(x) \qquad Q(r) \neq 0$$

Thus

$$P'(x) = (x - r)^{k-1}[kQ(x) + (x - r)Q'(x)]$$

and $P'(r) = 0$ provided $k > 1$. For $k = 1$, $P'(r) \neq 0$, so that $P(x)$ and $P'(x)$ have no zeros in common if $P(x)$ has no multiple zeros. In this latter case $P(x)$ and $P'(x)$ are relatively prime so that polynomials $A(x)$ and $B(x)$ exist such that

$$A(x)P(x) + B(x)P'(x) = 1 \tag{10.72}$$

To evaluate an integral of the type

$$\int \frac{C(x)}{[P(x)]^m}\,dx \tag{10.73}$$

provided $P(x)$ has no multiple zeros, we make use of (10.72). Thus

$$\frac{C(x)}{[P(x)]^m} = \frac{A(x)C(x)}{[P(x)]^{m-1}} + \frac{B(x)P'(x)C(x)}{[P(x)]^m}$$

and

$$\int \frac{C(x)}{[P(x)]^m}\,dx = \int \frac{A(x)C(x)}{[P(x)]^{m-1}}\,dx + \frac{1}{1-m}\,\frac{B(x)C(x)}{[P(x)]^{m-1}} + \frac{1}{m-1}\int \frac{[B(x)C'(x) + B'(x)C(x)]}{[P(x)]^{m-1}}\,dx \tag{10.74}$$

Since m has been reduced by 1, we can repeat this process until integrals of the type (10.71) occur.

Example 10.24. We consider

$$I_n(x) = \int_0^x \frac{dt}{(t^2+1)^n} = \int_0^x \frac{dt}{(t^2+1)^{n-1}} - \int_0^x \frac{(t/2)\cdot 2t}{(t^2+1)^n}\,dt$$

since $1 \cdot (t^2 + 1) - (t/2)(2t) \equiv 1$. Integration by parts yields

$$\begin{aligned} I_n(x) &= I_{n-1}(x) - \frac{t}{2}\,\frac{(t^2+1)^{1-n}}{1-n}\bigg|_0^x + \frac{\frac{1}{2}}{1-n}\int_0^x \frac{dt}{(t^2+1)^{n-1}} \\ &= \frac{2n-3}{2n-2}\,I_{n-1}(x) + \frac{x}{2(n-1)(x^2+1)^{n-1}} \end{aligned} \tag{10.75}$$

From $I_1(x) = \tan^{-1} x$ one has $I_2(x) = \frac{1}{2}\tan^{-1} x + \frac{1}{2}x/(x^2+1)$. Repeated application of (10.75) yields

$$I_n(x) = \frac{1\cdot 3\cdot 5 \cdot\cdot\cdot (2n-3)}{2\cdot 4\cdot 6 \cdot\cdot\cdot (2n-2)}\tan^{-1} x + R(x)$$

where $R(x)$ is a rational function of x.

If $F(x)$ is a polynomial, it can be written in the canonical form

$$F(x) = X_1 X_2^2 \cdots X_m^m$$

with $X_1, X_2, \ldots, X_m$, polynomials relatively prime to each other. Since X_1 is relatively prime to $X_2^2 \cdots X_m^m$, we have

$$A(x)X_1 + B(x)X_2^2 \cdots X_m^m = 1$$

so that
$$\frac{B(x)}{X_1} + \frac{A(x)}{X_2^2 X_3^3 \cdots X_m^m} = \frac{1}{X_1 X_2^2 \cdots X_m^m}$$

Continuing this process by noting that X_2^2 is relatively prime to $X_3^3 \cdots X_m^m$, one can write

$$\frac{f(x)}{F(x)} = \frac{A_1(x)}{X_1} + \frac{A_2(x)}{X_2^2} + \cdots + \frac{A_m(x)}{X_m^m}$$

if $f(x)$ is also a polynomial. The evaluation of

$$\int \frac{f(x)}{F(x)}\, dx$$

is reduced to an evaluation of integrals of the type given by (10.73), which in turn can be reduced to the simpler form given by (10.71).

Let $R(x, y)$ be a rational function of x and y,

$$R(x, y) = \frac{\sum_{j=0}^{n} \sum_{i=0}^{m} a_{ij} x^i y^j}{\sum_{j=0}^{q} \sum_{i=0}^{p} b_{ij} x^i y^j} \tag{10.76}$$

To evaluate

$$\int R(x, \sqrt{Ax + B})\, dx \tag{10.77}$$

with $y = \sqrt{Ax + B}$, one notes that the change of variable $t^2 = Ax + B$, $dx = (2t\, dt)/A$, $x = (t^2 - B)/A$, reduces (10.77) to an integral of the type discussed above.

Example 10.25. To evaluate

$$\int \frac{\sqrt{x + 1}}{x}\, dx \tag{10.78}$$

we let $t^2 = x + 1$, $2t\, dt = dx$, so that (10.78) becomes

$$\int \frac{2t^2\, dt}{t^2 - 1} = 2 \int \left(1 + \frac{1}{t^2 - 1}\right) dt$$
$$= 2 \int \left[1 + \frac{1}{2}\left(\frac{1}{t - 1} - \frac{1}{t + 1}\right)\right] dt = 2t + \ln \frac{t - 1}{t + 1} + C$$

and
$$\int \frac{\sqrt{x + 1}}{x}\, dx = 2\sqrt{x + 1} + \ln \frac{\sqrt{x + 1} - 1}{\sqrt{x + 1} + 1} + C$$

We can evaluate

$$\int R(x, \sqrt{ax^2 + bx + c})\, dx \tag{10.79}$$

provided R is a rational function. Let $x = \alpha$, $y = \beta$ satisfy

$$y^2 = Ax^2 + Bx + C$$

so that $\beta^2 = A\alpha^2 + B\alpha + C$. The equation of the straight line through the point (α, β) is $y - \beta = t(x - \alpha)$ with t the slope. The intersection of this straight line with the curve $y^2 = Ax^2 + Bx + C$ is easily seen to yield the coordinates

$$\begin{aligned} x &= \frac{B + A\alpha - 2\beta t + \alpha t^2}{t^2 - A} \\ y &= \beta + \frac{(B + 2A\alpha - 2\beta t)t}{t^2 - A} \end{aligned} \tag{10.80}$$

Since x, y, and dx are rational functions of t, one can integrate (10.79) by the methods discussed above.

Example 10.26. We evaluate $\int \frac{dx}{x\sqrt{x^2 + p^2}}$. We have $y^2 = x^2 + p^2$, $A = 1$, $B = 0$, $C = p^2$, and we choose $\alpha = 0$ so that $\beta = p$. From (10.80)

$$x = \frac{2pt}{1 - t^2} \qquad y = p\,\frac{1 + t^2}{1 - t^2} \qquad dx = \frac{2p(1 + t^2)}{(1 - t^2)^2}\, dt$$

so that

$$\int \frac{dx}{x\sqrt{x^2 + p^2}} = \frac{1}{p}\int \frac{dt}{t} = \frac{1}{p} \ln \frac{-p + \sqrt{x^2 + p^2}}{x} + C$$

The integral (10.79) is a special case of the following type of integral. Suppose $R(x, y)$ is a rational function of x and y, with y depending implicitly on x through the relation $f(x, y) = 0$. If the curve $f(x, y) = 0$ is *unicursal* in the sense that we can describe the curve parametrically by $x = \varphi(t)$, $y = \psi(t)$ with $\varphi(t)$ and $\psi(t)$ rational functions of t, then

$$\int R(x, y)\, dx = \int R(\varphi(t), \psi(t))\varphi'(t)\, dt \tag{10.81}$$

and this integral can be evaluated by the methods discussed above.

A rational function of the trigonometric functions can be integrated by the use of the following change of variable:

$$\begin{aligned} & & t &= \tan \frac{x}{2} = \sqrt{\frac{1 - \cos x}{1 + \cos x}} \\ x &= 2 \tan^{-1} t & & \\ dx &= \frac{2\, dt}{1 + t^2} & \cos x &= \frac{1 - t^2}{1 + t^2} \\ & & \sin x &= \frac{2t}{1 + t^2} \end{aligned} \tag{10.82}$$

For example,

$$\int \frac{dx}{\sin x} = \int \frac{2/(1+t^2)}{2t/(1+t^2)}\,dt = \ln t + C = \ln\left(\tan\frac{x}{2}\right) + C$$

The elliptic integral of the second kind appears in a natural manner if one attempts to find the arc length of an ellipse. Let $x = a\sin\varphi$, $y = b\cos\varphi$, $0 \leqq \varphi < 2\pi$ be the parametric representation of an ellipse. Then $ds^2 = dx^2 + dy^2 = (a^2\cos^2\varphi + b^2\sin^2\varphi)\,d\varphi^2$, so that

$$L = a\int_0^{2\pi} \sqrt{1 - e^2\sin^2\varphi}\,d\varphi$$

with $e^2 = (a^2 - b^2)/a^2 < 1$ if $b^2 < a^2$. The elliptic integral of the second kind is defined as

$$E_2(k, \varphi) = \int_0^{\varphi} \sqrt{1 - k^2\sin^2\varphi}\,d\varphi \qquad |k| < 1 \tag{10.83}$$

The period of a simple pendulum leads to the elliptic integral of the first kind, given by

$$E_1(k, \varphi) = \int_0^{\varphi} \frac{d\varphi}{\sqrt{1 - k^2\sin^2\varphi}} \qquad |k| < 1 \tag{10.84}$$

Extensive tables can be found in the literature for the evaluation of these important elliptic integrals.

Problems

Integrate the following:

1. $\int \frac{x^2\,dx}{(ax+b)^2}$
2. $\int \frac{dx}{x^2(ax+b)^2}$
3. $\int \frac{dx}{x^2(x-1)^3}$
4. $\int \frac{x\,dx}{(ax+b)^2(cx+d)}$
5. $\int \sqrt{ax^2+b}\,dx$
6. $\int \frac{\sqrt{x^2+1}}{x}\,dx$
7. $\int \frac{dx}{x\sqrt{ax^2+bx+c}}$ for $a > 0$, $a < 0$
8. $\int \frac{dx}{a + b\cos cx}$ for $a^2 > b^2$, $a^2 < b^2$
9. $\int \frac{dx}{a\sin cx + b\cos cx}$
10. Let α and β be real roots of $Ax^2 + Bx + C = 0$. Show that

$$\sqrt{Ax^2 + Bx + C} = (x - \beta)\sqrt{A\,\frac{x-\alpha}{x-\beta}}$$

and that $t = \sqrt{A \dfrac{x - \alpha}{x - \beta}}$ defines x as a rational function of t. Apply this result to evaluate $\int \sqrt{x^2 - 3x + 2}\, dx$.

11. Let $I_m = \int_0^{\pi/2} \sin^m x\, dx$. Show that

$$I_m = \frac{m-1}{m} I_{m-1} \qquad I_{2p} = \frac{(2p-1)!!}{(2p)!!}\frac{\pi}{2} \qquad I_{2p+1} = \frac{(2p)!!}{(2p+1)!!}$$

with $(2p)!! = 2 \cdot 4 \cdot 6 \cdots (2p)$, $(2p+1)!! = 1 \cdot 3 \cdot 5 \cdots (2p+1)$.

10.20. Sequences and Series. A sequence of constant terms

$$s_1, s_2, s_3, \ldots, s_n, \ldots \tag{10.85}$$

is simply a set of numbers which can be put into one-to-one correspondence with the positive integers. To completely determine the sequence, we must be given the specific rule for determining the nth term, $n = 1, 2, 3, \ldots$. We may also look upon the sequence, $\{s_n\}$, of (10.85) as a function defined only over the integers, so that $f(n) = s_n$, $n = 1, 2, 3, \ldots$.

DEFINITION 10.26. The sequence of terms $\{s_n\}$ is said to converge if a number S exists such that

$$\lim_{n\to\infty} s_n = S \tag{10.86}$$

Equation (10.86) states that for each $\epsilon > 0$ there exists an integer $N(\epsilon)$ such that $|S - s_n| < \epsilon$ for $n \geqq N(\epsilon)$. In general, the smaller ϵ is chosen, the greater is the corresponding N. The existence of an integer N for the given ϵ implies that the sequence becomes and remains within an ϵ distance of S.

Example 10.27. It is easy to show that the sequence $2, 1\frac{1}{2}, 1\frac{1}{3}, \ldots, 1 + 1/n, \ldots$ converges to the limit 1. We have $|s_n - 1| = |1/n| < \epsilon$ if $n > 1/\epsilon$, so that $N(\epsilon) = [1/\epsilon]$, where $[1/\epsilon]$ is the first integer greater than $1/\epsilon$.

In most cases we cannot readily determine the limit of the sequence even though the sequence converges. Cauchy obtained a criterion for the convergence of a sequence which does not depend on knowing the limit of the sequence.

Cauchy's Criterion. If, for any $\epsilon > 0$, an integer N exists such that $|s_{n+p} - s_n| < \epsilon$ for $n \geqq N$ and all $p \geqq 0$, the sequence $\{s_n\}$ converges to a unique limit S.

This criterion is certainly necessary, for if $\lim_{n\to\infty} s_n = S$, then $|S - s_n| < \epsilon/2$ for $n \geqq N$, $|S - s_{n+p}| < \epsilon/2$ for $n \geqq N$, $p \geqq 0$, so that $|s_{n+p} - s_n| < \epsilon$ for $n \geqq N$, $p \geqq 0$. Conversely, assume that a given sequence satisfies Cauchy's criterion. Choose $\epsilon = 1$, so that $|s_{n+p} - s_n| < 1$ for $n \geqq N_1$, $p \geqq 0$. Hence $|s_{N_1+p} - s_{N_1}| < 1$ for $p \geqq 0$, and all terms in the sequence following s_{N_1} are within a unit distance of s_{N_1}. There are only a finite number of terms preceding s_{N_1}, so that the sequence must be bounded. From the Weierstrass-Bolzano theorem at least one limit point exists for the infinite set of numbers $\{s_n\}$. Let us assume that two limit points exist, say, S and T, $S < T$. We choose $\epsilon = (T - S)/2$ and note that $|s_{n+p} - s_n| < \epsilon$ for $n \geqq N$, $p \geqq 0$. This means that after s_N the terms of the sequence are clustered within an ϵ distance of each other. Let the reader conclude that, if S is a limit point of the set $\{s_n\}$, then T cannot be a limit point, and conversely. Thus Cauchy's criterion guarantees that the sequence has a unique limit, S, with $\lim_{n\to\infty} s_n = S$. It should be understood that by first considering $\epsilon = 1$ we showed that the sequence was bounded. Then we deduced that only one limit point could exist.

Example 10.28. If we return to the sequence of Example 24, we note that

$$|s_{n+p} - s_n| = \left|\frac{p}{n(n+p)}\right| < \frac{1}{n} < \epsilon$$

if $n > 1/\epsilon$, $p \geqq 0$, since $p/(p+n) < 1$. Again $N = [1/\epsilon]$.

By applying the Weierstrass-Bolzano theorem it is very easy to prove that a bounded monotonic nondecreasing sequence of real numbers always converges. As an example of the use of this result, we consider the sequence

$$s_n = \left(1 + \frac{1}{n}\right)^n \qquad n = 1, 2, 3, \ldots$$

We show first that the sequence is bounded. We have from Newton's binomial expansion

$$\begin{aligned} s_n &= 1 + n\left(\frac{1}{n}\right) + \frac{n(n-1)}{2!}\frac{1}{n^2} + \frac{n(n-1)(n-2)}{3!}\frac{1}{n^3} + \cdots + \frac{1}{n^n} \\ &= 1 + 1 + \frac{1 - 1/n}{2!} + \frac{(1 - 1/n)(1 - 2/n)}{3!} + \cdots + \frac{1}{n^n} \end{aligned}$$

so that

$$s_n < 1 + 1 + \frac{1}{2!} + \frac{1}{3!} + \cdots < 1 + 1 + \frac{1}{2} + \frac{1}{2^2} + \frac{1}{2^3} + \cdots = 3$$

Every term of the sequence is less than 3. Next we show that $s_n > s_{n-1}$, so that the sequence is monotonic increasing. First let us notice that, if α and β are any two numbers such that $\alpha > \beta \geqq 0$, then from

$$\alpha^n - \beta^n = (\alpha - \beta)(\alpha^{n-1} + \alpha^{n-2}\beta + \cdots + \alpha\beta^{n-2} + \beta^{n-1})$$

we obtain

$$n\alpha^{n-1}(\alpha - \beta) > \alpha^n - \beta^n > n\beta^{n-1}(\alpha - \beta)$$

Now let $\alpha = 1 + 1/(n-1) > \beta = 1 + 1/n$, $n > 1$, so that

$$\beta^n > \alpha^n - n\alpha^{n-1}(\alpha - \beta) = \alpha^{n-1}[\alpha - n(\alpha - \beta)]$$

yields

$$s_n = \left(1 + \frac{1}{n}\right)^n > s_{n-1}\left(1 + \frac{1}{n-1} - \frac{1}{n-1}\right) = s_{n-1} \qquad n > 1$$

Thus the sequence converges to a unique limit, called e.

$$\lim_{n\to\infty}\left(1 + \frac{1}{n}\right)^n = e = 2.71828 \cdots$$

The infinite series

$$u_1 + u_2 + \cdots + u_n + \cdots \tag{10.87}$$

with u_n well defined for $n = 1, 2, \ldots$ can be given a clear meaning if we reduce the series to a sequence. We define $s_1 = u_1$, $s_2 = u_1 + u_2$, . . . ,

$$s_n = \sum_{i=1}^{n} u_i \qquad n = 1, 2, 3, \ldots \tag{10.88}$$

If the sequence $\{s_n\}$ converges to a unique limit S, we say that the series $\sum_{i=1}^{\infty} u_i$ converges to S and write

$$S = \sum_{i=1}^{\infty} u_i \tag{10.89}$$

If the sequence fails to converge, we say that the series diverges. In general, one of three things will occur when we consider the convergence of a series:

1. The series converges.
2. The series increases beyond bound, and so diverges properly.
3. The series oscillates and does not converge.

Example 10.29. The series $\sum_{j=0}^{\infty} m^j$, $m < 1$, converges to the limit $1/(1 - m)$ since $s_n = \sum_{j=0}^{n} m^j = (1 - m^{n+1})/(1 - m) \to \dfrac{1}{(1 - m)}$ as $n \to \infty$. The series $1 + 1 + 1 + \cdots + 1 + \cdots$ increases beyond bound since $s_n = n$. The series $1 - 1 + 1 - 1 + \cdots + (-1)^{n+1} + \cdots$ does not converge since $s_n = \frac{1}{2}[1 + (-1)^{n+1}]$ fails to converge. The terms of the sequence oscillate between 1 and zero.

If we apply the Cauchy criterion to the sequence $s_n = \sum_{j=1}^{n} u_j$, $n = 1, 2, \ldots$, we note that a necessary and sufficient condition for the convergence of the series $\sum_{j=1}^{\infty} u_j$ is that for any $\epsilon > 0$ an integer N exists such that

$$\left| \sum_{j=n+1}^{n+p} u_j \right| < \epsilon \tag{10.90}$$

for $n \geqq N$, all $p > 0$, since $s_{n+p} = \sum_{j=1}^{n+p} u_j$, $s_n = \sum_{j=1}^{n} u_j$. Formula (10.90) is equivalent to the statement that for any $\epsilon > 0$ an integer N exists such that

$$\left| \sum_{j=n+1}^{\infty} u_j \right| < \epsilon \qquad \text{for } n \geqq N$$

If we write

$$\sum_{j=1}^{\infty} u_j = \sum_{j=1}^{n} u_j + R_n$$

with $R_n = \sum_{j=n+1}^{\infty} u_j$, we note that the series converges if for any $\epsilon > 0$ we can find an integer N such that all remainders, R_n, are less than ϵ for $n \geqq N$.

If we apply (10.90) for a convergent series with $p = 1$, we note that for any $\epsilon > 0$ an integer N exists such that $|u_{n+1}| < \epsilon$ for $n \geqq N$. Thus a necessary condition that a series converge is that the nth term of the series tends to zero as n becomes infinite. However, the harmonic series $\sum_{n=1}^{\infty} \frac{1}{n}$ diverges even though the nth term tends to zero as n becomes infinite. That $\sum_{n=1}^{\infty} \frac{1}{n}$ diverges is seen by writing

$$\begin{aligned} 1 + \tfrac{1}{2} + \tfrac{1}{3} + \cdots + \frac{1}{n} + \cdots &= 1 + \tfrac{1}{2} + (\tfrac{1}{3} + \tfrac{1}{4}) \\ &\quad + (\tfrac{1}{5} + \tfrac{1}{6} + \tfrac{1}{7} + \tfrac{1}{8}) + (\tfrac{1}{9} + \cdots + \tfrac{1}{16}) + \cdots \\ &> 1 + \tfrac{1}{2} + \tfrac{1}{2} + \tfrac{1}{2} + \cdots + \tfrac{1}{2} + \cdots \end{aligned}$$

We now consider some practical tests to determine whether a series of *positive* terms converges or not.

1. Since $u_n \geqq 0$, we note that the sequence $s_n = \sum_{j=1}^{n} u_j$, $n = 1, 2, 3, \ldots$, is monotonic nondecreasing. If a constant, M, exists such that

$s_n = \sum_{j=1}^{n} u_j < M$ for all n, the sequence is bounded and hence converges.

Thus the series $\sum_{=1}^{\infty} \frac{1}{j!}$ converges since

$$s_n = \sum_{j=1}^{n} \frac{1}{j!} < 1 + \frac{1}{2} + \frac{1}{2^2} + \cdots = 2$$

for all n.

2. If $v_n \geqq u_n \geqq 0$ and $\sum_{n=1}^{\infty} v_n$ converges, then $\sum_{n=1}^{\infty} u_n$ converges. This is the *comparison test of Weierstrass*. The proof is trivial and is left as an exercise for the reader. Conversely, if $u_n \geqq v_n \geqq 0$, and if $\sum_{n=1}^{\infty} v_n$ diverges, then $\sum_{n=1}^{\infty} u_n$ diverges. The series $\sum_{n=1}^{\infty} \frac{1}{n^\alpha}$, $\alpha < 1$, diverges, since $1/n^\alpha > 1/n$ for $n = 1, 2, \ldots$.

3. *The Integral Test.* Let $\varphi(x) \geqq 0$ be a monotonic nonincreasing function defined for $x \geqq 1$. Let the reader show that

$$\varphi(1) + \varphi(2) + \cdots + \varphi(n-1) \geqq \int_1^n \varphi(x)\,dx \geqq \varphi(2) + \varphi(3) + \cdots + \varphi(n) \quad (10.91)$$

From (10.91) the reader can deduce that $\sum_{n=1}^{\infty} \varphi(n)$ converges or diverges according to whether $\int_1^\infty \varphi(x)\,dx$ exists or not.

The series $\sum_{n=1}^{\infty} \frac{1}{n^\alpha}$, $\alpha > 1$, is seen to converge since $\varphi(x) = \frac{1}{x^\alpha}$ is monotonic decreasing and $\lim_{n\to\infty} \int_1^n \frac{dx}{x^\alpha} = \frac{1}{\alpha - 1}$ for $\alpha > 1$.

4. *D'Alembert's Ratio Test.* Suppose that a constant d exists such that for $n \geqq N$ we have $a_{n+1}/a_n < d$. If $d < 1$, $\sum_{n=1}^{\infty} a_n$ converges. From $a_{N+1} < da_N$, $a_{N+2} < da_{N+1} < d^2 a_N, \ldots, a_{N+p} < d^p a_N, \ldots$ we note that $\sum_{n=N}^{\infty} a_n < a_N \sum_{r=0}^{\infty} d^r = a_N/(1-d)$. Thus $\sum_{n=1}^{\infty} a_n$ is bounded and so con-

verges. If $a_{n+1}/a_n > d \geqq 1$ for $n \geqq N$, the reader can show that $\sum_{n=1}^{\infty} a_n$ diverges. At times it may be useful to consider

$$\lim_{n\to\infty} \frac{a_{n+1}}{a_n} = d \tag{10.92}$$

If $d < 1$, then $a_{n+1}/a_n < d' < 1$ for $n \geqq N$. Why? Thus the series converges. If $d > 1$, the series can be shown to diverge. The case $d = 1$ is indeterminate since examples can be found for which both convergence and divergence exist.

5. *The Cauchy nth-root Test.* Assume that $\lim_{n\to\infty} \sqrt[n]{a_n} = k < 1$. Then for $n \geqq N$ we have $\sqrt[n]{a_n} < k' < 1$, so that $a_n < (k')^n$ for $n \geqq N$. Since $\sum_{n=N}^{\infty} (k')^n$ converges, of necessity $\sum_{n=1}^{\infty} a_n$ converges. The reader should show that, if $\lim_{n\to\infty} \sqrt[n]{a_n} = k > 1$, then $\sum_{n=1}^{\infty} a_n$ diverges. The series $\sum_{n=1}^{\infty} \frac{1}{2^n}$ converges since $\sqrt[n]{\frac{1}{2^n}} = \frac{1}{2} < 1$.

6. *Raabe's Test.* We consider the series $\sum_{n=1}^{\infty} a_n$, $a_n > 0$. If

$$\frac{a_n}{a_{n+1}} = 1 + \frac{\sigma}{n} + f(n) \tag{10.93}$$

with $\lim_{n\to\infty} nf(n) = 0$, then $\sum_{n=1}^{\infty} a_n$ converges or diverges according as $\sigma > 1$ or $\sigma \leqq 1$. From (10.93) we have

$$\lim_{n\to\infty} \left[n\frac{a_n}{a_{n+1}} - (n+1) \right] = \sigma - 1$$

Assume $\sigma > 1$ so that $\sigma - 1 = k > 0$. For $n \geqq N$ we have

$$n\frac{a_n}{a_{n+1}} - (n+1) > \frac{k}{2}$$

or

$$\frac{2}{k}[na_n - (n+1)a_{n+1}] > a_{n+1} \tag{10.94}$$

Let $n = N, N+1, N+2, \ldots, N+p-1$, in (10.94), and add. We obtain

$$\sum_{n=N}^{N+p-1} a_{n+1} < \frac{2}{k}[Na_N - (N+p)a_{N+p}] < \frac{2}{k} Na_N$$

Since $(2/k)Na_N$ is a fixed number, the series $\sum_{n=N}^{N+p-1} a_{n+1}$ is bounded for all p. Hence $\sum_{n=1}^{\infty} a_n$ is bounded and so converges. If $\sigma < 1$, then for $n \geqq N$ we have $a_{n+1} > \frac{n}{n+1} a_n$. Thus

$$a_{N+1} > \frac{N}{N+1} a_N, \quad a_{N+2} > \frac{N+1}{N+2} a_{N+1} > \frac{N}{N+2} a_N, \quad \ldots,$$
$$a_{N+p} > \frac{N}{N+p} a_N, \quad \ldots$$

so that

$$\sum_{p=1}^{\infty} a_{N+p} > Na_N \sum_{p=1}^{\infty} \frac{1}{N+p} = \infty$$

and $\sum_{n=1}^{\infty} a_n$ diverges. It can also be shown that $\sum_{n=1}^{\infty} a_n$ diverges if $\sigma = 1$.

As an example we consider $\sum_{n=1}^{\infty} \frac{1}{n^2}$. We have

$$\frac{a_n}{a_{n+1}} = \frac{(n+1)^2}{n^2} = \frac{n^2 + 2n + 1}{n^2} = 1 + \frac{2}{n} + \frac{1}{n^2}$$

Since $\lim_{n \to \infty} n \cdot \frac{1}{n^2} = 0$ and $\sigma = 2 > 1$, the series converges. The series $\sum_{n=1}^{\infty} \frac{1}{n}$ diverges since $\sigma = 1$.

Of particular interest are those series with alternating positive and negative terms. We consider

$$\sum_{n=1}^{\infty} (-1)^{n+1} a_n = a_1 - a_2 + a_3 - a_4 + \cdots + (-1)^{n+1} a_n + \cdots \tag{10.95}$$

In order that $\sum_{n=1}^{\infty} (-1)^{n+1} a_n$ converge, of necessity, $a_n \to 0$ as $n \to \infty$. Let us assume further that $a_{n+1} \leqq a_n$ for $n = 1, 2, \ldots$. We show that the series given by (10.95) converges. First let us note that

$$\begin{aligned} S_{2n} &= (a_1 - a_2) + (a_3 - a_4) + \cdots + (a_{2n-1} - a_{2n}) \geqq 0 \\ &= a_1 - (a_2 - a_3) - (a_4 - a_5) - \cdots - (a_{2n-2} - a_{2n-1}) - a_{2n} \leqq a_1 \end{aligned}$$

Hence the sequence $S_2, S_4, \ldots, S_{2n}, \ldots$ is monotonic nondecreasing and bounded, and thus converges to a unique limit S. Moreover $S_{2n+1} = S_{2n} + a_{2n+1}$, so that $\lim_{n\to\infty} S_{2n+1} = S$ since $\lim_{n\to\infty} a_{2n+1} = 0$ by assumption. Thus the alternating series converges to a unique limit $S \geqq 0$.

If $\sum_{n=1}^{\infty} (-1)^{n+1}a_n$ converges but $\sum_{n=1}^{\infty} a_n$ diverges, we say that the series is *conditionally* convergent. If $\sum_{n=1}^{\infty} a_n$ converges, we say that the alternating series is *absolutely* convergent.

Problems

1. Apply the integral test to the series $\sum_{k=2}^{\infty} \frac{1}{k(\ln k)^\alpha}$, and show that the series converges or diverges according as $\alpha > 1$ or $\alpha \leqq 1$.

2. Show that the series $1 - \frac{1}{2} + \frac{1}{3} - \frac{1}{4} + \cdots + (-1)^{n+1}\frac{1}{n} + \cdots$ converges.

3. Show that a necessary and sufficient condition for the convergence of $\sum_{n=1}^{\infty} a_n$ is that for any $\epsilon > 0$ an integer N exists such that $|a_{n+1} + a_{n+2} + \cdots + a_{m+1}| < \epsilon$ for $m \geqq n \geqq N$.

4. Show that, for $|r| < 1$, $\lim_{n\to\infty} r^n = 0$. *Hint:* $|r|^{n+1} = |r|\,|r|^n < |r|^n$.

5. Show that $\lim_{n\to\infty} \sqrt[n]{\frac{1}{n!}} = 0$ so that $\sum_{n=1}^{\infty} \frac{1}{n!}$ converges.

6. Show that the series

$$1 + \frac{\alpha\beta}{\gamma} + \frac{\alpha(\alpha+1)\beta(\beta+1)}{\gamma(\gamma+1)2!} + \frac{\alpha(\alpha+1)(\alpha+2)\beta(\beta+1)(\beta+2)}{\gamma(\gamma+1)(\gamma+2)3!} + \cdots$$

converges if $\gamma > \alpha + \beta$ and diverges otherwise.

7. Show that the series $\sum_{n=1}^{\infty} \frac{1}{n\sqrt[n]{n}}$ diverges.

8. Consider the convergence of $\sum_{n=1}^{\infty} \left[\frac{(2n-1)!!}{(2n)!!}\,\frac{4n+3}{2n+2}\right]^2$ with

$$(2n)!! = 2 \cdot 4 \cdot 6 \cdots (2n)$$

$(2n-1)!! = 1 \cdot 3 \cdot 5 \cdots (2n-1)$.

9. Assume $a_n \geqq a_{n+1} > 0$ for $n = 1, 2, 3, \ldots$, and further assume that $\sum_{n=1}^{\infty} a_n$ converges. Show that $\lim_{n\to\infty} na_n = 0$. Why does $\sum_{n=1}^{\infty} \frac{1}{n}$ diverge?

10. Let $\sum_{n=1}^{\infty} s_n$ converge to S, $\sum_{n=1}^{\infty} t_n$ converge to T. Define $\sigma_n = \left(\sum_{i=1}^{n} s_i\right)\left(\sum_{j=1}^{n} t_j\right)$, and show that $\lim_{n\to\infty} \sigma_n = ST$.

11. Show that the series $\sum_{n=1}^{\infty} \frac{(-1)^{n+1}}{n}$ is conditionally convergent. Is it possible to rearrange the terms of this sequence so that the resulting sequence would converge to π or any other number?

12. Consider the sequence $k_1, k_2, \ldots, k_n, \ldots$, defined by $k_{i+1} = 2\sqrt{k_i}/1 + k_i$, $0 < k_1 < 1$, $i = 1, 2, \ldots$. Show that $k_{i+1} > k_i$ and $0 < k_i < 1$ for $i = 1, 2, \ldots$. Then prove that $\lim_{i\to\infty} k_i = 1$.

13. Consider the sequence of complex numbers, $a_n + b_n i$, $n = 1, 2, 3, \ldots$. If the sequence $\{a_n\}$ converges to a and the sequence $\{b_n\}$ converges to b, show that the sequence $\{a_n + b_n i\}$ converges to $a + bi$ in the sense that for any $\epsilon > 0$ an integer $N(\epsilon)$ exists such that

$$|(a + bi) - (a_n + b_n i)| < \epsilon \qquad \text{for } n \geqq N$$

14. Show that the convergence of a sequence can be made to depend on the convergence of a series. *Hint:*

$$s_n = s_1 + (s_2 - s_1) + (s_3 - s_2) + \cdots + (s_n - s_{n-1})$$

10.21. Sequences and Series with Variable Terms. Let us consider a sequence of functions

$$f_1(x), f_2(x), \ldots, f_n(x), \ldots \tag{10.96}$$

with $f_n(x)$, $n = 1, 2, \ldots$, defined on the range $a \leqq x \leqq b$. For any number $x = c$ of the interval (a, b) we can investigate the convergence of the sequence of constant terms

$$f_1(c), f_2(c), \ldots, f_n(c), \ldots \tag{10.97}$$

If the sequence of (10.97) converges, we can write

$$\lim_{n\to\infty} f_n(c) = A_c \tag{10.98}$$

where A_c is a constant which obviously depends on the number $x = c$. If the sequence of (10.96) converges for all x on the interval (a, b), we obtain a set of numbers $\{A_x\}$ which defines a function of x on (a, b), for to each x there corresponds a unique A_x. We write

$$\lim_{n\to\infty} f_n(x) = f(x) \qquad a \leqq x \leqq b \tag{10.99}$$

with $f(x) = A_x$.

Example 10.30. We consider the sequence $\{f_n(x)\}$ with $f_n(x) = x/(1 + nx)$, $n = 1, 2, \ldots, 0 \leqq x \leqq 1$. At $x = 0$ we have $f_n(0) = 0$, $n = 1, 2, 3, \ldots$, so that $\lim_{n\to\infty} f_n(0) = 0$. For $x \neq 0$ it is obvious that $\lim_{n\to\infty} f_n(x) = \lim_{n\to\infty} \dfrac{x}{1 + nx} = 0$. The limiting function for this example is $f(x) \equiv 0$, $0 \leqq x \leqq 1$. If we consider the sequence $\{f_n(x)\}$ with $f_n(x) = 1/(1 + nx)$, $n = 1, 2, 3, \ldots, 0 \leqq x \leqq 1$, we note that

$$\lim_{n\to\infty} f_n(0) = \lim_{n\to\infty} 1 = 1$$

whereas, for $x \neq 0$, $\lim_{n\to\infty} f_n(x) = 0$. The limiting function in this latter case is defined by $f(x) = 0$ for $0 < x \leqq 1$, $f(0) = 1$. This function is discontinuous at $x = 0$. Let the reader graph a few terms of this sequence.

Example 10.31. Let us see how rapidly the sequence $\{x/(1 + nx)\}$ converges to $f(x) \equiv 0$, $0 \leqq x \leqq 1$. At $x = 0$ the convergence occurs immediately since $f_n(0) = 0$ for $n = 1, 2, 3, \ldots$. For $x \neq 0$ we have

$$|f_n(x) - f(x)| = \frac{x}{1 + nx} < \epsilon \tag{10.100}$$

$\epsilon > 0$ and arbitrary, provided $(1 + nx)/x < 1/\epsilon$ or $n > 1/\epsilon - 1/x$. Thus for $n > 1/\epsilon$ (10.100) will hold since $1/\epsilon > 1/\epsilon - 1/x$. If N is the first integer greater than $1/\epsilon$, we shall have $|f_n(x) - f(x)| < \epsilon$ for $n \geqq N$. It is important to note that an integer N can be found which is independent of x for $0 \leqq x \leqq 1$. We say that the sequence converges uniformly to its limiting value $f(x) \equiv 0$.

DEFINITION 10.27. The sequence $\{f_n(x)\}$ defined for $a \leqq x \leqq b$ is said to converge *uniformly* to $f(x)$ if for any $\epsilon > 0$ an integer N exists such that

$$|f_n(x) - f(x)| < \epsilon \tag{10.101}$$

holds for all x on (a, b) provided $n \geqq N$.

Uniform convergence is essentially the following: If one imagines that a circular tube of arbitrary radius $\epsilon > 0$ surrounds $f(x)$ throughout the length of the definition of $f(x)$, then uniform convergence guarantees that an integer N can be found such that, for $n \geqq N$, $f_n(x)$ will also lie inside the tube for all x on (a, b). In other words, all terms of the sequence from the Nth term onward lie inside the ϵ tube throughout the length of the tube. The value of N, in general, depends on ϵ. Since the curves $f_n(x)$, $n = 1, 2, 3, \ldots$, are two-dimensional, the tube need only be two-dimensional. It is not necessary that $f(x)$, and hence the tube, be continuous.

Example 10.32. Let us consider the sequence $\{1 - x^n\}$, $n = 1, 2, 3, \ldots$, defined for $-\frac{1}{2} \leqq x \leqq \frac{1}{2}$. From Prob. 4, Sec. 10.20, we have that $\lim_{n\to\infty} x^n = 0$ if $|x| \leqq \frac{1}{2}$.

Thus $f(x) = \lim_{n\to\infty} f_n(x) = \lim_{n\to\infty} (1 - x^n) = 1$ for $-\frac{1}{2} \leqq x \leqq \frac{1}{2}$. We wish to determine whether the convergence is uniform. We have $|f_n(x) - f(x)| = |x|^n < (\frac{2}{3})^n$, since $|x| \leqq \frac{1}{2}$ is our range of definition. Hence we can make $|f_n(x) - f(x)| < \epsilon$ if we choose n sufficiently large so that $(\frac{2}{3})^n < \epsilon$. This can be done if we choose $n > \ln \epsilon/(\ln 2 - \ln 3)$ for $0 < \epsilon < 1$. If $\epsilon \geqq 1$, we can choose $n \geqq 1$. Thus the sequence $\{f_n(x)\}$ converges uniformly to $f(x)$ on the range $-\frac{1}{2} \leqq x \leqq \frac{1}{2}$.

Example 10.33. We consider the sequence $\{nxe^{-nx}\}$ defined for $0 \leqq x \leqq 1$. Let the reader show that

$$f(x) = \lim_{n\to\infty} nxe^{-nx} = 0 \qquad 0 \leqq x \leqq 1$$

We show, however, that the sequence does not converge uniformly to $f(x)$. First we note that

$$|f_n(x) - f(x)| = nxe^{-nx} \qquad 0 \leqq x \leqq 1$$

If the convergence were uniform, then for $\epsilon = 0.01$ we would be able to find an integer N such that $nxe^{-nx} < 0.01$ for $n \geqq N$ and for all x on $0 \leqq x \leqq 1$. In particular Nxe^{-Nx} would be less than 0.01 for all x on $(0, 1)$. If we choose $x = 1/N$, we have $N \cdot (1/N)e^{-N\cdot(1/N)} = e^{-1} > 0.01$, a contradiction. Hence the sequence fails to converge uniformly to $f(x)$ on the range $0 \leqq x \leqq 1$.

The following theorems will emphasize the importance of uniform convergence:

THEOREM 10.29. Let $\{f_n(x)\}$ be a sequence of continuous functions defined for $a \leqq x \leqq b$. If the sequence converges uniformly to $f(x)$ on (a, b), then $f(x)$ is continuous on (a, b).

The proof is simple. Let $x = c$ be any point of (a, b), and choose any $\epsilon > 0$. From

$$f(c+h) - f(c) = [f(c+h) - f_n(c+h)] + [f_n(c+h) - f_n(c)] + [f_n(c) - f(c)]$$

we have

$$|f(c+h) - f(c)| \leqq |f(c+h) - f_n(c+h)| + |f_n(c+h) - f_n(c)| + |f_n(c) - f(c)| \qquad (10.102)$$

From uniform convergence we note that an integer N exists such that $|f_n(x) - f(x)| < \epsilon/3$ for $n \geqq N$ and for all x on (a, b). Applying this result to (10.102) yields

$$|f(c+h) - f(c)| < \tfrac{2}{3}\epsilon + |f_N(c+h) - f_N(c)|$$

Since $f_N(x)$ is continuous at $x = c$, we have $|f_N(c+h) - f_N(c)| < \epsilon/3$ for $|h| < \delta$, $\delta > 0$. Hence for any $\epsilon > 0$ a $\delta > 0$ exists such that $|f(c+h) - f(c)| < \epsilon$ for $|h| < \delta$. Q.E.D.

Example 10.33 shows that $f(x)$ can be continuous even though the convergence is not uniform. Uniform convergence is a sufficient condition for continuity to occur, but is not necessary.

THEOREM 10.30. Let $\{f_n(x)\}$ be a sequence of continuous functions defined for $a \leqq x \leqq b$. If the sequence converges uniformly to $f(x)$, then

$$\int_a^b f(x)\,dx = \int_a^b \lim_{n\to\infty} f_n(x)\,dx = \lim_{n\to\infty} \int_a^b f_n(x)\,dx \tag{10.103}$$

The proof of (10.103) proceeds as follows: Define $R_n(x)$ by

$$f(x) = f_n(x) + R_n(x) \tag{10.104}$$

Since $f(x)$ is continuous from the previous theorem, $R_n(x)$ is also continuous, for it is the difference of two continuous functions. From (10.104) we have

$$\int_a^b f(x)\,dx = \int_a^b f_n(x)\,dx + \int_a^b R_n(x)\,dx$$

or

$$\left|\int_a^b f(x)\,dx - \int_a^b f_n(x)\,dx\right| = \left|\int_a^b R_n(x)\,dx\right|$$

From uniform convergence we note that for any $\epsilon > 0$, and hence for $\epsilon/(b-a) > 0$, an integer N exists such that $|R_n(x)| < \epsilon/(b-a)$ for $n \geqq N$ and for all x on (a, b). Thus

$$\left|\int_a^b f(x)\,dx - \int_a^b f_n(x)\,dx\right| \leqq \int_a^b |R_n(x)|\,dx < \epsilon$$

for $n \geqq N$. This means that the sequence $\int_a^b f_n(x)\,dx$ converges to $\int_a^b f(x)\,dx$, so that (10.103) results. Q.E.D. It should be emphasized that Theorem 10.30 was shown to be true provided a and b are finite.

Example 10.34. We consider the sequence of Example 10.31. We have

$$\begin{aligned}\lim_{n\to\infty} \int_0^1 \frac{x}{1+nx}\,dx &= \lim_{n\to\infty} \frac{1}{n}\int_0^1 \left(1 - \frac{1}{1+nx}\right) dx \\ &= \lim_{n\to\infty} \frac{1}{n}\left[1 - \frac{\ln(1+n)}{n}\right] = 0\end{aligned}$$

Moreover $f(x) \equiv 0$ so that $\int_0^1 f(x)\,dx = 0$, which checks the results of Theorem 10.30 since the convergence was seen to be uniform.

Example 10.35. Let the reader show that the sequence $\{nxe^{-nx^2}\}$ does not converge uniformly on the range $0 \leqq x \leqq 1$ but does converge to $f(x) \equiv 0$. Now

$$\int_0^1 nxe^{-nx^2}\,dx = -\tfrac{1}{2}e^{-nx^2}\Big|_0^1 = \tfrac{1}{2}(1 - e^{-n})$$

so that

$$\lim_{n\to\infty} \int_0^1 nxe^{-nx^2}\,dx = \tfrac{1}{2} \neq \int_0^1 f(x)\,dx = 0$$

On the other hand, the sequence $\{nx/(1 + n^2x^2)\}$ does not converge uniformly on the range $0 \leqq x \leqq 1$ but does converge to $f(x) \equiv 0$, and

$$\lim_{n\to\infty} \int_0^1 \frac{nx}{1+n^2x^2}\,dx = \lim_{n\to\infty} \frac{\ln(1+n^2)}{2n} = 0 = \int_0^1 f(x)\,dx$$

Uniform convergence is a sufficient condition to apply (10.103), but it is not a necessary condition.

THEOREM 10.31. Let $\{f_n(x)\}$ be a sequence defined over $a \leqq x \leqq b$ which is known to converge to the constant $f(c)$ at $x = c$, $a \leqq c \leqq b$. Assume further that the sequence $\{f_n'(x)\}$ converges uniformly to $g(x)$ on (a, b) with $f_n'(x)$ continuous on (a, b) for $n = 1, 2, \ldots$. We show that the sequence $\{f_n(x)\}$ converges uniformly to a function $f(x)$ such that $f'(x) = g(x)$ for $a \leqq x \leqq b$, that is,

$$\lim_{n\to\infty} f_n'(x) = f'(x) = \frac{d}{dx}\left[\lim_{n\to\infty} f_n(x)\right] \tag{10.105}$$

Since the sequence $\{f_n'(x)\}$ converges uniformly to $g(x)$ we have from Theorem 10.30 that

$$\int_c^x g(x)\,dx = \lim_{n\to\infty}\int_c^x f_n'(x)\,dx = \lim_{n\to\infty}\left[f_n(x) - f_n(c)\right] \tag{10.106}$$

From (10.106) the reader can readily deduce that

$$\lim_{n\to\infty} f_n(x) = \int_c^x g(x)\,dx + f(c)$$

Thus $\lim\limits_{n\to\infty} f_n(x) = f(x)$ exists, and since $g(x)$ is continuous, we have $f'(x) = g(x)$. From $f'(x) = f_n'(x) + R_n(x)$ with $|R_n(x)| < \epsilon$ for $n \geqq N$ let the reader show that the sequence $\{f_n(x)\}$ converges to $f(x)$ uniformly.

If $f_n(x)$, $n = 1, 2, 3, \ldots$ is defined for $a \leqq x \leqq b$, we say that the series $\sum\limits_{n=1}^{\infty} f_n(x)$ converges at $x = c$ if the series of constant terms $\sum\limits_{n=1}^{\infty} f_n(c)$ converges, $a \leqq c \leqq b$. We write $f(c) = \sum\limits_{n=1}^{\infty} f_n(c)$, where

$$f(c) = \lim_{n\to\infty}\sum_{k=1}^{n} f_k(c)$$

If $\sum\limits_{n=1}^{\infty} f_n(x)$ converges for all x on (a, b), we write

$$f(x) = \sum_{n=1}^{\infty} f_n(x) = \lim_{n\to\infty}\sum_{k=1}^{n} f_k(x) \tag{10.107}$$

It is convenient to express $f(x)$ as a finite series plus a remainder which obviously contains an infinite number of terms,

$$f(x) = \sum_{k=1}^{n} f_k(x) + R_n(x) \tag{10.108}$$

with $R_n(x) = \sum_{k=n+1}^{\infty} f_k(x)$.

DEFINITION 10.28. The series $\sum_{n=1}^{\infty} f_n(x)$ is said to converge uniformly to $f(x)$ on (a, b) if for any $\epsilon > 0$ an integer N exists such that

$$|R_n(x)| < \epsilon \qquad \text{for } n \geqq N, \qquad a \leqq x \leqq b$$

In terms of the Cauchy criterion we note that uniform convergence exists if for any $\epsilon > 0$ an integer N exists such that

$$|s_{n+p}(x) - s_n(x)| < \epsilon \tag{10.109}$$

for $n \geqq N$, all $p > 0$, and for all x on (a, b), with $s_n(x) = \sum_{k=1}^{n} f_k(x)$.

The results concerning continuity, integration, and differentiation of uniformly convergent sequences apply equally well to any uniformly convergent series. To prove these results, one need only change the series into a sequence [see (10.88)].

Before constructing a few tests for determining the uniform convergence of a series we prove a result due to Abel. Let $u_1, u_2, \ldots, u_n$ be a monotonic nonincreasing sequence of positive terms, $u_{j+1} \leqq u_j$, $u_j > 0$. Let $a_1, a_2, \ldots, a_n$ be any set of numbers, with M and m the maximum and minimum, respectively, of the set

$$a_1, a_1 + a_2, a_1 + a_2 + a_3, \ldots, \sum_{i=1}^{n} a_i$$

We show that

$$mu_1 \leqq \sum_{j=1}^{n} a_j u_j \leqq Mu_1 \tag{10.110}$$

Proof. Let $s_1 = a_1$, $s_2 = a_1 + a_2, \ldots, s_n = \sum_{i=1}^{n} a_i$, so that $a_1 = s_1$, $a_2 = s_2 - s_1$, $a_3 = s_3 - s_2, \ldots, a_n = s_n - s_{n-1}$. We write

$$\begin{aligned}\sum_{j=1}^{n} a_j u_j &= \sum_{j=1}^{n} (s_j - s_{j-1}) u_j \qquad s_0 = 0 \\ &= s_1 u_1 + (s_2 - s_1) u_2 + (s_3 - s_2) u_3 + \cdots + (s_n - s_{n-1}) u_n \\ &= s_1(u_1 - u_2) + s_2(u_2 - u_3) + \cdots + s_{n-1}(u_{n-1} - u_n) + s_n u_n\end{aligned}$$

Since $u_{j+1} \leqq u_j$ for $j = 1, 2, \ldots, n - 1$, and since $u_n > 0$, we note that

$$\sum_{j=1}^{n} a_j u_j \leqq M(u_1 - u_2) + M(u_2 - u_3) + \cdots + M u_n = M u_1$$
$$\sum_{j=1}^{n} a_j u_j \geqq m(u_1 - u_2) + m(u_2 - u_3) + \cdots + m u_n = m u_1$$

which proves (10.110).

A. *Abel's Test for Uniform Convergence.* $\sum_{n=1}^{\infty} a_n u_n(x)$ converges uniformly on (a, b) if

1. $\sum_{n=1}^{\infty} a_n$ converges.
2. $u_n(x) > 0$ and $u_n(x) \geqq u_{n+1}(x)$, $a \leqq x \leqq b$, $n \geqq 1$.
3. $|u_1(x)| < k$ for all x on (a, b).

Proof. Since $\sum_{n=1}^{\infty} a_n$ converges, an integer N exists such that $\left| \sum_{n=N+1}^{N+p} a_n \right| < \epsilon/k$ for all $p > 0$. From (10.110), Abel's lemma, we have

$$\left| \sum_{n=N+1}^{N+p} a_n u_n(x) \right| < \frac{\epsilon}{k} u_{N+1}(x) \leqq \frac{\epsilon}{k} u_1(x) < \epsilon$$

for all $p > 0$. Q.E.D.

B. *The Weierstrass M Test.* Assume $|u_n(x)| \leqq M_n =$ constant, for $n = 1, 2, 3, \ldots, a \leqq x \leqq b$. If $\sum_{n=1}^{\infty} M_n$ converges, then $\sum_{n=1}^{\infty} u_n(x)$ converges uniformly on (a, b). *Proof.* For any $\epsilon > 0$ an integer N exists such that

$$\sum_{n=N+1}^{N+p} M_n < \epsilon \qquad \text{for all } p > 0$$

But $$\left| \sum_{n=N+1}^{N+p} u_n(x) \right| \leqq \sum_{n=N+1}^{N+p} |u_n(x)| \leqq \sum_{n=N+1}^{N+p} M_n < \epsilon \qquad \text{Q.E.D.}$$

C. $\sum_{n=1}^{\infty} u_n(x) v_n(x)$ is uniformly convergent on (a, b) if:

1. $\left|\sum_{i=1}^{n} u_i(x)\right| < M$ = constant, for all n and for all x on (a, b).

2. $\sum_{n=1}^{\infty} |v_{n+1}(x) - v_n(x)|$ is uniformly convergent on (a, b).

3. $v_n(x) \to 0$ uniformly as $n \to \infty$.

Proof. Let $s_n(x) = \sum_{i=1}^{n} u_i(x) \qquad |s_n(x)| < M$ from (1)

$$S_n(x) = \sum_{i=1}^{n} u_i(x)v_i(x)$$

Now

$$\begin{aligned} S_{n+p} - S_n &= u_{n+1}v_{n+1} + u_{n+2}v_{n+2} + \cdots + u_{n+p}v_{n+p} \\ &= (s_{n+1} - s_n)v_{n+1} + \cdots + (s_{n+p} - s_{n+p-1})v_{n+p} \\ &= -s_nv_{n+1} + s_{n+1}(v_{n+1} - v_{n+2}) + \cdots + s_{n+p}v_{n+p} \end{aligned}$$

$$\begin{aligned} |S_{n+p} - S_n| &\leqq |s_n|\,|v_{n+1}| + \sum_{r=n+1}^{n+p-1} |s_r|\,|v_r - v_{r+1}| + |s_{n+p}|\,|v_{n+p}| \\ &< M\left[|v_{n+1}| + |v_{n+p}| + \sum_{r=n+1}^{n+p-1} |v_r - v_{r+1}|\right] \end{aligned}$$

Since $v_n(x) \to 0$ uniformly as $n \to \infty$, we have $|v_{n+1}| < \epsilon/3M$, $|v_{n+p}| < \epsilon/3M$ for $n \geqq N_1$. From (2), $\sum_{r=n+1}^{n+p-1} |v_r - v_{r+1}| < \epsilon/3M$ for $n \geqq N_2$. For N equal to the larger of N_1, N_2 we have

$$|S_{n+p} - S_n| < \epsilon \qquad n \geqq N, \text{ all } p > 0 \qquad \text{Q.E.D.}$$

Example 10.36. We consider

$$\varphi(x) = 1^{-x} - 2^{-x} + 3^{-x} - 4^{-x} + \cdots + (-1)^{n-1}n^{-x} + \cdots$$

for $0 < \delta \leqq x \leqq R$. We have $\varphi(x) = \sum_{n=1}^{\infty} a_nv_n(x)$ with

$$(1)\ a_n = (-1)^{n-1} \qquad \left|\sum_{i=1}^{n} a_i\right| < 2$$

$$\begin{aligned} (2)\ |v_{n+1} - v_n| &= |(n+1)^{-x} - n^{-x}| = x\left|\int_n^{n+1} t^{-(x+1)}\,dt\right| \qquad x > 0 \\ &\leqq x\int_n^{n+1} t^{-(\delta+1)}\,dt < xn^{-(\delta+1)} \end{aligned}$$

$$\sum_{n=1}^{\infty} |v_{n+1} - v_n| < x\sum_{n=1}^{\infty} n^{-(\delta+1)} \leqq R\sum_{n=1}^{\infty} n^{-(\delta+1)}$$

Since $\sum_{n=1}^{\infty} n^{-(\delta+1)}$ converges, the Weierstrass M test states that $\sum_{n=1}^{\infty} |v_{n+1} - v_n|$ converges uniformly for $0 < \delta \leqq x \leqq R$.

(3) $n^{-x} \leqq n^{-\delta} \to 0$ so that $n^{-x} \to 0$ uniformly

From (C), $\varphi(x)$ converges uniformly for $0 < \delta \leqq x \leqq R$.

An important type of series is the power series

$$P(x) = \sum_{n=0}^{\infty} a_n x^n \tag{10.111}$$

If we apply the ratio test, we note that the series of (10.111) converges if

$$\lim_{n\to\infty} \left| \frac{a_{n+1}x^{n+1}}{a_n x^n} \right| < 1$$

or

$$|x| < \lim_{n\to\infty} \left| \frac{a_n}{a_{n+1}} \right| = R \tag{10.112}$$

We call R the radius of convergence of the series. The word "radius" comes into play if we consider the complex series $P(z) = \sum_{n=0}^{\infty} a_n z^n$, $z = x + iy$, which converges for $|z| < R$. If $P(x)$ converges for $|x| < R$, it follows that the series $Q(x) = \sum_{n=0}^{\infty} |a_n| x^n$ also converges for $|x| < R$ [see (10.112)]. The ratio test fails to tell us anything concerning the end points $x = -R$, $x = R$. The convergence of $\sum_{n=0}^{\infty} a_n R^n$ and $\sum_{n=0}^{\infty} a_n(-R)^n$ must be examined by the methods of Sec. 10.20 or by other means.

Example 10.37. (1) $\sum_{n=0}^{\infty} n!x^n$ converges only for $x = 0$. (2) $\sum_{n=0}^{\infty} x^n$ converges for $-1 < x < 1$. (3) $\sum_{n=0}^{\infty} \frac{x^n}{n!}$ converges for all finite x.

If $P(x) = \sum_{n=0}^{\infty} a_n x^n$ converges for $|x| < R$, it is very easy to show that $\sum_{n=0}^{\infty} a_n x^n$ converges uniformly for $|x| \leqq R_1 < R$. We have for $|x| \leqq R_1$

that $|a_nx^n| \leqq |a_n|R_1^n$, $n = 1, 2, 3, \ldots$, and since $\sum_{n=0}^{\infty} |a_n|R_1^n$ converges, the Weierstrass M test yields uniform convergence. Since each term of the series is continuous, $P(x)$ is continuous inside its region of convergence. Let the reader show that $P'(x)$ has the same radius of convergence as $P(x)$. Why is it that $P'(x) = \sum_{n=1}^{\infty} na_nx^{n-1}$?

A result due to Abel concerning power series can be stated as follows: If $\sum_{n=0}^{\infty} a_n$ converges to s, then $\sum_{n=0}^{\infty} a_nx^n$ is uniformly convergent for $0 \leqq x \leqq 1$, and $\lim_{x \to 1} \sum_{n=0}^{\infty} a_nx^n = s$. The proof depends on the result given by (10.110). Since $\sum_{n=0}^{\infty} a_n$ converges, we know that for any $\epsilon > 0$ an integer N exists such that $|a_n + a_{n+1} + \cdots + a_{n+p}| < \epsilon$ for $n \geqq N$, all $p \geqq 0$. For $0 \leqq x \leqq 1$ we know that x^n, $n = 0, 1, 2, 3, \ldots$, is a monotonic nonincreasing sequence, so that

$$\left| \sum_{i=n}^{n+p} a_ix^i \right| \leqq \epsilon x^n \leqq \epsilon$$

[see (10.110)]. Hence the series is uniformly convergent for $0 \leqq x \leqq 1$ so that $P(x) = \sum_{n=0}^{\infty} a_nx^n$ is continuous on the interval $0 \leqq x \leqq 1$. Hence $\lim_{x \to 1} P(x) = P(1) = \sum_{n=0}^{\infty} a_n = s$. For example, it is known that

$$\ln (1 + x) = \sum_{n=1}^{\infty} \frac{(-1)^{n+1}x^n}{n}$$

for $|x| < 1$. Since $\sum_{n=1}^{\infty} \frac{(-1)^{n+1}}{n}$ converges, we have $\ln 2 = \sum_{n=1}^{\infty} \frac{(-1)^{n+1}}{n}$. The converse of Abel's theorem is not true. If $P(x) = \sum_{n=0}^{\infty} a_nx^n$ converges to s as $x \to 1$, we cannot say that $\sum_{n=1}^{\infty} a_n$ converges to s. An example due to Tauber is as follows: $P(x) = \sum_{n=0}^{\infty} (-1)^n x^n = 1/(1+x)$, and $\lim_{x \to 1} P(x) = \frac{1}{2}$. However, $\sum_{n=0}^{\infty} (-1)^n$ is not convergent.

Problems

1. Show that $1/(1+t^2) = \sum_{n=0}^{\infty} (-1)^n t^{2n}$ converges uniformly for $0 \leqq t \leqq x < 1$. Integrate, and show that

$$\tan^{-1} x = \sum_{n=0}^{\infty} (-1)^n \frac{x^{2n+1}}{2n+1}$$

$$\pi = 4 \sum_{n=0}^{\infty} (-1)^n \frac{1}{2n+1}$$

2. If $\lim_{n \to \infty} \sqrt[n]{a_n} = r$ exists, show that $P(z) = \sum_{n=0}^{\infty} a_n (z - z_0)^n$ converges for $|z - z_0| < 1/r$.

3. The series $\sum_{n=0}^{\infty} u_n(x)$ is said to be absolutely convergent for $a \leqq x \leqq b$ if $\sum_{n=0}^{\infty} |u_n(x)|$ converges for $a \leqq x \leqq b$. Show that $S(x) = \sum_{n=1}^{\infty} \frac{x^2}{(1+x^2)^n}$ is absolutely convergent for $|x| \leqq 1$ and that $S(x)$ is not uniformly convergent for $|x| \leqq 1$.

4. Show that $\sum_{n=0}^{\infty} \frac{\sin (2n+1)x}{2n+1}$ is uniformly convergent for $\pi/4 \leqq x \leqq 3\pi/4$. By considering $x = \pi/2$ show that the series is not absolutely convergent for $\pi/4 \leqq x \leqq 3\pi/4$.

5. Show that

$$\int_0^1 \frac{\ln x}{1-x}\,dx = \sum_{n=0}^{\infty} \int_0^1 x^n \ln x \, dx = -\sum_{n=0}^{\infty} \frac{1}{(n+1)^2} = -\frac{\pi^2}{6}$$

6. Prove that the series $\sum_{n=1}^{\infty} \frac{x^{\frac{3}{2}}}{1+n^2x^2}$ is uniformly convergent for all x.

7. Prove that $\int_0^1 \ln \frac{1+x}{1-x} \frac{dx}{x} = \frac{\pi^2}{4}$.

8. Let $S(x) = \sum_{n=1}^{\infty} u_n(x)$ converge uniformly for $a \leqq x \leqq b$, and assume further that $\lim_{x \to c} u_n(x) = L_n$ for all n, $u_n(x)$ not necessarily continuous. If $\sum_{n=1}^{\infty} L_n$ converges, show that $\lim_{x \to c} S(x) = \sum_{n=1}^{\infty} L_n$.

9. Let $a_{mn} \geqq 0$ for $m = 1, 2, 3, \ldots, n = 1, 2, 3, \ldots$, and suppose a constant K exists such that

$$\sum_{n=1}^{N} \sum_{m=1}^{M} a_{mn} < K$$

for all M, N. Show that $\sum_{n=1}^{\infty} \sum_{m=1}^{\infty} a_{mn}$ exists and that

$$\sum_{n=1}^{\infty} \sum_{m=1}^{\infty} a_{mn} = \sum_{m=1}^{\infty} \sum_{n=1}^{\infty} a_{mn} \tag{10.113}$$

Write the elements $\{a_{mn}\}$ as a square array, and interpret (10.113).

10. If $P(z) = \sum_{n=0}^{\infty} a_n z^n$ and

$$\begin{aligned} P(z + w) &= \sum_{n=0}^{\infty} a_n (z + w)^n \\ &= \sum_{n=0}^{\infty} \sum_{r=0}^{n} a_n \binom{n}{r} w^r z^{n-r} \end{aligned}$$

converges absolutely, that is, $\sum_{n=0}^{\infty} \sum_{r=0}^{n} |a_n| \binom{n}{r} |w|^r |z|^{n-r}$ converges, show that

$$P(z + w) = \sum_{r=0}^{\infty} \sum_{s=0}^{\infty} \binom{s + r}{r} a_{s+r} z^s w^r \tag{10.114}$$

Apply (10.114) to $E(z) = \sum_{n=0}^{\infty} \frac{z^n}{n!}$, and show that $E(z + w) = E(z)E(w)$.

11. A series $\sum_{n=1}^{\infty} u_n(x)$ is said to be "boundedly convergent" for the interval $a \leqq x \leqq b$ if $\sum_{n=1}^{\infty} u_n(x)$ converges for all x on (a, b) and if a constant M exists such that $\left| \sum_{n=1}^{\infty} u_n(x) \right| < M$ for all x on (a, b). The series is said to be uniformly continuous on (a, b) except for the point c if the series converges uniformly on the intervals $a \leqq x \leqq c - \delta$, $c + \delta \leqq x \leqq b$, however small δ may be, $\delta > 0$. Show that if a series is uniformly convergent on $a \leqq x \leqq b$ except for a finite number of points and if the series is also boundedly convergent then the series may be integrated term by term.

12. For what ranges of x do the following series converge uniformly?

$$\sum_{n=1}^{\infty} \frac{1}{x^2 + n^2} \qquad \sum_{n=1}^{\infty} \frac{1}{x^2 n^2}$$

13. *Second Law of the Mean for Integrals (Bonnet).* Consider

$$\int_a^b f(x)\varphi(x)\,dx \approx \sum_{j=0}^{n-1} f(x_j)\varphi(x_j)(x_{j+1} - x_j)$$

Let $\varphi(x)$ be positive and monotonic nonincreasing on (a, b), and consider the set of numbers

$$\begin{aligned} S_0 &= f(a)(x_1 - a) \\ S_1 &= f(a)(x_1 - a) + f(x_1)(x_2 - x_1) \\ &\cdots\cdots\cdots\cdots\cdots \\ S_n &= \sum_{j=0}^{n-1} f(x_j)(x_{j+1} - x_j) \to \int_a^b f(x)\,dx \end{aligned}$$

Let A and B be the minimum and maximum values of the set S_i, $i = 0, 1, 2, \ldots, n$, apply Abel's result of (10.110), and show that

$$A\varphi(a) \leqq \int_a^b f(x)\varphi(x)\,dx \leqq B\varphi(a)$$

If $f(x)$ is continuous, show that

$$\int_a^b f(x)\varphi(x)\,dx = \varphi(a) \int_a^\xi f(x)\,dx \qquad a \leqq \xi \leqq b$$

Consider $\varphi(x) - \varphi(b) \neq 0$, $\varphi(x)$ monotonic nondecreasing and positive, and show that

$$\int_a^b f(x)\varphi(x)\,dx = \varphi(a) \int_a^\xi f(x)\,dx + \varphi(b) \int_\xi^b f(x)\,dx \tag{10.115}$$

10.22. Dini's Conditions. A simple example shows that (10.103) does not necessarily hold if one of the limits of integration is infinite. Let us consider the sequence $\{f_n(x)\}$ with $f_n(x) = (2x/n^2)e^{-x^2/n^2}$, $n = 1, 2, 3, \ldots$, with $x \geqq 0$. It is easy to see that $\lim_{n\to\infty} f_n(x) = f(x) \equiv 0$ for $x \geqq 0$. By setting $f_n'(x) = 0$ one easily shows that the maximum value of $f_n(x)$ for $x \geqq 0$ occurs at $x = n/\sqrt{2}$ so that $|f_n(x)| \leqq \sqrt{2/e}\,(1/n) < 1/n < \epsilon$ for $n \geqq N = 1 + [1/\epsilon]$. Hence the sequence converges uniformly to its limit $f(x) \equiv 0$ for $x \geqq 0$. On the other hand,

$$\begin{aligned} \lim_{n\to\infty} \int_0^\infty f_n(x)\,dx &= \lim_{n\to\infty} \int_0^\infty \frac{2x}{n^2} e^{-x^2/n^2}\,dx = \lim_{n\to\infty} 1 = 1 \\ &\neq \int_0^\infty \lim_{n\to\infty} f_n(x)\,dx = \int_0^\infty 0 \cdot dx = 0 \end{aligned}$$

We now state and prove a result due to Dini. We can write

$$\int_a^\infty \sum_{n=1}^\infty u_n(x)\,dx = \sum_{n=1}^\infty \int_a^\infty u_n(x)\,dx \tag{10.116}$$

if the following conditions are fulfilled:

1. $u_n(x)$ is continuous for $x \geqq a$, $n = 1, 2, 3, \ldots$.
2. $\lim_{x\to\infty} v_n(x) = v_n(\infty)$ exists for $n = 1, 2, 3, \ldots$, with

$$v_n(x) = \int_a^x u_n(t)\,dt$$

3. $\sum_{n=1}^\infty u_n(x)$ converges uniformly for $a \leqq x \leqq X$, X arbitrary but finite.
4. $\sum_{n=1}^\infty v_n(x)$ converges uniformly for $x \geqq a$.

Proof. From (3) and (2) we can write

$$\int_a^X \sum_{n=1}^\infty u_n(x)\,dx = \sum_{n=1}^\infty \int_a^X u_n(x)\,dx = \sum_{n=1}^\infty v_n(X)$$

Hence

$$\int_a^\infty \sum_{n=1}^\infty u_n(x)\,dx \equiv \lim_{X\to\infty} \int_a^X \sum_{n=1}^\infty u_n(x)\,dx = \lim_{X\to\infty} \sum_{n=1}^\infty v_n(X)$$

To prove (10.116), we must show that $\lim_{X\to\infty} \sum_{n=1}^\infty v_n(X) = \sum_{n=1}^\infty v_n(\infty)$, since $\sum_{n=1}^\infty v_n(\infty) = \sum_{n=1}^\infty \int_a^\infty u_n(x)\,dx$. It is first necessary to show that $\sum_{n=1}^\infty v_n(\infty)$ exists by making use of (2) and (4). From (4) an integer N can be found such that

$$\left| \sum_{n=1}^\infty v_n(X) - \sum_{n=1}^N v_n(X) \right| < \frac{\epsilon}{4} \qquad \text{for all } X$$

From (2), $\lim_{X\to\infty} v_n(X) = v_n(\infty)$ for $n = 1, 2, 3, \ldots, N$, so that $\sum_{n=1}^N [v_n(X) - v_n(\infty)] < \epsilon/4$ for $X \geqq X_0$, since the limit of a finite sum is the sum of the limits.

The identity

$$\sum_{n=1}^\infty v_n(X) - \sum_{n=1}^N v_n(\infty) = \left[\sum_{n=1}^\infty v_n(X) - \sum_{n=1}^N v_n(X) \right] + \sum_{n=1}^N [v_n(X) - v_n(\infty)]$$

shows that for any $\epsilon > 0$ we can find an integer N and then an X_0 such that

$$\left|\sum_{n=1}^{\infty} v_n(X) - \sum_{n=1}^{N} v_n(\infty)\right| < \frac{\epsilon}{2} \tag{10.117}$$

for $X \geqq X_0$. By the same reasoning we have

$$\left|\sum_{n=1}^{\infty} v_n(X) - \sum_{n=1}^{N+p} v_n(\infty)\right| < \frac{\epsilon}{2}$$

for $X \geqq X_1$ and p a positive fixed integer. Choosing any X larger than both X_0 and X_1 yields

$$\left|\sum_{n=1}^{N+p} v_n(\infty) - \sum_{n=1}^{N} v_n(\infty)\right| < \epsilon \tag{10.118}$$

Now (10.118) holds independent of any X since no X appears in (10.118), so that for any $\epsilon > 0$ an integer N exists such that (10.118) holds for all $p > 0$. This is exactly the Cauchy criterion for convergence, so that $\sum_{n=1}^{\infty} v_n(\infty)$ exists. Hence for $\epsilon/2 > 0$ an integer N_1 exists such that

$$\left|\sum_{n=1}^{\infty} v_n(\infty) - \sum_{n=1}^{N_1} v_n(\infty)\right| < \frac{\epsilon}{2} \tag{10.119}$$

Choosing the larger N of (10.117) and (10.119) and combining (10.117) with (10.119) yields

$$\left|\sum_{n=1}^{\infty} v_n(X) - \sum_{n=1}^{\infty} v_n(\infty)\right| < \epsilon \tag{10.120}$$

for $X \geqq X_0$. Formula (10.120) is just the statement that

$$\lim_{X\to\infty} \sum_{n=1}^{\infty} v_n(X) = \sum_{n=1}^{\infty} v_n(\infty) \qquad \text{Q.E.D.}$$

Example 10.38. We consider the Bessel function

$$J_0(x) = \sum_{n=0}^{\infty} \frac{(-1)^n(x/2)^{2n}}{n!n!}$$

and show that $\int_0^\infty e^{-x}J_0(x)\,dx = 1/\sqrt{2}$. The nth term of the series $e^{-x}J_0(x)$ is

$u_n(x) = e^{-x}\dfrac{(-1)^n(x/2)^{2n}}{n!n!}$, and we see that $u_n(x)$ is continuous for $x \geqq 0$. Also

$$v_n(x) = \int_0^x u_n(t)\,dt = \frac{(-1)^n}{2^{2n}n!n!}\int_0^x e^{-t}t^{2n}\,dt$$

and
$$v_n(\infty) = \lim_{x\to\infty} v_n(x) = \frac{(-1)^n}{2^{2n}n!n!}\int_0^\infty e^{-t}t^{2n}\,dt = \frac{(-1)^n(2n)!}{2^{2n}n!n!}$$

Let the reader show that $\lim_{n\to\infty} v_n(\infty) = 0$. By the ratio test it is easy to determine that $e^{-x}J_0(x)$ converges for $|x| \leqq X$, X arbitrary. Finally it remains to show that $\sum_{n=0}^{\infty} v_n(x)$ converges uniformly for $x \geqq 0$. If we write

$$\left|\sum_{n=0}^{\infty} v_n(x)\right| \leqq \sum_{n=0}^{\infty} \frac{(2n)!}{2^{2n}n!n!} \tag{10.121}$$

we cannot show uniform convergence since the series of constant terms in (10.121) diverges. However, through integration by parts the reader can readily verify that

$$\frac{1}{2^{2n+2}(n+2)!(n+2)!}\int_0^x e^{-t}t^{2n+2}\,dt < \frac{1}{2^{2n}n!n!}\int_0^x e^{-t}t^{2n}\,dt \qquad x > 0$$

so that $\sum_{n=0}^{\infty} v_n(x)$ is a series of alternating terms with $|v_{n+1}(x)| < |v_n(x)|$ for $x > 0$, $v_n(0) = 0$, $n = 0, 1, 2, \ldots$. For such a series the remainder after n terms of the series $\sum_{n=0}^{\infty} v_n(x)$ has the property that

$$|R_n(x)| \leqq |v_{n+1}(x)| \leqq \frac{(2n+2)!}{2^{2n+2}(n+1)!(n+1)!} \tag{10.122}$$

Since the right-hand side of (10.122) tends to zero as n becomes infinite, we note that $\sum_{n=0}^{\infty} v_n(x)$ converges uniformly for $x \geqq 0$. Applying (10.116) yields

$$\int_0^\infty e^{-x}J_0(x)\,dx = \sum_{n=0}^{\infty} \frac{(-1)^n(2n)!}{2^{2n}n!n!}$$

In the next section (Example 10.40) we shall show that

$$(1+x)^{-\frac{1}{2}} = \sum_{n=0}^{\infty} \frac{(-1)^n(2n)!}{2^{2n}n!n!}x^n \qquad -1 < x \leq 1 \tag{10.123}$$

For $x = 1$ we have $\int_0^\infty e^{-x}J_0(x)\,dx = 1/\sqrt{2}$.

Problems

1. Let $S(x) = \sum_{n=1}^{\infty} \frac{1}{(x+n)^3}$, $x \geqq 0$. Show that $\int_0^\infty S(x)\,dx = \frac{1}{2}\sum_{n=1}^{\infty}\frac{1}{n^2}$.

2. We wish to determine $f(x)$ such that

$$\int_0^\infty e^{-sx} f(x)\,dx = \frac{1}{\sqrt{1+s^2}} \qquad s \geqq 0$$

Assume $f(x) = \sum_{n=0}^{\infty} a_n x^n$, integrate formally, make use of (10.123), and show that $f(x) = J_0(x)$. Then justify your work.

3. Consider $f_n(x) = n^2 x e^{-nx}$, $n = 1, 2, 3, \ldots$, show that $\lim_{n\to\infty}\int_0^\infty f_n(x)\,dx \neq \int_0^\infty \lim_{n\to\infty} f_n(x)\,dx$, and determine which one of Dini's conditions is not fulfilled.

4. Consider $f_n(x) = nx/(1 + n^2x^2)$, $n = 1, 2, 3, \ldots$, and show that

$$\lim_{n\to\infty}\int_0^\infty f_n(x)\,dx = \int_0^\infty \lim_{n\to\infty} f_n(x)\,dx$$

Do Dini's conditions hold for this case?

10.23. Taylor Series. Let $f(x)$ be a function defined on $c \leqq x \leqq d$ such that $f'(x), f''(x), \ldots, f^{(n)}(x)$ exist on (c, d) with $f^{(n)}(x)$ R-integrable. We note that, for $c \leqq a \leqq x \leqq d$ or $c \leqq x \leqq a \leqq d$,

$$\int_a^x f^{(n)}(t)\,dt = f^{(n-1)}(x) - f^{(n-1)}(a)$$

$$\int_a^x dx \int_a^x f^{(n)}(t)\,dt = f^{(n-2)}(x) - f^{(n-2)}(a) - f^{(n-1)}(a)(x-a)$$

$$\int_a^x dx \int_a^x dx \int_a^x f^{(n)}(t)\,dt = f^{(n-3)}(x) - f^{(n-3)}(a) - f^{(n-2)}(a)(x-a) - f^{(n-1)}(a)\frac{(x-a)^2}{2!}$$

Continuing this integration process yields

$$\int_a^x dx \int_a^x dx \int_a^x \cdots \int_a^x f^{(n)}(t)\,dt = f(x) - f(a) - f'(a)(x-a) - f''(a)\frac{(x-a)^2}{2!} - f'''(a)\frac{(x-a)^3}{3!} - \cdots - f^{(n-1)}(a)\frac{(x-a)^{n-1}}{(n-1)!}$$

so that

$$f(x) = \sum_{r=0}^{n-1} f^{(r)}(a)\frac{(x-a)^r}{r!} + R_n(x) \tag{10.124}$$

$$R_n(x) = \int_a^x dx \int_a^x dx \int_a^x \cdots \int_a^x f^{(n)}(t)\,dt$$

If M is any bound of $|f^{(n)}(x)|$ on the interval (a, x), we have

$$|R_n(x)| \leq \int_a^x |dx| \int_a^x |dx| \int_a^x \cdots \int_a^x M|dt| = \frac{M|x-a|^n}{n!} \qquad (10.125)$$

Another form of $R_n(x)$ can be obtained if we note that $F(x)$ defined by

$$F(x) = \frac{1}{(n-1)!} \int_a^x f^{(n)}(t)(x-t)^{n-1}\, dt \qquad F(a) = 0$$

yields

$$F'(x) = \frac{1}{(n-2)!} \int_a^x f^{(n)}(t)(x-t)^{n-2}\, dt \qquad F'(a) = 0$$

$$F''(x) = \frac{1}{(n-3)!} \int_a^x f^{(n)}(t)(x-t)^{n-3}\, dt \qquad F''(a) = 0$$

$$\cdots\cdots\cdots\cdots\cdots\cdots\cdots\cdots\cdots\cdots$$

$$F^{(n)}(x) = f^{(n)}(x)$$

by applying the results of Prob. 4, Sec. 10.17. Integrating

$$F^{(n)}(x) = f^{(n)}(x)$$

n times over the range (a, x) yields $F(x) = R_n(x)$, so that

$$R_n(x) = \frac{1}{(n-1)!} \int_a^x f^{(n)}(t)(x-t)^{n-1}\, dt \qquad (10.126)$$

The inequality of (10.125) results immediately from (10.126).

If it is known that $f(x)$ has derivatives of all orders and if it can be shown that $\lim_{n\to\infty} R_n(x) = 0$, then (10.124) yields the important Taylor-series expansion of $f(x)$ about $x = a$,

$$f(x) = \sum_{r=0}^{\infty} f^{(r)}(a) \frac{(x-a)^r}{r!} \qquad (10.127)$$

It must be emphasized that (10.127) cannot be used as an expression for $f(x)$ unless $R_n(x) \to 0$ as $n \to \infty$ (see Example 10.41). A special case of (10.127) occurs if $a = 0$, with

$$f(x) = \sum_{r=0}^{\infty} f^{(r)}(0) \frac{x^r}{r!} \qquad (10.128)$$

Equation (10.128) is the Maclaurin-series expansion of $f(x)$ about the origin.

Example 10.39. Consider $f(x) = e^x$ with $a = 0$. All the derivatives of $f(x)$ exist for $|x| \leqq X$, X arbitrary. For $|x| \leqq X$ we have $|f^{(n)}(x)| = |e^x| \leqq e^X$, so that

$$|R_n(x)| \leqq \frac{e^X|x|^n}{n!} \leqq \frac{e^X X^n}{n!} \qquad |x| \leqq X$$

The reader can verify that $\lim_{n\to\infty} \frac{X^n}{n!} = 0$. A simple proof of this statement occurs if one considers the series, $\sum_{n=0}^{\infty} \frac{X^n}{n!}$, which is seen to converge by the ratio test. Hence the nth term, of necessity, must tend to zero. Thus $|R_n(x)| \to 0$ as $n \to \infty$ for all x, and $e^x = \sum_{r=0}^{\infty} \frac{x^r}{r!}$ since $f^{(r)}(0) = 1$ for $r = 0, 1, 2, \ldots$.

Example 10.40. Consider $f(x) = (1 + x)^{-\frac{1}{2}}$. The reader can verify that

$$f^{(n)}(0) = (-1)^n \frac{1 \cdot 3 \cdot 5 \cdots (2n-1)}{2^n} = (-1)^n \frac{(2n)!}{2^{2n}n!},$$

$n = 0, 1, 2, 3, \ldots$. Moreover

$$|f^{(n)}(x)| = \left| \frac{(2n)!}{2^{2n}n!} (1 + x)^{-(n+\frac{1}{2})} \right|$$

and for $|x| \leqq \sigma < 1$ we have

$$|R_n(x)| \leqq \frac{(2n)!}{2^{2n}n!n!} \frac{\sigma^n}{(1 - \sigma)^{n+\frac{1}{2}}} \tag{10.129}$$

Let the reader verify that $\lim_{n\to\infty} R_n(x) = 0$. Thus

$$(1 - x)^{-\frac{1}{2}} = \sum_{n=0}^{\infty} \frac{(-1)^n(2n)!}{2^{2n}n!n!} x^n \qquad |x| < 1 \tag{10.130}$$

The series of (10.130) converges for $x = 1$ and diverges for $x = -1$. The reader can verify directly that $\lim_{n\to\infty} R_n(1) = 0$.

Example 10.41. Consider $f(x) = e^{-1/x^2}$, $x \neq 0$, $f(0) = 0$. We have

$$f'(0) = \lim_{x\to 0} \frac{f(x) - f(0)}{x - 0} = \lim_{x\to 0} \frac{e^{-1/x^2}}{x} = 0$$

To compute $f''(0)$, we note that

$$f''(0) = \lim_{x\to 0} \frac{f'(x) - f'(0)}{x - 0} = \lim_{x\to 0} \frac{2e^{-1/x^2}}{x^4} = 0$$

It can be shown that $f^{(n)}(0) = 0$ for $n = 0, 1, 2, 3, \ldots$. Hence the Maclaurin series $\sum_{n=0}^{\infty} f^{(n)}(0) \frac{x^n}{n!}$ converges to zero for all values of x since every term of the series is zero. Returning to (10.124), we note that $f(x) = e^{-1/x^2} = R_n(x)$ for all n so that $R_n(x)$ does not tend to zero as n becomes infinite. The Maclaurin series of $f(x)$ does not converge to $f(x)$ for this example. We note that the convergence of a Taylor-series expansion of a function $f(x)$ does not guarantee that the series converges to $f(x)$. One must always investigate the remainder, $R_n(x)$.

Another form of the Taylor series can be obtained if we replace x by $a + h$, so that

$$f(a + h) = \sum_{r=0}^{\infty} f^{(r)}(a) \frac{h^r}{r!} \tag{10.131}$$

If we allow a to vary by replacing a by x, we obtain

$$f(x + h) = \sum_{r=0}^{\infty} f^{(r)}(x) \frac{h^r}{r!} \tag{10.132}$$

Equation (10.131) is very useful if we know $f^{(r)}(a)$ for all r and if we wish to find an approximate value of $f(a + h)$. If only a finite number of terms of the Taylor series are used in approximating $f(a + h)$, an estimate of the size of the error can be obtained from (10.125).

Example 10.42. In Example 10.39 we saw that $|R_n(x)| \leqq e^X|x|^n/n!$ for the Maclaurin-series expansion of e^x, $|x| \leqq X$. If we wish to find an approximate value of ϵ, we let $x = 1$ and note that $|R_n(1)| \leqq e/n! < 3/n!$. For $n = 8$ we have $|R_8(1)| < 3/8! < 0.0001$, so that $\sum_{r=0}^{7} \frac{1}{r!} \approx 2.718$ yields a value of e accurate to three places. Actually $e = 2.71828 \ldots$.

The Taylor-series extension to a function of more than one variable is not very difficult. Consider $f(x^1, x^2, \ldots, x^n)$ as a function of the n variables $x^1, x^2, \ldots, x^n$. Let $h^1, h^2, \ldots, h^n$ be any set of constants, and define $\varphi(t)$ by

$$\begin{aligned} \varphi(t) &= f(x^1 + h^1t, x^2 + h^2t, \ldots, x^n + h^nt) \\ &= f(u^1, u^2, \ldots, u^n) \qquad u^i = x^i + h^it \qquad i = 1, 2, \ldots, n \end{aligned} \tag{10.133}$$

We consider $x^1, x^2, \ldots, x^n$ as constants temporarily so that $\varphi(t)$ is looked upon as a function of the single variable t. Let us assume that $\varphi(t)$ has a Maclaurin-series expansion which converges for $t = 1$. We have

$$\varphi(t) = \sum_{n=0}^{\infty} \frac{d^n\varphi(0)}{dt^n} \frac{t^n}{n!}$$

Now $$\frac{d\varphi}{dt} = \frac{\partial f}{\partial u^\alpha} \frac{\partial u^\alpha}{\partial t} = \frac{\partial f}{\partial u^\alpha} h^\alpha$$

and, at $t = 0$,

$$\left(\frac{d\varphi}{dt}\right)_{t=0} = \frac{\partial f}{\partial x^\alpha} h^\alpha \qquad u^\alpha = x^\alpha \qquad \text{at } t = 0$$

Similarly
$$\frac{d^2\varphi}{dt^2} = \frac{\partial^2 f}{\partial u^\beta \, \partial u^\alpha} h^\alpha \frac{\partial u^\beta}{\partial t} = \frac{\partial^2 f}{\partial u^\beta \, \partial u^\alpha} h^\alpha h^\beta$$

and $\left(\frac{d^2\varphi}{dt^2}\right)_{t=0} = \frac{\partial^2 f}{\partial x^\alpha \, \partial x^\beta} h^\alpha h^\beta$. Continuing in this manner yields

$$\varphi(t) = f(x^1, x^2, \ldots, x^n) + \left(\frac{\partial f}{\partial x^\alpha} h^\alpha\right) t + \left(\frac{\partial^2 f}{\partial x^\beta \, \partial x^\alpha} h^\alpha h^\beta\right)\frac{t^2}{2!} + \cdots + \left(\frac{\partial^n f}{\partial x^\gamma \cdots \partial x^\beta \, \partial x^\alpha} h^\alpha h^\beta \cdots h^\gamma\right)\frac{t^n}{n!} + \cdots$$

For $t = 1$ we obtain

$$f(x^1 + h^1, x^2 + h^2, \ldots, x^n + h^n) = f(x^1, x^2, \ldots, x^n) + \frac{\partial f}{\partial x^\alpha} h^\alpha + \frac{1}{2!}\frac{\partial^2 f}{\partial x^\beta \, \partial x^\alpha} h^\alpha h^\beta + \cdots + \frac{1}{n!}\frac{\partial^n f}{\partial x^\gamma \cdots \partial x^\beta \, \partial x^\alpha} h^\alpha h^\beta \cdots h^\gamma + \cdots \tag{10.134}$$

For a function of two variables (10.134) becomes

$$f(x + h, y + k) = f(x, y) + \left(h\frac{\partial f}{\partial x} + k\frac{\partial f}{\partial y}\right) + \frac{1}{2!}\left(h\frac{\partial f}{\partial x} + k\frac{\partial f}{\partial y}\right)^{(2)} + \cdots + \frac{1}{n!}\left(h\frac{\partial f}{\partial x} + k\frac{\partial f}{\partial y}\right)^{(n)} + \cdots \tag{10.135}$$

where
$$\left(h\frac{\partial f}{\partial x} + k\frac{\partial f}{\partial y}\right)^{(n)} = \sum_{r=0}^{n} \binom{n}{r} h^r k^{n-r} \frac{\partial^n f}{\partial x^r \, \partial y^{n-r}}$$

Problems

1. Show that $\sin x = \sum_{n=0}^{\infty} \frac{(-1)^n x^{2n+1}}{(2n+1)!}$ holds for all values of x.

2. Show that $\cos x = \sum_{n=0}^{\infty} \frac{(-1)^n x^{2n}}{(2n)!}$ holds for all values of x.

3. Show that $\sinh x = \sum_{n=0}^{\infty} \frac{x^{2n+1}}{(2n+1)!}$ and $\cosh x = \sum_{n=0}^{\infty} \frac{x^{2n}}{(2n)!}$ hold for all values of x.

4. Show that $\ln(1 + x) = \sum_{n=1}^{\infty} \frac{(-1)^{n-1} x^n}{n}$ holds for $-1 < x \leq 1$.

5. Show that $\ln \frac{1+x}{1-x} = 2 \sum_{n=0}^{\infty} \frac{x^{2n+1}}{2n+1}$ holds for $|x| < 1$. If $N = \frac{1+x}{1-x} > 0$, show that $0 < x < 1$. Determine ln 5 accurate to four places.

6. How many terms in the Taylor-series expansion of $\sin x$ about $x = \pi/6$ are needed to determine sin 31° accurate to six places? Evaluate sin 31° accurate to six places.

7. By integrating $1/(1 + x^2)$ find the Maclaurin expansion of $\tan^{-1} x$.

8. Show that

$$\ln \frac{x+1}{x} = 2 \sum_{n=0}^{\infty} \frac{1}{(2n+1)(2x+1)^{2n+1}} \qquad x > 0$$

9. Find the first four terms of the Maclaurin-series expansion of $\ln \cos x$. For what range of x is the series valid?

10. Find the Taylor series of $\cos^2 x$ about $x = \pi/3$.

11. If $f^{(n)}(x)$ is continuous on (a, x), show that

$$\begin{aligned} R_n(x) &= f^{(n)}(a + \theta(x-a)) \frac{(x-a)^n}{n!} \qquad 0 \leqq \theta \leqq 1 \\ &= [f^{(n)}(a) + \epsilon] \frac{(x-a)^n}{n!} \end{aligned}$$

by making use of (10.126).

10.24. Extrema of Functions. The Lagrange Method of Multipliers. One says that $f(x)$ has an extremum at a point $x = c$ if an $\eta > 0$ exists such that $f(c + h) - f(c)$ has the same sign for $|h| < \eta$. If $f(c + h) \leqq f(c)$ for $|h| < \eta$, we say that $f(x)$ has a local maximum at $x = c$. If $f(c + h) \geqq f(c)$ for $|h| < \eta$, we say that $f(c)$ has a local minimum at $x = c$. If an extremum of $f(x)$ at $x = c$ exists and if $f(x)$ is differentiable at $x = c$, we have

$$\begin{aligned} f'(c) &= \lim_{h \to 0} \frac{f(c+h) - f(c)}{h} \leqq 0 \text{ or } \geqq 0 \\ f'(c) &= \lim_{h \to 0} \frac{f(c) - f(c-h)}{h} \geqq 0 \text{ or } \leqq 0 \end{aligned} \tag{10.136}$$

so that $f'(c) = 0$. Now assume $f'(c) = f''(c) = \cdots = f^{(n-1)}(c) = 0$, $f^{(n)}(c) \neq 0$, and further assume that $f^{(n)}(x)$ is continuous in a neighborhood of $x = c$. Applying (10.124) and the result of Prob. 11, Sec. 10.23, we have

$$\begin{aligned} f(x) &= f(c) + [f^{(n)}(c) + \epsilon] \frac{(x-c)^n}{n!} \\ f(c+h) - f(c) &= \frac{h^n}{n!} [f^{(n)}(c) + \epsilon] \end{aligned}$$

Since $f^{(n)}(c) \neq 0$ and since $\epsilon \to 0$ as $h \to 0$, we can choose h sufficiently small so that $f^{(n)}(c) + \epsilon$ is one-signed. Hence $f(c + h) - f(c)$ will be one-signed provided n is even (h can be positive and negative). Thus a necessary and sufficient condition that $f(x)$ have an extremum at $x = c$ is that $f'(c) = 0$ and the first nonvanishing derivative of $f(x)$ at $x = c$ be of even order. If $f^{(n)}(c) > 0$, a minimum occurs, while if $f^{(n)}(c) < 0$, a maximum occurs. Why?

For a function of two variables we have

$$f(x + h, y + k) - f(x, y) = \frac{\partial f}{\partial x} h + \frac{\partial f}{\partial y} k + \frac{1}{2!} (f_{xx}h^2 + 2f_{xy}hk + f_{yy}k^2) + \cdots$$

If we neglect the higher-order terms (h^3, h^2k, . . .), we note that $f(x + h, y + k) - f(x, y)$ will be one-signed provided $\frac{\partial f}{\partial x} = 0$, $\frac{\partial f}{\partial y} = 0$, and $(f_{xx}h^2 + 2f_{xy}hk + f_{yy}k^2)$ does not change sign for arbitrarily small positive and negative values of h and k. Let the reader deduce that $f(x, y)$ has an extremal at (x, y) provided $\frac{\partial f}{\partial x} = 0$, $\frac{\partial f}{\partial y} = 0$, and

$$\left(\frac{\partial^2 f}{\partial x\, \partial y}\right)^2 - \frac{\partial^2 f}{\partial x^2} \frac{\partial^2 f}{\partial y^2} < 0 \tag{10.137}$$

If $f_{xx} > 0$ or $f_{yy} > 0$, a minimum occurs, and if $f_{xx} < 0$ or $f_{yy} < 0$, a maximum occurs. Why?

Let us consider the function $z = f(x, y)$ to be extremalized under the condition $\varphi(x, y) =$ constant. If $\varphi(x, y)$ is a closed and bounded curve, and if $f(x, y)$ is continuous at every point of the curve $\varphi(x, y) = c$, we know that there will be a point P_0 on the curve $\varphi(x, y) = c$ such that $f(x, y)$ will take on its maximum value at P_0. Remember that we restrict $P(x, y)$ to lie on the curve $\varphi(x, y) = c$. The same statement applies if we consider the minimum value of $f(x, y)$. Now in the elementary calculus we would solve $\varphi(x, y) = c$ for y as a function of x, say, $y = \psi(x)$, substitute $y = \psi(x)$ into $z = f(x, y)$, to obtain $z = f(x, \psi(x))$, and then we would set $\frac{dz}{dx} = 0$. We also have

$$\frac{dz}{dx} = \frac{\partial f}{\partial x} + \frac{\partial f}{\partial y}\frac{dy}{dx} = \frac{\partial f}{\partial x} + \frac{\partial f}{\partial y}\left[\frac{-\partial\varphi/\partial x}{\partial\varphi/\partial y}\right] = 0 \tag{10.138}$$

since $\frac{\partial \varphi}{\partial x} + \frac{\partial \varphi}{\partial y}\frac{dy}{dx} = 0$. We can obtain (10.138) by another procedure due to Lagrange. Consider the new function

$$U = f(x, y) + \lambda\varphi(x, y) \tag{10.139}$$

We differentiate U with respect to x and y, assuming that x and y are independent variables, while λ is considered as a parameter. Setting $\frac{\partial U}{\partial x} = 0$, $\frac{\partial U}{\partial y} = 0$ yields

$$\begin{aligned} \frac{\partial f}{\partial x} + \lambda \frac{\partial \varphi}{\partial x} &= 0 \\ \frac{\partial f}{\partial y} + \lambda \frac{\partial \varphi}{\partial y} &= 0 \end{aligned} \tag{10.140}$$

Eliminating λ in (10.140) yields (10.138). This is Lagrange's method of multipliers.

More generally, if $y = f(x_1, x_2, \ldots, x_n)$ is to be extremalized subject to the conditions $\varphi_i(x_1, x_2, \ldots, x_n) = c_i$, $i = 1, 2, \ldots, m$, we form

$$U = f(x_1, x_2, \ldots, x_n) + \sum_{i=1}^{m} \lambda_i \varphi_i(x_1, x_2, \ldots, x_n)$$

and we set

$$\frac{\partial U}{\partial x_j} = \frac{\partial f}{\partial x_j} + \sum_{i=1}^{m} \lambda_i \frac{\partial \varphi_i}{\partial x_j} = 0 \qquad j = 1, 2, \ldots, n \tag{10.141}$$

The elimination of the λ_i, $i = 1, 2, \ldots, m$, along with the equations $\varphi_i(x_1, x_2, \ldots, x_n) = c_i$, $i = 1, 2, \ldots, m$, yields the values of $x_1, x_2, \ldots, x_n$, which extremalize y.

Example 10.43. We wish to find the ratio of altitude to radius of a cylindrical glass (open top) having a maximum volume for a fixed surface area. We have

$$V = \pi x^2 y \qquad S = 2\pi xy + \pi x^2 = \text{constant}$$

From $U = \pi x^2 y + \lambda(2\pi xy + \pi x^2)$ we have

$$\begin{aligned} \frac{\partial U}{\partial x} &= 2\pi[xy + \lambda(y + x)] = 0 \\ \frac{\partial U}{\partial y} &= \pi[x^2 + \lambda(2x)] = 0 \end{aligned}$$

Eliminating λ yields $y/x = 1$. It is easy to show that $y/x = 1$ yields a maximum volume.

Problems

1. Consider a cylindrical buoy with two conical ends. Let x be the radius of the cylinder, y the altitude of the cylinder, and z the altitude of the cones. Show that $x:y:z = \frac{\sqrt{5}}{2}:1:1$ yields a maximum volume for a fixed surface area.

2. Find the maximum distance from the origin to the curve $x^3 + y^3 - 3xy = 0$.

3. Derive (10.137) if $f_{xx}h^2 + 2f_{xy}hk + f_{yy}k^2$ does not change sign for arbitrary values of h and k.

4. A rectangular box has dimensions x, y, z. Show that $x:y:z = 1:1:1$ yields a minimum surface area for a fixed volume.

5. Consider the set of values $y_0, y_1, y_2, \ldots, y_n$ and the polynomial

$$y(x) = a_0 + a_1x + a_2x^2 + \cdots + a_mx^m$$

$m < n$. Show that the set of a_i, $i = 0, 1, 2, \ldots, m$, which extremalize

$$S = \sum_{j=0}^{n} (y_j - \Sigma a_i x_j^i)^2$$

satisfy $\sum_{i=0}^{m} s_{i+k}a_i = t_k$ with $s_k = \sum_{j=0}^{n} x_j^k$, $t_k = \sum_{j=0}^{n} y_j x_j^k$.

10.25. Numerical Methods. Let us assume that we wish to find a root of $f(x) = 0$, and let us further assume that $f'(x)$ exists. We can plot a rough graph of $y = f(x)$ and attempt to read the value of x at which the curve crosses the x axis, $y = 0$ (see Fig. 10.4).

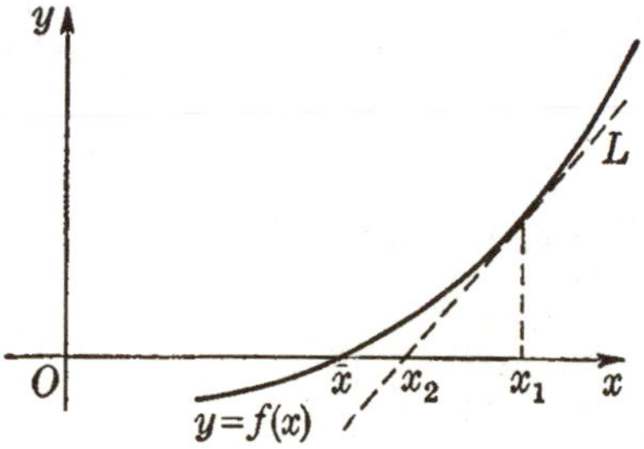

FIG. 10.4

If we choose a point x_1 not far removed from $\bar{x}$, $f(\bar{x}) = 0$, we note that the equation of the tangent line at $(x_1, f(x_1))$ is $y - f(x_1) = f'(x_1)(x - x_1)$. This line, L, intersects the x axis at the point $x_2 = x_1 - f(x_1)/f'(x_1)$, $f'(x_1) \neq 0$, obtained by setting $y = 0$. This process can be continued, with

$$x_{n+1} = x_n - \frac{f(x_n)}{f'(x_n)} \qquad f'(x_n) \neq 0 \tag{10.142}$$

If the sequence $x_1, x_2, \ldots, x_n, \ldots$ can be shown to converge to a limit c, then

$$\lim_{n\to\infty} x_{n+1} = \lim_{n\to\infty} x_n - \lim_{n\to\infty} \frac{f(x_n)}{f'(x_n)}$$

and $c = c - f(c)/f'(c)$ so that $f(c) = 0$, provided $f(x)$ and $f'(x)$ are continuous at $x = c$, $f'(c) \neq 0$. Thus $c = \bar{x} = \lim_{n\to\infty} x_n$. It is apparent that difficulties will occur if $f'(x)$ is small near $x = \bar{x}$.

Example 10.44. For $f(x) = x^2 - 2 = 0$ we have $f'(x) = 2x$, and (10.142) becomes

$$x_{n+1} = x_n - \frac{x_n^2 - 2}{2x_n} = \frac{x_n^2 + 2}{2x_n} \tag{10.143}$$

From (10.143) we note that $0 < x_{n+1} < x_n$ provided $x_n^2 > 2$, $x_n > 0$. Let the reader deduce that $x_{n+1}^2 > 2$ if $x_n^2 > 2$. Thus if we choose $x_1 = 2$, (10.143) will yield a monotonic decreasing sequence bounded below by zero. The sequence must converge to a solution of $x^2 - 2 = 0$. The computation may be arranged as follows:

n	x_n	$x_n^2 + 2$	$2x_n$	$x_{n+1} = \dfrac{x_n^2 + 2}{2x_n}$
1	2	6	4	1.7
2	1.7	4.89	3.4	1.44
3	1.44	4.0736	2.88	1.414
4	1.414	3.999396	2.828	1.4142
5	1.4142	3.99996164	2.8284	1.41412

The desired root to five decimal places is $x = 1.41412$.

An iteration process may be used to find a root of $x - \varphi(x) = 0$ or $x = \varphi(x)$. We shall assume that $|\varphi'(x)| < A < 1$. We define x_{n+1} by

$$x_{n+1} = \varphi(x_n) \qquad n = 1, 2, 3, \ldots \tag{10.144}$$

with x_1 arbitrary. Thus

$$\begin{aligned} x_{n+1} - x_n &= \varphi(x_n) - \varphi(x_{n-1}) \\ &= (x_n - x_{n-1})\varphi'(\xi_n) \end{aligned}$$

with ξ_n lying between x_{n-1} and x_n, by applying the law of the mean. Thus

$$|x_{n+1} - x_n| = |x_n - x_{n-1}| \cdot |\varphi'(\xi_n)| < |x_n - x_{n-1}|A \tag{10.145}$$

Now $\quad x_n = x_0 + (x_1 - x_0) + (x_2 - x_1) + \cdots + (x_n - x_{n-1})$

and $x_0 + \sum_{k=1}^{\infty} |x_k - x_{k-1}|$ converges by the ratio test since $\dfrac{|x_{n+1} - x_n|}{|x_n - x_{n-1}|} < A < 1$, from (10.145). Thus

$$\lim_{n\to\infty} x_n = x_0 + \lim_{n\to\infty} \sum_{k=1}^{n} (x_k - x_{k-1}) = \bar{x}$$

exists. From (10.144) and the continuity of $\varphi(x)$ we have

$$\lim_{n\to\infty} x_n = \bar{x} = \lim_{n\to\infty} \varphi(x_{n-1}) = \varphi(\bar{x})$$

and $\bar{x}$ is the desired root of $x = \varphi(x)$.

Example 10.45. We find a root of $x = \frac{1}{2}\sin x + 1$, $\varphi(x) = \frac{1}{2}\sin x + 1$,

$$|\varphi'(x)| = \tfrac{1}{2}|\cos x| \leq \tfrac{1}{2}$$

The computation is arranged as follows:

n	x_n	$\sin x_n$	$x_{n+1} = \frac{1}{2}\sin x_n + 1$
1	0	0	1
2	1	0.8415	1.4208
3	1.4208	0.9888	1.4944
4	1.4944	0.9971	1.4986
5	1.4986	0.9974	1.4987
6	1.4987	0.9974	1.4987

We started with $x_1 = 0$ and obtained the desired root, $x = 1.4987$, accurate to four decimal places. Had we graphed $y = x$ and $y = \frac{1}{2}\sin x + 1$, we would have noticed that $x_1 = 1.5$ would be a good starting point.

It is often useful to replace a given function by a polynomial which approximates the given function to a high degree of accuracy. Such a polynomial is called an interpolating function. Let $f(x)$ be a function defined for $x_0 \leqq x \leqq x_n$, and let us assume that we have evaluated $f(x)$ at $x_0, x_1, x_2, \ldots, x_n$, obtaining the $n + 1$ values $y_i = f(x_i)$, $i = 0, 1, 2, \ldots, n$. A simple way of constructing a polynomial of degree n, $L_n(x)$, which satisfies $L_n(x_i) = y_i$, $i = 0, 1, 2, \ldots, n$, is due to Lagrange. Let us note that the polynomial

$$y_0 \frac{(x - x_1)(x - x_2) \cdots (x - x_n)}{(x_0 - x_1)(x_0 - x_2) \cdots (x_0 - x_n)}$$

has the value zero at $x = x_1, x_2, \ldots, x_n$ and has the value y_0 at $x = x_0$. A sum of such polynomials yields *Lagrange's interpolating polynomial*,

$$\begin{aligned} L_n(x) = y_0 &\frac{(x - x_1)(x - x_2) \cdots (x - x_n)}{(x_0 - x_1)(x_0 - x_2) \cdots (x_0 - x_n)} \\ &+ y_1 \frac{(x - x_0)(x - x_2) \cdots (x - x_n)}{(x_1 - x_0)(x_1 - x_2) \cdots (x_1 - x_n)} + \cdots \\ &\quad + y_n \frac{(x - x_0)(x - x_1) \cdots (x - x_{n-1})}{(x_n - x_0)(x_n - x_1) \cdots (x_n - x_{n-1})} \end{aligned} \tag{10.146}$$

One naturally inquires how closely $L_n(x)$ approximates $f(x)$ for $x \neq x_i$, $i = 0, 1, 2, \ldots, n$. To answer this question, we construct $F(x)$ defined by

$$F(x) = f(x) - L_n(x) - R(x - x_0)(x - x_1) \cdots (x - x_n) \tag{10.147}$$

with R a constant. Let us note that $F(x)$ vanishes for $x = x_0, x_1, x_2, \ldots, x_n$, since $f(x_i) = L(x_i)$, $i = 0, 1, 2, \ldots, n$. Now let us choose any point $\bar{x}$ of the interval $x_0 \leqq x \leqq x_n$ other than $x_0, x_1, x_2, \ldots, x_n$ and determine R so that $F(\bar{x}) = 0$, that is,

$$f(\bar{x}) - L_n(\bar{x}) - R(\bar{x} - x_0)(\bar{x} - x_1) \cdots (\bar{x} - x_n) = 0 \tag{10.148}$$

Up to the present no restriction was placed on $f(x)$. If we assume that $f(x)$ is differentiable at every point of $x_0 \leqq x \leqq x_n$, then $F(x)$ is differentiable on this range. If we apply the law of the mean (or Rolle's theorem) to the intervals (x_0, x_1), (x_1, x_2), . . . , $(x_i, \bar{x})$, $(\bar{x}, x_{i+1})$, . . . , (x_{n-1}, x_n), we note that $F'(x)$ will vanish at $n + 1$ points. If we further assume that $f''(x)$, $f'''(x)$, . . . , $f^{(n+1)}(x)$ exist on $x_0 \leqq x \leqq x_n$, we note that $F^{(n+1)}(x)$ will vanish at a point $x = s$, $x_0 \leqq s \leqq x_n$. Since $L_n(x)$ is a polynomial of degree n, we have $\dfrac{d^{n+1}L_n(x)}{dx^{n+1}} = 0$. From (10.147) we obtain

$$F^{(n+1)}(s) = 0 = f^{(n+1)}(s) - R(n+1)! \qquad x_0 \leqq s \leqq x_n$$

since $\dfrac{d^{n+1}}{dx^{n+1}} [(x - x_0)(x - x_1) \cdots (x - x_n)] = (n + 1)!$. Substituting $R = f^{(n+1)}(s)/(n + 1)!$ into (10.148) yields

$$f(\bar{x}) - L_n(\bar{x}) = \frac{f^{(n+1)}(s)}{(n+1)!} (\bar{x} - x_0)(\bar{x} - x_1) \cdots (\bar{x} - x_n) \qquad (10.149)$$

Equation (10.149) holds for any value of $\bar{x}$ in the interval $x_0 \leqq x \leqq x_n$ although (10.149) was obtained by choosing $x \neq x_i$, $i = 0, 1, 2, \ldots, n$, for one notes that $f(x_i) = L_n(x_i)$ is consistent with (10.149).

If M is any upper bound of $|f^{(n+1)}(x)|$ on $x_0 \leqq x \leqq x_1$, we have

$$|f(x) - L_n(x)| \leqq \frac{M}{(n+1)!} |x - x_0|\,|x - x_1| \cdots |x - x_n| \qquad (10.150)$$

The inequality (10.150) is useful in finding an upper bound to the difference between $f(x)$ and $L_n(x)$.

Example 10.46. The third-degree polynomial passing through the points (0, −5), (1, −3), (2, −1), (3, 7) is

$$\begin{aligned} L_3(x) &= -5 \frac{(x-1)(x-2)(x-3)}{(-1)(-2)(-3)} + (-3) \frac{x(x-2)(x-3)}{1(-1)(-2)} \\ &\quad + (-1) \frac{x(x-1)(x-3)}{2 \cdot 1 \cdot (-1)} + 7 \frac{x(x-1)(x-2)}{3 \cdot 2 \cdot 1} \\ &= x^3 - 3x^2 + 4x - 5 \end{aligned}$$

Example 10.47. Let $L_6(x)$ be the polynomial which coincides with $\sin x$ at $x = 0$, $\pi/6$, $\pi/3$, $\pi/2$, $\frac{2}{3}\pi$, $\frac{5}{6}\pi$, π. Since $\left|\dfrac{d^7 \sin x}{dx^7}\right| \leq 1 = M$, we have

$$|\sin x - L_6(x)| \leqq \frac{\left| x\left(x - \frac{\pi}{6}\right)\left(x - \frac{\pi}{3}\right)\left(x - \frac{\pi}{2}\right)\left(x - \frac{2}{3}\pi\right)\left(x - \frac{5}{6}\pi\right)(x - \pi)\right|}{7!}$$

For $x = 40°$ or $\frac{2}{9}\pi$ we have

$$\begin{aligned} |\sin 40° - L_6(\tfrac{2}{9}\pi)| &\leq \frac{\pi^7}{7!}\left(\frac{2}{9}\right)\left(\frac{1}{18}\right)\left(\frac{1}{9}\right)\left(\frac{5}{18}\right)\left(\frac{4}{9}\right)\left(\frac{11}{18}\right)\left(\frac{7}{9}\right) = \frac{1}{7!}\left(\frac{\pi}{9}\right)^7 385 \\ &< 0.00005 \end{aligned}$$

Thus the use of $L_6(\frac{2}{9}\pi)$ would yield a value of $\sin 40°$ accurate to four decimal places.

Let us consider a function $f(x)$ defined on $-h \leqq x \leqq h$. We construct the polynomial $L_2(x)$, which coincides with $f(x)$ at $x = -h, 0, h$. From (10.146) we have

$$L_2(x) = f(-h)\,\frac{x(x-h)}{2h^2} + f(0)\,\frac{(x+h)(x-h)}{-h^2} + f(h)\,\frac{(x+h)x}{2h^2}$$

and
$$\int_{-h}^{h} L_2(x)\,dx = \frac{h}{3}\,[f(-h) + 4f(0) + f(h)] \tag{10.151}$$

We may well inquire how the value of (10.151) differs from $\int_{-h}^{h} f(x)\,dx$. This difference is

$$\varphi(h) = \int_{-h}^{h} f(x)\,dx - \frac{h}{3}\,[f(-h) + 4f(0) + f(h)]$$

We note that $\varphi(0) = 0$. Let us assume that $f'(x)$, $f''(x)$, $f'''(x)$, and $f^{(iv)}(x)$ exist. Differentiating $\varphi(h)$ with respect to h yields

$$\begin{aligned}\varphi'(h) &= f(h) + f(-h) - \frac{1}{3}\,[f(-h) + 4f(0) + f(h)] - \frac{h}{3}\,[-f'(-h) + f'(h)]\\ &= \frac{2}{3}\,[f(h) + f(-h)] - \tfrac{4}{3}f(0) - \frac{h}{3}\,[f'(h) - f'(-h)]\end{aligned}$$

and $\varphi'(0) = 0$. Differentiating again yields

$$\begin{aligned}\varphi''(h) &= \frac{2}{3}\,[f'(h) - f'(-h)] - \frac{1}{3}\,[f'(h) - f'(-h)] - \frac{h}{3}\,[f''(h) + f''(-h)]\\ &= \frac{1}{3}\,[f'(h) - f'(-h)] - \frac{h}{3}\,[f''(h) + f''(-h)]\end{aligned}$$

so that $\varphi''(0) = 0$. Differentiating once more yields

$$\begin{aligned}\varphi'''(h) &= \frac{1}{3}[f''(h) + f''(-h)] - \frac{1}{3}[f''(h) + f''(-h)] - \frac{h}{3}[f'''(h) - f'''(-h)]\\ &= -\frac{h}{3}\,[f'''(h) - f'''(-h)]\\ &= -\frac{2h^2}{3}\,f^{(iv)}(\bar{h}) \qquad -h \leqq \bar{h} \leqq h\end{aligned}$$

Now

$$\varphi(h) = \int_0^h dh \int_0^h dh \int_0^h \varphi'''(h)\,dh = \int_0^h dh \int_0^h dh \int_0^h (-\tfrac{2}{3})h^2 f^{(iv)}(\bar{h})\,dh$$

If M is any upper bound of $|f^{(iv)}(x)|$ on $-h \leqq x \leqq h$, we have

$$|\varphi(h)| \leqq M \int_0^h |dh| \int_0^h |dh| \int_0^h \tfrac{2}{3}h^2\,|dh| = \frac{M|h|^5}{90} \tag{10.152}$$

The inequality (10.152) yields an estimate of the error if we use *Simpson's* rule as given by (10.151) to approximate $\int_{-h}^{h} f(x)\,dx$. We note that Simpson's rule is exact if $f(x)$ is a polynomial of degree less than or equal to 3 since $f^{(iv)}(x) \equiv 0$. Since (10.151) depends only on the values of $f(x)$ at the equally spaced points $-h$, 0, h, we can extend (10.151) to the interval $a \leqq x \leqq b$ by subdividing (a, b) into an even number of intervals

$$a, a + h, a + 2h, a + 3h, \ldots, a + (n - 1)h, a + nh = b$$

applying Simpson's rule to the intervals

$$(a, a + 2h), (a + 2h, a + 4h), \ldots, (a + (n - 2)h, a + nh)$$

and adding. This yields

$$\begin{aligned}\int_a^b f(x)\,dx &= \frac{h}{3}[f(a) + 4f(a + h) + f(a + 2h)] \\ &\quad + \frac{h}{3}[f(a + 2h) + 4f(a + 3h) + f(a + 4h)] + \cdots \\ &\quad + \frac{h}{3}[f(a + (n - 2)h) + 4f(a + (n - 1)h) + f(b)] + E \\ &= \frac{h}{3}[f(a) + 4f(a + h) + 2f(a + 2h) + 4f(a + 3h) \\ &\quad + 2f(a + 4h) + \cdots + 4f(b - h) + f(b)] + E \qquad (10.153)\end{aligned}$$

with
$$|E| \leqq \frac{n}{2}\left[\frac{Mh^5}{90}\right] = \frac{Mh^4}{180} nh = \frac{Mh^4(b - a)}{180} \qquad (10.154)$$

M is any upper bound of $|f^{(iv)}(x)|$ on the interval $a \leqq x \leqq b$. The greater the number of subdivisions chosen and hence the smaller the value of h, the more accurate becomes Simpson's rule for approximating an integral.

Example 10.48. If we apply Simpson's rule to $\ln 2 = \int_1^2 \frac{dx}{x}$, $n = 10$, we note that $|E| \leqq \frac{(0.1)^4}{180}\frac{24}{x^5} \leqq \frac{24}{180}(0.1)^4 < 0.0002$, so that we can expect to obtain a value of ln 2 accurate to at least three decimal places. Applying (10.153) yields

$$\begin{aligned}\ln 2 &\approx \frac{1}{30}\left(1 + \frac{4}{1.1} + \frac{2}{1.2} + \frac{4}{1.3} + \frac{2}{1.4} + \frac{4}{1.5} + \frac{2}{1.6} + \frac{4}{1.7} + \frac{2}{1.8} + \frac{4}{1.9} + \frac{1}{2}\right) \\ &\approx 0.69314\end{aligned}$$

In the tables one notes that $\ln 2 = 0.69315$ accurate to five decimal places.

We conclude this section with a discussion of the Euler-Maclaurin sum formula. Let us suppose that $f(x)$ has continuous derivatives at least to the order r on the interval $0 \leqq x \leqq 1$. If we integrate by parts, we obtain

$$\begin{aligned}\int_0^1 f(x)\,dx &= xf(x)\Big|_0^1 - \int_0^1 xf'(x)\,dx \\ &= f(1) - \int_0^1 xf'(x)\,dx \\ &= \tfrac{1}{2}[f(0) + f(1)] + \tfrac{1}{2}[f(1) - f(0)] - \int_0^1 xf'(x)\,dx \\ &= \tfrac{1}{2}[f(0) + f(1)] + \tfrac{1}{2}\int_0^1 f'(x)\,dx - \int_0^1 xf'(x)\,dx \\ &= \tfrac{1}{2}[f(0) + f(1)] - \int_0^1 (x - \tfrac{1}{2})f'(x)\,dx \qquad (10.155)\end{aligned}$$

We define $B_2(x)$ such that $B_2'(x) = x - \frac{1}{2}$, $B_2(1) = 0$. Integration yields $B_2(x) = \frac{1}{2}(x^2 - x)$. Replacing $x - \frac{1}{2}$ by $B_2'(x)$ in (10.155) and integrating by parts yields

$$\int_0^1 f(x)\,dx = \tfrac{1}{2}[f(0) + f(1)] + \int_0^1 B_2(x)f''(x)\,dx \qquad (10.156)$$

We now define $B_3(x)$ such that $B_3'(x) = B_2(x) + b_2$ with the stipulation that $B_3(0) = B_3(1) = 0$. Thus $B_3(x) = x^3/6 - x^2/4 + b_2x + c$, and $c = 0$, $b_2 = \frac{1}{12}$. From $B_2(x) = B_3'(x) - b_2$ we have upon integration by parts

$$\int_0^1 f(x)\,dx = \tfrac{1}{2}[f(0) + f(1)] - b_2[f'(1) - f'(0)] - \int_0^1 B_3(x)f'''(x)\,dx$$

This process is continued so that a sequence of polynomials (Bernoulli) and a sequence of constants (Bernoulli) are generated. We have

$$\begin{aligned} B_k'(x) &= B_{k-1}(x) + b_{k-1} \qquad k \geqq 3 \\ B_k(0) &= B_k(1) = 0 \end{aligned} \qquad (10.157)$$

and

$$\begin{aligned}\int_0^1 f(x)\,dx = \tfrac{1}{2}[f(0) + f(1)] - b_2[f'(1) - f'(0)] + b_3[f''(1) - f''(0)] \\ - b_4[f'''(1) - f'''(0)] + \cdots + (-1)^{r-1}b_r[f^{(r-1)}(1) - f^{(r-1)}(0)] \\ + (-1)^r \int_0^1 B_{r+1}'(x)f^{(r)}(x)\,dx \qquad (10.158)\end{aligned}$$

In the generation of the Bernoulli polynomials and constants no mention was made of B_0, B_1, b_1, b_0. For convenience we choose $B_0(x) \equiv 0$, $B_1(x) \equiv x$, $b_0 = 1$, $b_1 = 0$.

Let us see whether it is possible to find two functions, $\Phi(x, t)$, $\varphi(t)$, such that

$$\begin{aligned}\Phi(x, t) &= \sum_{r=0}^{\infty} B_r(x)t^r \\ \varphi(t) &= \sum_{r=0}^{\infty} b_r t^r\end{aligned} \qquad (10.159)$$

If $\Phi(x, t)$ and $\varphi(t)$ exist satisfying (10.159), we call them the generating functions of the Bernoulli polynomials and Bernoulli numbers, respectively. The series expansions are simply the Taylor series expansions (see Sec. 10.23), so that $b_r = \varphi^{(r)}(0)/r!$, $B_r(x) = \frac{1}{r!}\left(\frac{\partial^r \Phi(x, t)}{\partial t^r}\right)_{t=0}$. Formally we have

$$\begin{aligned}
\frac{\partial \Phi(x, t)}{\partial x} &= \sum_{r=0}^{\infty} B_r'(x)t^r \\
&= B_1'(x)t + B_2'(x)t^2 + \sum_{r=3}^{\infty} B_r'(x)t^r \\
&= t + (x - \tfrac{1}{2})t^2 + \sum_{r=3}^{\infty} [B_{r-1}(x) + b_{r-1}]t^r \\
&= t + (x - \tfrac{1}{2})t^2 + t \sum_{s=2}^{\infty} (B_s(x) + b_s)t^s \\
&= t + (x - \tfrac{1}{2})t^2 + t \sum_{s=0}^{\infty} (B_s(x) + b_s)t^s - t(1 + xt) \\
&= t\Phi(x, t) + t\varphi(t) - \tfrac{1}{2}t^2
\end{aligned} \tag{10.160}$$

so that
$$\frac{\partial \Phi(x, t)}{\partial x} - t\Phi(x, t) = t\varphi(t) - \tfrac{1}{2}t^2$$

Integrating with respect to x yields

$$\Phi(x, t) = A(t)e^{xt} - \varphi(t) + \tfrac{1}{2}t$$

with $A(t)$ an arbitrary function of t. Now $\Phi(0, t) = 0$ since $B_r(0) = 0$, so that $A(t) = \varphi(t) - \frac{1}{2}t$, and

$$\Phi(x, t) = [\varphi(t) - \tfrac{1}{2}t](e^{xt} - 1)$$

Finally $\Phi(1, t) = \sum_{r=0}^{\infty} B_r(1)t^r = t$ since $B_1(1) = 1$, $B_r(1) = 0$, $r \neq 1$. Hence $[\varphi(t) - \frac{1}{2}t](e^t - 1) = 1$, and

$$\begin{aligned}
\varphi(t) &= \frac{t}{2} + \frac{t}{e^t - 1} \\
\Phi(x, t) &= t\,\frac{e^{xt} - 1}{e^t - 1}
\end{aligned} \tag{10.161}$$

One can now consider $\varphi(t)$ and $\Phi(x, t)$ of (10.161) and justify the series expansions (10.159) along with the term-by-term differentiation given in

(10.160). One then easily shows that (10.157) is valid and that $B_0(x) = 0$, $B_1(x) = x$, $b_0 = 1$, $b_1 = 0$. We omit proof of this fact. An important property of the Bernoulli constants is discernible if we note that

$$\varphi(-t) = -\frac{t}{2} - \frac{t}{e^{-t} - 1} = -\frac{t}{2} + \frac{te^t}{e^t - 1}$$
$$= -\frac{t}{2} + t + \frac{t}{e^t - 1} = \frac{t}{2} + \frac{t}{e^t - 1} = \varphi(t)$$

Hence $\varphi(t)$ is an even function so that $b_{2r+1} = 0$ for all r [see (10.159)]. If we take $r = 2s + 1$ and apply (10.157) to (10.158), we obtain

$$\int_0^1 f(x)\,dx = \tfrac{1}{2}[f(0) + f(1)] - b_2[f'(1) - f'(0)] - b_4[f'''(1) - f'''(0)] - \cdots - b_{2s}[f^{(2s-1)}(1) - f^{(2s-1)}(0)] - \int_0^1 B_{2s+1}(x)f^{(2s+1)}(x)\,dx \quad (10.162)$$

If we apply (10.162) to $g(x) = f(x + 1)$, we obtain

$$\int_0^1 g(x)\,dx = \int_0^1 f(x + 1)\,dx = \int_1^2 f(x)\,dx$$
$$= \tfrac{1}{2}[f(1) + f(2)] - b_2[f'(2) - f'(1)] - \cdots - \int_1^2 B_{2s+1}(x)f^{(2s+1)}(x)\,dx$$

Continuing this process for the intervals (2, 3), (3, 4), . . . , $(n - 1, n)$ and adding yields the *Euler-Maclaurin formula*,

$$\int_0^n f(x)\,dx = \tfrac{1}{2}f(0) + f(1) + f(2) + \cdots + f(n - 1) + \tfrac{1}{2}f(n) - b_2[f'(n) - f'(0)] - b_4[f'''(n) - f'''(0)] - \cdots - b_{2s}[f^{(2s-1)}(n) - f^{(2s-1)}(0)] - \sum_{m=0}^{n-1} \int_m^{m+1} B_{2s+1}(x - m)f^{(2s+1)}(x)\,dx \quad (10.163)$$

The Bernoulli polynomials and constants of (10.163) can be calculated from (10.157) or (10.161). It can be shown that if $f^{(2s+1)}(x)$ is monotonic decreasing for $0 \leqq x \leqq n$ and if the odd derivatives of $f(x)$ have the same sign on $0 \leqq x \leqq n$, then the true value of $\int_0^n f(x)\,dx$ always lies between the sums of s and $s + 1$ terms of the series following $\frac{1}{2}f(0) + f(1) + \cdots + f(n - 1) + \frac{1}{2}f(n)$ in (10.163). For this case we need not be concerned with

$$\sum_{m=0}^{n-1} \int_m^{m+1} B_{2s+1}(x - m)f^{(2s+1)}(x)\,dx$$

Example 10.49. *Stirling's Formula.* For $f(x) = \ln (x + 1)$ we note that

$$f^{(2s+1)}(x) = \frac{(2s)!}{(x + 1)^{2s+1}} > 0$$

for $0 \leqq x \leqq n$ and $f^{(2s+1)}(x)$ is monotonic decreasing on $0 \leqq x \leqq n$. Integration by parts yields

$$(n+1)\ln(n+1) - n = \int_0^n \ln(x+1)\,dx = \tfrac{1}{2}\ln 1 + \ln 2 + \ln 3 + \cdots$$
$$+ \ln(n-1) + \tfrac{1}{2}\ln n - b_2\left(\frac{1}{n+1} - 1\right) - b_4\left(\frac{2}{(n+1)^3} - 2\right) - \cdots \quad (10.164)$$

Now

$$(n+1)\ln(n+1) = (n+1)\ln\left[n\left(1+\frac{1}{n}\right)\right] = (n+1)\ln n + (n+1)\ln\left(1+\frac{1}{n}\right)$$

and, for large n, $(n+1)\ln(n+1) \approx (n+1)\ln n$, since

$$\lim_{n\to\infty}(n+1)\ln\left(1+\frac{1}{n}\right) = 1$$

Neglecting those terms which tend to zero as $n \to \infty$, we note from (10.164) that

$$(n+\tfrac{1}{2})\ln n - n \approx \ln n! + C \quad (10.165)$$

where C is the sum of the constant terms in (10.164). Let the reader apply (4.47), with (10.165) to show that $C = -\frac{1}{2}\ln 2\pi$. Thus

$$\ln \frac{n!}{n^n\sqrt{2\pi n}} \approx -n$$
$$n! \approx \sqrt{2\pi n}\, n^n e^{-n} = \sqrt{2\pi n}\left(\frac{n}{e}\right)^n \quad (10.166)$$

Formula (10.166) is Stirling's formula for approximating $n!$ for n large. The true meaning of (10.166) is that

$$\lim_{n\to\infty}\frac{n!}{\sqrt{2\pi n}\left(\dfrac{n}{e}\right)^n} = 1$$

Problems

1. Find a root of $x^3 - 3x + 1 = 0$ lying between $x = 0$ and $x = 1$, accurate to six decimal places.

2. Find a root of $x = \ln(x+1) + 1$, accurate to five decimal places.

3. Find a polynomial $p(x)$ such that $p(0) = 0$, $p(1) = 1$, $p(2) = 2$, $p(3) = 3$, $p(4) = 11$.

4. Apply Simpson's rule with 10 subdivisions to approximate $\int_0^1 \frac{dx}{1+x^2}$. Do the same for $\int_0^1 \frac{dx}{\sqrt{2 - \sin^2 x}}$. How accurate can you expect your answers to be?

5. Show that $b_4 = -\frac{1}{720}$, $B_4(x) = \frac{1}{4!}(x^4 - 2x^3 + x^2)$.

6. Show that $C = -\frac{1}{2}\ln 2\pi$ [see (10.165)].

10.26. The Lebesgue Integral. It is rather obvious that not all functions are Riemann-integrable (R-integrable). We consider the interval $0 \leqq x \leqq 1$ and define $f(x) = 1$ if x is irrational, $f(x) = 2$ if x is rational. The upper Darboux integral is seen to be 2, whereas the lower Darboux integral is 1 (see Sec. 10.16). Thus $f(x)$ is not R-integrable on $0 \leqq x \leqq 1$. However, let us consider the following: It is known that the rationals are countable (see Sec. 10.11), so that we can enumerate the rationals on the interval $0 \leqq x \leqq 1$, and we designate the rationals by

$$r_1, r_2, r_3, \ldots, r_n, \ldots$$

We define $f_1(x)$ on $0 \leqq x \leqq 1$ by $f_1(x) = 1$ for $x \neq r_1$, $f_1(r_1) = 2$. We define $f_2(x)$ by $f_2(x) = 1, x \neq r_1, r_2, f(r_1) = f(r_2) = 2$. Proceeding in this manner, we construct a sequence of functions

$$f_1(x), f_2(x), \ldots, f_n(x), \ldots \tag{10.167}$$

with $f_n(x) = 1$ except at $x = r_1, r_2, \ldots, r_n$, and at these rational points, $f_n(x) = 2$. It is apparent that

$$\lim_{n\to\infty} f_n(x) = f(x) = \begin{matrix} 1 & \text{for } x \text{ irrational} \\ 2 & \text{for } x \text{ rational} \end{matrix}$$

Moreover the sequence $\{f_n(x)\}$ is monotonic nondecreasing as n increases, and we write $f_n(x) \nearrow f(x)$. The reader can easily prove that every function of the sequence $\{f_n(x)\}$ is R-integrable, with $\int_0^1 f_n(x)\,dx = 1$ for all n. We define the Lebesgue integral of $f(x)$ as

$$L(f) = \int_0^1 f(x)\,dx \equiv \lim_{n\to\infty} \int_0^1 f_n(x)\,dx = \lim_{n\to\infty} 1 = 1 \tag{10.168}$$

Even though $f(x)$ is not R-integrable, it is Lebesgue integrable (L-integrable), and the value of its Lebesgue integral is defined by (10.168).

More generally, we have the following situation: Let C be the class of all R-integrable functions on the range $a \leqq x \leqq b$, and let $f(x)$ be the limit of a monotonic nondecreasing sequence, $\{f_n(x)\}$, of functions belonging to C. We define the Lebesgue integral of $f(x)$ by

$$L(f) = \int_a^b f(x)\,dx = \lim_{n\to\infty} \int_a^b f_n(x)\,dx \tag{10.169}$$

provided the limit exists. Let $\mathbf{C}$ denote the class of all such functions which are L-integrable. Any function which is R-integrable is automatically L-integrable since $f(x) = f_n(x) \nearrow f(x)$ and

$$\lim_{n\to\infty} \int_a^b f(x)\,dx = \int_a^b f(x)\,dx$$

Thus C is a proper subset of $\mathbf{C}$. Any function which is R-integrable is L-integrable, and the values of the two integrals are the same. It can be shown that if $f(x)$ is a limit of a monotonic nondecreasing sequence of functions belonging to $\mathbf{C}$ then $f(x)$ also belongs to $\mathbf{C}$. An analogous situation occurs in the real-number system. We define the irrational numbers as the limits of bounded sequences of rational numbers. The limit of a bounded sequence of irrational and/or rational numbers is again a real number, rational or irrational.

To extend the class **C** in order to embrace a larger class of functions, we consider the following: A function $f(x)$ is said to belong to the class $\mathbf{C}_1$ if for any $\epsilon > 0$ we can find two functions, $g(x)$ and $h(x)$, belonging to **C** such that

$$g(x) \leqq f(x) \leqq h(x) \qquad \text{for } a \leqq x \leqq b$$

and

$$L(h) - L(g) < \epsilon \tag{10.170}$$

If $f(x)$ belongs to $\mathbf{C}_1$ we define its Lebesgue integral by

$$\mathfrak{L}(f) \equiv \int_a^b f(x)\,dx = \sup_{g \epsilon \mathbf{C}} L(g) = \inf_{h \epsilon \mathbf{C}} L(h) \tag{10.171}$$

for all $g(x)$ and $h(x)$ satisfying (10.170). The reader can easily verify that if $f(x)$ belongs to **C** then $f(x)$ belongs to $\mathbf{C}_1$ and $L(f) = \mathfrak{L}(f)$. The set $\mathbf{C}_1$ contains all Lebesgue-integrable functions.

The ideas contained above lead to the concepts of measurable functions and measurable sets. In most texts one begins with the definition of the measure of a set, and afterward the Lebesgue integral is defined. We omit a discussion of measure, but we do emphasize that the theory of measure is of prime importance in modern probability theory. The Lebesgue integral plays an important role in modern mathematical analysis with its applications to the Fourier integral, probability theory, ergodic theory, and other fields.

REFERENCES

Burington, R. S., and C. C. Torrance: "Higher Mathematics," McGraw-Hill Book Company, Inc., New York, 1939.

Courant, R.: "Differential and Integral Calculus," Interscience Publishers, Inc., New York, 1947.

Franklin, P.: "A Treatise on Advanced Calculus," John Wiley & Sons, Inc., New York, 1940.

Goursat, E.: "A Course in Mathematical Analysis," Ginn & Company, Boston, 1904.

Hildebrand, F. B.: "Introduction to Numerical Analysis," McGraw-Hill Book Company, Inc., New York, 1956.

Householder, A. S.: "Principles of Numerical Analysis," McGraw-Hill Book Company, Inc., New York, 1953.

Kamke, E.: "Theory of Sets," Dover Publications, New York, 1950.

Milne, W. E.: "Numerical Calculus," Princeton University Press, Princeton, N. J., 1949.

Munroe, M. E.: "Introduction to Measure and Integration," Addison-Wesley Publishing Company, Cambridge, Mass., 1953.

Newman, M. H. A.: "Topology of Plane Sets," Cambridge University Press, New York, 1939.

Titchmarsh, M. A.: "The Theory of Functions," Oxford University Press, New York, 1939.

Widder, D. V.: "Advanced Calculus," Prentice-Hall, Inc., New York, 1947.

Wilson, E. B.: "Advanced Calculus," Ginn & Company, Boston, 1912.

INDEX

A CATALOG OF SELECTED

DOVER BOOKS

IN SCIENCE AND MATHEMATICS

Mathematics

FUNCTIONAL ANALYSIS (Second Corrected Edition), George Bachman and Lawrence Narici. Excellent treatment of subject geared toward students with background in linear algebra, advanced calculus, physics and engineering. Text covers introduction to inner-product spaces, normed, metric spaces, and topological spaces; complete orthonormal sets, the Hahn-Banach Theorem and its consequences, and many other related subjects. 1966 ed. 544pp. 6⅛ x 9¼. 0-486-40251-7

DIFFERENTIAL MANIFOLDS, Antoni A. Kosinski. Introductory text for advanced undergraduates and graduate students presents systematic study of the topological structure of smooth manifolds, starting with elements of theory and concluding with method of surgery. 1993 edition. 288pp. 5⅜ x 8½. 0-486-46244-7

VECTOR AND TENSOR ANALYSIS WITH APPLICATIONS, A. I. Borisenko and I. E. Tarapov. Concise introduction. Worked-out problems, solutions, exercises. 257pp. 5⅝ x 8¼. 0-486-63833-2

AN INTRODUCTION TO ORDINARY DIFFERENTIAL EQUATIONS, Earl A. Coddington. A thorough and systematic first course in elementary differential equations for undergraduates in mathematics and science, with many exercises and problems (with answers). Index. 304pp. 5⅜ x 8½. 0-486-65942-9

FOURIER SERIES AND ORTHOGONAL FUNCTIONS, Harry F. Davis. An incisive text combining theory and practical example to introduce Fourier series, orthogonal functions and applications of the Fourier method to boundary-value problems. 570 exercises. Answers and notes. 416pp. 5⅜ x 8½. 0-486-65973-9

COMPUTABILITY AND UNSOLVABILITY, Martin Davis. Classic graduate-level introduction to theory of computability, usually referred to as theory of recurrent functions. New preface and appendix. 288pp. 5⅜ x 8½. 0-486-61471-9

AN INTRODUCTION TO MATHEMATICAL ANALYSIS, Robert A. Rankin. Dealing chiefly with functions of a single real variable, this text by a distinguished educator introduces limits, continuity, differentiability, integration, convergence of infinite series, double series, and infinite products. 1963 edition. 624pp. 5⅜ x 8½. 0-486-46251-X

METHODS OF NUMERICAL INTEGRATION (SECOND EDITION), Philip J. Davis and Philip Rabinowitz. Requiring only a background in calculus, this text covers approximate integration over finite and infinite intervals, error analysis, approximate integration in two or more dimensions, and automatic integration. 1984 edition. 624pp. 5⅜ x 8½. 0-486-45339-1

INTRODUCTION TO LINEAR ALGEBRA AND DIFFERENTIAL EQUATIONS, John W. Dettman. Excellent text covers complex numbers, determinants, orthonormal bases, Laplace transforms, much more. Exercises with solutions. Undergraduate level. 416pp. 5⅜ x 8½. 0-486-65191-6

RIEMANN'S ZETA FUNCTION, H. M. Edwards. Superb, high-level study of landmark 1859 publication entitled "On the Number of Primes Less Than a Given Magnitude" traces developments in mathematical theory that it inspired. xiv+315pp. 5⅜ x 8½. 0-486-41740-9

Math–Decision Theory, Statistics, Probability

INTRODUCTION TO PROBABILITY, John E. Freund. Featured topics include permutations and factorials, probabilities and odds, frequency interpretation, mathematical expectation, decision-making, postulates of probability, rule of elimination, much more. Exercises with some solutions. Summary. 1973 edition. 247pp. 5⅜ x 8½. 0-486-67549-1

STATISTICAL AND INDUCTIVE PROBABILITIES, Hugues Leblanc. This treatment addresses a decades-old dispute among probability theorists, asserting that both statistical and inductive probabilities may be treated as sentence-theoretic measurements, and that the latter qualify as estimates of the former. 1962 edition. 160pp. 5⅜ x 8½. 0-486-44980-7

APPLIED MULTIVARIATE ANALYSIS: Using Bayesian and Frequentist Methods of Inference, Second Edition, S. James Press. This two-part treatment deals with foundations as well as models and applications. Topics include continuous multivariate distributions; regression and analysis of variance; factor analysis and latent structure analysis; and structuring multivariate populations. 1982 edition. 692pp. 5⅜ x 8½. 0-486-44236-5

LINEAR PROGRAMMING AND ECONOMIC ANALYSIS, Robert Dorfman, Paul A. Samuelson and Robert M. Solow. First comprehensive treatment of linear programming in standard economic analysis. Game theory, modern welfare economics, Leontief input-output, more. 525pp. 5⅜ x 8½. 0-486-65491-5

PROBABILITY: AN INTRODUCTION, Samuel Goldberg. Excellent basic text covers set theory, probability theory for finite sample spaces, binomial theorem, much more. 360 problems. Bibliographies. 322pp. 5⅜ x 8½. 0-486-65252-1

GAMES AND DECISIONS: INTRODUCTION AND CRITICAL SURVEY, R. Duncan Luce and Howard Raiffa. Superb nontechnical introduction to game theory, primarily applied to social sciences. Utility theory, zero-sum games, n-person games, decision-making, much more. Bibliography. 509pp. 5⅜ x 8½. 0-486-65943-7

INTRODUCTION TO THE THEORY OF GAMES, J. C. C. McKinsey. This comprehensive overview of the mathematical theory of games illustrates applications to situations involving conflicts of interest, including economic, social, political, and military contexts. Appropriate for advanced undergraduate and graduate courses; advanced calculus a prerequisite. 1952 ed. x+372pp. 5⅜ x 8½. 0-486-42811-7

FIFTY CHALLENGING PROBLEMS IN PROBABILITY WITH SOLUTIONS, Frederick Mosteller. Remarkable puzzlers, graded in difficulty, illustrate elementary and advanced aspects of probability. Detailed solutions. 88pp. 5⅜ x 8½. 0-486-65355-2

PROBABILITY THEORY: A CONCISE COURSE, Y. A. Rozanov. Highly readable, self-contained introduction covers combination of events, dependent events, Bernoulli trials, etc. 148pp. 5⅝ x 8¼. 0-486-63544-9

THE STATISTICAL ANALYSIS OF EXPERIMENTAL DATA, John Mandel. First half of book presents fundamental mathematical definitions, concepts and facts while remaining half deals with statistics primarily as an interpretive tool. Well-written text, numerous worked examples with step-by-step presentation. Includes 116 tables. 448pp. 5⅜ x 8½. 0-486-64666-1

Math–Geometry and Topology

ELEMENTARY CONCEPTS OF TOPOLOGY, Paul Alexandroff. Elegant, intuitive approach to topology from set-theoretic topology to Betti groups; how concepts of topology are useful in math and physics. 25 figures. 57pp. 5⅜ x 8½. 0-486-60747-X

A LONG WAY FROM EUCLID, Constance Reid. Lively guide by a prominent historian focuses on the role of Euclid's Elements in subsequent mathematical developments. Elementary algebra and plane geometry are sole prerequisites. 80 drawings. 1963 edition. 304pp. 5⅜ x 8½. 0-486-43613-6

EXPERIMENTS IN TOPOLOGY, Stephen Barr. Classic, lively explanation of one of the byways of mathematics. Klein bottles, Moebius strips, projective planes, map coloring, problem of the Koenigsberg bridges, much more, described with clarity and wit. 43 figures. 210pp. 5⅜ x 8½. 0-486-25933-1

THE GEOMETRY OF RENÉ DESCARTES, René Descartes. The great work founded analytical geometry. Original French text, Descartes's own diagrams, together with definitive Smith-Latham translation. 244pp. 5⅜ x 8½. 0-486-60068-8

EUCLIDEAN GEOMETRY AND TRANSFORMATIONS, Clayton W. Dodge. This introduction to Euclidean geometry emphasizes transformations, particularly isometries and similarities. Suitable for undergraduate courses, it includes numerous examples, many with detailed answers. 1972 ed. viii+296pp. 6⅛ x 9¼. 0-486-43476-1

EXCURSIONS IN GEOMETRY, C. Stanley Ogilvy. A straightedge, compass, and a little thought are all that's needed to discover the intellectual excitement of geometry. Harmonic division and Apollonian circles, inversive geometry, hexlet, Golden Section, more. 132 illustrations. 192pp. 5⅜ x 8½. 0-486-26530-7

THE THIRTEEN BOOKS OF EUCLID'S ELEMENTS, translated with introduction and commentary by Sir Thomas L. Heath. Definitive edition. Textual and linguistic notes, mathematical analysis. 2,500 years of critical commentary. Unabridged. 1,414pp. 5⅜ x 8½.. Three-vol. set.
Vol. I: 0-486-60088-2 Vol. II: 0-486-60089-0 Vol. III: 0-486-60090-4

SPACE AND GEOMETRY: IN THE LIGHT OF PHYSIOLOGICAL, PSYCHOLOGICAL AND PHYSICAL INQUIRY, Ernst Mach. Three essays by an eminent philosopher and scientist explore the nature, origin, and development of our concepts of space, with a distinctness and precision suitable for undergraduate students and other readers. 1906 ed. vi+148pp. 5⅜ x 8½. 0-486-43909-7

GEOMETRY OF COMPLEX NUMBERS, Hans Schwerdtfeger. Illuminating, widely praised book on analytic geometry of circles, the Moebius transformation, and two-dimensional non-Euclidean geometries. 200pp. 5⅝ x 8¼. 0-486-63830-8

DIFFERENTIAL GEOMETRY, Heinrich W. Guggenheimer. Local differential geometry as an application of advanced calculus and linear algebra. Curvature, transformation groups, surfaces, more. Exercises. 62 figures. 378pp. 5⅜ x 8½.
0-486-63433-7